Imidazole Dipeptides

Chemistry, Analysis, Function and Effects

Food and Nutritional Components in Focus

Series Editor:
Professor Victor R Preedy, *School of Medicine, King's College London, UK*

Titles in the Series:
1: Vitamin A and Carotenoids: Chemistry, Analysis, Function and Effects
2: Caffeine: Chemistry, Analysis, Function and Effects
3: Dietary Sugars: Chemistry, Analysis, Function and Effects
4: B Vitamins and Folate: Chemistry, Analysis, Function and Effects
5: Isoflavones: Chemistry, Analysis, Function and Effects
6: Fluorine: Chemistry, Analysis, Function and Effects
7: Betaine: Chemistry, Analysis, Function and Effects
8: Imidazole Dipeptides: Chemistry, Analysis, Function and Effects

How to obtain future titles on publication:
A standing order plan is available for this series. A standing order will bring delivery of each new volume immediately on publication.

For further information please contact:
Book Sales Department, Royal Society of Chemistry, Thomas Graham House, Science Park, Milton Road, Cambridge, CB4 0WF, UK
Telephone: +44 (0)1223 420066, Fax: +44 (0)1223 420247
Email: booksales@rsc.org
Visit our website at www.rsc.org/books

Imidazole Dipeptides

Chemistry, Analysis, Function and Effects

Edited by

Victor R Preedy
School of Medicine, King's College London, UK
Email: editor@publicationeditor.org.uk

THE QUEEN'S AWARDS
FOR ENTERPRISE:
INTERNATIONAL TRADE
2013

Food and Nutritional Components in Focus No. 8

Print ISBN: 978-1-84973-890-3
PDF eISBN: 978-1-78262-261-1
ISSN: 2045-1695

A catalogue record for this book is available from the British Library

Published by The Royal Society of Chemistry,
Thomas Graham House, Science Park, Milton Road,
Cambridge CB4 0WF, UK

Registered Charity Number 207890

For further information see our web site at www.rsc.org

Preface

Recently, there have been major advances in our understanding of the chemistry and function of nutritional components. This has been enhanced by rapid developments in analytical techniques and instrumentation. Chemists, food scientists and nutritionists are however separated by divergent skills, and professional disciplines. Hitherto, this transdisciplinary divide has been difficult to bridge.

The series *Food and Nutritional Components in Focus* aims to cover in a single volume the chemistry, analysis, function and effects of single components in the diet or its food matrix. Its aim is to link scientific disciplines so that information becomes more meaningful and applicable to health in general.

The series *Food and Nutritional Components in Focus* covers the latest knowledge base and has a structured format with major subsections covering

- Compounds in Context
- Chemistry
- Analysis
- Function and Effects

In some books the section on Chemistry is also linked with Biochemistry. Each chapter has a novel cohort of features namely by containing:

- Summary Points
- Key Facts (areas of focus explained for the lay person)
- Definitions of Words and Terms

The series covers numerous classes of dietary components including, for example, minerals, vitamins, food additives, and so on. The chapters are written by national or international experts, specialists and leaders in the field.

Food and Nutritional Components in Focus No. 8
Imidazole Dipeptides: Chemistry, Analysis, Function and Effects
Edited by Victor R Preedy

Published by the Royal Society of Chemistry, www.rsc.org

Imidazole Dipeptides has the following material: Section 1 Imidazole Dipeptides in Context has material on histidine-containing dipeptides, and carnosine and its homologs in foods. Section 2 Chemistry and Biochemistry has chapters on the chemistry of imidazole dipeptides, carnosine derivatives, chiral carnosine, enzymes producing carnosine and anserine, anserine- and carnosine-degrading enzymes, protein covalent modifications, reactive carbonyl species, antioxidant activities, DNA and protein degradation. Section 3 Analysis has a chapter on liquid chromatography and tandem mass spectrometry, while Section 4 Function and Effects has additional information on antioxidant activity, as well as material on inflammaging, blood glucose, muscle, exercise, homeostasis, muscle contractility, fatigue, wound healing, ischemia/reperfusion, antigenotoxic effects, cancers, colon cancer, neuroprotection, vascular dementia, Alzheimer's disease, behavior, kidney diseases, liver injury the blood–retinal barrier and gastric protection.

Imidazole Dipeptides is specifically designed for chemists, analytical scientists, forensic scientists, food scientists, dieticians, nutritionists, food scientists, health professionals and research academics. The series is suitable for lecturers and teachers in food and nutritional sciences. Importantly, the series will be a valuable resource for college or university libraries as a reference guide.

Professor Victor R Preedy
King's College London

Contents

Imidazole Dipeptides in Context

Chapter 1 Carnosine in the Context of Histidine-Containing Dipeptides 3
Giulio Vistoli

1.1 Introduction 3
1.2 Histidine: Reactivity, Metabolism and Physiological Roles 4
1.2.1 Histidine's Reactivity: The Key Role of the Imidazole Ring 4
1.2.2 Histidine Metabolism and Physiological Roles 6
1.3 Proteinogenic Histidine Dipeptides 7
1.3.1 Specific Activity Profiles 7
1.3.2 Carbonyl-Quencher Activity 9
1.4 Nonproteinogenic Histidine Dipeptides: Carnosine and Its Derivatives 11
1.4.1 Optimizing the Carnosine's Carbonyl Quenching: A Mechanistic Perspective 12
1.4.2 Optimizing the Carnosine Pharmacokinetic Profile: A Balance Game 14
1.5 Summary 17
Summary Points 17
Key Facts of Carbonyl Quenching 18
Definitions of Words and Terms 18
List of Abbreviations 19
References 19

Food and Nutritional Components in Focus No. 8
Imidazole Dipeptides: Chemistry, Analysis, Function and Effects
Edited by Victor R Preedy

Published by the Royal Society of Chemistry, www.rsc.org

Chapter 2 Carnosine and Its Homologs in Foods 23
Pier Giorgio Peiretti and Giorgia Meineri

2.1 Main Functions of Carnosine and Its Homologs for the Protection of Human Health 23
2.2 Diet and Human Health: Benefits of Carnosine and Its Homologs Contained in Meat Products 24
2.3 Influence of Carnosine and Its Homologs on the Quality of Meat for Human Consumption 25
2.4 Content of Carnosine and Its Homologs in Animal Tissues 28
2.5 Human Nutritional Intakes of Carnosine and Its Homologs in Meat 31
Summary Points 33
Key Facts of Imidazole Dipeptides Occurrence in Food 33
Definitions of Words and Terms 34
References 35

Chemistry and Biochemistry

Chapter 3 The Chemistry of Imidazole Dipeptides 43
Daniel Perrone, Mariana Monteiro, and Vanessa Naciuk Castelo-Branco

3.1 Introduction 43
3.2 Physicochemical and Spectroscopic Data 45
3.3 Food Sources 50
3.4 Reactivity and Synthesis 53
3.5 Chemical Basis of Imidazole Dipeptides Physiological Roles 54
3.5.1 Buffering Capacity 54
3.5.2 Antioxidant Activity 55
Summary Points 55
Keys Facts of Muscle Physiology 56
Definitions and Explanation of Key Terms 56
List of Abbreviations 57
References 58

Chapter 4 The Chemistry of Carnosine Derivatives 61
Francesco Bellia, Enrico Rizzarelli, and Graziella Vecchio

4.1 Introduction 61
4.2 L-Carnosine Derivatives 63
4.2.1 Derivatization of the Carnosine Amino Group 63
4.2.2 Derivatization of the Carnosine Carboxylic Group 70
4.2.3 Double Functionalization of Carboxylic and Amino Groups 71

4.3 Derivatives of D-Carnosine 72
4.4 Metal Complexes of L- and D-Carnosine Derivatives 73
4.5 Nanosystems Based on Carnosine 76
4.6 Conclusions 77
Summary Points 77
Key Facts of Chemical Modifications of Natural Products (Carnosine being among them) 78
Definitions of Words and Terms 78
List of Abbreviations 80
Acknowledgements 80
References 80

Chapter 5 Chiral Carnosine **84**
Dan Su and Yonggui Song

5.1 Structure, Physical and Chemical Properties 84
5.2 Extraction and Synthesis 84
5.3 Physiological Activities 85
5.3.1 L-Carnosine 85
5.3.2 D-Carnosine 86
5.4 Pharmacological Effects and Applications 87
5.4.1 L-Carnosine 87
5.4.2 D-Carnosine 88
5.5 Determination Methods 89
5.5.1 HPLC Method 89
5.5.2 HPLC-MS/MS Method 90
5.5.3 CE Method 90
5.5.4 Separation of Optical Isomers of Carnosine 90
5.6 Pharmacokinetics of L-carnosine and D-carnosine 91
5.7 Concluding Remarks 91
Summary Points 92
Key Facts of Chiral Carnosine 92
Definitions and Explanation of Key Terms 92
List of Abbreviations 93
References 93

Chapter 6 The Biochemistry of Enzymes Producing Carnosine and Anserine **99**
Jakub Drozak, Emile Van Schaftingen, and Maria Veiga-da-Cunha

6.1 Introduction 99
6.2 Carnosine Synthase 100
6.2.1 Brief History of Research 100
6.2.2 Molecular Identification and Species Distribution 101
6.2.3 Enzyme Specificity and Metabolite Repair 102

6.2.4 Tissue Distribution 104
6.2.5 Regulation of Skeletal Muscle Carnosine Synthase Activity 105
6.2.6 Perspectives 106
6.3 Carnosine *N*-Methyltransferase 106
6.3.1 Brief History of Research 106
6.3.2 Molecular Identity and Biochemical Properties of the Chicken Enzyme 107
6.3.3 Hnmt-Like Gene Expression 109
6.3.4 Phylogeny 109
6.3.5 Perspectives 110
Summary Points 112
Key Facts about the Importance of Molecular Identification of Enzymes 112
Definitions of Words and Terms 113
List of Abbreviations 113
References 114

Chapter 7 The Biochemistry of Anserine and Carnosine Degrading Enzymes 118
Shoji Yamada

7.1 Introduction 118
7.2 The Enzymes Named "Carnosinase" 119
7.2.1 Carnosine Dipeptidase II 121
7.2.2 Carnosine Dipeptidase I 124
7.3 The Enzyme Named "Anserinase" or "Acetylhistidine Deacetylase" 127
7.3.1 Xaa–Methyl-His Dipeptidase 128
7.4 The Phylogenetic History of Anserine and Carnosine-Degrading Enzymes 130
7.5 Conclusions 131
Summary Points 132
Key Facts 133
Definitions of Words and Terms 133
List of Abbreviations 134
Acknowledgements 134
References 134

Chapter 8 Carnosine and Derivatives as Inhibitors of Protein Covalent Modifications Induced by Reactive Carbonyl Species 139
Mara Colzani, Davide Garzon, and Giancarlo Aldini

8.1 Reactive Carbonyl Species and Protein Covalent Modification 139
8.1.1 α,β-Unsaturated Aldehydes 140

8.1.2 Dialdehdyes 141
8.1.3 Keto-Aldehydes 142
8.2 Biological Implications of RCS-Induced Protein Covalent Modifications 145
8.2.1 Protein Function Derangement 145
8.2.2 The Antigenic Properties of RCS Protein Adducts 146
8.2.3 The Damaging AGEs–RAGE Axis 146
8.2.4 The Amyloidogenic Properties of RCS Protein Adducts 146
8.3 Endogenous Detoxification of RCS 147
8.3.1 Phase I Metabolism 147
8.3.2 Phase II Metabolism 147
8.4 Carnosine and Histidine Dipeptides as RCS-Sequestering Agents 149
8.4.1 *In Vitro* Studies 150
8.4.2 *In Vivo* Studies 154
8.5 Carnosine Derivatives as a Novel Class of Bioactive Compounds 159
8.6 Conclusion 161
Summary Points 162
Definitions of Words and Terms 162
List of Abbreviations 163
References 164

Chapter 9 Antioxidant Activity of Imidazole Dipeptides – Prevention of DNA and Protein Degradation **170**
Masayoshi Takahashi, Shoji Hagiwara, and Hiroshi Nabetani

9.1 Introduction 170
9.2 EMADD Procedure 172
9.2.1 Principle 172
9.2.2 Reagents 173
9.2.3 Preparation of ROS Solutions 173
9.2.4 Measurement of the Effect of Preventing ROS-Induced DNA Degradation 174
9.3 Determination of Antioxidant Effects 177
9.4 Results 177
9.4.1 Degradation of DNA by Various ROS 177
9.4.2 Comparison of Naturally Occurring Antioxidants against ROS-Induced DNA Degradation 179
9.4.3 Comparison of Various Antioxidant Activity Assays 180
9.5 Discussion 182
9.6 Perspectives 184
Summary Points 185

Key Facts 186
Definitions of Words and Terms (Abbreviations) 186
Acknowledgements 187
References 187

Analysis

Chapter 10 Liquid Chromatography Coupled to Tandem Mass Spectrometry to Analyze Imidazole Dipeptides **191**
Maria-José Motilva, Alba Macià, Alba Domínguez, and Agustín Labrador

10.1 Introduction 191
10.2 Analysis of Imidazole Dipeptides 192
10.2.1 Chromatographic Conditions 192
10.2.2 MS/MS as the Detector System 200
10.3 Instrumental Quality Parameters 201
10.4 Applications of LC-MS/MS for the Analysis of Imidazole Dipeptides 207
10.4.1 Meat Samples 207
10.4.2 Biological Samples 208
10.5 Conclusions 209
Summary Points 209
Key Facts of Mass Spectrometry 210
Definition of Words and Terms 210
List of Abbreviations 211
Acknowledgments 211
References 211

Function and Effects

Chapter 11 Antioxidant Activity of Imidazole Dipeptides **217**
Ewa Biazik and Wiesław Kopeć

11.1 Introduction 217
11.2 Modification of Oxidation Processes in Liposome Model Systems *via* Imidazole Dipeptides 219
11.3 Radical-Scavenging Activity of Imidazole Dipeptides 221
11.4 Aldehyde-Quenching Activity of Imidazole Dipeptides 223
11.5 Buffering Activity of Imidazole Dipeptides 225
11.6 Antioxidant Activity of Imidazole Dipeptides in Other Systems 226
11.6.1 Protein Model System 226
11.6.2 Tissue Model System 227
11.7 Antioxidant Activity of Pseudopeptides 227

11.8 Antioxidant Activity of *N*-Acyl Imidazole Dipeptide Derivatives and Adducts 229
Summary Points 232
Key Facts 233
Definitions of Words and Terms 233
List of Abbreviations 234
References 235

Chapter 12 Inflammaging, Oxidative Stress and Carnosine: Role of Hormetic Vitagenes 238
Vittorio Calabrese, Sergio Davinelli, Maria Luca, Davide Zella, Edward J. Calabrese, and Giovanni Scapagnini

12.1 Introduction 238
12.2 Hormesis 241
12.3 Cellular Stress Response, HSF Biology and the Vitagene Network 244
12.4 Carnosine as an Inducer of Vitagenes in Oxi-Inflammaging 248
Summary Points 250
Definitions of Words and Terms 250
List of Abbreviations 250
References 251

Chapter 13 Carnosine and the Control of Blood Glucose 257
Nobuaki Okumura, Mamoru Tanida, Akira Niijima, Nobuo Tsuruoka, Yoshinobu Kiso, Yuko Horii, and Katsuya Nagai

13.1 Introduction 257
13.2 Effect of Carnosine on Blood Glucose 259
13.2.1 Suppression of 2-Deoxy-D-Glucose-Induced Hyperglycemia by Carnosine 259
13.2.2 Dose Dependence of Carnosine Action on 2DG-Hyperglycemia 259
13.2.3 Effects of Carnosine on Plasma Insulin and Glucagon Levels 260
13.2.4 Effects of Carnosine on Sympathetic and Parasympathetic Nerve Activities 261
13.2.5 Effects of Histamine and L-Histidine on 2DG-Hyperglycemia 262
13.2.6 Involvement of Histamine Receptors in Carnosine Action on 2DG-Hyperglycemia 262
13.3 Carnosine Actions on Blood Pressure and Autonomic Regulation of Peripheral Tissues 263
13.3.1 Effects of Carnosine on Blood Pressure and Renal Sympathetic Nerve Activity 263

13.3.2 Effect of Carnosine on Lipid Metabolism in Adipose Tissues and Cancer 265
13.4 Carnosine Synthesis, Release, and Hydrolysis 265
13.4.1 Carnosine Biosynthesis by Carnosine Synthetase 265
13.4.2 Carnosine Concentration in the Blood 266
13.4.3 Carnosine Hydrolysis by Carnosine Dipeptidases 266
13.4.4 Tissue Distribution of CN1 and CN2 267
13.5 Hypothalamic Regulation of Autonomic Nervous System and Carnosine Action 268
13.5.1 Involvement of Histamine Neurons in the TMN of the Hypothalamus 268
13.5.2 Involvement of the SCN, the Master Circadian Clock 269
13.6 Effects of Anserine on Blood Glucose and Autonomic Nervous System 270
13.7 Conclusions 270
Summary Points 271
Key Facts 271
Definitions and Explanations of Key Terms 272
List of Abbreviations 273
References 273

Chapter 14 β-Alanine, Muscle Carnosine and Exercise 277
Bryan Saunders, Guilherme G. Artioli, Craig Sale, and Bruno Gualano

14.1 Introduction 277
14.2 β-Alanine and Muscle Carnosine 278
14.3 β-Alanine Supplementation and Exercise 280
14.3.1 Single-Bout High-Intensity Exercise 281
14.3.2 High-Intensity Intermittent and Repeated-Bout High-Intensity Exercise 284
14.3.3 Exercise and Training 285
14.4 Conclusions 286
14.5 Practical Applications 287
Summary Points 288
Key Facts of High-Intensity Exercise 288
Key Facts of Increased Muscle Carnosine and Exercise 288
Definitions of Words and Terms 289
List of Abbreviations 289
Acknowledgements 290
References 290

Chapter 15 Human Skeletal Muscle Carnosine and Its Function: A Focus on Homeostasis, Muscle Contractility and pH 295
I. Everaert and W. Derave

15.1 Introduction 295
15.2 Determinants of Muscle-Carnosine Content 296
15.2.1 Muscle Fiber-Type Composition 296
15.2.2 Gender, Age and Androgens 297
15.2.3 Diet 298
15.3 Physiological Functions of Carnosine in Contracting Muscle 299
15.3.1 Intramyocellular Buffering 301
15.3.2 Calcium Release of Sarcoplasmic Reticulum 302
15.3.3 Calcium Sensitivity of the Contractile Apparatus 304
15.3.4 Release 305
15.3.5 Carnosine–Histidine–Histamine Pathway 305
Summary Points 305
Key Facts 306
Key Terms 306
List of Abbreviations 307
Acknowledgements 308
References 308

Chapter 16 Anserine as a Suppressor of Fatigue 313
Yoshinori Takahashi

16.1 Introduction 313
16.2 Effect of Single Administration of Anserine-Containing Salmon Muscle Extracts on Physical Fatigue 314
16.3 Effect of Anserine-Containing Salmon Muscle Extracts on Exercise Performance and Kinetic Visual Acuity among Athletes 317
16.4 Effect of Anserine-Containing Salmon Muscle Extracts on Asthenopia 318
16.5 Conclusions 319
Summary Points 321
Key Facts 321
Key Facts Regarding Asthenopia 321
Definitions of Words and Terms 322
List of Abbreviations 323
References 323

Chapter 17 Carnosine Promotion of Wound Healing **325**
I. Ansurudeen, E. Forsberg, and K. Brismar

17.1 Wounds 325
17.1.1 Type of Wounds 325
17.1.2 Wound Healing 326
17.1.3 Factors Involved in Wound Healing 328
17.1.4 Abnormal Wound Healing 329
17.2 Carnosine 329
17.2.1 Carnosine in Wound Healing 329
17.3 Conclusions 335
Summary Points 336
Key Facts 336
Key Terms 337
List of Abbreviations 337
Acknowledgement 338
References 338

Chapter 18 Carnosine and Ischemia/Reperfusion Injury **341**
Latif Abbasoğlu, Semra Doğru-Abbasoğlu, and Müjdat Uysal

18.1 Ischemia/Reperfusion Injury 341
18.1.1 Carnosine and Ischemia/Reperfusion Injury 344
18.2 Conclusions 350
Summary Points 350
Key Facts of I/R Injury 351
Definitions of Words and Terms 351
List of Abbreviations 352
References 352

Chapter 19 Antigenotoxic Effects of Carnosine **356**
Ahmet Güner, and Lokman Alpsoy

19.1 Carnosine 356
19.2 Antigenotoxic Effects of Carnosine 358
19.2.1 Protective Role of Carnosine on Mutagenicity (AMES) 359
19.2.2 Protective Role of Carnosine on DNA Damage 359
19.2.3 Protective Role of Carnosine on Chromosome Damage 361
19.3 Antigenotoxic Mechanism of Carnosine 363

19.3.1 Carnosine Inhibits Glycalization 363
19.3.2 Carnosine has Metal-Ion-Chelator and Free-Radical-Scavenger Properties 365
19.3.3 Carnosine Activates Antioxidative Enzymes and Increases the Level of Nonenzymatic Antioxidative Compounds 365
Summary Points 366
Key Facts 367
Definitions of Words and Terms 367
List of Abbreviations 368
References 368

Chapter 20 Carnosine and Cancer **372**
Frank Gaunitz, Henry Oppermann, and Alan Hipkiss

20.1 Carnosine and Cancer – Experimental Evidence 372
20.2 Possible Mechanisms for Carnosine's Effects on Cancer 373
20.2.1 Gene Regulation and Cell Signaling 373
20.3 Cell Cycle 376
20.4 Apoptosis 377
20.5 Metastasis 377
20.6 Glycolysis 377
20.6.1 Regulation of Glycolysis 378
20.7 Carnosine and Induction of PDK4 mRNA Expression 379
20.8 Carnosine as a Drug for Cancer Treatment 381
20.8.1 Bioavailability and Mode of Administration 381
20.8.2 Transport and Delivery 382
20.8.3 Carnosinase Activity and Plasma Levels of Carnosine 382
20.8.4 Dosage and Side Effects 382
20.8.5 Derivatives of Carnosine and Related Compounds for Cancer Treatment 383
20.9 Conclusions 383
Summary Points 383
Key Facts of Signal Transduction Pathways 384
Key Facts of Pyruvate Dehydrogenase Kinase 384
Key Facts of the Warburg Effect 385
Definition of Words and Terms 385
List of Abbreviations 388
References 389

Chapter 21 **L-Carnosine and Human Colon Cancer** **393**
Barbara Iovine, Mariangela Garofalo, Maria Orefice, and Maria Assunta Bevilacqua

21.1 Colon Cancer 393
21.1.1 A Comprehensive Summary of Colorectal Cancers 394
21.1.2 Colorectal Carcinoma Genetics 395
21.2 What are the Main Causes of Colon Cancer? 397
21.2.1 Diet and Colorectal Cancer 397
21.2.2 Functional Relationship between Inflammation and Colon Cancer 398
21.3 The Dipeptide L-Carnosine in Human Colon Cancer 399
21.3.1 L-Carnosine Inhibits the Inflammatory Response of Cancer Intestinal Epithelial Cells 399
21.3.2 L-Carnosine Induces Cell-Cycle Arrest in Colon Cancer Cells 401
21.4 Conclusions 404
21.4.1 Critical Molecular Pathways Involved in the L-Carnosine Effect on Colon Cancer 404
Summary Points 405
Key Facts 406
Key Features of Signal Transduction Mediated by the Ras/Raf/MEK/ERK Pathway 406
Definitions of Words and Terms 407
List of Abbreviations 408
References 409

Chapter 22 **Carnosine-Induced Neuroprotection** **412**
S. M. Fayaz, Arshad Majid, and G. K. Rajanikant

22.1 Introduction 412
22.2 Pleiotropic Properties of Carnosine 413
22.2.1 Carnosine as an Antioxidant 413
22.2.2 Carnosine as Antiglycating Agent 416
22.2.3 Carnosine as Metal-Ion Chelator 416
22.2.4 Carnosine and Altered Proteins 417
22.3 Role of Carnosine in Neurological Disorders 417
22.3.1 Cerebral Ischemia 417
22.3.2 Alzheimer's Disease 419
22.3.3 Parkinson's Disease (PD) 421
22.3.4 Brain Tumor 423
22.3.5 Other Neurological Disorders 425
22.4 Conclusions 426
Summary Points 427
Key Facts of Oxidative Stress 427

Definitions of Words and Terms 428
List of Abbreviations 428
References 429

Chapter 23 Carnosine and Vascular Dementia **432**
Jing Ma and Zhong Chen

23.1 Carnosine 432
23.2 Carnosine in Nervous System Disease 434
23.3 Hypothetical Functions of Carnosine in the Nervous System 435
23.4 Vascular Dementia 435
23.4.1 The Background of Vascular Dementia 435
23.4.2 The Cellular Pathology of VaD 437
23.4.3 The Animal Model of VaD 438
23.5 Carnosine in VaD 438
23.5.1 The Effect of Carnosine on VaD 438
23.5.2 Mechanisms of Carnosine in VaD 440
23.6 Conclusions 444
Summary Points 444
Key Facts 444
Key Facts of Vascular Dementia 444
Key Facts of Alzheimer's Dementia 445
Definitions of Words and Terms 445
List of Abbreviations 447
References 448

Chapter 24 Carnosine and Alzheimer's Disease-Related Fibril Formation **451**
Alessandra Aloisi and Ross Rinaldi

24.1 Carnosine and Alzheimer's Disease-Related Fibril Formation 451
24.2 Amyloid Proteins and AD 452
24.2.1 Cross-β Structure of Aβ Amyloid Fibrils 452
24.3 Screening for Negative Modulators and Carnosine Activity 453
24.3.1 Carnosine and β-Amyloid Peptide: Interaction Dynamics 456
24.4 Critical Points in Perspective 461
Summary Points 462
Key Facts 462
Key Features of Amyloid Fibrils 462
Key Features of Alzheimer's Disease (AD) 463
Key Features of Protein Folding into Functional Protein 464
Key Features of Chaperones in Proteostasis 464

Definitions of Words and Terms 464
Key Terms 464
List of Abbreviations 466
References 466

Chapter 25 The Function of Carnosine and Its Homologs on Behavior 471
Mitsuhiro Furuse

25.1 The Presence of Carnosine in Animals 471
25.2 The Behavioral Effects of the Central Administration of Carnosine and Its Constituents 472
25.3 Nitric Oxide is Involved in Carnosine-Induced Hyperactivity 476
25.4 The Effect of Replacing the Amino Acid Residue of Carnosine 481
25.4.1 The Central Administration of Dipeptides in the Form of β-Alanyl Branched Chain Amino Acids Induces Hyperactivity 481
25.4.2 The Effect of the Substitution of Amino Terminal β-Ala on Carnosine-Induced Sedative Effects Under an Acute Stressful Condition in Chicks 482
25.4.3 The Reverse Structure of Carnosine Induced Sedative and Hypnotic Effects Under Acute Stress Conditions 482
Summary Points 483
Key Facts 486
Definitions of Words and Terms 486
List of Abbreviations 487
References 488

Chapter 26 Carnosine and Kidney Diseases 493
Katarzyna Kiliś-Pstrusińska

26.1 Introduction 493
26.2 Biological Activity of Carnosine in Relation to Kidneys 494
26.3 Carnosine and Kidney Diseases 495
26.3.1 Hypertension 495
26.3.2 Acute Kidney Injury 497
26.3.3 Diabetic Nephropathy 498
26.3.4 Others 500
26.4 Carnosinase and Kidney 500
26.5 Summary 503
Summary Points 503

Key Facts 504
Key Facts of Renoprotection 504
Key Facts of Importance of Carnosine in Kidney Diseases 504
Definitions of Words and Terms 505
List of Abbreviations 505
References 506

Chapter 27 Carnosine Protection Against Liver Injury 510
Müjdat Uysal, Necla Koçak-Toker, and Semra Doğru-Abbasoğlu

27.1 Introduction 510
27.2 Liver Injuries and Carnosine 513
27.2.1 Ischemia–Reperfusion-Induced Liver Injury and Carnosine 514
27.2.2 Acetaminophen-Induced Liver Injury and Carnosine 514
27.2.3 Thioacetamide-Induced Liver Injury and Carnosine 515
27.2.4 Cadmium-Induced Liver Injury and Carnosine 515
27.2.5 Sepsis-Induced Liver Injury and Carnosine 516
27.2.6 Stress-Induced Liver Injury and Carnosine 516
27.2.7 Age-Induced Liver Injury and Carnosine 517
27.2.8 Alcoholic Liver Injury and Carnosine 518
27.2.9 Nonalcoholic Fatty Liver Disease and Carnosine 520
27.3 Liver Disease and Polaprezinc 521
Summary Points 523
Key Facts of Liver Injury 523
Definition of Words and Terms 523
List of Abbreviations 524
References 525

Chapter 28 The Blood–Retinal Barrier and Carnosine 528
Yoshiyuki Kubo, Shin-Ichi Akanuma, and Ken-Ichi Hosoya

28.1 Introduction 528
28.2 Blood–Retinal Barrier (BRB) and Membrane Transporters 529
28.3 Carrier-Mediated L-Carnosine Transport in the Different Tissues 530
28.4 Oligopeptide Transporters (SLC15A) in the Retina 531
28.5 Minor Contribution of Dipeptide Transport at the Inner and Outer BRB 532

28.6 Is L-Carnosine Synthesized from β-alanine and L-histidine in the Retina? 534
28.7 Influx Transport of β-alanine and L-Histidine across the BRB 534
28.8 Carrier-Mediated Influx Transport of β-Alanine at the Inner BRB 536
28.9 Carrier-Mediated Influx Transport of L-Histidine across the Inner BRB 536
28.10 TAUT and LAT1 are Involved in Retinal L-Carnosine Synthesis 538
28.11 β-Alanine and L-Histidine Transport at the Outer BRB 539
28.12 Concluding Remarks 539
Summary Points 540
Key Facts 540
Key Features of Dipeptide Transport at the Inner and Outer BRB 540
Key Features of β-Alanine and L-Histidine Transport at the BRB 541
Key Features of L-histidine Transport at the Inner BRB 541
Key Facts about β-Alanine Transport at the Inner BRB 541
Key Facts about Transporter Involved in the Supply of L-Carnosine to the Retina 542
Definition of Words and Terms 542
List of Abbreviations 543
References 544

Chapter 29 L-Carnosine and Zinc in Gastric Protection 548
Hyuk Soon Choi, Eun Sun Kim, Bora Keum, Hoon Jai Chun, and Mi-Kyung Sung

29.1 Introduction 548
29.2 L-Carnosine and Zinc 549
29.3 Mechanism Underlying Gastric Mucosal Damage 551
29.4 Defense and Repair Mechanisms in the Gastric Mucosa 553
29.4.1 Bicarbonate Secretion 554
29.4.2 Prostaglandin E Synthase 554
29.4.3 Mucin Glycoproteins 554
29.4.4 Growth Factors 555
29.5 Mechanisms Underlying Zinc L-Carnosine Action 556
29.5.1 Antiulcer Effect 556
29.5.2 Anti-Inflammatory Action 556

29.5.3 Antioxidant Action 556
29.5.4 Growth Factors 557
29.5.5 Other Potential Cytoprotective Factors 557
29.6 Clinical Evidence of, and Future Research of Zinc L-Carnosine in Gastric Mucosal Protection 558
Summary Points 560
Key Facts 561
Definitions of Words and Terms 561
References 563

Subject Index **566**

Section I
Imidazole Dipeptides in Context

CHAPTER 1

Carnosine in the Context of Histidine-Containing Dipeptides

GIULIO VISTOLI*[a]

[a]Dipartimento di Scienze Farmaceutiche, Università degli Studi di Milano, Via Mangiagalli, 25, I-20133 Milan, Italy
*E-mail: giulio.vistoli@unimi.it

1.1 Introduction

Among the proteinogenic amino acids, histidine plays manifold roles mostly ascribable to the reactivity of the imidazole ring that characterizes its side chain. While being one of the least abundant residues (representing about 2% of the total protein residues), histidine plays key roles in protein folding and functions. Indeed, the histidine imidazole ring is the coordinating group in many metalloproteins and is involved in several enzymatic mechanisms such as in the catalytic triad in which histidine acts as a general base increasing the nucleophilicity of surrounding serine, threonine, or cysteine residues. Thus, histidine is able to shuttle protons by abstracting them through its basic nitrogen atom as seen in carbonic anhydrases, where it shuttles protons away from the zinc-bound water molecule to restore the active form of the enzyme (Fisher *et al.*, 2005). Moreover, histidine is involved in regulating protein functions due to its post-translational modification by the addition of a phosphate

Food and Nutritional Components in Focus No. 8
Imidazole Dipeptides: Chemistry, Analysis, Function and Effects
Edited by Victor R Preedy

Published by the Royal Society of Chemistry, www.rsc.org

group to yield the 1-phosphohistidine or the 3-phosphohistidine adducts (Attwood *et al.*, 2007). Although an exhaustive characterization of this reversible modification in human cells is still lacking, mainly due to the instability of the phosphoramidate bond, accumulating evidence emphasizes the crucial role of histidine phosphorylation in cell signaling, as exemplified by its role in regulating the flux through the K^+ channel KCa3.1 caused by histidine phosphorylation within the protein C-terminus (Beckman-Sundh *et al.*, 2011).

When considering the roles of histidine in proteins, its key presence also in several peptides comes as no surprise. A relevant example of histidine-rich peptides is offered by the salivary histatins, which are metal-binding peptides endowed with potent antibacterial, antifungal and wound-healing activities (Kavanagh and Dowd, 2004). These few examples emphasize the role of histidine in the manifold protein world and represent an indirect but convincing explanation of the remarkable role played also by small histidine-containing dipeptides (HCD) among which carnosine and its congeners represent the most important group. Based on these premises, the chapter is focused on histidine dipeptides and starts from an introductory analysis of histidine as a single amino acid, discussing its reactivity, metabolism and physiological roles to move to histidine dipeptides as grouped into proteinogenic and non-proteinogenic dipeptides. Among the latter, the chapter analyzes indepth the biological profiles for carnosine and related dipeptides. While discussing the multifaceted properties of the histidine-containing dipeptides, the chapter is focused on their carbonyl-quenching activity detailing mechanisms, reactivity and structure–activity relationships.

1.2 Histidine: Reactivity, Metabolism and Physiological Roles

1.2.1 Histidine's Reactivity: The Key Role of the Imidazole Ring

The reactivity of the histidine residue is mostly ascribable to the physicochemical properties of its imidazole ring that can be seen as a combination of those of pyridine and pyrrole (Richaud *et al.*, 2011). As shown in Figure 1.1, the N–H nitrogen atom has an acid character (pK = 14.52) which brings to mind (and surpasses) that of pyrrole (pK = 17.51), while the multiply bound nitrogen atom resembles that of pyridine, while being more basic (pK = 7.01 *vs.* pK = 5.20 for pyridine). Besides its amphoteric nature, tautomerism is another feature characterizing the imidazole ring that indeed exists in two identical tautomers showing such rapid equilibria that they cannot be separated or isolated. The imidazole basicity renders it an optimal buffer at physiological pH and several studies showed that the buffering capacity in human tissues is mostly related to the histidine concentration (Li and Hong, 2011).

With regard to the coordination abilities, imidazole is considered as a moderate σ-donor due to the lone pair located at the basic nitrogen atom and

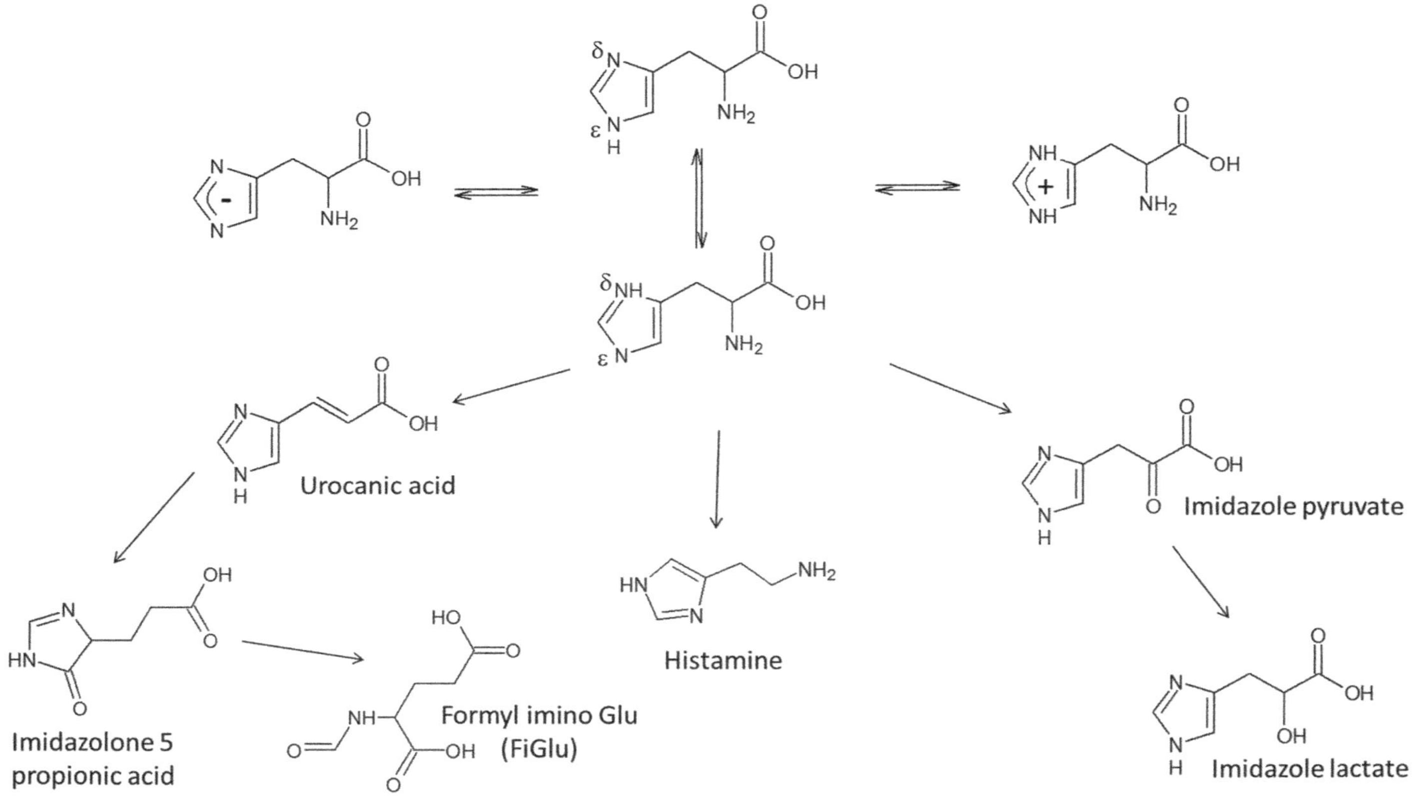

Figure 1.1 Main equilibria characterizing protonation and tautomerization of histidine. The two possible tautomers are called τ tautomer when the Nε atom is protonated and π tautomer when the Nδ atom is protonated. The three major catabolic pathways of histidine are also reported. They are catalyzed by histidinase, histidine decarboxylase and histidine transaminase.

a weak π-acceptor, albeit its π-excessive nature. Hence, it prevalently forms η^1–(σ-N) complexes that are favored with electron-rich metals in low-oxidation states with high-energy d-orbitals although in rare cases imidazole was found to behave as a π-donor.

Due to its electronic properties, imidazole undergoes electrophilic substitutions which can occur at the basic nitrogen atom and at the electron-rich C4 and C5 atoms. The imidazole reactivity towards electrophilic reagents is greater than in benzene but lower than in the five-membered heterocycles with only one heteroatom. The reactivity decreases for reactions requiring acidic conditions and is increased by electron-donating substituents. Nucleophilic substitutions can occur only when an appropriate leaving group is located at the electron deficient C2 atom. The imidazole ring is moderately susceptible towards nucleophilic substitutions and its reactivity is enhanced by electron-withdrawing substituents (Bhatnagar *et al.*, 2011).

With regard to the Michael addition (a reaction that plays a pivotal role for the carbonyl quenching of the histidine dipeptides), imidazole can act as a donor due to the nucleophilicity of its basic nitrogen atom (the so-called aza-Michael addition). However, the imidazole ring is not nucleophilic enough to react with α,β unsaturated carbonyls at neutral pH values and requires the presence of catalysts, as seen in the organocatalytic mechanisms where the Michael addition is promoted by the initial formation of more reactive enamines (Lakhdar *et al.*, 2012).

As seen for other heterocycles, imidazole can directly react with the hydroxyl radical giving the hydroxyimidazolyl radical anion that then undergoes a slow elimination of water to yield the dehydroimidazolyl radical that is reasonably more stable than the starting hydroxyl radical. Such a reaction can explain antioxidant effects of the imidazole-based compounds as seen in urocanic acid and imidazole-4-acetic acid that come from histidine metabolism (Hartman *et al.*, 1990).

1.2.2 Histidine Metabolism and Physiological Roles

Histidine is an essential amino acid; its *de novo* biosynthesis is highly conserved from archea to plants. While children have to obtain histidine from diet, adults are independent on the dietary sources due to both the low daily requirement and the protein turnover. Dairy, meat, poultry and fish are good sources of histidine as well as rice, wheat and rye. Furthermore, the intestinal flora can produce absorbable histidine.

As schematized in Figure 1.1, the histidine catabolism involves three major pathways. The first is based on the histidinase enzyme that converts histidine into urocanic acid by catalyzing the irreversible elimination of ammonia. By hydration and rearrangement, urocanic acid is transformed into imidazolone-5-propionate that in turn can yield glutamate *via* the *N*-formimino glutamate intermediate (FiGlu) and successive transformylation. Since the last step involves the transfer of the formyl group to tetrahydrofolate, a folate deficiency can be detected by the urinary accumulation of the FiGlu intermediate (Cooperman and Lopez, 2002). The second pathway plays a key role in signal transduction since it involves decarboxylation to give histamine by histidine decarboxylase (Ohtsu, 2010). Lastly, transaminase enzymes can convert histidine into imidazole pyruvate and then into imidazole lactate.

The major disorders in histidine metabolism are histidinemia and urocanase deficiency. The former is a benign recessive autosomal disorder due to a deficiency of histidinase resulting in a limited histidine metabolism and consequent accumulation in liver and blood. The latter is a rare normally benign recessive autosomal disorder caused by a deficiency of urocanase and involving a block in conversion of the urocanic acid and its consequent accumulation, while histidine levels remain substantially unchanged (Bender, 2012).

1.3 Proteinogenic Histidine Dipeptides

While considering the greater relevance of nonproteinogenic histidine-containing dipeptides (see Section 1.4), this section is focused on some important proteinogenic histidine dipeptides to frame carnosine and its derivatives in the context as broadly as possible.

For completeness, this section analyzes dipeptides in which histidine is in both the C-terminal and N-terminal position. Notably, the former usually exist in their linear form, while the latter are often present as cyclic diketopiperazines, which are rather stable derivatives, although either of the lactamic groups can hydrolyze resulting in two different dipeptides (Wang *et al.*, 2013). First, the specific roles of the proteinogenic dipeptides are discussed (see Section 1.3.1), while their common carbonyl quencher activity is analyzed in the following paragraph (see Section 1.3.2).

1.3.1 Specific Activity Profiles

The activity profiles of the histidine-containing dipeptides can be roughly subdivided into two major groups since they can be related to the discussed imidazole reactivity or to specific receptor binding. Besides the specific roles discussed below, it is worth remembering that all hydrolyzable histidine containing (di)peptides can be seen as reservoirs of histamine. Similarly, all histidine (di)peptides act as proton buffers due to the imidazole basicity.

With regard to the first group, almost all histidine dipeptides conserve the capacity to chelate metal ions. This feature is particularly relevant for the His–His dipeptide that chelates Cu(I) ions in a very favorable geometry (Himes *et al.*, 2007). Again, several studies investigated the capacity of Gly–His (Figure 1.2, **8**) to form complexes with metal ions, among which gold-based complexes have attracted great interest because of their potential use as anticancer agents. Notably, the metal chelation is often combined with the imidazole-based radical scavenging and determines the antioxidant profile exhibited by several histidine-containing dipeptides (Hartman *et al.*, 1990).

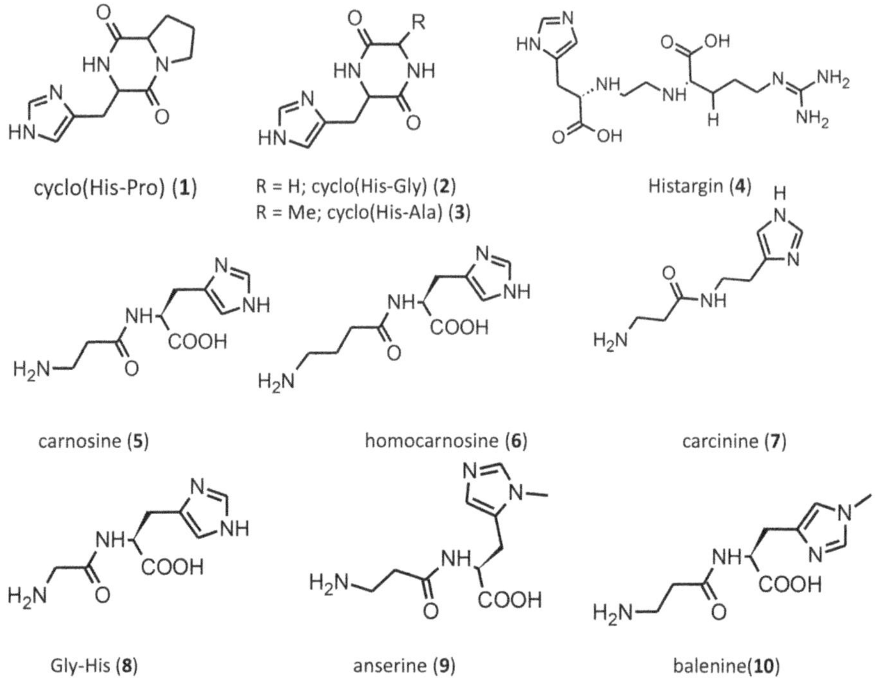

Figure 1.2 Molecular structure of some selected histidine-containing dipeptides.

Another feature connected to the imidazole reactivity and characterizing some histidine dipeptides is related to its catalytic role as a general base. Accordingly, Ser–His, and less markedly Ala–His, were found to catalyze the formation of a RNA phosphodiester bond, thus promoting the RNA polymerization. Interestingly, this feature might explain the nonenzymatic prebiotic formation of catalytic nucleic acids that are believed to have had a key role in the origin of life. Ser–His was also found to catalyze the formation of peptide bonds, as seen in short dipeptides and also this feature may have played a role for the origin of biomacromolecules (Wieczorek *et al.*, 2013).

The second group collects those dipeptides the biological profile of which is related to their interaction with well-defined biological targets. A first set of histidine-containing dipeptides was found to possess significant anxiolytic and sedative activities. For example, the dipeptide Met–His (aka, wheylin-1), which arises from the digestion of β-lactoglobulin by termolysin, was found to exert an anxiolytic-like activity in mice that is ascribable to activation of the $GABA_A$ receptor (Yamada *et al.*, 2014). Moreover, several cyclodipeptides were found to induce a significant prolongation of the pentobarbital narcosis due to a $GABA_A$ potentiation, as demonstrated for cyclo(His–Pro) (Suzuki *et al.*, 1981).

Cyclo(His–Pro) (Figure 1.2, **1**) is an endogenous dipeptide structurally related to a tyreotropin-releasing hormone. Besides the already mentioned $GABA_A$ potentiation, it elicits in the central nervous system multiple biological activities, which can be related to a presynaptic dopaminergic mechanism and involve a leptin-like function. At the gastrointestinal level, cyclo(His–Pro) was found to act as a gut peptide influencing the entero-insular axis and improving glycemic control (Minelli *et al.*, 2008). Notably, other histidine-containing dipeptides (such as His–Ala and His–Leu) were found to influence the glucose metabolism mainly through inhibition of the intestinal dipeptidyl peptidase-4 (DPP-4) (Nongonierma *et al.*, 2013).

Cyclo(His–Pro) has recently attracted great interest for its neuroprotective role that was first observed in traumatic injuries of the spinal cord and then confirmed in other models of experimental injuries of the nervous system. The involved mechanism still remains unclear even though recent studies suggest the contribution of the expression of small heat-shock proteins.

Another set of histidine-containing dipeptides were found to exert biological activities at the cardiovascular level. Among them, the Trp–His dipeptide was found to possess a vasorelaxant activity that is ascribable to its ability to decrease the intracellular [Ca^{++}] concentration by blocking a voltage-dependent l-type Ca^{++} channel (VDCC) (Wang *et al.*, 2010). Moreover, Trp–His was found to prevent atherosclerosis in apo E deficient mice by multiple (and still poorly understood) mechanisms that do not include the regulation of lipid metabolism (Matsui *et al.*, 2010). Furthermore, some dipeptides such as cyclo(His–Gly) (Figure 1.2, **2**) were found to inhibit thrombin-induced platelet aggregation.

Moreover, some histidine-containing dipeptides showed anticancer activity. For example, cyclo(His–Phe) displays a remarkable antitumor activity, as demonstrated by a great reduction of cell viability in cervical carcinoma cells. Again, cyclo(His–Gly) and, more markedly, cyclo(His–Ala) (Figure 1.2, **3**) show an inhibition of HeLa cells comparable to cisplatin at the same concentration, while the greater activity of the latter can be explained by increased lipophilicity that promotes permeation into cells and enhances its effects (Lucietto *et al.*, 2006).

Similarly, some histidine-containing dipeptides were found to exert significant antimicrobial activities. Among them, Trp–His and His–Arg represented relevant frameworks to design analogs that were found to be active against several Gram-negative and Gram-positive bacterial strains as well as against a fungal strain and to be devoid of cytotoxic effects (Sharma *et al.*, 2009). Finally, some histidine-containing dipeptides, such as γ-Glu–His, are kokumi molecules that enhance complex taste of the matured cheese (Toelstede *et al.*, 2009).

1.3.2 Carbonyl-Quencher Activity

As mentioned in Section 1.1, this chapter is mostly focused on the quenching activity of histidine-containing dipeptides towards reactive carbonyl species (RCS). Regardless of their source, RCS can be subdivided into the two major groups, namely α,β unsaturated carbonyls (*e.g.*, 4-hydroxynonenal and acrolein) and dicarbonyls (*e.g.*, glyoxal and methylglyoxal).

The enhanced reactivity of the former is mainly ascribable to their capacity to yield Michael-type adducts due to the marked electrophilicity of the β carbon atom that is combined with that of the carbonyl group, which maintains its ability to condense with suitable nucleophilic molecules. As seen with carnosine, there can be a crosstalk mechanism between the two electrophilic groups since a nucleophilic molecule can first react with the carbonyl group and the so-obtained intermediate can promote the Michael addition on the β carbon atom through an organocatalytic mechanism.

The notable reactivity of the α dicarbonyls is ascribable to the powerful activation of one electron-withdrawing carbonyl group on the other that renders them able to condense at the same time with more nucleophilic centers yielding crosslinked products as exemplified by the imidazolium-based adducts. Hence, RCS can be seen as strongly electrophilic compounds that can react with suitable nucleophilic compounds generating stable condensation products. The rational design of RCS scavengers should carefully take into account the reactivity of such carbonyl species with a view to maximizing their quenching, while avoiding unselective (and undesired) condensation reactions with physiological carbonyl compounds (Vistoli *et al.*, 2013a).

The first study analyzing the quenching activity of the histidine-containing dipeptides towards α,β unsaturated carbonyls was dated 1999 by Zhou and Decker (Zhou and Decker, 1999). They analyzed the consumption of *trans*-2-hexenal when incubated with a set of nucleophilic peptides. Albeit without determining the precise quenching mechanism, the study first revealed the remarkable quenching activity of carnosine and some analogs. The obtained results evidenced that the tested nonhistidine-containing dipeptides are understandably devoid of quenching activity, while the proteinogenic histidine-containing dipeptides show a clear quenching activity although, on average, it is less pronounced than that of carnosine. Again, single amino acids (histidine included) lack a significant quenching activity and the same holds true also for the imidazole ring, thus confirming that it is not nucleophilic enough to yield aza-Michael additions alone.

The same authors also examined the reactivity of the histidine-containing dipeptides towards hexanal, a saturated aldehyde that originates from lipid catabolism and can be seen as a prototype of endogenous unreactive carbonyl species useful in assessing the quenching selectivity towards RCS. The results revealed that carnosine is unable to stably trap hexanal, while some proteinogenic dipeptides showed a significant quenching activity, which is related to the lipophilicity of the dipeptide, the higher the lipophilicity the greater the reactivity (Zhou and Decker, 1999).

Since then, the quenching activity of the proteinogenic histidine-containing dipeptides has been rarely investigated, even though a recent study provided a rather systematic analysis of their quenching activity towards HNE and pyridoxal chosen as example of physiological nonreactive carbonyl compound (Vistoli *et al.*, 2013b). The obtained results confirmed that proteinogenic dipeptides are less reactive than carnosine although the study shed light on some dipeptides that are endowed with a significant reactivity combined with a notable selectivity since they do not trap pyridoxal (*e.g.* L-Lys–L-His–OMe and L-Tyr–D-His–OMe). The obtained results were also amenable to QSAR analyses that evidenced that flexibility and polarity play key roles in determining quenching activity and selectivity.

1.4 Nonproteinogenic Histidine Dipeptides: Carnosine and Its Derivatives

Among the nonproteinogenic histidine-containing dipeptides, carnosine and its analogs represent by far the most important group although other dipeptides have been reported as exemplified by histidinomethylalanine that was isolated in milk products (Walter *et al.*, 1994) or by histargin (Figure 1.2, **4**) which is produced by actinomyteces and is able to selectively inhibit carboxypeptidase B (Umezawa *et al.*, 1984). As an aside, diphthamide, the key diphtheria toxin effector, can also be seen as an atypical dipeptide in which the two residues are linked through the central imidazole ring.

Although a complete understanding of their physiological and biochemical role is still missing, the reported activities of carnosine (βAla–His, Figure 1.2, **5**) and its derivatives can be mostly related to imidazole reactivity and involve proton buffering, metal chelation, antioxidant and carbonyl quenching (Boldyrev *et al.*, 2013). Regarding the carbonyl quenching, carnosine shows a remarkable and selective activity towards α,β unsaturated carbonyls and this can be explained by its concerted quenching mechanism (see below). By contrast, the quenching activity of carnosine towards α dicarbonyls is still debated and appears to be mostly ascribable to the reactivity of its primary amine without involving specific and concerted mechanisms.

Nevertheless, carnosine cannot have concrete medicinal applications due to its unfavorable pharmacokinetic profile. Indeed, it is actively absorbed at the gastrointestinal level by the hPepT1 peptide transporter and is rapidly hydrolyzed by serum carnosinase (CN2, EC 3.4.13.20), a specific dipeptidase found in plasma and in the brain (Bellia *et al.*, 2014). Starting from the two constituent amino acids, carnosine can be resynthesized only in those tissues possessing carnosine synthase (EC 6.3.2.11), an ATP-dependent enzyme belonging to the ATPGD1 (ATP-Grasp domain containing protein 1) superfamily. This implies that only skeletal and heart muscle and certain brain regions (such as the olfactory bulb) can contain carnosine at significant concentrations.

When considering such an unsuitable pharmacokinetic profile, the extended set of carnosine derivatives proposed over the last few years come as no surprise. For simplicity, the attention here is focused on the natural carnosine derivatives which are exploited to derive the key principles that can guide the rational design of bioavailable, efficient and selective RCS quenchers by optimizing both quenching activity (Section 1.4.1) and pharmacokinetic profile (Section 1.4.2).

1.4.1 Optimizing the Carnosine's Carbonyl Quenching: A Mechanistic Perspective

The mechanism by which carnosine reacts with α,β unsaturated carbonyls is a clear example of multistep concerted mechanisms (Aldini *et al.*, 2013). The first step involves the reversible condensation between the carbonyl group and the carnosine primary amine to yield the corresponding unsaturated imino intermediate. Only α,β unsaturated carbonyls can then undergo the second step involving the intramolecular Michael addition between the imidazole ring and the acceptor β-carbon atom. Due to the reversibility of the initial imino adducts, carnosine cannot stably quench physiological carbonyl, thus reaching the required selectivity. The substantial inactivity of the histidine alone (see above) emphasizes the catalytic role of the imino intermediate that increases the electrophilic character of the acceptor β-carbon atom and approaches the two reactive centers constraining them in a pose conducive to the addition.

With regard to carnosine derivatives (see Table 1.1), *N*-acetyl carnosine is present in mammalian cardiac and skeletal muscles as well as in the central nervous system of vertebrates at significant concentrations. *N*-Acetyl carnosine is resistant to the effect of carnosinase hydrolysis and should liberate carnosine with difficulty due to its marked polarity. When considering the described quenching mechanism, it is not surprising that the *N*-acetylation completely removes the quenching activity towards HNE and this finding can be seen as a further confirmation of the role played by the imino intermediate. By contrast, *N*-acetyl carnosine maintains the chemical properties directly related to the imidazole ring, namely antioxidant, proton-buffering, and metal-chelating capabilities and as such it can have moderate protective effects, as confirmed by some animal models (Boldyrev *et al.*, 2013).

Homocarnosine (GABA–His, Figure 1.2, **6**) is an endogenous dipeptide distributed in cerebral regions and cerebrospinal fluid where it can act as a reservoir for the neurotransmitter GABA. Homocarnosine is recognized by both peptide transporters and human serum carnosinase for which it behaves

Table 1.1 Comparison of some biochemical properties of carnosine and derivatives.

Compound	p*K* amine	HNE quenching[a]	CN2 hydrolysis	hPepT1 transport
Carnosine	9.33	***	+	+
N-Acetylcarnosine	—	—	–	n.d.
Homocarnosine	9.89	*	+	+
Gly–His	8.05	*	n.d.	+
Anserine	9.37	**	+	+
Carcinine	n.d.	**	–	n.d.
D-Carnosine	9.33	***	–	–

[a]Marks summarize quenching activity: *** equal carnosine; ** slightly lower; * markedly lower.

as a dose-dependent inhibitor (Peters *et al.*, 2010). Besides the properties directly related to the imidazole ring and exemplified by its capacity to form stable copper(II) complexes, it maintains a quenching activity that is about one half that of carnosine (Boldyrev *et al.*, 2013). Such a decrease can be, at least partly, explained by the higher basicity of its primary amine due to the increased distance from the peptide bond. Indeed, the imino intermediate can be formed only by the neutral amino group, thus suggesting that the initial intermediate could be promoted by lowering the amino basicity. Nevertheless, such an approach should be cautiously pursued since weakly basic amino groups tend to give stable imino adducts thus depleting physiological carbonyls. This suggests that amine basicity has to remain in a well-defined range that optimizes the quenching activity without yielding irreversible Schiff bases.

Carcinine (βAla–Histamine, Figure 1.2, **6**) is present in the cardiac muscle of several mammals and can be seen as an inactive conjugate of histamine, although pharmacological studies revealed that it behaves as a selective H3 antagonist that might be useful in the treatment of neuronal diseases, such as epilepsy and locomotor or cognitive deficits. Moreover, the lack of the carboxyl group renders carcinine resistant to carnosinase, while its recognition by peptide transporters has not yet been determined. Besides the properties related to the imidazole ring, carcinine maintains a good quenching activity towards HNE, as evidenced by its capacity to inhibit and even reverse the carbonylation processes of retinal proteins, thus preventing vision loss in progressive retinopathies (Marchette *et al.*, 2012). This finding suggests that the carboxyl group, which is not involved in the quenching mechanism, might be modified to improve the pharmacokinetic profile.

As seen with many proteinogenic dipeptides, the already described nor-derivative Gly–His dipeptide (Figure 1.2, **8**) is less active than carnosine and this can be explained in terms of reduced flexibility that characterizes the imino intermediate and hampers the productive approaching of the imidazole ring to the β-carbon atom. Flexibility indeed plays a pivotal role in the mechanism of the HNE quenching by allowing the imino intermediate to assume catalytically productive conformations. This suggests that, while conserving the same reactive centers, the quenching activity could be improved by modifications that favor a more folded conformational profile. This prompted the design of aryl derivatives of carnosine that should favor the folded conformations by π–π stacking. Among them, *O*-methyl-β3-homotyrosine–D-His showed an increased folding degree and a quenching activity about three times that of carnosine (Vistoli *et al.*, 2012).

Moreover, the activity profile of Gly–His allows another interesting consideration since it shows a reduced basicity compared to carnosine but a selectivity comparable with that of carnosine (Vistoli *et al.*, 2013b). This means that its primary amine is not weakly basic enough to generate stable imino adducts and this suggests that it should be possible to design carnosine derivatives that possess less basic and more reactive amino groups without undermining the required selectivity.

The two *N*-methylated carnosine derivatives (anserine, βAla–3MethylHis, Figure 1.2, **9** and balenine, βAla–1MethylHis, Figure 1.2, **10**) show a different distribution, since anserine is found in nearly all mammalian orders while balenine is significantly present only in marine mammals (*Cetacea*). In both cases, they are present in skeletal and cardiac muscle where they contribute to the buffering activity, while in the central nervous system they are neuroprotective and putative neurotransmitters, even though their precise role is still unclear. They are synthesized by specific carnosine *N*-methyl transferases (EC 2.1.1.22) (Drozak *et al.*, 2013) and are hydrolyzed by serum carnosinase. Besides the properties directly related to the imidazole reactivity, they are able to quench HNE, as experimentally confirmed only for anserine that shows an activity comparable with that of carnosine, a result easily explainable since the *N*-methylation does not involve the basic nitrogen atom (Boldyrev *et al.*, 2013).

The recently investigated D-carnosine enantiomer (Gly–D-His) showed a quenching activity superimposable to that of carnosine, an expected result when considering that two enantiomers have to possess an identical reactivity. More importantly, D-carnosine is completely resistant to the carnosinase effect, but it is no longer recognized by peptide transporters and is too polar to be passively absorbed. This prompted the design of lipophilic prodrugs among which D-carnosine octyl ester was found to be the most promising derivative, as demonstrated in animal studies (Aldini *et al.*, 2014). D-Carnosine represents a meaningful example of the challenges posed by the achievement of bioavailable derivatives that render their rational design a sort of balance game.

1.4.2 Optimizing the Carnosine Pharmacokinetic Profile: A Balance Game

A bioavailable carnosine derivative has to lose the capacity to interact with serum carnosinase while conserving the recognition by peptide transporters and this is clearly a difficult task since these biological targets stabilize rather similar interaction patterns. To better clarify such recognition processes, human serum carnosinase and hPepT1 were modeled by homology techniques and utilized to simulate their interaction with carnosine. Figure 1.3 shows the putative complex between carnosine and serum carnosinase and reveals that both substrate-charged termini are involved in key ionic contacts (Vistoli *et al.*, 2006), while the nitrogen atoms of the imidazole ring stabilize H-bonds which explain the specificity of the enzyme towards histidine-containing dipeptides (Bellia *et al.*, 2014). Apart from modifying the peptide bond (a strategy rarely pursued until now), this result suggests that a stable carnosine derivative could be obtained by masking the carboxyl as confirmed by carcinine. The precise interaction pattern stabilized by carnosine can also explain why D-carnosine was found to be resistant.

Figure 1.4 depicts the putative complex between carnosine and hPepT1 and evidences the polar interactions stabilized by both charged termini, while the

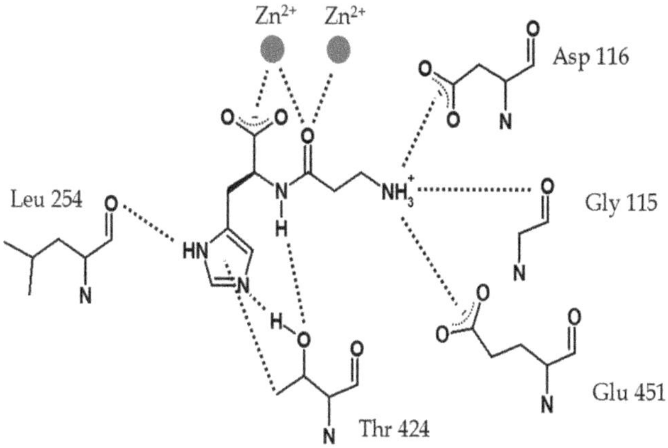

Figure 1.3 Main interactions stabilizing the putative complex between carnosine and human serum carnosinase. It is worth noting the key interactions stabilized by the zinc ions with both the carboxylate and the carbonyl group, the ammonium interacts with Asp116 and Glu451, while the imidazole ring elicits H-bonds with Leu254 and Thr424. The figure is reprinted from Vistoli *et al.* (2006) with kind permission from ACS.

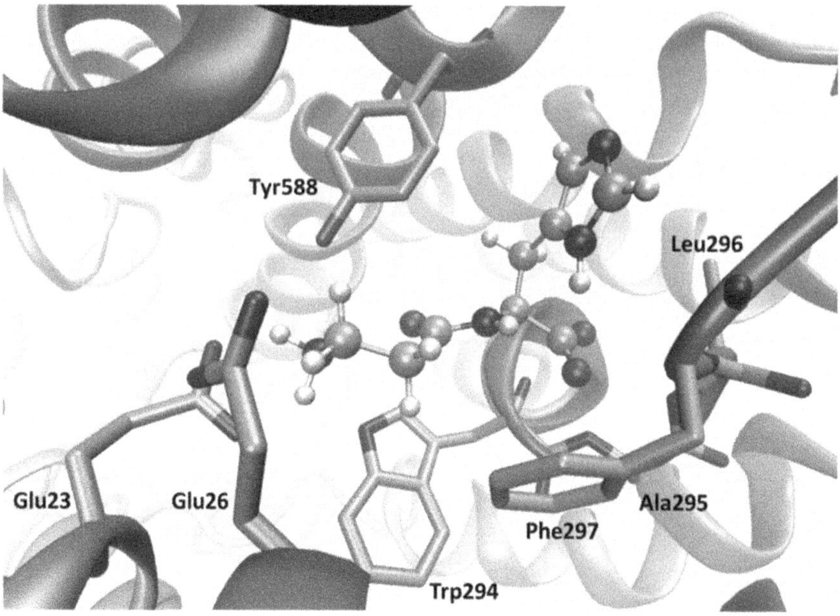

Figure 1.4 Main interactions stabilizing the putative complex between carnosine and human peptide transporter hPepT1. One may note the interactions stabilized by the ammonium head that involve a strong H-bond with Tyr588 plus ion pairs with Glu23 and Glu26. The carboxyl terminus elicits H-bonds with the backbone atoms of Ala295, Leu296 and Phe297 and the imidazole ring stabilizes π–π interactions with Tyr588. (Adapted from Vistoli *et al.* (2012), with permission from Springer Science + Business Media.)

imidazole ring mostly stabilizes hydrophobic contacts (Pedretti *et al.*, 2008). This result rationalizes the enantioselectivity of the binding and underlines the limited role of the backbone groups that appear not to be mandatory for the transport. Modifications of the carboxyl group are better tolerated than those of the amine and this suggests that carnosine derivatives modified at the C-terminus might yet be recognized by hPepT1.

Taken together, the structure–activity relationships derived by the carnosine derivatives and concerning both their quenching activity and their pharmacokinetic profile can be schematized by Figure 1.5 and summarized as follows (Vistoli *et al.*, 2012):

(1) The amino group and the imidazole ring are required for quenching activity and could be replaced by more reactive moieties provided that the obtained derivatives maintain an optimal selectivity without depleting physiological carbonyls.
(2) The peptide bond and the carboxyl group can be vastly modified without altering the quenching activity. Moreover, such modifications should prevent the hydrolysis by serum carnosinase without affecting the active transport.
(3) Unfortunately the inversion of the histidine chiral center has a markedly detrimental effect on the recognition by peptide transporters.
(4) The β-alanine skeleton can be modified to favor an optimal conformational profile of the imino intermediate. The available data for homocarnosine and proteinogenic dipeptides suggest that the ethylene linker represents a good balance to optimize basicity and flexibility.

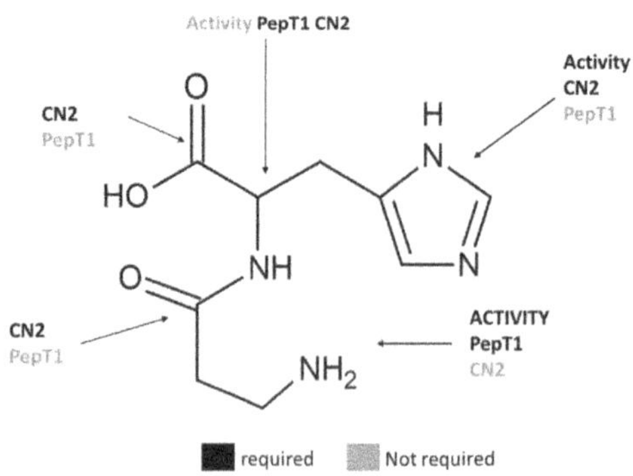

Figure 1.5 Schematic representation of the major structure–activity relationships (SARs) as derived by analyzing the reported carnosine derivatives. (Adapted from Vistoli *et al.* (2012), with permission from Springer Science + Business Media.)

1.5 Summary

As discussed in an excellent recent review (Boldyrev *et al.*, 2013), the comparison of the biochemical roles of the endogenous histidine-containing dipeptides reveals that most of them (*e.g.* metal-chelating, buffering, radical scavenging) are connected with the reactivity of the imidazole ring and are shared by any histidine-containing peptide and even by the single histidine amino acid. Therefore, carbonyl quenching represents the sole carnosine property that involves the imidazole ring but also requires the crucial contribution of the β-alanine residue and cannot be elicited by histidine alone.

Such a consideration possesses two relevant implications. First, carbonyl quenching is the only feature that shows marked differences among the natural histidine-containing dipeptides and allows structure–activity relationships to be derived. Secondly and although the complete biological role of carnosine is not yet clarified, the relevance of carbonyl quenching might explain why a complex and energy-consuming molecular system has evolved to synthesize, modulate, transport, and degrade carnosine (and its derivatives) across the organism. Future studies might clarify whether carnosine possesses additional (receptor-mediated?) biochemical properties that can further justify why the carnosine system evolved.

Summary Points

- The chapter is focused on the role of histidine and histidine-containing dipeptides.
- The reactivity of histidine is mostly ascribable to its imidazole ring and includes proton-buffering, radical scavenging and metal chelation.
- In proteins, histidines are often involved in enzymatic machineries as in the well-known catalytic triad.
- The histidine-containing dipeptides can be subdivided into proteinogenic and nonproteinogenic dipeptides.
- The proteinogenic dipeptides show multifaceted biological activities including sedative, neuroprotective, vasorelaxant, antiaggregant, antitumoral and antimicrobial properties.
- Among the nonproteinogenic dipeptides, carnosine plays a key role, although its biological profile is still debated.
- Among the investigated biological properties of carnosine, the capacity to scavenge reactive carbonyl species appears to only one requiring the contribution of β-alanine.
- Carnosine shows a selective quenching towards α,β unsaturated carbonyls that are trapped by a concerted mechanism that culminates in the Michael addition.
- Carnosine derivatives show different quenching activity profiles that allow structure–activity relationships to be developed.

Key Facts of Carbonyl Quenching

1. Reactive carbonyl species (RCS) coming from the oxidative catabolism of sugar and lipids can condense with suitable nucleophilic endogenous compounds generating the advanced glycoxidation end products (AGEs) and the advanced lipoxidation end products (ALEs).
2. Several studies emphasize the correlations and the pathogenic mechanisms between the amount of AGEs/ALEs and the progression of different diseases. Similarly, the beneficial effects of compounds able to neutralize AGE/ALE are now well-described.
3. Among the possible inhibition mechanisms, direct scavenging appears the most effective one since it converts RCS into nonreactive and easily excretable products.
4. Among the known carbonyl quenchers, carnosine appears the most promising compound due to its remarkable activity combined with a satisfactory selectivity since it is unable to deplete physiological carbonyls.
5. When considering the unsuitable pharmacokinetic profile of carnosine, the notable number of derivatives reported in the literature in the last few years comes as no surprise.

Definitions of Words and Terms

Tautomers are isomers of an organic molecule that dynamically interconvert by a chemical reaction called tautomerization. They differ only for thc position of a proton combined with a switch between adjacent single and double bonds. Due to the rapid interconversion, tautomers cannot be separated or isolated.

Antioxidant is a molecule able to donate an electron to a rampaging free radical and neutralize it, thus delaying or inhibiting the so-induced cellular damage.

Proton buffering involves ionizable compounds with ionization constants reasonably close to the physiological pH that can act as buffers in *in vivo* conditions. Thus, the imidazole ring with its pK around 7 renders histidine the major buffering system *in vivo*.

Metal chelation involves the formation of two or more coordinate bonds between a ligand and a metal ion. The imidazole ring shows a significant capacity to elicit coordinate bonds and histidine was found to chelate metal ions in simple peptides and in complex metalloproteins.

Reactive carbonyl species (RCS) are electrophilic compounds that come from the oxidative catabolism of sugars and lipids. They can be subdivided into α,β unsaturated carbonyls (*e.g.* acrolein or hydroxynonenal) and dicarbonyls (*e.g.* glyoxal or malondialdehyde). RCS can condense with nucleophilic sites within proteins, nucleic acids or phospholipids forming covalent adducts named AGE or ALE, depending on their biochemical source.

AGE Covalent adducts formed by RCS coming from the oxidative catabolism of sugars. They can be also generated by the direct condensation between reducing sugars and nucleophilic residues (glycation) *via* the so-called Maillard reaction.

ALE Covalent adducts formed by RCS coming from the oxidative catabolism of lipids. With regard to α,β unsaturated carbonyls, condensation reactions can involve both the carbonyl group and the unsaturated β carbon atom that can give Micheal additions with suitable nucleophilic groups.

Carnosine is a nonproteinogenic histidine-containing dipeptide. Among its various biological activities, the carbonyl quenching appears to be the most interesting one and leads to the generation of many derivatives designed to be selective, bioavailable and effective RCS scavengers.

hPepT1 is a peptide transporter belonging to the POT (proton oligopeptide transporter) family. It modulates intestinal absorption and renal reabsorption of di- and tripeptides in a proton-dependent way by a co-transport mechanism.

Serum carnosinase is a hydrolytic metalloenzyme found in plasma and is responsible for the rapid and selective digestion of the absorbed carnosine. It is expressed also in CNS where it can hydrolyze homocarnosine, thus modulating the release of GABA in the brain.

List of Abbreviations

AGE	Advanced glycoxidation end-products
ALE	Advanced lipoxidation end-products
ATPGD1	ATP-Grasp domain containing protein 1
CN2	Serum carnosinase
DPP-4	Dipeptidyl peptidase-4
FiGlu	*N*-Formimino glutamate
GABA	Gamma amino butyric acid
HCD	Histidine-containing dipeptide
HNE	(±)-4-Hydroxy-2-*trans*-nonenal
hPepT1	Human peptide transporter 1
QSAR	Quantitative structure–activity relationship models
RCS	Reactive carbonyl species
VDCC	Voltage-dependent l-type Ca^{++} channel

References

Aldini, G., Carini, M., Yeum, K. J. and Vistoli, G., 2014. Novel molecular approaches for improving enzymatic and nonenzymatic detoxification of 4-hydroxynonenal: toward the discovery of a novel class of bioactive compounds. *Free Radical Biology & Medicine*. 69: 145–156.

Aldini, G., Vistoli, G., Stefek, M., Chondrogianni, N., Grune, T., Sereikaite, J., Sadowska-Bartosz, I. and Bartosz, G., 2013. Molecular strategies to prevent, inhibit, and degrade advanced glycoxidation and advanced lipoxidation end products. *Free Radical Research*. 47(suppl. 1): 93–137.

Attwood, P. V., Piggott, M. J., Zu, X. L. and Besant, P. G., 2007. Focus on phosphohistidine. *Amino Acids*. 32: 145–156.

Bellia, F., Vecchio, G. and Rizzarelli, E., 2014. Carnosinases, their substrates and diseases. *Molecules*. 19: 2299–2329.

Beckman-Sundh, U., Ek, B., Zetterqvist, O. and Ek, P., 2011. A screening method for phosphohistidine phosphatase 1 activity. *Upsala Journal of Medical Sciences*. 116: 161–168.

Bender, D., 2012. *Amino Acid Metabolism*. John Wiley & Sons.

Bhatnagar, A., Sharma, P. and Kumar, N., 2011. A review on imidazoles: their chemistry and pharmacological potentials. *International Journal of PharmTech Research*. 3: 268–282.

Boldyrev, A. A., Aldini, G. and Derave, W., 2013. Physiology and pathophysiology of carnosine. *Physiological Reviews*. 93: 1803–1845.

Cooperman, J. M. and Lopez, R., 2002. The role of histidine in the anemia of folate deficiency. *Experimental Biology and Medicine*. 227: 998–1000.

Drozak, J., Chrobok, L., Poleszak, O., Jagielski, A. K. and Derlacz, R., 2013. Molecular identification of carnosine *N*-methyltransferase as chicken histamine *N*-methyltransferase-like protein (hnmt-like). *PLoS One*. 8: e64805.

Fisher, Z., Hernandez Prada, J. A., Tu, C., Duda, D., Yoshioka, C., An, H., Govindasamy, L., Silverman, D. N. and McKenna, R., 2005. Structural and kinetic characterization of active-site histidine as a proton shuttle in catalysis by human carbonic anhydrase II. *Biochemistry*. 44: 1097–1105.

Hartman, P. E., Hartman, Z. and Ault, K. T., 1990. Scavenging of singlet molecular oxygen by imidazole compounds: high and sustained activities of carboxy terminal histidine dipeptides and exceptional activity of imidazole-4-acetic acid. *Photochemistry and Photobiology*. 51: 59–66.

Himes, R. A., Park, G. Y., Barry, A. N., Blackburn, N. J. and Karlin, K. D., 2007. Synthesis and X-ray absorption spectroscopy structural studies of Cu(I) complexes of histidylhistidine peptides: the predominance of linear 2-coordinate geometry. *Journal of the American Chemical Society*. 129: 5352–5353.

Kavanagh, K. and Dowd, S., 2004. Histatins: antimicrobial peptides with therapeutic potential. *Journal of Pharmacy and Pharmacology*. 56: 285–289.

Lakhdar, S., Baidya, M. and Mayr, H., 2012. Kinetics and mechanism of organocatalytic aza-Michael additions: direct observation of enamine intermediates. *Chemical Communications*. 48: 4504–4506.

Li, S. and Hong, M., 2011. Protonation, tautomerization, and rotameric structure of histidine: a comprehensive study by magic-angle-spinning solid-state NMR. *Journal of the American Chemical Society*. 133: 1534–1544.

Lucietto, F. R., Milne, P. J., Kilian, G., Frost, C. L. and Van De Venter, M., 2006. The biological activity of the histidine-containing diketopiperazines cyclo(His–Ala) and cyclo(His–Gly). *Peptides*. 27: 2706–2714.

Marchette, L. D., Wang, H., Li, F., Babizhayev, M. A. and Kasus-Jacobi, A., 2012. Carcinine has 4-hydroxynonenal scavenging property and neuroprotective effect in mouse retina. *Investigative Ophthalmology & Visual Science*. 53: 3572–3583.

Matsui, T., Sato, M., Tanaka, M., Yamada, Y., Watanabe, S., Fujimoto, Y., Imaizumi, K. and Matsumoto, K., 2010. Vasodilating dipeptide Trp–His can prevent atherosclerosis in apo E-deficient mice. *British Journal of Nutrition*. 103: 309–313.

Minelli, A., Bellezza, I., Grottelli, S. and Galli, F., 2008. Focus on cyclo(His–Pro): history and perspectives as antioxidant peptide. *Amino Acids*. 35: 283–289.

Nongonierma, A. B., Mooney, C., Shields, D. C. and Fitzgerald, R. J., 2013. Inhibition of dipeptidyl peptidase IV and xanthine oxidase by amino acids and dipeptides. *Food Chemistry*. 141: 644–653.

Ohtsu, H., 2010. Histamine synthesis and lessons learned from histidine decarboxylase deficient mice. *Advances in Experimental Medicine and Biology*. 709: 21–31.

Pedretti, A., De Luca, L., Marconi, C., Negrisoli, G., Aldini, G. and Vistoli, G., 2008. Modeling of the intestinal peptide transporter hPepT1 and analysis of its transport capacities by docking and pharmacophore mapping. *ChemMedChem*. 3: 1913–1921.

Peters, V., Kebbewar, M., Jansen, E. W., Jakobs, C., Riedl, E., Koeppel, H., Frey, D., Adelmann, K., Klingbeil, K., Mack, M., Hoffmann, G. F., Janssen, B., Zschocke, J. and Yard, B. A., 2010. Relevance of allosteric conformations and homocarnosine concentration on carnosinase activity. *Amino Acids*. 38: 1607–1615.

Richaud, A., Barba-Behrens, N. and Méndez, F., 2011. Chemical reactivity of the imidazole: a semblance of pyridine and pyrrole? *Organic Letters*. 13: 972–975.

Sharma, R. K., Reddy, R. P., Tegge, W. and Jain, R., 2009. Discovery of Trp–His and His–Arg analogues as new structural classes of short antimicrobial peptides. *Journal of Medicinal Chemistry*. 52: 7421–7431.

Suzuki, T., Sakurada, T., Tadano, T., Kisara, K., Sasaki, Y. and Suzuki, K., 1981. Effect of cyclic dipeptides containing histidine on pentobarbital narcosis. *Journal of Pharmacobiodynamics*. 4: 377–379.

Toelstede, S., Dunkel, A. and Hofmann, T., 2009. A series of kokumi peptides impart the long-lasting mouthfulness of matured Gouda cheese. *Journal of Agricultural and Food Chemistry*. 57: 1440–1448.

Umezawa, H., Aoyagi, T., Ogawa, K., Iinuma, H., Naganawa, H., Hamada, M. and Takeuchi, T., 1984. Histargin, a new inhibitor of carboxypeptidase B, produced by actinomycetes. *Journal of Antibiotics (Tokyo)*. 37: 1088–1090.

Vistoli, G., Pedretti, A., Cattaneo, M., Aldini, G. and Testa, B., 2006. Homology modeling of human serum carnosinase, a potential medicinal target, and MD simulations of its allosteric activation by citrate. *Journal of Medicinal Chemistry*. 49: 3269–3277.

Vistoli, G., Carini, M. and Aldini, G., 2012. Transforming dietary peptides in promising lead compounds: the case of bioavailable carnosine analogs. *Amino Acids*. 43: 111–126.

Vistoli, G., De Maddis, D., Cipak, A., Zarkovic, N., Carini, M. and Aldini, G., 2013a. Advanced glycoxidation and lipoxidation end products (AGEs and ALEs): an overview of their mechanisms of formation. *Free Radical Research*. 47(suppl. 1): 3–27.

Vistoli, G., De Maddis, D., Straniero, V., Pedretti, A., Pallavicini, M., Valoti, E., Carini, M., Testa, B. and Aldini, G., 2013b. Exploring the space of histidine containing dipeptides in search of novel efficient RCS sequestering agents. *European Journal of Medicinal Chemistry*. 66: 153–160.

Walter, A. W., Henle, T., Haessner, R. and Klostermeyer, H., 1994. Studies on the formation of lysinomethylalanine and histidinomethylalanine in milk products. *Zeitschrift fuer Lebensmittel-Untersuchung und -Forschung*. 199: 243–247.

Wang, Y., Wang, P., Ma, H. and Zhu, W., 2013. Developments around the bioactive diketopiperazines: a patent review. *Expert Opinion on Therapeutic Patents*. 23: 1415–1433.

Wang, Z., Watanabe, S., Kobayashi, Y., Tanaka, M. and Matsui, T., 2010. Trp–His, a vasorelaxant di-peptide, can inhibit extracellular Ca^{2+} entry to rat vascular smooth muscle cells through blockade of dihydropyridine-like L-type Ca^{2+} channels. *Peptides*. 31: 2060–2066.

Wieczorek, R., Dörr, M., Chotera, A., Luisi, P. L. and Monnard, P. A., 2013. Formation of RNA phosphodiester bond by histidine-containing dipeptides. *ChemBioChem*. 14: 217–223.

Yamada, A., Mizushige, T., Kanamoto, R. and Ohinata, K., 2014. Identification of novel β-lactoglobulin-derived peptides, wheylin-1 and -2, having anxiolytic-like activity in mice. *Molecular Nutrition & Food Research*. 58: 353–358.

Zhou, S. and Decker, E. A., 1999. Ability of carnosine and other skeletal muscle components to quench unsaturated aldehydic lipid oxidation products. *Journal of Agricultural and Food Chemistry*. 47: 51–55.

Zhou, S. and Decker, E. A., 1999. Ability of amino acids, dipeptides, polyamines, and sulfhydryls to quench hexanal, a saturated aldehydic lipid oxidation product. *Journal of Agricultural and Food Chemistry*. 47: 1932–1936.

CHAPTER 2

Carnosine and Its Homologs in Foods

PIER GIORGIO PEIRETTI*[a] AND GIORGIA MEINERI[b]

[a]Institute of Science of Food Production, National Research Council, Via L. da Vinci 44, 10095 Grugliasco, Italy; [b]Department of Veterinary Sciences, University of Torino, Via L. da Vinci 44, 10095 Grugliasco, Italy
*E-mail: piergiorgio.peiretti@ispa.cnr.it

2.1 Main Functions of Carnosine and Its Homologs for the Protection of Human Health

The interest in carnosine (β-alanyl-L-histidine) and other imidazole dipeptides, such as anserine (β-alanyl-L-3-methyl-histidine), homocarnosine (γ-aminobutyryl-L-histidine) and balenine or ophidine (β-alanyl-L-3-methylhistidine), has recently increased.

Some studies have shown that carnosine and its related molecules are capable of acting as antioxidants, able to counter the particularly harmful chemical phenomena that occur in the body such as oxidation and glycoxidation (Quinn *et al.*, 1992).

The rate and extent of lipid oxidation in muscle tissue is governed by a number of factors, the most important being the level of polyunsaturated fatty acids that are found in the intramuscular fat tissue, in the cell membranes and in the intracellular lipid fraction (Buckley *et al.*, 1995).

The chemical molecules that can be considered as an index of glycoxidative stress are pentosidine, glycated hemoglobin and carbonyl groups (Singh

Food and Nutritional Components in Focus No. 8
Imidazole Dipeptides: Chemistry, Analysis, Function and Effects
Edited by Victor R Preedy

Published by the Royal Society of Chemistry, www.rsc.org

et al., 2001). However, the animal organism has some defense mechanisms against oxidation and glycation: one of these is represented by carnosine, a dipeptide present primarily in the skeletal muscle tissue of mammals (Babizhayev *et al.*, 2001).

Yen *et al.* (2002) reported a highly effective antioxidant of carnosine in models of liposomes and deoxyribose. They found that carnosine in high quantities (50 mg ml^{-1}) was effective in protecting against protein oxidation.

As well as in muscle tissue, carnosine and imidazole dipeptides are also found in great concentrations in the central nervous system; homocarnosine is mainly found in the cerebrospinal fluid. These imidazole dipeptides have been shown to be effective in preventing the damage and/or death of neurons caused by exposure to oxidative stress. Some studies have shown that tissue levels of carnosine decrease with age (Stuerenburg, 2000) and that the concentration of homocarnosine in human cerebrospinal fluid can decrease between 4- and 10-fold with advancing age (Huang *et al.*, 2005). Given the decline of carnosine during the aging processes, diets rich in carnosine are likely to become increasingly important for human aging.

The dietary intake of carnosine and its homologs may also be important in the presence of neuropathology; Ahmed *et al.* (2005) in fact have recently reported an association between Alzheimer disease and increased levels of protein glycation products in cerebrospinal fluid. Given that homocarnosine suppresses protein glycation, it can be assumed that a progressive decline in the concentration of this dipeptide could increase glycation in the cerebrospinal fluid.

The additional imidazole dipeptides are considered to be devoid of side effects; indeed, recent studies show that its antisenescent effects may enable carnosine to be used therapeutically in the treatment of generative diseases of the eye and in as yet poorly studied diseases, such as autism (Tsoi *et al.*, 2011). Tsai *et al.* (2010) reported that carnosine could provide antioxidant and anti-inflammatory defenses against the development of Parkinson's disease. Carnosine is responsible for a series of activities related to detoxification of the body from free radicals and byproducts of peroxidation, especially at the level of cell membranes, but recent studies have shown that these imidazole dipeptides show important physiological properties such as buffering capacity, protection of the cell membrane, the formation of inactive complexes with transition metals, the immune function of macrophage adjustment, anti-inflammatory and antibacterial activity, as well as acting as neurotransmitters or vasodilators, preventing ischemic cardiovascular damage, and finally modulating some enzymes (Aristoy *et al.*, 2004).

2.2 Diet and Human Health: Benefits of Carnosine and Its Homologs Contained in Meat Products

Carnosine and other imidazole dipeptides are of great nutritional interest in human diets as they are the major antioxidant molecules found in meat products for human consumption.

The carnosine present in the human body comes both from endogenous and from exogenous sources. The endogenous part is synthesized from the essential amino acids, alanines and histidines through the action of a specific synthetase, carnosinase, while diet provides all exogenous carnosine, and is capable of introducing approximately 50/250 mg per day (when at least one portion of beef, pork or chicken is consumed per day).

Carnosine is the major imidazole compound in mammalian tissues (Abe, 1995). Many studies have reported that the skeletal muscle tissue of most vertebrates contains high levels of carnosine and this has been determined in the commercial meat products (Davey, 1960; Abe and Okuma, 1995).

Meat and its byproducts is a potential source of carnosine for use as a food antioxidant. It has been suggested that molecular oxidation and glycation, as well as the pathologies associated with the consumption of sugars, deleterious aldehydes, and ketones, could be prevented and improved by carnivorous diets containing carnosine, anserine and homocarnosine (Brownlee, 2001).

Wu *et al.* (2003), studying the antioxidant activities of carnosine, anserine, some free amino acids and their combinations, found that anserine, at concentrations from 2.5 to 40 mM, exhibited the greatest reducing power among all the studied compounds. Their results revealed that anserine and carnosine are electron donors and could react with free radicals to convert them to more stable products and terminate radical chain reactions.

Boldyrev *et al.* (1988) demonstrated that carnosine and anserine could decrease membrane lipid oxidation by measuring the level of thiobarbituric acid reactive substances.

Carnosine and anserine have been found in the skeletal muscle tissue of most vertebrates (Davey, 1960), while carnosine is not available in plants; indeed, carnosine is practically absent from vegetarian diets. Krajcovicova-Kudlackova *et al.* (2002) found that levels of advanced glycation end products were higher in the plasma of vegetarians than in those of omnivorous people, while Hipkiss (2005) suggested that carnivorous diets contain a potential antiglycating agent, *i.e.* carnosine, whilst vegetarians may lack the intake of the dipeptide. The discovery of potentially protective components of meat is still poorly known among public opinion; indeed, vegetarian diets are usually considered to be very healthy. However, many doctors report that their patients with diabetes who eat meat seem to control the secondary complications of diabetes better than exclusively vegetarian patients.

2.3 Influence of Carnosine and Its Homologs on the Quality of Meat for Human Consumption

Carnosine may serve to promote the quality of meat for human consumption, preserving it from oxidation and favoring its organoleptic properties (color, taste). As mentioned in the previous paragraphs, skeletal muscle is susceptible to oxidative deterioration (Buckley *et al.*, 1995).

The process of lipid peroxidation is caused by a direct reaction between body fats and free radicals derived from molecular oxygen (Hsieh and Kinsella, 1989). The rate and extent of lipid oxidation in muscle tissue is governed by a number of factors, the most important being the level of polyunsaturated fatty acids, found in intramuscular fat tissue, in cell membranes and in the intracellular lipid fraction (Buckley *et al.*, 1995).

One of the most important causes of the deterioration of meat is lipid oxidation and the thiobarbituric acid reactive substances test seems to be an appropriate method for the routine assessment of secondary oxidation in meat samples (Fernández *et al.*, 1997).

Lipid peroxidation lowers the nutritional value of meat and adversely affects the flavor and taste of the meat (Addis and Warner, 1991). In fact, quality changes in pork and poultry meat are revealed through adverse changes in flavor, color, nutritional value and through the production of toxic and carcinogenic compounds (Jensen *et al.*, 1998).

In addition, the terminal oxidation products (malondialdehyde, 4-hydroxy-*trans*-2-nonenal and other advanced glycation end products) that are formed as a result of the lipid oxidation process in animal tissues have cytotoxic, mutagenic and carcinogenic effects (Sayre *et al.*, 2006). The lipid oxidation of meat, especially during their processing can be inhibited by the addition of antioxidants.

To prevent or retard oxidation reactions, different antioxidant systems are found in muscle tissue. These include α-tocopherol, and antioxidant enzymes such as glutathione peroxidase, superoxide dismutase, catalase and dipeptides containing histidine such as carnosine.

The contribution of α-tocopherol for the oxidative stability of skeletal muscle is largely influenced by diet. Dietary supplementation of tocopherol increases the muscle concentrations of α-tocopherol and inhibits both the oxidation of lipids and the deterioration of color (Meineri *et al.*, 2013).

Recently, it has been shown that the oxidative stability of skeletal muscle is largely influenced by dipeptides containing histidine, such as carnosine and anserine. These dipeptides, which are found in meat, are regarded as bioactive food components (Gardner *et al.*, 1991).

In recent years, interest in the potential antioxidant activity of carnosine in meat has increased considerably (Calvert and Decker, 1992; O'Neill *et al.*, 1999). These antioxidants, which are naturally present in skeletal muscle, may contribute to the intrinsic antioxidant potential of the muscle and may act as a natural antioxidant in processed meat products (Decker *et al.*, 1992). Numerous studies have shown that carnosine (0.5–1.5%) is able to inhibit lipid oxidation in turkey meat (Calvert and Decker, 1992), in pork (Decker and Crum, 1991), and in beef (Shantha and Decker, 1995). Moreover, extracts of muscle beef containing carnosine are able to inhibit the oxidation of lipids *in vitro* (Decker, 1995).

Imidazole dipeptides control and prevent lipid oxidation by inactivating catalysts and/or free radicals in the cell cytosol of muscle fibers, and their concentrations vary widely according to species and muscle type, as will be discussed in the next section. Zhou and Decker (1999) studied the ability of carnosine and various other related compounds to protect biomolecules from lipid oxidation products and to prevent the formation of aldehydes from monounsaturated and polyunsaturated fatty acids. Some researchers suggest that carnosine may be important to reduce the toxicity of the lipid oxidation products in biological systems and to minimize rancidity in meat products. Carnosine, indeed, inhibits the catalysis of lipid oxidation by pro-oxidant molecules of the aqueous phase such as iron and copper, and interferes with the propagation steps in lipid oxidation by peroxyl radicals (Chan and Decker, 1994; Decker, 1995).

Besides, in meat, imidazole dipeptides are linked to sensory perception and to the generation of some aromatic compounds in meat (Pereira-Lima *et al.*, 2000). Their function is to reduce the rancid taste and improve color stability (Jiménez-Colmenero *et al.*, 2010). Carnosine is highly efficient at maintaining an acceptable red color in ground beef during storage and has therefore been suggested as a useful antioxidant in foodstuffs of animal origin (Badr, 2007).

Regarding the effect of carnosine on the preservation of meat, Das *et al.* (2006) concluded that the shelf life of buffalo meat treated with 1.0% carnosine could be extended to 8 days under refrigerated storage, without undesirable changes in color or smell, while the control sample can be stored for a maximum of only 6 days. So carnosine, either as a component of meat or as an additive, may be useful in preventing the formation of anomalous flavors and has also been shown to increase the shelf life of meat and meat products when compared with untreated groups of meat products (Decker, 1995).

O'Neill *et al.* (1999) showed that a combination in the diet of α-tocopherol and carnosine results in greater antioxidant protection than α-tocopherol antioxidant alone. These antioxidants have also shown a synergistic effect on the inhibition of cholesterol oxidation and oxidation accelerated by salt; these results confirm that carnosine shows good potential as an antioxidant in poultry meat.

Djenane *et al.* (2004) have shown that the combination of carnosine with ascorbic acid has provided the best antioxidant protection in fresh meat products stored in a modified atmosphere.

Regarding any possible mechanisms of deterioration and decreases in carnosine content in meat, a recent study has shown that the level of carnosine is not affected by retention time (Bauchart *et al.*, 2006), but can be altered by cooking processes. The concentrations of these bioactive compounds may decrease with cooking, through the loss of water-soluble compounds such as taurine, carnosine and creatine in broth (Purchas *et al.*, 2004).

Peiretti *et al.* (2012) studied the effects of cooking methods on carnosine and its counterparts in beef and turkey meat. They observed that beef subjected to boiling suffers an approximately 50% loss of carnosine, probably due to the high water solubility of carnosine and its homologs; microwave cooking causes an average antioxidant loss of about 20%; cooking by microwave/grill reduced the amount of carnosine by about 10%. Regarding the anserine and homocarnosine content, the greater loss was observed for boiling (about 70%), while for the other cooking methods the value ranged from 30% to 70%. The data fluctuated more for turkey meat: reduction in carnosine was observed (25%) for microwave/grill and grill cooking at high temperatures. Similarly, anserine and homocarnosine content decreased slightly in the case of microwave/grill and cooking on the grill at high temperatures (2–7%) and by 10–30% in other cases. In the case of beef, it can be seen that boiling resulted in an average loss of 50% of carnosine on the same sample of raw meat. One probable cause of this depletion was the high water solubility of carnosine and related compounds, as reported by Nielsen *et al.* (2002). Studies showed that covalent adducts between carnosine and polypeptides are found in preparations of beef broth (Bauchart *et al.*, 2007).

In beef stew, a decrease of up to 30% in carnosine concentration has been reported (Kansci *et al.*, 1993).

Bauchart *et al.* (2006) studied the postmortem presence of low molecular weight peptides (<5 kDa) in the pectoral muscle of cattle after 14 days of storage at 4 °C and vacuum cooking for 90 min at 75 °C. They found that 89% of the amino acids in fresh muscle was made up of carnosine, anserine and glutathione, and also showed that the levels of these compounds were lower in cooked meat compared to fresh muscle.

A meat extract with a high carnosine and poor pro-oxidant content may have a potential as a natural antioxidant (Chan *et al.*, 1993) and as an alternative to the use of synthetic antioxidants such as butylated hydroxytoluene and butylated hydroxyanisole that, at high concentrations, have toxic properties and cannot therefore always be recommended (Kansci *et al.*, 1993). In conclusion, we can say that carnosine and anserine are endogenous antioxidants in the meat and have potential efficacy as food additives (Chan and Decker, 1994).

2.4 Content of Carnosine and Its Homologs in Animal Tissues

Imidazole dipeptides are widely distributed in tissues of vertebrate organisms: in particular, carnosine is an endogenous peptide present in high concentrations in muscle tissue, but also in other organs such as brain, liver, kidneys and skin (Quinn *et al.*, 1992). These compounds are also present in cardiac muscle, although at values 50–100 times lower than in skeletal muscle (O'Dowd *et al.*, 1988).

The metabolites of carnosine, such as anserine, balenine and carcinine are molecules generated by specific enzymes (Boldyrev and Abe, 1999). Velíšek *et al.* (2006) reported a brief examination of the main ways leading to the biosynthesis of most of the important peptides that occur in food.

The level of carnosine in the tissues is controlled by a number of enzymes that transform carnosine into related compounds, such as carcinine (by decarboxylation), *N*-acetyl (by acetylation), and anserine or balenine (by methylation) or in the corresponding amino acids by hydrolysis (Boldyrev and Abe, 1999). The hydrolysis is mainly due to tissue carnosinase and anserinase. These enzymes act on carnosine, anserine, and homocarnosine, and to a lesser extent on some other aminoacyl-L-histidine dipeptides (Pegova *et al.*, 2000).

Yamada *et al.* (2009) found the imidazole compound-related, named *N*-α-acetylhistidine, as well as histidine, carnosine, anserine and balenine, that are the major nonprotein nitrogenous constituents in the skeletal muscle of vertebrate animals.

Regarding the difference between the different species, mammals and birds have large amounts of carnosine and anserine and a small amount of balenine (Abe, 2000). Among terrestrial mammals, the carnosine content is greatest in cattle, pigs, equines, and many other mammals (see Table 2.1).

Some mammals, such as rabbits, sheep and goats, deer, and kangaroos, which show jumping locomotion, have higher levels of anserine than carnosine (Abe, 1995).

The contributions of anserine, carnosine and balenine were determined in various muscles and other tissues of pigs by Carnegie *et al.* (1982). Because the anserine/carnosine/balenine ratio in pork is very different from that of other species, it is possible to distinguish between original and imitation hams.

Wu and Shiau (2002) studied carnosine content in the muscles of six types of commercial chicken. The carnosine content ranged from 8 to 162 mg per 100 g of muscle, indicating that a lower carnosine intake could be obtained by replacing the edible meat with chicken bones.

Aquatic mammals, such as whales and dolphins, mainly contain balenine, a molecule that is rarely found in terrestrial mammals.

The class Aves (birds) shows both carnosine and anserine, though the latter compound is predominant. It is well known that the antioxidant capacity of anserine is about five times higher than that of carnosine and this may explain why birds live longer than mammals of similar size (Wu *et al.*, 2003).

Table 2.1 Concentration of imidazole dipeptides in the muscle of terrestrial mammals and birds. (Adapted from Abe and Okuma (1995).)

	Carnosine (μmol g^{-1} wet weight)	Anserine (μmol g^{-1} wet weight)	Balenine (μmol g^{-1} wet weight)
Ox leg	26.1	5.94	0.103
Pig leg	29.5	1.42	1.77
Horse leg	42.6	0.176	0.019
Deer leg	3.35	13.9	3.91
Chicken leg	5.70	17.1	0.055
Chicken breast	10.4	32.0	0.197
Turkey leg	4.53	20.5	0.077
Turkey breast	11.2	46.0	0.810

In fishes, van Waarde (1988) has classified the following five distribution patterns of imidazole dipeptides in skeletal muscle:

(1) migratory pelagic fish (tuna family) contain large amounts of histidine and anserine;
(2) the *Anguillidae* (eels) and *Acipenseridae* (sturgeon) families contain higher quantities of carnosine;
(3) salmonids (salmon) and *Gadidae* (cod), and the subclass *Elasmobranchii* (sharks and rays) contain mainly anserine;
(4) *Clupeidae* (herring and sardines) and cyprinids (carp) contain only histidine at high concentrations;
(5) *Pleuronectidae* (sole) and *Percidae* (perches) contain very low levels of imidazole compounds.

Carnosine is also contained in large quantities in the skin of frogs belonging to *Eleutherodactylus* and *Leptodactylus* species (Daly and Heatwole, 2007).

Among the reptiles, only those belonging to the suborder *Serpentes* (snakes) contain large amounts of balenine in tissues.

The turtles (order *Testudines*) contain higher amounts of carnosine in skeletal muscle, while the suborders *Sauria* (skinks), and *Crocodilia* (crocodiles) contain higher amounts of anserine.

Carnosine levels and sources of imidazole dipeptides also vary within animals of the same species, depending on the breed, sex (Abe and Okuma, 1995), localization of muscle tissue (Tian *et al.*, 2007), and type of muscle fibers (Rao and Gault, 1989).

As for the difference between animal breeds, Tian *et al.* (2007) showed that the carnosine content in the muscles of the Black-Bone Silky Fowl chicken breed was significantly higher than that of White Plymouth Rocks, indicating that Black-Bone Silky Fowl would be a better breed of chicken for carnosine supplements.

As regards gender comparisons, it has been found that the carnosine content in the flesh of females is higher than in that of males belonging to the same species (Chan and Decker, 1994).

Regarding the anatomical location of the muscle tissue, Maikhunthod and Intarapichet (2005) studied the content of carnosine in chicken breast and thigh muscles. They found that fresh chicken breast had a carnosine content twice to four times higher than the thigh.

There is also a correlation, within the same species, between the type of fibers and the carnosine content in the various muscles studied (Rao and Gault, 1989). Carnosine and anserine are found in higher concentrations in the "white", so-called "fast muscle fibers"; indeed, these muscle fibers in chicken contain over five times more anserine and carnosine than "red muscle fibers", considered "slow fibers" (Davey, 1960). This difference can be explained by the fact that the carnosine content differs according to the type of muscle metabolism; muscles with a generally anaerobic metabolism (such as *longissimus dorsi*) have a higher carnosine content than those with an oxidative metabolism (Rao and Gault, 1989). The carnosine content is significantly higher in muscles with white fibers than in muscles with red fibers, while no significant difference is seen in the anserine content. This result agreed with previous studies (Decker, 1995).

2.5 Human Nutritional Intakes of Carnosine and Its Homologs in Meat

The major nutritional source of carnosine for humans is the meat of livestock (Park *et al.*, 2005). In the literature, there are many studies that report the amount of carnosine and imidazole dipeptides in commercially fresh meat and meat products (Rao and Gault, 1989; Abe and Okuma, 1995).

Peiretti *et al.* (2011) found that the total carnosine-related content (carnosine + anserine + homocarnosine) in meat was as follows, in ascending order: beef < rabbit < pig < horse < chicken < turkey.

Regarding the anserine content, an almost inverse tendency was observed; meats with the highest carnosine content were those with the lowest levels of anserine (Table 2.2).

Anserine is specific to the skeletal muscles of some birds (particularly abundant in the muscle of chicken and turkey), fish and marine mammals, but is virtually absent in mammalian tissues, with the exception of the rabbit, where almost twice as much anserine is found compared to carnosine (Arihara, 2006).

Other studies confirm these data: indeed, Gil-Agusti *et al.* (2008) developed a procedure for liquid chromatography to determine the content of carnosine and anserine in samples of chicken meat (hamburger), turkey (breast and hamburger), pork (minced) and beef (hamburger and minced meat). They have shown that while the carnosine concentration was lowest in chicken and turkey (0.67 and 0.66 mg g^{-1}, respectively), the anserine content was highest (2.88 and 4.20 mg g^{-1}, respectively). Minced pork meat contained the least anserine and the highest concentration of carnosine (respectively 0.05 and 2.44 mg g^{-1}), while the anserine (0.20 and 0.32 mg g^{-1}, respectively) and carnosine content (1.75 and 1.96 mg g^{-1}, respectively) was intermediate in hamburger and minced beef (Table 2.3).

In rabbit, turkey, chicken, beef and horse meat, homocarnosine is present in a small percentage compared to anserine, while in pork loin it is more than twice that of anserine (Table 2.2).

Table 2.2 Concentration of imidazole dipeptides in the raw meats obtained from different species. (Adapted from Peiretti *et al.* (2011).)

	Carnosine (%w/w)	Anserine (%w/w)	Homocarnosine (%w/w)
Horse fillet	1.73–2.54	0.010–0.020	0.0037–0.0065
Pork loin	1.57–1.80	0.041–0.059	0.14–0.18
Beef fillet	1.22–1.49	0.098–0.280	0.014–0.024
Rabbit hind leg	0.36–0.46	1.12–1.45	0.027–0.054
Chicken breast	0.70–0.92	1.46–1.70	0.093–0.160
Turkey breast	0.36–0.79	2.11–2.46	0.12–0.19

Balenine is usually present in low amounts in the muscles of different species of mammals and chicken, while in marine mammals, such as whales, it is the predominant histidine dipeptide (Aristoy *et al.*, 2004).

Jones *et al.* (2011) analyzed 10 foods commonly consumed in the British diet (turkey, chicken, tuna, beef, lamb, pork, mackerel, rainbow trout and two types of shrimp). They showed that there can be a difference of 17 times in the imidazole dipeptide content of different foods; indeed, the highest anserine value was found in shrimps and the lowest values in pork, whereas the highest carnosine content was found in beef and the lowest in rainbow trout fillet (Table 2.4).

In conclusion, carnosine and anserine have gained importance for human nutrition due to their presence and availability in meat products (Hilton and Mizuma, 1993). An interesting study has shown the effect of dietary supplementation of imidazole dipeptides, increasing muscle carnosine content (Maynard *et al.*, 2001).

The concentration of carnosine in human tissues can thus be increased by dietary supplementation; indeed, carnosine has been shown to be absorbed apparently intact in an orally ingested meal (Gardner *et al.*, 1991). A recent study

Table 2.3 Concentration of imidazole dipeptides in commercial meat samples. (Adapted from Gil-Agusti *et al.* (2008).)

	Carnosine (mg g^{-1})	Anserine (mg g^{-1})
Chicken–turkey hamburger	0.67	2.88
Beef hamburger	1.75	0.20
Minced beef	1.96	0.32
Minced pork	2.44	0.05
Turkey breast	0.66	4.20

Table 2.4 Concentration of imidazole dipeptides in the dietary sources commonly consumed within the British diet. (Adapted from Jones *et al.* (2011).)

	Carnosine (mmol kg^{-1} DM)	Anserine (mmol kg^{-1} DM)	Total imidazole dipeptides (mmol kg^{-1} DM)
Honduran prawns whole	40.9	321.8	362.8
North Atlantic prawns whole	51.5	287.2	338.6
Turkey pectoral	46.7	176.3	223.0
Chicken pectoral	54.0	144.2	198.1
Tuna midfillet	4.7	137.4	142.1
Beef rump	70.4	19.3	89.7
Lamb rump	33.6	47.1	80.8
Pork rump	55.4	1.1	56.5
Mackerel midfillet	34.6	2.0	36.6
Rainbow trout midfillet	0.5	21.2	21.6

also demonstrated the bioavailability of carnosine in human plasma (Park *et al.*, 2005) after ingestion of beef, while Bauchart *et al.* (2007) investigated the occurrence of carnosine in the intestinal digesta of pigs after a meat meal. Chan and Decker (1994) showed that anserine and carnosine could potentially act as a bioactive antioxidant, since it has been suggested to be absorbed intact (without hydrolysis of the peptide bonds) through the intestine into the blood.

Furthermore, carnosine also affects the content of the digestive processes. Purchas *et al.* (2006) found that carnosine levels increase by 20% in cooked samples of meat as a result of pepsin digestion, and a further 8% following pancreatic digestion and suggest that the protein breakdown occurring during digestion results in the formation of dipeptide-like carnosine.

Summary Points

- This chapter focuses on imidazole dipeptides (carnosine, anserine, homocarnosine, balenine or ophidine) occurrence in food.
- Carnosine and its homologs are present in skeletal muscle cells and other tissues.
- These imidazole dipeptides have been extensively studied because of their physiological effects, including antioxidant properties or buffering activity and their potential therapeutic applications.
- Carnosine is responsible for a variety of activities related to the body detoxification from free-radical species.
- Total carnosine-related content (carnosine + anserine + homocarnosine) in meat was as follows, in ascending order: beef < rabbit < pig < horse < chicken < turkey.
- The contents of carnosine and its homologs also vary within animals of the same species, depending on the breed, sex, production system, localization of muscle tissue, and type of muscle fibers.
- The influence of carnosine and its homologs on the quality of meat for human consumption will also be considered.

Key Facts of Imidazole Dipeptides Occurrence in Food

1. The animal organism has some defense mechanisms against oxidation and glycation.
2. Carnosine and its related molecules are capable of acting as antioxidants.
3. Carnosine might act as a natural scavenger of dangerous reactive aldehydes and of endogenous molecules.
4. Carnosine, anserine, homocarnosine, balenine or ophidine are present primarily in the skeletal muscle tissue and in the central nervous system, but also in other organs such as brain, liver, kidneys and skin of vertebrate organisms.
5. The level of carnosine in the tissues is controlled by a number of enzymes.

6. Carnosine may serve to promote the quality of meat for human consumption, preserving it from oxidation and favoring its organoleptic properties.
7. The concentration of carnosine in human tissues can be increased by dietary supplementation.

Definitions of Words and Terms

Additive. Food additives are substances added to food to preserve flavor or enhance its taste and appearance; may be either of animal and vegetable origin.

Free radicals. Free radicals are atoms or groups of atoms with an odd (unpaired) number of electrons and can be formed when oxygen interacts with certain molecules. Once formed these highly reactive radicals can start a chain reaction, like dominoes. Their danger comes from the damage they can do when they react with important cellular components or the cell membrane.

Glycation. Glycation is a reaction that takes place when simple sugar molecules, such as fructose or glucose, become attached to proteins or lipid fats without the moderation of an enzyme. This process is called a Maillard reaction. The process of glycation results in the formation of early glycation products, such Shiff bases and Amadori products, which over time, undergo slow and complex rearrangements until the formation of advanced glycation end products.

Livestock. Livestock is a term used to defined domesticated animals raised in an agricultural setting to produce food or for work.

Molecular oxygen. Molecular oxygen is a molecule that is composed of two oxygen atoms, O_2, that has no color, odor or taste. The solar radiation with wavelengths less than 242 nm can break the molecular oxygen into two oxygen atoms, $O_2 \rightarrow O + O$.

Oxidative stress. Oxidative stress is a phenomenon that reflects an imbalance between the systemic manifestations of reactive oxygen species and a biological system's ability to detoxify or repair the resulting damage. Disturbances in the normal redox state of the cells may cause toxic effects through the production of peroxides and free radicals that damage all cell components, including proteins, lipids and nucleic acids.

Polyunsaturated fatty acids. Fatty acids that contain more than one double bond per molecule. This class includes many important compounds, such as some essential fatty acids.

Rancidity. Rancidity is the chemical decomposition of fats, oils and other lipids. When these processes occur in food, undesirable odors and flavors can result. Rancidification can also detract from the nutritional value of the food as some vitamins are highly sensitive to degradation.

Reactive aldehydes. Aldehydes are organic compounds that incorporate a carbonyl functional group and they are among the most reactive compounds of organic chemistry.

Thiobarbituric acid reactive substances. Thiobarbituric acid reactive substances are molecules formed as a byproduct of lipid peroxidation (*i.e.* as degradation products of fats), which can be detected by an assay using thiobarbituric acid as a reagent. Because reactive oxygen species have extremely short half-lives, they are difficult to measure directly. Instead, what can be measured are several products of the damage produced by oxidative stress, such as thiobarbituric acid reactive substances.

References

Abe, H., 1995. Histidine-related dipeptides: distribution, metabolism, and physiological function. In: Hochachka, P. W. and Mommsen, T. P. (ed.) *Biochemistry and Molecular Biology of Fishes Metabolic Biochemistry*. Elsevier, Amsterdam, vol. 4, pp. 309–333.

Abe, H. and Okuma, E., 1995. Discrimination of meat species in processed meat products based on the ratio of histidine dipeptides. *Nippon Shokuhin Kagaku Kogaku Kaishi*. 42: 827–834.

Abe, H., 2000. Role of histidine-related compounds as intracellular proton buffering constituents in vertebrate muscle. *Biochemistry (Moscow)*. 65: 757–765.

Addis, P. B. and Warner, G. L., 1991. The potential herbal aspects of lipid oxidation products in food. In: Aruoma, O. I. and Halliwell, B. (ed.) *Free radicals and Food Additives*. Taylor and Francis, London, pp. 77–119.

Ahmed, N., Ahmed, U., Thornalley, P. J., Hager, K., Fleischer, G. and Munch, G., 2005. Protein glycation, oxidation and nitration adduct residues and free adducts of cerebrospinal fluid in Alzheimer's disease and link to cognitive impairment. *Journal of Neurochemistry*. 92: 255–263.

Arihara, K., 2006. Strategies for designing novel functional meat products. *Meat Science*. 74: 219–229.

Aristoy, M. C., Soler, C. and Toldra, F., 2004. A simple, fast and reliable methodology for the analysis of histidine dipeptides as markers of the presence of animal origin proteins in feeds for ruminants. *Food Chemistry*. 84: 485–491.

Babizhayev, M. A., Deyev, A. I., Yermakova, V. N., Semiletov, Y. A., Davydova, N. G., Kurysheva, N. I., Zhukotskii, A. V. and Goldman, I. M., 2001. *N*-Acetylcarnosine, a natural histidine-containing dipeptide, as a potent ophthalmic drug in treatment of human cataracts. *Peptides*. 22: 979–994.

Badr, H. M., 2007. Antioxidative activity of carnosine in gamma irradiated ground beef and beef patties. *Food Chemistry*. 104: 665–679.

Bauchart, C., Rémond, D., Chambon, C., Patureau Mirand, P., Savary-Auzeloux, I., Reynès, C. and Morzel, M., 2006. Small peptides (<5 kDa) found in ready-to-eat beef meat. *Meat Science*. 74: 658–666.

Bauchart, C., Savary-Auzeloux, I., Patureau Mirand, P., Thomas, E., Morzel, M. and Rémond, D., 2007. Carnosine concentration of ingested meat affects carnosine net release into the portal vein of minipigs. *Journal of Nutrition*. 137: 589–593.

Boldyrev, A. A., Dupin, A. M., Pindel, E. V. and Severin, S. E., 1988. Antioxidative properties of histidine-containing dipeptides from skeletal muscles of vertebrates. *Comparative Biochemistry and Physiology, Part B: Biochemistry & Molecular Biology*. 89: 245–250.

Boldyrev, A. and Abe, H., 1999. Metabolic transformation of neuropeptide carnosine modifies its biological activity. *Cellular and Molecular Neurobiology*. 19: 163–175.

Brownlee, M., 2001. Biochemistry and molecular cell biology of diabetic complications. *Nature*. 414: 813–820.

Buckley, D. J., Morrissey, P. A. and Gray, J. I., 1995. Influence of dietary vitamin E on the oxidative stability and quality of pig meat. *Journal of Animal Science*. 73: 3122–3130.

Calvert, J. T. and Decker, E. A., 1992. Inhibition of lipid oxidation by combinations of carnosine and various antioxidants in ground turkey. *Journal of Food Quality*. 15: 423–433.

Carnegie, P. R., Hee, K. P. and Bell, A. W., 1982. Ophidine (β-alanyl-L-3-methylhistidine, 'balenine') and other histidine dipeptides in pig muscles and tinned hams. *Journal of the Science of Food and Agriculture*. 33: 795–801.

Chan, K. M., Decker, E. A. and Means, W. J., 1993. Extraction and activity of carnosine, a naturally occurring antioxidant in beef muscle. *Journal of Food Science*. 58: 1–4.

Chan, K. M. and Decker, E. A., 1994. Endogenous skeletal muscle antioxidants. *Critical Reviews in Food Science and Nutrition*. 34: 403–426.

Daly, J. W. and Heatwole, H., 2007. Occurrence of carnosine in skin of certain leptodactylid frog. *Cellular and Molecular Life Sciences*. 22: 764–765.

Das, A. K., Anjaneyulu, A. S. R. and Biswas, S., 2006. Effect of carnosine preblending on the quality of ground buffalo meat. *Food Chemistry*. 97: 531–538.

Davey, C. L., 1960. The effects of carnosine and anserine on glycolytic reactions in skeletal muscle. *Archives of Biochemistry and Biophysics*. 89: 296–302.

Decker, E. A. and Crum, A. D., 1991. Inhibition of oxidative rancidity in salted ground pork by carnosine. *Journal of Food Science*. 56: 1179–1181.

Decker, E. A., Crum, A. D. and Calvert, J. T., 1992. Differences in the antioxidant mechanism of carnosine in the presence of copper and iron. *Journal of Agricultural and Food Chemistry*. 40: 756–759.

Decker, E. A., 1995. The Role of phenolics, conjugated linoleic acid, carnosine, and pyrroloquinoline quinone as nonessential dietary antioxidants. *Nutrition Reviews*. 53: 49–58.

Djenane, D., Martínez, L., Sánchez-Escalante, A., Beltrán, J. A. and Roncalés, P., 2004. Antioxidant effect of carnosine and carnitine in fresh beef steaks stored under modified atmosphere. *Food Chemistry*. 85: 453–459.

Fernández, J., Pérez-Álvarez, J. P. and Fernández-López, J. P., 1997. Thiobarbituric acid test for monitoring lipid oxidation in meat. *Food Chemistry*. 59: 345–353.

Gardner, M. L. G., Illingworth, K. M., Kelleher, J. and Wood, D., 1991. Intestinal absorption of the intact peptide carnosine in man, and comparison with intestinal permeability to lactulose. *Journal of Physiology*. 431: 411–422.

Gil-Agusti, M., Esteve-Romero, J. and Carda-Broch, S., 2008. Anserine and carnosine determination in meat samples by pure micellar liquid chromatography. *Journal of Chromatography A*. 1189: 444–450.

Hilton, C. W. and Mizuma, H., 1993. Bioactive peptides in food. *Annals of Medicine*. 25: 427–428.

Hipkiss, A. R., 2005. Glycation, ageing and carnosine: are carnivorous diets beneficial? *Mechanisms of Ageing and Development*. 126: 1034–1039.

Hsieh, R. J. and Kinsella, J. E., 1989. Oxidation of polyunsaturated fatty acids: mechanisms, products and inhibition with emphasis on fish. *Advances in Food and Nutrition Research*. 33: 233–341.

Huang, Y., Duan, J., Chen, H., Chen, M. and Chen, G., 2005. Separation and determination of carnosine-related peptides using capillary electrophoresis with laser-induced fluorescence detection. *Electrophoresis*. 26: 593–599.

Kansci, G., Genot, C., Meynier, A. and Gandemer, G., 1993. The antioxidant activity of carnosine and its consequences on the volatile profiles of liposomes during iron/ascorbate induced phospholipid oxidation. *Food Chemistry*. 60: 165–175.

Krajcovicova-Kudlackova, M., Sebekova, K., Schinzel, R. and Klvanova, J., 2002. Advanced glycation end products and nutrition. *Physiological Research*. 51: 313–316.

Jensen, C., Lauridsen, C. and Bertelsen, G., 1998. Dietary vitamin E: quality and storage stability of pork and poultry. *Trends in Food Science and Technology*. 9: 62–72.

Jiménez-Colmenero, F., Ventanas, J. and Toldrá, F., 2010. Nutritional composition of dry cured ham and its role in a healthy diet. *Meat Science*. 84: 585–593.

Jones, G., Smith, M. and Harris, R., 2011. Imidazole dipeptide content of dietary sources commonly consumed within the British diet. *Proceedings of the Nutrition Society*. 70(OCE6): E363.

Maikhunthod, B. and Intarapichet, K. O., 2005. Heat and ultrafiltration extraction of broiler meat carnosine and its antioxidant activity. *Meat Science*. 71: 364–374.

Maynard, L. M., Boissonneault, G. A., Chow, L. K. and Bruckner, G. G., 2001. High levels of dietary carnosine are associated with increased concentrations of carnosine and histidine in rat soleus muscle. *Journal of Nutrition*. 131: 287–290.

Meineri, G., Medana, C., Giancotti, V., Visentin, S. and Peiretti, P. G., 2013. Effect of dietary supplementation of vitamin E in pigs to prevent the formation of carcinogenic substances in meat products. *Journal of Food Composition and Analysis*. 30: 67–72.

Nielsen, C. U., Supuran, C. T., Scozzafava, A., Frokiaer, S., Steffansen, B. and Brodin, B., 2002. Transport characteristics of L-carnosine and the anticancer derivative 4-toluenesulfonylureido-carnosine in a human epithelial cell line. *Pharmaceutical Research*. 19: 1337–1344.

O'Dowd, J. J., Robins, D. J. and Miller, D. J., 1988. Detection, characterisation, and quantification of carnosine and other histidyl derivatives in cardiac and skeletal muscle. *Biochimica et Biophysica Acta*. 967: 241–249.

O'Neill, L. M., Galvin, K., Morrissey, P. A. and Buckley, D. J., 1999. Effect of carnosine, salt and dietary vitamin E on the oxidative stability of chicken meat. *Meat Science*. 52: 89–94.

Park, Y. J., Volpe, S. L. and Decker, E. A., 2005. Quantitation of carnosine in human plasma after dietary consumption of beef. *Journal of Agricultural and Food Chemistry*. 53: 4736–4739.

Pegova, A., Abe, H. and Boldyrev, A., 2000. Hydrolysis of carnosine and related compounds by mammalian carnosinases. *Comparative Biochemistry and Physiology, Part B: Biochemistry & Molecular Biology*. 127: 443–446.

Peiretti, P. G., Medana, C., Visentin, S., Giancotti, V., Zunino, V. and Meineri, G., 2011. Determination of carnosine, anserine, homocarnosine, pentosidine and thiobarbituric acid reactive substances contents in meat from different animal species. *Food Chemistry*. 126: 1939–1947.

Peiretti, P. G., Medana, C., Visentin, S., Dal Bello, F. and Meineri, G., 2012. Effect of cooking method on carnosine and its homologs, pentosidine and thiobarbituric acid-reactive substance contents in beef and turkey meat. *Food Chemistry*. 132: 80–85.

Pereira-Lima, C. I., Ordoñez, J. A., García de Fernando, G. D. and Cambero, M. I., 2000. Influence of heat treatment on carnosine, anserine and free amino acid composition of beef broth and its role in flavour development. *European Food Research and Technology*. 210: 165–172.

Purchas, R. W., Rutherfurd, S. M., Pearce, P. D., Vather, R. and Wilkinson, B. H. P., 2004. Cooking temperature effects on the forms of iron and levels of several other compounds in beef semitendinosus muscle. *Meat Science*. 68: 201–207.

Purchas, R. W., Busboom, J. R. and Wilkinson, B. H. P., 2006. Changes in the forms of iron and in concentrations of taurine, carnosine, coenzyme Q10, and creatine in beef *longissimus* muscle with cooking and simulated stomach and duodenal digestion. *Meat Science*. 74: 443–449.

Quinn, P. J., Boldyrev, A. and Formazuyk, V. E., 1992. Carnosine its properties, function and potential therapeutic applications. *Molecular Aspects of Medicine*. 13: 379–444.

Rao, M. V. and Gault, N. F. S., 1989. The influence of fibre-type composition and associated biochemical characteristics on the acid buffering capacities of several beef muscles. *Meat Science*. 26: 5–18.

Sayre, L. M., Lin, D., Yuan, Q., Zhu, X. and Tang, X., 2006. Protein Adducts Generated from Products of Lipid Oxidation: focus on HNE and ONE. *Drug Metabolism Reviews*. 38: 651–675.

Shantha, N. C. and Decker, E. A., 1995. Conjugated linoleic acid concentrations in cooked beef containing antioxidants and hydrogen donors. *Journal of Food Lipids*. 2: 57–64.

Singh, R., Barden, A., Mori, T. and Beilin, L., 2001. Advanced glycation end-products: a review. *Diabetologia*. 44: 129–146.

Stuerenburg, H. J., 2000. The roles of carnosine in aging of skeletal muscle and in neuromuscular diseases. *Biochemistry (Moscow)*. 65: 862–865.

Tian, Y., Xie, M., Wang, W., Wu, H., Fu, Z. and Lin, L., 2007. Determination of carnosine in Black-Bone Silky Fowl (*Gallus gallus domesticus* Brisson) and common chicken by HPLC. *European Food Research and Technology*. 226: 311–314.

Tsai, S. J., Kuo, W. W., Liu, W. H. and Yin, M. C., 2010. Antioxidative and anti-inflammatory protection from carnosine in the striatum of MPTP-treated mice. *Journal of Agricultural and Food Chemistry*. 58: 11510–11516.

Tsoi, B., He, R. R., Yang, D. H., Li, Y. F., Li, X. D., Li, W. X., Abe, K. and Kurihara, H., 2011. Carnosine ameliorates stress-induced glucose metabolism disorder in restrained mice. *Journal of Pharmacological Sciences*. 117: 223–229.

Velíšek, J., Kubec, R. and Cejpek, K., 2006. Biosynthesis of food constituents: peptides – a review. *Czech Journal of Food Sciences*. 24: 149–155.

van Waarde, A., 1988. Biochemistry of non-protein nitrogenous compounds in fish, including the use of amino acids for anaerobic energy production. *Comparative Biochemistry and Physiology, Part B: Biochemistry & Molecular Biology*. 91: 207–228.

Wu, H. C. and Shiau, C. Y., 2002. Proximate composition, free amino acids and peptides contents in commercial chicken and other meat essences. *Journal of Food and Drug Analysis*. 10: 170–177.

Wu, H. C., Shiau, C. Y., Chen, H. M. and Chiou, T. K., 2003. Antioxidant activities of carnosine, anserine, some free amino acids and their combination. *Journal of Food and Drug Analysis*. 11: 148–153.

Yamada, S., Kawashima, K., Baba, K., Oku, T. and Ando, S., 2009. Occurrence of a novel acetylated amino acid, *N*-α-acetylhistidine, in skeletal muscle of freshwater fish and other ectothermic vertebrates. *Comparative Biochemistry and Physiology, Part B: Biochemistry & Molecular Biology*. 152: 282–286.

Yen, W. J., Chang, L. W., Lee, C. P. and Duh, P. D., 2002. Inhibition of lipid peroxidation and nonlipid oxidative damage by carnosine. *Journal of the American Oil Chemists' Society*. 79: 329–333.

Zhou, S. and Decker, E. A., 1999. Ability of carnosine and other skeletal muscle components to quench unsaturated aldehydic lipid oxidation products. *Journal of Agricultural and Food Chemistry*. 47: 51–55.

Section II
Chemistry and Biochemistry

CHAPTER 3

The Chemistry of Imidazole Dipeptides

DANIEL PERRONE*[a], MARIANA MONTEIRO[b], AND VANESSA NACIUK CASTELO-BRANCO[c]

[a]Laboratório de Bioquímica Nutricional e de Alimentos, Departamento de Bioquímica, Instituto de Química, Universidade Federal do Rio de Janeiro, Av. Athos da Silveira Ramos 149, CT, Bloco A, sala 528A, 21941-909, Rio de Janeiro, Brazil; [b]Departamento de Nutrição Básica e Experimental, Instituto de Nutrição Josué de Castro, Universidade Federal do Rio de Janeiro, Av. Carlos Chagas Filho, 373, CCS, bloco J, 2° andar, sala 16, 21941-902, Rio de Janeiro, Brazil; [c]Departamento de Bromatologia, Faculdade de Farmácia, Universidade Federal Fluminense, Rua Mário Viana, 523, 24241-000, Niteroi, Brazil
*E-mail: danielperrone@iq.ufrj.br

3.1 Introduction

Histidine-derived dipeptides, also know as imidazole dipeptides, have long been known to constitute animal muscle tissue. So far, four imidazole dipeptides have been described: carnosine (β-alanyl-L-histidine), anserine (β-alanyl-*N*-π-methyl-L-histidine), balenine (β-alanyl-*N*-τ-methyl-L-histidine) and homocarnosine (γ-aminobutyryl-L-histidine) (Figure 3.1). In addition to these, L-histidine methyl derivatives have also been reported.

Carnosine was isolated more than one hundred years ago from a meat extract by Gulewitsch and Amiradzibi (1900) and its structure was determined

Food and Nutritional Components in Focus No. 8
Imidazole Dipeptides: Chemistry, Analysis, Function and Effects
Edited by Victor R Preedy

Published by the Royal Society of Chemistry, www.rsc.org

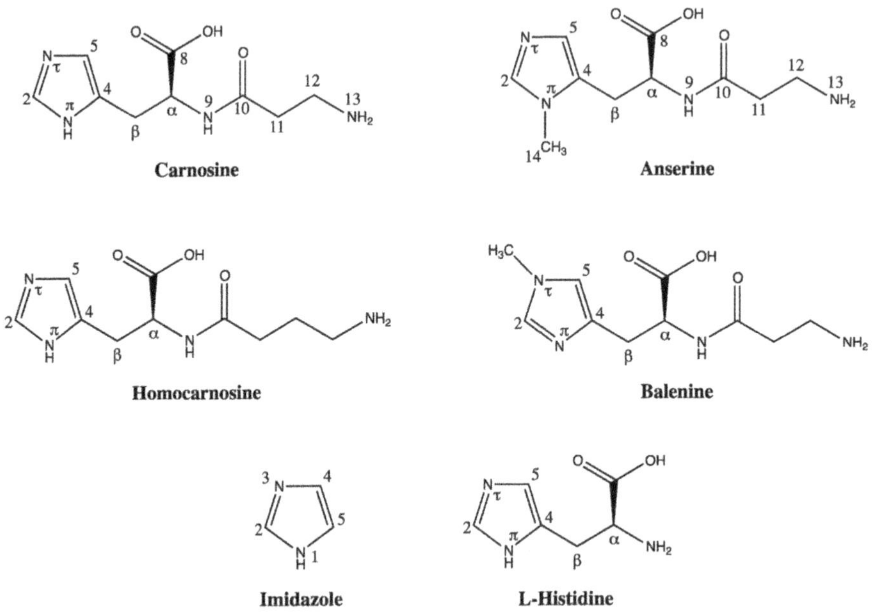

Figure 3.1 Chemical structures of imidazole dipeptides, imidazole and L-histidine.

almost two decades later by Baumann and Ingvaldsen (1918). Anserine was the second imidazole dipeptide described and its name derives from the subfamily *Anserinae* that includes birds such as goose, from which meat this compound was first isolated by Ackermann *et al.* (1929). Balenine isolation and structure determination provoked some confusion. This dipeptide was first isolated by Pocchiari *et al.* (1962) from a baleen whale (*Balaenoptera* family) meat sample, which explains its name. However, many years before, Imamura (1939) isolated a dipeptide from snake (*Ophiophagus* genus) muscle that was named ophidine and wrongly identified as 5-alanyl-2-methyl-L-histidine. The correct structure of this dipeptide was only resolved much later by Nakai and Tsujigado (1965) and found to be identical to that of balenine, which was the final consensus name. Nevertheless, some authors still use the term ophidine or, even worse, wrongly identify this peptide as 5-alanyl-2-methyl-L-histidine.

The nomenclature of L-histidine methyl derivatives has also been the subject of controversy. According to IUPAC rules, the nitrogen atoms of the imidazole ring of histidine are denoted by *pros* ("near", abbreviated π) and *tele* ("far", abbreviated τ) in relation to their proximity to the side chain. The carbon atom between the two ring nitrogen atoms is numbered 2 (as in imidazole), and the carbon atom next to the τ nitrogen is numbered 5. The carbon atoms of the aliphatic chain are designated α and β, following general amino acids nomenclature rules. This recommendation by IUPAC aimed to homogenize two different imidazole ring-numbering systems that had been simultaneously used for a long time by biochemists, who generally

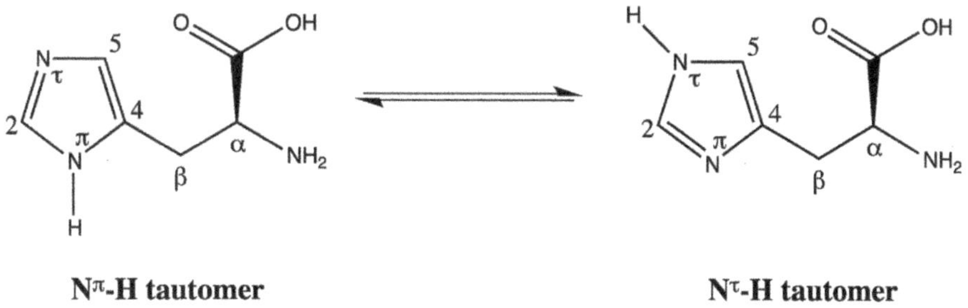

Figure 3.2 Tautomerism of the imidazole ring of L-histidine.

numbered as 1 the nitrogen atom adjacent to the side chain, and by organic chemists, who designated it as 3. This numbering should also be used for histamine, the decarboxylation product of L-histidine, as well as substituted L-histidine derivatives.

The imidazole ring of L-histidine and its derivatives may exist as an equilibrium of two tautomeric structures, N^{τ}—H and N^{π}—H (Giralt *et al.*, 1986), as shown in Figure 3.2. The tautomeric equilibrium of free L-histidine and most L-histidine containing peptides and proteins is shifted towards the N^{τ}—H tautomer, with a few exceptions. The existence of these constitutionally tautomers was confirmed by ^{13}C and ^{15}N NMR and Raman spectroscopy studies (Tanokura, 1983) and may play an important role on the biological action of imidazole dipeptides, specially for those that exhibit hormonal activity.

3.2 Physicochemical and Spectroscopic Data

The chemical and physical properties of imidazole dipeptides are given in Table 3.1.

The electronic absorption spectrum of 0.88 mM carnosine is shown in Figure 3.3. The absorbance bands observed at 264.5, 214 and 209 nm can be assigned to n–π*, π–π*, and π–π* electronic transitions, respectively. At low concentrations (0.07 mM) the peak at 264.5 nm virtually disappears leaving only the broad band around 214 nm. Other imidazole derivatives show similar UV-Vis absorption spectra.

The vibrational FT-IR spectrum of carnosine is shown in Figure 3.4. The most important vibrations in the high-frequency region include asymmetric stretching modes for the terminal amine (3237 and 3049 cm^{-1}, NH_2, N13). In intermediate frequencies, the most important peaks include stretching vibrations of several bonds, such as the amide carbonyl (1643 cm^{-1}, C10), the imidazole ring carbons (1563 and 1461 cm^{-1}, C4 and C5), the carboxyl carbonyl (1400 cm^{-1}, C8), the carbon–nitrogen bonds (1335 and 1312 cm^{-1}, C2, C5 and Cα) and the nitrogen–carbon–nitrogen imidazole ring bond (1227 and 1162 cm^{-1}, Nπ, C2 and Nτ). Peaks corresponding to inplane bending of pyrrolic and methylene groups (1432 cm^{-1}, Nτ, C11 and C12) and to breathing of the imidazole ring (1335, 1312 and 1270 cm^{-1}) can also be assigned in this region

Table 3.1 Chemical and physical data of carnosine, anserine, balenine and homocarnosine.

Compound	Molecular formula	Molecular weight (g mol^{-1})	Melting point (°C)	p*K* value of imidazole group	Water solubility at 25 °C (g L^{-1})
Carnosine	$C_9H_{14}N_4O_3$	226.23	253	7.01	10.6
Anserine	$C_{10}H_{16}N_4O_3$	240.26	238	7.15	4.0
Balenine	$C_{10}H_{16}N_4O_3$	240.26	323	6.93	4.1
Homocarnosine	$C_{10}H_{16}N_4O_3$	240.26	345	7.10	7.4

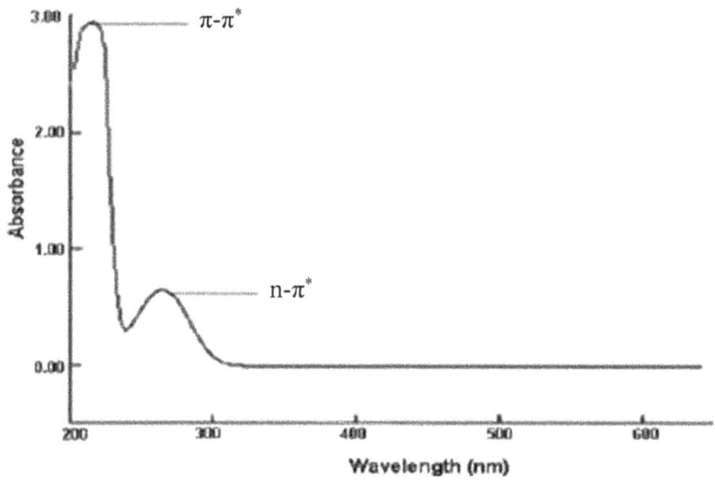

Figure 3.3 UV-Vis absorption spectrum of carnosine, showing the n–π* and π–π* electronic transitions. (Reproduced with permission from Branham *et al.* (2011).)

of the spectrum. In the low-frequency region, the most important bands correspond to inplane deformation of the methine (1095 and 979 cm^{-1}, Cα, C2 and C5), the imidazole ring (859 cm^{-1}), the terminal amine (838 cm^{-1}, N13) and carboxyl (625 cm^{-1}, C8).

Carnosine has been the subject of numerous NMR investigations, even being detected with ^{1}H NMR in intact muscle (Arús *et al.*, 1984). The ^{1}H and ^{13}C resonance signals of carnosine were unambiguously assigned using a variety of NMR techniques including proton-carbon chemical shift correlations, titrations of NMR chemical shifts, coupling constants, and isotope shifts (Friedrich and Wasylishen, 1986). More recently, Branham *et al.* (2011) conducted two-dimensional NMR experiments, including ^{1}H correlation spectroscopy (COSY), ^{13}C gradient heteronuclear multiple-bond correlation spectroscopy (gHMBC), and ^{15}N gHMBC. The NMR assignments proposed by these authors are given in Table 3.2.

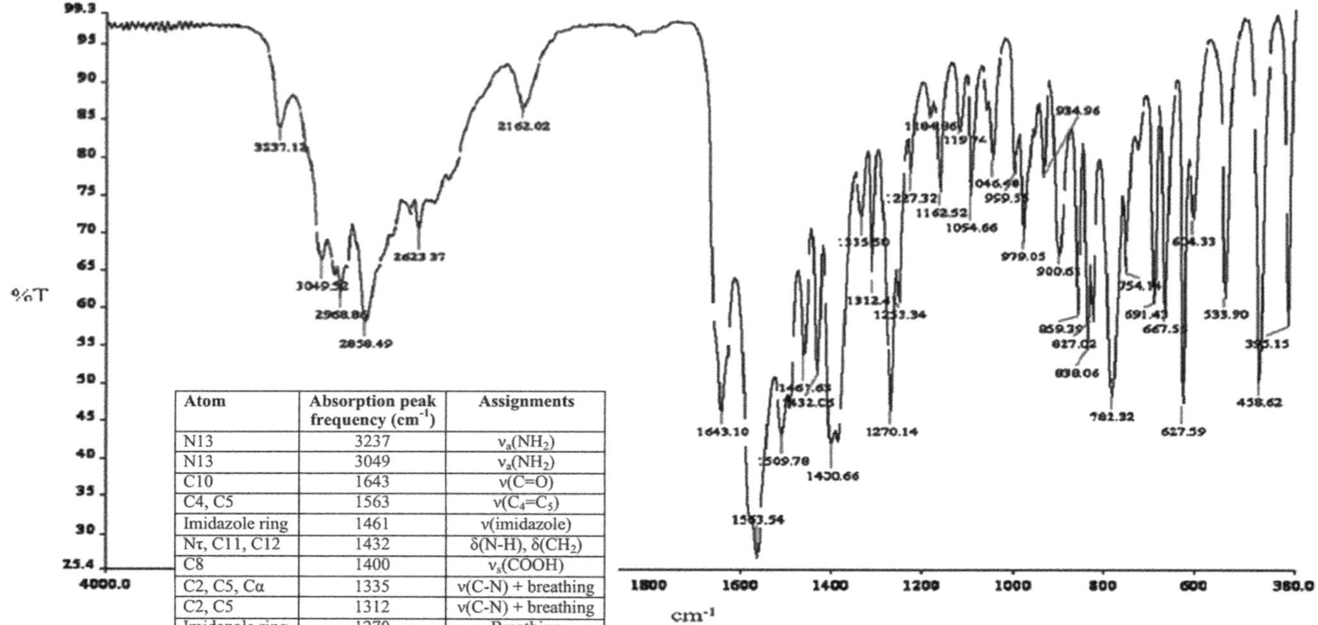

Atom	Absorption peak frequency (cm^{-1})	Assignments
N13	3237	$v_a(NH_2)$
N13	3049	$v_a(NH_2)$
C10	1643	$v(C{=}O)$
C4, C5	1563	$v(C_4{=}C_5)$
Imidazole ring	1461	v(imidazole)
Nτ, C11, C12	1432	δ(N-H), $\delta(CH_2)$
C8	1400	$v_s(COOH)$
C2, C5, Cα	1335	v(C-N) + breathing
C2, C5	1312	v(C-N) + breathing
Imidazole ring	1270	Breathing
Nπ, C2, Nτ	1227	v(NCN) + δ(N-H)
Nπ, C2, Nτ	1162	v(NCN) + δ(N-H)
Cα	1095	δ(C-H)
C2, C5	979	δ(C-H)
Imidazole ring	859	δ(imidazole)
N13	838	$\delta(NH_2)$
C8	625	δ(COOH)

Figure 3.4 FT-IR spectrum of carnosine with vibration assignments. (Adapted with permission from Branham *et al.* (2011).)

Table 3.2 Proposed ^{1}H, ^{13}C and ^{15}N assignments for carnosine. (Adapted with permission from Branham *et al.* (2011).)

Atom	Chemical shift relative to TMS (ppm)	Assignments
H12	2.60	β-Ala-methylene
Hβ	2.92; 3.08	His-methylene
H11	3.16	α-Ala-methylene
Hα	4.40	α-His-methine
H5	6.89	Ring methane
H2	7.66	Imine H
H9	7.89	Amide N—H
Cβ	29.0	β-His-methine
C12	32.5	β-Ala-methylene
C11	36.1	α-Ala-methylene
Cα	55.3	α-His-methine
C4	117.6	Ring C
C5	133.2	Ring methine C
C2	135.9	Ring imine C
C10	171.7	Amide carbonyl
C8	177.9	Carboxylate
N13	30.9	NH_3
N9	129.5	Amide N
Nτ	188.5	Pyrrole N
Nπ	220.7	Imine N

The proton NMR COSY spectrum of carnosine is shown in Figure 3.5. Multiplet peaks at δ 2.60 and 3.16 are representative of the two protons attached each at alanine C12 and C11, respectively. The two protons attached to histidine Cβ result in two double doublet peaks at δ 2.92 and 3.08, while the proton attached to histidine Cα results in a double doublet peak at δ 4.40. A singlet peak at δ 6.89 and a doublet peak at δ 7.66 are representative of imidazole protons attached at C5 and C2, respectively. The amide proton results in a small broad singlet at δ 7.89 due to hydrogen bonding.

The ^{13}C gHMBC spectrum of carnosine is shown in Figure 3.6. Triplet peaks at δ 29.0, 32.5 and 36.1 are representative of histidine Cβ and alanine C12 and C11, respectively. Imidazole ring carbons result in peaks at δ 117.6 (C4, doublet), 133.2 (C5, singlet) and 135.9 (C2, doublet). Singlet peaks at δ 171.7 and 177.9 correspond to amide and carboxylate carbons.

The ^{15}N gHMBC spectrum of carnosine is shown in Figure 3.7. Peaks at δ 30.9, 129.5, 188.5 and 220.7 are representative of terminal amine, amide and imidazole pyrrole (Nτ) and imine (Nπ) nitrogen atoms, respectively.

Spectroscopic data for imidazole dipeptides other than carnosine is scarce. Quantitative ^{13}C NMR spectrum of minced white muscle meat of Atlantic salmon was obtained without chemical pretreatment of the sample and gave rise to sufficiently resolved signals to allow estimation of anserine content (Aursand *et al.*, 1995). In the aliphatic region of the obtained spectrum (Figure 3.8a), peaks at δ 26.0 and 53.4 corresponds to histidine Cβ and Cα, respectively. Peaks at δ 33.2 and 35.8 are also observed in this region of the

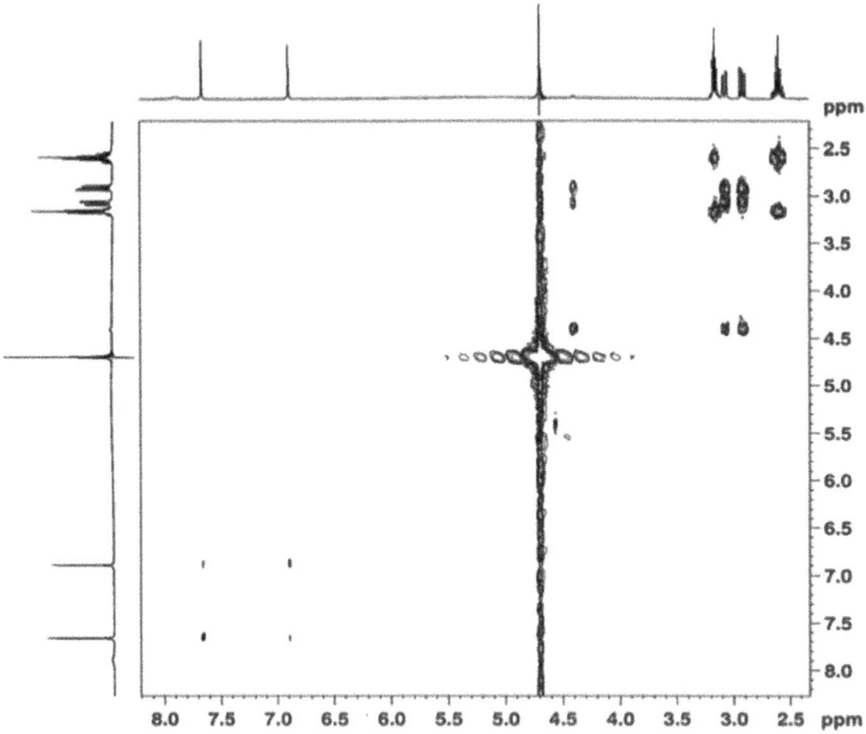

Figure 3.5 ^{1}H-NMR COSY spectrum of carnosine (500 MHz). Chemical shift reported in ppm (δ) downfield from tetramethylsilane (TMS). (Reproduced with permission from Branham *et al.* (2011).)

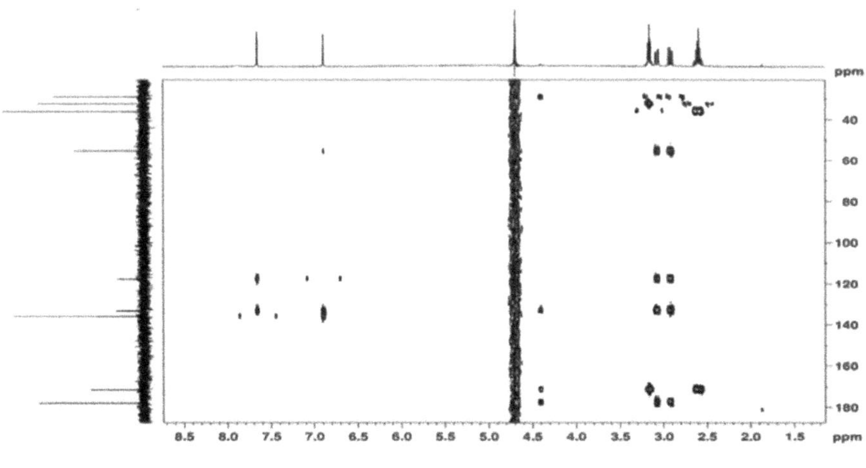

Figure 3.6 ^{1}H–^{13}C NMR gHMBC spectrum of carnosine (125 MHz). Chemical shift reported in ppm (δ) downfield from tetramethylsilane (TMS). (Reproduced with permission from Branham *et al.* (2011).)

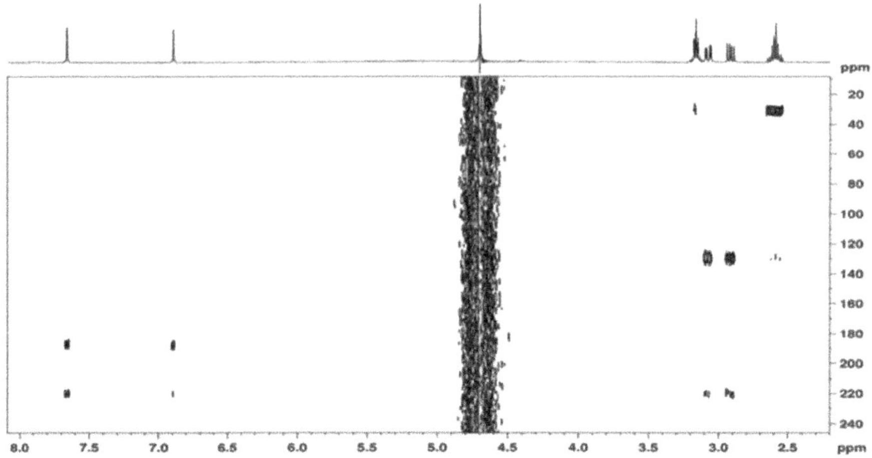

Figure 3.7 $^{1}H-^{15}N$ NMR gHMBC spectrum of carnosine (50 MHz). Chemical shift reported in ppm (δ) downfield from liquid ammonia. (Reproduced with permission from Branham *et al.* (2011).)

spectrum and represent the methyl group attached to Nπ and the C12 methylene, respectively. Imidazole ring carbons result in well-separated peaks located in the olefinic region of the NMR spectrum (Figure 3.8b) at δ 118.2, 131.1 and 135.5, representative of C5 (methine), C4 and C2 (imine). The carboxylate carbon results in a peak at δ 176.5, located in the carbonyl region of the spectrum (Figure 3.8c). The quantification of total imidazole dipeptides is based on the signals from C2 and C5, common to all compounds. Anserine-specific quantification is based on the unique peak of the methyl group attached to imidazole Nπ.

3.3 Food Sources

The naturally occurring imidazole dipeptides carnosine, anserine, homocarnosine and balenine are found exclusively in animal tissue, being particularly abundant in skeletal muscle and nervous tissue (Gil-Agustí *et al.*, 2008). The contents of the major imidazole dipeptides, carnosine and anserine, in various meat types is presented in Table 3.3. While carnosine and anserine are broadly distributed in different kinds of meat, balenine and homocarnosine are found in meat pork (Schöenherr, 2002; Peiretti *et al.*, 2011) at low concentrations (<0.5 g/100 g).

The distribution of these dipeptides has been studied in many vertebrates, but few invertebrates. Meats with high carnosine content are usually those that have low anserine levels and *vice versa* (Gil-Agustí *et al.*, 2008). Thus, skeletal muscles of terrestrial mammals and amphibians have mostly carnosine, fish and birds primarily anserine, and reptiles and cetaceans are rich in balenine (Abe, 2000). Invertebrates contain small amounts of either

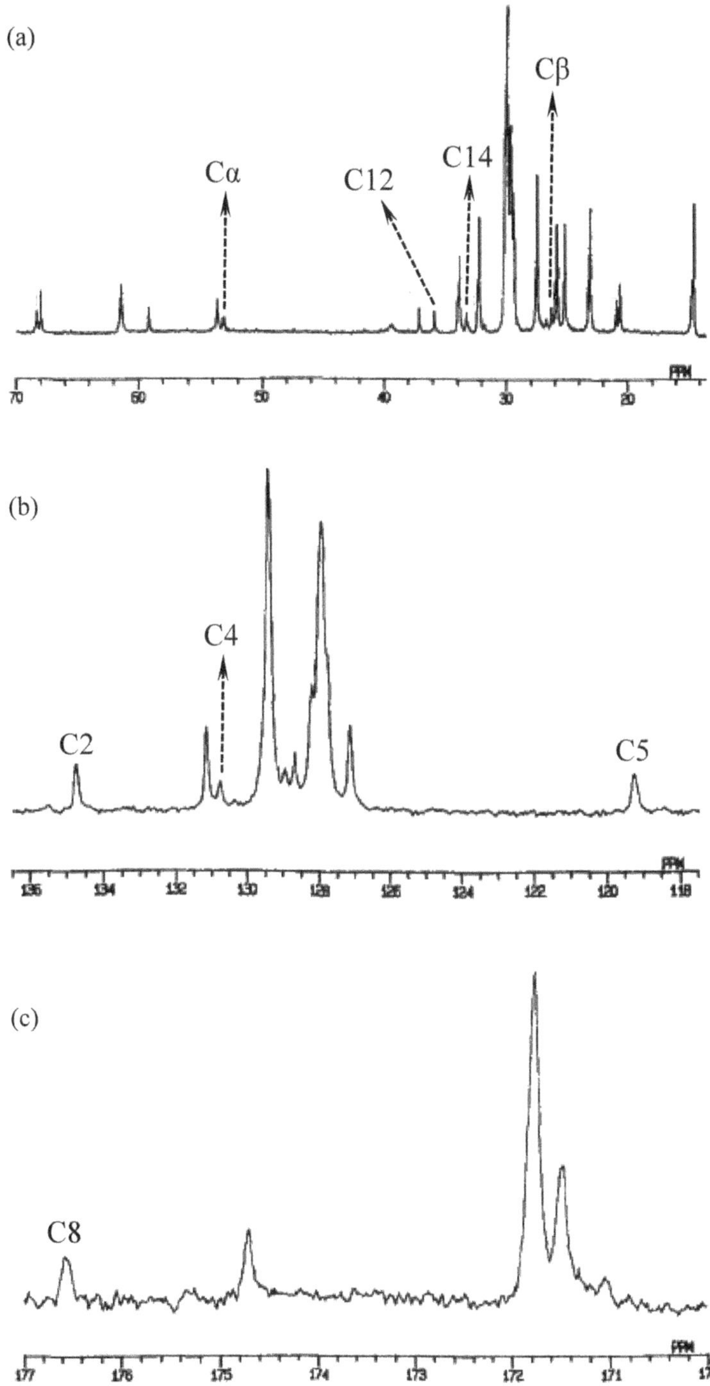

Figure 3.8 Expansion of the aliphatic (a), olefinic (b) and carbonyl (c) regions of the ^{13}C NMR spectrum of minced white muscle from Atlantic salmon (100 MHz), showing the signal assignments for the anserine carbons. (Adapted with permission from Aursand *et al.* (1995).)

Table 3.3 Imidazole dipeptide contents of commonly consumed dietary sources.

Animal/species	Sample	Carnosine (mg g^{-1})	Anserine (mg g^{-1})	References
Chicken	Breast	0.6–1.2	34.6	Jones *et al.* (2011)
	Thigh	0.3–0.4	n.d.[a]	Intarapichet and Maikhu-nthod (2005)
Turkey	Breast	0.7–5.9	4.2–22.5	Gil-Augustí *et al.* (2008)
Beef	Fillet	13.8	1.7	
	Hamburger	1.7	0.2	Peiretti *et al.* (2011)
Pork	Loin	17.0	0.5	
Rabbit	Leg	4.0	12.7	
Horse	Fillet	28.0	0.1	
Tuna	Midfillet	1.1	33.0	Jones *et al.* (2011)
Mackerel	Midfillet	7.8	0.5	
Rainbow trout	Midfillet	0.1	5.1	
Prawns		9.3–11.6	69.0–77.3	

[a]Not determined.

carnosine (mollusks and crustaceans) or anserine (oyster) (Lukton and Olcott, 1958). Pork and beef meat tends to present higher percentages (>80%) of carnosine in relation to the total imidazole dipeptides contents than turkey and chicken meat (<30%) (Peiretti *et al.*, 2011).

The content of imidazole dipeptides, mainly carnosine, differs according to genotype, gender, age, breeding conditions and muscle type of animals (glycolytic or oxidative muscles). Fresh chicken breast meat has 2- to 4-fold higher carnosine contents compared to thigh meats. The carnosine content of female meat is significantly higher than that of male meat. For instance, the carnosine contents of female chicken breast and thigh (1.04 and 0.35 mg g^{-1}, respectively) are higher in comparison to the same parts of male animals (0.71 and 0.30 mg g^{-1}, respectively) (Intarapichet and Maikhunthod, 2005). Regarding seafood, prawns show high contents of imidazole dipeptides (~85 mg g^{-1}), mainly anserine (~70 mg g^{-1}), while fish show lower contents, ranging from 5 to 35 mg g^{-1} (anserine is the major dipeptide found in tuna and rainbow trout, while mackerel is richer in carnosine) (Jones *et al.*, 2011).

Additionally to intrinsic factors, cooking might cause degradation of imidazole dipeptides. Cooking beef and turkey meat by boiling and broiling causes an average loss of 50% of carnosine in comparison to the corresponding raw meat samples. In contrast, microwave cooking causes a smaller loss (~13%) of imidazole dipeptides, for both anserine and homocarnosine (Peiretti *et al.*, 2012). Therefore, the use of microwave oven guarantees low carnosine and related-compounds degradation.

Heating by cooking, roasting or other processes enhances the formation of Maillard reactions products, including Amadori compounds, dicarbonyls melanoidins and advanced glycation end products (AGEs). AGEs are late products of glucose–protein or glucose–lipid interactions that lead to the formation of unstable reactive intermediates such as glycoxidation products, which might be associated with the pathogenesis of the degenerative processes related to aging, diabetes, kidney disease and Alzheimer's disease (Koschinsky *et al.*, 1997; Ahmed *et al.*, 2005). It has been suggested that carnosine, and possibly anserine and homocarnosine, could be potential antiglycating agents (Hipkiss, 2005). Thus, a carnosine-rich diet might be an important factor to reduce the risk of the development of these deleterious effects.

Imidazole dipeptides are potential biomarkers of meat intake, especially anserine and its urinary metabolite 1-methylhistidine. The excretion level of anserine in urine may serve as a marker of recent meat, chicken, or fish intake, whereas 1-methylhistidine may serve as a marker of intake within the last few days. However, these markers cannot be used to discriminate between red and white meat ingestion (Dragsted, 2010).

3.4 Reactivity and Synthesis

Imidazole can be considered as having properties similar to both pyrrole and pyridine. Electrophilic reagents would attack the unshared electron pair on N-3, but not that on the "pyrrole" nitrogen since it is the part of the aromatic sextet. While the imidazole ring is fairly susceptible to electrophilic attack on an annular carbon, it is much less likely to undergo a nucleophilic substitution reaction unless a strongly electron-withdrawing group is substituted elsewhere in the ring. In the absence of such activation the position most prone to nucleophilic attack is C-2. The overall reactivity of imidazole is related to sets of resonance structures in which dipolar contributors have finite importance. These predict electrophilic attack in imidazole at N-3 or any ring carbon atom, nucleophilic attack at C-2 or C-1 and also the amphoteric nature of the molecule (Bhatnagar *et al.*, 2011).

The synthesis of peptides may be accomplished by chemical or enzymatic routes (Guzmán *et al.*, 2007). Chemical routes may occur either in solution or solid phase and are used for the routine production of peptides of up to 50 amino acid residues (Kimmerlin and Seebach, 2005). A simple synthetic route of carnosine was described by Turner (1953) entailing the condensation of phthaloyl-β-alanyle chloride with L-histidine in the presence of triethylamine at low temperature that yields phthaloyl-β-alanyle histidine. Carnosine is isolated from this compound after detachment of the phthaloyl group with hydrazine. Related imidazole peptides may also be prepared using this straightforward method. Synthesis of carnosine by trimethylsilyl protection of functional groups was reported more recently by Vezenkov and Yanachkov (1991), as well as the synthesis of imidazole dipeptides phosphoanalogs (Khomutov, 1990).

The enzymatic synthesis of small peptides, such as imidazole dipeptides, is a useful approach to overcome limitations in chemical synthesis, such as lack of specificity and the high environmental burden (Guzmán *et al.*, 2007). Heyland *et al.* (2010) developed a whole-cell biocatalyst using a recently identified enzyme class of β-aminopeptidases for the synthesis of carnosine with yields of up to 71%.

3.5 Chemical Basis of Imidazole Dipeptides Physiological Roles

Many physiological functions have been ascribed to imidazole dipeptides, such as direct activators of some glycolytic enzymes; enhancers for systems sensitive to Ca^{2+}; oxygen- and copper-chelating capacity; antioxidant activity by scavenging of free radicals; buffering capacity (Abe, 1995).

3.5.1 Buffering Capacity

During the anaerobic high-intensity burst involved in the movement of animals for long periods, the maintenance of intracellular pH of skeletal muscle is critically important. Proton accumulation due to the hydrolysis of ATP to ADP leads to an intracellular pH drop that, in turn, inactivates glycolytic enzymes such as phosphofructokinase and hence decrease glycolytic flux. This phenomenon is also observed during prolonged anoxia of perfused rat heart and it is known to affect the rate of other metabolic functions. From an evolutionary point of view, the improvement of cellular buffering systems enhanced the capability for anaerobic exercise performance or anoxia tolerance. In addition to classic buffering systems, the role of nonbicarbonate buffering of vertebrate muscle has been recognized as paramount to muscular pH balance. The imidazole group of L-histidine is the main contributor to the intracellular buffering capacity in vertebrate muscle. L-Histidine occurs as free residues on the polypeptide chain of proteins, as free L-histidine, and as imidazole dipeptides (Abe, 2000). Imidazole dipeptides correspond to 40% or more of the buffering in muscle of terrestrial animals (Davey, 1960), while in fish muscle this ranges from 7% in trout to over 60% in marlin (Abe *et al.*, 1985).

Because the pK values of these imidazole groups are close to intracellular pH (Table 3.1), one of the two nitrogen atoms of the imidazole ring can be protonated in the physiological range of pH, acting as potent proton buffering constituents. The ability of imidazole groups in keeping the physiological pH close to their pK values is known as the "alphastat regulation", a theory proposed by Reeves (1972). According to this theory, imidazole groups have a degree of dissociation (referred to as α) of 0.55 in the intracellular fluid and this remains constant despite changes in temperature. Other candidates are organic phosphate compounds such as nucleotides and sugar phosphates and several organic acids such as citrate and succinate, only when they exist in large amounts in muscle cytosol.

Proton-buffering capacity (termed the β value) due to nonbicarbonate buffering compounds is defined as the μmoles of NaOH or HCl required to change the pH by one unit per gram tissue. This is measured as the "slyke" unit based on the original technique of Van Slyke. Marine diving mammals and warm-bodied tuna show high β values (47.5 to 136 μmol NaOH/pH), on average higher than terrestrial mammals (50.2 to 82.8 μmol NaOH/pH). This may relate to the anaerobic capacity for prolonged breath-hold diving. The buffering capacity of mammalian muscle strongly correlates with myoglobin content, which is found in higher contents in muscle of marine mammals in comparison to terrestrial ones, as well as with the activity of lactate dehydrogenase (Castellini and Somero, 1981). Moreover, an inverse correlation between imidazole dipeptides content and aerobic metabolism has been observed, suggesting that these compounds are more important in muscles that are used mostly for burst activity and that rely upon glycolytic and lactic acid metabolisms (Abe *et al.*, 1985).

3.5.2 Antioxidant Activity

Oxidative reactions are implicated in the development of food deterioration, altering color, flavor, texture and nutritional values. Moreover, toxic substances could be formed during these reactions, increasing the risk of development of numerous diseases including atherosclerosis and cancer (Choe and Min, 2009). Carnosine and related dipeptides have been postulated to have numerous actions in muscle systems, including inhibition of oxidative reactions (Decker *et al.*, 2000).

Carnosine shows antioxidant activity through two mechanisms: metal chelating and free-radical scavenging (Decker and Crum, 1993; Peiretti *et al.*, 2011; Manhiani *et al.*, 2013). Consequently, carnosine is able to reduce concentrations of thiobarbituric acid reactive substances (TBARs, secondary oxidation products) due to its capacity to interact with aldehydic lipid oxidation products through Schiff base or Michael addition-type reactions (Decker *et al.*, 2000). Carnosine might also react with a variety of aldehydes, including saturated, α,β-monounsaturated, polyunsaturated, and hydroxylated aldehydes, which could be potentially toxic to biological tissues (Hipkiss, 2005).

An inverse association between the total content of carnosine, anserine and homocarnosine and the formation of TBARs in meat from different animal species has been demonstrated (Peiretti *et al.*, 2011). Besides, carnosine is more efficient in inhibiting TBARs formation than butylated hydroxytoluene (BHT) and α-tocopherol during refrigerated storage of cooked salted and unsalted ground pork (Decker and Crum, 1993). Anserine may also contribute to decrease TBARs levels in cooked turkey meat (Peiretti *et al.*, 2012). Therefore, carnosine and related compounds might prevent the formation of off-flavor and potentially toxic compounds during meat processing and storage.

Summary Points

- Imidazole dipeptides are histidine-derived dipeptides present only in animal tissue, mainly skeletal muscle.
- The main compounds are carnosine (β-alanyl-L-histidine), anserine (β-alanyl-*N*-π-methyl-L-histidine), balenine (β-alanyl-*N*-τ-methyl-L-histidine) and homocarnosine (γ-aminobutyryl-L-histidine).

- Imidazole dipeptides may exist as an equilibrium of two tautomeric structures, which play an important role on their biological action.
- The imidazole ring of histidine-derived dipeptides is prone to electrophilic attack at the imine nitrogen.
- Synthesis of imidazole dipeptides may be done through chemical or enzymatic routes.
- Imidazole dipeptides content and distribution in animal tissue depend on genotype, gender, age, breeding conditions and muscle type of animals.
- Imidazole dipeptides are potential biomarkers of meat intake.
- Imidazole dipeptides might act as activators of some glycolytic enzymes, enhancers for systems sensitive to Ca^{2+}, chelators of oxygen and copper, antioxidants and buffering agents.

Keys Facts of Muscle Physiology

1. Muscle contraction consumes large amounts of energy (ATP) and is precisely regulated so that contraction/relaxation cycles produce motility.
2. Contraction occurs when muscles are stimulated by an electrical nervous impulse (depolarization).
3. Proteins involved in the physical process of contraction are myosin and actin, which form the actomyosin complex during contraction/relaxation cycles.
4. Calcium is released during depolarization from the sarcoplasmic reticulum to the cytoplasm, binds to troponin C and stimulates the formation of the actomyosin complex.
5. Muscles fibers are classified as fast-twitch (white muscles) or slow-twitch (red muscles).
6. Aerobic glycolytic system is predominant in red muscle, while anaerobic glycolytic system is predominant in white muscle.
7. ATP production in anaerobic glycolysis is much less efficient than in aerobic respiration and also results in incomplete substrate oxidation and accumulation of lactate.

Definitions and Explanation of Key Terms

Advanced glycation end products. Unstable reactive intermediate products of the Maillard reaction, also called glycotoxins, with the ability to promote oxidative stress and inflammation, increasing the risk of chronic disease development.

Buffering. Property of weak acids and their salts to neutralize the action of acid or alkaline agents maintaining constant medium pH.

Chemical shift. Frequency of the resonance expressed with reference to a standard compound, applied to describe the relative position of the resonance in a nuclear magnetic resonance (NMR) spectrum.

Electrophilicity. Ability of an electron-poor chemical species to accept two electrons forming a covalent bond with an electron-rich species, known as a nucleophile.

Fourier transform infrared spectroscopy. Technique used for the measurement of high quality infrared spectra, suitable to determine bonding mechanisms of a solid, liquid or gas.

Free-radical scavenging. Capacity of molecules to donate a hydrogen atom to stabilize free radicals, such as superoxide anion ($^{\bullet}O_2^-$), hydroxyl radical ($HO^{\bullet}$) and peroxyl radical ($ROO^{\bullet}$).

Inplane deformation. Type of bending vibration mode characterized by changes in bond angle between bonds with a common atom or the movement of a group of the atoms.

Maillard reaction. Nonenzymatic reaction between reducing sugars (*e.g.* glucose) and amino groups of proteins, lipids and nucleic acids, forming Schiff bases and Amadori compounds and the characteristic yellow-brown color of thermally processed food.

Metal chelating. Capacity of chemical compounds to bind metallic ions, improving their stability and inhibiting metal interactions with other metals and nutrients.

Nucleophilicity. Ability of an electron-rich chemical species to donate two electrons forming a covalent bond with a positive nuclear charge of an electron poor species, known as an electrophile.

Nuclear magnetic resonance. Phenomenon that occurs when the nuclei of certain atoms are immersed in a static magnetic field and exposed to a second oscillating magnetic field.

p*K* value. Acid dissociation constant that refers to a quantitative measure of the strength of an acid in solution.

Stretching. Symmetric or asymmetric vibrational modes of covalent bonds in molecules that influences the absorption in the infrared region of the electromagnetic spectrum.

Tautomerism. Chemical isomerism characterized by relatively easy interconversion of isomeric forms in equilibrium.

List of Abbreviations

ADP	Adenosine diphosphate
AGEs	Advanced glycation end products
ATP	Adenosine triphosphate
BHT	Butylated hydroxytoluene
COSY	Correlation spectroscopy
FT-IR	Fourier transform infrared spectroscopy
gHMBC	Gradient heteronuclear multiple-bond correlation spectroscopy
IUPAC	International Union of Pure and Applied Chemistry
NMR	Nuclear magnetic resonance
TBARs	Thiobarbituric acid reactive substances
TMS	Tetramethylsilane

References

Abe, H., Dobson, G. P., Hoeger, U. and Parkhouse, W. S., 1985. Role of histidine-related compounds to intracellular buffering in fish skeletal muscle. *American Journal of Physiology*. 249: R449–R454.

Abe, H., 1995. Histidine-related dipeptides: distribution, metabolism, and physiological function. In: Hochachka, P. W. and Mommsen, T. P. (ed.) *Biochemistry and Molecular Biology of Fishes*. Elsevier Science, Amsterdam, The Netherlands, vol. 4, pp. 309–333.

Abe, H., 2000. Role of histidine-related compounds as intracellular proton buffering constituents in vertebrate muscle. *Biochemistry*. 65(7): 757–765.

Ackermann, D., Timpe, O. and Poller, K., 1929. Über das Anserin, einen neuen Bestandteil der Vogelmuskulatur. *Zeitschrift für Physiologische Chemie*. 183: 1–10.

Ahmed, N., Ahmed, U., Thornalley, P. J., Hager, K., Fleischer, G. and Münch, G., 2005. Protein glycation, oxidation and nitration adduct residues and free adducts of cerebrospinal fluid in Alzheimer's disease and link to cognitive impairment. *Journal of Neurochemistry*. 92: 255–263.

Arús, C., Bárány, M., Westler, W. M. and Markley, J. L., 1984. ^{1}H NMR of Intact Tissues ate 11.1 T. *Journal of Magnetic Resonance*. 57: 519–525.

Aursand, M., Jørgensen, L. and Grasdalent, H., 1995. Quantitative high-resolution ^{13}C nuclear magnetic resonance of anserine and lactate in white muscle of Atlantic salmon (Salmo salar). *Comparative Biochemistry and Physiology, Part B: Biochemistry & Molecular Biology*. 112B(2): 315–321.

Baumann, L. and Ingvaldesn, T., 1918. Concerning histidine and carnosine. The synthesis of carnosine. *Journal of Biological Chemistry*. 35: 263–276.

Bhatnagar, A., Sharma, P. K. and Kumar, N., 2011. A review on "imidazoles": their chemistry and pharmacological potentials. *International Journal of PharmTech Research*. 3(1): 268–282.

Branham, M. L., Singh, P., Bisetty, K., Sabela, M. and Govender, T., 2011. Preparation, spectrochemical, and computational analysis of L-carnosine (2-[(3-aminopropanoyl)amino]-3-(1*H*-imidazol-5-yl)propanoic acid) and its ruthenium(II) coordination complexes in aqueous solution. *Molecules*. 16: 10269–10291.

Castellini, M. A. and Somero, G. N., 1981. Buffering capacity of vertebrate muscle: correlations with potentials for anaerobic function. *Journal of Comparative Physiology B*. 143: 191–198.

Choe, E. and Min, D. B., 2009. Mechanisms of antioxidants in the oxidation of foods. *Comprehensive Reviews in Food Science and Food Safety*. 8: 345–358.

Davey, C. L., 1960. The significance of carnosine and anserine in striated skeletal muscle. *Archives of Biochemistry and Biophysics*. 89: 303–308.

Decker, E. A. and Crum, A. D., 1993. Antioxidant activity of carnosine in cooked ground pork. *Meat Science*. 34: 245–253.

Decker, E. A., Livisay, S. A. and Zhou, S., 2000. A re-evaluation of the antioxidant activity of purified carnosine. *Biochemistry*. 65: 766–770.

Dragsted, L. O., 2010. Biomarkers of meat intake and the application of nutrigenomics. *Meat Science*. 84: 301–307.

Friedrich, J. O. and Wasylishe, R. E., 1986. A ^{1}H and ^{13}C nuclear magnetic resonance study of carnosine. *Canadian Journal of Chemistry*. 64: 2132–2138.

Gil-Agustí, M., Esteve-Romero, J. and Carda-Broch, S., 2008. Anserine and carnosine determination in meat samples by pure micellar liquid chromatography. *Journal of Chromatography A*. 1189: 444–450.

Giralt, E., Ludevid, M. D. and Pedroso, E., 1986. The relevance of imidazole tautomerism for hormonal activity of histidine-containing peptides. *Bioorganic Chemistry*. 14: 405–416.

Gulewitsch, W. and Amiradzibi, S., 1900. Über das Carnosine, eine neue organische Base des Fleischextraktes. *Berichte der Deutschen Chemischen Gesellschaft*. 33: 1902–1908.

Guzmán, F., Barberis, S. and Illanes, A., 2007. Peptide synthesis: chemical or enzymatic. *Electronic Journal of Biotechnology*. 10(2): 279–314.

Heyland, J., Antweiler, N., Lutz, J., Heck, T., Geueke, B., Kohler, H. P. E., Blank, L. M. and Schmid, A., 2010. Simple enzymatic procedure for L-carnosine synthesis: whole-cell biocatalysis and efficient biocatalyst recycling. *Microbial Biotechnology*. 3(1): 74–83.

Hipkiss, A. R., 2005. Glycation, ageing and carnosine: are carnivorous diets beneficial? *Mechanisms of Ageing and Development*. 126: 1034–1039.

Imamura, H., 1939. Chemie der Schlangen – I. Über die *N*-haltigen Extraktivstoffe der Schlangenmuskeln. *Journal of Biochemistry*. 30: 479–490.

Intarapichet, K. O. and Maikhunthod, B., 2005. Genotype and gender differences in carnosine extracts and antioxidant activities of chicken breast and thigh meats. *Meat Science*. 71: 634–642.

Jones, G., Smith, M. and Harris, R., 2011. Imidazole dipeptide content of dietary sources commonly consumed within the British diet. *Proceedings of the Nutrition Society*. 70(OCE6): E363.

Khomutov, A. R., 1990. Synthesis of phosphoanalogues of histidine and carnosine. *Journal of the American Chemical Society*. 75: 2388–2390.

Kimmerlin, T. and Seebach, D., 2005. '100 years of peptide synthesis': ligation methods for peptide and protein synthesis with applications to β-peptide assemblies. *Journal of Peptide Research*. 65(2): 229–260.

Koschinsky, T., He, C. J., Mitsuhashi, T., Bucala, R., Liu, C., Buenting, C., Heitmann, K. and Vlassara, H., 1997. Orally absorbed reactive glycation products (glycotoxins): an environmental risk factor in diabetic nephropathy. *Proceedings of the National Academy of Sciences of the United States of America*. 94: 6474–6479.

Lukton, A. and Olcott, H. S., 1958. Content of free imidazole compounds in muscle tissue of aquatic animals. *Food Research*. 23: 611–618.

Manhiani, P. S., Northcutt, J. K., Han, I., Bridges, W. C. and Dawson, P. L., 2013. Antioxidant activity of carnosine extracted from various poultry tissues. *Poultry Science*. 92(2): 444–453.

Nakai, T. and Tsujigado, N., 1965. Beta-alanyl dipeptide preparations from whale muscles made by several workers. *Journal of Biochemistry*. 57(6): 812–814.

Peiretti, P. G., Medana, C., Visentin, S., Dal Bello, F. and Meineri, G., 2012. Effect of cooking method on carnosine and its homologues, pentosidine and thiobarbituric acid-reactive substance contents in beef and turkey meat. *Food Chemistry*. 132: 80–85.

Peiretti, P. G., Medana, C., Visentin, S., Giancotti, V., Zunino, V. and Meineri, G., 2011. Determination of carnosine, anserine, homocarnosine, pentosidine and thiobarbituric acid reactive substances contents in meat from different animal species. *Food Chemistry*. 126: 1939–1947.

Pocchiari, F., Tentori, L. and Vivaldi, G., 1962. The presence of the dipeptide β-alanyl-3-methylhistidine in whale meat extract. *Scientific Reports of the Istituto Superiore di Sanita*. 2: 188–194.

Reeves, R. B., 1972. An imidazole alphastat hypothesis for vertebrate acid-base regulation: tissue carbon dioxide content and body temperature in bullfrogs. *Respiration Physiology*. 14: 219–236.

Schöenherr, J., 2002. Analysis of products of animal origin in feeds by determination of carnosine and related dipeptides by high-performance liquid chromatography. *Journal of Agriculture and Food Chemistry*. 50: 1945–1950.

Tanokura, M., 1983. ^{1}NMR study on the tautomerism of the imidazole ring residues. I. microscopic p*K* values and molar ratios of tautomers in histidine-containing peptides. *Biochimica et Biophysica Acta*. 742: 576–585.

Turner, R. A., 1953. Synthesis of carnosine and related peptides by the phthaloyl method. *Journal of the American Chemical Society*. 75: 2388–2390.

Vezenkov, L. T. and Yanachkov, O. B., 1991. Synthesis of carnosine by trimethylsilyl protection. *Doklady Bolgarskoi Akademii Nauk*. 44: 53–56.

CHAPTER 4

The Chemistry of Carnosine Derivatives

FRANCESCO BELLIA[b], ENRICO RIZZARELLI[b], AND GRAZIELLA VECCHIO*[a]

[a]Dipartimento di Scienze Chimiche, University of Catania, V.le A.Doria 6, 95125 Catania, Italy; [b]Institute of Biostructure and Bioimaging, CNR, V.le A.Doria 6, 95125 Catania, Italy
*E-mail: gr.vecchio@unict.it

4.1 Introduction

Carnosine (β-alanyl-L-histidine, L-Car) is a natural multifunctional dipeptide (Derave and Sale, 2012). It is present in various animal species and has high concentration (millimolar range) in skeletal muscles and olfactory bulb of mammalians (Boldyrev *et al.*, 2013). A number of biological activities have been proposed for Car such as physiological buffering, metal-chelating, antioxidant, and free-radical scavenger activity. L-Car has been shown to trap reactive carbonyl species (RCS) preventing the formation of advanced lipoxidation and glycation end products and this quenching ability is considered one of the main protection mechanisms *in vivo*. Favorable effects of carnosine on survival and learning ability of animals under ischemic injury were also shown (Dobrota *et al.*, 2005). In the context of neurodegenerative disorders, L-Car is able to reduce the formation of Aβ aggregates *in vitro* (Fu *et al.*, 2008). Moreover, it has been reported that Car has a strong effect

Food and Nutritional Components in Focus No. 8
Imidazole Dipeptides: Chemistry, Analysis, Function and Effects
Edited by Victor R Preedy

Published by the Royal Society of Chemistry, www.rsc.org

in restoring mitochondrial functioning and in counteracting amyloid pathology in triple-transgenic Alzheimer's disease model mice. Very recently, a proteomic approach revealed that L-Car affects tumor cell growth by causing interference with protein folding/processing and hypoxia-inducible factor 1-α (HIF-1α) signaling in gliobastomas (Asperger *et al.*, 2011).

In addition, the metal-binding ability of L-Car especially for copper(II) and zinc(II) ions, has extensively been studied *in vitro* (Baran, 2000). The zinc(II)–carnosine complex or polaprezin (Promac®, Zeria Pharmaceutical Co., Ltd.) is a drug effective for the gastric ulcers and other lesions in the alimentary tract (Katayama *et al.*, 2000). The copper- and zinc-mediated neurotoxicity involved in several pathologies, such as amyotrophic lateral sclerosis, Alzheimer's and Parkinson's diseases, might be reduced or prevented by endogenous metal-chelating agents, such as L-Car (Barnham and Bush, 2008).

Biological properties, such as antiaging and antioxidant, make L-Car extensively used for nutraceutical or cosmetic applications (Bellia *et al.*, 2011).

As for all biological species, carnosine homeostasis in humans is carefully regulated by a specific synthase and by carnosinases. Because of the high activity of the serum-circulating carnosinase isoform in humans, the therapeutic uses of carnosine are very limited. The presence of serum carnosinases in humans has been defined as a "paradox", not yet fully understood (Boldyrev *et al.*, 2013). Several attempts have been made to overcome this limitation and exploit the potential therapeutic action of L-Car. The chemical modification of natural carnosine seems to be a very promising approach.

The simplest carnosine derivative reported is *N*-acetyl-carnosine (NAC). This is a naturally occurring Car derivative partially resistant to carnosinase hydrolysis and it has been reported as a carnosine prodrug (Babizhayev *et al.*, 2009). It is commercially available and used as an ophthalmic prodrug for the prevention of cellular senescence and the management of cataracts.

After the simple acetyl carnosine a number of derivatives have been synthesized and characterized.

Generally speaking, the conjugation of carnosine with several types of organic molecules has the main purpose of increasing the blood permanence time of carnosine. In addition, the multifunctional activity of Car can be improved, because of the potential beneficial role of the conjugating moiety and the synergism with the peptide properties as well. Finally, the functionalizing moiety could aim at delivery to a specific target.

D-Carnosine (β-alanyl-D-histidine, D-Car), the enantiomer of the natural dipeptide, has been reported as a very promising compound because it maintains many activities of L-enantiomer, but it is not hydrolyzed by carnosinase and so high concentration could be maintained in the serum (Vistoli *et al.*, 2009). The beneficial effect of D-Car due to the carnosinase resistance in the treatment of acute spinal cord injury has been recently reported. However, D-Car is less bioavailable than L-Car, because it is not recognized by hPepT1, a transporter responsible for the uptake of a broad array of small peptides in the colon.

In this chapter, we will review the data reported on L- and D-Car derivatives and their potential applications.

4.2 L-Carnosine Derivatives

A number of carnosine derivatives have been synthesized with different aims. Most of them have been synthesized with the purpose of obtaining carnosinase-resistant compounds. Derivatives of the amino or carboxylic group have been mainly obtained using different chemical strategies. A summary of the carnosine derivatives is reported in Tables 4.1 and 4.2.

Here, we discuss these derivatives according to the different functionalization site of carnosine. It is important to highlight that the transporter hPepT1 can recognize carnosine derivatives if the carboxylic group is not modified (Bai *et al.*, 1991). On the contrary, the amino group has been tested not to be essential for the recognition of the transporter. Otherwise, the prevention role of glycation reactions into biological systems is related to the free amino group of carnosine.

4.2.1 Derivatization of the Carnosine Amino Group

A number of derivatives have been synthesized by modifying the amino group of the dipeptide (Table 4.1). A largely used synthetic procedure consists in the alkylation of the Car amino group by a nucleophilic substitution reaction using tosylate, iodide or bromide of the wished molecule, as reported in Figure 4.1a.

In other cases, amide derivatives have been synthesized by an amide-coupling reaction between the amino group of carnosine and a selected carboxylic acid, in the presence of typically used activating agents (Figure 4.1b). Carnosine is used as an ester (methyl, ethyl or benzyl) to protect the carboxylic group. The final product obtained as an ester is easily hydrolyzed with bases.

The amino derivatives of Car do not have the primary amino group and this makes this class of derivatives not able to react with reactive carbonyl species (RCS). These derivatives are reported in Table 4.1.

The amino group of carnosine has been modified with antioxidant compounds, vitamins and sugars.

4.2.1.1 Derivatives with Antioxidant and Protective Activities

The carnosine conjugates of *R* or *S*-Trolox ((*S*,*S*)-6-hydroxy-2,5,7,8-tetramethylchroman-2-carboxylic acid) (Table 4.1, **1*R*, 1*S***), the water-soluble analog of vitamin E, have been designed to exploit the cooperative effect of the biological activities shown by the two constituents. **1*R*** and **1*S*** are stable against human serum carnosinase and may circulate longer in the blood stream (Stvolinsky *et al.*, 2012). Their antioxidant activities have been tested by

Table 4.1 Amino derivatives of L-carnosine. The table summarizes derivatives of carnosine synthesized modifying the amino group of the dipeptide. A largely used synthetic procedure consists in the alkylation of the Car amino group by a nucleophilic substitution reaction using tosylate, iodide or bromide (RX) of the desired molecule.

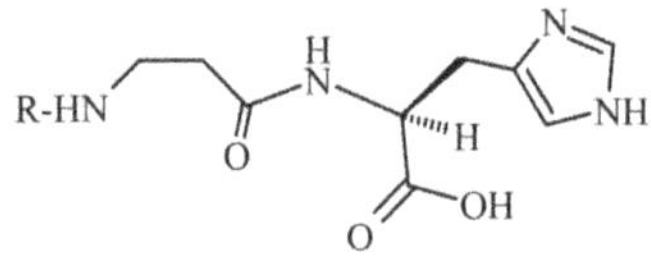

R	Abbreviation
CH_3CO	NAC
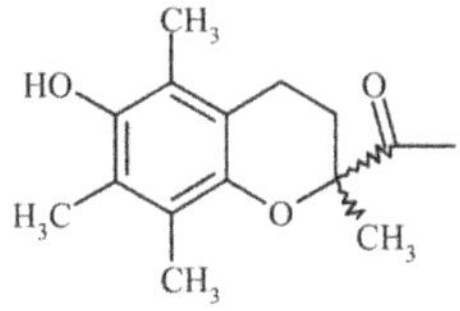 R- or S-Trolox	**1*R*** or **1*S***
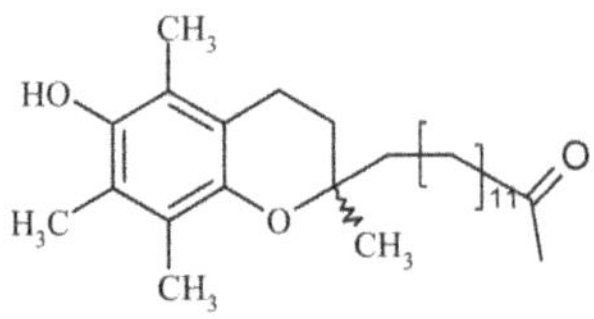 Vitamin E	**2**
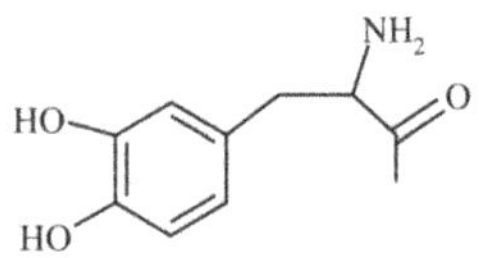 L-Dopa	**3**
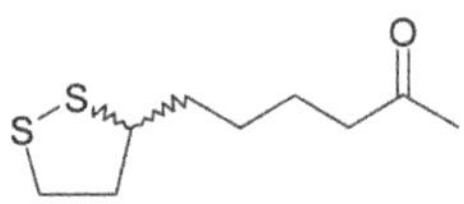 Lipoic acid *R* or *S*	**4*R*** or **4*S***
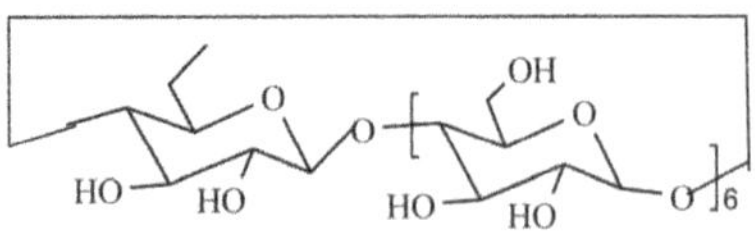 β-Cyclodextrin	**5**
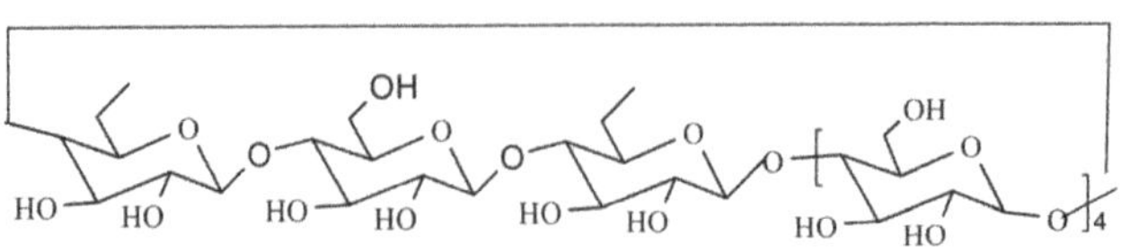 β-Cyclodextrin	**6**

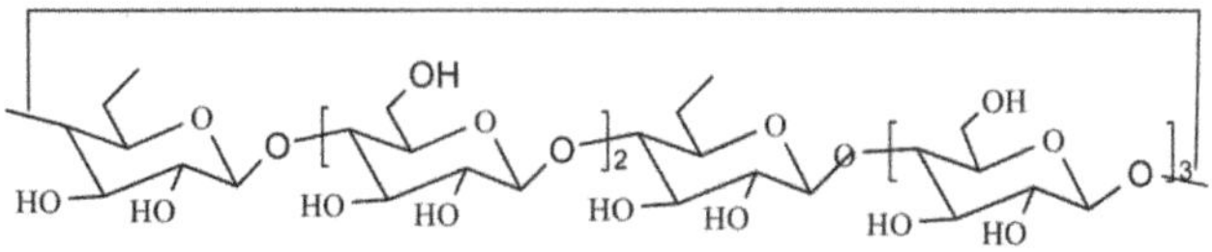

β-Cyclodextrin

7

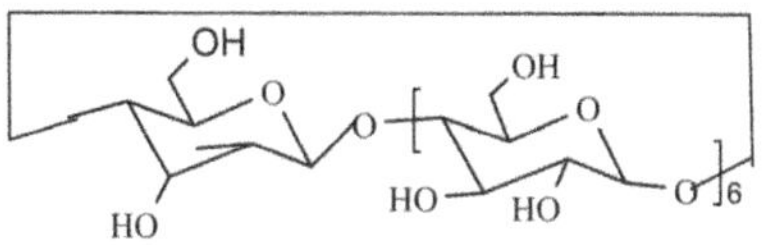

β-Monoaltrocyclodextrin

8

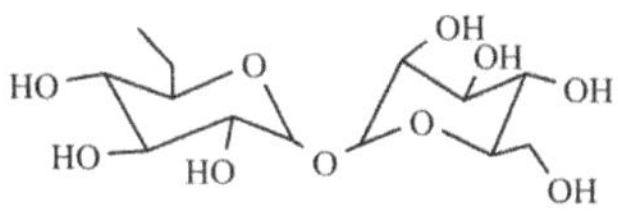

9

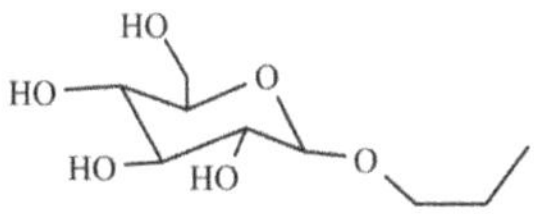

Glucose

10

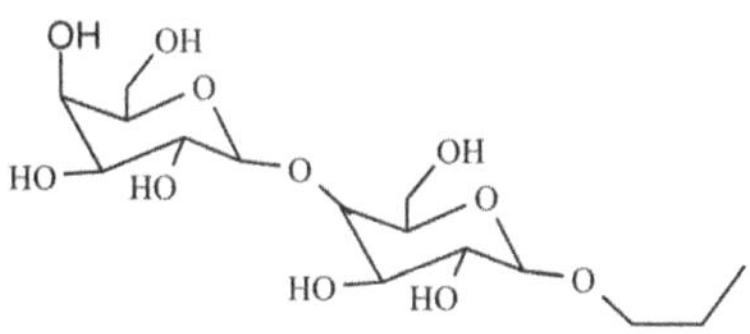

Lactose

11

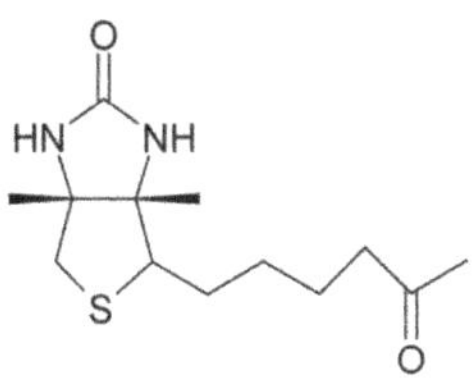

Biotin

12

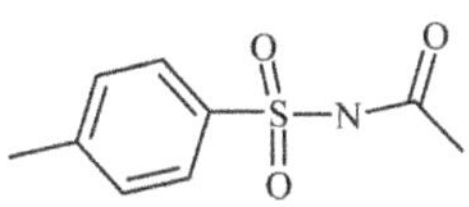

Tosylureido

13

Table 4.2 Carboxylic and amino derivatives of L-carnosine. The table summarizes derivatives of carnosine synthesized modifying the carboxylic group. The primary amide of carnosine is the simplest example of this family of derivatives. The most used synthetic strategy is the amide-coupling reaction starting from the two constituent amino acids such as β-alanine protected with *tert*-butyloxycarbonyl group (Boc) with histidine amide in the presence of typically used activating agents.

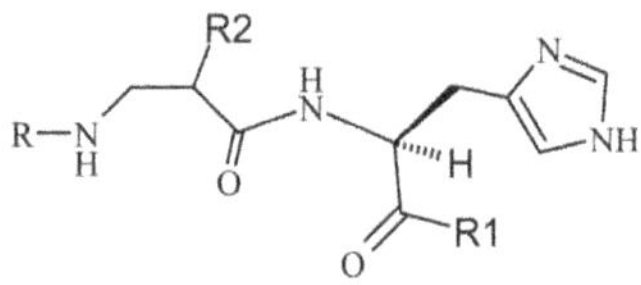

R	R_1	R_2	Abbreviation
H	NH_2	H	**14**
H	$NHCH_3$	H	**15**
H	$NHCH_2CH_3$	H	**16**
H	$NHCH_2{-}C_6H_5$	H	**17**
H	$NHCH_2{-}CH_2{-}CH_3$	H	**18**
H	$NHCH_2{-}CH_2{-}CH_2{-}CH_3$	H	**19**
H	$NHCH_2{-}(CH)_n{-}CH_3$ $n = 4$	H	**20**
H	$NHCH_2{-}(CH)_n{-}CH_3$ $n = 6$	H	**21**
H	$NHCH_2{-}(CH)_n{-}CH_3$ $n = 8$	H	**22**
H	$NHCH_2{-}(CH)_n{-}CH_3$ $n = 10$	H	**23**
H	OCH_2CH_3	H	**24**
H	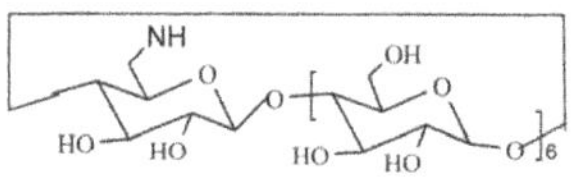 β-Cyclodextrin	H	**25**
H	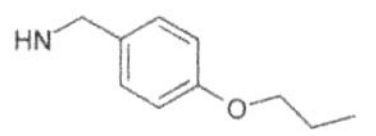	H	**26**
H		H	**27**
H	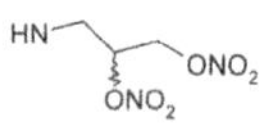 *R* or *S*	H	**28*R*** or **28*S***
H	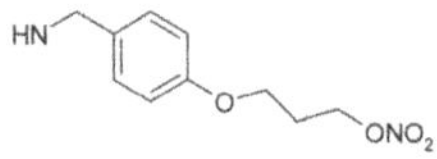	H	**29**
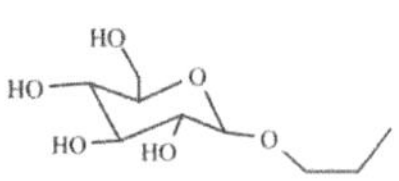 Glucose	NH_2	H	**30**

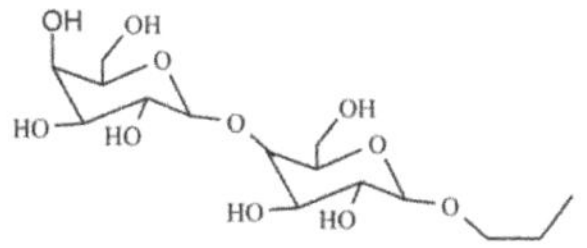 Lactose	NH_2	H	**31**
CH_3CONH	OH	NH_2	**32**
CH_3CONH	OH	CH_3CONH	**33**
$(CH_3)_3$–C–O–CO–*tert*-boc	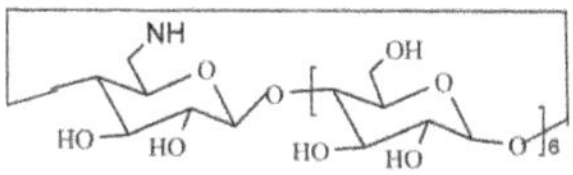 β-Cyclodextrin	H	**34**

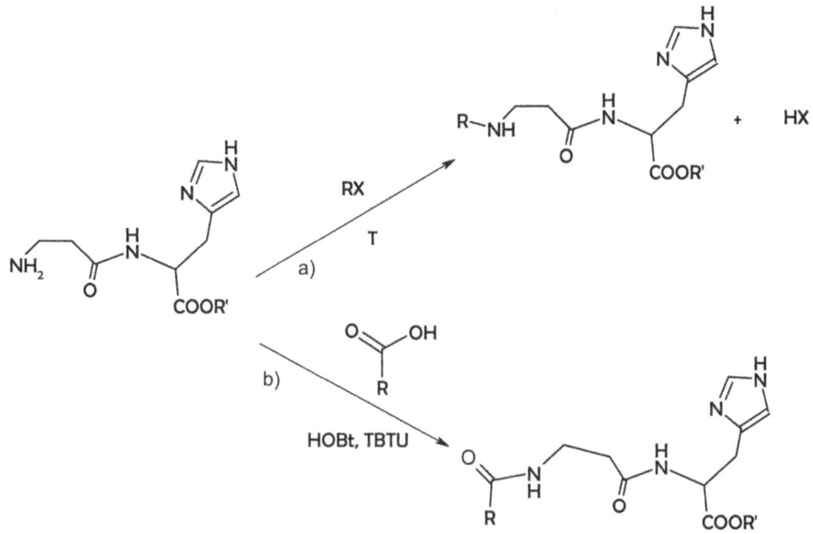

Figure 4.1 Derivatization of the carnosine amino group. The figure summarizes the reaction schemes of carnosine derivatization by modifying the amino group of the dipeptide. (a) Nucleophilic substitution reaction using tosylate, iodide or bromide (RX) of the desired molecule. (R is reported in Table 4.1). (b) Amide coupling reaction between the amino group of carnosine and a carboxylic acid (HOBt is 1-hydroxybenzotriazole, TBTU is *N*,*N*,*N'*,*N'*-tetramethyl-*O*-(benzotriazol-1-yl)uronium tetrafluoroborate). Carnosine methyl, ethyl or benzyl ester is used.

means of several assays (red blood cell hemolysis, DPPH scavenger, lipoprotein oxidation). The conjugates generally show a good activity, in some cases higher than that of both Trolox and Car. A higher quenching activity against the DPPH radical has been reported for the conjugates with respect to that of the constituents. In this assay, the activity has been shown to depend on the configuration of Trolox: the *R*-Trolox conjugate (Table 4.1, **1*R***) is more active than the epimer with *S*-Trolox (Table 4.1, **1*S***). A similar trend has been shown in the protection of rat neurons from oxidative stress: **1*R*** > **1*S*** > L-Car. The antioxidant effects of Trolox conjugates have also been tested on the lifespan of the fruit fly, *Drosophila melanogaster*. In this model, **1*S*** is more active than Car, whereas **1*R*** is less active than the dipeptide. In addition, **1*S*** caused an increase in the life span of both male and female fruit flies, whereas carnosine increased life span only in male fruit flies. These data suggest the presence of additional cellular targets in *D. melanogaster* for *S*-Trolox in comparison with simple neuronal cells. In combination with more hydrophobic nature of Trolox moiety one can expect that **1*S*** or **1*R*** can more easily penetrate the membrane barrier and to be more active as a membrane protector and antistressing agent compared to carnosine. The novel synthetic derivative **1*S*** has been tested *in vivo* and it increases the resistance of rats to acute hypobaric hypoxia by protecting the brain from oxidative-stress damage. This effect is also accompanied by the ability to improve the long-term memory better than carnosine does in Morris water maze test, probably due to the increased efficiency of the brain antioxidant system. The results suggest that **1*S*** is a potential drug in neurologic diseases and in aging (Stvolinsky *et al.*, 2012).

Vitamin E has been also conjugated to Car (Table 4.1, **2**) through a very demanding synthetic procedure (Astete *et al.*, 2013). The main function of vitamin E moiety in the new conjugate **2** is to act as a bioactive shuttle using the tocopherol transfer protein to transport carnosine to different tissues.

L-Car has also been functionalized with L-Dopa (L-3,4-dihydroxyphenylalanine) (Table 4.1, **3**) as a potential prodrug for Parkinson's disease, with the aim of reducing the reactive oxygen species (ROS) generated by the L-Dopa treatment in Parkinson's disease (Sozio *et al.*, 2008). Interestingly, the sustained release of L-Dopa in the rat striatum after the administration of the new compounds has been shown. Anyway, the conjugate **3** has not shown significant antioxidant activity *in vivo*.

New conjugates of *R*- or *S*-α-lipoic acid and Car have been synthesized (Nativi *et al.*, 2013) (Table 4.1, **4*R***, **4*S***). The two diastereoisomers have been isolated but only their mixture (**4**) has been studied. **4** shows very good antioxidant activity. In addition, **4** has been investigated as an antihyperalgesic agent *in vivo*, inducing neurophatic pain through oxaliplatinum. Neither lipoic acid and Car shows antihyperalgesic activity and so their conjugation produces a new activity in **4**. The action of **4** has been related to its ability to act as an antagonist of a transient receptor potential 1 (TRPA1) ion channel. The disulfide bridge of the lipoic acid makes this molecule able to form a covalent bond with cysteine residues of TRPA1 (Wang *et al.*, 2011). The molecule is bonded to TRPA1 and oriented in order to be able to complex Ca^{2+} ion through the carnosine moiety, thus blocking TRPA1, which is regulated by Ca^{2+}. **4** is an antagonist of TRPA1 unlike L-Car and lipoic acid. This compound does not show cytotoxicity toward astrocyte cell viability or cardiotoxicity *in vivo* and for these reasons is a very promising drug for neuropathic pain control (Nativi *et al.*, 2013).

Another class of promising carnosine derivatives are its glycoconjugates. Car has been functionalized with oligosaccharides, such as β-cyclodextrin, in different positions of the sugar (Table 4.1, **5**–**8**). Cyclodextrins are particularly used in pharmaceutical science for their ability to include and stabilize drugs (Bellia *et al.*, 2009). β-Cyclodextrin has been functionalized at the upper rim (primary OHs) with one (**5**) (Bellia *et al.*, 2008) or two Car moieties (Mineo *et al.*, 2004) (**6**, **7**) and monofunctionalized at the lower rim (secondary OHs) (La Mendola *et al.*, 2002) (**8**). The conjugation with cyclodextrins stabilizes the dipeptide against the carnosinase hydrolysis and improves antioxidant activity in comparison to that of the natural dipeptide. The antioxidant activity has been determined by indirect and direct methods, such as the pulse radiolysis technique.

It is worth noting that the prevention of low density lipoprotein (LDL) oxidation induced by copper(II) ions has also been observed with **5** and **8** at concentrations 10–20 times lower than that reported for L-Car (Bellia *et al.*, 2008).

Other Car conjugates of monosaccharides and disaccharides have been synthesized (Lanza *et al.*, 2011) (Table 4.1, **9**, **10**, **11**). The trehalose derivative of carnosine (**9**) has also been synthesized to combine the properties of trehalose and Car (Grasso *et al.*, 2011a). Trehalose is a sugar distributed in many living systems with protective and moisturising functions. Its ability to protect proteins against the denaturation process and conformational changes has been focused on and related to potential application in the treatment of Huntington's disease (Tanaka *et al.*, 2004). The Car–trehalose conjugate has been tested *in vitro* (LDL assay). As in the case of the cyclodextrin moiety, the trehalose increased the antioxidant properties and protected carnosine from the degradation by carnosinase.

4.2.1.2 Delivery Agents

With the aim of selectively addressing carnosine and its antioxidant function, glucose and lactose functionalizations have been carried out (**10**, **11**). An important physiological role of the conjugating moiety in the carnosine derivatization is enhancing the bioavailability of the dipeptide by facilitating the site-specific transport to different tissues. In recent years, it has been shown that some animal lectins, galectins, are important mediators in inflammatory diseases. The key role that lectins play in recognition processes of sugar residues has prompted efforts to synthesize the glycoconjugates of small molecules or proteins to be specifically bound to a selected lectin. In keeping with the behavior typically shown by carnosine conjugates, these kinds of derivatives are stable to carnosinases (Lanza *et al.*, 2011).

A new Car derivative with biotin (H vitamin) (**12**) has been obtained in a one-pot synthesis in order to exploit the ability of biotin to be recognized by streptavidin or avidin. As typically found, the derivative is resistant to the carnosinase action (Bellia *et al.*, 2013).

4-Toluensulfonylureido derivatives of amines have been conjugated to Car (Table 4.1, **13**) and tested as a target moiety for the delivery to tumor cells (Nielsen *et al.*, 2002). Compounds containing this aromatic moiety have been shown to act as anticancer agents for their ability to inhibit the carbonic anhydrase in tumor cells (Supuran *et al.*, 2001). 4-Toluensulfonylureido carnosine (**13**) has been shown to be stable to serum carnosinase and to have good affinity for the hPetT1 transporter. However, *trans*-epithelial transport of **13** was not very effective, thus excluding such an application for carnosine-tosylate.

4.2.2 Derivatization of the Carnosine Carboxylic Group

The carboxylic group has been modified as an ester or an amide (Lanza *et al.*, 2011; Bertinaria *et al.*, 2011) (Table 4.2, **14–34**). The primary amide of Car is the simplest example of this family of derivatives (**14**). The most used synthetic strategy due to the good yield is the amide coupling reaction starting from the two constituent amino acids, such as β-alanine protected with *tert*-butyloxycarbonyl group (Boc) and histidine amide, in the presence of typically used activating agents (Figure 4.2).

In some cases, imidazole nitrogen has also been protected with groups other than Boc.

Amide bond formation seems to be a very promising functionalization because it deeply modifies the resistance of carnosine *versus* human serum carnosinase, while maintaining some important biological functions of the dipeptide. It has been reported that the presence of a free carboxylic group is important in the recognition of the carnosinase enzymes. Therefore, the conversion of carboxyl group into amide makes the derivatives very stable in serum. Furthermore, this class of derivatives has the primary amine group available and so the antiglycating activity is maintained.

In addition to simple amido-carnosine **14**, other secondary amides have been synthesized (Table 4.2, **14–29**).

A recent study reports on the comparative investigation in dependence of the amide side chain lipophilicity (Bertinaria *et al.*, 2011) (Table 4.2, **14–23**). The antioxidant capability of these derivatives has been tested in comparison to that of carnosine and it is not favored by the lipophilicity. Interestingly, the

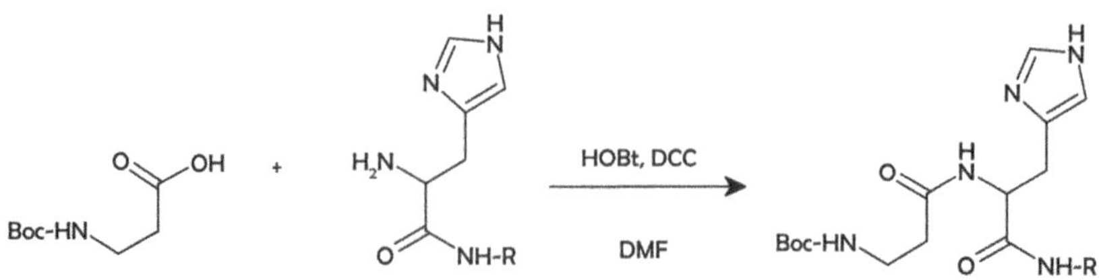

Figure 4.2 Synthetic scheme for carnosine amide derivatives. The figure summarizes an amide-coupling reaction between the amino group of histidine amide and β-alanine protected with *tert*-butyloxycarbonyl group. HOBt is 1-hydroxybenzotriazole, DCC = *N*,*N*'-dicyclohexylcarbodiimide.

modification of the carboxyl group maintains the quenching activity of HNE (4-hydroxy-*trans*-2-nonenal) reported for carnosine, though the derivatives show a lower activity than that of the dipeptide. The amido derivative **18**, which is moderately more hydrophobic with respect to **14**, is able to protect primary hippocampus neurons against HNE-induced death, showing a very significant increase in comparison to L-Car. In fact, carnosine is hydrolyzed by carnosinases and it is not able to exploit any protective activity. Derivative **18** is also able to cross the blood–brain barrier (BBB) and to concentrate in the rat brain after intravenous administration. This finding makes **18** very promising as a neuroprotective agent.

Amino-β-cyclodextrin has also been used to prepare the amide of carnosine (Table 4.2, **25**) and its stability in the human serum has been reported (Bellia *et al.*, 2008). In addition, radical scavenger ability against reactive oxygen species (ROS) has been evaluated for **25**.

4.2.2.1 NO-Donors

Very recently, carnosine has also been conjugated with NO-donor molecules (**26–29**). Microvascular disorders, related to aging or oxidative stress, have been associated to a reduced capacity of cells to produce NO. Furthermore, chronic mild long-lasting cerebrovascular dysfunction, involving basal NO deficit, significantly contributes to plaque development in Alzheimer's disease (AD). The NO-donor carnosine conjugates (Table 4.2, **27–29**) and their analogs without NO-donor moiety (Table 4.2, **18**, **26**) display carnosine-like properties, differing from carnosine for their high serum stability. These compounds are able to complex copper(II) ions at physiological pH with good stability constants of the complexes and protect LDL from oxidation induced by copper(II) ions. All products show moderately potent HNE quenching activity. The NO-donor compounds relaxed rat aorta strips *via* a NO-dependent mechanism. **28*R*** has been selected for an *in vivo* evaluation of organ protection in a model of cerebral ischaemia/reperfusion injury. **18** has been investigated for comparison. Both derivatives protect from hypoxia-induced brain damage at lower concentrations than carnosine; **28*R*** also decreased serum TNF-α levels. This class of NO-donor based on carnosine is worthy of further studies as potential tools for treating chronic vascular and neurodegenerative diseases related to the reduced NO bioavailability (Bertinaria *et al.*, 2012).

4.2.3 Double Functionalization of Carboxylic and Amino Groups

Other derivatives with double functionalization or modified in the ethylenic chain have also been reported. For instance, the amide of carnosine has also been glycoconjugated to an amino group in order to study the metal complexing ability (Lanza *et al.*, 2011) (Table 4.2, **30**, **31**). As expected, these derivatives were resistant to the carnosinases.

Carnosine derivatives whose side chain has been introduced on the ethylenic chain of β-alanine have also been reported (Table 4.2, **32**, **33**). Carnosinase resistance was higher in the case of **33**. Compound **32** showed only partial carnosinase resistance. Derivative **33** was also able to inhibit carnosinase. These compounds have shown antioxidant activity quenching ˙OH and peroxynitrite *in vitro* (Cacciatore *et al.*, 2005).

A cyclodextrin conjugate of L-carnosine protected by Boc (**34**) has been synthesized as an intermediate in the synthesis of **5**. This system has been characterized by NMR and a self-inclusion of the Boc chain has been reported (Vecchio *et al.*, 2001).

4.3 Derivatives of D-Carnosine

The design and synthesis of D-Car prodrugs have been inspired by the fact that the potential applications of L-Car are very limited by its instability in human plasma. Recently, increasing attention has been paid to the functionalization of D-Car and several compounds have been synthesized with the aim of increasing the bioavailability of D-Car. D-Car is poorly recognized by the intestinal hPepT1 transporter and the chemical functionalization has the role to modify D-Car in order to be absorbed at the intestinal level (Orioli *et al.*, 2011). Some of these derivatives have been patented by Flamma S.p.A.

D-Car prodrugs with different computed lipophilicity have been designed (Table 4.3, **35–48**) (Orioli *et al.*, 2011). Derivatives with both amino and carboxyl groups modified were synthesized in order to study the hydrolysis mechanism in rat plasma. Several protective actions have been observed, such as the reduction of carbonyl stress markers, such as advanced glycoxidation product (AGE), of hyperlipidemias and the prevention of renal and vascular injuries. Derivatives that can rapidly release D-Car after ester hydrolysis have been considered for an *in vivo* investigation. The D-Car octyl ester (**47**) has been selected to undergo an extensive evaluation *in vivo* in the Zucker rat.

Mice treated with **47** showed reduced lesion area and a more stable plaque phenotype compared to untreated animals, with reduced foam-cell accumulation, inflammation and apoptosis and increased clearance of apoptotic bodies and collagen deposition, resulting in a decreased necrotic core formation. Likewise, renal lesions were attenuated in D-carnosine octyl ester-treated *versus* untreated mice, with lower inflammation, apoptosis and fibrosis. This was associated with increased urinary levels of HNE–carnosine adducts, reduced protein carbonylation and oxidative stress (Menini *et al.*, 2012).

The D-Car derivatives have also been synthesized to compare the stereochemical differences of the L and D-Car conjugates. Trehalose (**49**) and cyclodextrin (**50**, **51**) conjugates of D-Car have been structurally investigated and their ability to complex metal ions has been correlated to the chirality of histidine (Vecchio *et al.*, 2001; Grasso *et al.*, 2011a,b), These systems are very interesting examples in the field of stereochemistry, but their potentialities as therapeutic agents have not been investigated yet.

Table 4.3 D-Carnosine derivatives. The table summarizes derivatives of D-carnosine. Recently, increasing attention has been paid on the functionalization of D-Car and several compounds have been synthesized with the aim of increasing the bioavailability of D-Car.

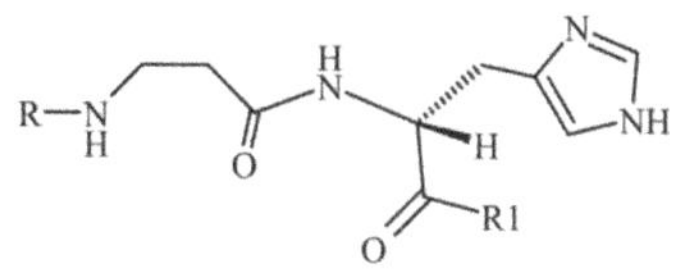

R	R_1	Abbreviation
CH_3CO	OH	**35**
CH_3CO	OCH_3	**36**
CH_3CO	OCH_2CH_3	**37**
CH_3CO	$O(CH_2)_7CH_3$	**38**
$CH_3CH_2O(CO)$	CH_3	**39**
$CH_3CH_2O(CO)$	$OCH(CH_3)_2$	**40**
$CH_3CH_2O(CO)$	$O(CH_2)_3CH_3$	**41**
$C_6H_5CH_2O(CO)$	OH	**42**
$C_6H_5CH_2O(CO)$	CH_3	**43**
$C_6H_5CH_2O(CO)$	OCH_2CH_3	**44**
H	OCH_2CH_3	**45**
H	$OCH(CH_3)_2$	**46**
H	$O(CH_2)_7CH_3$	**47**
H	$O(CH_2)_{15}CH_3$	**48**
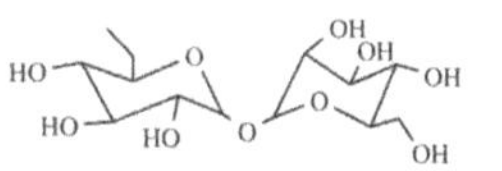 Trehalose	OH	**49**
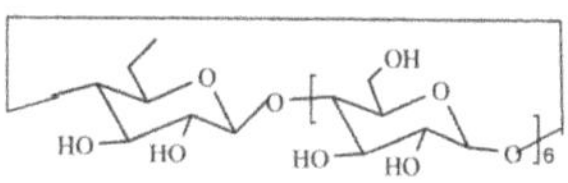 β-Cyclodextrin	OH	**50**
$(CH_3)_3$–C–O–CO–*tert*-boc	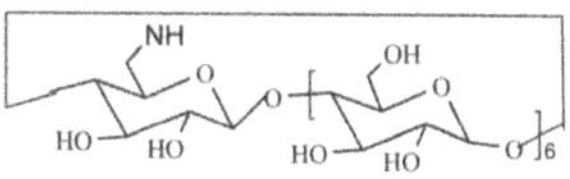	**51**

4.4 Metal Complexes of L- and D-Carnosine Derivatives

The metal-binding properties of L-Car have also been widely investigated due to the hypothesized physiological role of Car as a metal chelator. In addition, the knowledge and the modulation of the complexing ability of carnosine derivatives could be of interest in the development of clinical approaches for the regulation of metal-ion homeostasis. Dyshomeostasis of redox metal ions, such as iron and copper ions, is highly correlated to a number of serious diseases: such as Alzheimer's and Parkinson's diseases, cancer and aging (Barnham and Bush, 2008).

Although L-Car is able to complex various transition-metal ions, the copper(II) complexes have been the most widely studied due to their thermodynamic stability. The copper(II)–L-Car system has largely been investigated through different techniques and a detailed thermodynamic characterization of the complex species has been reported. The stability constants of copper(II) complexes of carnosine derivatives are very similar to those of carnosine (Bellia *et al.*, 2012).

The main species formed at physiological pH and millimolar concentrations of both carnosine and copper(II) is the monomeric species [Cu(L-Car)H_{-1}], together with a secondary dimeric species [Cu_2(L-Car)$_2H_{-2}$] (Figure 4.3). These species have been determined in the most of the copper(II) complexes with amino-functionalized carnosine derivatives.

The copper(II) complexes of L-Car–cyclodextrin conjugate **5** (Table 4.1) are the first example ever reported of carnosine derivative complexes (Bonomo *et al.*, 2003). In spite of the steric hindrance due to the β-cyclodextrin bound to the amino group, these L-Car derivatives show a similar metal-binding behavior of the natural dipeptide in terms of the stoichiometry of the copper complex species and their thermodynamic stabilities. A full characterization of the first member of the glycoconjugate family (**5**) and its copper(II) complexes by means of potentiometric and calorimetric measurements, ESI-MS, EPR and CD spectroscopy has been reported. The cyclodextrin–L-Car conjugate (**8**) forms the monomeric species involving a secondary OH of cyclodextrin in the metal-ion coordination.

β-Cyclodextrins di-functionalized with L-Car **6** and **7** (Table 4.1) have also been characterized for their abilities to form metal complexes (Figure 4.4). Interestingly, from the point of view of supramolecular chemistry, it has been found that in the case of the 6A,6C L-Car–CD conjugate (Table 4.1, **6**), the formation of hexameric complex species is induced by the coordination

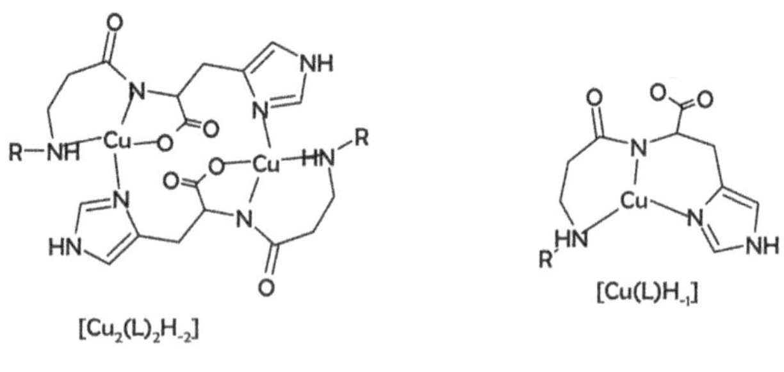

Figure 4.3 Schematic structures of copper(II) complexes of carnosine derivatives. L is a generic derivative of carnosine. In the case of carnosine R = H. The main species formed at physiological pH and millimolar concentrations of both carnosine and copper(II) is the monomeric species [Cu(L-Car)H_{-1}], together with a secondary dimeric species [Cu_2(L-Car)$_2H_{-2}$]. These species have been determined in most of the copper(II) complexes with amino-functionalized carnosine derivatives.

of copper(II) with the L-Car units (Figure 4.4a). The regioisomer 6A,6D (Figure 4.4b) forms only dimeric species involving two ligands and four metal ions.

Copper(II) complexes of D-Car conjugates with β-cyclodextrin (Grasso *et al.*, 2011a) (Table 4.2, **16**) or trehalose (Grasso *et al.*, 2011b) (**15**) have also been reported. In these cases, the dimeric species showed a significantly lower stability constant than those reported for the analogs of L-Car derivatives. This stereoselectivity is driven by noncovalent interactions such as

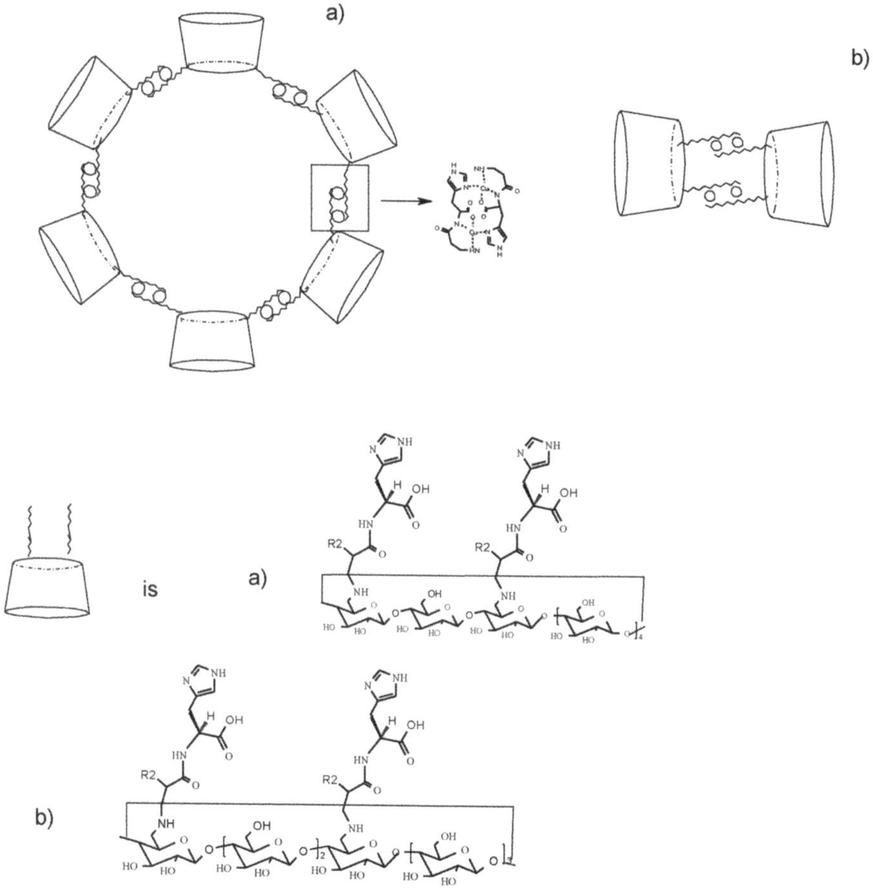

Figure 4.4 Schematic structures of copper(II) complexes of cyclodextrin conjugates with carnosine. (a) Hexameric complexes of β-cyclodextrin A,C-difunctionalized with carnosine; (b) dimeric complexes of β-cyclodextrin A,D-difunctionalized with carnosine. β-Cyclodextrins 6-difunctionalized with L-Car form copper(II) metal complexes. In the case of the L-Car conjugates at 6A,6C positions of cyclodextrins (a) hexameric complex species are formed by the coordination of copper(II) with the L-Car units. The regioisomer 6A,6D (b) forms only dimeric species involving two ligands and four metal ions.

hydrogen bonds, CH–π interactions, hydrophobic and steric effects involving the chiral moieties cyclodextrin or trehalose.

The copper(II) complex of L-Car shows superoxide dismutase (SOD) activity. The ability to catalyze the dismutation of the superoxide anion is also found for the copper complexes with the L-Car–cyclodextrin derivatives. Moreover, it is interesting to note that the carnosine derivatization with β-cyclodextrin induces an increase of the SOD activity in indirect assays with respect to that of the L-Car complexes.

The ester (**24**) and amide derivatives of L-carnosine (Table 4.2, **14** and **23**) have also shown the ability to bind copper(II) ion, forming complex species very similar to those of the copper(II)–L-Car system (Table 4.1). Only amide derivatives are able to form dimeric species (Lanza *et al.*, 2011; Bertinaria *et al.*, 2012), though with lower thermodynamic stability than carnosine. The amide group is able to coordinate the metal ion in the deprotonated form. L-Carnosine ethyl ester is not able to form dimeric species due to the lack of carboxylate, differently from L-Car.

Copper(II) stability constants have been also determined for the NO donors, which maintain similar ability to complex copper(II) as carnosine. It is interesting to note that the diastereisomeric complexes of copper(II) with **28*R*** or **28*S*** (Table 4.2) show a significant difference in the ability to complex copper(II) in forming the dimeric species.

4.5 Nanosystems Based on Carnosine

Nanoparticles (NPs) are a new class of emerging therapeutics in nanomedicine (Doane and Burda, 2012).

Recently, the interest for NPs has been increasing in different fields and also NPs based on carnosine have been prepared. NPs of ferrite have been coated with carnosine, exploiting the ability of carboxylic group to bind iron oxide. These magnetic NPs have been fully characterized.

In order to coat gold nanoparticles with carnosine, different approaches have been used. Carnosine functionalized with L-lipoic acid (Table 4.1, **4**) has been used to exploit reactivity of sulfur with gold (Saada *et al.*, 2011). These NPs have been investigated as activators of carbonic anhydrase. In this system the role of carnosine has been related to the presence of histidine, as the activity of carnosine nanoparticles is very similar to that of histidine nanoparticles.

Recently, the biotin carnosine conjugates (Table 4.1, **12**) have been used to prepare gold NPs using a noncovalent approach (Bellia *et al.*, 2013). The Gold NPs have been coated with streptavidin (SA) or avidin (AV). SA and AV have a very high affinity to biotin and its derivatives with association constants about 10^{14} M^{-1}. This strong bonding has been exploited to introduce carnosine molecules on the NP surface.

The chemical derivatization of carnosine with a bulky hydrophobic aromatic unit *N*-(fluorenyl-9-methoxycarbonyl) produced an amphiphilic compound able to form fibrillar structures based on antiparallel β-sheets.

Moreover, the chelation of zinc(II) by the carnosine moiety leads to a transition from twisted fibrils into wide nanotapes. For these reasons, this system might have applications in diagnostic or delivery systems for healthcare. The formation of nanotapes was also induced by a C_{16} chain linked to the carnosine moiety (Castelletto *et al.*, 2011).

4.6 Conclusions

Nature has been a source of medicinal products for millennia, with many useful drugs developed especially from plant sources. Chemical modification of natural compounds together with combinatorial chemistry has shifted the focus of drug-discovery efforts from Nature to the laboratory bench.

The natural molecule carnosine shows interesting potentiality in a number of diseases together with the big limitation of its very short life in human serum. Chemical modification is very effective in obtaining new carnosine derivatives with improved potentiality. A large number of carnosine derivatives have been synthesized and structurally characterized, and some of them have been investigated *in vivo*.

The main features of the carnosine derivatives are the stability in the serum against carnosinase and the synergic activity of carnosine with the functionalizing moiety that often enhances the beneficial effects of the dipeptide. In addition, the functionalizing moiety can improve the lipophilicity of the molecule to aid the BBB crossing.

The carnosinase resistance has been reported for most of the carnosine conjugates, whether the dipeptide is derivatized through the amine or the carboxyl group. This is in keeping with the fact that both the functional groups are essential for the carnosinase recognition. The scavenger abilities against ROS, RCS and HNE have been the most studied activities of the carnosine derivatives in a number of different models and, in several cases, it has been improved with respect to that of the natural dipeptide. Derivatives have been tested by very different assays and for this reason, a structure–activity relationship cannot be derived for all the derivatives synthesized.

Finally, the carnosine ability to bind several metal ions, mainly copper(II) and zinc(II), has been exploited to design several conjugates able to form metal complexes with different stoichiometry and stability.

On the basis of the wide interest in carnosine, these derivatives represent very promising systems. Among them, a system having the main important therapeutic actions of carnosine improved by the resistance to carnosinase will probably be selected in the future for clinical trials.

Summary Points

- This chapter focuses on carnosine derivatives.
- A number of carnosine derivatives have been synthesized by different chemical procedures.

- Carnosine derivatives are differently designed to conjugate the biological activity of the carnosine with that of the functionalizing moiety.
- The chemical functionalization has the aim of stabilizing carnosine against the carnosinase enzymes.
- The chemical functionalization is a very promising approach in improving the potential therapeutic uses of carnosine.
- Due to the reactive oxygen species (ROS) and reactive carbonyl species (RCS) scavenger action of carnosine, its chemical modification can have applications in the case of neurodegenerative diseases.
- In the last years carnosine derivatives have been tested *in vivo* with very successful results.
- Until now there have been no application of these derivatives *in vivo* in human beings.

Key Facts of Chemical Modifications of Natural Products (Carnosine being among them)

1. Natural products traditionally have played an important role in drug discovery and were the basis of most early medicines.
2. A number of currently used drugs are derivatives of natural products.
3. Important drugs for infectious immunological, cardiovascular, neurological, inflammatory and cancer diseases have been discovered starting from natural products.
4. Aspirin, one of the most famous drugs, which is an example of a naturally derived product was discovered in the late 1890s and is still being used as an analgesic and anti-inflammatory drug.
5. Penicillin created in interest in discovering other compounds from natural sources for the treatment of bacterial infections, resulting in nearly all novel classes of antibiotics being from natural product sourced scaffolds until 1962.

Definitions of Words and Terms

Biotin. Biotin, vitamin H or vitamin B7 is a water-soluble vitamin. It is a coenzyme for the synthesis of fatty acids, some amino acids and carbohydrates.

Cyclodextrin. Cyclodextrins (CDs) are cyclic oligomers of D(+)-glucopyranosyl units linked by a-1,4-glycosidic bonds. The most common CDs are α-, β- and γ-CD with six, seven or eight glucose units, respectively. These molecules are water soluble and have a truncated cone shape and a hydrophobic cavity. They are widely used in cosmetics, drug formulations, and chromatography.

Drug delivery. Drug delivery is the method of administering pharmaceuticals. Drug delivery is finalized to improve the bioavailability of the drug.

Glycation. It is the nonenzymatic reaction of biological molecules with reducing sugars. *In vivo* the glycated products are known as advanced glycation end products, or AGES. AGES are mediators of various complications of diabetes and aging.

HNE (4-hydroxy-(*E*)-2-nonenal). HNE is a reactive carbonyl species produced by lipid peroxidation, often tested to assay the reactive carbonyl species (RCS) scavenger ability of compounds.

Lectin. Lectins are a family of proteins able to bind carbohydrates in a selective manner. Lectins are involved in a number of important biological process through specific carbohydrate–lectin recognition.

TRPA1. Transient receptor potential cation channel, subfamily A, member 1 (TRPA1) is an ion channel in the plasma membrane of animal cells. It acts as a sensor for pain and cold.

Nanoparticle (NP). NPs are systems of nanometric dimension, used with different aims. A large number of nanoparticles based on organic systems, such as lipids, polymers, peptides, and polysaccharides have been synthesized. In some cases, organic molecules have been used to functionalize inorganic NPs of gold, ferrite, silica and others to conjugate their properties to the inorganic core properties. Some of them have been applied in the medical field for the treatment of cancer as nanotherapeutics.

NO-donor. This is a molecule able to release NO. Some $^{\cdot}NO$ donors have potential therapeutic value. In fact, NO is a highly reactive radical and plays an important role in the regulation of a wide range of physiological processes, including cellular immunity, angiogenesis, neurotransmission, and platelet aggregation. $^{\cdot}NO$ is synthesized by the nitric oxide synthases (NOS) from the amino acid L-arginine.

Prodrug. This is a molecule able to release the drug *in vivo* for enzymatic action or other metabolic processes. Prodrug is inactive and is used to improve the bioavailability of the drug and to reduce its toxicity.

Reactive carbonyl species (RCS). Saturated carbonyls can react with amino groups to form a Schiff base. In addition to this reactivity, RCS have high electrophilicity due to the α,β-unsaturated bond and hence form Michael adducts with nucleophilic targets such as thiol- and amino groups. Some RCS such as the 4-hydroxy-(*E*)-2-nonenal (HNE) and 4-oxo-(*E*)-2-nonenal are very reactive and toxic.

Reactive oxygen species (ROS). ROS (including $O_2^{-\cdot}$, H_2O_2, $^{\cdot}OH$) are constitutively produced in aerobic cells as byproducts of the cellular respiration. ROS are important messengers in normal cell signaling. Various environmental and pathological stresses lead to a rise the ROS level in cells. A variety of diseases have been associated with ROS dyshomeostasis.

Trehalose. Trehalose is a disaccharide formed by α-glucose units 1,1′ linked. It is a nonreducing sugar. It is found in different species. In prokaryotes, trehalose is thought to serve as a carbon source and a structural component. In some bacteria such as *Escherichia coli*, trehalose is synthesized in response to high osmolality. In the animal kingdom, trehalose has been found in very high amounts. Trehalose has not been found in higher vertebrates. Maintaining the three-dimensional structure and stabilizing biomolecules is one of the most important properties of trehalose.

List of Abbreviations

AV	Avidin
BBB	Blood–brain barrier
Car	Carnosine
CD	Cyclodextrin
DPPH	2,2-Diphenylpicrylhydrazyl or di(phenyl)-(2,4,6-trinitrophenyl) iminoazanium
ESI MS	Electrospray ionization mass spectrometry
EPR	Electron paramagnetic resonance
HIF-1α	Hypoxia-inducible factor 1-α
HNE	4-Hydroxy-(*E*)-2-nonenal
hPepT1	Human peptide transporter 1
LDL	Low-density lipoprotein
NAC	*N*-Acetyl carnosine
NP	Nanoparticle
RCS	Reactive carbonyl species
ROS	Reactive oxygen species
SA	Streptavidin
SOD	Superoxide dismutase
TRPA1	Transient receptor potential cation channel, A1

Acknowledgements

We thank all the coworkers, whose names are listed in the references for their contributions to this research. We thank MIUR (2008R23Z7K, PRIN 2008F5A3AF_005, FIRB RINAME) for financial support.

References

Asperger, A., Renner, C., Menzel, M., Gebhardt, R., Meixensberger, J. and Gaunitz, F., 2011. Identification of factors involved in the anti-tumor activity of carnosine on glioblastomas using a proteomics approach. *Cancer Investigation*. 29: 272–281.

Astete, C. E., Songe Meador, D., Spivak, D. and Sabliov, C., 2013. Synthesis of vitamin E-carnosine (VECAR): new antioxidant molecule with potential application in atherosclerosis. *Synthetic Communications*. 43: 1299–1313.

Babizhayev, M. A., Micans, P., Guiotto, A. and Kasus-Jacobi, A., 2009. *N*-Acetylcarnosine lubricant eyedrops possess all-in-one universal antioxidant protective effects of L-carnosine in aqueous and lipid membrane environments, aldehyde scavenging, and transglycation activities inherent to cataracts: A clinical study of the new vision-saving drug *N*-acetylcarnosine eyedrop therapy in a database population of over 50 500 patients. *American Journal of Therapeutics*. 16: 517–533.

Bai, F. P., Subramaniam, P., Mosberg, H. I. and Amidon, G. L., 1991. Structural requirement for the intestinal mucosal-cell peptide transporter: the need for N-terminal alpha-amino group. *Pharmaceutical Research*. 8: 593–599.

Baran, E. J., 2000. Metal complexes of carnosine. *Biochemistry (Mosc).* 65: 789–797.

Barnham, K. J. and Bush, A. I., 2008. Metals in Alzheimer's and Parkinson's diseases. *Current Opinion in Chemical Biology*. 12: 222–228.

Bellia, F., Amorini, A. M., La Mendola, D., Vecchio, G., Tavazzi, B., Giardina, B., Di Pietro, V., Lazzarino, G. and Rizzarelli, E., 2008. New glycosidic derivatives of histidine-containing dipeptides with antioxidant properties and resistant to carnosinase activity. *European Journal of Medicinal Chemistry.* 43: 373–380.

Bellia, F., La Mendola, D., Pedone, C., Rizzarelli, E., Saviano, M. and Vecchio, G., 2009. Selectively functionalized cyclodextrins and their metal complexes. *Chemical Society Reviews*. 38: 2756–2781.

Bellia, F., Vecchio, G., Cuzzocrea, S., Calabrese, V. and Rizzarelli, E., 2011. Neuroprotective features of carnosine in oxidative driven diseases. *Molecular Aspects of Medicine*. 32: 258–266.

Bellia, F., Vecchio, G. and Rizzarelli, E., 2012. Carnosine derivatives: new multifunctional drug-like molecules. *Amino Acids*. 43: 153–163.

Bellia, F., Oliveri, V., Rizzarelli, E. and Vecchio, G., 2013. New derivative of carnosine for nanoparticle assemblies. *European Journal of Medicinal Chemistry*. 70: 225–236.

Bertinaria, M., Rolando, B., Giorgis, M., Montanaro, G., Guglielmo, S., Buonsanti, M. F., Carabelli, V., Gavello, D., Daniele, P. G., Fruttero, R. and Gasco, A., 2011. Synthesis, physicochemical characterization, and biological activities of new carnosine derivatives stable in human serum as potential neuroprotective agents. *Journal of Medicinal Chemistry*. 54: 611–621.

Bertinaria, M., Rolando, B., Giorgis, M., Montanaro, G., Marini, E., Collino, M., Benetti, E., Daniele, P. G., Fruttero, R. and Gasco, A., 2012. Carnosine analogues containing NO-donor substructures: synthesis, physico-chemical characterization and preliminary pharmacological profile. *European Journal of Medicinal Chemistry*. 54: 103–112.

Bonomo, R. P., Bruno, V., Conte, E., De Guidi, G., La Mendola, D., Maccarrone, G., Nicoletti, F., Rizzarelli, E., Sortino, S. and Vecchio, G., 2003. Potentiometric, spectroscopic and antioxidant activity studies of SOD mimics containing carnosine. *Dalton Transactions*. 4406–4415.

Boldyrev, A. A., Aldini, G. and Derave, W., 2013. Physiology and pathophysiology of carnosine. *Physiological Reviews*. 93: 1803–1843 and references therein.

Boldyrev, A., 2005. Carnosine protects the brain of rats and Mongolian gerbils against ischemic injury: after-stroke-effect. *Neurochemical Research.* 30: 1283–1288.

Cacciatore, I., Cocco, A., Costa, M., Fontana, M., Lucente, G., Pecci, L. and Pinnen, F., 2005. Biochemical properties of new synthetic carnosine analogues containing the residue of 2,3-diaminopropionic acid: the effect of *N*-acetylation. *Amino Acids*. 28: 77–83.

Castelletto, V., Cheng, G., Greenland, B. W., Hamley, I. W. and Harris, P. J. F., 2011. Tuning the self-assembly of the bioactive dipeptide l -carnosine by incorporation of a bulky aromatic substituent. *Langmuir*. 27: 2980–2988.

Derave, W. and Sale, C., 2012. Carnosine in exercise and disease: introduction to the International Congress held at Ghent University, Belgium, July 2011. *Amino Acids*. 43: 1–4.

Doane, T. L. and Burda, C., 2012. The unique role of nanoparticles in nanomedicine: Imaging, drug delivery and therapy. *Chemical Society Reviews*. 41: 2885–2911.

Dobrota, D., Fedorova, T., Stvolinsky, S., Babusikova, E., Likavcanova, K., Drgova, A., Strapkova, A. and Boldyrev, A., 2005. Carnosine protects the brain of rats and Mongolian gerbils against ischemic injury: after-stroke-effect. *Neurochemical Research*. 30: 1283–1288.

Fu, Q., Dai, H., Hu, W., Fan, Y., Shen, Y., Zhang, W. and Chen, Z., 2008. Carnosine protects against Abeta42-induced neurotoxicity in differentiated rat PC12 cells. *Cellular and Molecular Neurobiology*. 28: 307–316.

La Mendola, D., Sortino, S., Vecchio, G. and Rizzarelli, E., 2002. Synthesis of new carnosine derivatives of β-cyclodextrin and their hydroxyl radical scavenger ability. *Helvetica Chimica Acta*. 85: 1633–1643.

Lanza, V., Bellia, F., D'Agata, R., Grasso, G., Rizzarelli, E. and Vecchio, G., 2011. New glycoside derivatives of carnosine and analogs resistant to carnosinase hydrolysis: synthesis and characterization of their copper(II) complexes. *Journal of Inorganic Biochemistry*. 105: 181–188.

Grasso, G. I., Arena, G., Bellia, F., Maccarrone, G., Parrinello, M., Pietropaolo, A., Vecchio, G. and Rizzarelli, E., 2011a. Intramolecular weak interactions in the thermodynamic stereoselectivity of copper(II) complexes with carnosine–trehalose conjugates. *Chemistry – A European Journal*. 17: 9448–9455.

Grasso, G. I., Bellia, F., Arena, G., Vecchio, G. and Rizzarelli, E., 2011b. Noncovalent interaction-driven stereoselectivity of copper(II) complexes with cyclodextrin derivatives of L- and D-carnosine. *Inorganic Chemistry*. 50: 4917–4924.

Katayama, S., Nishizawa, K., Hirano, M., Yamamura, S. and Momose, Y., 2000. Effect of polaprezinc on healing of acetic acid-induced stomatitis in hamsters. *Journal of Pharmacy & Pharmaceutical Sciences*. 3: 114–117.

Menini, S., Iacobini, C., Ricci, C., Scipioni, A., Blasetti Fantauzzi, C., Giaccari, A., Salomone, E., Canevotti, R., Lapolla, A., Orioli, M., Aldini, G. and Pugliese, G. D., 2012. Carnosine octylester attenuates atherosclerosis and renal disease in ApoE null mice fed a Western diet through reduction of carbonyl stress and inflammation. *British Journal of Pharmacology*. 166: 1344–1356.

Mineo, P., Vitalini, D., La Mendola, D., Rizzarelli, E., Scamporrino, E. and Vecchio, G., 2004. Coordination features of difunctionalized β-cyclodextrins with carnosine: ESI-MS and spectroscopic investigations on 6A,6D-di(β-alanyl-L-histidine)-6A,6D-dideoxy-*b*-cyclodextrin and 6A,6C-di(β-alanyl-L-histidine)-6A,6C-dideoxy-β-cyclodextrin and their copper(II) complexes. *Journal of Inorganic Biochemistry*. 98: 254–265.

Nativi, C., Gualdani, G., Dragoni, E., Cesare, Di, Mannelli, L., Sostegni, S., Norcini, M., Gabrielli, G., La Marca, G., Richichi, B., Francesconi, O., Moncelli, M. R., Ghelardini, C. and Roelens, S., 2013. A TRPA1 antagonist reverts oxaliplatin-induced neuropathic pain. *Scientific Reports*. 3: 2005. DOI: 10.1038/srep02005.

Nielsen, C. U., Supuran, C. T., Scozzafava, A., Frokjaer, S., Steffansen, B. and Brodin, B., 2002. Transport characteristics of L-carnosine and the anticancer derivative 4-toluenesulfonylureido-carnosine in a human epithelial cell line. *Pharmaceutical Research*. 19: 1337–1344.

Orioli, M., Vistoli, G., Regazzoni, L., Pedretti, A., Lapolla, A., Rossoni, G., Canevotti, R., Gamberoni, L., Previtali, M., Carini, M. and Aldini, G., 2011. Design, synthesis, ADME properties, and pharmacological activities of beta-alanyl-D-histidine (D-carnosine) prodrugs with improved bioavailability. *ChemMedChem*. 6: 1269–1282.

Saada, M. C., Montero, J. L., Vullo, D., Scozzafava, A., Winum, J. Y. and Supuran, C. T., 2011. Carbonic anhydrase activators: gold nanoparticles coated with derivatized histamine, histidine, and carnosine show enhanced activatory effects on several mammalian isoforms. *Journal of Medicinal Chemistry*. 54: 1170–1177.

Sozio, P., Iannitelli, A., Cerasa, L. S., Cacciatore, I., Cornacchia, C., Giorgioni, G., Ricciutelli, M., Nasuti, C., Cantalamessa, F. and Di Stefano, A., 2008. New L-dopa codrugs as potential antiparkinson agents. *Archiv der Pharmazie*. 341: 412–417.

Stvolinsky, S., Toropova, K., Gordeeva, M., Kazey, V., Sato, T., Meguro, K. and Boldyrev, A., 2012. Carnosine and its (*S*)-Trolox™ derivative protect animals against oxidative stress. *Amino Acids*. 43: 165–170.

Supuran, C. T., Briganti, F., Tilli, S., Chegwidden, W. R. and Scozzafava, A., 2001. Carbonic anhydrase inhibitors: sulfonamides as antitumor agents?. *Bioorganic & Medicinal Chemistry*. 9: 703–714.

Tanaka, M., Machida, Y., Niu, S., Ikeda, T., Jana, N. R., Doi, H., Kurosawa, M., Nekooki, M. and Nukina, N., 2004. Trehalose alleviates polyglutamine-mediated pathology in a mouse model of Huntington disease. *Nature Medicine*. 10: 148–154.

Vecchio, G., La Mendola, D. and Rizzarelli, E., 2001. The synthesis and conformation of β-cyclodextrins functionalized with enantiomers of Boc-carnosine. *Journal of Supramolecular Chemistry*. 1: 87–95.

Vistoli, G., Orioli, M., Pedretti, A., Regazzoni, L., Canevotti, R., Negrisoli, G., Carini, M. and Aldini, G., 2009. Design, synthesis, and evaluation of carnosine derivatives as selective and efficient sequestering agents of cytotoxic reactive carbonyl species. *ChemMedChem*. 4: 967–975.

Wang, Y. Y., Chang, R. B., Allgood, S. D., Silver, W. L. and Liman, E. R., 2011. A TRPA1-dependent mechanism for the pungent sensation of weak acids. *Journal of General Physiology*. 137: 493–505.

CHAPTER 5

Chiral Carnosine

DAN SU[a] AND YONGGUI SONG*[a]

[a]The National Pharmaceutical Engineering Center for Solid Preparation in Chinese Herbal Medicine, Jiangxi University of Traditional Chinese Medicine, 56 Yangming Road, Nanchang, 330006, China
*E-mail: songkang9925@163.com

5.1 Structure, Physical and Chemical Properties

The chemical name of carnosine is 2-(3-aminopro-panoylamino)-3-(3*H*-imidazol-4-yl)propanoic acid (IUPAC-*IUB* Commission on Biological Nomenclature, 1971; Barrans *et al.*, 1976). The formula of carnosine is $C_9H_{14}N_4O_3$, and the relative molecular mass is 226.24. As shown in Figure 5.1, L-carnosine contains an *S*-configuration chiral carbon atom. Both are white crystalline powder, odorless. L-, and D-Carnosine (Figure 5.1) are soluble in water, 0.1 mol L^{-1} HCl, 0.1 mol L^{-1} NaOH, and almost insoluble in methanol, acetonitrile, ethanol, chloroform, and acetone. The melting point of them is 251.0–253.2 °C. The specific rotation of L-carnosine is $[\alpha]_D^{20}$ + 20.6 (Sui *et al.*, 2009).

5.2 Extraction and Synthesis

Natural L-carnosine is initially extracted from animal muscles. Now, there are many papers and patents published regarding various synthetic pathways for L-carnosine. High-purity D-carnosine can also be prepared by chemical synthesis (King and Kidd, 1949; Turner, 1953; Rinderknecht *et al.*, 1964; Vinick,

Food and Nutritional Components in Focus No. 8
Imidazole Dipeptides: Chemistry, Analysis, Function and Effects
Edited by Victor R Preedy

Published by the Royal Society of Chemistry, www.rsc.org

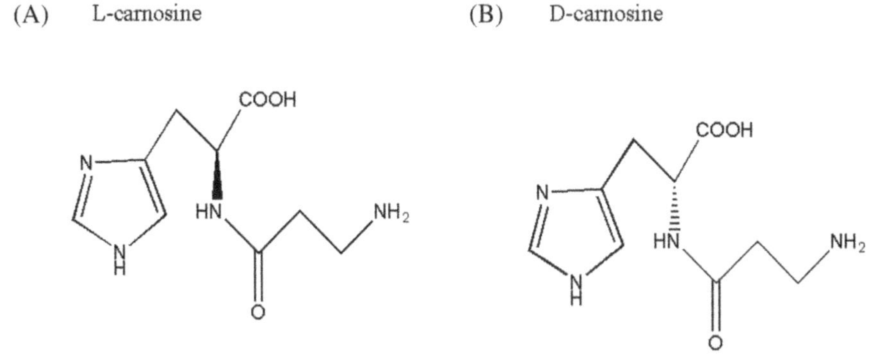

Figure 5.1 Structure of L-carnosine (A), and D-carnosine (B).

1982; Skoblik *et al.*, 1995; Stafan *et al.*, 2001; Yamaoka *et al.*, 2005; Lee *et al.*, 2007; Shasad *et al.*, 2007; Sui *et al.*, 2009; Albinati *et al.*, 2010).

5.3 Physiological Activities

5.3.1 L-Carnosine

In mammals, carnosine synthesized by carnosine synthetase is widely distributed in various tissues, especially abundant in muscle, brain, and eye lens up to 20 mmol L^{-1} (Pegova *et al.*, 2000). As a natural dipeptide, most physiological activities of L-carnosine has basically been a consensus such as buffering physiological pH, chelating metal ions, scavenging free radicals and antilipid peroxidation.

5.3.1.1 Buffering Physiological pH

The adduction chemistry of carnosine involves the amino group, carboxyl group and the imidazole ring of L-histidine. This composition contributes to buffer physiological pH *in vivo* for receiving or giving protons. Consequently, L-carnosine can reduce lipid peroxidation caused by pH fluctuations. Abe revealed that proton buffering was the main physiological function of histidine-related compounds (Abe, 2000). In addition, the buffering capacities of carnosine in the pH range of 6.5–7.5 were high and relatively insensitive to changes in temperature within the range from 5 to 40 °C (Abe and Okuma, 1991).

5.3.1.2 Chelating Metal Ions

The imidazole ring from amino acid residues of L-carnosine has the ability to chelate metal ions such as copper, zinc, cobalt, ferrous ions, *etc.* L-Carnosine was found to form a complex with copper that decreases its catalytic activity and slows the lipid peroxidation (Decker *et al.*, 1992). As a hydrogen donor, L-carnosine maintains the iron ion divalent to prevent free-radical generation by electron transfer with oxygen (Kohen *et al.*, 1988). The property of chelating copper or zinc ions makes L-carnosine exhibit neuroprotection (Boldyrev and Abe, 1999).

5.3.1.3 *Scavenging Free Radicals*

Carnosine is an effective free-radical scavenger, because the histidine residues on its side chain can serve as a hydrogen acceptor to capture hydroxyl radicals, elemental oxygen and hydrogen-peroxide radicals, and can also inhibit fat oxidation induced by whether metal or nonmetal ions. Stvolinsky revealed carnosine can combine with $O_2^{\cdot}$ to form charge-transfer complexes to prevent the conversion of H_2O_2 to $HO^{\cdot}$, which is more toxic, and thereby mitigate the damage to the brain tissue caused by the reactive oxygen species (Stvolinsky *et al.*, 2000). Stecchini reported that it could clear the O_2^-, OH^-, peroxide radicals, and thus could inhibit the oxidation of lipids. Therefore, carnosine can reduce the oxidation of biological macromolecules and protect normal biological functions of cells (Stecchini *et al.*, 1998).

5.3.2 D-Carnosine

D-Carnosine is a non-naturally occurring dipeptide. In recent years, people have been increasingly concerned about its physical activities.

5.3.2.1 *Reaction with Cytotoxic Aldehyde*

Jurgen *et al.* (1996) found that the histidine-related dipeptide such as L-carnosine, anserine, balenine, homocarnosine and homoanserine had an equivalent effect in the treatment of diabetes to D-carnosine. Also, the latter can be used to react with dihydroxyacetone and the reaction is stable, the reaction rate is consistent with L-carnosine. The findings of Glanacarlo *et al.* (2008) indicated that like L-carnosine, D-carnosine could react with cytotoxic unsaturated aldehydes, particularly HNE (4-hydroxy-*trans*-2-nonenal), and acrolein. After a 1 h reaction, the inhibition rates of L- and D-carnosine against HNE were 44% and 49.2%, respectively, and after 24 h were 96.4% and 98.3%, respectively. After the 12 week gavage of L- and D-carnosine in obese Zucker rats with metabolic disorders, the increments of D-carnosine in rat plasma and urine, were 3 and 9 times, respectively, greater than those of L-carnosine. This showed that D-carnosine has better metabolic stability than L-carnosine, and can reach the blood and kidneys with prototype form to exert its pharmacological activity, and has a role in carnosinase hydrolysis resistance, thereby D-carnosine can react with cytotoxic aldehyde effectively and quickly.

5.3.2.2 *Inhibition of Cytotoxics*

The study of Vistoli *et al.* (2009) showed that D-carnosine, as well as its enantiomer L-carnosine, has a role in cytotoxin (produced by unsaturated fatty acid lipid peroxidation) inhibition.

5.3.2.3 Inhibition of the Formation of Crystal Fibrils

Attanasio *et al.* (2009) revealed that D- and L-carnosine could inhibit the formation of eye α-crystalline fibrils, and almost completely restore the lost activity of heated or denaturing agents added in chaperones.

5.4 Pharmacological Effects and Applications

5.4.1 L-Carnosine

Since the amazing peptide was found in 1990, it has been recognized as the most potential natural antioxygen. Its pharmacological effects reported mainly in the following areas:

5.4.1.1 Antioxidant Effects

L-Carnosine inhibits nonenzymic glycosylation and crosslinking of proteins induced by reactive aldehydes. It significantly delays the oxidation of lipids and protects human low-density lipoprotein against oxidation and glycation.

5.4.1.2 Antiaging Effects

This function is closely linked with antioxidants. L-Carnosine delays aging mainly through the DNA and protein protection since oxidative DNA damage is a major cause of cellular aging.

5.4.1.3 Anticancer Effects

Many studies of L-carnosine and cancer have been developed over several years. It have been shown to reduce proliferation of Sarcoma-180, expressing Her2/neu, and HCT116 tumor cells (Nagai and Suda, 1985; Renner *et al.*, 2010; Iovine *et al.*, 2012). When it is combined with vitamin E, mammary cancer could be suppressed efficiently (Boissonneault *et al.*, 1998). Recently, the possible use of L-carnosine for malignant brain tumors also been discussed (Gaunitz and Hipkiss, 2012).

5.4.1.4 Neuroprotection Effects

As an antioxidant and a free-radical scavenger, L-carnosine could protect the nervous system resisting protein oxidation and radicals (Bellia *et al.*, 2011; Davinelli *et al.*, 2013; Ji *et al.*, 2014).

5.4.1.5 Prevention of Alzheimer's Disease (AD)

During the processing of AD, L-carnosine could directly reduce the congestion and deposition of β-amyloid peptide (β-A) or indirectly decrease these by inhibiting AGEs (Seidler and Yeargans, 2002; Boldyrev *et al.*, 2004).

5.4.1.6 Promotion of Wound Healing

L-Carnosine can accelerate cellular healing of damaged tissue (Kalman *et al.*, 1998; Roberts *et al.*, 1998).

5.4.1.7 Prevention of Chronic Diseases

For its reaction with cytotoxics, which is caused by unsaturated fatty acids oxidation, L-carnosine could treat chronic diseases as potential drug targets (Aldini *et al.*, 2006).

5.4.1.8 Reduction of Atherosclerosis and Amelioration of Dyslipidaemia

L-Carnosine was found to greatly reduced obese-related diseases in obese Zucker rats, by significantly restraining the development of dyslipidaemia, hypertension and renal injury (Aldini *et al.*, 2011). L-Carnosine could protect endothelial cells by its antioxidant effects and facilitate aldehyde removal from atherosclerotic lesions, it may be an important determinant of atherosclerotic lesion formation (Barski *et al.*, 2013).

5.4.1.9 Diabetes Regulation

Recently, many studies have revealed that advanced glycoxidation end products (AGEs) were closely related with complications caused by diabetes. The inhibition of AGEs and amelioration of glucose metabolism effectively (Peters *et al.*, 2012) made L-carnosine a perspective drug treatment for diabetes (Jurgen *et al.*, 1996).

5.4.1.10 Ocular Treatment

Like imidazole-containing dipeptides, L-carnosine can enhance α-crystallin chaperone activity, in consequence to prevent, delay or dissolve the human cataract (Babizhayev, 2010; Babizhayev, 2012). The Pharmacological Committee of the Ministry of Health and Medical Industry of the Russian Federation permitted 5% carnosine eye drops for medical use (IuF *et al.*, 1996). L-Carnosine combination with dexpanthenol and/or hyaluronic acid could be prepared for phthalmic solutions, including contact lens care solutions and eye drops (Wagenaar, 2010).

Besides, L-carnosine could prolong survival from heart failure (Gary and Pamela, 1996), treat muscular fatigue states and improve athletic performance (Nagai, 1984).

5.4.2 D-Carnosine

Compared to L-carnosine, the pharmacological effects researches of the D-form are less. The published literature concerning three aspects is as follows.

5.4.2.1 Amelioration of Dyslipidaemia

As for L-carnosine, the D-form was also found to greatly reduced obese-related diseases in obese Zucker rats, by significantly restraining the development of dyslipidaemia, hypertension and renal injury (Aldini *et al.*, 2011).

5.4.2.2 Treatment of Acute Spinal Cord Injury

D-Carnosine treatment reduces the development of inflammation and tissue injury associated with spinal cord trauma (Di Paola *et al.*, 2011).

5.4.2.3 Reduction of Atherosclerosis

D-Carnosine octylester attenuates atherosclerosis and renal disease through reduction of carbonyl stress and inflammation (Menini *et al.*, 2012).

5.4.2.4 Treatment of Cataract Disease

As for L-carnosine, the D-form also shows protective effects on α-crystallin amyloid fibril formation (Attanasio *et al.*, 2009).

In addition, lipophilic derivatives of D-carnosine for their greater bioavailability than L-form and distributed in the lung were proposed to detoxify the cytotoxic carbonyl compounds induced by cigarette smoke (Negrisoli *et al.*, 2009). D-Carnosine was also prepared as a medicament or diet supplement for the treatment or prevention of oxidative stress-induced disorders (Aldini *et al.*, 2005). D-Carnosine is effective for treatment of disorders involving highly reactive and cytotoxic carbonyl derivatives, especially Alzheimer's disease, atherosclerosis, disorders secondary to diabetes, and age-associated tissue degeneration (Glanacarlo *et al.*, 2008).

5.5 Determination Methods

The development of various determination methods for carnosine has always attracted researchers for its difficulties caused by being highly polar and having a low molecular weight. The major methods documented were high performance liquid chromatography (HPLC), high performance liquid chromatography/tandem mass spectrometry (HPLC-MS), capillary electrophoresis (CE), *etc.*

5.5.1 HPLC Method

Ion-exchange chromatography, precolumn derivatization, and reverse-phase chromatography using NH_2 column are often applied in HPLC methods. Aristoy *et al.* (2004) analyzed histidine dipeptides for ruminants combining phthalaldehyde precolumn derivatization with cation exchange chromatography (Aristoy *et al.*, 2004). In addition, cation-exchange chromatography was directly applied for determination of L-carnosine in commercial canned soups and natural meat extracts (Sri *et al.*, 2000). Isothiocyanate (PITC), 2,4-dinitrochlorobenzene (DNCB), phthalaldehyde and 2-mercaptoethanol are common derivatizing reagents adopted for analysis of L-carnosine (Cornet and Bousset, 1999; Intarapichet and Maikhunthod, 2005; Hu *et al.*, 2009). The novel derivatizing reagent, 4-(5,6-dimethoxy-2-phthalimidinyl)-2-methoxyphenylsulfonyl chloride as a fluorescent labeling reagent is used in sensitive determination of carnosine in urine (Tsuruta *et al.*, 2010). The NH_2 column was also adopted to determine the contents of L-carnosine directly (Han *et al.*, 2009). In addition, the applications of pure micellar liquid chromatography and hydrophilic chromatographic system (HILIC) have been reported (Gil-Agusti *et al.*, 2008; Mora *et al.*, 2008).

5.5.2 HPLC-MS/MS Method

Orioli *et al.* (2005) determined L-carnosine in rat muscles by HPLC-MS/MS using heptafluorobutyric anhydride (HFBA) as an ion-pairing reagent. The contents of histidine-containing dipeptides in rat tissues were also reported using liquid chromatography/electrospray ionization tandem mass spectrometry as the analysis method (Aldini *et al.*, 2004). Recently, an ultraperformance liquid chromatography (UPLC) coupled to tandem mass spectrometry (MS/MS) method for the determination of the bioactive dipeptides, carnosine and anserine was developed and applied to analyze them in chicken broth (Macià *et al.*, 2012).

5.5.3 CE Method

Huang *et al.* (2005) separated and determined carnosine-related peptides using capillary electrophoresis with laser-induced fluorescence detection. In addition, microchip electrophoresis with chemiluminescence was reported in quantification of carnosine-related peptides (Zhao *et al.*, 2009).

5.5.4 Separation of Optical Isomers of Carnosine

The derivating agents, phthalaldehyde (OPA) and *N*-acetyl-L-aminothiopropionic acid were adopted to react with L-, and D-carnosine individually to produce a pair of nonenantiomer derivatives, and then realized chiral separation by reversed-phase liquid chromatography. In addition, chiral ligand exchange chromatography successfully satisfied enantiomeric separation of L-, and D-carnosine, which was carried out by a chiral chromatographic column with a mobile phase composed of copper sulfate as a chiral ligand (Su and Song, 2013).

5.6 Pharmacokinetics of L-carnosine and D-carnosine

The pharmacokinetics data clearly showed that the stereoselective disparation between L- and D-carnosine was unequivocal after intragastrical administration both L-, and D-carnosine. It was demonstrated that L-carnosine was preferred for the active transport process, and was absorbed more quickly. The concentration–time profile of elimination phase showed that L-carnosine was eliminated more quickly than D-carnosine. L-Carnosine showed better gastrointestinal absorption (Su *et al.*, 2010).

After administration of L- or D-carnosine separately, the metabolism in rat of L-CAR and D-CAR also exhibited significant isomeric selectivity (the disparity was mainly attributed to the highly isomer selectivity). D-Carnosine was resistant to hydrolysis by the carnosinase, and was therefore more stable in the plasma. The mutual conversion was not observed for both L-, and D-carnosine. It was different from some D-amino acids transformed to L-type by related amino acid racemase *in vivo* (Su and Song, 2011).

In chronic treatment, D- and L-carnosine were administered with the diet water for 12 weeks at a dose of 30 mg kg^{-1}. D-Carnosine was found in plasma to be far more prevalent than the L-form, urine and kidney, suggesting better metabolic stability (Glanacarlo *et al.* 2008).

Recently, D-car is thought to be less bioavailable than L-carnosine because it is not recognized by Pept1 of the colon. Thus, several hydrophilic (Grasso *et al.*, 2011a,b) and lipophilic molecules have been conjugated to D-carnosine with the aim of increasing the bioavailability of D-carnosine (Negrisoli *et al.*, 2009; Orioli *et al.*, 2011).

5.7 Concluding Remarks

Until now, the exploration on the biological activity and mechanism of L-carnosine is still a hot topic. Meanwhile, the research on D-carnosine is increasingly gaining attention. Recently, some literature has focused on the mechanism of stereoselectivity of L-, and D-carnosine comprising the key role of the contacts with metal ions (Vistoli *et al.*, 2010; Grasso *et al.*, 2011a,b; Pietropaolo, 2012).

It is worth noting that we cannot make a simple judgment for L- or D- being better just from the experiments *in vitro*. For L-carnosine, the therapeutic use is strongly limited by the mechanism governing its homeostasis. For D-carnosine, the pharmacology studies *in vivo* are still lacking. Therefore, the field needs to be explored further for both L- and D-carnosine. In addition, more extensive studies would be valuable into the pharmacokinetics properties of D-carnosine including absolute bioavailability, tissues distribution, metabolism and excretion, which would be completely different from L-carnosine. The results of the above studies would be especially helpful to evaluate and design appropriate formulations for D-carnosine to obtain better bioavailability. It would be an advisable approach to improve its pharmacokinetics profile except for designing and synthesizing derivatives of D-carnosine.

Summary Points

- The basic information such structure, physical and chemical properties of chiral carnosine have been described.
- The major ways to extract or synthesis L-carnosine and D-carnosine have been summarized.
- The key physiological activities between L-, and D-carnosine have been explained.
- The pharmacological effects and applications for both L-, and D-carnosine have been reviewed.
- The main determination methods for L-carnosine, D-carnosine and the both optical isomers simultaneously have been described.
- The pharmacokinetics profiles of L-carnosine and D-carnosine have been explained.
- What also needs to be explored for chiral carnosine has been discussed.

Key Facts of Chiral Carnosine

1. L-Carnosine and D-carnosine are optical enantiomers.
2. Unlike the L-form, D-carnosine is a non-naturally occurring dipeptide and is obtained by synthesis.
3. The physiological activities of L-carnosine were studied in more depth, such as buffering physiological pH, chelating metal ions, scavenging free radicals and antilipid peroxidation.
4. D-Carnosine has been found to react with cytotoxic aldehyde, inhibit cytotoxics, and restrain the formation of crystal fibrils.
5. L-Carnosine has very broad applications owing to its various physiological activities and pharmacological effects.
6. Compared to L-carnosine, studies of the pharmacological effects researches and applications of D-form are still less.
7. The differential pharmacokinetics between L-carnosine and D-carnosine revealed the disparity of disposal mode *in vivo.*

Definitions and Explanation of Key Terms

Chiral molecules. Chiral molecules have point chirality at a single stereogenic atom, which has four different substituents. The two enantiomers of such compounds are said to have different absolute configurations at this center. This center is thus stereogenic.
Dipeptide. A dipeptide is sometimes ambiguously designated as two classes of organic compounds: Its molecules contain either two amino acids joined by a single peptide bond or one amino acid with two peptide bonds.
Physiological pH. This is the part of human homeostasis concerning the proper balance between acids and bases, also called body pH. The body is very sensitive to its pH level, so strong mechanisms exist to maintain it.

Outside the acceptable range of pH, proteins are denatured and digested, enzymes lose their ability to function, and death may occur.

HNE. 4-Hydroxynonenal, or 4-hydroxy-2-nonenal or 4-HNE or HNE, is an α,β-unsaturated hydroxyalkenal that is produced by lipid peroxidation in cells. 4-HNE is the primary α,β-unsaturated hydroxyalkenal formed in this process.

β-Amyloid peptide. A glycoprotein associated with Alzheimer's disease, and derived from a precursor, A4, capable of several forms through alternative splicing.

AGEs. In human nutrition and biology, advanced glycation end products, known as AGEs, are substances that can be a factor in the development or worsening of many degenerative diseases, such as diabetes, atherosclerosis and chronic renal failure.

PepT1. Peptide transporter 1 (PepT 1), also known as solute carrier family 15 member 1 (SLC15A1), is a protein that in humans is encoded by *SLC15A1* gene. PepT 1 is a solute carrier for oligopeptides. It functions in renal oligopeptide reabsorption and in the intestines in a proton-dependent way, hence acting like a cotransporter.

Bioavailability. The degree to which or rate at which a drug or other substance is absorbed or becomes available at the site of physiological activity after administration.

List of Abbreviations

HNE 4-Hydroxy-*trans*-2-nonenal
DNA Deoxyribonucleic acid
AD Alzheimer's disease
β-A β-Amyloid peptide
AGEs Advanced glycoxidation end products
PITC Isothiocyanate
DNCB 2,4-Dinitrochlorobenzene
HFBA Heptafluorobutyric anhydride
OPA Phthalaldehyde
NAC *N*-Acetyl-L-aminothiopropionic acid
CAR Carnosine
Pept1 Peptide transporter 1

References

Abe, H. and Okuma, E., 1991. Effect of temperature on the buffering capacities of histidine-related compounds and fish skeletal muscle. *Nippon Suisan Gakkais*. 57: 2101–2107.

Abe, H., 2000. Role of histidine-related compounds as intracellular proton buffering constituents in vertebrate muscle. *Biochemistry (Moscow)*. 65: 891–900.

Albinati, A., Cesarotti, E., Mason, S. A., Rimoldi, I., Rizzato, S. and Zerla, D., 2010. Histidine and deuterium-labelled histidine by asymmetric catalytic reduction and assignment of the absolute stereochemistry by neutron diffraction. *Tetrahedron: Asymmetry*. 21: 1162–1165.
Aldini, G., Orioli, M., Carini, M. and Facino, R. M., 2004. Profiling histidine-containing dipeptides in rat tissues by liquid chromatography/electrospray ionization tandem mass spectrometry. *Journal of Mass Spectrometry*. 39: 1417–1428.
Aldini, G., Canevotti, R. and Negrisoli, G., 2005. Compositions containing D-carnosine, *U.S. Pat.*, No. 20080194493 A1.
Aldini, G., Dalle-Donne, I., Colombo, R., Maffei Facino, R., Milzani, A. and Carini, M., 2006. Lipoxidation-derived reactive carbonyl species as potential drug targets in preventing protein carbonylation and related cellular dysfunction. *ChemMedChem*. 1: 1045–1058.
Aldini, G., Orioli, M., Rossoni, G., Savi, F., Braidotti, P., Vistoli, G., Yeum, K. J., Negrisoli, G. and Carini, M. J., 2011. The carbonyl scavenger carnosine ameliorates dyslipidaemia and renal function in Zucker obese rats. *Journal of Cellular and Molecular Medicine*. 15: 1339–1354.
Aristoy, M. C., Soler, C. and Toldra, F., 2004. A simple, fast and reliable methodology for the analysis of histidine dipeptides as markers of the presence of animal origin proteins in feeds for ruminants. *Food Chemistry*. 84: 485–491.
Attanasio, F., Cataldo, S., Fisichella, S., Nicoletti, S., Nicoletti, V. G., Pignataro, B., Savarino, A. and Rizzarelli, E., 2009. Protective effects of L- and D-carnosine on α-crystallin amyloid fibril formation: implications for cataract disease. *Biochemistry*. 48: 6522–6531.
Babizhayev, M. A., 2010. Method for topical treatment of eye disease and composition and device for said treatment, *U.S. Pat.*, No. 7795203 B2.
Babizhayev, M. A., 2012. Structural and functional properties, chaperone activity and posttranslational modifications of alpha-crystallin and its related subunits in the crystalline lens: *n*-acetylcarnosine, carnosine and carcinine act as alpha-crystallin/small heat shock protein enhancers in prevention and dissolution of cataract in ocular drug delivery formulations of novel therapeutic agents. *Recent Patents on Drug Delivery & Formulation*. 6: 107–148.
Barrans, Y., Bellocq, A. M., Cotrait, M. and Richard, H., 1976. Structure cristalline et conformation en solution aqueuse de la carnosine ([beta]-alanyl-L-histidine). *Journal of Molecular Structure*. 30: 225–242.
Barski, O. A., Xie, Z., Baba, S. P., Sithu, S. D., Agarwal, A., Cai, J., Bhatnagar, A. and Srivastava, S., 2013. Dietary carnosine prevents early atherosclerotic lesion formation in apolipoprotein E-null mice. *Arteriosclerosis Thrombosis and Vascular Biology*. 33: 1162–1170.
Bellia, F., Vecchio, G., Cuzzocrea, S., Calabrese, V. and Rizzarelli, E., 2011. Neuroprotective features of carnosine in oxidative driven diseases. *Molecular Aspects of Medicine*. 32: 258–266.
Boissonneault, G. A., Hardwick, T. A., Bogardus, S. L., Chan, W. K., Tatum, V., Glauert, H. P., Chow, C. K. and Decker, E. A., 1998. Interactions between

carnosine and vitamin E in mammary cancer risk determination. *Nutrition Research*. 18: 723–733.

Boldyrev, A. and Abe, H., 1999. Metabolic transformation of neuropeptide carnosine modifies its biological activity. *Cellular and Molecular Neurobiology*. 19: 163–175.

Boldyrev, A., Koudinov, A., Berezov, T. and Carpenter, D. O., 2004. Amyloid-ß induced cell death is independent of free radicals. *Journal of Alzheimer's Disease*. 6: 633–638.

Cornet, M. and Bousset, J., 1999. Free amino acids and dipeptides in porcine muscles: differences between "red" and "white" muscles. *Meat Science*. 51: 215–219.

Davinelli, S., Di Marco, R., Bracale, R., Quattrone, A., Zella, D. and Scapagnini, G., 2013. Synergistic effect of L-carnosine and EGCG in the prevention of physiological brain aging. *Current Pharmaceutical Design*. 19: 2722–2727.

Decker, A. E., Crum, A. D. and Calvert, J. T., 1992. Differences in the antioxidant mechanism of carnosine in the presence of copper and iron. *Journal of Agricultural and Food Chemistry*. 40: 756–759.

Di Paola, R., Impellizzeri, D., Salinaro, A. T., Mazzon, E., Bellia, F., Cavallaro, M., Cornelius, C., Vecchio, G., Calabrese, V., Rizzarellic, E. and Cuzzocrea, S., 2011. Administration of carnosine in the treatment of acute spinal cord injury. *Biochemical Pharmacology*. 82: 1478–1489.

Gary, P. Z. and Pamela, R., 1996. Method for prolonging survival from heart failure by administration of carnosine, *U.S. Pat.*, No. 5585396.

Gaunitz, F. and Hipkiss, A. R., 2012. Carnosine and cancer: a perspective. *Amino Acids*. 43: 135–142.

Gil-Agusti, M., Esteve-Romero, J. and Carda-Broch, S., 2008. Anserine and carnosine determination in meat samples by pure micellar liquid chromatography. *Journal of Chromatography, A*. 1189: 444–450.

Glanacarlo, A., Renato, C. and Gianpaolo, N., 2008. Compositions containing D-carnosine, *U.S. Pat.*, No. 20080194493.

Grasso, G. I., Arena, G., Bellia, F., Maccarrone, G., Parrinello, M., Pietropaolo, A., Vecchio, G. and Rizzarelli, E., 2011a. Intramolecular weak interactions in the thermodynamic stereoselectivity of copper(II) complexes with carnosine–trehalose conjugates. *Chemistry – A European Journal*. 17: 9448–9455.

Grasso, G. I., Bellia, F., Arena, G., Vecchio, G. and Rizzarelli, E., 2011b. Noncovalent interaction-driven stereoselectivity of copper(II) complexes with cyclodextrin derivatives of L- and D-carnosine. *Inorganic Chemistry*. 50: 4917–4924.

Han, M., Chen, X., Ji, X. and Bi, K., 2009. Determination of L-carnosine and its related substances by HPLC. *Chinese Pharmaceutical Journal*. 44: 1111–1113.

Hu, Q., Chen, X., Bi, K. and Yao, J., 2009. Pre-column derivatization determination of carnosine by RP-HPLC. *Journal of Shenyang Pharmaceutical University*. 36: 376–378.

Huang, Y., Duan, J., Chen, H., Chen, M. and Chen, G., 2005. Separation and determination of carnosine-related peptides using capillary electrophoresis with laser-induced fluorescence detection. *Electrophoresis*. 26: 593–599.

Intarapichet, K. O. and Maikhunthod, B., 2005. Genotype and gender differences in carnosine extracts and antioxidant activities of chicken breast and thigh meats. *Meat Science*. 71: 634–642.

Iovine, B., Iannella, M. L., Nocella, F., Pricolo, M. R. and Bevilacqua, M. A., 2012. Carnosine inhibits KRAS-mediated HCT116 proliferation by affecting ATP and ROS production. *Cancer Letters*. 315: 122–128.

IuF, M., Formaziuk, V. E. and Sergienko, V. I., 1996. Development of carnosine eyedrops and assessing their efficacy in corneal diseases. *Vestnik Oftalmologia*. 113: 27–31.

IUPAC-IUB Commission on Biological Nomenclature, 1971. *Biochimica et Biophysica Acta*. 229: 1–17.

Ji, Y. S., Park, J. W., Heo, H., Park, J. S. and Park, S. W., 2014. The neuroprotective effect of carnosine (β-alanyl-L-histidine) on retinal ganglion cell following ischemia–reperfusion injury. *Current Eye Research*. 39: 634–641.

Jurgen, M., Alan, R. H. and Sianna, P., 1996. Method for the treatment of the complications and pathology of diabetes, *U.S. Pat.*, No. 5561110.

Kalman, D., Colker, C. M., Stark, R., Minsch, A., Wilets, I. and Antonio, J., 1998. Effect of pyruvate supplementation on body composition and mood. *Current Therapeutic Research*. 59: 793–802.

King, K. E. and Kidd, D. A. A., 1949. A new synthesis of glutamine and of γ-dipeptides of glutamic acid from phthalylated intermediates[J]. *Journal of the Chemical Society*. 1949: 3315–3319.

Kohen, R., Yamamoto, Y., Cundy, K. C. and Ames, B. N., 1988. Antioxidant activity of carnosine, homocarnosine, and anserine present in muscle and brain. *Proceedings of the National Academy of Sciences of the United States of America*. 85: 3175–3179.

Lee, T. S., Yook, J. S., Yook, C. H., Lee, J. C. and Lee, C. M., 2007. Method for simple preparing L-carnosine and thereof whit high yield and high purity without purification step, *KR. Pat.*, No. 2007050319.

Macià, A., Motilva, M. J., Romero, M. P., Labradorb, A., Domínguezb, A. and Peirob, L., 2012. Improved liquid-chromatography tandem mass spectrometry method for the determination of the bioactive dipeptides, carnosine and anserine: application to analysis in chicken broth. *Talanta*. 93: 293–300.

Menini, S., Iacobini, C., Ricci, C., Scipioni, A., Fantauzzi, C. B., Giaccari, A., Salomone, E., Canevotti, R., Lapolla, A., Orioli, M., Aldini, G. and Pugliese, G., 2012. D-Carnosine octylester attenuates atherosclerosis and renal disease in ApoE null mice fed a Western diet through reduction of carbonyl stress and inflammation. *British Journal of Pharmacology*. 166: 1344–1356.

Mora, L., Sentandreu, M. Á. and Toldrá, F., 2008. Contents of creatine, creatinine and carnosine in porcine muscles of different metabolic types. *Meat Science*. 79: 709–715.

Nagai, K., 1984. Verwendrug von L-carnosin zur Tumorbehandlung, *D.E. Pat.*, No. 3424781 A1.

Nagai, K. and Suda, T., 1985. Antineoplastic effects of carnosine and betaalanine-physiological considerations of its antineoplastic effects. *Journal of the Physiological Society of Japan*. 48: 741–747.

Negrisoli, G., Canevotti, R. and Previstali, M., 2009. Dipeptide compounds containing D-histidine, *U.S. Pat.*, No. 20090306165 A1.

Orioli, M., Aldini, G., Beretta, G., Facino, R. M. and Carini, M., 2005. LC-ESI-MS/MS determination of 4-hydroxy-*trans*-2-nonenal Michael adducts with cysteine and histidine-containing peptides as early markers of oxidative stress in excitable tissues. *Journal of Chromatography B: Analytical Technologies in the Biomedical and Life Sciences*. 827: 109–118.

Orioli, M., Vistoli, G., Regazzoni, L., Pedretti, A., Lapolla, A., Rossoni, G., Canevotti, R., Gamberoni, L., Previtali, M., Carini, M. and Aldini, G., 2011. Design, synthesis, ADME properties, and pharmacological activities of beta-alanyl-D-histidine (D-carnosine) prodrugs with improved bioavailability. *ChemMedChem*. 6: 1269–1282.

Pegova, A., Abe, H. and Boldyrev, A., 2000. Hydrolysis of carnosine and related compounds by mammalian carnosinases. *Comparative Biochemistry and Physiology, Part B: Biochemistry & Molecular Biology*. 127: 443–446.

Peters, V., Schmitt, C. P., Zschocke, J., Gross, M. L., Brismar, K. and Forsberg, E., 2012. Carnosine treatment largely prevents alterations of renal carnosine metabolism in diabetic mice. *Amino Acids*. 42: 2411–2416.

Pietropaolo, A., 2012. Exploring metal-driven stereoselectivity of glycopeptides by free-energy calculations. *Pure and Applied Chemistry*. 84: 1919–1930.

Renner, C., Zemitzsch, N., Fuchs, B., Geiger, K. D., Hermes, M., Hengstler, J., Gebhardt, R., Meixensberger, J. and Gaunitz, F., 2010. Carnosine retards tumor growth in vivo in an NIH3T3-HER2/neu mouse model. *Molecular Cancer Research*. 9: 1–7.

Rinderknecht, H., Rebane, T. and Ma, V., 1964. Synthesis of carnosine, anserine and isoanserine. *Journal of Organic Chemistry*. 29: 1968–1970.

Roberts, P. R., Black, K. W., Santamauro, J. T. and Zaloga, G. P., 1998. Dietary Peptides Improve wound healing following surgery. *Nutrition*. 14: 266–269.

Seidler, N. W. and Yeargans, G. S., 2002. Effects of thermal denaturation on protein glycation. *Life Sciences*. 70: 1789–1799.

Shasad, S., Devid, S., Michoel, D. T. and Hans, H. L., 2007. Coupling of functional hydrogen bonds in pyridoxal-5-phosphate-enzyme model systems observed by solid-state NMR spectroscopy. *Journal of the American Chemical Society*. 129: 4440–4455.

Skoblik, T. I., Markin, A. Y., Zaborina, T. G., Filippovich, E. I., Zheltukhina, G. A., Erstigneeva, R. P., Babizhaev, M. A. and Sharkov, V. Y., 1995. Method for preparation of carnosine, *R.U. Pat.*, No. 2030422.

Sri Kantha, S., Takeuchi, M., Watabe, S. and Ochi, H., 2000. HPLC determination of carnosine in commercial canned soups and natural meat extracts. *LWT-Food Science and Technology*. 33: 60–62.

Stafan, H., Thomas, R. and Dario, V., 2001. Preparation of β-alanimamide dipeptides from N-unprotected amino acids oresters and cyanoacetic acid derivatives, *W.O. Pat.*, No. 0164638.

Stecchini, M. L., TORRE, M., Sarais, I., Fuochi, P. G., Tubaro, F. and Ursini, F., 1998. Carnosine increases irradiation resistance of *Aeromonas hydrophila* in minced turkey meat. *Journal of Food Science*. 63: 147–149.

Stvolinsky, S., Kukley, M., Dobrota, D., Mezesova, V. and Boldyrev, A., 2000. Carnosine protects rats under global ischemia. *Brain Research Bulletin*. 53: 445–448.

Su, D., Bi, K., Zhou, C., Song, Y., Wei, B., Geng, L., Liu, W. and Chen, X., 2010. Enantioselective Separation and determination of carnosine in rat plasma using a fluorescence HPLC method for stereoselective pharmacokinetic studies. *Chromatographia*. 71: 603–608.

Su, D. and Song, Y., 2011. Chiral pharmacokinetics of carnosine in rats. *Chinese Journal of Hospital Pharmacy*. 31: 1874–1877.

Su, D. and Song, Y., 2013. Enantioseparation of carnosine by HPLC with chiral ligand-exchange method. *Chinese Journal of Modern Applied Pharmacy*. 30: 75–79.

Sui, X., Chen, X., Fang, L., Wei, Y. and Bi, K., 2009. Spectroscopic data and structure confirmation of the drug L-carnosine. *Pharmacy Today*. 19: 29–31.

Tsuruta, Y., Maruyama, K., Inoue, H., Kosha, K., Date, Y., Okamura, N., Eto, S. and Kojima, E., 2010. Sensitive determination of carnosine in urine by high-performance liquid chromatography using 4-(5,6-dimethoxy-2-phthalimidinyl)-2-methoxyphenylsulfonyl chloride as a fluorescent labeling reagent. *Journal of Chromatography B: Analytical Technologies in the Biomedical and Life Sciences*. 878: 327–332.

Turner, R. A., 1953. Synthesis of carnosine and related peptides by the phthaloyl method. *Journal of the American Chemical Society*. 75: 2388–2390.

Vinick, F. J., 1982. Process for preparing L-carnosine, *U.S. Pat.*, No. 4359416.

Vistoli, G., Orioli, M., Pedretti, A., Regazzoni, L., Canevotti, R., Negrisoli, G., Carini, M. and Aldini, G., 2009. Design, synthesis, and evaluation of carnosine derivatives as selective and efficient sequestering agents of cytotoxic reactive carbonyl species. *ChemMedChem*. 4: 967–975.

Vistoli, G., Pedretti, A., Marconi, C. and Aldini, G., 2010 Combined in silico approaches for drug design and pharmacokinetic optimization of a set of carnosine analogues as potent and selective carbonyl quenchers. Convegno Nazionale della Divisione di Chimica Farmaceutica della Società Chimica Italiana. Available at: http://159.149.85.2/cms/downloads/posters/Poster_2010_Abano.pdf. Accessed September 2010.

Wagenaar, L. J., 2010. Ophthalmic solutions, including contact lens care solutions and eye drops comprising carnosine, preferably in combination with dexpanthenol and/or hyaluronic acid, *U.S. Pat.*, No. 20110230424 A1.

Yamaoka, N., Egawa, H., Takenaka, S. and Moriwaki, H., 2005. Preparation of carnosine, *J.P. Pat.*, No. 2005306782.

Zhao, S., Huang, Y., Shi, M., Huang, J. and Liu, Y., 2009. Quantification of carnosine-related peptides by microchip electrophoresis with chemiluminescence detection. *Analytical Biochemistry*. 393: 105–110.

CHAPTER 6

The Biochemistry of Enzymes Producing Carnosine and Anserine

JAKUB DROZAK*[a], EMILE VAN SCHAFTINGEN[b], AND MARIA VEIGA-DA-CUNHA*[b]

[a]Department of Metabolic Regulation, Institute of Biochemistry, Faculty of Biology, University of Warsaw, Miecznikowa 1, 02-096 Warsaw, Poland; [b]Welbio and Laboratory of Physiological Chemistry, de Duve Institute, Université catholique de Louvain, Avenue Hippocrate 75, B1.75.08, B-1200 Brussels, Belgium
*E-mail: jdrozak@biol.uw.edu.pl, maria.veiga@uclouvain.be

6.1 Introduction

Although about 2000 articles in the PubMed describe a whole variety of biochemical properties, potential physiological roles, and therapeutic uses of carnosine, anserine and related imidazole dipeptides, only a few of them are devoted to the mechanisms of formation of these compounds. Furthermore, most of the reports on the synthesizing enzymes were published at least 30 years ago, at a time when molecular identification of proteins was nearly out of reach. Results describing the molecular identity of carnosine synthase and carnosine *N*-methyltransferase have only recently become available, giving new insights into the biosynthesis of these dipeptides (see Figure 6.1 for

Food and Nutritional Components in Focus No. 8
Imidazole Dipeptides: Chemistry, Analysis, Function and Effects
Edited by Victor R Preedy

Published by the Royal Society of Chemistry, www.rsc.org

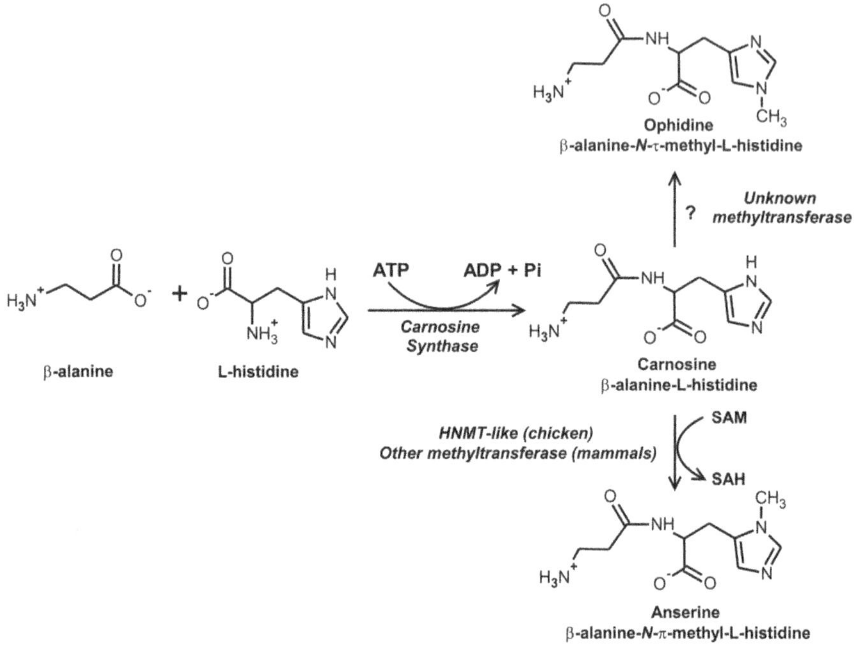

Figure 6.1 Pathway for the synthesis of carnosine, anserine and ophidine. Vertebrate carnosine synthase catalyzes the synthesis of carnosine. Hnmt-like and, in mammals, an unidentified methyltransferase convert carnosine to anserine. The formation of ophidine is presumably also catalyzed by a methyltransferase. SAM, *S*-adenosyl-L-methionine; SAH, *S*-adenosyl-L-homocysteine.

an overview of carnosine and methylcarnosine biosynthetic pathways). The purpose of this chapter is to summarize recent advances in our knowledge on the properties of these enzymes and to discuss perspectives for further research, particularly in the context of health and disease.

6.2 Carnosine Synthase

6.2.1 Brief History of Research

The presence of an enzyme able to catalyze the ATP-dependent synthesis of carnosine was first reported about 60 years ago by the groups of Winnick and Meister (Kalyankar and Meister, 1959; Stenesh and Winnick, 1960). The enzyme from chicken pectoral muscle, an abundant source, has been mostly investigated, but later studies have also been performed on enzymes from other sources, such as rat brain (Skaper *et al.*, 1973), mouse olfactory bulb (Horinishi *et al.*, 1978), and human brain (Kish *et al.*, 1979). Carnosine synthase was shown to be ATP dependent and to catalyze the synthesis not only of carnosine, but also, with lower activity, of homocarnosine, the other physiological histidine-containing dipeptide. Hence, the enzyme is sometimes named carnosine/homocarnosine synthase. Its reaction mechanism was proposed to involve the formation of a β-alanyl-adenylate intermediate and result in the hydrolysis of ATP to AMP and inorganic pyrophosphate (Kalyankar and Meister, 1959). However, these conclusions were not based on the direct demonstration of the reaction stoichiometry, and their validity has been questioned (Bauer, 2005).

It should be noted that a different enzyme, able to catalyze the synthesis of carnosine and anserine from β-alanine and (methyl)-histidine *in the absence of ATP*, has been purified from the Japanese eel muscle (Tsubone *et al.*, 2007). This enzyme also shows a high *hydrolytic* activity with carnosine and anserine, and its synthetic activity is demonstrated at concentrations of amino acids (0.25 M β-alanine, 0.16 M *N*-π-methyl-L-histidine) that are about 3 orders of magnitude higher than their physiological intracellular concentrations. Such concentrations are indeed expected to allow reversal of the hydrolytic reaction. For thermodynamic reasons, it is unlikely that this enzyme is involved *in vivo* in the synthesis of carnosine or related dipeptides. This enzyme is therefore not further described in this chapter.

6.2.2 Molecular Identification and Species Distribution

Carnosine synthase is now known to be the product of the ATPGD1 (ATP-grasp domain containing 1) gene (Drozak *et al.*, 2010), recently renamed CARNS1. This enzyme is a member of the superfamily of ATP-grasp proteins, which includes enzymes catalyzing the ATP-assisted ligation of a carboxylic acid to an amino or thiol group-containing molecule (usually *via* the formation of an acylphosphate intermediate) with the associated production of ADP and Pi (Galperin and Koonin, 1997; Fawaz *et al.*, 2011).

This identification was accomplished following a ≈1500-fold purification of carnosine synthase from chicken pectoral muscle, until only two protein bands of 90 and 100 kDa could still be visualized by SDS-PAGE. Investigation of the reaction stoichiometry with this preparation showed that the ligation of β-alanine and L-histidine is accompanied by the hydrolysis of ATP to ADP and inorganic phosphate, rather than to AMP and inorganic pyrophosphate as had long been admitted on the basis of indirect evidence (Kalyankar and Meister, 1959).

Mass spectrometry analysis of the purified proteins did not initially disclose a logical candidate, *i.e.* a putative ligase, and the identification largely relied on database searches (Drozak *et al.*, 2010). Most of the ligases that convert ATP to ADP and inorganic phosphate belong to the ATP-grasp family (Galperin and Koonin, 1997; Fawaz *et al.*, 2011). By searching the Pfam database (Punta *et al.*, 2012) for members of this family with unknown function, it turned out that ATPGD1 was the only mouse or human protein with a subunit size (≈100 kDa) and tissue distribution (preferential expression in muscle and brain) that were compatible with carnosine synthase. In 2009, at the time this research was being done, the corresponding chicken protein and gene were absent from databases, but a few chicken expression sequence tags (ESTs) showed similarity with mammalian ATPGD1, suggesting that this protein is also present in this species. This was confirmed by amplification and sequencing of the full-length chicken cDNA, and demonstration that the deduced protein sequence corresponded to that of the 100 kDa polypeptide present in the carnosine/homocarnosine synthase preparation that was purified from chicken pectoral muscle. The final confirmation came from establishing that both mouse and human recombinant ATPGD1 catalyze the ATP-dependent synthesis of carnosine and homocarnosine (Drozak *et al.*, 2010).

Carnosine synthase is about twice as large as other members of the ATP-grasp family. This is because it contains two distinct ATP-grasp domains, a C-terminal one, which is highly conserved (65% identity between human and chicken) and presumably includes the catalytically active part, and a N-terminal one, which is much less conserved (47% identity between human and chicken) and can only be revealed as a member of the ATP-grasp family by performing iterative psi BLAST searches using the NCBI/Blast public website (http://www.blast.ncbi.nlm.nih.gov/Blast.cgi). We speculated previously that the two domains may have been involved in the synthesis of a tripeptide in a primitive form of the protein (Drozak *et al.*, 2010). Interestingly, the bivalve *Corbicula japonica* contains a tripeptide named acorbine, whose sequence is β-alanyl-ornithyl-ornithine (Uchisawa *et al.*, 2007). Because of the ability of carnosine synthase to catalyze the synthesis of β-alanyl-ornithine, it is tempting to speculate that the synthesis of the tripeptide is carried out by a *C. japonica* homolog of carnosine synthase. Unfortunately, no genome sequence is available for *C. japonica* or related species.

Homologs of carnosine synthase are found in mammals, birds, and many other vertebrate species, including a teleost fish (*Misgurnus anguillicaudatus*) where ESTs derived from the olfactory epithelium encode a protein closely related to carnosine synthase. Other fishes, whose genomes have been extensively sequenced (*e.g. Takifugu rubripes*, *Oreochromis niloticus*), have no detectable CARNS1 homolog. Homologs of the better conserved C-terminal half of carnosine synthase are also found in the green algae *Volvox carteri*, *Chlamydomonas reinhardtii* and *Chlorella variabilis* and in the oyster parasite *Perkinsus marinus*, suggesting that these species may contain carnosine or a related peptide.

6.2.3 Enzyme Specificity and Metabolite Repair

Carnosine synthase is a ligase with a rather broad specificity (Winnick and Winnick, 1958; Kalyankar and Meister, 1959; Horinishi *et al.*, 1978; Drozak *et al.*, 2010). It is indeed able to ligate histidine, not only to β-alanine, but also to γ-aminobutyrate and its isomer, β-aminobutyrate, though not to glycine, α-alanine, or α-aminobutyrate. Kinetic studies indicate that carnosine synthase shows a marked preference for β-alanine compared to γ-aminobutyrate, the catalytic efficiency (k_{cat}/K_M) being about 20 times higher for the former than for the latter substrate (Drozak *et al.*, 2010). This dual specificity allows carnosine synthase to synthesize carnosine in muscle, where β-alanine, though not γ-aminobutyrate is available, but mostly homocarnosine in brain, which contains γ-aminobutyrate concentrations in the millimolar range, *i.e.* orders of magnitude higher than the concentration of β-alanine. It is, however, worth remarking that the specificity of carnosine synthase means that it will not use the abundant α-amino acid glycine and L-alanine.

β-alanine and γ-aminobutyrate are ligated by carnosine synthase not only to L-histidine (though not to D-histidine), but also, in some cases, to methylhistidine, L-ornithine, L-lysine, and L-arginine (Kalyankar and Meister, 1959; Horinishi *et al.*, 1978; Drozak *et al.*, 2010). Figure 6.2 shows that the mouse enzyme is quite good at using ornithine and the chicken enzyme, *N*-π-methyl-L-histidine, whereas the human enzyme appears to be more specific for histidine (Drozak *et al.*, 2010). Taking into account the rather high concentration of lysine compared to histidine (≈2-fold higher in muscle and brain tissues; Perry *et al.*, 1971a; Divino Filho *et al.*, 1997; Mero, 1999), one would

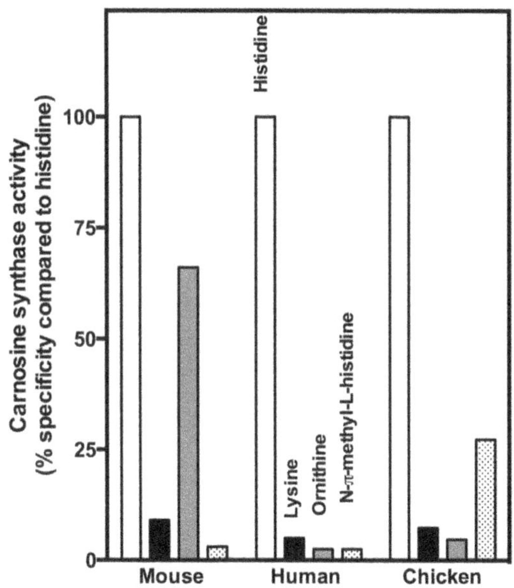

Figure 6.2 Substrate specificity of carnosine synthase from various species. The substrate specificity of mouse, human and chicken carnosine synthase is shown as a percentage of their catalytic efficiency (K_{cat}/K_M value) for histidine, which was systematically set to 100% for all three enzymes. This research was originally published in The Journal of Biological Chemistry. Drozak *et al.*, 2010. Molecular identification of carnosine synthase as ATP-grasp domain-containing protein 1 (ATPGD1). *Journal of Biological Chemistry*. 285: 9346–9356 © The American Society for Biochemistry and Molecular Biology.

expect that the concentration of β-alanyllysine and γ-aminobutyryllysine would amount to ≈10% of that of carnosine and homocarnosine, respectively. However, the β-alanyllysine concentration in muscle (Matsuoka *et al.*, 1969) and the γ-aminobutyryllysine concentration in bovine (Kumon *et al.*, 1970a) and other mammalians brain (Nakajima *et al.*, 1969) are much lower than this value, possibly because hydrolysis of the basic dipeptides by a specific peptidase (Kumon *et al.*, 1970b). Recent data from our laboratory indicate that this is indeed due to the presence of a peptidase, which we have purified and molecularly identified as a protein of unknown function. This enzyme hydrolyzes the β-alanine and γ-aminobutyrate derivatives of ornithine, lysine, and arginine, but does not act significantly (catalytic efficiency more than 3000-fold lower) on carnosine and homocarnosine (Veiga-da-Cunha and Van Schaftingen unpublished results). Thus, carnosine synthase and the β-alanyllysine dipeptidase are a further illustration of metabolite repair. According to this emerging concept, enzymes are often less specific than usually assumed, they make abnormal metabolites, but these abnormal metabolites may be destroyed by specific repair enzymes (Linster *et al.*, 2013; Van Schaftingen *et al.*, 2013).

6.2.4 Tissue Distribution

Carnosine or its methylated derivatives anserine or ophidine accumulate in a range of magnitudes and different ratios in the skeletal muscle of numerous species (Boldyrev *et al.*, 2013), but there is little quantitative information regarding the actual expression of carnosine synthase in the skeletal muscle of these animals.

Brain appears to be particularly rich in homocarnosine in humans, with values of the order of 0.5–1 μmol g^{-1}, compared to 0.1 μmol g^{-1} in rabbit, 0.05 μmol g^{-1} in rats and even lower values in dog and cat (Abraham *et al.*, 1962; Kanazawa and Sano, 1967; Perry *et al.*, 1971a,b). This difference in homocarnosine concentration agrees with the more than 10-fold higher specific activity of carnosine synthase found in human brain biopsies (≈23 ± 10 nmol h^{-1} g^{-1} for cerebral cortex; Kish e*t al.*, 1979) compared to rat brain (≈0.2–0.4 nmol h^{-1} g^{-1}; Ng and Marshall, 1978). Interestingly, several studies indicate that the region of the brain that has the highest carnosine/homocarnosine synthase activity is the olfactory bulb (Ng and Marshall, 1978), the only brain region where carnosine is abundant.

Expression data for human CARNS1 available at BioGPS database (Su *et al.*, 2004; Wu *et al.*, 2009), indicate that carnosine synthase mRNA is most abundant in brain and spinal cord, and less abundant in skeletal muscle. Predominant expression of CARNS1 mRNA (KIAA-1394) in human brain was also reported by Nagase *et al.* (2000), in the study that first reported the sequence of the ATPGD1 cDNA among a series of other human cDNAs encoding large proteins. In mouse, the BioGPS data base shows that the highest levels of expression are found in skeletal muscle and spinal cord, with lower levels in various regions of the brain and the cerebellum (Lattin *et al.*, 2008).

In humans, the BioGPS data also shows that carnosine synthase mRNA is practically absent from fetal brain, in agreement with the observation that homocarnosine is undetectable in fetal brain (Van Regemorter *et al.*, 1972).

Bauer and collaborators (Bauer *et al.*, 1979; Bakardjiev and Bauer, 2000; De Marchis *et al.*, 2000) studied carnosine biosynthesis in rat primary cell cultures of skeletal muscle and central nervous system (CNS) tissue and demonstrated that it was restricted to muscle cells and to the oligodendrocytes derived from brain and spinal cord. This last finding is consistent with BioGPS data showing that human CARNS1 mRNA is absent from astrocytes and barely detectable in neurons. It would be most interesting to investigate the localization of the CARNS1 mRNA in the olfactory bulb, to help resolve the controversy on the localization of the synthase in neurons (Biffo *et al.*, 1990; De Marchis, 2000) or in the ensheathing cells (Bakardjiev, 1997).

6.2.5 Regulation of Skeletal Muscle Carnosine Synthase Activity

As for the activity of any enzyme, what determines carnosine synthesis in tissues is the amount of carnosine synthase, the concentration of its substrates and the concentration of other molecules that regulate its activity. Three substrates need to be taken into consideration: ATP, histidine and β-alanine. The kinetic properties of carnosine synthase indicate that the K_M for ATP-Mg (0.2–0.6 mM depending on the species; Drozak *et al.*, 2010) is much lower than the prevailing concentration of ATP in muscle at rest, suggesting that the enzyme is virtually saturated by this substrate.

The concentration of histidine in muscle (0.36–0.6 mM in humans; Divino Filho *et al.*, 1997; Mero, 1999) is slightly above the K_M (0.36 mM for human carnosine synthase; Drozak *et al.*, 2010). This signifies that a doubling of the histidine concentration in muscle would, at best, cause a 33% increase in the rate of carnosine synthesis, suggesting that there is little room for increasing carnosine synthesis in this way.

It is difficult to find values for the intracellular concentration of β-alanine in skeletal muscle (or in other human tissues), but it is likely that it is much lower than the K_M value of carnosine synthase for this amino acid (0.46 mM and 0.09 mM for mouse and human, respectively). In mouse, the concentration of β-alanine is 20 μM in testis and 50 μM in brain (Rzem and Van Schaftingen, unpublished results; Murakami *et al.*, 2009) and it is reasonable to assume that the skeletal muscle concentration is of the same order of magnitude. This suggests that raising the muscle concentration of β-alanine could greatly affect carnosine synthesis, providing that the transport does not become limiting.

What may limit the accumulation of carnosine is the feedback inhibition exerted by carnosine on its own synthesis. Carnosine is a competitive inhibitor (*versus* β-alanine), for which K_i values of 14.8 ± 1.15 and 7.4 ± 0.6 mM have been observed for the human and mouse enzymes, respectively

(Veiga-da-Cunha and Van Schaftingen, unpublished results). These concentrations are of the same order of magnitude as the muscle concentration of carnosine, suggesting that a significant inhibition may be exerted by the reaction product.

6.2.6 Perspectives

The sustained growth in the number of deciphered genomes will lead to a better appreciation of the distribution of carnosine synthase in Nature. It will be most interesting to compare this information with the biochemical demonstration of the presence of carnosine or related peptides in the relevant organisms. Due to the rather broad specificity of carnosine synthase, it may be that in some organisms or some cells, its true physiological significance would be to synthesize peptides other than carnosine, such as *e.g.* β-alanyllysine or β-alanylornithine. It may also be that the N-terminal ATP-grasp domain present in CARNS1 is active in some species, leading to the formation of a tripeptide.

Numerous questions still remain on the regulation of the synthesis of carnosine. Is the feedback inhibition seen *in vitro* similar in physiological conditions? Does this feedback inhibition involve also anserine and ophidine? Could the N-terminal degenerated ATP-grasp domain play a role in regulation of carnosine synthase?

Until now, no carnosine or homocarnosine deficiency has been described, although a decrease of homocarnosine in the cerebrospinal fluid of two subjects (with normal concentration of γ-aminobutyrate) has been reported (Gerrits *et al.*, 1993). It is therefore presently unknown if the absence of carnosine or homocarnosine has major consequences on health. The creation of an animal model for these deficiencies and the detection of patients with mutations in carnosine synthase will be critical to gain a better appreciation of the physiological importance of these two dipeptides.

6.3 Carnosine *N*-Methyltransferase

6.3.1 Brief History of Research

Although the initial research on anserine formation dates back to the early 1940s, the matter has only been occasionally investigated during the following six decades, meaning that the information on the anserine-producing enzyme(s) is still fragmentary. Using deuterium-labeled methionine, Schenck *et al.* (1943) showed that the *N*-methyl group of anserine could be derived from methionine in rabbit muscles. This suggested that the methyl group of methionine was transferred on the imidazole ring either of free L-histidine or of carnosine. Martignoni and Winnick (1953) were probably the first to postulate that, in chicken, anserine could indeed be produced by one of these two mechanisms, and the use of cell-free enzyme preparations of vertebrate muscles in radiotracer experiments supported their suggestion. In a series of papers, Kalyankar and Meister (1959), Winnick and Winnick (1960), and Stenesh and Winnick (1960) showed that a partially purified chicken carnosine synthase (EC 6.3.2.11) promotes the condensation of β-alanine with either L-histidine or *N*-π-methyl-L-histidine, yielding carnosine or anserine, respectively. Due to its broad specificity towards histidine analogs, chicken carnosine synthase was then named "carnosine-anserine synthase", despite being a lot less active with *N*-π-methyl-L-histidine than with L-histidine (Stenesh and Winnick, 1960). Furthermore, *N*-π-methyl-L-histidine was found to be a very poor substrate for mouse and rat carnosine synthase, suggesting that this enzyme does not produce anserine in these mammalian species (Skaper *et al.*, 1973; Horinishi *et al.*, 1978).

In the meantime, Winnick and Winnick (1959) and McManus (1962) reported the presence, in chicken muscle, of carnosine *N*-methyltransferase (EC 2.1.1.22), an enzyme that catalyzes the formation of anserine by transferring the methyl group of *S*-adenosyl-L-methionine (SAM) to carnosine, and suggested this activity to be an important source of anserine in chicken muscle. Further evidence confirming the significance of the *N*-methyltransferase in the production of the dipeptide came later from studies on the time-course of its appearance in developing chicken muscle (McManus and Benson, 1967), as well as in primary cultures of chicken muscle cells (Bauer and Schulz, 1994). The authors of these reports concluded that anserine is formed secondarily to carnosine, emphasizing the importance of carnosine *N*-methyltransferase (Figure 6.1). Furthermore, the activity of the latter enzyme appeared to be induced when the concentration of carnosine reached a sufficient level in the cell.

The presence of carnosine *N*-methyltransferase in rabbit muscle was described by Raghavan *et al.* (1992). These authors discovered this activity while studying actin *N*-methyltransferase, an enzyme that methylates histidine residues in this protein. Unlike the avian enzyme (see below), this mammalian enzyme is not yet molecularly identified.

6.3.2 Molecular Identity and Biochemical Properties of the Chicken Enzyme

In order to identify the gene encoding carnosine *N*-methyltransferase, this enzyme was purified 640-fold from chicken pectoral muscle. Tryptic fragments were analyzed by mass spectrometry, leading to the finding that one of the proteins present in the preparation was a putative methyltransferase designated "histamine *N*-methyltransferase-like (HNMT-like)" (Drozak *et al.*, 2013). The identity of carnosine *N*-methyltransferase with this protein was confirmed by expression, purification and analysis of the properties of recombinant chicken HNMT-like protein.

Chicken HNMT-like protein comprises 322 amino acids and behaves as a monomer (≈36 kDa) in its native form. It shares a high degree of amino acid conservation (49% identity and 69% similarity, Figure 6.3) with chicken

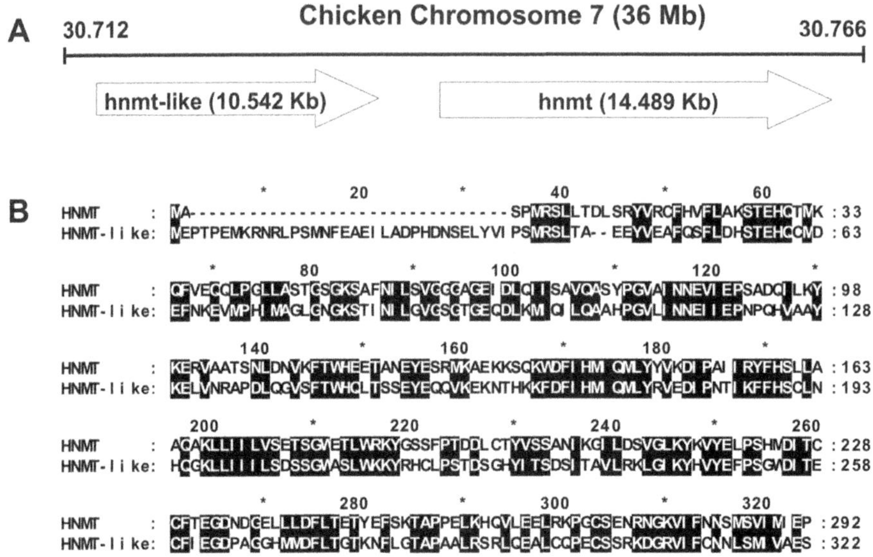

Figure 6.3 Chromosomal localization of chicken histamine *N*-methyltransferase (hnmt) and histamine *N*-methyltransferase-like (hnmt-like) genes and alignment of their amino acid sequences. (A) The hnmt and hnmt-like genes are in tandem repeat on chromosome 7. (B) Chicken HNMT (XP_422143.2) and HNMT-like protein (XP_001234740.1), which share 49% identity, have been aligned. Similar amino acid residues are highlighted by black boxes.

histamine *N*-methyltransferase (HNMT), an enzyme that catalyzes the formation of *N*-τ-methyl-L-histamine by the transfer of the methyl group of SAM to histamine inactivating it. Interestingly, the HNMT-encoding gene (*hnmt*) is located on chicken chromosome 7, in tandem repeat with the *hnmt-like* (LOC771456) gene, suggesting that both genes are paralogs that resulted from a gene-duplication event that happened during the evolution of vertebrates (*cf.* Section 6.3.4).

Intriguingly, histamine *N*-methyltransferase methylates the *N*-τ atom of the imidazole moiety of histamine, whereas the *N*-π atom is methylated in anserine. It was therefore of importance to verify that HNMT-like methylated the appropriate nitrogen of carnosine. This was done by analyzing the product formed by HNMT-like and showing (1) that its fragmentation pattern in tandem mass spectrometry corresponded with that of anserine (Drozak *et al.*, 2013) and (2) that acid hydrolysis of this product yielded *N*-π-methyl-L-histidine and not *N*-τ-methyl-L-histidine (Drozak *et al.*, unpublished results). These findings suggest the imidazole moiety of histamine and carnosine is oriented differently in the catalytic sites of HNMT and HNMT-like, respectively.

As recently reported (Drozak *et al.*, 2013), both chicken purified carnosine *N*-methyltransferase and recombinant HNMT-like protein display K_M values for SAM (11 and 14 μM, respectively) and for carnosine (1.6 mM) that are consistent with the results reported by McManus (1962) (K_M = 9 μM for SAM and 4 mM for carnosine). Taking into account the intracellular concentrations of SAM and carnosine in chicken muscles (about 30 μM and 10 mM, respectively), carnosine *N*-methyltransferase appears to work with close to saturating concentrations of substrates *in vivo*. It is therefore intriguing that chicken muscles contain both carnosine and anserine at millimolar concentrations, despite the fact that the activity of carnosine *N*-methyltransferase (3.2 pmol min^{-1} mg^{-1} protein; Drozak *et al.*, 2013) is several fold higher than the activity of carnosine synthase (0.74 pmol min^{-1} mg^{-1} protein; Drozak *et al.*, 2010). This observation suggests that there is an unidentified endogenous factor limiting the activity of the methyltransferase in chicken muscles. One possibility is anserine, the product of methyltransferase reaction, which could act as a feedback inhibitor.

Chicken carnosine *N*-methyltransferase, whether recombinant or purified from tissue, is very specific for its substrate, as is also histamine methyltransferase (Maslinski and Fogel, 1991; Drozak *et al.*, 2013). Several compounds structurally related to carnosine, including L-histidine and histamine, were tested as potential methyl group acceptors and none of them turned out to be a substrate for the enzyme (Drozak *et al.*, 2013; McManus, 1962), though methylation of certain natural analogs of carnosine such as homocarnosine (γ-aminobutyryl-L-histidine) has not been investigated thus far.

6.3.3 Hnmt-Like Gene Expression

Expression data deposited in the NCBI Gene Expression Omnibus (GEO) repository show that the level of *hnmt-like* transcripts in chicken retinal pigment epithelium rapidly increases during embryogenesis, reaching its maximal level just before hatching (GEO Profiles: GDS2964[ACCN] LOC771456). In view of these data it is likely that carnosine *N*-methyltransferase might be particularly active in avian retina. Indeed, the retinas of birds were reported to be rich in anserine (Margolis and Grillo, 1984), though the importance of this phenomenon remains to be elucidated.

6.3.4 Phylogeny

Phylogenetic analysis suggests that the *hnmt* (encoding histamine *N*-methyltransferase) and *hnmt-like* (encoding carnosine *N*-methyltransferase) genes arose through the duplication of an ancestral gene (Figure 6.4). This event of duplication probably happened rather late during the evolution of vertebrates, in the course of radiation of the Sauropsida (reptilians and birds), and led to the presence of both *hnmt* and *hnmt-like* genes in birds, in the lizard (*Anolis carolinensis*) and the alligator. Unexpectedly, *hnmt-like* genes are not detected in genomes of mammals and amphibians (*i.e. Xenopus* sp.), despite the fact that anserine is present in most mammalian species (though not in humans) (Boldyrev *et al.*, 2013). These findings suggest that the *hnmt-like* gene has been lost in mammalians and that the carnosine *N*-methyltransferase that is present in the muscle of several mammalian species (Crush, 1970), including rabbit (Raghavan *et al.*, 1992) and rat (Drozak *et al.*, unpublished results), must be a completely different enzyme from "HNMT-like". Thus, convergent evolution took place in the emergence of the anserine-synthesizing enzymes, which is not uncommon among methyltransferases (Schubert *et al.*, 2003).

Intriguingly, several fish genomes comprise two "*hnmt-like*" genes (*e.g. Oreochromis niloticus*, GenBank gene symbols LOC100703530 and LOC100704419), which are in tandem repeat and encode proteins that show a lower degree of identity between them (≈41%) than found between histamine-methyltransferase and carnosine methyltransferase (≈49%). This suggests that a more ancient gene duplication occurred during evolution of fishes, leading to two methyltransferases, one of which, confusingly designated HNMT-like 1, is closely related to both histamine- and carnosine *N*-methyltransferases of tetrapods and is most likely a histamine-methyltransferase. The function of the other homolog (HNMT-like 2) is unknown. Since none of the genomes of the three fishes mentioned in Figure 6.4 appears to comprise a carnosine synthase homolog, we do not expect to find a real carnosine *N*-methyltransferase in these species. The second homolog of unknown function was presumably lost during evolution before the duplication leading to the separation of histamine- and carnosine *N*-methyltransferases.

6.3.5 Perspectives

Many suggestions have been made over the years concerning the role of anserine, though no convincing theory has been provided yet, especially in view of its co-occurrence with carnosine. The identification of carnosine *N*-methyltransferase opens up a new area of research, and there are several reasons to believe that its exploration will facilitate progress in the understanding of the physiological function of anserine. For example, analysis of *hnmt-like* gene expression will lead to a precise identification of anserine-producing cells and tissues, while the use of specific antibodies raised against homogenous recombinant enzyme will allow one to immunolocalize minute amounts of the methyltransferase in tissues. The results of these explorations will in turn be useful for the identification of factors regulating or affecting the expression of the *hnmt-like* gene and the activity of the enzyme during the ontogenetic development of animals.

As mentioned above, further work is needed to identify the methyltransferase that produces anserine in mammals. This information will produce a better understanding of the situation in humans. It may turn out that the gene encoding the human enzyme is absent or is mutated. Alternatively, the absence of anserine in humans may be more apparent than real. Anserine might indeed be formed only at a particular stage of human ontogeny, for example. If the activity of carnosine *N*-methyltransferase is indeed present in

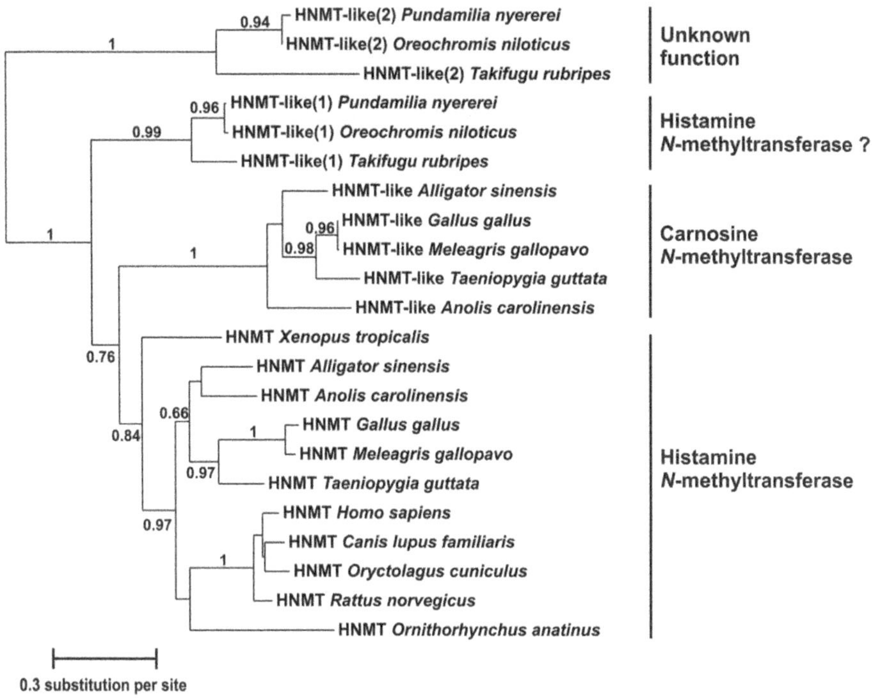

Figure 6.4 Phylogenetic tree of vertebrate histamine *N*-methyltransferase (HNMT) and histamine *N*-methyltransferase-like (HNMT-like) proteins. Protein sequences were aligned using T-Coffee (Notredame *et al.*, 2000) and the tree was inferred with the use of PhyML (Guindon and Gascuel, 2003). Brunch support values assessed using the aLRT test are indicated (Anisimova and Gascuel, 2006). Proved or hypothetical enzyme function of shown proteins are indicated on the right. The following sequences were used: **HNMT**: *A. sinensis*, XP_006018452.1; *A. carolinensis*, XP_003215142.1; *C. lupus familiaris*, (XP_533340.1); *G. gallus*, XP_422143.2; *H. sapiens*, NP_008826.1; *M. gallopavo*, XP_003207782.1, *O. anatinus*, XP_001510925.2; *O. cuniculus* (XP_002712173.1); *R. norvegicus*, NP_112306.1; *T. guttata*, XP_002194327.1; X. tropicalis, NP_001072355.1. **HNMT-like**: *A. sinensis*, XP_006018451.1; *A. carolinensis*, XP_003215153.1; *G. gallus*, XP_001234740.1; *M. gallopavo*, XP_003207778.1; *T. guttata*, XP_002194298.1. **HNMT-like(1)**: *O. niloticus*, XP_003445374.1; *P. nyererei*, XP_005745752.1; *T. rubripes*, XP_003962122.1; **HNMT-like(2)**: *O. niloticus*, XP_003445450.1; *P. nyererei*, XP_005745755.1; *T. rubripes*, XP_003962121.1. Note that HNMT-like sequence (XP_006018451.1) of *Alligator sinensis* is incorrectly described as "histamine *N*-methyltransferase (HNMT)" in GenBank database.

humans, we will need to find in which cells and consider that anserine deficiency might result in disorders, most likely those in which oxidative stress and oxidative damage play an important role. Hence, the identification of the gene encoding mammalian carnosine *N*-methyltransferase would also be of considerable diagnostic importance.

Finally, future work should also aim at identifying the methyltransferase(s) that produce(s) ophidine in several species (see Figure 6.1). Considering that this compound occurs in widely different organisms (*e.g.* Japanese striped snake, red deer, fin whale), it would not be surprising if "ophidine-synthesizing enzymes" would have appeared several times independently during evolution.

Summary Points

- Carnosine synthase is the enzyme that synthesizes carnosine (β-alanine-L-histidine) and homocarnosine (γ-aminobutyrate-L-histidine), while carnosine *N*-methyltransferase catalyzes the *N*-methylation of carnosine, yielding anserine (β-alanine-*N*-π-methyl-L-histidine).
- Recently, carnosine synthase has been molecularly identified as vertebrate ATP-grasp domain containing protein 1 (ATPGD1) and carnosine *N*-methyltransferase (in chicken) as histamine *N*-methyltransferase-like protein (HNMT-like), giving a new insight into the biosynthesis of carnosine and anserine.
- Carnosine synthase has ≈950 amino acids, it is twice as long as most of the proteins of the ATP-grasp family and it contains a highly conserved C-terminal half that clearly belongs to the ATP-grasp family and is most likely responsible for the catalytic activity.
- The *N*-terminal half of carnosine synthase is much less conserved but is also homologous to proteins of the ATP-grasp family.
- Chicken carnosine *N*-methyltransferase (HNMT-like) comprises 322 amino acids and is found as a monomer (≈36 kDa) in its native form.
- Chicken carnosine *N*-methyltransferase is specific for carnosine, and has a K_M for carnosine and *S*-adenosyl-L-methionine equal to 1.6 mM and 14 μM, respectively, which are well below intracellular concentrations of these metabolites.
- Avian HNMT-like protein is closely related to histamine *N*-methyltransferase (HNMT) and both proteins are most likely paralogs resulting from a gene duplication event that happened once during the evolution of vertebrates.
- Since the gene encoding HNMT-like protein is absent from the genomes of mammals, the anserine-producing enzyme must be a very different protein in these vertebrates.

Key Facts about the Importance of Molecular Identification of Enzymes

1. Enzymes are biocatalysts that allow metabolic reactions to proceed efficiently.
2. Most enzymes are proteins that are encoded by genes.

3. Identification of genes that encode enzymes (*i.e.* molecular identification of enzymes) provides crucial information on biological properties of enzymes and allows transgenic animal models to be created where the gene of interest can be knocked out to help understanding its physiological importance.
4. Molecular identification of enzymes is thus essential to fully understand their physiological functions and may have an immediate practical application in the comprehension of disease, for example.

Definitions of Words and Terms

BLAST. Computer program allowing the search of homologous proteins, DNA or RNA in databases and their alignment.

Catalytic efficiency. It is defined as the k_{cat}/K_M ratio. It is useful to compare different substrates of a single enzyme: the best substrates are those with the highest catalytic efficiency.

Expression sequence tag (EST). Is a short cDNA sequence that covers only part of the whole cDNA and can be used to identify gene transcripts.

$\mathbf{k_{cat}}$. Catalytic constant of an enzyme, indicating its "turnover number", *i.e.* the maximal number of product molecules it can form per unit of time and per catalytic site. Its value equals the V_{max} divided by the enzyme concentration.

$\mathbf{K_M}$. Michaelis–Menten constant. In enzyme kinetics, K_M indicates the concentration of substrate at which an enzyme reaction proceeds at half the maximum velocity (V_{max}).

Ontogeny (ontogenesis). The process of the development of an individual organism from a fertilized egg to an adult form.

Orthologs. Homologous proteins (or genes) that exert the same function in different species.

Paralogs. Homologous proteins (or genes) that exert different functions.

Phylogeny (phylogenesis). The process of the evolutionary development of a species or a group of organisms.

$\mathbf{V_{max}}$. The maximal rate (maximal velocity) of an enzyme. It is reached when the catalytic site of an enzyme is fully occupied by the substrate(s).

List of Abbreviations

ATPGD1	ATP-grasp domain containing protein 1
CARNS1	Carnosine synthase
HNMT-like	Histamine *N*-methyltransferase-like protein
HNMT	Histamine *N*-methyltransferase protein
SAH	*S*-Adenosyl-L-homocysteine
SAM	*S*-Adenosyl-L-methionine
GEO	Gene Expression Omnibus
NCBI	The National Center for Biotechnology Information
EST	Expression sequence tag

References

Anisimova, M. and Gascuel, O., 2006. Approximate likelihood-ratio test for branches: A fast, accurate, and powerful alternative. *Systematic Biology*. 55: 539–552.

Abraham, D., Pisano, J. J. and Udenfriend, S., 1962. The distribution of homocarnosine in mammals. *Archives of Biochemistry and Biophysics*. 99: 210–213.

Bakardjiev, A., 1997. Biosynthesis of carnosine in primary cultures of rat olfactory bulb. *Neuroscience Letters*. 227: 115–118.

Bakardjiev, A. and Bauer, K., 2000. Biosynthesis, release, and uptake of carnosine in primary cultures. *Biochemistry (Mosc)*. 65: 779–782.

Bauer, K., 2005. Carnosine and homocarnosine, the forgotten, enigmatic peptides of the brain. *Neurochemical Research*. 30: 1339–1345.

Bauer, K., Salnikow, J., de Vitry, F., Tixier-Vidal, A. and Kleinkauf, H., 1979. Characterization and biosynthesis of omega-aminoacyl amino acids from rat brain and the C-6 glioma cell line. *Journal of Biological Chemistry*. 254: 6402–6407.

Bauer, K. and Schulz, M., 1994. Biosynthesis of carnosine and related peptides by skeletal muscle cells in primary culture. *European Journal of Biochemistry*. 219: 43–47.

Biffo, S., Grillo, M. and Margolis, F. L., 1990. Cellular localization of carnosine-like and anserine-like immunoreactivities in rodent and avian central nervous system. *Neuroscience*. 35: 637–651.

Boldyrev, A. A., Aldini, G. and Derave, W., 2013. Physiology and pathophysiology of carnosine. *Physiological Reviews*. 93: 1803–1845.

Crush, K. G., 1970. Carnosine and related substances in animal tissues. *Comparative Biochemistry and Physiology*. 34: 3–30.

De Marchis, S., Modena, C., Peretto, P., Migheli, A., Margolis, F. L. and Fasolo, A., 2000. Carnosine-related dipeptides in neurons and glia. *Biochemistry (Mosc)*. 65: 824–833.

Divino Filho, J. C., Bergström, P., Stehle, P. and Fürst, P., 1997. Simultaneous measurements of free amino acid patterns of plasma, muscle and erythrocytes in healthy human subjects. *Clinical Nutrition*. 16: 299–305.

Drozak, J., Chrobok, L., Poleszak, O., Jagielski, A. K. and Derlacz, R., 2013. Molecular identification of carnosine *N*-methyltransferase as chicken histamine *N*-methyltransferase-like protein (hnmt-like). *PLoS One*. 8: e64805.

Drozak, J., Veiga-da-Cunha, M., Vertommen, D., Stroobant, V. and Van Schaftingen, E., 2010. Molecular identification of carnosine synthase as ATP-grasp domain-containing protein 1 (ATPGD1). *Journal of Biological Chemistry*. 285: 9346–9356.

Fawaz, M. V., Topper, M. E. and Firestine, S. M., 2011. The ATP-grasp enzymes. *Bioorganic Chemistry*. 39: 185–191.

Galperin, M. Y. and Koonin, E. V., 1997. A diverse superfamily of enzymes with ATP-dependent carboxylate-amine/thiol ligase activity. *Protein Science*. 6: 2639–2643.

Gerrits, G. P., Monnens, L. A., Gabreëls, F. J., De Abreu, R. A., Koster, A. and Trijbels, J. M., 1993. Cerebrospinal fluid amino acids, purines and pyrimidines as a tool in the study of metabolic brain diseases. *Journal of Inherited Metabolic Disease*. 16: 670–675.

Guindon, S. and Gascuel, O., 2003. A simple, fast, and accurate algorithm to estimate large phylogenies by maximum likelihood. *Systematic Biology*. 52: 696–704.

Horinishi, H., Grillo, M. and Margolis, F. L., 1978. Purification and characterization of carnosine synthetase from mouse olfactory bulbs. *Journal of Neurochemistry*. 31: 909–919.

Kalyankar, G. D. and Meister, A., 1959. Enzymatic synthesis of carnosine and related beta-alanyl and gamma-aminobutyryl peptides. *Journal of Biological Chemistry*. 234: 3210–3218.

Kanazawa, A. and Sano, I., 1967. A method of determination of homocarnosine and its distribution in mammalian tissues. *Journal of Neurochemistry*. 14: 211–214.

Kish, S. J., Perry, T. L. and Hansen, S., 1979. Regional distribution of homocarnosine, homocarnosine-carnosine synthetase and homocarnosinase in human brain. *Journal of Neurochemistry*. 32: 1629–1636.

Kumon, A., Matsuoka, Y., Nakajima, T., Kakimoto, Y., Imaoka, N. and Sano, I., 1970a. Isolation and indentification of *N*-alpha-(beta-alanyl)lysine and *N*-alpha-(gamma-aminobutyryl)lysine from bovine brain. *Biochimica et Biophysica Acta*. 200: 170–171.

Kumon, A., Matsuoka, Y., Kakimoto, Y., Nakajima, T. and Sano, I., 1970b. A peptidase that hydrolyzes Na-(gamma-aminobutyryl)lysine. *Biochimica et Biophysica Acta*. 200: 466–474.

Lattin, J. E., Schroder, K., Su, A. I., Walker, J. R., Zhang, J., Wiltshire, T., Saijo, K., Glass, C. K., Hume, D. A., Kellie, S. and Sweet, M. J., 2008. Expression analysis of G protein-coupled receptors in mouse macrophages. *Immunome Research*. 4: 5.

Linster, C. L., Van Schaftingen, E. and Hanson, A. D., 2013. Metabolite damage and its repair or pre-emption. *Nature Chemical Biology*. 9: 72–80.

Margolis, F. L. and Grillo, M., 1984. Carnosine, homocarnosine and anserine in vertebrate retinas. *Neurochemistry International*. 6: 207–209.

Martignoni, P. and Winnick, T., 1954. Biosynthesis of carnosine and anserine in the chick. *Journal of Biological Chemistry*. 208: 251–261.

Maslinski, C. and Fogel, W. A., 1991. Catabolism of histamine, In: Uvnäs, B. (ed.) *Handbook of Experimental Pharmacology*. Histamine and histamine antagonists. Springer, Berlin-Heidelberg, vol. 97, pp. 165–189.

Matsuoka, M., Nakajima, T. and Sano, I., 1969. Identification of alpha-(beta-alanyl)-lysine in rabbit muscle. *Biochimica et Biophysica Acta*. 177: 169–171.

McManus, I. R., 1962. Enzymatic synthesis of anserine in skeletal muscle by *N*-methylation of carnosine. *Journal of Biological Chemistry*. 237: 1207–1211.

McManus, I. R. and Benson, M. S., 1967. Studies on the formation of carnosine and anserine in pectoral muscle of the developing chick. *Archives of Biochemistry and Biophysics*. 119: 444–453.

Mero, A., 1999. Leucine supplementation and intensive training. *Sports Medicine*. 27: 347–358.

Murakami, T., Yamane, H., Tomonaga, S. and Furuse, M., 2009. Forced swimming and imipramine modify plasma and brain amino acid concentrations in mice. *European Journal of Pharmacology*. 602: 73–77.

Nakajima, T., Kakimoto, Y., Kumon, A., Matsuoka, M. and Sano, I., 1969. Alpha-(gamma-aminobutyryl)-lysine in mammalian brain: its identification and distribution. *Journal of Neurochemistry*. 16: 417–422.

Nagase, T., Kikuno, R., Ishikawa, K. I., Hirosawa, M. and Ohara, O., 2000. Prediction of the coding sequences of unidentified human genes. XVI. The complete sequences of 150 new cDNA clones from brain which code for large proteins in vitro. *DNA Research*. 7: 65–73.

Ng, R. H. and Marshall, F. D., 1978. Regional and subcellular distribution of homocarnosine-carnosine synthetase in the central nervous system of rats. *Journal of Neurochemistry*. 30: 87–90.

Notredame, C., Higgins, D. G. and Heringa, J., 2000. T-Coffee: a novel method for fast and accurate multiple sequence alignment. *Journal of Molecular Biology*. 302: 205–217.

Perry, T. L., Berry, K., Hansen, S., Diamond, S. and Mok, C., 1971a. Regional distribution of amino acids in human brain obtained at autopsy. *Journal of Neurochemistry*. 18: 513–519.

Perry, T. L., Hansen, S., Berry, K., Mok, C. and Lesk, D., 1971b. Free amino acids and related compounds in biopsies of human brain. *Journal of Neurochemistry*. 18: 521–528.

Punta, M., Coggill, P. C., Eberhardt, R. Y., Mistry, J., Tate, J., Boursnell, C., Pang, N., Forslund, K., Ceric, G., Clements, J., Heger, A., Holm, L., Sonnhammer, E. L., Eddy, S. R., Bateman, A. and Finn, R. D., 2012. The Pfam protein families database. *Nucleic Acids Research*. 40(Database issue): D290–D301.

Raghavan, M., Lindberg, U. and Schutt, C., 1992. The use of alternative substrates in the characterization of actin-methylating and carnosine-methylating enzymes. *European Journal of Biochemistry*. 210: 311–318.

Schenck, J. R., Simmonds, S., Cohn, M., Stevens, C. M. and du Vigneaud, V., 1943. The relation of transmethylation to anserine. *Journal of Biological Chemistry*. 149: 355–359.

Schubert, H. L., Blumenthal, R. M. and Cheng, X., 2003. Many paths to methyltransfer: a chronicle of convergence. *Trends in Biochemical Sciences*. 28: 329–335.

Skaper, S. D., Das, S. and Marshall, F. D., 1973. Some properties of a homocarnosine-carnosine synthetase isolated from rat brain. *Journal of Neurochemistry*. 6: 1429–1445.

Stenesh, J. J. and Winnick, T., 1960. Carnosine-anserine synthetase of muscle. 4. Partial purification of the enzyme and further studies of beta-alanyl peptide synthesis. *Biochemical Journal*. 77: 575–581.

Su, A. I., Wiltshire, T., Batalov, S., Lapp, H., Ching, K. A., Block, D., Zhang, J., Soden, R., Hayakawa, M., Kreiman, G., Cooke, M. P., Walker, J. R. and

Hogenesch, J. B., 2004. A gene atlas of the mouse and human protein-encoding transcriptomes. *Proceedings of the National Academy of Sciences of the United States of America*. 101: 6062–6067.

Tsubone, S., Yoshikawa, N., Okada, S. and Abe, H., 2007. Purification and characterization of a novel imidazole dipeptide synthase from the muscle of the Japanese eel Anguilla japonica. *Comparative Biochemistry and Physiology Part B: Biochemistry and Molecular Biology*. 146: 560–567.

Uchisawa, H., Naraoka, T. and Ono, T., 2007. A novel ornithine-containing tripeptide isolated from the extract of the brackish-water bivalve *Corbicula japonica*. *Biochimica et Biophysica Acta*. 1770: 790–796.

Van Regemorter, N., Mardens, Y., Lowenthal, A., Van Sande, M. and Lavinha, F., 1972. Distribution of two peptides of γ-aminobutyric acid in human brain and CSF. *Clinica Chimica Acta*. 38: 59–65.

Van Schaftingen, E., Rzem, R., Marbaix, A., Collard, F., Veiga-da-Cunha, M. and Linster, C. L., 2013. Metabolite proofreading, a neglected aspect of intermediary metabolism. *Journal of Inherited Metabolic Disease*. 36: 427–434.

Winnick, R. E. and Winnick, T., 1958. Carnosine-anserine synthetase of muscle. II. Specificity and inhibition studies with amino acid analogs. *Bulletin de la Société de Chimie Biologique (Paris)*. 40: 1727–1735.

Winnick, R. E. and Winnick, T., 1959. Pathways and the physiological site of anserine formation. *Nature*. 183: 1466–1468.

Winnick, R. E. and Winnick, T., 1960. Carnosine-anserine synthetase of muscle. III. Experiments on the mechanism of peptide bond formation. *Biochimica et Biophysica Acta*. 37: 214–220.

Wu, C., Orozco, C., Boyer, J., Leglise, M., Goodale, J., Batalov, S., Hodge, C. L., Haase, J., Janes, J., Huss, III, J. W. and Su, A. I., 2009. BioGPS: an extensible and customizable portal for querying and organizing gene annotation resources. *Genome Biology*. 10: R130.

CHAPTER 7

The Biochemistry of Anserine and Carnosine Degrading Enzymes

SHOJI YAMADA*[a]

[a]Laboratory of Marine Biochemistry, Faculty of Fisheries, Kagoshima University, 4-50-20 Shimoarata, Kagoshima, Japan
*E-mail: yamada@fish.kagoshima-u.ac.jp

7.1 Introduction

Structurally unusual imidazole dipeptides, such as carnosine (β-alanyl-L-His), anserine (β-alanyl-*N*-π-methyl-L-His), ophidine (or balenine; β-alanyl-*N*-τ-methyl-L-His), and homocarnosine (γ-aminobutyryl-L-His), containing a non-α-amino acid (*i.e.* β-alanine or γ-aminobutyric acid) at the N-terminus and an imidazole-related amino acid (L-His, *N*-π-methyl-L-His, or *N*-τ-methyl-L-His) at the C-terminus occur widely in skeletal muscles and central nervous systems of vertebrates (Crush, 1970; Chan and Decker, 1994; Bonfanti, 1999). Moreover, *N*-α-acetyl-L-His (NAH) has been found as one of the naturally occurring imidazole-related compounds in ectothermic vertebrates (Anastasi *et al.*, 1964; Baslow, 1965; Yamada *et al.*, 2009). The discovery of these imidazole-related compounds accompanied that of several corresponding degrading enzymes, while classification of these enzymes has been confused over the past half-century.

Food and Nutritional Components in Focus No. 8
Imidazole Dipeptides: Chemistry, Analysis, Function and Effects
Edited by Victor R Preedy

Published by the Royal Society of Chemistry, www.rsc.org

Peptidases are of great relevance to life science, and so the MEROPS peptidase database (http://www.merops.sanger.ac.uk) was constructed to integrate sources of information about them (Rawlings *et al.*, 2012). The organizational principle of the database is a hierarchical classification in which homologous sets of the proteins of interest are grouped in families and the homologous families are grouped in clans. The MEROPS database (release 9.10) includes 54 clans, nearly 250 families and more than 2000 codes for peptidases. A few of the families contain two or more rather distinct groups of peptidases, as shown by a deep divergence in the dendrogram. For these, subfamilies are recognized. While the EC List of the Nomenclature Committee of the International Union of Biochemistry and Molecular Biology (NC-IUBMB) is a classification based on reaction catalyzed, the MEROPS peptidase database is a classification of individual peptidases according to molecular structure and homology.

According to the MEROPS database, there are six enzymes that can catalyze degradation of imidazole dipeptides (Table 7.1). Three of these enzymes, peptidase V (MEROPS ID: M20.004), Xaa–His dipeptidase (M20.007), and BapF peptidase (*Pseudomonas* sp.) (P01.003), are known to exist only in bacterial species. The other three, carnosine dipeptidase II (M20.005), carnosine dipeptidase I (M20.006), and Xaa–methyl-His dipeptidase (M20.013), are distributed in vertebrates. All the vertebrate enzymes are highly homologous, belonging to the family M20 subfamily F of the metallopeptidase H clan, and can catalyze the hydrolysis of anserine, carnosine, homocarnosine, and/or NAH as follows:

$$\text{anserine} + H_2O \rightarrow N\text{-}\pi\text{-methyl-L-His} + \beta\text{-alanine}$$

$$\text{carnosine} + H_2O \rightarrow \text{L-His} + \beta\text{-alanine}$$

$$\text{homocarnosine} + H_2O \rightarrow \text{L-His} + \gamma\text{-aminobutyric acid}$$

$$\text{NAH} + H_2O \rightarrow \text{L-His} + \text{acetate}$$

The nomenclature cytosol nonspecific dipeptidase (EC 3.4.13.18), β-Ala–His dipeptidase (EC 3.4.13.20) and Xaa–methyl-His dipeptidase (EC 3.4.13.5), is accepted for enzymes as the enzymes that can hydrolyze these compounds by the NC-IUBMB. Recent studies demonstrate that these enzymes correspond to carnosine dipeptidase II, carnosine dipeptidase I, and Xaa–methyl-His dipeptidase in the MEROPS database, respectively.

This review is, therefore, intended to give a detailed account of the current knowledge of these three vertebrate enzymes, which are frequently named "anserinase"or "carnosinase". The bacterial enzymes including peptidase V, Xaa–His dipeptidase, and BapF peptidase (*Pseudomonas* sp.) will not be discussed.

7.2 The Enzymes Named "Carnosinase"

The enzymes named "carnosinase" have a long history of classification. In 1949, Hanson and Smith (1949) first described a carnosine degrading enzyme from hog kidney and named the enzyme "carnosinase", since its properties

Table 7.1 The six enzymes that catalyze the degradation of the imidazole dipeptides in the MEROPS peptidase database (release 9.10).[a]

MEROPS					NC-IUBMB[b]		Sources		Gene symbols
Family	Subfamily	ID	Name	Other names	Accepted name	EC number	Organisms	Tissues	
M20	A	M20.004	Peptidase V	β-Ala–His dipeptidase			Bacteria		pepV
M20	C	M20.007	Xaa–His dipeptidase	Peptidase D			Bacteria		pepD
M20	F	M20.005	Carnosine dipeptidase II	Tissue carnosinase, prolinase, glycyl-leucine dipeptidase	Cytosol nonspecific dipeptidase[c]	3.4.13.18[c]	Animals	All tissues	CNDP2
M20	F	M20.006	Carnosine dipeptidase I	Serum carnosinase	β-Ala–His dipeptidase[c]	3.4.13.20[c]	Vertebrates	Brain, serum, kidney, skeletal muscle	CNDP1
M20	F	M20.013	Xaa–methyl-His dipeptidase	Anserinase, acetylhistidine deacetylase	Xaa–methyl-His dipeptidase[d]	3.4.13.5[d]	Ectothermic vertebrates	Brain, ocular fluids, skeletal muscle	ANSN
P1	(None)	P01.003	BapF peptidase (*Pseudomonas* sp.)				*Pseudomonas aeruginosa*		

[a]http://merops.sanger.ac.uk.
[b]The Nomenclature Committee of the International Union of Biochemistry and Molecular Biology.
[c]Based on Teufel *et al.* (2003).
[d]Based on Yamada *et al.* (2005).

were distinct from those of any previously described protease. The enzyme was partially purified and was shown by inhibition and activation studies to be a metalloenzyme like other exopeptidases. Rosenberg (1960) purified the enzyme from the same tissue by heating, ammonium sulfate and acetone fractionation, and zone electrophoresis, and described several enzymatic properties. Thereafter, studies by several research groups using different tissues (serum, kidney, and brain) of several mammal species, including human, rat, and mouse, provided clear evidence for the existence of two different kinds of carnosine degrading enzymes (Perry *et al.*, 1968; Murphey *et al.*, 1972; Lenney, 1976; Lenney *et al.*, 1977; Wolos *et al.*, 1978; Margolis *et al.*, 1979; Lenney *et al.*, 1982; Margolis *et al.*, 1983; Lenney *et al.*, 1985; Kunze *et al.*, 1986; Lenney, 1990b; Jackson *et al.*, 1991).

Of the two enzymes, one enzyme was given multiple names, such as "manganese-dependent carnosinase" by Margolis *et al.* (1979), "human tissue carnosinase" by Lenney *et al.* (1985), "β-Ala–Arg hydrolase" by Kunze *et al.* (1986), and "human cytosolic nonspecific dipeptidase" by Lenney (1990a). This enzyme actually is not a carnosine-specific enzyme but an enzyme with broad substrate specificity against dipeptides, exhibiting characteristics identical to another enzyme, which was already known as "cytosol nonspecific dipeptidase" (EC 3.4.13.18). Cytosol nonspecific dipeptidase is also known as "prolinase", "glycyl-leucine dipeptidase", "glycyl-glycine dipeptidase", and the MEROPS name carnosine dipeptidase II.

Another carnosine-degrading enzyme described as 'serum carnosinase' by Murphey *et al.* (1973), as "manganese-independent carnosinase" by Margolis *et al.* (1979), as "carnosinase" by Kunze *et al.* (1986), and as "human serum carnosinase" by Jackson *et al.* (1991), is an imidazole dipeptide-specific enzyme with narrow substrate specificity. In 1992, the enzyme was newly registered as "β-Ala–His dipeptidase" (EC 3.4.13.20) by the NC-IUBMB. β-Ala–His dipeptidase was given the MEROPS name carnosine dipeptidase I. Cytosol nonspecific dipeptidase (*i.e.*, carnosine dipeptidase II) and β-Ala–His dipeptidase (*i.e.*, carnosine dipeptidase I) were finally distinguished by molecular identification. Teufel *et al.* (2003) identified two distinct clones belonging to a metallopeptidase family in the Human Genome Sciences human brain cDNA libraries: one is CNDP2 encoding carnosine dipeptidase II; the other is CNDP1 encoding carnosine dipeptidase I. Although the two enzymes were formerly given the same name "carnosinase", those properties and biological roles are absolutely different.

7.2.1 Carnosine Dipeptidase II

7.2.1.1 Distribution

In human, the activity of carnosine dipeptidase II was found in each of the 17 tissues assayed, with kidney, liver, spleen and brain being the richest sources of this enzyme (Lenney *et al.*, 1985). Very little was found in the human uterus, although this organ was the richest source of the enzyme in rat (Lenney, 1976). Butterworth and Priestman (1982) also reported that the activity of carnosine dipeptidase II, which they described as "prolinase" at that time, occurred widely in nearly every human tissue examined. The CNDP2 mRNAs of two fish species Nile tilapia (*Oreochromis niloticus*) (Yamada *et al.*, 2005) and Japanese eel (*Anguilla japonica*) (Oku *et al.*, 2012), and that of human as well (Teufel *et al.*, 2003), were ubiquitously expressed in every tissue examined. All the enzymes from rat (kidney and liver), mouse (liver), and human (placenta and kidney) tissues were found only in the cytosolic fraction by subcellular localization (Lenney *et al.*, 1985).

7.2.1.2 Enzymatic Properties

Historically, this enzyme has been described by many researchers under various names, such as glycyl-glycine dipeptidase (formerly EC 3.4.13.11), glycyl-leucine dipeptidase (formerly EC 3.4.13.2), microsomal dipeptidase (renal dipeptidase; formerly EC. 3.4.13.11), and prolyl dipeptidase (prolinase; formerly EC 3.4.13.8), in various tissues of mammals. Carnosine dipeptidase II has a broader specificity than carnosine dipeptidase I. The enzyme can hydrolyze many ordinary dipeptides, such as Gly–L-Leu, L-Pro–L-Leu, Gly–Gly, Gly–L-His, L-Ala–L-Ala, L-Pro–L-Ala, L-His–L-Ala, L-Leu–L-Leu, and L-Ala–L-Leu, as well as carnosine. There is, however, no reliable evidence that this enzyme hydrolyzes anserine and homocarnosine.

This enzyme is extremely sensitive to bestatin (K_i = 0.5 nM) (Peppers and Lenney, 1988), which is a very potent competitive inhibitor of leucyl aminopeptidase and aminopeptidase B. This inhibitor can be used to distinguish carnosine dipeptidase II from carnosine dipeptidase I and Xaa–methyl-His dipeptidase. Since the enzyme is stabilized and activated by dithiothreitol or 2-mercaptoethanol, such a reducing agent should be added to a purification buffer for high recovery. Metal-chelating agents strongly inhibit carnosine dipeptidase II, and Mn^{2+} ions are required for the full enzymatic activity (Lenney *et al.*, 1985; Kunze *et al.*, 1986).

Teufel *et al.* (2003) revealed that the gene CNDP2 located on human chromosome 18 encodes a protein with the characteristics of the enzyme. They reported that the purified recombinant human CNDP2 protein migrated at 54 kDa on SDS-PAGE, and molecular weight determined by MALDI-TOF mass spectrometry was in agreement with its predicted value deduced from the amino acid sequence (52.7 kDa). The human CNDP2 protein apparently forms a homodimer, judging from gel filtration. Unno *et al.* (2008) first presented the crystal structures of mouse carnosine dipeptidase II complexed with bestatin, as a nonhydrolyzable substrate analog, together with Zn^{2+} at a resolution of 1.7 A and that with Mn^{2+} at 2.3 A. The mouse enzyme is a homodimer in a noncrystallographic asymmetric unit, and each subunit is composed of two domains; domain A that is complexed with bestatin and the two metal ions, and domain B that provides the major interface for dimer formation. They indicated that the bestatin molecule bound to domain A interacts with several residues of domain B of the other subunit, and these interactions are likely to be essential for enzyme activity. Figure 7.1 shows the deduced amino-acid sequence of carnosine dipeptidase II (CNDP2) from Japanese eel, aligned with those of carnosine dipeptidase I (CNDP1) and Xaa–methyl-His dipeptidase (ANSN).

7.2.1.3 Biological Roles

Carnosine dipeptidase II may be mainly involved in intracellular catabolism of dipeptidic substrates as a housekeeping enzyme. It is thought that this enzyme hydrolyzes many dipeptides of the type Pro-Xaa during proline-rich collagen breakdown (Bauer, 2004). Moreover, it has been suggested that the role of the enzyme is to provide the amino acids L-His and β-alanine, the products of carnosine, to the amino acid pool for protein synthesis in the cells that secrete nasal mucus and in olfactory receptor neurons (Chen *et al.*, 1994).

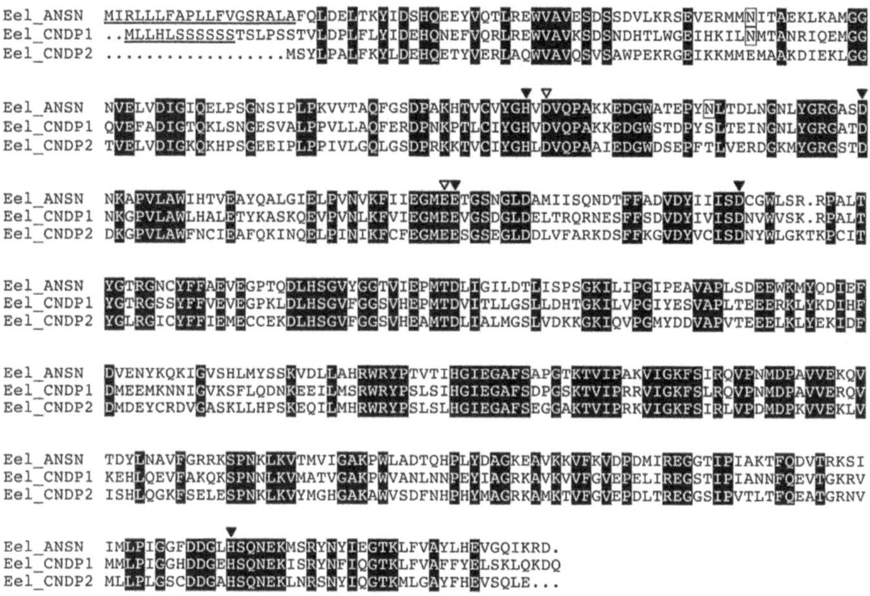

Figure 7.1 The deduced amino-acid sequences of carnosine dipeptidase II (CNDP2), carnosine dipeptidase I (CNDP1), and Xaa–methyl-His dipeptidase (ANSN) of Japanese eel (*Anguilla japonica*). The nucleotide accession numbers are AB219567 (CNDP2), AB207245 (CNDP1), and AB207246. (ANSN) (Oku *et al.*, 2012). Identical amino acids are highlighted in black. Gaps inserted into sequences are indicated by dots. The N-terminal signal peptide regions are underlined. The potential *N*-glycosylation sites are surrounded by squares. The active-site and metal-binding-site residues are marked with white triangles and black triangles, respectively.

7.2.2 Carnosine Dipeptidase I

7.2.2.1 Distribution

Perry *et al.* (1968) reported that human serum hydrolyzes carnosine and anserine, and named the enzyme "serum carnosinase". It was the first report to describe carnosine dipeptidase I. Carnosinuria was reported in patients with severe retardation and neurological disturbances, and this was associated with reduced serum carnosinase activity (Perry *et al.*, 1968; van Heeswijk *et al.*, 1969). Lenney *et al.* (1982) reported that human serum carnosinase is a different enzyme from kidney cellular carnosinase (that is carnosine dipeptidase II), and predicted that these enzymes are the products of two different genes. Jackson *et al.* (1991) investigated distribution of the activity of carnosine dipeptidase I in serum of mammals. Six higher primates including human, chimpanzee, gibbon, gorilla, orangutan, and pygmy chimpanzee were found to have enzyme activity. No activity was detected in the serum of nonprimate mammals such as dog, horse, calf, hog, rat, mouse, rabbit, baboon, guinea pig, armadillo, and Chinese hamster. Carnosine dipeptidase I is, therefore, considered to be a primate-specific enzyme in serum.

Teufel *et al.* (2003) reported that the human gene CNDP1 encoding carnosine dipeptidase I is expressed in the brain and liver. In contrast, the mouse and rat orthologs of CNDP1 were detected exclusively in the kidney, and not in the brain. This supports previous results describing the existence of a mouse kidney enzyme that resembles human serum carnosinase (Margolis *et al.*, 1979; Margolis *et al.*, 1983). Immunocytochemistry with an antiserum against the mouse kidney enzyme demonstrated that the enzyme is localized in the proximal tubules of kidney, glandular cells of uterus, nasal olfactory mucosa, and in vomeronasal and certain other nerve pathways. The human enzyme is localized in several layers of the retina and in certain neuronal tracts of the cerebral cortex, cerebellar cortex, olfactory bulb, hippocampus, and in disseminated tracts presumably from the internal capsule, interspersed among the basal ganglia (Jackson *et al.*, 1994). The enzyme was also present in the epithelial cells of the choroid plexus and in corpora amylacea, seen in many regions of the human CNS. Janssen *et al.* (2005) reported the CNDP1 mRNA and its protein expression in the glomerulus of the human kidney. Taken together, it can be concluded that the human gene CNDP1 is expressed in the CNS, including retina, liver, and kidney, and with the corresponding enzyme present in the serum.

The enzyme activity of carnosine dipeptidase I has not been detected in nonmammalian species. Although Partmann (1976) reported carnosine-degrading enzyme activity in the skeletal muscle of European eel (*Anguilla anguilla*), which contains a large amount of carnosine, its enzymatic properties have remained unknown until recently. Oku *et al.* (2012) reported the purification and molecular identification of carnosine dipeptidase I from the skeletal muscle of

Japanese eel. This was the first report describing the isolation of the dipeptidase encoded by CNDP1 from a nonmammalian species. They showed that the mRNA of CNDP1 is expressed in the liver and kidney of the fish, but not in the brain. Database searches revealed that an ortholog of CNDP1 is universally distributed in mammals, birds, reptiles and amphibians; interestingly, the gene is restricted to only two taxa, salmonids and eels, among the ray-finned fishes. Other ray-finned fishes, such as zebrafish, fugu, and medaka, whose genomes have been completely sequenced, do not possess the gene encoding carnosine dipeptidase I.

7.2.2.2 *Enzymatic Properties*

Unlike carnosine dipeptidase II with broad substrate specificity, carnosine dipeptidase I shows narrow specificity only on the imidazole dipeptides and some dipeptides of Xaa–His. When multiple dipeptides were tested as possible substrates, using a highly purified human serum enzyme preparation, carnosine and anserine were the two most rapidly hydrolyzed substrates (Jackson *et al.*, 1991). L-Ala–L-His and Gly–L-His, containing histidine at the carboxyl end, also were rapidly hydrolyzed. Homocarnosine was relatively slowly hydrolyzed by the human native enzyme. Contrary to the native enzyme, the recombinant human enzyme did not degrade anserine (Teufel *et al.*, 2003). For carnosine dipeptidase I purified from mouse kidney, anserine was a poor substrate and homocarnosine a nonsubstrate (Margolis *et al.*, 1983). In contrast, carnosine dipeptidase I purified from Japanese eel muscle showed strong activities against L-Ala–L-His and anserine (Table 7.2). Carnosine and homocarnosine were poor substrates for the fish enzyme. The eel carnosine dipeptidase I can be easily distinguished from both carnosine dipeptidase II and Xaa–methyl-His dipeptidase by its inability to hydrolyze ordinary dipeptides such as Gly–L-Leu and L-Pro–Gly, and *N*-α-acetyl amino acids such as *N*-α-acetyl-L-Met and NAH. The pH optimum for activity against carnosine is 8.0–8.5 (Lenney *et al.*, 1982) or 7.5–8.5 (Teufel *et al.*, 2003). Activity against carnosine is enhanced in the presence of Cd^{2+} (Lenney *et al.*, 1982).

Carnosine dipeptidase I is a *N*-glycosylated protein (Jackson *et al.*, 1991). The human and Japanese eel enzymes have a typical N-terminal signal peptide region and *N*-glycosylation site(s) (Teufel *et al.*, 2003; Oku *et al.*, 2012), indicating that both proteins are secretory. Indeed, the human enzyme exists in the extracellular serum.

The coding region of human CNDP1 is translated into 508 amino acids, which includes a signal peptide of 29 amino acids containing a Leu-repeat and three *N*-glycosylation sites. On the other hand, Japanese eel CNDP1 encodes 492 amino acids, which includes a signal peptide of 11 amino acids containing a Ser-repeat and one *N*-glycosylation site (Figure 7.1). Both mature enzymes from human and Japanese eel form a homodimer. Although crystal-structure analysis has not yet been performed, a 3D model of human carnosine dipeptidase I has been generated by homology modeling techniques,

Table 7.2 Substrate specificity of carnosine dipeptidase I and Xaa–methyl-His dipeptidase of Japanese eel (*Anguilla japonica*).[a]

	Relative activity (L-Ala–L-His = 100%)[b]	
Substrate	Carnosine dipeptidase I	Xaa–methyl-His dipeptidase
L-Ala–L-His	100	100
Carnosine	7	9
Anserine	78	33
Homocarnosine	4	14
Gly–L-Leu	2	352
Gly–D-Leu	0	15
Gly–L-His	6	91
Gly–Gly	24	306
L-Cys–Gly	0	1
L-His–Gly	3	7
L-Pro–Gly	2	180
L-Leu–Gly	0	69
L-Ala–L-Pro	0	0
L-His–L-Leu	1	1
N-α-Chloroacetyl-L-Leu	9	501
N-α-Acetyl-L-Leu	1	53
N-α-Acetyl-L-Met	1	80
N-α-Acetyl-L-Glu	0	8
N-α-Acetyl-L-Asp	0	3
N-α-Acetyl-L-His	2	43
L-Ala-*p*-nitroanilide	0	0
L-Leu-*p*-nitroanilide	1	1
L-Leu-β-naphthylamide	0	0
Benzyloxycarbonyl-Gly–L-Leu	0	9

[a]Data from Oku *et al.* (2012).
[b]The purified enzymes were incubated at 37 °C for 1 h with 2 mM substrate in 100 mM *N*-ethylmorpholine–HCl buffer, pH 8.0, containing 0.1 mM $CoSO_4$ and 1 mg mL^{-1} bovine serum albumin.

and indicates the molecular mechanism by which citrate ions act as allosteric activators of the enzyme (Vistoli *et al.* 2006).

7.2.2.3 Biological Roles

The physiological function of carnosine dipeptidase I has not been elucidated. It has been postulated that one possible role of the enzyme is to release the inhibitory neurotransmitter γ-aminobutyric acid from homocarnosine in the higher primate CNS, including that of humans, in which the dipeptide is found in high concentration (Jackson *et al.*, 1994). The enzyme may catalyze a second metabolic reaction for the formation of free γ-aminobutyric acid in the CNS, in addition to glutamate decarboxylation.

Many studies have focused on genetic association of carnosine dipeptidase I with human diabetic nephropathy in the past several years, after the gene

CNDP1 encoding the enzyme was confirmed as a susceptibility locus in 2005. Janssen *et al.* (2005) detected significant evidence for association between diabetic nephropathy and a Leu-repeat in the signal peptide of CNDP1. An allelic variant with a lower number (5× Leu) of a Leu-repeat in exon 2 of the gene on human chromosome 18, termed "Mannheim allele", is observed at greater frequency in diabetic individuals without kidney disease in some populations, and is also associated with a low secretion of the enzyme in serum (Janssen *et al.*, 2005; Freedman *et al.*, 2007; Mooyaart *et al.*, 2009). It has been hypothesized that diabetic individuals with a lower number of Leu-repeat in the signal peptide have lower serum enzyme activity levels, and are protected against diabetic nephropathy, because carnosine, a physiological substrate for carnosine dipeptidase I, acts as a protective factor against adverse effects of high glucose levels on renal cells. Meta-analyses on association between the CNDP1 Mannheim allele and diabetic nephropathy susceptibility using fixed and random effects models have also confirmed that Mannheim allele microsatellite polymorphism is associated with diabetic nephropathy susceptibility, especially in the type-2 diabetic nephropathy and in the Caucasian population (Zhu *et al.*, 2013). Moreover, *N*-glycosylation of the enzyme was found to be essential for appropriate secretion and enzyme activity, in addition to the Leu-repeat polymorphism of CNDP1 (Riedl *et al.*, 2010).

7.3 The Enzyme Named "Anserinase" or "Acetylhistidine Deacetylase"

Fishes belonging to the family Gadidae, as well as the family Salmonidae, contain anserine in their skeletal muscles as a major imidazole-related compound. In 1955, Jones (1955) found that this imidazole dipeptide in the skeletal muscle of Atlantic cod (*Gadus morhua*) was hydrolyzed by an endogenous enzyme, with liberation of β-alanine and *N*-π-methyl-L-His in roughly stoichiometric proportion, and originally named the enzyme "anserinase". This was the first study to demonstrate the activity of Xaa–methyl-His dipeptidase. Jones (1956) also reported that this cod enzyme is a metal-activated dipeptidase with low substrate specificity. The enzyme was registered as EC 3.4.13.5 by the IUB in 1972. In a separate line of research, Baslow (1963) reported the postmortem degradation of NAH (termed "neurosine" in the article) to L-His in the brain of killifish (*Fundulus heteroclitus*). Subsequently, partial purification of the NAH degrading enzyme from the brain of skipjack tuna (*Katsuwonus pelamis*) was reported (Baslow and Lenney, 1967). The name acetylhistidine deacetylase with EC 3.5.1.34 was recommended for the enzyme by the IUB in 1972. However, both "anserinase" and "acetylhistidine deacetylase" were judged to be the same by the IUB in 1981, on the basis of a study indicating similarity of the skipjack tuna brain enzyme and the cod skeletal muscle enzyme reported by Lenney *et al.* (1978). The enzyme is now incorporated into a single entry by the accepted name "Xaa–methyl-His dipeptidase" (EC. 3.4.13.5).

7.3.1 Xaa–Methyl-His Dipeptidase

7.3.1.1 Distribution

Xaa–methyl-His dipeptidase, unlike carnosine dipeptidase II and carnosine dipeptidase I, shows a limited distribution in vertebrates. This enzyme is present only in ectothermic vertebrates, including ray-finned fishes, amphibians and reptiles (Yamada *et al.*, 2005). The enzyme is absent from endothermic vertebrates, including birds and mammals, and from the primitive fishes such as agnathans. Therefore, it can be said that Xaa–methyl-His dipeptidase is a specific enzyme for ectothermic-jawed vertebrates.

There are two kinds of possible physiological substrates for Xaa–methyl-His dipeptidase in ectothermic vertebrates: one is NAH (in brain and eye) and the other is anserine (in skeletal muscle). The strong activity of the enzyme colocalizes with NAH ubiquitously in the brain and eye of ectothermic vertebrates. In the fish eye, the enzyme exists in the ocular fluids (vitreous body and aqueous humor), and the substrate exists in the lens. Exceptionally, some fish species, such as Atlantic cod, haddock (*Melanogrammus aeglefinus*), and Japanese eel, contain significant activity in their skeletal muscles (Jones, 1955; Perez and Jones, 1962; Oku *et al.*, 2012). On the contrary, the skeletal muscles of rainbow trout (*Oncorhynchus mykiss*) and Nile tilapia have trace or no activity (Yamada *et al.*, 1991; Yamada *et al.*, 1994). The gene ANSN encoding Xaa–methyl-His dipeptidase is expressed strongly in the brain, retina, skeletal muscle, and kidney, and slightly in the spleen, but not expressed in the lens, intestine, gill, heart, and liver of Nile tilapia (Yamada *et al.*, 2005). Whilst Nile tilapia skeletal muscle expresses the gene, this tissue does not show detectable activity. The mRNA of ANSN is more strongly expressed in the liver than in the brain of Japanese eel (Oku *et al.*, 2012). In the central nervous system of rainbow trout, the mesencephalon is the richest source of the enzyme, while the enzyme activity of the spinal cord is about one-fifteenth that of the mesencephalon (Yamada *et al.*, 1993). In subcellular fractionation of the fish brain, 90% of the total activity was recovered from the soluble supernatant fraction, whereas 40% of the total NAH was recovered from the crude mitochondrial pellet, suggesting subcellular compartmentalization of the enzyme and the substrate.

7.3.1.2 Enzymatic Properties

In contrast with carnosine dipeptidase I, Xaa–methyl-His dipeptidase shows broad specificity on dipeptides and *N*-α-acetylated amino acids (Table 7.2). The dipeptidase can hydrolyze not only imidazole dipeptides, such as anserine, carnosine, and homocarnosine, but also ordinary dipeptides, such as L-Ala–L-His, Gly–L-Leu, Gly–Gly, and L-Pro–Gly. This enzyme also shows strong deacetylating activity against *N*-α-acetylated amino acids, such as NAH, *N*-α-acetyl-L-Met, *N*-α-acetyl-L-Leu, and *N*-α-chloroacetyl-L-Leu. However, the enzyme is unable to hydrolyze the following *N*-α-acetylated compounds: *N*-α-acetyl-Gly, *N*-α-acetyl-L-Ala, *N*-α-acetyl-L-Val, *N*-α-acetyl-L-Cys, *N*-α-acetyl-L-Asp, *N*-α-acetyl-L-Tyr, *N*-α-acetyl-L-ornithine, and *N*-ω-acetylhistamine (Baslow and Lenney, 1967; Yamada *et al.*, 2005). Tripeptides including Gly–Gly–L-Leu, Gly–L-Leu–L-Tyr, and L-Leu–Gly–Gly are not substrates for Xaa–methyl-His dipeptidase.

The enzyme is strongly inhibited by metal-chelating reagents like EDTA and 1,10-phenanthroline. The enzyme purified from the brain of rainbow trout was inactivated by incubation with 1 mM EDTA (Yamada *et al.*, 1993). When the inactivated enzyme was incubated with 2 mM divalent metal ion salts, the activity was restored completely only by Co^{2+}. Addition of Zn^{2+} and Mn^{2+} produced a partial reactivation, while Mg^{2+}, Sr^{2+}, and Ca^{2+} produced no restoration. The rainbow trout enzyme is also inhibited by the thiol-reactive agent *p*-chloromercuribenzoate (1 mM), but not by iodoacetate and *N*-ethylmaleimide. High concentrations (0.1–0.5 M) of phosphate buffer are also inhibitory, possibly by complexing the metal ion required for activity (Lenney *et al.*, 1978). The enzyme is hardly affected by bestain (10 μM), which is a very potent competitive inhibitor of carnosine dipeptidase II.

Xaa–methyl-His dipeptidase, as well as carnosine dipeptidase II and carnosine dipeptidase I, forms a homodimer, and contains seven conserved residues (two active-site residues and five metal-binding-site residues) as shown in Figure 7.1. The enzyme is a secretory *N*-glycoprotein.

7.3.1.3 Biological Roles

In ectothermic vertebrates, the lens of the eye has high levels of NAH, which is synthesized from L-His and acetyl coenzyme A, as catalyzed by histidine *N*-acetyltransferase inside the lens (Baslow, 1966; Yamada and Arikawa, 2014). Xaa–methyl-His dipeptidase exists in the ocular fluids outside the lens. Therefore, there is a complete compartmentalization of the synthetic and hydrolytic enzymes. From these findings, Baslow (1998) has proposed that NAH functions as a metabolically recyclable molecular water pump for water homeostasis. Water influx or generation of metabolic water in the lens may serve as the trigger mechanism to open a Ca-dependent gate for the release of NAH down its gradient, along with its associated water. The released NAH is immediately hydrolyzed by Xaa–methyl-His dipeptidase in the ocular fluids. L-His is then transported into the lens, where NAH is again synthesized by histidine *N*-acetyltransferase.

Since NAH is present in high concentration in the brain of ectothermic vertebrates as well as the lens, it is obvious that this compound is the physiological substrate for Xaa–methyl-His dipeptidase. Based on the physiological substrate for the enzyme in the brain and eye, "acetylhistidine deacetylase", rather than "Xaa–methyl-His dipeptidase", might be a suitable name for the enzyme.

7.3.1.4 Decomposition of Anserine in Seafood

Considerable increases in both *N*-π-methyl-L-His and β-alanine contents of Atlantic cod skeletal muscle (containing a large amount of anserine) were found to occur during storage in ice before the onset of bacterial invasion of the tissue (Jones, 1955), and the breakdown of anserine was catalyzed by Xaa–methyl-His dipeptidase as described above. Similar observations were reported by Ruiz-Capillas and Moral (2001) in hake (*Merluccius merluccius*) muscle during ice storage. They suggested that the loss of anserine during storage was attributed to "anserinase" activity, and proposed that anserine, as well as the two degradation compounds, could be used as quality control indices for chilled hake. Although no enzymological studies are available for hake, the endogenous enzyme may be Xaa–methyl-His dipeptidase because hake belongs to the same order, Gadiformes, as Atlantic cod and haddock, whose skeletal muscles are known to contain the enzyme. In contrast, Bauchart *et al.* (2007) reported that the anserine concentration in rainbow trout muscle (also containing a large amount of anserine) was almost unaffected by ice storage. As the skeletal muscle of rainbow trout has no activity of Xaa–methyl-His dipeptidase (Yamada *et al.*, 1991), it is presumed that muscle anserine is stable in salmonids, unlike cod and its allied species, during ice storage.

7.4 The Phylogenetic History of Anserine and Carnosine Degrading Enzymes

All three enzymes that catalyze the degradation of the imidazole dipeptides, including carnosine dipeptidase II, carnosine dipeptidase I, and Xaa–methyl-His dipeptidase, belong to the same evolutionary metallopeptidase subfamily M20F, but are distinctly separated from each other in a molecular phylogenetic tree (Figure 7.2). Two main clades exist: non-secretory type (including carnosine dipeptidase II) and secretory type (including carnosine dipeptidase I and Xaa–methyl-His dipeptidase). The proposed molecular evolution of these enzymes in vertebrates is shown in Figure 7.3 (Yamada *et al.*, 2005; Oku *et al.*, 2011). Since carnosine dipeptidase II is ubiquitously distributed in modern species of vertebrates as a housekeeping enzyme, it can be estimated that this gene was the original gene for duplication, resulting in carnosine dipeptidase I and Xaa–methyl-His dipeptidase.

The first gene duplication from carnosine dipeptidase II occurred after the divergence between the Urochordata (tunicates) and Craniata (animals with skulls). One copy of the gene preserved the ancestral function (that is carnosine dipeptidase II), while the other copy was free from selective pressure, resulting in a primitive secretory type. The latter has been inherited by some modern species of the jawless fishes, such as Far Eastern brook lamprey (*Lethenteron reissneri*) and Arctic lamprey (*Lethenteron camtschaticum*) (Oku *et al.*, 2011).

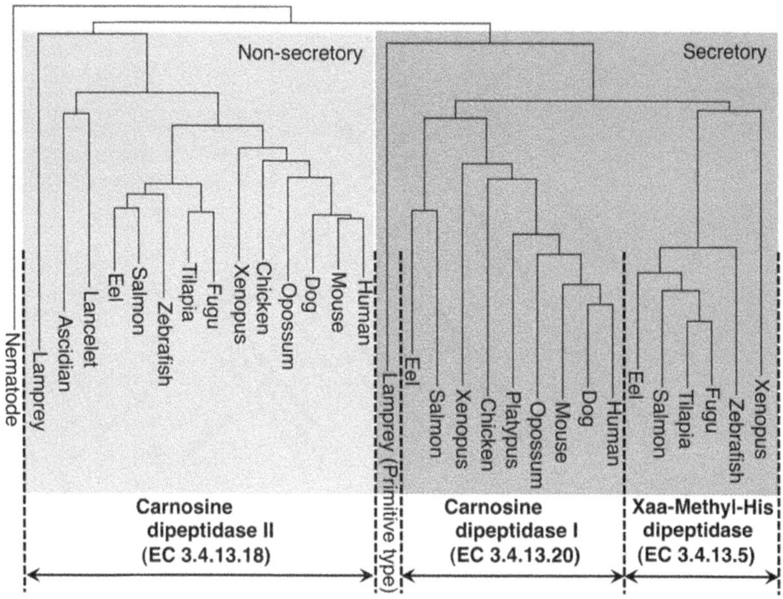

Figure 7.2 The phylogenetic tree of carnosine dipeptidase II, carnosine dipeptidase I, and Xaa–methyl-His dipeptidase in vertebrates. The phylogenetic tree based on the amino-acid sequences was constructed by the neighbor-joining method of the CLUSTAL W program. The homologous genes were obtained from public databases. The light- and dark-shaded areas indicate nonsecretory and secretory proteins, respectively.

The second duplication event of the primitive secretory-type gene occurred in the Gnathostomata (jawed vertebrates) lineage. Before the divergence between the Actinopterygii (ray-finned fishes) and Sarcopterygii (lobe-finned fishes and four-legged vertebrates), one copy may have evolved to the carnosine dipeptidase I gene, and the other copy to the Xaa–methyl His dipeptidase gene. While the carnosine dipeptidase I gene has been extensively distributed from ectotherms to endotherms, the Xaa–methyl-His dipeptidase gene has been lost in modern species of both endothermic birds and mammals. Further studies are needed to hypothesize why the Xaa–methyl-His dipeptidase gene (together with its substrate, NAH) has been lost not only in mammals but also in birds, despite the inference that modern birds and reptiles have diverged from a common ancestor.

7.5 Conclusions

There is no doubt that recent molecular identification of the imidazole dipeptide degrading enzymes, called "carnosinase" and "anserinase", will accelerate research into the functions of bioactive imidazole dipeptides in human health. Needless to say, this acceleration will depend on many scientists' tireless pursuit of truth in the past.

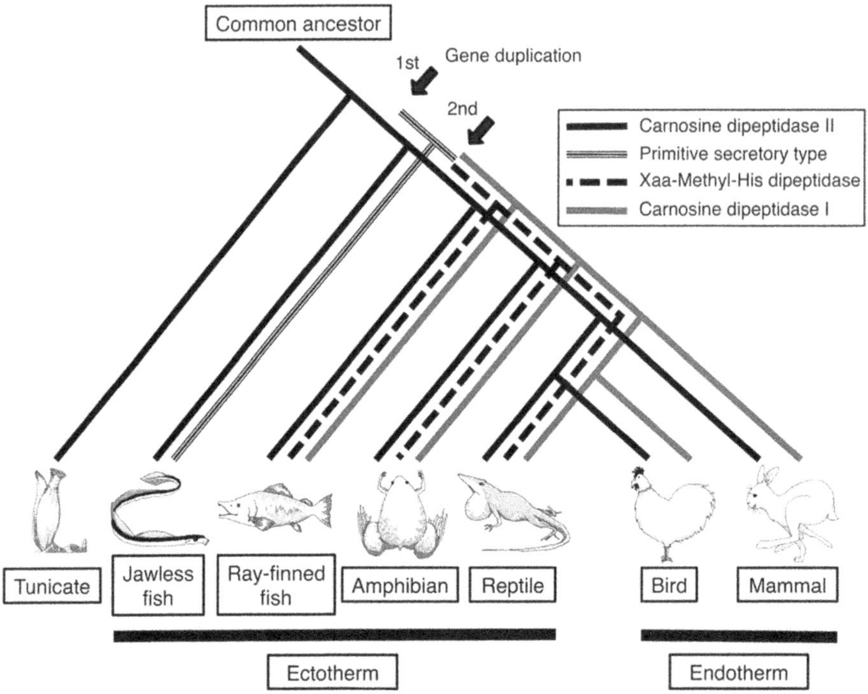

Figure 7.3 The molecular evolution and distribution of carnosine dipeptidase II, carnosine dipeptidase I, and Xaa-methyl-His dipeptidase in vertebrates. Source: Yamada, S., 2013. Xaa–methyl-His dipeptidase. In: Rawlings, N. D. and Salvesen, G. S. (ed.) Handbook of Proteolytic Enzymes, 3rd edn. Academic Press, Oxford, UK, pp. 1603–1608.

Summary Points

- This chapter focuses on three kinds of vertebrate enzymes catalyzing the hydrolysis of the imidazole dipeptides.
- All the vertebrate enzymes belong to the family M20 subfamily F of the metallopeptidase H clan.
- Carnosine dipeptidase II (EC 3.4.13.18) with broad specificity is a cytosolic housekeeping enzyme.
- Carnosine dipeptidase I (EC 3.4.13.20) with narrow specificity is specifically secreted into human serum.
- Xaa–methyl-His dipeptidase (EC 3.4.13.5) with broad specificity is an ectotherm-specific enzyme deacetylating *N*-α-acetyl-L-His in the brain and eye.
- Phylogenetic analyses indicate carnosine dipeptidase I and Xaa–methyl-His dipeptidase have been evolved from carnosine dipeptidase II after two gene-duplication events.

Key Facts

Key features of dipeptidase classification. This table lists the key facts of the classification of the enzymes belonging to "dipeptidases" by the Nomenclature Committee of the International Union of Biochemistry and Molecular. EC: Enzyme Commission.

1. All known enzymes are divided into six main groups: "oxidoreductases (EC 1)", "transferases (EC 2)", "hydrolases (EC 3)", "lyases (EC 4)", "isomerases (EC 5)", and "ligases (EC 6)".
2. The enzymes belonging to EC 3 catalyze the hydrolysis of various bonds.
3. The enzymes acting on peptide bond belong to EC 3.1 and are called "peptidases".
4. The substrates for the enzymes "dipeptidases" with EC 3.4.13 are dipeptides.
5. At present, eleven "dipeptidases" including three kinds of the imidazole dipeptide degrading enzymes (EC 3.4.13.5, EC 3.4.13.18, and EC 3.4.13.20) are registered.

Definitions of Words and Terms

Carnosinuria. A condition in which carnosine is excreted in excessive amounts in the urine. This aminoaciduria occurs in carnosinemia (an inherited metabolic disorder of carnosine dipeptidase I deficiency) or by meat ingestion.

Competitive inhibitor. A chemical compound that closely resembles the structure of the substrate. The inhibitor competes with the substrate by binding at the active site of the target enzyme.

Ectotherm. A cold-blooded animal that regulates its body temperature by exchanging heat with its environment. It is also called a poikilotherm.

Endotherm. A warm-blooded animal that is able to maintain a constant body temperature. It is also called a homeotherm.

***N*-α-Acetylhistidine.** A natural imidazole compound found in very high concentration ubiquitously in the central nervous system of ectothermic vertebrates, including reptiles, amphibians and bony fish. The compound is absent from homothermic vertebrates, including mammals and birds. *N*-α-Acetylhistidine is synthesized from L-His and acetyl-CoA by histidine *N*-acetyltransferase (EC 2.3.1.33), and hydrolyzed to L-His and acetate by Xaa–methyl-His dipeptidase (EC 3.4.13.5). Little is known of the biological role.

***N*-Glycosylation.** Post-translational modification of proteins. A process by which sugars are chemically attached to a nitrogen atom of an asparagine residue in a protein. It is also termed *N*-linked glycosylation.

Homodimer. A protein composed of two identical protein subunits (polypeptide chains).

Housekeeping enzyme. An enzyme that is needed for the maintenance of basic cellular function and is found in all cells of an organism.

Peptidase. An enzyme that hydrolyzes peptide bonds of peptides or proteins, resulting in amino acids or small peptides. Also termed protease, proteinase and proteolytic enzyme.
Signal peptide. A short peptide present at the N-terminus of a protein. It is a significant factor determining the destination to which that protein is delivered. Enzymes known as signal peptidases cleave off a signal peptide of a protein, once its final destination has been reached.

List of Abbreviations

NAH *N*-α-Acetyl-L-His

Acknowledgements

I greatly thank Dr. Steven M. Plakas (FDA, Dauphin Island, AL., USA) for the review of this manuscript. I also thank Yoko Yamada for help in some figure preparation.

References

Anastasi, A., Correale, P. and Erspamer, V., 1964. Occurrence of *N*-acetylhistidine in the central nervous system and the eye of *Rana esculenta*. *Journal of Neurochemistry*. 11: 63–66.
Baslow, M. H., 1963. The enzymatic degradation of neurosine as an index of fish quality. *American Zoologist*. 3: 536(Abstract No. 266).
Baslow, M. H., 1965. Neurosine, its identification with *N*-acetyl-L-histidine and distribution in aquatic vertebrates. *Zoologica*. 50: 63–66.
Baslow, M. H., 1966. *N*-Acetyl-L-histidine synthetase activity from the brain of the killifish. *Brain Research*. 3: 210–213.
Baslow, M. H., 1998. Function of the *N*-acetyl-L-histidine system in the vertebrate eye. Evidence in support of a role as a molecular water pump. *Journal of Molecular Neuroscience*. 10: 193–208.
Baslow, M. H. and Lenney, J. F., 1967. α-*N*-Acetyl-L-histidine amidohydrolase activity from the brain of the skipjack tuna *Katsuwonus pelamis*. *Canadian Journal of Biochemistry*. 45: 337–340.
Bauchart, C., Chambon, C., Mirand, P. P., Savary-Auzeloux, I., Remond, D. and Morzel, M., 2007. Peptides in rainbow trout (*Oncorhynchus mykiss*) muscle subjected to ice storage and cooking. *Food Chemistry*. 100: 1566–1572.
Bauer, K., 2004. Cytosol nonspecific dipeptidase. In: Barrett, A. J. Rawlings, N. D. and Woessner, J. F. (ed.) *Handbook of Proteolytic Enzymes*, 2nd edn. Elsevier, London, UK, pp. 1020–1022.
Bonfanti, L., Peretto, P., De Marchis, S. and Fasolo, A., 1999. Carnosine-related dipeptides in the mammalian brain. *Progress in Neurobiology*. 59: 333–353.
Butterworth, J. and Priestman, D., 1982. Fluorimetric assay for prolinase and partial characterisation in cultured skin fibroblasts. *Clinica Chimica Acta*. 122: 51–60.

Chan, K. M. and Decker, E. A., 1994. Endogenous skeletal muscle antioxidants. *Critical Reviews in Food Science and Nutrition*. 34: 403–426.

Chen, Y., Getchell, T. V., Sparks, D. L. and Getchell, M. L., 1994. Cellular localization of carnosinase in the human nasal mucosa. *Acta Oto-Laryngologica*. 114: 193–198.

Crush, K. G., 1970. Carnosine and related substances in animal tissues. *Comparative Biochemistry and Physiology*. 34: 3–30.

Freedman, B. I., Hicks, P. J., Sale, M. M., Pierson, E. D., Langefeld, C. D., Rich, S. S., Xu, J., McDonough, C., Janssen, B., Yard, B. A., van der Woude, F. J. and Bowden, D. W., 2007. A leucine repeat in the carnosinase gene CNDP1 is associated with diabetic end-stage renal disease in European Americans. *Nephrology, Dialysis, Transplantation*. 22: 1131–1135.

Hanson, H. T. and Smith, E. L., 1949. Carnosinase: an enzyme of swine kidney. *Journal of Biological Chemistry*. 179: 789–801.

van Heeswijk, P. J., Trijbels, J. M., Schretlen, E. D., van Munster, P. J. and Monnens, L. A., 1969. A patient with a deficiency of serum-carnosinase activity. *Acta Paediatrica Scandinavica*. 58: 584–592.

Jackson, M. C., Kucera, C. M. and Lenney, J. F., 1991. Purification and properties of human serum carnosinase. *Clinica Chimica Acta*. 196: 193–205.

Jackson, M. C., Scollard, D. M., Mack, R. J. and Lenney, J. F., 1994. Localization of a novel pathway for the liberation of GABA in the human CNS. *Brain Research Bulletin*. 33: 379–385.

Janssen, B., Hohenadel, D., Brinkkoetter, P., Peters, V., Rind, N., Fischer, C., Rychlik, I., Cerna, M., Romzova, M., de Heer, E., Baelde, H., Bakker, S. J. L., Zirie, M., Rondeau, E., Mathieson, P., Saleem, M. A., Meyer, J., Koppel, H., Sauerhoefer, S., Bartram, C. R., Nawroth, P., Hammes, H. P., Yard, B. A., Zschocke, J. and van der Woude, F. J., 2005. Carnosine as a protective factor in diabetic nephropathy – Association with a leucine repeat of the carnosinase gene CNDP1. *Diabetes*. 54: 2320–2327.

Jones, N. R., 1955. The free amino acids of fish; 1-methylhistidine and β-alanine liberation by skeletal muscle anserinase of codling (*Gadus callarias*). *Biochemical Journal*. 60: 81–87.

Jones, N. R., 1956. Anserinase and other dipeptidase activity in skeletal muscle of codling (*Gadus callarias*). In: *Abstract of the 352nd Meeting of the Biochemical Society (Proceedings of the Biochemical Society)*, p. 20.

Kunze, N., Kleinkauf, H. and Bauer, K., 1986. Characterization of two carnosine-degrading enzymes from rat brain. Partial purification and characterization of a carnosinase and a β-alanyl-arginine hydrolase. *European Journal of Biochemistry*. 160: 605–613.

Lenney, J. F., 1976. Specificity and distribution of mammalian carnosinase. *Biochimica et Biophysica Acta*. 429: 214–219.

Lenney, J. F., 1990a. Human cytosolic carnosinase: evidence of identity with prolinase, a non-specific dipeptidase. *Biological Chemistry Hoppe-Seyler*. 371: 167–171.

Lenney, J. F., 1990b. Separation and characterization of two carnosine-splitting cytosolic dipeptidases from hog kidney (carnosinase and non-specific dipeptidase). *Biological Chemistry Hoppe-Seyler*. 371: 433–440.

Lenney, J. F., Baslow, M. H. and Sugiyama, G. H., 1978. Similarity of tuna *N*-acetylhistidine deacetylase and cod fish anserinase. *Comparative Biochemistry and Physiology, Part B: Biochemistry & Molecular Biology*. 61: 253–258.

Lenney, J. F., George, R. P., Weiss, A. M., Kucera, C. M., Chan, P. W. and Rinzler, G. S., 1982. Human serum carnosinase: characterization, distinction from cellular carnosinase, and activation by cadmium. *Clinica Chimica Acta*. 123: 221–231.

Lenney, J. F., Kan, S. C., Siu, K. and Sugiyama, G. H., 1977. Homocarnosinase: a hog kidney dipeptidase with a broader specificity than carnosinase. *Archives of Biochemistry and Biophysics*. 184: 257–266.

Lenney, J. F., Peppers, S. C., Kucera-Orallo, C. M. and George, R. P., 1985. Characterization of human tissue carnosinase. *Biochemical Journal*. 228: 653–660.

Margolis, F. L., Grillo, M., Brown, C. E., Williams, T. H., Pitcher, R. G. and Elgar, G. J., 1979. Enzymatic and immunological evidence for two forms of carnosinase in the mouse. *Biochimica et Biophysica Acta*. 570: 311–323.

Margolis, F. L., Grillo, M., Grannot-Reisfeld, N. and Farbman, A. I., 1983. Purification, characterization and immunocytochemical localization of mouse kidney carnosinase. *Biochimica et Biophysica Acta*. 744: 237–248.

Mooyaart, A. L., van Valkengoed, I. G., Shaw, P. K., Peters, V., Baelde, H. J., Rabelink, T. J., Bruijn, J. A., Stronks, K. and de Heer, E., 2009. Lower frequency of the 5/5 homozygous CNDP1 genotype in South Asian Surinamese. *Diabetes Research and Clinical Practice*. 85: 272–278.

Murphey, W. H., Lindmark, D. G., Patchen, L. I., Housler, M. E., Harrod, E. K. and Mosovich, L., 1973. Serum carnosinase deficiency concomitant with mental retardation. *Pediatric Research*. 7: 601–606.

Murphey, W. H., Patchen, L. and Lindmark, D. G., 1972. Carnosinase: a fluorometric assay and demonstration of two electrophoretic forms in human tissue extracts. *Clinica Chimica Acta*. 42: 309–314.

Oku, T., Ando, S., Hayakawa, T., Baba, K., Nishi, R., Shiozaki, K. and Yamada, S., 2011. Purification and identification of a novel primitive secretory enzyme catalyzing the hydrolysis of imidazole-related dipeptides in the jawless vertebrate *Lethenteron reissneri*. *Peptides*. 32: 648–655.

Oku, T., Ando, S., Tsai, H.-C., Yamashita, Y., Ueno, H., Shiozaki, K., Nishi, R. and Yamada, S., 2012. Purification and identification of two carnosine-cleaving enzymes, carnosine dipeptidase I and Xaa–methyl-His dipeptidase, from Japanese eel (*Anguilla japonica*). *Biochimie*. 94: 1281–1290.

Partmann, W., 1976. A carnosine splitting enzyme activity in skeletal muscle of eels (in German). *Archiv fuer Fischereiwissenschaft*. 27: 55–62.

Peppers, S. C. and Lenney, J. F., 1988. Bestatin inhibition of human tissue carnosinase, a non-specific cytosolic dipeptidase. *Biological Chemistry Hoppe-Seyler*. 369: 1281–1286.

Perez, B. S. and Jones, N. R., 1962. Effects of tetracycline antibiotics on the products of anserinase action in chill stored haddock (*Gadus aeglefinus*) muscle. *Journal of Food Science*. 27: 69–72.

Perry, T. L., Hansen, S. and Love, D. L., 1968. Serum-carnosinase deficiency in carnosinaemia. *Lancet*. 1: 1229–1230.

Rawlings, N. D., Barrett, A. J. and Bateman, A., 2012. MEROPS: the database of proteolytic enzymes, their substrates and inhibitors. *Nucleic Acids Research*. 40: D343–D350.

Riedl, E., Koeppel, H., Pfister, F., Peters, V., Sauerhoefer, S., Sternik, P., Brinkkoetter, P., Zentgraf, H., Navis, G., Henning, R. H., Van Den Born, J., Bakker, S. J. L., Janssen, B., van der Woude, F. J. and Yard, B. A., 2010. *N*-Glycosylation of carnosinase influences protein secretion and enzyme activity implications for hyperglycemia. *Diabetes*. 59: 1984–1990.

Rosenberg, A., 1960. Purification and some properties of carnosinase of swine kidney. *Archives of Biochemistry and Biophysics*. 88: 83–93.

Ruiz-Capillas, C. and Moral, A., 2001. Changes in free amino acids during chilled storage of hake (*Merluccius merluccius* L.) in controlled atmospheres and their use as a quality control index. *European Food Research and Technology*. 212: 302–307.

Teufel, M., Saudek, V., Ledig, J. P., Bernhardt, A., Boularand, S., Carreau, A., Cairns, N. J., Carter, C., Cowley, D. J., Duverger, D., Ganzhorn, A. J., Guenet, C., Heintzelmann, B., Laucher, V., Sauvage, C. and Smirnova, T., 2003. Sequence identification and characterization of human carnosinase and a closely related non-specific dipeptidase. *Journal of Biological Chemistry*. 278: 6521–6531.

Unno, H., Yamashita, T., Ujita, S., Okumura, N., Otani, H., Okumura, A., Nagai, K. and Kusunoki, M., 2008. Structural basis for substrate recognition and hydrolysis by mouse carnosinase CN2. *Journal of Biological Chemistry*. 283: 27289–27299.

Vistoli, G., Pedretti, A., Cattaneo, M., Aldini, G. and Testa, B., 2006. Homology modeling of human serum carnosinase, a potential medicinal target, and MD simulations of its allosteric activation by citrate. *Journal of Medicinal Chemistry*. 49: 3269–3277.

Wolos, A., Piekarska, K., Glogowski, J. and Konieczka, I., 1978. Two molecular forms of swine kidney carnosinase. *International Journal of Biochemistry*. 9: 57–62.

Yamada, S. and Arikawa, S., 2014. An ectotherm homologue of human predicted gene NAT16 encodes histidine *N*-acetyltransferase responsible for *N*-α-acetylhistidine synthesis. *Biochimica et Biophysica Acta*. 1840: 434–442.

Yamada, S., Kawashima, K., Baba, K., Oku, T. and Ando, S., 2009. Occurrence of a novel acetylated amino acid, *N*α-acetylhistidine, in skeletal muscle of freshwater fish and other ectothermic vertebrates. *Comparative Biochemistry and Physiology, Part B: Biochemistry & Molecular Biology*. 152: 282–286.

Yamada, S., Tanaka, Y. and Ando, S., 2005. Purification and sequence identification of anserinase. *FEBS Journal*. 272: 6001–6013.

Yamada, S., Tanaka, Y., Sameshima, M. and Furuichi, M., 1991. *N*-α-Acetylhistine metabolism in fish. 2. Distribution of *N*-α-acetylhistidine-deacetylating enzyme in tissues of rainbow trout. *Nippon Suisan Gakkaishi*. 57: 1601.

Yamada, S., Tanaka, Y., Sameshima, M. and Furuichi, M., 1993. Properties of *N*-α-acetylhistidine deacetylase in brain of rainbow trout *Oncorhynchus mykiss*. *Comparative Biochemistry and Physiology, Part B: Biochemistry & Molecular Biology*. 106: 309–315.

Yamada, S., Tanaka, Y., Sameshima, M. and Furuichi, M., 1994. Effects of starvation and feeding on tissue *N*-α-acetylhistidine levels in Nile tilapia *Oreochromis niloticus*. *Comparative Biochemistry and Physiology, Part A: Molecular & Integrative Physiology*. 109: 277–283.

Zhu, J. M., Wang, B., Li, J., Chen, G. M., Fan, Y. G., Feng, C. C., Pan, H. F. and Ye, D. Q., 2013. D18S880 microsatellite polymorphism of carnosinase gene and diabetic nephropathy: a meta-analysis. *Genetic Testing and Molecular Biomarkers*. 17: 289–294.

CHAPTER 8

Carnosine and Derivatives as Inhibitors of Protein Covalent Modifications Induced by Reactive Carbonyl Species

MARA COLZANI[a], DAVIDE GARZON[a], AND GIANCARLO ALDINI*[a]

[a]Department of Pharmaceutical Sciences, Università degli Studi di Milano, Via Mangiagalli 25, 20133 Milan, Italy
*E-mail: giancarlo.aldini@unimi.it

8.1 Reactive Carbonyl Species and Protein Covalent Modification

Reactive carbonyl species (RCS) are a class of exogenous and endogenous reactive compounds that can covalently react with nucleophilic targets such as proteins, phospholipids and nucleic acids, forming damaging adducts (Aldini *et al.*, 2007b), see Table 8.1. Proteins represent the most studied target of RCS and the corresponding reaction products are named advanced lipoxidation end products (ALEs) when the modifying moiety derives from lipids, and advanced glycation end products (AGEs) when it comes from sugar.

From a chemical point of view, the RCS so far reported to react *in vivo* with proteins are quite heterogeneous and can be divided into the following

Food and Nutritional Components in Focus No. 8
Imidazole Dipeptides: Chemistry, Analysis, Function and Effects
Edited by Victor R Preedy

Published by the Royal Society of Chemistry, www.rsc.org

Table 8.1 Key facts of reactive carbonyl species (RCS). This table lists the key facts or reactive carbonyl species (RCS) which are reactive by-products of lipids and sugars oxidation.

- Reactive carbonyl species are cytotoxic compounds generated by oxidative damage.
- Reactive carbonyl species can be formed either by lipid or sugar oxidation.
- Reactive carbonyl species react with proteins forming protein adducts that can induce cell damage.
- Reactive carbonyl species are involved in the pathogenesis of some oxidative-based diseases.
- A strict correlation between the amount of RCS, of the corresponding protein adducts in tissues and fluids and disease states has been found in both animal and human subjects.
- A substantial amount of literature is now available reporting the molecular and cellular pathogenic mechanisms for the involvement of reactive carbonyl species in the onset and progression of various diseases, including atherosclerosis, diabetes and some neurological disorders.
- Reactive carbonyl species are detoxified by phase I and II metabolism forming unreactive metabolites.
- Carnosine reacts with reactive carbonyl species and is believed to take part to their detoxification.
- Compounds effective as inhibitors of reactive carbonyl species or able to block their biological effects have been found to significantly ameliorate different oxidative based diseases.
- Reactive carbonyl species are now considered as promising drug targets.

groups: (1) α,β-unsaturated aldehydes; (2) dialdehydes; (3) keto-aldehydes, see Figure 8.1. The following paragraphs briefly describe the sources of RCS and their mechanism of reactions with proteins (please refer to the review by Vistoli *et al.* (2013) for a more detailed description of RCS).

8.1.1 α,β-Unsaturated Aldehydes

α,β-Unsaturated aldehydes include hydroxylated (*i.e.* 4-hydroxynonenal [HNE], 4-hydroxyhexenal [HHE]) and nonhydroxylated (acrolein [ACR], crotonaldehdye, nonenal, hexenal) derivatives. They can have either endogenous or exogenous sources and different mechanisms of formation, although the lipid-peroxidation cascade involving free PUFAs or phospholipids represents the main pathway (Vistoli *et al.*, 2013). Acrolein is also generated in inflammatory conditions by the degradation of threonine and of spermine and spermidine by myeloperoxidase and amine oxidase, respectively (Seiler, 2004; Stevens and Maier, 2008).

The reactivity of the α,β-unsaturated aldehydes towards the nucleophilic sites of proteins (Lys, His, Cys, Arg and the terminal amino group) is regulated by their electrophilic nature, due to the carbonyl function which acts as an electron-withdrawing group, decreasing the electron richness of the conjugated alkene group and making the alkene β-carbon atom electron

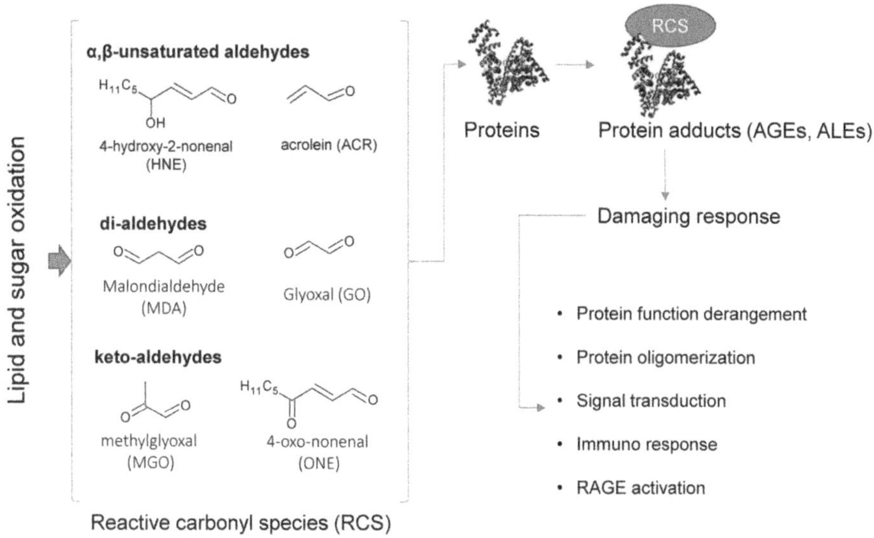

Figure 8.1 Major classes and chemical structures of the most studied reactive carbonyl species (RCS). RCS are breakdown products formed by the oxidative damage of lipids and sugars, belonging to different chemical classes including α,β-unsaturated aldehydes, dialdehydes and keto-aldehydes. RCS are electrophilic compounds able to form covalent adducts with nucleophilic substrates such as proteins, forming adducts (AGEs and ALEs) that can elicit a damaging response.

deficient (Lopachin *et al.*, 2009). The Michael adduct represents the main reaction product and it is formed by the covalent reaction between the electrophilic β-carbon atom of the unsaturated aldehydes and the nucleophilic (electron-rich species) sites of the protein such as the thiolate anion of cysteines, the ε-amino group of lysines and the imidazolic ring of histidines (Figure 8.2). The α,β-unsaturated aldehydes can also generate a reversible Schiff base through the dehydration reaction between the aldehydic carbonyl group and the primary amino group of the protein (the ε-terminal group of Lys or the amino-terminus of the protein) (Figure 8.2).

8.1.2 Dialdehdyes

Malondialdehyde (MDA) and glyoxal (GO) are the two most studied dialdehydes involved in the formation of protein adducts. The former is generated by the lipid peroxidation cascade involving PUFA and it reacts with both Lys and Arg, forming several protein adducts and crosslinks. The protein covalent adduction of MDA is due to the nondissociated enolate form (the β-hydroxyacrolein) that, by reacting with the ε-amino group of Lys, forms the *N*-ε-(2-propenal)lysine (*N*-propenal-Lys), a reactive intermediate that can further react forming additional products (Figure 8.3).

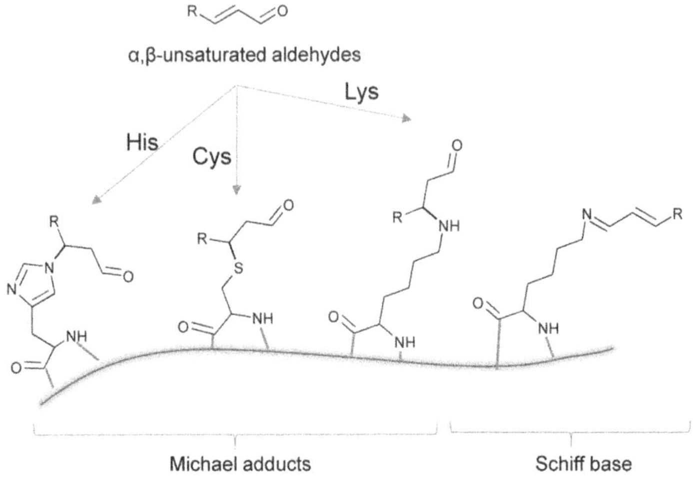

Figure 8.2 Covalent adduction of α,β-unsaturated aldehydes with nucleophilic sites of proteins. C3 of α,β-unsaturated aldehydes react with the nucleophilic sites of proteins (His, Cys and Lys) forming Michael adducts and with the ε-amino group of Lys forming the corresponding Schiff base.

Glyoxal has both exogenous and endogenous sources. Exogenous sources include cooked foods (Degen *et al.*, 2012) and the environment (combustion) (Fu *et al.*, 2008). Exogenous glyoxal is easily absorbed. At the endogenous level, glyoxal can be generated by the oxidation of both lipids and sugars (O'Brien *et al.*, 2005) as well as by the degradation of the Amadori products (Thornalley *et al.*, 1999).

N-ε-(Carboxymethyl)lysine (CML) and *S*-(carboxymethyl)cysteine (CMC) represent the two main reaction products of GO with proteins. Glyoxal can also react with Arg forming imidazolinone derivatives and *N*-(carboxymethyl) arginine. Due to the presence of the two carbonyl groups, GO can also react with two nucleophilic sites forming crosslinks named GOLD and GODIC, formed by the reaction of one molecule of glyoxal with two Lys or with one Lys and one Arg, respectively (Figure 8.3).

8.1.3 Keto-Aldehydes

Methylglyoxal (MGO), 4-oxo-nonenal (ONE) and isoketals (also called levuglandins) are the most studied keto-aldehydes responsible for protein covalent modifications.

MGO has different exogenous and endogenous sources and several mechanisms of formation. At the endogenous level, the main source is represented by the involvement of the enzymatic or nonenzymatic degradation of the triose phosphate intermediates originating from the glycolytic processes (Kalapos, 1999). MGO reacts primarily with Arg and Lys forming imidazolinone derivatives and carboxyethyl-lysine (CEL) adduct, respectively.

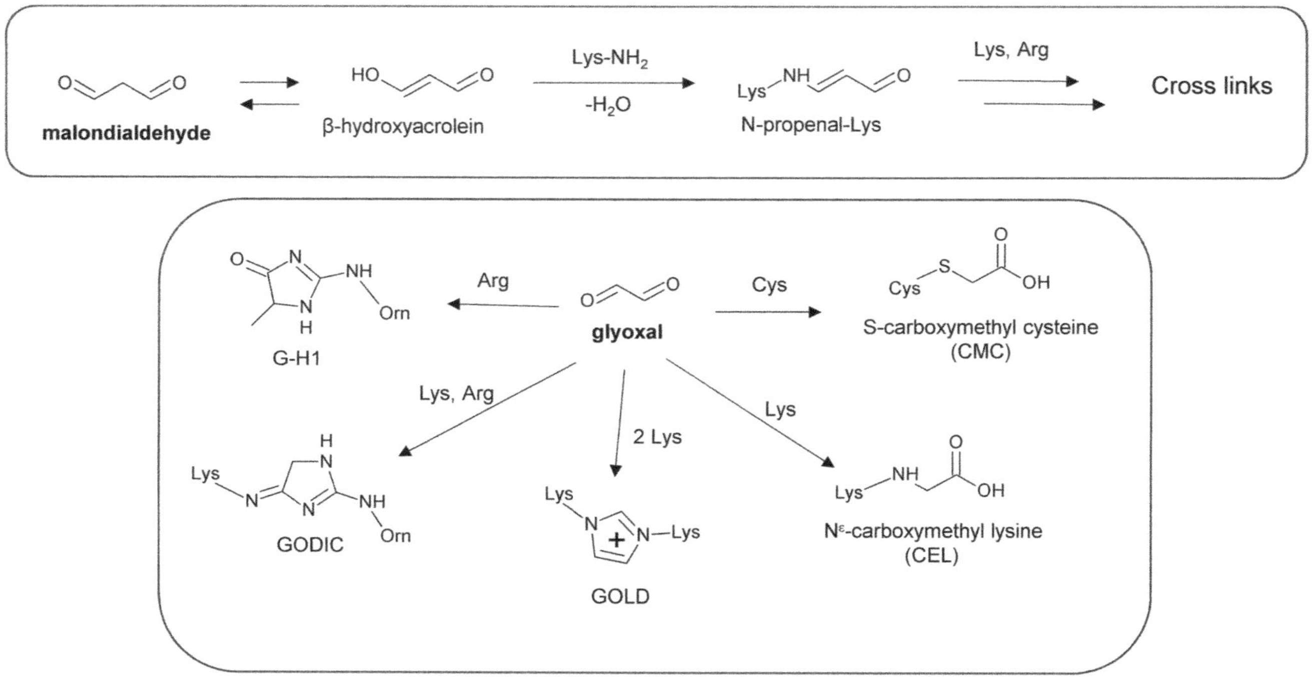

Figure 8.3 Reaction mechanisms of formation for the malondialdehyde- (upper panel) and glyoxal-based adducts with nucleophilic sites. Upper panel – MDA, through the nondissociated enolate form (the β-hydroxyacrolein) is able to react with the ε-amino group of Lys, forming the *N*-propenal-Lys, a reactive intermediate that can further react forming additional products. Lower panel – Glyoxal can react with Lys and Cys forming *N*-ε-(carboxymethyl)lysine (CML) and *S*-(carboxymethyl)cysteine (CMC), which represent the two main reaction products of GO with proteins. Glyoxal can also react with Arg forming imidazolinone derivatives and *N*-(carboxymethyl) arginine. Due to the presence of the two carbonyl groups, GO can also react with two nucleophilic sites forming crosslinks named GOLD and GODIC, formed by the reaction of one molecule of glyoxal with two Lys or with one Lys and one Arg, respectively.

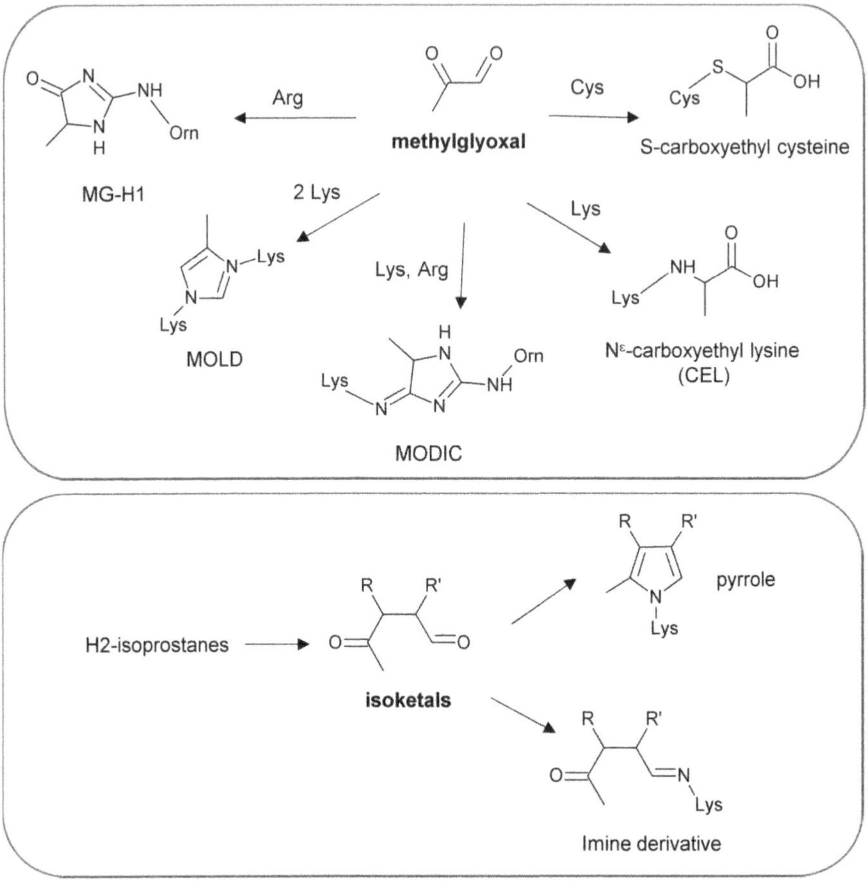

Figure 8.4 Reaction mechanisms of formation for the methylglyoxal- (upper panel) and isoketals-based adducts with nucleophilic sites. Upper panel: reaction pathway of MGO. MGO reacts primarily with Arg and Lys forming imidazolinone derivatives and carboxyethyl-lysine (CEL) adduct, respectively. Similarly to GO, it can also form crosslinks involving Lys and Arg residues. Lower panel: reaction pathway of isoketals. Isoketals are highly reactive γ-keto-aldehydes able to react with the ε-amino group of Lys forming a reversible imine derivative or an irreversible pyrrole adduct through a pyrrolidine intermediate.

Similarly to GO, it can also form crosslinks involving Lys and Arg residues (Figure 8.4).

4-oxo-Nonenal (ONE) is generated by the lipid-oxidation and it is more reactive than HNE towards nucleophilic sites. It forms Michael adducts by the reaction of C2 and C3 with His, Lys and Cys; moreover, it can form crosslinks and Schiff base by reacting with Lys.

Isoketals are highly reactive γ-keto-aldehydes generated by the free-radical-mediated lipid peroxidation of H2-isoprostanes. Isoketals react with the ε-amino group of Lys forming a reversible imine derivative or an irreversible pyrrol adduct through a pyrrolidine intermediate (Davies *et al.*, 2004) (Figure 8.4).

8.2 Biological Implications of RCS-Induced Protein Covalent Modifications

In the past, RCS such as HNE and MDA and the corresponding protein adducts were mainly considered as biomarkers of oxidative damage (Janero, 1990; Requena *et al.*, 1996; Moore and Roberts, 1998). Several assays for their measurement have been set up and used extensively; among the most widespread, there is the TBARS assay, based on thiobarbituric acid, for measuring MDA and the assay based on the derivatizing agent 2,4-dinitrophenylhydrazine, to quantify protein carbonyl (Dalle-Donne *et al.*, 2006). Later, several studies demonstrated that RCS and the corresponding protein adducts are also involved in the onset and progression of several oxidative-based diseases such as diabetes, atherosclerosis and some neurological disorders, therefore nowadays, RCS are widely studied as pathogenetic factors as well as drug targets (Aldini *et al.*, 2006; Aldini *et al.*, 2007b). Several mechanisms have been considered to explain the biological effects induced by RCS and protein adducts, as summarized below.

8.2.1 Protein Function Derangement

When RCS react with a target protein, the function of the protein itself can be deranged due to several mechanisms, such as a conformational change or the impairment of the interaction with the natural substrate when the RCS bind the active site of the protein. Proteomic studies have identified several proteins highly susceptible to RCS modification and undergoing function derangement (Colzani *et al.*, 2013; Fedorova *et al.*, 2014). In general, most of the target proteins are characterized by nucleophilic sites that are not only solvent accessible, but also characterized by a peculiar amino acid surrounding environment that stabilizes the most nucleophilic forms (the thiolate anion in the case of Cys or the uncharged ε-amino group in the case of Lys). Accordingly, as an example, we found that actin is a very susceptible cellular target of electrophilic agents due to the presence of several accessible nucleophilic targets; among these Cys374 was found to be the most reactive (Aldini *et al.*, 2005a). The peculiar reactivity of Cys374 can be explained not only by considering its accessibility, but also by the presence of surrounding amino acids that stabilize the thiolate anion (Aldini *et al.*, 2005a). By using MS strategies we found that actin is the protein target of several electrophilic agents including HNE, ACR, (Dalle-Donne *et al.*, 2007) and 15-deoxy-delta 12,14-prostaglandin J2 (15d-PGJ_2) (Aldini *et al.*, 2007a). In the case of quite bulky electrophiles, such as 15d-PGJ_2, the function of actin was found to be compromised also at the cellular level, due to a conformational change that induces a depolymerization process of fibrillar actin as well as the inhibition of actin polymerization, leading to cytoskeletal damage and cell death (Aldini *et al.*, 2007a).

The liver fatty acid-binding protein (L-FABP) is another interesting example regarding the derangement of protein function induced by the covalent modification generated by RCS. L-FABP adduction by HNE, that has been found in several pathological conditions such as alcoholic liver disease, impairs its stability and ligand-binding ability (Smathers *et al.*, 2012).

Other proteins whose functions are altered by RCS covalent adducts have recently been reviewed (Madian and Regnier, 2010; Madian *et al.*, 2011).

8.2.2 The Antigenic Properties of RCS Protein Adducts

The antigenic properties of RCS protein adducts and the consequent inflammatory response represents an additional mechanism of oxidative damage. Circulating antibodies against epitopes formed by the reaction of RCS with proteins were found in the sera of patients affected by chronic obstructive pulmonary disease (COPD), as well as in a murine model of chronic ozone exposure (Kirkham *et al.*, 2011). HNE protein adduction was also proposed as a possible antigenic stimulus for systemic lupus erythematosus (SLE) autoantibodies. Alzolibani *et al.* recently demonstrated that the structural perturbation of histone-H2A by HNE makes the protein immunogenic and that the neo-epitope might play a role in the induction of circulating autoantibodies in SLE (Alzolibani *et al.*, 2013). Human serum albumin is another protein that, once modified by RCS such as HNE, increases the production of autoantibodies (Khatoon *et al.*, 2012).

8.2.3 The Damaging AGEs–RAGE Axis

More recently, the role of the receptor RAGE has been widely considered as an additional mechanism of oxidative damage. RAGE is a multiligand transmembrane receptor consisting of an extracellular region, which is composed of one ligand-binding V-type Ig domain and two Ig-like C-domains (C1 and C2), a short hydrophobic transmembrane-spanning region, and a signal-transducing cytoplasmic domain (Fritz, 2011). The RAGE receptor is activated by AGEs, in particular those generated by GO and MGO (Xue *et al.*, 2011). RAGE activation induces the transcription factor NFk-B, leading to a proinflammatory and profibrotic response (Kierdorf and Fritz, 2013). The involvement of RAGE activation for some AGEs has been well documented in some animal models, although a clear structure–affinity relationship between AGEs and RAGE is still missing.

8.2.4 The Amyloidogenic Properties of RCS Protein Adducts

RCS can induce oligomerization and aggregation of some target proteins. As an example, HNE covalently reacts with amyloid-β peptides forming crosslinks and promoting Alzheimer protofibril formation (Siegel *et al.*, 2007).

α-Synuclein was reported as another RCS target; its covalent modification has been found to induce protein oligomerization (Näsström *et al.*, 2011) and an increase of dopaminergic toxicity (Xiang *et al.*, 2013).

8.3 Endogenous Detoxification of RCS

RCS are efficiently detoxified to unreactive derivatives by both phase I and II metabolic pathways. Phase I metabolism is mainly mediated by aldehyde-oxidizing enzymes and by aldehyde-reducing enzymes, while phase II is mediated by glutathione-dependent enzymes, as reviewed by O'Brien *et al.* (2005).

8.3.1 Phase I Metabolism

The oxidation of the carbonyl group to the corresponding carboxylic function is usually catalyzed by aldehyde dehydrogenase enzymes (ALDH) and NAD^+ as cofactor. Different aldehyde-oxidizing enzymes, characterized by different levels of affinity towards RCS, have so far been reported. For example, ALDH2 efficiently oxidizes HNE (Siems and Grune, 2003), but not MGO, while α-oxoaldehyde dehydrogenase oxidizes MGO to pyruvate, but not glyoxal nor 3-deoxyglucosone (Izaguirre *et al.*, 1998).

The reduction of RCS to the corresponding aldehydes is catalyzed by enzymes belonging to three groups of enzymes: the alcohol dehydrogenase family, the aldo-keto reductase superfamily and the short-chain dehydrogenase/reductase family (O'Brien *et al.*, 2005).

8.3.2 Phase II Metabolism

Two main GSH-dependent enzymes are involved in the phase II metabolism of RCS: the glutathione transferases and the glyoxalase system.

The glutathione transferases (GSTs) catalyze the conjugation of GSH with the C3 of alkenals and hydroxyalkenals forming the corresponding Michael adducts. The adducts still maintain the aldehydic function and can be further metabolized by ALDH or ADH to form the corresponding acid or alcohol derivatives. In the kidney, the GSH adducts are then hydrolyzed and acetylated forming the mercapturic acid derivatives, which are then excreted in the urine (Alary *et al.*, 2003) (Figure 8.5).

The glyoxalase system detoxifies oxoaldehydes, such as glyoxal and methylglyoxal, into the corresponding α-hydroxy acid; it consists of two enzymes, glyoxalase 1 and glyoxalase 2 and of GSH as a cofactor. As an example, the metabolic detoxification of MG starts with the GSH-methylglyoxal hemithioacetal that is formed by the nonenzymatic reaction between glutathione and MG; the metabolite is then isomerized to *S*-D-lactoylglutathione by glyoxalase 1 and is then converted to D-lactate and GSH by the enzyme glyoxalase 2 (Thornalley, 1990) (Figure 8.5).

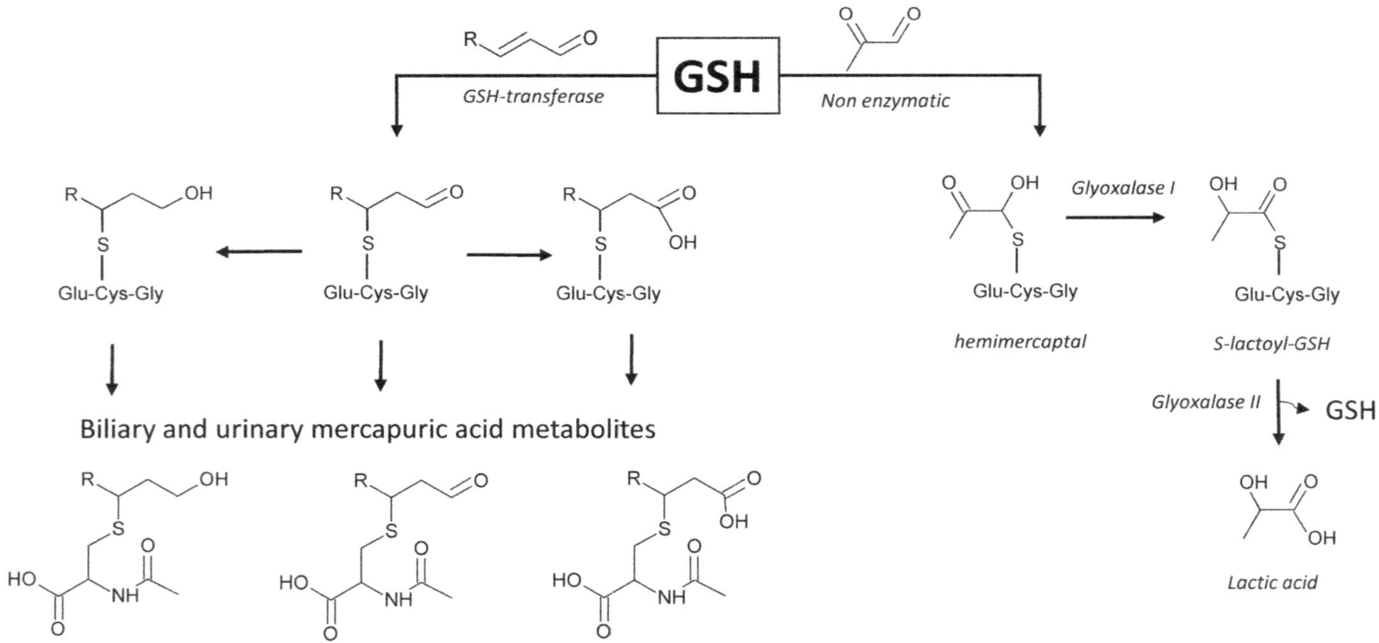

Figure 8.5 GSH-dependent metabolic pathways for the detoxification of α,β-unsaturated aldehydes (left) and oxoaldehydes (right). GSH-dependent detoxification of α,β-unsaturated aldehydes. The glutathione (GSH) transferases catalyze the conjugation of GSH with the C3 of α,β-unsaturated alkenals and hydroxyalkenals forming the corresponding Michael adducts that are further metabolized by ALDH or ADH to form the corresponding acid or alcohol derivatives. In the kidney, the GSH adducts are then hydrolyzed and acetylated forming the mercapturic acid derivatives, which are then excreted in the urine. MG detoxification – the metabolic detoxification of MG starts with the GSH-methylglyoxal hemithioacetal that is formed by the nonenzymatic reaction between glutathione and MG; the metabolite is then isomerized to *S*-D-lactoylglutathione by glyoxalase 1 and is then converted to D-lactate and GSH by the enzyme glyoxalase 2.

8.4 Carnosine and Histidine Dipeptides as RCS-Sequestering Agents

Carnosine (β-alanyl-L-histidine) is a physiological dipeptide present in large amounts in muscular and nervous tissues of vertebrate animals. Interest in this endogenous compound is due not only to its high concentrations in some tissues (spanning the millimolar range in the skeletal muscle), but also to the fact that its synthesis, modulation, transport, and degradation is regulated by a quite complex and energy-consuming molecular system (Boldyrev *et al.*, 2013). After its discovery at the beginning of the 20th century, its biological role remained obscure for many decades, although several hypotheses have been proposed: among others, carnosine and derivatives such as anserine, balenine and homocarnosine (Figure 8.6) have been proposed as possible buffering agents, metal-ion chelators, antioxidants and neuromodulators, as recently reviewed by Boldyrev *et al.* (2013).

Moreover, carnosine was found to inhibit protein oxidation, glycation and covalent adduction induced by RCS. In this regard, the quenching activity (ability of a compound to react with RCS forming unreactive adducted metabolites) of carnosine and related histidine dipeptides towards the different classes of RCS has been studied *in vitro* and *in vivo* since the late 1990s. Overall, these studies proposed carnosine, together with GSH, as the most important RCS-sequestering agents in vertebrate organisms, and suggested

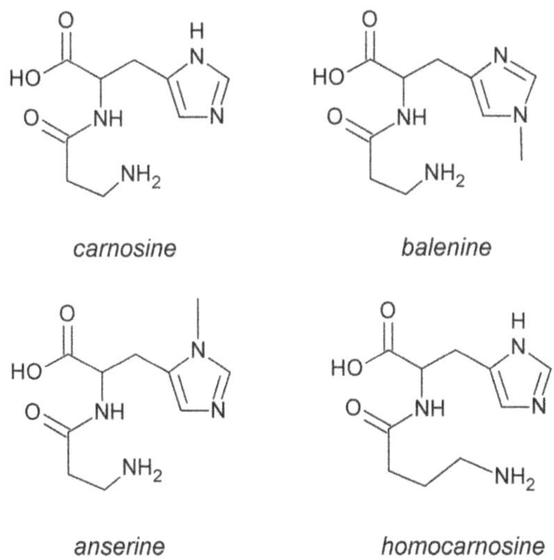

Figure 8.6 Chemical structures of carnosine and its naturally occurring derivatives. Carnosine is a dipeptide, composed of β-alanine and L-histidine. Carnosine is the archetype of a family of naturally occurring derivatives including anserine, ophidine (balenine), the L-histidine methylated, and homocarnosine, in which β-alanine is replaced by γ-aminobutyric acid (GABA).

its role in preventing the formation of AGEs and ALEs (Aldini *et al.*, 2005b). Such a mechanism has been tested in several *in vitro* and *in vivo* studies conducted by different research groups and besides contributing towards a better understanding of the biological role of the endogenous dipeptide, it also spurred a growing interest towards carnosine derivatives as a novel class of bioactive compounds (Vistoli *et al.*, 2012). The *in vitro* and *in vivo* studies demonstrating the quenching activity of carnosine towards different RCS are summarized below.

8.4.1 *In Vitro* Studies

The first experiments demonstrating the *in vitro* ability of carnosine to react with RCS were performed by incubating the molecule in the presence of different RCS at physiological pH (generally pH 7.4) and temperature (37–40 °C); the reactivity was then estimated by measuring the RCS disappearance by HPLC (Zhou and Decker, 1999). More complex experiments were then carried out, consisting of incubating pure proteins with RCS, and then estimating the effect of carnosine on the formation of protein carbonyls by observing the modulation of protein crosslinks and oligomers by electrophoresis (Hipkiss and Chana, 1998; Hipkiss *et al.*, 1998). Later, molecular studies on the reaction between carnosine and α-β-unsaturated aldehydes were performed not only to characterize their reaction products, but also to elucidate their reaction mechanism (Aldini *et al.*, 2002a; Liu *et al.*, 2003). It should be noted that analyses aiming at the study of the reaction mechanisms of carnosine towards other classes of RCS, such as dialdehydes and keto-aldehydes, are still missing; also the comparison of the quenching efficiency of carnosine in respect to the well-known RCS quenchers, such as aminoguanidine, hydralazine and pyridoxamine has not yet been reported.

8.4.1.1 α-β-Unsaturated Aldehydes

Carnosine was reported to quench *trans*-2-hexenal and 4-hydroxy-2-*trans*-nonenal (HNE) by measuring by HPLC their disappearance upon incubation with carnosine (Zhou and Decker, 1999). The quenching activity of carnosine was higher in respect to its constituent amino acids, β-alanine and histidine, as well as of other histidine-containing dipeptides such as leucyl-histidine, isoleucyl-histidine and valyl-histidine. Polyunsaturated aldehydes (*trans*,*trans*-2,4-decadienal and *trans*,*trans*-2,4-hexadienal) and the nonconjugated monounsaturated aldehyde *trans*-4-decenal reacted weakly with carnosine.

Further studies confirmed carnosine as an effective quencher of HNE and identified and characterized the reaction products by MS (Aldini *et al.*, 2002a). In particular, two reaction products were identified by incubating HNE with carnosine: an imine macrocyclic derivative (**2a** in Figure 8.7) and a Michael adduct, stabilized as a 5-member cyclic hemi-acetal (**5a**). The suggested adduction chemistry presents multiple steps: it begins with the

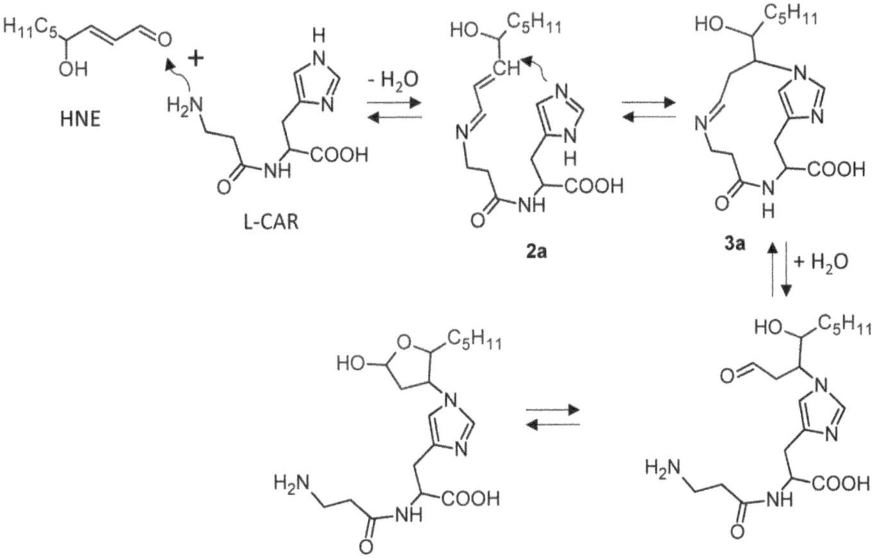

Figure 8.7 Reaction mechanism of carnosine with HNE. With permission from Aldini, G., M. Carini, *et al.* (2002a). Carnosine reacts with HNE trougth a two-step mechanism. The reaction starts with the formation of a reversible Schiff base (an α,β-unsaturated imine, **2a**) to yield the macrocyclic adduct **3a** through an intramolecular Michael addition, which hydrolyzes to form the stable hemiacetal derivative **5a**.

formation of a reversible Schiff base (an α,β-unsaturated imine, **2a**) to yield the macrocyclic adduct **3a** through an intramolecular Michael addition, which hydrolyzes to form the stable hemiacetal derivative **5a** (Aldini *et al.*, 2002a; Liu *et al.*, 2003). Structural analyses were based on NMR and mass spectrometry and were performed on carnosine and its analogs *N*-acetyl-carnosine and anserine (a *N*-π-methyl derivative of carnosine). *N*-Acetylation of the β-alanine amine group significantly reduced the quenching activity; this, together with the observation on the limited quenching capability of the two constitutive amino acids, suggested that both the amino group of the β-alanyl residue and the imidazole ring of L-histidine synergistically trap HNE. The two-step reaction of carnosine with HNE also explains the selectivity of the dipeptide towards α,β-unsaturated aldehydes and that carnosine does not crossreact with physiological aldehydes, such as pyridoxal phosphate, a damaging mechanism quite common to well-known RCS scavenging agents such as hydralazine, aminoguanidine and edaravone (Aldini *et al.*, 2010).

The fact that HNE reacts with proteins mainly on Lys and His residues (Nadkarni and Sayre, 1995) suggests that carnosine may act as a detoxifying agent by mimicking the target amino acids of proteins.

Such quenching activity was then demonstrated in tissue by detecting the carnosine-HNE Michael adduct **5a** in rat skeletal muscle homogenate

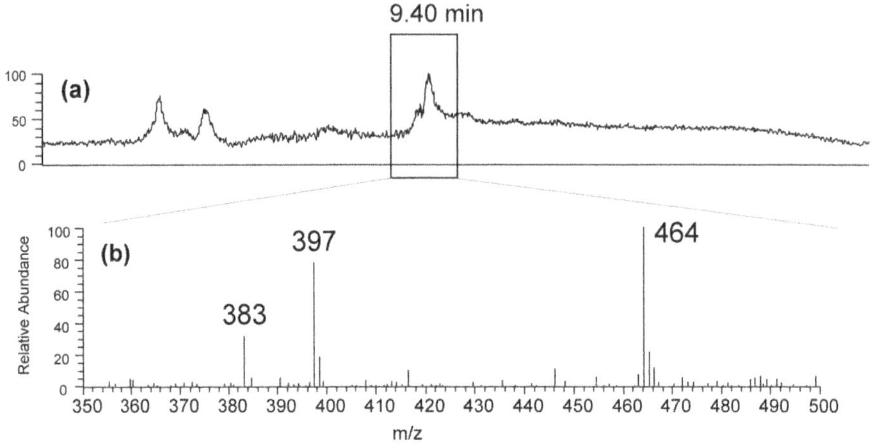

Figure 8.8 LC/MS analysis (positive-ion mode) of skeletal muscle homogenate incubated with HNE (5 nmole mg^{-1} protein). The upper panel (a) shows the total ion current (TIC); the lower panel (b) shows the MS spectrum of the peak eluted at of 9.4 min. The ions at *m/z* 383, *m/z* 397 and *m/z* 464 correspond to Michael adducts of HNE with carnosine, anserine and GSH, respectively. The figure is adapted from Aldini, G., P. Granata, *et al.* (2002b).

(Figure 8.8) upon *in vitro* spontaneous oxidation in aerobic conditions (Aldini *et al.*, 2002a) (Aldini *et al.*, 2002b; Orioli *et al.*, 2005), thereby confirming the RCS scavenging of carnosine in biological media and that carnosine-HNE can be considered as a biomarker of lipid peroxidation. In these studies, the anserine-HNE Michael adduct was observed for the first time in oxidized rat muscle, too (Figure 8.8). This finding is consistent with the fact that anserine is also present at high concentration in rat muscle and exerts a quenching activity similar to carnosine *in vitro*.

Recently, the ability of carnosine and other carbonyl quenchers (hydralazine, aminoguanidine and pyridoxamine) to quench HNE was compared, using ubiquitin as a model of protein undergoing RCS-induced carbonylation (Colzani *et al.*, 2014). HNE covalently modified ubiquitin on lysine and histidine residues, to form Schiff base and Michael adducts: all tested compounds were found to dose-dependently inhibit ubiquitin carbonylation, but carnosine and hydralazine were the most effective quenchers.

In addition to HNE, the reaction between carnosine and acrolein was also described in detail (Carini *et al.*, 2003). The reaction products obtained *in vitro* were analyzed by mass spectrometry; several molecular species were detected, including macrocyclics deriving from the formation of the iminic bond between the terminal amino group followed by intramolecular Michael addition of the C3 of acrolein to histidine (Figure 8.9, **1c**). Additional reaction products were the *N*-β-3-formyl-3,4-dehydropiperidinos (Figure 8.9, **2c**) deriving from the Michael addition of two acrolein molecules to the amino group of β-alanine, followed by aldol condensation and dehydration.

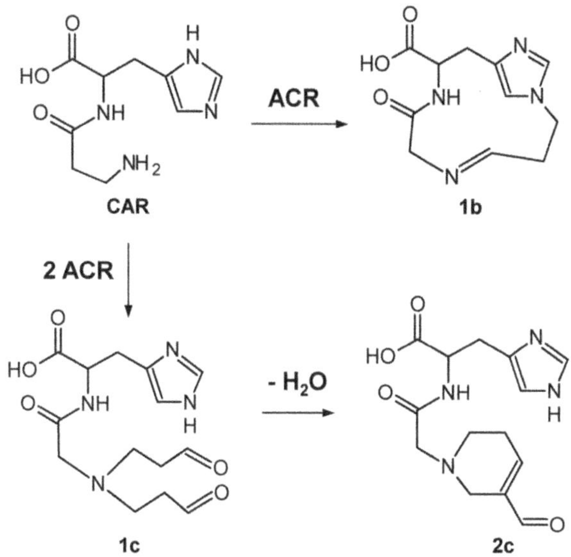

Figure 8.9 Predominant reaction pathways of CAR with ACR. With permission from Carini *et al.* (2003). Carnosine reacts with ACR forming several molecular species, among which the predominant are: the 14-membered macrocyclic derivatives, deriving from the formation of the iminic bond between the terminal amino group followed by intramolecular Michael addition of the C(3) of the ACR moiety to histidine (structure **1b**); the *N*-β-(3-formyl-3,4-dehydropiperidino) derivatives arising from the Michael addition of two acrolein molecules to the amino group of β-alanine (**1c**), followed by an aldol condensation and dehydration (**2c**).

8.4.1.2 Dialdehydes and Keto-Aldehydes

Besides α,β-unsaturated aldehydes, carnosine was reported to directly react with other reactive RCS (Hipkiss *et al.*, 2001), such as the dialdehydes MDA and GO, the keto-aldehyde MGO as well as with reducing sugars (Hipkiss *et al.*, 1995). Several analytical and biochemical methods were used to demonstrate the RCS quenching effect, although the reaction mechanisms as well as the reaction products have not yet been elucidated. For example, the reaction obtained *in vitro* between carnosine and methylglyoxal was observed by spectrophotometry: carnosine was found to inhibit methylglyoxal-mediated modification of ovalbumin, chosen as model protein (Hipkiss and Chana, 1998). Carnosine also reacts with MDA (Hipkiss *et al.*, 1998), as demonstrated by its effect in preventing MDA-induced protein crosslink and oligomerization. Carnosine was also found to protect cultured human and rat cells against the effects of toxic RCS such as formaldehyde, acetaldehyde, sugar-deriving AGEs and MDA itself. The *in vitro* studies carried out by Hipkiss (2009) mainly focused on the RCS quenching ability of carnosine and had a pioneering role in highlighting the protective role of carnosine against protein carbonylation in cells, promoting carnosine and related analogs as therapeutic targets for the treatment of pathologies involving noxious RCS, such as secondary diabetic complications, inflammation and possibly Alzheimer's disease.

8.4.2 *In Vivo* Studies

GSH-dependent conjugation represents the most efficient phase II detoxification reaction towards electrophilic compounds, including RCS, leading to urinary mercapturic acid metabolites. As reported above, *in vitro* studies carried out either in solution or in skeletal muscle homogenates indicate that carnosine and analogs can also act as nucleophilic agents able to detoxify RCS (in particular α,β-unsaturated aldehydes), forming unreactive conjugated metabolites. Such studies suggest that besides GSH, carnosine could also represent an important detoxification pathway of RCS, acting in particular areas, such as skeletal muscle, or when GSH is depleted. This important issue has been investigated, although not conclusively, in several *in vivo* studies, as reported in the present chapter.

One of the first pieces of evidence on the *in vivo* formation of covalent adducts between toxic endogenous aldehydes and histidine dipeptides was reported by Orioli *et al.* (2007). An analytical LC-MS method was developed and validated for the determination of RCS adducts with mercapturic acid and carnosine in the urine of obese Zucker rats. The Zucker rat was selected because it represents a well-established animal model of metabolic syndrome, since it is characterized by obesity, hyperlipidaemia, insulin resistance and renal injury; in this animal, the oxidative stress and lipid hydroperoxides lead to RCS and ultimately to protein carbonylation processes that are involved in the onset and progression of tissue damage (Dominguez *et al.*, 2006). The mass spectrometry method applied in this study was set to scan the molecules able to generate the immonium ion of histidine (*m*/*z* 110), which was used as a diagnostic marker of carnosine products; this method, based on the precursor ion scan (PIS) mode, allowed identifying the urinary metabolites of carnosine.

The method was applied to compare the urine of young lean rats (4 weeks old), old lean rats (6 months old) and Zucker rats. As reported in Table 8.2, three distinct precursor ions at *m*/*z* 314, 328 and 383 were identified in Zucker rats; none of them was identified in young lean rats, while two of them (*m*/*z* 314 and 383) were detected in old lean rats. The characterization of the three carnosine products by MS/MS led to the identification of His-1,4-dihydroxynonane (His-DHN, at *m*/*z* 314), His-1,4-hydroxynonanoic acid (His-HNA, at *m*/*z* 328) and the Michael adduct carnosine-HNE (CAR-HNE, at *m*/*z* 383). These results confirm that RCS – and in particular HNE – is generated in physiological conditions, and that its formation is associated with aging and with the metabolic syndrome. Once identified and characterized, a validated multiple reaction monitoring (MRM) method was set up for their quantitation and to measure HNE-mercapturic acid metabolites, such as the 1,4-dihydroxynonane mercapturic acid (DHN-MA), with a good sensitivity.

Table 8.2 Urinary RCS metabolites containing histidine. RCS adducts metabolites containing histidine found in rat urine by precursor and product ion scan analyses. Two adducts were found in the old lean rats, three in the Zucker rats and none of the adducts were found in the young lean rats. Data are adapted from Orioli *et al.* (2007).

His-containing metabolites		Urine		
		Lean rats 4 week old	Lean rats 6 months old	Zucker rats
His-DHN (m/z 314)	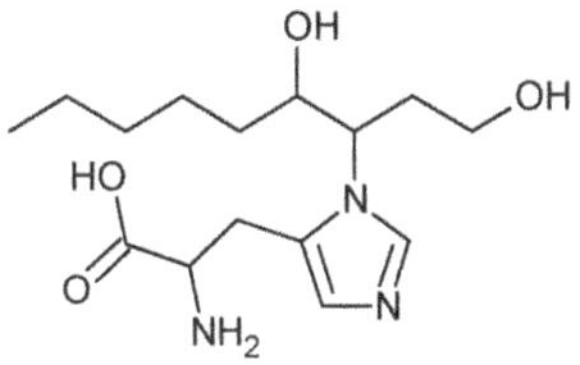	Not found	Found	Found
His-HNA (m/z 328)	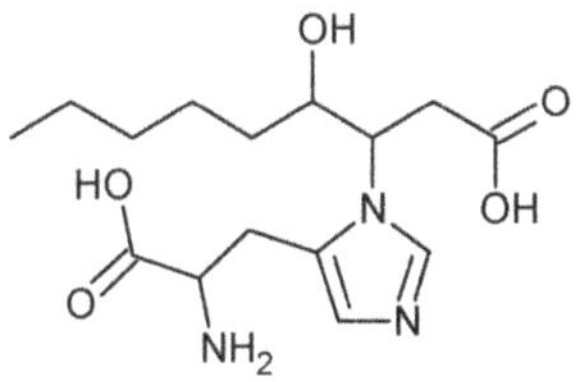	Not found	Not found	Found
CAR-HNE (m/z 383)	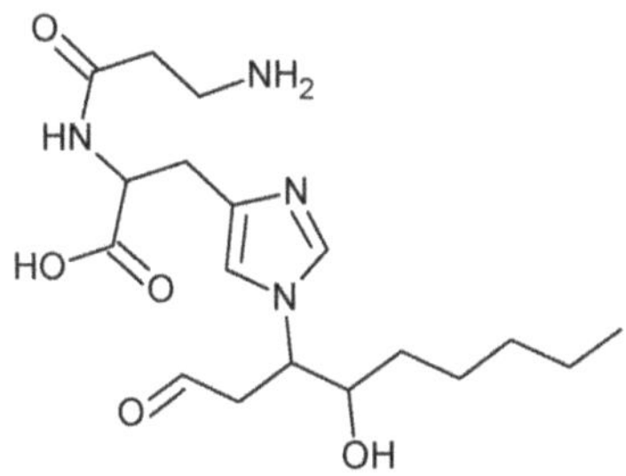	Not found	Found	Found

Table 8.3 Quantitative analysis of urinary RCS metabolites containing histidine. Quantitative analysis of RCS metabolites containing His in lean and Zucker rat urine. Data are from Orioli *et al.* (2007).

His-containing metabolites	Urine	
	Lean rats (nmole per 24 h)	Zucker rats (nmole per 24 h)
His-DHN (m/z 314)	0.9	3.8
DHN-MA (m/z 322)	16	60
CAR-HNE (m/z 383)	0.8	2

The results, summarized in Table 8.3, well indicate that GSH and carnosine act as nucleophilic substrates in the detoxification of HNE, and that GSH is more reactive, highlighting the role of GSH as first-line defense against cytotoxic HNE. Upon a significant oxidative stress, GSH is likely to be consumed, therefore carnosine, due to the fact that it is quite stable towards oxidation (it has a poor antioxidant activity in respect to GSH), might acquire a crucial role in detoxifying cytotoxic RCS: this mechanism would suggest an interesting cooperation between the histidine dipeptides and GSH in RCS metabolism. It is still not clear whether the His-HNE metabolites detected in Zucker rats derive from carnosine-HNE adducts or from other sources, such as HNE-protein digestion.

The ability of carnosine to form covalent adducts with HNE has been confirmed by different studies, which show that the formation of carnosine-HNE metabolites correlate with the inhibition of protein carbonyls formation and with a protective effect as well, suggesting that the pharmacological activity of carnosine can be mediated by the RCS detoxifying activity and by the prevention of ALEs formation. In particular, the chronic administration of L-carnosine and of its enantiomer, D-carnosine (β-alanyl-D-histidine), to Zucker obese rat significantly reduced the development of dyslipidemia, hypertension and renal injury, as demonstrated by urinary parameters and by microscope analysis of renal tissue biopsies (Aldini *et al.*, 2011). Such a pharmacological effect was associated with an improved excretion of HNE, as carnosine-HNE and histidine-HNE, thereby reducing the accumulation of AGEs and protein carbonyls. These data suggest that the biological effect of endogenous L-carnosine is not based on a prohistaminic effect, since D-carnosine does not undergo peptidic hydrolysis and therefore it is not a histidine precursor; conversely, a direct carbonyl-quenching mechanism can explain, at least partially, the pharmacological effects of carnosine (Aldini *et al.*, 2011).

The protective and HNE-quenching effect of carnosine was then confirmed in apoE null mice feeding a western and proatherogenic diet (Menini *et al.*, 2012). Treatment with D-carnosine octylester (a D-carnosine prodrug, see Chapter 1) attenuated atherosclerosis and renal diseases, an effect that was associated with increased urinary levels of HNE-carnosine adducts and reduced protein carbonylation, circulating and tissue ALEs, expression of receptors for these products, and systemic and tissue oxidative stress (Menini *et al.*, 2012).

The RCS-quenching ability and the antiatherogenic effect of D-carnosine (given as octyl-ester) was confirmed and further characterized by an independent research group (Barski *et al.*, 2013). Oral treatment with the drug decreased atherosclerotic lesion formation in the aortic valves of apolipoprotein E-null mice. Such pharmacological activity was associated with a RCS-scavenging effect of D-carnosine, determined by a decrease in the accumulation of the adducts protein–acrolein, protein-4-hydroxyhexenal and protein-4-hydroxynonenal in atherosclerotic lesions and by a concomitant increase of urinary excretion of aldehydes as carnosine conjugates (Barski *et al.*, 2013) (Table 8.4).

The first paper reporting the identification of carnosine-RCS metabolites in humans was published by Baba *et al.* (2013). Urine samples from healthy, nonsmoker adults were analyzed by LC/MS with the aim of identifying carnosine-aldehyde metabolites. The most abundant metabolites identified by this study were carnosine-propanol (*m*/*z* 285), carnosine-propanal (*m*/*z* 283),

Table 8.4 Quantitative analysis of RCS adducted metabolites containing histidine in healthy humans. Acrolein adducts are the most abundant ones, in particular 3-HPMA, carnosine-propanol, histidine propanol. The concentration is reported as pmole mg^{-1} creatine. The data are from (Baba *et al.*, 2013).

Adduct urine concentration (pmole mg^{-1} creatine)		Urine
		Healthy humans
Carnosine-propanal (*m*/*z* 283)	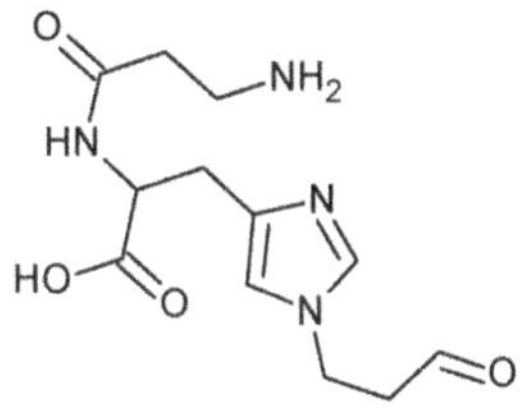	66 ± 17
Carnosine-propanol (*m*/*z* 285)	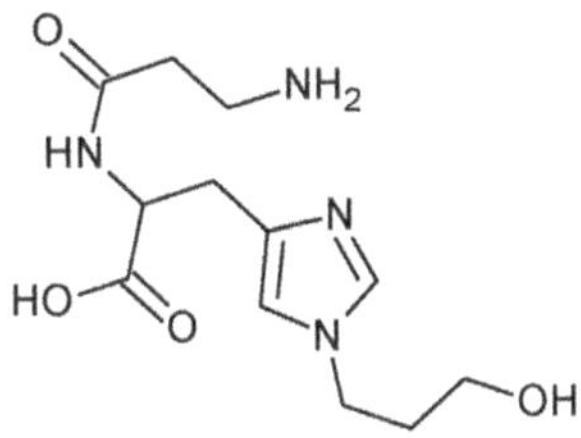	200 ± 75
Histidine-propanal (*m*/*z* 212)	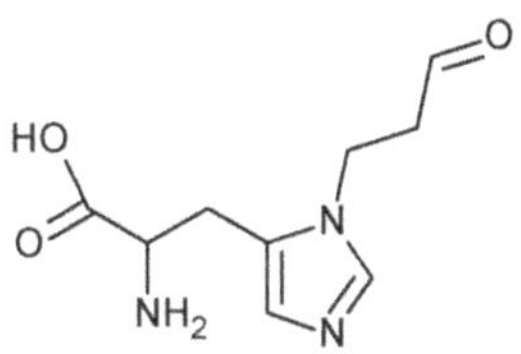	227 ± 150
Histidine-propanol (*m*/*z* 214)	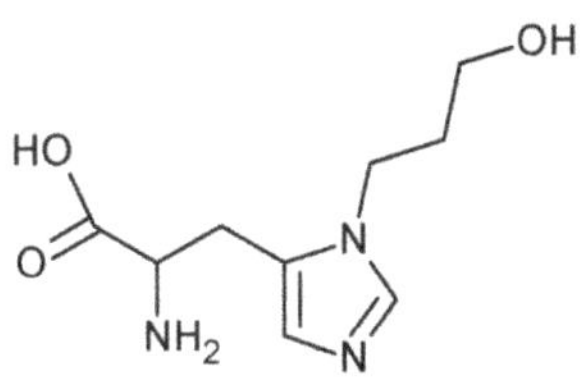	127 ± 84
3-HPMA (*m*/*z* 220)	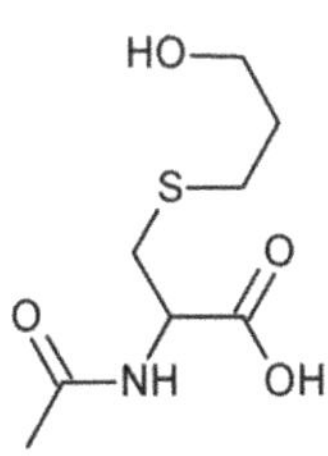	695 ± 213

(continued)

Table 8.4 (*continued*)

Adduct urine concentration (pmole mg^{-1} creatine)		Urine
		Healthy humans
Carnosine-HNE (*m*/*z* 383)	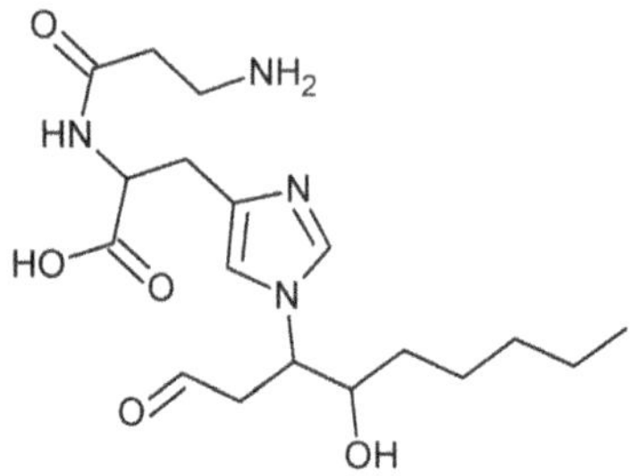	Found at trace levels
Carnosine-DHN (*m*/*z* 385)	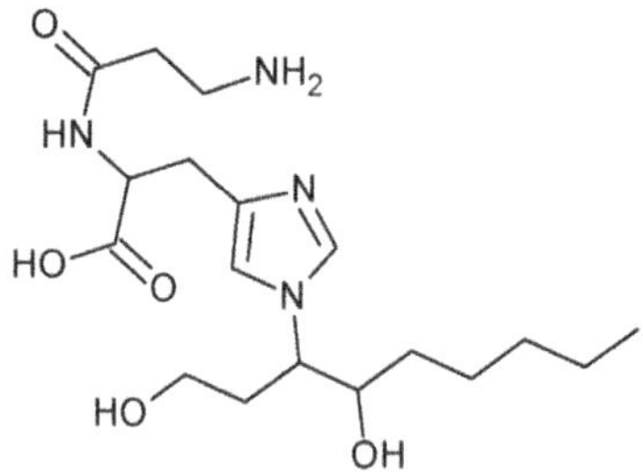	Found at trace levels
His-HNE (*m*/*z* 312)	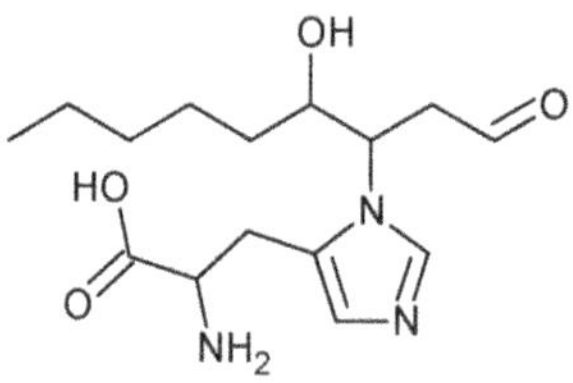	2.7 ± 1
His-DHN (*m*/*z* 314)	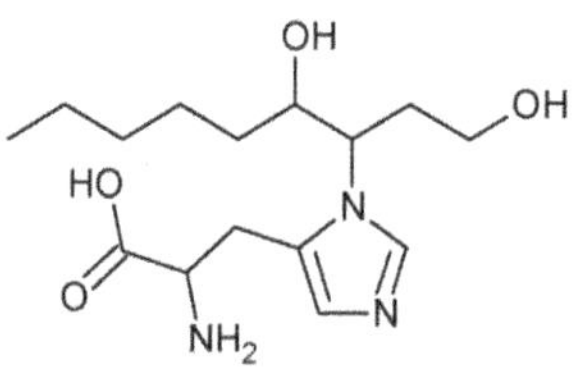	2.9 ± 1

histidine-propanol (*m*/*z* 214) and histidine-propanal (*m*/*z* 212), resulting from the acrolein adduction by carnosine followed by reduction and hydrolysis. The majority of the metabolites (75%) were in their reduced form, suggesting that the reduction of carnosine/histidine conjugates is an important step of their metabolism. The origin of the histidine conjugates is not clear, since they can arise either from the hydrolysis of the adducted carnosine, from the direct reaction between histidine and acrolein or from the hydrolysis of adducted histidines in proteins. HNE adducts were found as well, but in lower amounts than ACR metabolites; for example, the urinary concentration of His-HNE (*m*/*z* 310) was 75-fold lower than His-propanal concentration.

Moreover, only trace levels of carnosine-HNE (*m/z* 383) and its reduced form carnosine-DHN (*m/z* 385) were detected. It is important to note that these observations were made on samples obtained from healthy volunteers, while no data have been reported for pathological conditions, where a significant increase of HNE is expected.

Taken together, the above mentioned studies indicate as follows: (1) carnosine and its analogs, together with GSH, represent an efficient endogenous detoxifying system for RCS and in particular for α,β-unsaturated aldehydes; (2) the metabolites arising from the adduction of carnosine/histidine and RCS are excreted in the urine mainly in their reduced form, both in healthy humans and rodents; (3) the carnosine adduction pathway seems a relevant metabolic fate of acrolein's overall metabolism; and finally that (4) in healthy rat and human urine only a low amount of HNE adducts was found: this can be explained by considering that in physiological conditions HNE is mainly detoxified by GSH and/or that HNE is degraded to generate other metabolites.

Concerning the reduction of RCS-carnosine adducts, Baba *et al.* (2013) found that aldose reductase is the enzyme involved in the reducing reaction and that it is able to catalyze the reduction not only of carnosine-propanals but also of other histidine peptides conjugates such as homocarnosine and anserine. Studies performed on mice lacking aldose reductase (KO mice) confirmed this observation, since no reduced metabolites were detected in the urine of these animals. The enzyme showed a limited catalytic activity on HNE conjugates; this can explain the low levels of the CAR-DHN adduct, in respect to its nonreduced form, in both rodent and human urines. Such limited activity was explained by considering that in HNE conjugates, HNE is in an hemiacetal form so that the aldehyde group is not available for the reducing reaction.

8.5 Carnosine Derivatives as a Novel Class of Bioactive Compounds

Carnosine and the methyl derivatives anserine and ophidine (balenine) are dietary dipeptides found in significant amounts in red and white meat and in some fish such as tuna, salmon and trout.

The dipeptides are absorbed in intact form, primarily in the jejunum, through an active transport, which is accomplished by the oligopeptide transporter 1 (PEPT1). In humans, once absorbed, carnosine and derivatives are efficiently hydrolyzed by specific hydrolytic enzymes, named carnosinase (Figure 8.10). Two forms of carnosinase have been identified so far: CN1, or serum carnosinase, and CN2, also named tissue carnosinase or cytosolic non-specific dipeptidase (Teufel *et al.*, 2003). CN1 is highly expressed in the serum of humans and causes the limited bioavailability of carnosine, as observed in intervention studies (Yeum *et al.*, 2010). The constitutive amino acids histidine and β-alanine are then absorbed from the serum to the tissues, where

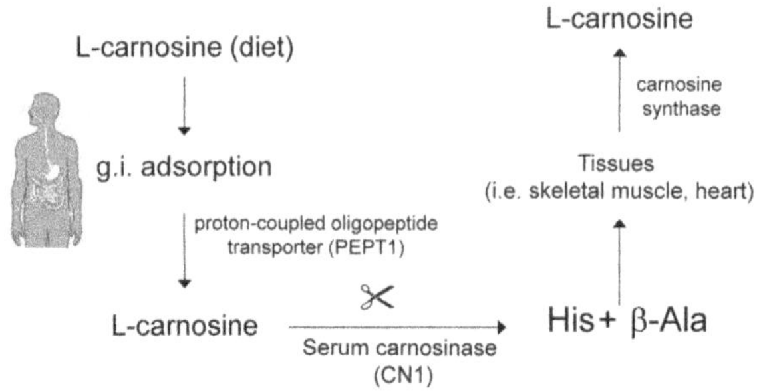

Figure 8.10 Metabolic fate of carnosine orally ingested in humans. Carnosine is absorbed in intact form, primarily in the jejunum, through an active transport, which is accomplished by the oligopeptide transporter 1 (PEPT1). Once absorbed, carnosine is hydrolyzed by serum carnosinase. The constitutive amino acids histidine and β-alanine are then absorbed from the serum to the tissues, where they are the substrates of the carnosine synthase that produces carnosine.

they are the substrates of the carnosine synthase that produces carnosine (Drozak *et al.*, 2010). By contrast CN1 is absent is nonprimate mammals; this explains the bioavailability of such dipeptides in rodent blood.

Considering the involvement of RCS in the pathogenetic mechanisms of several human diseases based on oxidative stress, serum RCS represents an important drug target for the discovery of bioactive compounds able to detoxify RCS (Aldini *et al.*, 2006; Aldini *et al.*, 2007b). Such an innovative drug target has been validated by testing the efficacy of carnosine in several animal models of oxidative-stress-based diseases, such as metabolic syndrome (Aldini *et al.*, 2011) and atherosclerosis (Menini *et al.*, 2012; Barski *et al.*, 2013) – where RCS act as pathogenetic factors. The observed pharmacological activity of carnosine in animal models can be explained by considering that the absence of serum carnosinase in rodents allows carnosine to reach a plasma concentration able to trap circulating RCS, as evidenced by the detection of its adducts in plasma and urine. In humans, carnosine probably acts as an endogenous detoxifying agent of RCS in those tissues where it is present in a high concentration such as skeletal muscle, brain and heart, but not in the blood stream due to the presence of carnosinase, even following treatment with oral doses of 60 mg kg^{-1} body (Yeum *et al.*, 2010; Everaert *et al.*, 2012).

Evidence, although indirect, of the potential activity of carnosine, when bioavailable in serum, is given by genetic studies. Janssen *et al.* first reported that diabetic patients with a decreased risk of developing diabetic nephropathy are characterized by a polymorphism of the CNDP1 gene, leading to a reduced secretion of serum carnosinase (Janssen *et al.*, 2005). The disease-protecting effect of lower serum carnosinase activity probably relies on the higher stability of carnosine in the blood stream or to a local effect of enhanced carnosine availability in the kidney, as commented by Boldyrev *et al.* (2013).

All the evidence reported above prompted the design of carnosine derivatives that, on the one hand, maintain or improve the RCS quenching efficacy and selectivity of carnosine, and on the other hand are characterized by improved bioavailability, as a result of being recognized by PEPT1, but not by carnosinase (Vistoli *et al.*, 2012). The first strategy used to obtain stable derivatives involved the inversion of the chiral center of histidine leading to the enantiomer D-carnosine (Aldini *et al.*, 2011). D-CAR was found to be stable in the plasma, since it is not recognized by carnosinase; its HNE-quenching efficacy and selectivity were superimposable on those of L-carnosine. However, D-CAR is less recognized by PEPT-1 in respect to L-CAR, resulting in a lower absorption, thus reducing bioavailability despite its carnosinase stability. With the aim of improving the bioavailability of D-carnosine, a set of lipophilic prodrugs have been designed and synthesized to be absorbed through a passive absorption (Orioli *et al.*, 2011). Among these, the octyl ester of D-CAR was selected for *in vivo* studies; its pharmacokinetic profile confirmed the improved bioavailability and its efficacy was tested in several animal models, including obese Zucker rats and ApoE null mice fed with a Western diet (see above) (Menini *et al.*, 2012; Barski *et al.*, 2013).

8.6 Conclusion

Several pieces of evidence indicate that carnosine, together with GSH, is an endogenous nucleophilic peptide able to detoxify cytotoxic RCS and in particular α,β-unsaturated aldehydes, such as HNE and ACR.

Michael adducts between carnosine and α,β-unsaturated aldehydes, in both reduced and nonreduced form, represent the main metabolites so far identified in both human and rodent urine and plasma. However, several questions remain open; in particular it is still not clear whether the carnosine detoxification pathway is catalyzed by enzymes. Recent evidence indicates that aldose reductase is involved in reducing the carbonyl group of the Michael adducts, thus forming more stable and unreactive adducts (Baba *et al.*, 2013). However, the existence of an enzyme catalyzing the formation of the Michael adduct and the biological function of these metabolites is still unexplored.

It is now well established that carnosine-mediated detoxification of RCS is associated with a protective effect in some animal models of oxidative-stress-based diseases including atherosclerosis, diabetes, ischemia/reperfusion damage and metabolic syndrome. Such promising effects can be explained in rodents, where carnosine, due to the absence of serum carnosinase, is bioavailable and acts as detoxifying agent of circulating RCS.

The effect of carnosine in humans is not yet known, due to the absence of controlled intervention studies; what is well established is the lack of carnosine bioavailability in serum due to the presence of carnosinase (Yeum *et al.*, 2010; Everaert *et al.*, 2012).

Overall, taking into consideration the biological effect of carnosine as RCS detoxifying agent, its lack of bioavailability and the pathogenetic effects of circulating RCS, the design of novel carnosine derivatives stable to carnosinase represent a promising therapeutic approach (Vistoli *et al.*, 2012; Aldini *et al.*, 2014).

Summary Points

- The chapter focuses on carnosine as a detoxifying agent of reactive carbonyl species.
- Reactive carbonyl species (RCS) are generated by lipid and sugar oxidation and belong to different chemical classes including α,β-unsaturated aldehydes, dialdehydes and keto-aldehydes.
- RCS and the corresponding reaction products with proteins (AGEs and ALEs) are involved in the pathogenic mechanisms of several oxidative-based pathologies, such as atherosclerosis and diabetes-related diseases.
- Several mechanisms have been considered to explain the damaging effects of RCS and of the corresponding adducts, among these the protein function derangement, the antigenic properties of RCS protein adducts, the damaging AGEs–RAGE axis and the amyloidogenic properties of RCS protein adducts.
- RCS are recognized as important pathogenetic mediators and as innovative drug targets;
- RCS are detoxified by phase I and II metabolism.
- Carnosine is an endogenous histidine dipeptide of dietary source, presents in mM concentrations in some tissues but not in serum due to the presence of carnosinases that catalytically hydrolyze carnosine to its constitutive amino acids.
- Carnosine was found to be a selective RCS sequestering agent, in particular towards α,β-unsaturated aldehydes.
- Carnosine and its analogs, together with GSH, represent an efficient endogenous detoxifying system for α,β-unsaturated aldehydes.
- The metabolites arising from the adduction of carnosine with α,β-unsaturated aldehydes are excreted in urine mainly in their reduced form, both in healthy humans and rodents;
- The carnosine adduction pathway seems to be a relevant metabolic fate of acrolein's overall metabolism.
- By considering the reduced bioavailability of carnosine in human serum due to the presence of serum carnosinase, carnosine analogs resistant to carnosinases have been proposed and tested in animal models as detoxifying agents of circulating RCS.
- Carnosine derivatives represent a novel class of bioactive compounds.

Definitions of Words and Terms

Aldehyde dehydrogenases (ALDHs) are a group of phase I enzymes that catalyze the oxidation (dehydrogenation) of aldehydes to the corresponding carboxylic acids through a $NAD(P)^+$-dependent reaction.

AGEs. Acronym for advanced glycation end products. AGEs indicates a heterogeneous class of covalent adducts arising from the reaction between proteins and RCS generated from the oxidative degradation of sugars. AGEs can also be generated by the direct condensation between reducing sugars and Lys residues (also called glycation) *via* the so-called Maillard reactions.

ALEs. Acronym for advanced lipoxidation end products. ALEs indicates the covalent adducts arising from the reaction between protein and RCS generated from the oxidation degradation of lipids.

Carnosine is a β-alanyl-L-histidine dipeptide discovered in 1900 as an abundant nonprotein nitrogen-containing compound of meat. The dipeptide is not only found in skeletal muscle, but also in other excitable tissues. Although the biological role of carnosine is still unknown, several biological activities have been proposed, among which is the carbonyl quenching activity.

D-Carnosine octylester is a new chemical entity synthesized to act as a prodrug of D-carnosine. D-carnosine (β-alanyl-D-histidine) is the enantiomer of carnosine, designed to be stable to carnosinases. D-carnosine octylester is absorbed through passive diffusion.

Glutathione (GSH) is an endogenous tripeptide (γ-L-glutamyl-L-cysteinylglycine) acting as a substrate in both conjugation reactions (phase II reactions) and reduction reactions. GSH is the main endogenous antioxidant able to reduce oxidants and forming the oxidized form (GSSG). GSSG is reduced to GSH by glutathione reductase and using NADPH as an electron donor.

4-Hydroxy-*trans*-nonenal (HNE) is an α,β-unsaturated aldehyde generated by the lipid-peroxidation of ω-6 polyunsaturated fatty acids such as arachidonic and linoleic acids. HNE is an electrophilic by-product able to form covalent adducts with nucleophilic substrates such as proteins, nucleic acids and phospholipids. It is metabolized through phase I and II reactions.

hPepT1 is a peptide transporter belonging to the POT (proton oligopeptide transporter) family. It modulates intestinal absorption and renal reabsorption of di- and tripeptides in a proton-dependent way, hence acting like as a cotransporter.

Reactive carbonyl species (RCS) are breakdown products formed by the oxidative damage to lipids and sugars. RCS are electrophilic compounds able to form covalent adducts with nucleophilic substrates and in particular with protein, nucleic acids and phospholipids.

Serum carnosinase (CN2) is a hydrolytic metallo-enzyme found in plasma and responsible for the rapid and selective digestion of the absorbed carnosine. It is expressed also in CNS, where it can hydrolyze homocarnosine thus modulating the release of GABA in the brain.

List of Abbreviations

15d-PGJ$_2$	15-Deoxy-delta 12,14-prostaglandin J2
ACR	Acrolein
AGEs	Advanced glycation end products
ALDH	Aldehyde dehydrogenase enzymes
ALEs	Advanced lipoxidation end products
CEL	*N*-ε-(Carboxyethyl)lysine

CN1	Serum carnosinase
CN2	Tissue carnosinase
CMC	*S*-(Carboxymethyl)cysteine
CML	*N*-ε-(Carboxymethyl)lysine
COPD	Chronic obstructive pulmonary disease
GO	Glyoxal
GSTs	Glutahione transferases
HNE	4-Hydroxynonenal
L-FABP	Liver fatty acid-binding protein
MGO	Methylglyoxal
MDA	Malondialdehyde
ONE	4-Oxononenal
PEPT1	Oligopeptide transporter 1
RCS	Reactive carbonyl species

References

Alary, J., Guéraud, F. and Cravedi, J. P., 2003. Fate of 4-hydroxynonenal in vivo: disposition and metabolic pathways. *Molecular Aspects of Medicine*. 24: 177–187.

Aldini, G., Carini, M., Beretta, G., Bradamante, S. and Facino, R. M., 2002a. Carnosine is a quencher of 4-hydroxy-nonenal: through what mechanism of reaction? *Biochemical and Biophysical Research Communications*. 298: 699–706.

Aldini, G., Carini, M., Vistoli, G., Shibata, T., Kusano, Y., Gamberoni, L., Dalle-Donne, I., Milzani, A. and Uchida, K., 2007a. Identification of actin as a 15-deoxy-delta(12,14)-prostaglandin J(2) target in neuroblastoma cells: mass spectrometric, computational, and functional approaches to investigate the effect on cytoskeletal derangement. *Biochemistry*. 46: 2707–2718.

Aldini, G., Carini, M., Yeum, K. J. and Vistoli, G., 2014. Novel molecular approaches for improving enzymatic and nonenzymatic detoxification of 4-hydroxynonenal: toward the discovery of a novel class of bioactive compounds. *Free Radical Biology & Medicine*. 69: 145–156.

Aldini, G., Dalle-Donne, I., Colombo, R., Facino, R. M., Milzani, A. and Carini, M., 2006. Lipoxidation-derived reactive carbonyl species as potential drug targets in preventing protein carbonylation and related cellular dysfunction. *ChemMedChem*. 1: 1045–1058.

Aldini, G., Dalle-Donne, I., Facino, R. M., Milzani, A. and Carini, M., 2007b. Intervention strategies to inhibit protein carbonylation by lipoxidation-derived reactive carbonyls. *Medicinal Research Reviews*. 27: 817–868.

Aldini, G., Dalle-Donne, I., Vistoli, G., Facino, R. M. and Carini, M., 2005a. Covalent modification of actin by 4-hydroxy-trans-2-nonenal (HNE): LC-ESI-MS/MS evidence for Cys374 Michael adduction. *Journal of Mass Spectrometry*. 40: 946–954.

Aldini, G., Facino, R. M., Beretta, G. and Carini, M., 2005b. Carnosine and related dipeptides as quenchers of reactive carbonyl species: from structural studies to therapeutic perspectives. *BioFactors*. 24: 77–87.

Aldini, G., Granata, P. and Carini, M., 2002b. Detoxification of cytotoxic alpha,beta-unsaturated aldehydes by carnosine: characterization of conjugated adducts by electrospray ionization tandem mass spectrometry and detection by liquid chromatography/mass spectrometry in rat skeletal muscle. *Journal of Mass Spectrometry*. 37: 1219–1228.

Aldini, G., Orioli, M., Rossoni, G., Savi, F., Braidotti, P., Vistoli, G., Yeum, K. J., Negrisoli, G. and Carini, M., 2011. The carbonyl scavenger carnosine ameliorates dyslipidemia and renal function in Zucker obese rats. *Journal of Cellular and Molecular Medicine*. 15: 1339–1354.

Aldini, G., Vistoli, G., Regazzoni, L., Benfatto, M. C., Bettinelli, I. and Carini, M., 2010. Edaravone Inhibits Protein Carbonylation by a Direct Carbonyl-Scavenging Mechanism: focus on reactivity, selectivity, and reaction mechanisms. *Antioxidants & Redox Signaling*. 12: 381–392.

Alzolibani, A. A., Al Robaee, A. A., Al-Shobaili, H. A. and Rasheed, Z., 2013. 4-Hydroxy-2-nonenal modified histone-H2A: a possible antigenic stimulus for systemic lupus erythematosus autoantibodies. *Cellular Immunology*. 284: 154–162.

Baba, S. P., Hoetker, J. D., Merchant, M., Klein, J. B., Cai, J., Barski, O. A., Conklin, D. J. and Bhatnagar, A., 2013. Role of aldose reductase in the metabolism and detoxification of carnosine-acrolein conjugates. *Journal of Biological Chemistry*. 288: 28163–28179.

Barski, O. A., Xie, Z., Baba, S. P., Sithu, S. D., Agarwal, A., Cai, J., Bhatnagar, A. and Srivastava, S., 2013. Dietary carnosine prevents early atherosclerotic lesion formation in apolipoprotein E-null mice. *Arteriosclerosis, Thrombosis, and Vascular Biology*. 33: 1162–1170.

Boldyrev, A. A., Aldini, G. and Derave, W., 2013. Physiology and pathophysiology of carnosine. *Physiological Reviews*. 93: 1803–1845.

Carini, M., Aldini, G., Beretta, G., Arlandini, E. and Facino, R. M., 2003. Acrolein-sequestering ability of endogenous dipeptides: characterization of carnosine and homocarnosine/acrolein adducts by electrospray ionization tandem mass spectrometry. *Journal of Mass Spectrometry*. 38: 996–1006.

Colzani, M., Aldini, G. and Carini, M., 2013. Mass spectrometric approaches for the identification and quantification of reactive carbonyl species protein adducts. *Journal of Proteomics*. 92: 28–50.

Colzani, M., Criscuolo, A., De Maddis, D., Garzon, D., Yeum, K. J., Vistoli, G., Carini, M. and Aldini, G., 2014. A novel high resolution MS approach for the screening of 4-hydroxy-*trans*-2-nonenal sequestering agents. *Journal of Pharmaceutical and Biomedical Analysis*. 91C: 108–118.

Dalle-Donne, I., Carini, M., Vistoli, G., Gamberoni, L., Giustarini, D., Colombo, R., Facino, R. M., Rossi, R., Milzani, A. and Aldini, G., 2007. Actin Cys374 as a nucleophilic target of alpha,beta-unsaturated aldehydes. *Free Radicals in Biology and Medicine*. 42: 583–598.

Dalle-Donne, I., Rossi, R., Colombo, R., Giustarini, D. and Milzani, A., 2006. Biomarkers of oxidative damage in human disease. *Clinical Chemistry*. 52: 601–623.

Davies, S. S., Amarnath, V. and Roberts, L. J., 2004. Isoketals: highly reactive gamma-ketoaldehydes formed from the H-2-isoprostane pathway. *Chemistry and Physics of Lipids*. 128: 85–99.

Degen, J., Hellwig, M. and Henle, T., 2012. 1,2-Dicarbonyl Compounds in Commonly Consumed Foods. *Journal of Agricultural and Food Chemistry*. 60: 7071–7079.

Dominguez, J. H., Wu, P., Hawes, J. W., Deeg, M., Walsh, J., Packer, S. C., Nagase, M., Temm, C., Goss, E. and Peterson, R., 2006. Renal injury: similarities and differences in male and female rats with the metabolic syndrome. *Kidney International*. 69: 1969–1976.

Drozak, J., Veiga-Da-Cunha, M., Vertommen, D., Stroobant, V. and Van Schaftingen, E., 2010. Molecular identification of carnosine synthase as ATP-grasp domain-containing protein 1 (ATPGD1). *Journal of Biological Chemistry*. 285: 9346–9356.

Everaert, I., Taes, Y., De Heer, E., Baelde, H., Zutinic, A., Yard, B., Sauerhofer, S., Vanhee, L., Delanghe, J., Aldini, G. and Derave, W., 2012. Low plasma carnosinase activity promotes carnosinemia after carnosine ingestion in humans. *American Journal of Physiology-Renal Physiology*. 302: F1537–F1544.

Fedorova, M., Bollineni, R. C. and Hoffmann, R., 2014. Protein carbonylation as a major hallmark of oxidative damage: update of analytical strategies. *Mass Spectrometry Reviews*. 33: 79–97.

Fritz, G., 2011. RAGE: a single receptor fits multiple ligands. *Trends in Biochemical Sciences*. 36: 625–632.

Fu, T. M., Jacob, D. J., Wittrock, F., Burrows, J. P., Vrekoussis, M. and Henze, D. K., 2008. Global budgets of atmospheric glyoxal and methylglyoxal, and implications for formation of secondary organic aerosols. *Journal of Geophysical Research: Atmospheres*. 113: D15303.

Hipkiss, A. R., 2009. Carnosine and its possible roles in nutrition and health. *Advances in Food and Nutrition Research*. 57: 87–154.

Hipkiss, A. R., Brownson, C. and Carrier, M. J., 2001. Carnosine, the anti-ageing, anti-oxidant dipeptide, may react with protein carbonyl groups. *Mechanisms of Ageing and Development*. 122: 1431–1445.

Hipkiss, A. R. and Chana, H., 1998. Carnosine protects proteins against methylglyoxal-mediated modifications. *Biochemical and Biophysical Research Communications*. 248: 28–32.

Hipkiss, A. R., Michaelis, J. and Syrris, P., 1995. Non-enzymatic glycosylation of the dipeptide L-carnosine, a potential anti-protein-cross-linking agent. *FEBS Letters*. 371: 81–85.

Hipkiss, A. R., Worthington, V. C., Himsworth, D. T. and Herwig, W., 1998. Protective effects of carnosine against protein modification mediated by malondialdehyde and hypochlorite. *Biochimica et Biophysica Acta*. 1380: 46–54.

Izaguirre, G., Kikonyogo, A. and Pietruszko, R., 1998. Methylglyoxal as substrate and inhibitor of human aldehyde dehydrogenase: comparison of kinetic properties among the three isozymes. *Comparative Biochemistry and Physiology, Part B: Biochemistry & Molecular Biology*. 119: 747–754.

Janero, D. R., 1990. Malondialdehyde and thiobarbituric acid-reactivity as diagnostic indices of lipid peroxidation and peroxidative tissue injury. *Free Radical Biology and Medicine*. 9: 515–540.

Janssen, B., Hohenadel, D., Brinkkoetter, P., Peters, V., Rind, N., Fischer, C., Rychlik, I., Cerna, M., Romzova, M., De Heer, E., Baelde, H., Bakker, S. J., Zirie, M., Rondeau, E., Mathieson, P., Saleem, M. A., Meyer, J., Köppel, H., Sauerhoefer, S., Bartram, C. R., Nawroth, P., Hammes, H. P., Yard, B. A., Zschocke, J. and Van Der Woude, F. J., 2005. Carnosine as a protective factor in diabetic nephropathy: association with a leucine repeat of the carnosinase gene CNDP1. *Diabetes*. 54: 2320–2327.

Kalapos, M. P., 1999. Methylglyoxal in living organisms – chemistry, biochemistry, toxicology and biological implications. *Toxicology Letters*. 110: 145–175.

Khatoon, F., Moinuddin, Alam, K. and Ali, A., 2012. Physicochemical and immunological studies on 4-hydroxynonenal modified HSA: implications of protein damage by lipid peroxidation products in the etiopathogenesis of SLE. *Human Immunology*. 73: 1132–1139.

Kierdorf, K. and Fritz, G., 2013. RAGE regulation and signaling in inflammation and beyond. *Journal of Leukocyte Biology*. 94: 55–68.

Kirkham, P. A., Caramori, G., Casolari, P., Papi, A. A., Edwards, M., Shamji, B., Triantaphyllopoulos, K., Hussain, F., Pinart, M., Khan, Y., Heinemann, L., Stevens, L., Yeadon, M., Barnes, P. J., Chung, K. F. and Adcock, I. M., 2011. Oxidative stress-induced antibodies to carbonyl-modified protein correlate with severity of chronic obstructive pulmonary disease. *American Journal of Respiratory and Critical Care Medicine*. 184: 796–802.

Liu, Y., Xu, G. and Sayre, L. M., 2003. Carnosine inhibits (*E*)-4-hydroxy-2-nonenal-induced protein cross-linking: structural characterization of carnosine-HNE adducts. *Chemical Research in Toxicology*. 16: 1589–1597.

Lopachin, R. M., Gavin, T., Petersen, D. R. and Barber, D. S., 2009. Molecular mechanisms of 4-hydroxy-2-nonenal and acrolein toxicity: nucleophilic targets and adduct formation. *Chemical Research in Toxicology*. 22: 1499–1508.

Madian, A. G., Myracle, A. D., Diaz-Maldonado, N., Rochelle, N. S., Janle, E. M. and Regnier, F. E., 2011. Differential carbonylation of proteins as a function of in vivo oxidative stress. *Journal of Proteome Research*. 10: 3959–3972.

Madian, A. G. and Regnier, F. E., 2010. Profiling carbonylated proteins in human plasma. *Journal of Proteome Research*. 9: 1330–1343.

Menini, S., Iacobini, C., Ricci, C., Scipioni, A., Fantauzzi, C. B., Giaccari, A., Salomone, E., Canevotti, R., Lapolla, A., Orioli, M., Aldini, G. and Pugliese, G., 2012. D-Carnosine octylester attenuates atherosclerosis and renal disease in ApoE null mice fed a Western diet through reduction of carbonyl stress and inflammation. *British Journal of Pharmacology*. 166: 1344–1356.

Moore, K. and Roberts, L. J., 1998. Measurement of lipid peroxidation. *Free Radical Research*. 28: 659–671.

Nadkarni, D. V. and Sayre, L. M., 1995. Structural definition of early lysine and histidine adduction chemistry of 4-hydroxynonenal. *Chemical Research in Toxicology*. 8: 284–291.

Näsström, T., Fagerqvist, T., Barbu, M., Karlsson, M., Nikolajeff, F., Kasrayan, A., Ekberg, M., Lannfelt, L., Ingelsson, M. and Bergström, J., 2011. The lipid peroxidation products 4-oxo-2-nonenal and 4-hydroxy-2-nonenal promote the formation of α-synuclein oligomers with distinct biochemical, morphological, and functional properties. *Free Radical Biology and Medicine*. 50: 428–437.

O'Brien, P. J., Siraki, A. G. and Shangari, N., 2005. Aldehyde sources, metabolism, molecular toxicity mechanisms, and possible effects on human health. *Critical Reviews in Toxicology*. 35: 609–662.

Orioli, M., Aldini, G., Benfatto, M. C., Facino, R. M. and Carini, M., 2007. HNE Michael adducts to histidine and histidine-containing peptides as biomarkers of lipid-derived carbonyl stress in urines: LC-MS/MS profiling in Zucker obese rats. *Analytical Chemistry*. 79: 9174–9184.

Orioli, M., Aldini, G., Beretta, G., Facino, R. M. and Carini, M., 2005. LC-ESI-MS/MS determination of 4-hydroxy-*trans*-2-nonenal Michael adducts with cysteine and histidine-containing peptides as early markers of oxidative stress in excitable tissues. *Journal of Chromatography B: Analytical Technologies in the Biomedical and Life Sciences*. 827: 109–118.

Orioli, M., Vistoli, G., Regazzoni, L., Pedretti, A., Lapolla, A., Rossoni, G., Canevotti, R., Gamberoni, L., Previtali, M., Carini, M. and Aldini, G., 2011. Design, synthesis, ADME properties, and pharmacological activities of beta-alanyl-D-histidine (D-carnosine) prodrugs with improved bioavailability. *ChemMedChem*. 6: 1269–1282.

Requena, J. R., Fu, M. X., Ahmed, M. U., Jenkins, A. J., Lyons, T. J. and Thorpe, S. R., 1996. Lipoxidation products as biomarkers of oxidative damage to proteins during lipid peroxidation reactions. *Nephrology Dialysis Transplantation*. 11(suppl. 5): 48–53.

Seiler, N., 2004. Catabolism of polyamines. *Amino Acids*. 26: 217–233.

Siegel, S. J., Bieschke, J., Powers, E. T. and Kelly, J. W., 2007. The oxidative stress metabolite 4-hydroxynonenal promotes Alzheimer protofibril formation. *Biochemistry*. 46: 1503–1510.

Siems, W. and Grune, T., 2003. Intracellular metabolism of 4-hydroxynonenal. *Molecular Aspects of Medicine*. 24: 167–175.

Smathers, R. L., Fritz, K. S., Galligan, J. J., Shearn, C. T., Reigan, P., Marks, M. J. and Petersen, D. R., 2012. Characterization of 4-HNE modified L-FABP reveals alterations in structural and functional dynamics. *PLoS One*. 7: e38459.

Stevens, J. F. and Maier, C. S., 2008. Acrolein: sources, metabolism, and biomolecular interactions relevant to human health and disease. *Molecular Nutrition & Food Research*. 52: 7–25.

Teufel, M., Saudek, V., Ledig, J. P., Bernhardt, A., Boularand, S., Carreau, A., Cairns, N. J., Carter, C., Cowley, D. J., Duverger, D., Ganzhorn, A. J., Guenet, C., Heintzelmann, B., Laucher, V., Sauvage, C. and Smirnova, T., 2003. Sequence identification and characterization of human carnosinase and a closely related non-specific dipeptidase. *Journal of Biological Chemistry*. 278: 6521–6531.

Thornalley, P. J., 1990. The glyoxalase system: new developments towards functional characterization of a metabolic pathway fundamental to biological life. *Biochemical Journal*. 269: 1–11.

Thornalley, P. J., Langborg, A. and Minhas, H. S., 1999. Formation of glyoxal, methylglyoxal and 3-deoxyglucosone in the glycation of proteins by glucose. *Biochemical Journal*. 344: Pt 1, 109–116.

Vistoli, G., Carini, M. and Aldini, G., 2012. Transforming dietary peptides in promising lead compounds: the case of bioavailable carnosine analogs. *Amino Acids*. 43: 111–126.

Vistoli, G., De Maddis, D., Cipak, A., Zarkovic, N., Carini, M. and Aldini, G., 2013. Advanced glycoxidation and lipoxidation end products (AGEs and ALEs): an overview of their mechanisms of formation. *Free Radical Research*. 47(suppl. 1): 3–27.

Xiang, W., Schlachetzki, J. C., Helling, S., Bussmann, J. C., Berlinghof, M., Schäffer, T. E., Marcus, K., Winkler, J., Klucken, J. and Becker, C. M., 2013. Oxidative stress-induced posttranslational modifications of alpha-synuclein: specific modification of alpha-synuclein by 4-hydroxy-2-nonenal increases dopaminergic toxicity. *Molecular and Cellular Neuroscience*. 54: 71–83.

Xue, J., Rai, V., Singer, D., Chabierski, S., Xie, J., Reverdatto, S., Burz, D. S., Schmidt, A. M., Hoffmann, R. and Shekhtman, A., 2011. Advanced glycation end product recognition by the receptor for AGEs. *Structure*. 19: 722–732.

Yeum, K. J., Orioli, M., Regazzoni, L., Carini, M., Rasmussen, H., Russell, R. M. and Aldini, G., 2010. Profiling histidine dipeptides in plasma and urine after ingesting beef, chicken or chicken broth in humans. *Amino Acids*. 38: 847–858.

Zhou, S. and Decker, E. A., 1999. Ability of carnosine and other skeletal muscle components to quench unsaturated aldehydic lipid oxidation products. *Journal of Agricultural and Food Chemistry*. 47: 51–55.

CHAPTER 9

Antioxidant Activity of Imidazole Dipeptides – Prevention of DNA and Protein Degradation

MASAYOSHI TAKAHASHI*[a], SHOJI HAGIWARA[a], AND HIROSHI NABETANI[b]

[a]Reaction and Separation Engineering Lab, Food Engineering Division, National Food Research Institute, National Agriculture and Food Research Organization, 2-1-12 Kannondai, Tsukuba City, Ibaraki Prefecture, 305-8642, Japan; [b]Director of Food Engineering Division, National Food Research Institute, National Agriculture and Food Research Organization, 2-1-12 Kannondai, Tsukuba, Ibaraki, 305-0031, Japan
*E-mail: mtjpbr@yahoo.co.jp

9.1 Introduction

The representative of imidazole dipeptide (IDP) carnosine (Car) was initially isolated from bovine meat by the Russian chemist V. S. Gulevitch in 1900, and its chemical formula was determined as a dipeptide of β-alanine and L-histidine. Its methylated derivatives, anserine (Ans) and balenine, were discovered in goose and whale meat. After the discovery of these compounds, it was assumed that IDPs function in osmotic control and pH buffering,

Food and Nutritional Components in Focus No. 8
Imidazole Dipeptides: Chemistry, Analysis, Function and Effects
Edited by Victor R Preedy

Published by the Royal Society of Chemistry, www.rsc.org

thereby helping to maintain muscles during active movement. It has also been demonstrated that carnosine is capable of reducing fatigue in experiments with mice; however, the biochemical basis of this effect remains unclear. Additionally, the metal-chelating activity and antioxidant activity of IDPs has been studied. In Russia, the antioxidant activity of IDPs was utilized in treating patients with disorders caused by exposure to strong radiation (Boldyrev, 2012).

Since IDPs are abundant in muscles undergoing strenuous activity, they are thought to be key substances involved in chemical processes that reduce fatigue. Excess free radicals accumulate during mitochondrial electron transfer, which produces the energy required for biochemical activity in living cells. It is assumed that the majority of reactive oxygen species (ROS) are generated subsequent to the initial production of hydrogen peroxide, particularly in the muscles of birds or large fish that engage in long-distance migration. ROS transfer energy to chlorine, nitrogen and hydrogen oxide, thereby dramatically increasing their reactivity. It is thought that fatigue arises as a consequence of dysfunction and structural failure of muscle-cell components due to ROS generation. Fatigue is but one of many diseases that are caused by the long-term leakage of the radicals. The diverse effects of ROS are likely to include the destruction of the structural components of the inner wall of blood vessels, arteriosclerosis, renal dysfunction and dysfunctional gene expression, such as in certain cancers (Yanai *et al.*, 2008). We followed the method previously described by Yanai *et al.* (2004) and discussed the pros and cons of the use of ROS-induced protein degradation as an indication of cellular damage, and subsequently reported a novel method for determining DNA degradation to the Japanese Society for Food Science and Technology. DNA degradation was chosen as the main indicator for this method, which is a key factor for cellular viability.

In this chapter, we will introduce a novel method for estimating naturally occurring antioxidants in foods. Samples subjected to this method are assessed for the major types of ROS that cause cellular damage. Among the various biological macromolecules, DNA is a very important target of ROS that causes a variety of diseases when damaged.

We will demonstrate the principle and the implementation of the above-described method: estimation method for antioxidants with DNA degradation (EMADD). Using this method, we investigated the effect of various representative antioxidant nutrients on ROS levels, and it was found that IDPs provide the most comprehensive protective effects among the antioxidant nutrient assayed. Hence, we have determined the basis of its multifaceted characteristic.

Cell death can occur due to a variety of reasons, and Harman first proposed that one of the major causes of cell death involves damage to cellular components, such as lipids, genomic DNA and proteins, arising from endogenously generated reactive oxygen species (ROS) (Harman, 1956; Beckman and Ames, 1998). Research into the role of ROS in the development of diseases and in aging is active and continuing (Dröge, 2002; Seifried *et al.*, 2007).

The recent attention that ROS has been receiving is reflected in the popularity of foods or supplements that are rich in antioxidants. These antioxidant foods and supplements are taken in an effort to counteract the effects of lifestyle-related diseases of aging in Japan and the United States (Devasagayam *et al.*, 2004; Atalay *et al.*, 2006).

Previously, Yanai *et al.* reported that there are differences in the effects of dietary antioxidants toward endogenous ROS and the artificially generated radicals used in antioxidant assays, such as ORAC score, which varied depending on the ROS species. Furthermore, they compared the effect of various antioxidants against endogenous ROSs using their original assay with protein degradation (Yanai *et al.*, 2004). This study demonstrated that the ORAC system reports that hydrophobic antioxidants, such as polyphenols, produce larger effects compared to hydrophilic antioxidants, such as IDP mixtures of carnosine and anserine (ACmix). In contrast, when these antioxidants were assayed for their ability to suppress endogenous ROS-induced protein degradation, each chemical species of naturally occurring antioxidant exhibited effects that were specific to the type of ROS. This report detailed the effects on the primary endogenous radicals (peroxynitrite, $ONOO^{\cdot}$; hydroxide, $OH^{\cdot}$; and hypochlorous, $ClO^{\cdot}$); however, hydrogen peroxide (H_2O_2) was not examined due to its weak ability to destroy protein polymers.

Therefore, in this study, we investigated the ability of various antioxidants to inhibit DNA degradation induced by four ROS species. As a result, when the DNA was used as the biological target, it was confirmed that all four ROS produced DNA damage *in vivo* in a concentration-dependent manner. The effects of various antioxidants were also observed using the protein-degradation system, with each of the antioxidants exhibiting specific activity profiles toward the ROS species examined. However, using ACmix purified from chicken extract, the results obtained using protein degradation differed from the DNA degradation data for the four endogenous ROS.

In this chapter, we will compare the results obtained using our new method with data from the protein degradation and ORAC methods, and discuss the effects of naturally occurring antioxidants, including ACmix.

9.2 EMADD Procedure

9.2.1 Principle

Initially, we determined the minimum time and concentration required for *in vitro* ROS-induced degradation of DNA. Antioxidant concentrations should be similar to the concentrations observed in the blood, and should be added to the assay at least 3 min prior to the onset of degradation. After sample treatment with an appropriate incubation time for each ROS at

37 °C, the samples are mixed with a loading buffer that contains a reductive stopping agent, applied to agarose gel electrophoresis the DNA bands are stained and visualized. The protective effects of the antioxidants are determined as the degree of inhibition of DNA degradation. The results of this method provide an estimation of the actual endogenous antioxidant capacity.

9.2.2 Reagents

9.2.2.1 Antioxidants

L-Carnosine (Car)	5.0 mM
L-Anserine–HNO_3(Ans)	5.0 mM
L-Ascorbic acid	5.0 mM
Na L-ascorbate (V.C) for DNA	5.0 mM
Catechin	0.5 mM
Glutathione, reductive (GSH)	2.5 mM
α-Tocopherol, water soluble (V.E)	1.0 mM
Astaxanthin, water soluble (Ast)	0.1 mM
CoenzymeQ10 (CoQ10)	1.0 mM

9.2.2.2 Preparation of Anserine and Carnosine Mixture from Chicken Extract

Chicken extracts were obtained from culled chickens with whole-body using hot-water extraction (95 °C, 4 h), and a mixture of anserine and carnosine (ACmix) was obtained following purification by ultrafiltration (Asahi Kasei, SEP-1013; a molecular weight cut-off of 3000 Daltons) and ion-exchange chromatography with Dowex50-X8 (Sigma), according to a published method (Yanai *et al.*, 2004). The extract composition was assessed using an automatic analyzer for the determination of amino acids, and the ratio of Ans and Car content were found to be 2.5 : 1.0, with total IDP purity greater than 92%.

9.2.3 Preparation of ROS Solutions

$ClO^{\cdot}$

Hypochlorite radical solution was prepared by diluting sodium hypochlorite (Wako, Japan) with distilled water (pH ~ 6.5). The concentration is to be adjusted as a series from 100 mM.

H_2O_2

The hydrogen peroxide solution was also prepared by dilution with distilled water (pH ~ 6.5). The concentration is to be adjusted as a series from 2.0 M.

$OH^{\cdot}$

The hydroxide radical solution was prepared using the Fenton reaction with ferric chloride. One hundred µl of stock solutions of 0.1 M $FeCl_3$, 0.1 M ascorbic acid and 0.1 M ethylene-diamine-tetra acetic acid (EDTA) were mixed in a test tube, to which 1000 µL of 0.13 M hydrogen peroxide was added to generate hydroxide radicals (total 1300 µL). The mixture was added at an adequate volume to the samples and the reaction was conducted for 60 min at 37 °C. All reagents were freshly prepared just prior to use, except for EDTA. The $OH^{\cdot}$ concentration is provided as mM the concentration of hydrogen peroxide added and the concentration is to be adjusted as a series from 100 mM (Yanai *et al.*, 2008).

$ONOO^{\cdot}$

The peroxynitrite radical solution was prepared using a modifying version of the method described by Radi *et al.* (1991), using a quenching reactor. The three solutions were concomitantly mixed by injection into a silicon tube and the mixture was subsequently poured into a beaker containing sodium hydroxide. One g of manganese oxide powder was then added to the solution to catalyze the removal of excess hydrogen peroxide. Following removal of the manganese oxide by filtration with filter paper, the absorbance (302 nm) of the yellow reaction solution was measured and the concentration of $ONOO^{\cdot}$ was calculated using the molecular absorption coefficient (1670/M). The concentration of $ONOO^{\cdot}$ obtained was 53.0 mM. The solution was aliquoted into 1 mL serum tubes and stored at −80 °C until use (Yanai *et al.*, 2004). The concentration is to be adjusted as a series from 40 mM.

9.2.4 Measurement of the Effect of Preventing ROS-Induced DNA Degradation

The antioxidant activity was measured in terms of the suppression of ROS-induced DNA degradation. DNA integrity was assessed using agarose gel electrophoresis and visualized with the DNA stain Mupid-Stain EYE (Mupid, USA). DNA manipulation was performed according to the "Molecular Cloning" manual (Sambrook and Russell, 2001).

9.2.4.1 Preparation of DNA: EDTA Removal

The procedure is outlined in the following flow chart:

Combine 100 μL of λ phage DNA solution (TOYOBO Co., Ltd.)
+5 μL of 3 M sodium acetate (pH 5.5),
+210 μL ethanol
↓ (15000 × *g*) centrifuge at 10 °C for 10 min
Discard the supernatant.

↓

Wash with 500 μL of 70% ethanol, chill on ice
* ↓, (15 000 × *g*) centrifuge at 10 °C for 5 min
* Discard the supernatant.
(* Repeat steps twice.)

↓

Subject to a vacuum of < 1 mPa for 10 min in a freeze dryer

↓

Add PBS (phosphate-buffered saline, Sigma) to dried lambda phage DNA to a concentration of 0.5 μg / μL^{-1}

↓

Store in a refrigerator (10 °C) overnight

↓

Ensure that the DNA is resuspended in PBS by slow pipetting, while avoiding shearing damage

9.2.4.2 DNA-Degradation Reaction

The DNA-degradation reactions were conducted in a total volume of 50 μL by adding 5 μL of the antioxidants solution and 5 μL of ROS in PBS to 3 μg of lambda phage DNA in 40 μL PBS. Each antioxidant was mixed with the DNA solution and incubated for 3 min at room temperature. Next, individual ROS were added to separate reactions and incubated at 37 °C. For $ClO^{\cdot}$ and $OH^{\cdot}$, reaction times were 60 min, whereas 120 min was used for $ONOO^{\cdot}$ and H_2O_2 (Yanai *et al.*, 2008).

Glass test tubes were employed since the H_2O_2 reaction is easily influenced by the cleanliness and quality of plastic material of the test tubes. The glass test tubes were immersed in 10 mM hypochlorous acid solution for 30 min after washing with an alkaline detergent, and subsequently washed thoroughly with tap water, and finally with distilled water to remove any trace contamination of reductive substances. Termination of the reaction was carried out by mixing 3.5 μL of stopping solution in agarose gel electrophoresis running buffer (6× concentration) to 17.5 μL of each reaction at the designated termination time. The composition of the stopping solution is as follows: 0.024% bromophenol blue, 300 mM dithiothreitol, 45.5% glycerol, 50% TAE. The 50% TAE was prepared by adding 1% volume of 50× TAE, as described in the molecular cloning manual (Sambrook and Russell, 2001).

9.2.4.3 *Agarose Gel Electrophoresis*

A 0.8% agarose gels were immersed in an electrophoresis device filled with 1× TAE buffer, and the samples (6 μL per well) were loaded in the wells and electrophoresis was performed at 100 V for 30 min. Lambda phage DNA was digested with EcoRI and Hind III and used as a molecular weight marker. After electrophoresis, the gel was soaked in distilled water for 1 min to remove the buffer components, and was subsequently stained in Mupid-Stain EYE solution (Mupid, USA) for 1 min. The gel was then placed in a volume of water that is 3× the gel volume, and incubated on a shaker for 20 min. Finally, the gel was transferred to fresh water of 5× the gel volume, and was incubated with shaking for 2 to 5 h to reduce the background color.

9.2.4.4 *Criteria of Estimation*

The degree of antioxidant activity is expressed as follows:

Completely prevention of DNA degradation, resulting in a single clear band at 23.1 kbp:	+++
Smeared band between 4.4 to 23.1 kbp:	++
Smeared bands > 2.0 kbp:	+
Trace bands < 2.0 kbp, ≈degradation control:	–
Degradation in excess of control:	– –

9.3 Determination of Antioxidant Effects

The effects of various antioxidants were measured using commonly used assays of antioxidant activity and compared with results obtained with the EMADD method. The AAPH method was carried out according to the procedure of Hirayama's chemical luminescence method (Hirayama *et al.*, 1997). For OH radical determination, AAPH was used instead of OH˙ prepared according to the Hirayama method described above, and measurements were performed using the chemical luminescence method in the same manner as described by Hirayama *et al.* (1997). Since the concentrations of antioxidants that suppress ROS oxidation activity by 50% are measured with these methods, smaller values indicate stronger antioxidant effects. In the ORAC method, the measured values are expressed (Trolox eq M/M) as equivalent values of Trolox, the antioxidant standard (Prior *et al.*, 2003).

9.4 Results

9.4.1 Degradation of DNA by Various ROS

The four major ROS constitutively produced *in vivo* (ClO˙, OH˙, ONOO˙ and H_2O_2) induced degradation of lambda phage DNA in a concentration-dependent manner. The degree of DNA degradation mediated by each ROS was compared. ClO˙ produced obvious DNA degradation at 0.2 mM (37 °C for 60 min), and complete degradation (bands < 0.6 kb) occurred at 2 mM (Figure 9.1). DNA degradation induced by 1 mM OH˙ (37 °C for 60 min) resulted in a high molecular weight smear (Figure 9.2). On the other hand, 3 mM ONOO˙ (37 °C for 120 min) induced marked DNA degradation, with the band disappearing at 4 mM (Figure 9.3). In comparison, H_2O_2-induced DNA degradation was first observed at 50 mM (37 °C for 120 min), and proceeded in a concentration-dependent manner up to 200 mM (Figure 9.4).

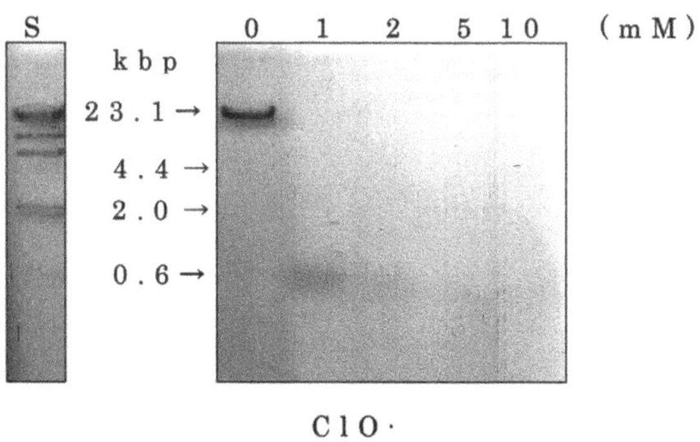

Figure 9.1 Dose-dependent degradation of λ-phage DNA with hypochloric acid. The reactions were performed at 37 °C with ClO˙ for 30 min. The numbers above the image indicate concentrations in mM of each ROS. S: Standard molecular marker (EcoRI, HindIII/λ-phage DNA).

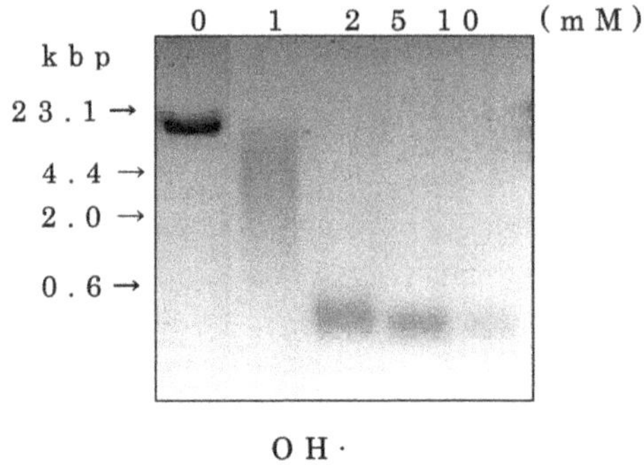

Figure 9.2 Dose-dependent degradation of λ-phage DNA with hydroxy radicals. The reactions were performed at 37 °C with OH˙ for 30 min. The numbers above the image indicate concentration in mM of each ROS.

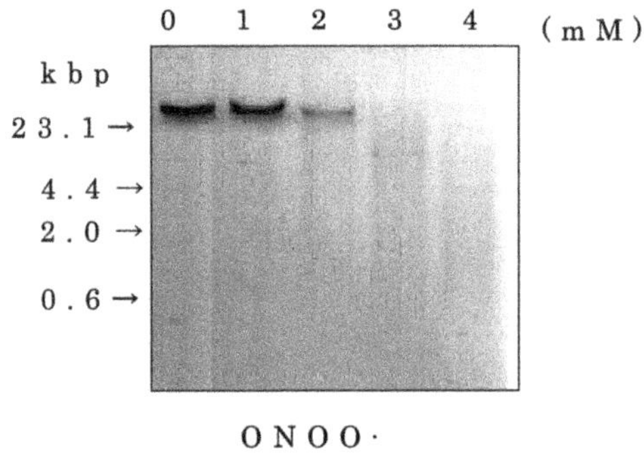

Figure 9.3 Dose-dependent degradation of λ-phage DNA with peroxynitrite radicals. The reactions were performed at 37 °C with ONOO˙ for 120 min. The numbers above the image indicate concentration in mM of each ROS.

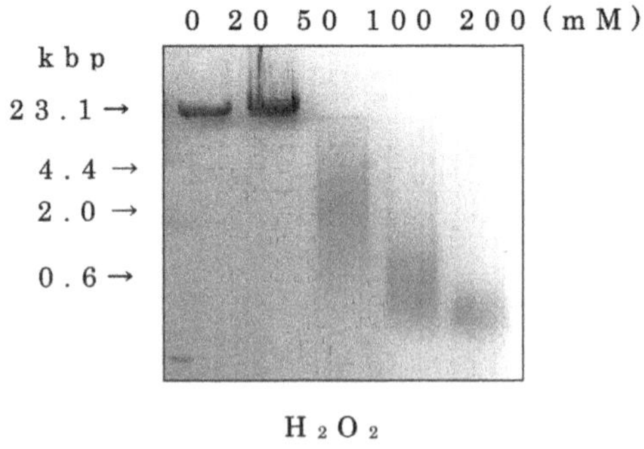

Figure 9.4 Dose-dependent degradation of λ-phage DNA with hydrogen peroxide. The reactions were performed at 37 °C with H_2O_2 for 120 min. The numbers above the image indicate concentrations in mM of each ROS.

9.4.2 Comparison of Naturally Occurring Antioxidants against ROS-Induced DNA Degradation

Three of the ROS ($ClO^{\cdot}$, $OH^{\cdot}$ and $ONOO^{\cdot}$) have been assessed in protein degradation assays, including ACmix, ferulic acid, and V.C, with the later showing the strongest antioxidant effect for each of the three ROSs (Yanai *et al.*, 2008). We compared the inhibitory effect of these antioxidants on ROS-induced (including H_2O_2, $ClO^{\cdot}$, $OH^{\cdot}$ and $ONOO^{\cdot}$) DNA degradation.

For $ClO^{\cdot}$ (5 mM) induced DNA degradation, ACmix and V.C showed inhibitory effects comparable to their effects in the protein degradation assay. However, ferulic acid did not produce an inhibitory effect (Figure 9.5). For $OH^{\cdot}$ (5 mM) induced DNA degradation, ACmix and ferulic acid showed strong inhibitory effects, whereas V.C inhibition was minimal (Figure 9.6).

For $ONOO^{\cdot}$ (3 mM) induced DNA degradation, ACmix and V.C exhibited strong inhibition, whereas ferulic acid produced only slight suppression (Figure 9.7). For H_2O_2 (50 mM) induced DNA degradation, which could not be estimated using the protein degradation system, ferulic acid strongly suppressed degradation. On the other hand, V.C showed no inhibitory effect and was observed to potentiate degradation (Figure 9.8).

Other naturally occurring antioxidants were assessed for potential inhibitory effects in the ROS-induced DNA degradation assay and the results are shown in table (Table 9.1), along with the results obtained with ACmix, V.C, and ferulic acid. These compounds showed inhibitory effects on DNA degradation were similar to previous observations using the protein degradation assay. Peptide-type antioxidants, such as GSH and ACmix, showed strong inhibitory effects against by $ClO^{\cdot}$-induced DNA and protein degradation, whereas hydrophobic antioxidants showed significant inhibitory effects against $OH^{\cdot}$-induced DNA and protein degradation. Only ACmix and VC showed antioxidant effects against $ONOO^{\cdot}$-induced DNA degradation.

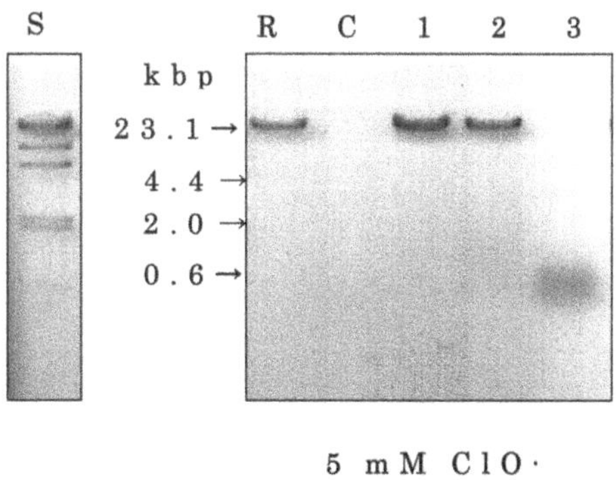

Figure 9.5 Antioxidant compatibility with hypochloric acid using DNA degradation. The reactions were performed at 37 °C with $ClO^{\cdot}$ for 30 min. S: Standard molecular marker (EcoRI, HindIII/λ-phage DNA), R: reference, λ-phage DNA without ROS or antioxidant, C; positive control with $ClO^{\cdot}$, 1: reaction as with C plus 5 mM anserine–carnosine, 2; with 5 mM vitamin C and 3; with 0.5 mM ferulic acid.

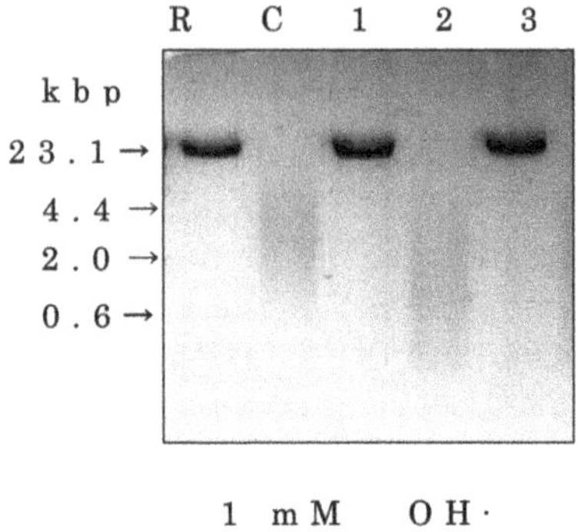

Figure 9.6 Antioxidant compatibility with hydroxy radicals using DNA degradation. The reactions were performed at 37 °C with $OH^{\cdot}$ for 30 min. R: reference, λ-phage DNA without ROS or antioxidant, C; positive control with $OH^{\cdot}$, 1: reaction as with C plus 5 mM anserine–carnosine, 2; with 5 mM vitamin C and 3; with 0.5 mM ferulic acid.

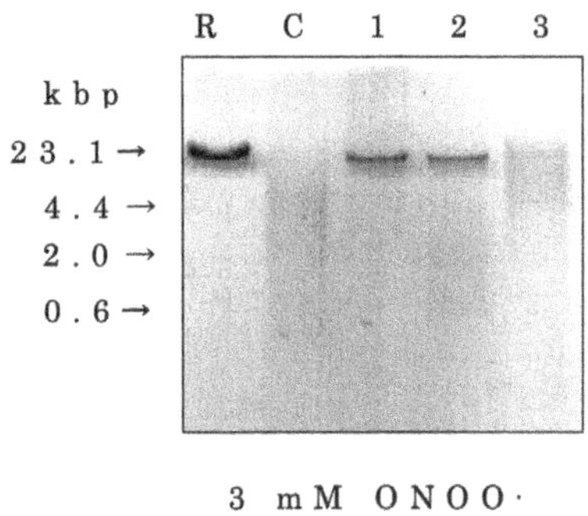

Figure 9.7 Antioxidant compatibility with peroxynitrite radicals using DNA degradation. The reactions were performed at 37 °C with $ONOO^{\cdot}$ for 120 min. R: reference, λ-phage DNA without ROS or antioxidant, C; positive control with $ONOO^{\cdot}$, 1: reaction as with C plus 5 mM anserine–carnosine, 2; with 5 mM vitamin C and 3; with 0.5 mM ferulic acid.

For H_2O_2-induced DNA degradation, strong suppressive effects were observed with GSH ACmix, astaxanthin and chlorogenic acid, while hydrophobic compounds were less effective and catechin, ferulic acid and CoQ10 showed no inhibitory effects.

ACmix showed inhibitory effects against all four ROS. The components of ACmix, Car and Ans, were tested individually and the results revealed that Car does not inhibit $OH^{\cdot}$- or $ONOO^{\cdot}$-induced DNA degradation, while Ans does.

9.4.3 Comparison of Various Antioxidant Activity Assays

Antioxidant activities of nutrients contained in foods were measured with the commonly used assays and the results were compared with those of the DNA-degradation method (Table 9.2). The polyphenol ferulic acid and the

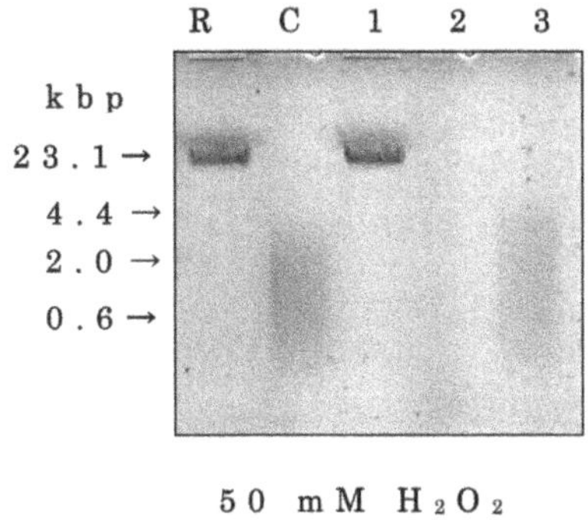

Figure 9.8 Antioxidant compatibility with hydrogen peroxide using DNA degradation. The reactions were performed at 37 °C with H_2O_2 for 120 min. R: reference, λ-phage DNA without ROS or antioxidant, C; positive control with H_2O_2, 1: reaction as with C plus 5 mM anserine–carnosine, 2; with 5 mM vitamin C and 3; with 0.5 mM ferulic acid.

Table 9.1 Comparison of the activity of naturally occurring antioxidants against four ROS using the DNA degradation assay.[a]

Antioxidants		ClO radical	OH radical	ONOO radical	H_2O_2
Anserine–NO_3	(5 mM)	+++	++	+	+++
Carnosine	(5 mM)	+++	−	−	+++
AC mix	(5 mM)	+++	++	++	+++
VC	(5 mM)	++	−−	++	−−
VE	(1 mM)	−	+	−	−
Coenzyme Q10	(1 mM)	−	++	−	−
Astaxantin	(0.1 mM)	−	+++	−	++
Catechin	(0.5 mM)	+++	++	−	−
Chlorogenic acid	(0.5 mM)	+	++	−	+++
Ferulic acid	(0.5 mM)	−	+++	−	−
Glutathione (GSH)	(2.5 mM)	+++	+	−	+++

[a]Protective efficiency was determined by densitometry analysis and is expressed as completely intact 23.1 kbp band (+++), mostly intact with smearing less than 4.4 kbp from the intact band (++), partially intact with smearing bands greater than 2.0 kbp that is obviously protected better than degradation control (+), not protected, similar to degradation control with smearing bands (−), exacerbated destruction, molecular weight bands less than 2 kbp or no staining (−−).

carotenoid astaxanthin, which are highly effective against $OH^{\cdot}$, were measured using the ORC method and produced Trolox equivalent values (Trolox eq. M/M) of 4.1 and 0.7, respectively, indicating that these compounds have strong antioxidant effects against AAPH. However, the values obtained for IDP were 0.3 or lower. The $OH^{\cdot}$ scavenging activity of IDP measured with the chemiluminescence method was as low as 3000 mM. In comparison, GSH showed strong antioxidant activity in the ORAC method and strong $OH^{\cdot}$ scavenging ability with the chemiluminescence method. However, GSH had only a weak effect, compared to other antioxidants, in inhibiting $OH^{\cdot}$-induced DNA degradation. Strong positive correlations were observed between the methods based on AAPH or $OH^{\cdot}$. The correlation coefficient between the AAPH

Table 9.2 Comparison of analytical methods of antioxidant activity.[a]

	OH radical	AAPH	ORAC	Protein	DNA
Antioxidants	IC_{50} (mM)	IC_{50} (μM)	(Trolox eq M/M)	(OH˙)	(OH˙)
Anserine–HNO_3	>3000	5030.0	0.1	–	++
Carnosine	>3000	4570.0	0.3	–	–
AC mix	2810	6952.0	0.1	±	++
VC	0.29	0.1	0.5	–	––
VE	4.56	30	1.0	++	+
Coenzyme Q10	26.10	890	0.6	+	++
Astaxantin	1.34	3	0.7	++	+++
Catechin	0.09	0.2	12.5	–	++
Chlorogenic acid	0.13	0.1	4.0	±	++
Ferulic acid	0.01	0.02	4.1	+++	+++
Glutathione (GSH)	1.40	20	12.3	±	+

[a]Explanation for each experiment was described in materials and methods of this paper. Lower values indicate higher antioxidant activity in experiments using OH radical and AAPH. Higher values or more + signs indicate higher antioxidant activity in the other experiments.

method and the OH˙ scavenging method was $R = 0.93$, and the correlation coefficient between the ORAC method and the OH˙ scavenging method was $R = 0.95$. On the other hand, ACmix showed only weak antioxidant activities against OH˙ in the chemiluminescence as well as in the protein degradation, AAPH and ORAC methods. However, ACmix showed moderate effect against OH˙ in the DNA degradation method. It was confirmed that this was due to the effect of Ans in ACmix. Evaluation of the IDPs by the ORAC method did not correlate with the inhibitory effects on DNA degradation.

9.5 Discussion

The induction of cell death by reactive oxygen species (ROS) is thought to be a causative factor in aging-related cellular senescence. ROS do not originate from external sources (*i.e.* ingestion or inhalation), but rather are constitutively produced during metabolic energy production, or by the immune system for host defenses. That is, when O_2 is utilized in the mitochondrial electron transport system, ROS (such as the superoxide anion, O_2^-) are produced as a natural byproduct (MacCord, 2002). H_2O_2 is produced in the process of O_2^- detoxification and is one of the four main types of endogenously produced ROS, along with OH˙, ClO˙ and ONOO˙.

In previous reports, estimations of carnosine antioxidant activity, which is one of the IDPs in ACmix, differed depending on the assay method employed, likely because of its metal chelating activity (Brown, 1981).

The following are examples of the problems associated with IDP:

IDP suppresses the production of OH˙ during the decomposition of H_2O_2 by the catalytic action of transition metal ions (Fenton reaction).

It is an antioxidizing agent for singlet oxygen (Kohen *et al.*, 1988).

It is an antioxidant against OH˙, but not to ClO˙ or H_2O_2 (Aruoma *et al.*, 1989).

In this study, estimation of antioxidant activity of various naturally occurring antioxidants, based on the ROS-induced DNA-degradation assay, produced results that were similar to those obtained with a protein degradation assay, as previously reported (Yanai *et al.*, 2008). Specifically, ACmix, which is a mixture of hydrophilic imidazole dipeptides present in animal tissues, showed a strong antioxidant activity against $ClO^{\cdot}$ that was comparable to V.C. On the other hand, the protein-degradation assay indicated strong inhibitory activities for polyphenols against $OH^{\cdot}$, and hydrophilic V.C against $ONOO^{\cdot}$ (Table 9.2). ACmix showed the strongest antioxidant activity against H_2O_2-induced DNA degradation, which is a substrate for biologically produced ROS that has not been previously studied. However, when the IDP components were individually assayed, it was revealed that the broad-spectrum antioxidative effect in ACmix is attributable to Ans, with Car having strong antioxidant activities against $ClO^{\cdot}$ and H_2O_2 only.

The results of our method may lead to changing viewpoints towards other antioxidants. Namely, while V.C exhibits strong antioxidant activities against $ClO^{\cdot}$ or $ONOO^{\cdot}$, our result showed that it potentiates DNA degradation following exposure to H_2O_2 or $OH^{\cdot}$. In addition, the polyphenolic antioxidant ferulic acid strongly inhibits $OH^{\cdot}$ in the protein degradation assay, as well as in the DNA degradation assay.

To summarize, we have shown here the activity of various naturally occurring antioxidants in inhibiting the effects of four ROS on DNA degradation (Table 9.1).

Among the antioxidants examined, ACmix was the only antioxidant that could suppress DNA degradation induced by all four ROS. The result showed marked differences when compared to protein degradation with respect to $ClO^{\cdot}$ quenching (Table 9.2).

In comparing the results of the common methods used for measuring antioxidant activity in foods, it was shown that carnosine and anserine have weak scavenging activity against hydroxyl radicals and its derivatives, whereas they have a strong protective function in suppressing damage to biological components (protein or DNA) caused by endogenous ROS. This is contrary to estimations using the ORAC method. This should be considered when using ORAC data as an indicator of food-derived antioxidant intake for the purpose of suppressing oxidative stress to maintain health.

In addition, with respect to the antioxidant effect of IDPs against H_2O_2 shown in this chapter, some compounds have been reported to exhibit similar properties:

Ferritin and H_2O_2 blocked lipid oxidation (Yoon *et al.*, 2010).

H_2O_2 inhibited decomposition catalyzed by the superoxide dismutase, Cu,Zn-SOD (Choi *et al.*, 1999).

These findings were consistent with the antioxidant effects of Ans and Car in ACmix toward H_2O_2 observed in this study. Furthermore, it was reported that H_2O_2, the starting substance of ROS metabolisms *in vivo*, promotes the induction of a senescent phenotype in cultured cells (Chen and Ames, 1994; von Zglinicki *et al.*, 1995; Hossain *et al.*, 2006), which can theoretically be reversed by IDPs. Additionally, IDPs have been reported to suppress the secretion of inflammatory cytokines (IL-8) that can cause oxidative stress *in vivo* (Son *et al.*, 2008). It has also been reported that IDPs suppress the formation of advanced glycation end product (AGE) (Hipkiss *et al.*, 2001). IDPs have a clear effect on preventing the induction of ROS-induced cell death and can be utilized to maintain health. We expect that they will assume a more important position in the food science of antioxidants.

9.6 Perspectives

Since there are several kinds of radicals that cause oxidative damage in the body, ignoring the specificity of antioxidative nutrients towards specific ROS could result in important findings being overlooked, with respect to preventing disease with nutrients. For example, there is statistical data that shows that there is risk associated with excessive vitamin C intake, which should therefore be avoided (Food and Nutrition Board, Institute of Medicine, 2002). In addition, the nutrition data laboratory (NDL) of the United States Department of Agriculture deleted the ORAC database of selected foods from their homepage in 2012, because there is no correlation between the effects of specific biologically active compounds (ORAC values) and human health, including polyphenols (USDA, 2012). When we assayed IDPs using the same method, we confirmed that they show very little activity. On the other hand, using the method presented herein, IDPs showed significant effects on the radical of chlorine, which is rarely observed with other nutrients. Furthermore, IDPs were also shown to have effects on a broad range of radicals.

Possible reasons that few researchers utilize EMADD to study antioxidants may include that this method requires the careful purification of phage DNA (removal of EDTA, which is a strong antioxidant) and is a more complicated procedure that requires electrophoresis and gel staining, rather than simple chemical methods. Initially, we used a 0.8% agarose gel with moderate hardness; however, electrophoretic separation could be optimized with harder gels of higher than 1.5% agarose, or a polyacrylamide gel, for visualization of shorter-chain plasmid DNA. The staining solution used in this study is a commercial reagent that does not require UV detection and has a safety profile amenable to food-development laboratories. However, by using ethidium bromide (EtBr), which is a carcinogen, or other similar chemical detection reagents, we can increase the sensitivity of the assay. As an alternative to gel electrophoresis, gel filtration chromatography can be implemented in EMADD, which could enable an automated high-performance process that can analyze large numbers of samples.

As the EMADD method is reliant the analysis of double-stranded DNA length, it is insensitive to nucleotide nicking. Hence, a new method employing electrophoresis of single-stranded DNA could improve the sensitivity of the assay. To further develop the field, more technological advances are required.

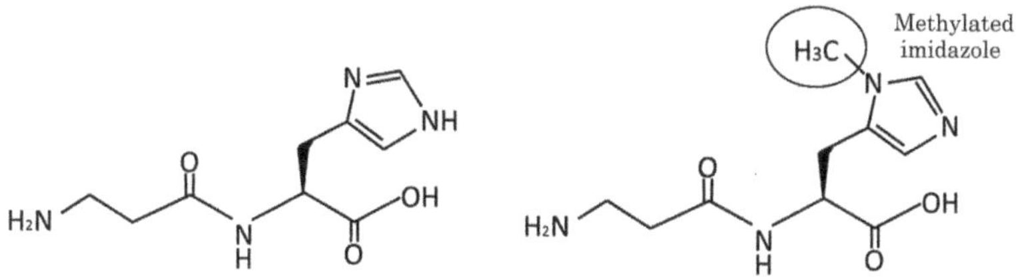

Figure 9.9 Structural formulas of carnosine and anserine. The difference between carnosine (left) and anserine (right) is the methylated imidazole heterocyclic group of β-alanyl-L-histidine dipeptides.

The results presented in this chapter showed the characterization of particular nutrients, to establish a basis for IDPs having comprehensive protective effects against ROS, as well as a scientific basis for traditional, proverbial recommendations that are orally communicated in many tribal cultures around the world.

The fact that anserine is simply methylated carnosine (Figure 9.9) may provide insight its antioxidative mechanism, as this minor modification results in vastly altered antioxidative characteristics. To further our understanding of the molecular orbits of these dipeptides, it is expected that subsequent studies will utilize quantum chemistry techniques to study the reaction intermediate of their metabolites.

Summary Points

- This chapter focuses on imidazole dipeptides (IDP) and a novel analytical method for antioxidants.
- The present method (EMADD) has been designed to determine the effect of compounds in preventing reactive oxygen species (ROSs)-induced DNA degradation. Individual antioxidants have a specific effect against each type of ROS.
- IDPs are naturally occurring antioxidants that are present in animal muscle and can prevent DNA degradation caused by a variety of ROS, including H_2O_2, $ClO^{\cdot}$, $OH^{\cdot}$ and $ONOO^{\cdot}$.
- Chicken meat extract is rich in two IDPs, carnosine and anserine, which have broad inhibitory effects on all four ROS.
- In recent years there has been a unification of methods used to estimation the effects of antioxidants, *e.g.* ORAC, resulting in evaluations being preferentially biased to hydrophobic over hydrophilic antioxidants. Therefore, the antioxidant effect of IDPs was estimated to be lower than their actual biological function.
- We argue that it is unsuitable to utilize a single standardized estimation method, such as ORAC, to formulate recommendations regarding nutrient intake.

Key Facts

1. The novel method for estimating the antioxidant activity of nutrients can distinguish their capacity to prevent endogenous radical-induced DNA degradation. EMADD: estimation method for antioxidants with DNA degradation.
2. EMADD can be applied for determining protective effects against $OH^{\cdot}$, $ONOO^{\cdot}$, $ClO^{\cdot}$, and hydrogen oxide.
3. Imidazole dipeptides, *e.g.* carnosine and anserine, have suppressive effects against a broad range of reactive oxygen species (ROS), including $ClO^{\cdot}$ and hydrogen oxide, and anserine in particular can prevent DNA degradation by all four ROS.
4. The methylated imidazole of Ans differs from Car in that it has a much broader range of antioxidant effects against various ROS.
5. Ferulic acid has the strongest effect against the $OH^{\cdot}$ radical, while vitamin C and the imidazole dipeptides are the most potent against the $ONOO^{\cdot}$ radical.
6. The consumption of various antioxidant ingredients in foods is recommended. Only IDP can be used alone from the viewpoint of antioxidant science.
7. Consumption of high doses of vitamin C could cause DNA degradation.

Definitions of Words and Terms (Abbreviations)

ACmix	Mixture of carnosine and anserine purified from chicken extract
Ans	Anserine
Ast	Astaxanthin
Car	Carnosine
$ClO^{\cdot}$	Hypochlorous radical of ROS
CoQ10	Coenzyme Q10
EMADD	Estimation method for antioxidants with DNA degradation
EDTA	Ethylenediaminetetraacetic acid
GSH	Glutathione, reductive form
H_2O_2	Hydrogen peroxide of ROS
IDP	Imidazole dipeptides
Kbp: Kilo base pairs:	One thousand DNA nucleotides
M	Molar, moles per liter
min	Minutes
mL	Milliliter
mPa	Millipascal, 0.001 N m^{-2}
NaOH	Sodium hydroxide
$ONOO^{\cdot}$	Peroxynitrite radical of ROS
$OH^{\cdot}$	Hydroxide radical of ROS
PBS	Phosphate-buffered saline, *i.e.* Dulbecco's buffered saline

ROS	Reactive oxygen species
V.C	Vitamin C
V.E	Vitamin E
μL	Microliter

Acknowledgements

This work was supported in part by research grants for a contracted research project "Development of evaluation and management methods for supply of safe, reliable, and functional food and farm produce." of the Ministry of Agriculture, Forestry and Fisheries of Japan. It was also carried out in collaboration with Tokai Bussan Co., Ltd.

References

Aruoma, O. I., Laughton, M. J. and Halliwell, B., 1989. Carnosine, homocarnosine and anserine: could they act as antioxidants in vivo? *Biochemistry Journal*. 264: 863–869.

Atalay, M., Lappalainen, J. and Sen, C. K., 2006. Dietary antioxidants for the athlete. *Current Sports Medicine Reports*. 5: 182–186.

Beckman, K. B. and Ames, B. N., 1998. The free radical theory of aging matures. *Physiological Reviews*. 78: 547–581.

Boldyrev, A. A., 2012. Carnosine: new concept for the function of an old molecule: a review. *Biochemistry (Moscow)*. 77: 313–326.

Brown, C. E., 1981. Interactions among carnosine, anserine, ophidine and copper in biochemical adaptation. *Journal of Theoretical Biology*. 88: 245–256.

Chen, Q. and Ames, B. N., 1994. Senescence-like growth arrest induced by hydrogen peroxide in human diploid fibroblast F65 cells. *Proceedings of the National Academy of Sciences of the United States of America*. 91: 4130–4134.

Choi, S. Y., Kwon, H. Y., Kwon, O. B. and Kang, J. H., 1999. Hydrogen peroxide-mediated Cu, Zn-superoxide dismutase fragmentation: protection by carnosine, homocarnosine and anserine. *Biochimica et Biophysica Acta*. 1472: 651–657.

Devasagayam, T. P., Tilak, J. C., Boloor, K. K., Sane, K. S., Ghaskadbi, S. S. and Lele, R. D., 2004. Free radicals and antioxidants in human health: current status and future prospects. *Journal of the Association of Physicians of India*. 52: 794–804.

Dröge, W., 2002. Free radicals in the physiological control of cell function. *Physiological Reviews*. 82: 47–95.

Food and Nutrition Board, Institute of Medicine, 2002. "Dietary Reference Intakes (DRIs): Estimated Average Requirements" (PDF). United States National Academy of Sciences. Available at: http://dsld.nlm.nih.gov/dsld/docs/Dietary_Reference_Intakes_Recommended_Intakes_for_Individuals.pdf. Accessed 27 November 2013.

Harman, D., 1956. Aging: a theory based on free radical and radiation chemistry. *Journals of Gerontology*. 11: 298–300.

Hipkiss, A. R., Brownson, C. and Carrier, M. J., 2001. Carnosine, the anti-ageing, anti-oxidant dipeptide, may react with protein carbonyl groups: a review. *Mechanisms of Ageing and Development*. 122: 1431–1445.

Hirayama, O., Takagi, M., Hukumoto, K. and Katoh, S., 1997. Evaluation of antioxidant activity by chemiluminescence. *Analytical Biochemistry*. 247: 237–241.

Hossain, M. N., Sakemura, R., Fujii, M. and Ayusawa, D., 2006. G-protein gamma subunit GNG11 strongly regulates cellular senescence. *Biochemical and Biophysical Research Communications*. 351: 645–650.

Kohen, R., Yamamoto, Y., Cundy, K. C. and Ames, B. N., 1988. Antioxidant activity of carnosine, homocarnosine, and anserine present in muscle and brain. *Proceedings of the National Academy of Sciences USA*. 85: 3175–3179.

MacCord, J. M., 2002. Superoxide dismutase in aging and disease: an overview. *Methods in Enzymology*. 349: 331–341.

Prior, R. L., Hoang, H., Gu, L., Wu, X., Bacchiocca, M., Howard, L., Hampsch-Woodill, M., Huang, D., Ou, B. and Jacob, R., 2003. Assays for hydrophilic and lipophilic antioxidant capacity (oxygen radical absorbance capacity (ORAC(FL))) of plasma and other biological and food samples. *Journal of Agricultural and Food Chemistry*. 51: 3273–3279.

Radi, R., Beckman, J. S., Bush, K. M. and Freeman, B. A., 1991. Peroxynitrite oxidation of sulhydryls. The cytotoxic potential of superoxide and nitric oxide. *Journal of Biological Chemistry*. 266: 4244–4259.

Sambrook, J. and Russell, D. W., 2001. *Molecular Cloning: A Laboratory Manual*, 3rd edn. (3 volume set). Cold Spring Harbor Laboratory Press, New York, USA.

Seifried, H. E., Anderson, D. E., Fisher, E. I. and Milner, J. A., 2007. A review of the interaction among dietary antioxidants and reactive oxygen species. *Journal of Nutritional Biochemistry*. 18: 567–579.

Son, D. O., Satsu, H., Kiso, Y., Totsuka, M. and Shimizu, M., 2008. Inhibitory effect of carnosine on interleukin-8 production in intestinal epithelial cells through translational regulation. *Cytokine*. 42: 265–276.

USDA: United States Department of Agriculture, Agricultural Research Service, 2012. Oxygen Radical Absorbance Capacity (ORAC) of Selected Foods, Release 2. Available at: http://www.ars.usda.gov/services/docs.htm?docid=15866. Accessed 13 October 2013.

Yanai, N., Shiotani, S., MIzuno, M., Nabetani, H. and Nakajima, M., 2004. Characteristics of anti-oxidative activity of carnosine and anserine mixture isolated from chicken extract: comparison with other botanical antioxidants. *Nippon Shokuhin Kagaku Kogaku Kaishi*. 51: 238–246.

Yanai, N., Shiotani, S., Hagiwara, S., Nabetani, H. and Nakajima, M., 2008. Antioxidant combination inhibits reactive oxygen species mediated damage. *Bioscience, Biotechnology, and Biochemistry*. 72: 3100–3106.

Yoon, J. H., Lee, M. S. and Kang, J. H., 2010. Reaction of ferritin with hydrogen peroxide induces lipid peroxidation. *BMB Reports*. 43: 219–224.

von Zglinicki, T., Saretzki, G., Döcke, W. and Lotze, C., 1995. Mild hyperoxia shortens telomeres and inhibits proliferation of fibroblasts: a model for senescence? *Experimental Cell Research*. 220: 186–193.

Section III
Analysis

CHAPTER 10

Liquid Chromatography Coupled to Tandem Mass Spectrometry to Analyze Imidazole Dipeptides

MARIA-JOSÉ MOTILVA*[a], ALBA MACIÀ[a], ALBA DOMÍNGUEZ[b], AND AGUSTÍN LABRADOR[b]

[a]Food Technology Department, Agrotecnio Research Center, Escola Tecnica Superior d'Enginyeria Agraria (ETSEA), University of Lleida, Av. Alcalde Rovira Roure 191, 25198 Lleida, Spain; [b]Preparados Alimenticios S.A. subsidiary of Gallina Blanca Star Group. Plaza Europa 41, 08908 L'Hospitalet de Llobregat, Barcelona, Spain
*E-mail: motilva@tecal.udl.cat

10.1 Introduction

Imidazole dipeptides, such as carnosine (β-alanyl-L-histidine) and the related derivatives, such as anserine (*N*-β-alanyl-3-methyl-L-histidine), homocarnosine (γ-amino-butyryl-histidine) and balenine (*N*-β-alanyl-1-methyl-histidine) are known as bioactive dipeptides due to their potential health benefits (Kovacs-Nolan and Mine, 2010).

These dipeptides have an imidazole ring and they have a pK_a values very close to the physiological pH. These compounds have low octanol–water

Food and Nutritional Components in Focus No. 8
Imidazole Dipeptides: Chemistry, Analysis, Function and Effects
Edited by Victor R Preedy

Published by the Royal Society of Chemistry, www.rsc.org

partition coefficient (log$P$, o/w), and this demonstrates the hydrophilicity of these compounds (United States National Library of Medicine, 2012). In humans, the diet is the main source of carnosine and its derivatives, which are found in significant amounts in beef, chicken and pork, as the red and white meats, and in fish (Gil-Agustí *et al.*, 2008).

Liquid chromatography (LC) coupled to tandem mass spectrometry (MS/MS) are a powerful analytical separation technique and detector system, respectively, that allows the analysis of compounds with good peak resolution, with high selectivity and sensitivity. The potential of this coupling for the analysis of imidazole dipeptides was first demonstrated in 2004 by Aldini *et al.* (2004). These authors used MS/MS with quadrupole-ion trap, as the high-resolution MS, for the determination of these compounds (carnosine, anserine, homocarnosine and balenine) in biological matrices, such as plasma and tissues from rats. Since that first study, the application of LC-MS/MS to the analysis of these dipeptides has been increased and different LC technologies, chromatographic modes and analyzers have been reported for the determination of these dipeptides in different matrices, such as meat samples and biological samples (Jansen *et al.*, 2006; Chen *et al.*, 2007; Yeum *et al.*, 2010; Peiretti *et al.*, 2011; Su *et al.*, 2011; Waters™, 2011; Macià *et al.*, 2012; Waters™, 2013).

In this chapter, the published literature for the analysis of imidazole dipeptides by LC coupled to MS/MS is reviewed. Special attention is paid to the different analyzers used. Additionally, the determination of these dipeptides in different matrices, such as meat and biological samples, is also focused upon, and the sample pretreatments needed in each sample matrix are examined.

10.2 Analysis of Imidazole Dipeptides

Table 10.1 shows the methods reported in the literature for the analysis of imidazole dipeptides by LC coupled to MS/MS. Figure 10.1 shows a schematic representation of these methodologies for the determination of imidazole dipeptides in meat and biological samples, and the chemical structures of these compounds. In the next two sections, the chromatographic conditions, the analytical columns and the mobile phases, and the MS/MS as the detector system used for the analysis of imidazole dipeptides are reviewed.

10.2.1 Chromatographic Conditions

10.2.1.1 Analytical Column

The reversed-phase (RP) is the chromatographic mode mainly used for the analysis of imidazole dipeptides by LC-MS/MS (Aldini *et al.*, 2004; Jansen *et al.*, 2006; Chen *et al.*, 2007; Yeum *et al.*, 2010; Peiretti *et al.*, 2011; Su *et al.*, 2011). In this chromatographic mode, the dipeptides are resolved according to their degree of hydrophobicity (apolarity). Therefore, first eluted anserine and

Table 10.1 Sample pretreatment and LC-MS/MS conditions for the analysis of imidazole dipeptides. The different sample pretreatment and LC-MS/MS conditions reported in the literature for the analysis of imidazole dipeptides.

Imidazole dipeptides	LC chromatographic mode	Stationary phase	Mobile phase	LC-detector system	Application	Sample pre-treatment	Extraction recovery (%R)	References
Meat samples								
Carnosine and anserine	Reversed-phase	Superspher RP-18 (150 × 2 mm, 4 µm)	Gradient mode (A) 0.1% formic acid (B) ACN + 0.1% formic acid	HPLC-MS/MS (TQD)	Chicken essences	Derivatization	—	Chen *et al.* (2007).
Carnosine, anserine and homo-carnosine	Reversed-phase	Sinergy polar-RP (150 × 2.1 mm, 3 µm)	Gradient mode (A) 5 mM HFBA (B) ACN	HPLC-MS (ion trap–orbitrap)	Meat	Protein pre-cipitation	—	Peiretti *et al.* (2011).
Carnosine and anserine	HILIC	BEH HILIC (100 × 2.1 mm, 1.7 µm)	Gradient mode (A) 0.65 mM ammonium acetate with 75% ACN (B) 4.55 mM ammonium acetate with 30% ACN	UPLC-MS/MS (TQD)	Chicken broth	Off-line SPE Off-line µSPE	80–90	Macià *et al.* (2012).

(*continued*)

Table 10.1 (*continued*)

Imidazole dipeptides	LC chromatographic mode	Stationary phase	Mobile phase	LC-detector system	Application	Sample pre-treatment	Extraction recovery (%R)	References
Biological samples								
Homocarnosine	Reversed-phase	Symmetry C_{18} (150 × 3.9 mm, 5 µm)	Isocratic mode 25 mM ammonium acetate in water/ MeOH (50/50, v/v)	HPLC-MS/MS (TQD)	Cerebrospinal fluid	Protein precipitation derivatization	105	Jansen *et al.* (2006).
Carnosine, anserine, balenine and homocarnosine	Reversed-phase	Sinergy polar-RP C_{18}	Isocratic mode Water/ACN/HFBA (90/10/0.1, v/v/v)	LC-MS/MS (quadrupole-ion trap)	Rat plasma Rat tissues	Protein precipitation	96.2–101.3	Aldini *et al.* (2004).
Carnosine and anserine	Reversed-phase	Sinergy polar-RP C_{18} (150 × 2 mm, 4 µm)	Isocratic mode Water/ACN/HFBA (90/10/0.1, v/v/v)	HPLC-MS/MS (TQD)	Plasma Urine	Protein precipitation	96–101	Yeum *et al.* (2010).
Carnosine	Reversed-phase	BEH C_{18} (50 × 2.1 mm, 1.7 µm)	Isocratic mode 0.2% formic acid/ACN (90/10, v/v)	UPLC-MS/MS (TQD)	Rabbit aqueous humor	Microdialysis	61	Su *et al.* (2011).

No specific application								
Carnosine and anserine	HILIC	BEH HILIC (50 × 2.1 mm, 1.7 µm)	Gradient mode (A) 10 mM ammonium formate with 0.125% formic acid at pH 3/ACN (5/95, v/v) (B) 10 mM ammonium formate with 0.125% formic acid at pH 3/ACN (50/50, v/v)	UPLC-MS/MS (TQD)	—	—	—	Waters™ (2011)
Carnosine and anserine	HILIC	CORTECS HILIC (50 × 2.1 mm, 1.6 µm)	Gradient mode (A) 10 mM ammonium formate with 0.125% formic acid/ACN (5/95, v/v) (B) 10 mM ammonium formate with 0.125% formic acid/ACN (50/50, v/v)	UPLC-MS/MS (TQD)	—	—	—	Waters™ (2013)

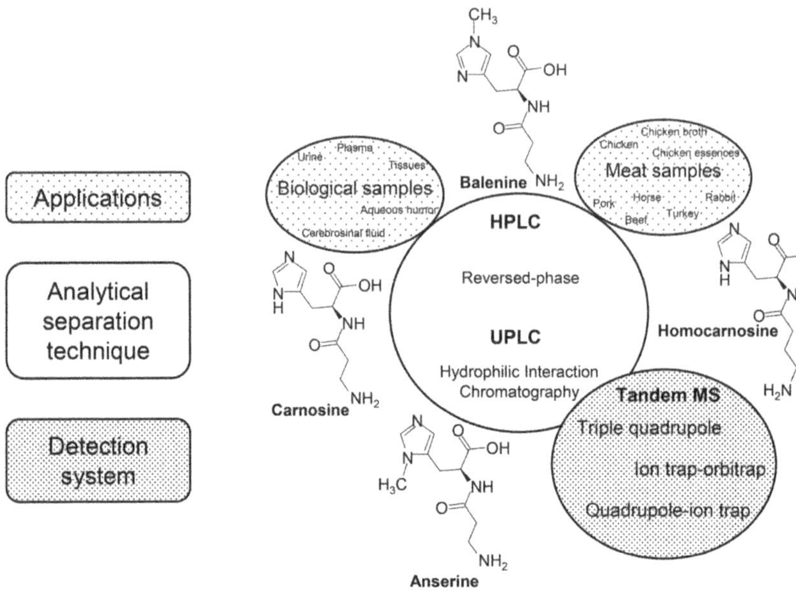

Figure 10.1 Schematic representation of the LC-MS/MS methods reported in the literature for the determination of imidazole dipeptides in meat and biological samples. The figure shows a schematic representation of the different methodologies reported in the literature for the analysis of imidazole dipeptides. These dipeptides have been analysed by high-performance liquid chromatography (HPLC) and ultraperformance liquid chromatography (UPLC) as the analytical separation technique. Reversed-phase and hydrophilic interaction chromatography are the chromatographic modes used by HPLC and UPLC, respectively. Triple quadrupole, ion trap–orbitrap and quadrupole-ion trap are the analyzers used in MS detection system. These methodologies have been applied for the analysis of biological and meat samples.

then carnosine, because anserine is more polar than carnosine, see Table 10.2. This elution order was observed when the stationary phase was RP-18 and the reported analytical columns were Superspher RP-18 (Chen *et al.*, 2007), Symmetry C_{18} (Jansen *et al.*, 2006) and Bridged ethylene hydride (BEH) C_{18} (Su *et al.*, 2011). Imidazole dipeptides are hydrophilic compounds that have low retention when this RP-18 stationary phase is used. This drawback is overcome with the addition of a derivatization reagent previous to the chromatographic separation (precolumn derivatization). Derivatization reagents play an important role in increasing the hydrophobicity of compounds that have a low retention in RP, such as imidazole dipeptides. As derivatization agents, butanol with hydrochloric acid (Chen *et al.*, 2007) and dabsyl-Cl (Jansen *et al.*, 2006) were reported for the analysis of carnosine and anserine (Chen *et al.*, 2007) and homocarnosine (Jansen *et al.*, 2006) in order to increase the retention of these dipeptides in the stationary phase RP-18. The main derivatives of anserine and carnosine with the derivatization reagent dabsyl-Cl

Table 10.2 Elution chromatographic order of imidazole dipeptides. A schematic summary of the elution chromatographic order of imidazole dipeptides reported in the literature for the analysis of these compounds by UPLC-MS/MS.

Chromatographic mode	Stationary phase	Characteristics	Elution order	Analytical columns	Reference
Reversed-phase (RP)	RP-18	Eluted according to their hydrophobicity (apolarity)	Anserine > carnosine > balenine > homocarnosine	Superspher RP-18 Symmetry C_{18} Bridged ethylene hybrid (BEH) C_{18}	Chen *et al.* (2007). Jansen *et al.* (2006). Su *et al.* (2011).
	Polar-RP	Increased retention of polar compounds	Carnosine > anserine > balenine > homocarnosine	Synergy polar-RP	Peiretti *et al.* (2011). Aldini *et al.* (2004). Yeum *et al.* (2010).
Hydrophilic interaction chromatography (HILIC)	Silica	Improved retention of the polar compounds	Carnosine > balenine > anserine	BEH HILIC	Macià *et al.* (2012). Waters™ (2011)
		Eluted according to their polarity		CORTECS HILIC	Waters™ (2013)

were generated to react the primary and secondary amines of these dipeptides with one and two dabsyl groups, respectively (Jansen *et al.*, 2006).

Another strategy reported to increase the retention of these compounds by RP is the use of stationary phases with polar groups, and the reported analytical column is Sinergy polar-RP C_{18} (Aldini *et al.*, 2004; Yeum *et al.*, 2010; Peiretti *et al.*, 2011). The addition of polar groups to the stationary phase allowed the retention of polar compounds, such as imidazole dipeptides, to be increased. As a result, the elution order of these compounds was changed in comparison with the use of RP-18 columns, and the compounds were eluted according to their degree of polarity. So, carnosine eluted first and then anserine (see Table 10.2).

All of these RP methodologies were carried out by conventional LC, high-performance LC (HPLC), and the stationary phases had a particle size of 3 μm (Peiretti *et al.*, 2011), 4 μm (Aldini *et al.*, 2004; Chen *et al.*, 2007; Yeum *et al.*, 2010), or 5 μm (Jansen *et al.*, 2006), an inner diameter of 4.6 mm and a column length of 150 or 250 mm. An exception was the proposed method by Su *et al.* (2011), who used the ultraperformance LC (UPLC) chromatography for the analysis of carnosine, and the analytical column was BEH C_{18} (50 × 2.1 mm, 1.7 μm). The chromatographic technique of UPLC has been developed in the last few years as a result of the improvements made in the packing chromatographic material of the analytical columns. This LC technology uses columns packed with smaller size particles, sub-2 μm, the length is shorter (from 50 to 100 mm), and the inner diameter is smaller (2.1 mm) than those reported in conventional HPLC. The fact of using these analytical columns packed with smaller size particles (sub-2 μm), resulted in higher peak efficiency and peak resolution, shorter analysis time and lower solvent consumption than the conventional LC. Nevertheless, this technique needs special chromatographic equipment because it must operate at high backpressure, up to 15 000 psi (Mazzeo *et al.*, 2005; Swartz, 2005*a*; Swartz, 2005*b*).

Although RP has a great powerful separation, this chromatographic mode is limited for the analysis of polar compounds, such as imidazole dipeptides, due to its low chromatographic retention, as has been commented before. The chromatographic mode named hydrophilic interaction chromatography (HILIC) has gained in popularity in the last decade because it has been reported to be an interesting alternative to RP, ion-exchange and normal-phase mode for the analysis of polar compounds. HILIC mode uses stationary phases of silica (Grumbach *et al.*, 2004; Roberto-*Gama*, 2012).

Recently, the HILIC mode has been reported for the analysis of imidazole dipeptides by using the UPLC technology, and the analytical columns used were BEH HILIC (100 × 2.1 mm, 1.7 μm) (Macià *et al.*, 2012) and (50 × 2.1 mm, 1.7 μm) (Waters™, 2011), and CORTECS HILIC (100 × 2.1 mm, 1.7 μm) (Waters™, 2013). By using these analytical columns, these dipeptides were eluted according to their degree of polarity or hydrophilicity, and thus, carnosine eluted first and then anserine (see Table 10.2).

An important characteristic of the almost stationary phases reported in the literature for the analysis of imidazole dipeptides in RP and HILIC mode is that they are packed with totally porous silica particles. Recently, a new generation of analytical columns has been developed and they are known as core–shell columns. These columns have a solid core and a porous shell (Guiochon and Gritti, 2011; Fekete *et al.*, 2012). In the literature, one methodology has been reported for the analysis of imidazole dipeptides (carnosine and anserine) by using this type of column, and in addition, the HILIC mode and UPLC technology were used (Waters™, 2013). The column was named CORTECS HILIC (50 × 2.1 mm, 1.6 µm). With the use of these columns (core–shell), the longitudinal diffusion (*B* coefficient), and the eddy diffusion (*A* coefficient) of the Van Deemter equation, are reduced compared to fully porous silica particles. The reductions of these coefficients have been reported to improve the peak resolution, sensitivity and peak capacities (Guiochon and Gritti, 2011; Fekete *et al.*, 2012).

10.2.1.2 Mobile Phase

For the analysis of imidazole dipeptides, the mobile phase is composed by acidified water, as eluent A, and organic modifier, as eluent B. The pH of the eluent A of the mobile phase is normally kept below 7, which is the pK_a values of the imidazole dipeptides. This pH is achieved by the addition of a small amount of acetate buffer (Jansen *et al.*, 2006), formate buffer (Waters™, 2011, Waters™, 2013), formic acid (Chen *et al.*, 2007; Su *et al.*, 2011), and perfluorinated carboxylic acid with longer *n*-alkyl chains, such as heptafluorobutyric acid (HFBA) (Aldini *et al.*, 2004; Yeum *et al.*, 2010; Peiretti *et al.*, 2011). The HFBA is an ion-pairing agent, and it has been reported to be a satisfactory alternative to trifluoroacetic acid (TFA) for the analysis of polar peptides (Pearson and McCroskey, 1996). The concentration of these acidic modifiers was from 0.5 to 25 mM for the acetate and formate buffers, and from 0.1 to 0.2% for the acids. It is important to keep the concentration of these acidic modifiers as low as possible in order to ensure satisfactory ionization of analytes (imidazole dipeptides).

As organic modifier (eluent B), methanol (Jansen *et al.*, 2006) and acetonitrle (Aldini *et al.*, 2004; Chen *et al.*, 2007; Yeum *et al.*, 2010; Peiretti *et al.*, 2011; Su *et al.*, 2011; Waters™, 2011; Macià *et al.*, 2012; Waters™, 2013) have been reported, with acetonitrile the most used. The separation of imidazole dipeptides was accomplished by both, isocratic (Aldini *et al.*, 2004; Jansen *et al.*, 2006; Yeum *et al.*, 2010; Su *et al.*, 2011) or gradient elution (Chen *et al.*, 2007; Peiretti *et al.*, 2011; Waters™, 2011; Macià *et al.*, 2012; Waters™, 2013).

When the HILIC mode was used, the mobile phase was made up of 0.65–10 mM ammonium acetate or formiate with a high amount of acetonitrile, as the eluent A; and 4.55–10 mM ammonium acetate or formiate with low amount of acetonitrile, as eluent B. In this chromatographic mode, water is introduced as the stronger eluting solvent instead of polar organic phase as in RP.

One of the more interesting advantages of this chromatographic mode (HILIC) is that it allows the retention of polar compounds, by using an environmentally friendly mobile phase, instead of normal-phase mode, which requires the use of toxic chlorinated solvents that are considered an ecological hazard. In addition, the fact of using acetonitrile-rich mobile phases provide favorable conditions for efficient droplet formation and desolvation within the MS source, resulting in improved sensitivity compared to the highly aqueous mobile phases used in RP (Grumbach *et al.*, 2004).

10.2.2 MS/MS as the Detector System

In the last few years, the use of MS/MS as the detector system for the analysis of imidazole dipeptides by LC has been considerably increased. These methodologies have been reported by using both the conventional HPLC (Aldini *et al.*, 2004; Jansen *et al.*, 2006; Chen *et al.*, 2007; Yeum *et al.*, 2010; Peiretti *et al.*, 2011) and the improved UPLC (Su *et al.*, 2011; Waters™, 2011; Macià *et al.*, 2012; Waters™, 2013) chromatographic systems.

The MS/MS detector system separates the analytes according to their *m*/*z* ratio, allowing the identification of individual compounds from a mixture in complexes matrices, such as meat and biological samples. Therefore, the use of this detector gives a high sensitivity and selectivity in comparison with others systems, such as UV, fluorescence or electrochemical detectors.

For the imidazole dipeptides analysis the ionization technique used has been electrospray (ESI) in the positive mode. In ESI, which is a soft ionization technique, the precursor or parent ion is obtained, and this ion is *m*/*z* 227 for carnosine, and *m*/*z* 241 for anserine, homocarnosine and balenine. These three dipeptides have the same nominal precursor ion and they are isobaric compounds.

Aldini *et al.* (2004) were the first authors that reported a LC-MS/MS method for the analysis of imidazole dipeptides. In this study, carnosine, anserine, homocarnosine and balenine were analyzed in rat plasma and rat tissues. Since then, different studies have been reported for the analysis of these compounds by using MS/MS as the detector system. These studies are based on a triple quadrupole (TQD) (Jansen *et al.*, 2006; Chen *et al.*, 2007; Yeum *et al.*, 2010; Peiretti *et al.*, 2011; Su *et al.*, 2011; Waters™, 2011; Macià *et al.*, 2012; Waters™, 2013), which uses the same two analyzers, and quadrupole-ion trap (Aldini *et al.*, 2004) and linear ion trap (LTQ)–orbitrap (Peiretti *et al.*, 2011), as the hybrid systems.

Among the different MS/MS techniques reported in the literature for the analysis of imidazole dipeptides, triple quadrupole (TQD) instruments are the most widely employed (Jansen *et al.*, 2006; Chen *et al.*, 2007; Yeum *et al.*, 2010; Su *et al.*, 2011; Waters™, 2011; Macià *et al.*, 2012; Waters™, 2013). This instrument uses two quadrupoles (Q_1 and Q_3) as the analyzers, and they are united in series. The second quadrupole (Q_2) is used as a collision cell and it contains an inert gas, such as argon. Then, in this cell, fragment ions or product ions can be generated by collision between the target analytes with the argon gas molecules. This system can be run in one of four modes for a variety of experiments, according to if the first or second quadrupole runs in full-scan mode or selected ion monitoring (SIM) mode. These experiments are daughter scan mode, parent scan mode, neutral scan mode, and the most sensitive selective reaction monitoring (SRM). The last one is used for quantification purposes and the others are used to fragment the molecule and thus identify an analyte in complex matrices, such as meat and biological samples.

On the other hand, when the two analyzers are not the same, the system is named a hybrid instruments. Aldini *et al.* (2004) and Peiretti *et al.* (2011) used quadrupole-ion trap and linear ion trap–orbitrap, respectively, as the hybrid instruments, for the analysis of imidazole dipeptides in biological and meat samples, respectively. These instruments combine the advantages of both different mass analyzers, and offer a great potential for the screening, the selectivity, the confirmation, and above all, the structural elucidation of unknown compounds in complex matrices, because an accurate mass of a characteristic fragment is given. Due to its great potential, these techniques are known as high-resolution (HR) MS techniques, and these have emerged as a perfect tool for profiling complex samples, such as foods and biological samples. In the studies reported by Aldini *et al.* (2004) and Peiretti *et al.* (2011), who used the HR-MS and the hybrid systems quadrupole-ion trap and linear ion trap–orbitrap, respectively, the isobaric compounds homocarnosine, anserine and balenine could be successfully resolved. The identification of the isobaric compounds anserine and homocarnosine was unequivocally achieved with the MS^2 fragmentation (Aldini *et al.*, 2004; Peiretti *et al.*, 2011). Nevertheless, MS^3 fragmentation was required for the resolution of the isobaric compounds anserine and balenine (Aldini *et al.*, 2004). Figure 10.2 shows a schematic representation for the analysis of carnosine, and for the identification of the isobaric compounds anserine, homocarnosine and balenine by using MS^n experiments. The chemical structures of the precursor ion and the different generated product ions reported in the literature for carnosine, anserine, homocarnosine and balenine are shown in Figure 10.3 (Aldini *et al.*, 2004; Yeum *et al.*, 2010; Peiretti *et al.*, 2011; Su *et al.*, 2011; Macià *et al.*, 2012).

As a conclusion, the use of the HR-MS techniques, such as quadrupole-ion trap and linear ion trap–orbitrap, is a good strategy for the analysis of isobaric compounds, their identification being often a difficult task. This strategy is much more reliable than the methodologies that use low-resolution MS, such as a single quadrupole, or above all, UV and fluorescence, as the detection systems. When low-resolution MS with a quadrupole is used for the analysis of isobaric compounds, it is essential to obtain a good peak resolution in order to select correctly the precursor ion. If these compounds are not well resolved in the first analyzer, mixed product ions (fragments) are obtained and this leads to incorrect conclusions and incorrect quantifications.

10.3 Instrumental Quality Parameters

Almost all of the developed LC-MS/MS methods reported in the literature for the analysis of imidazole dipeptides have been validated, and the quality parameters of the method been studied. These are linearity, precision,

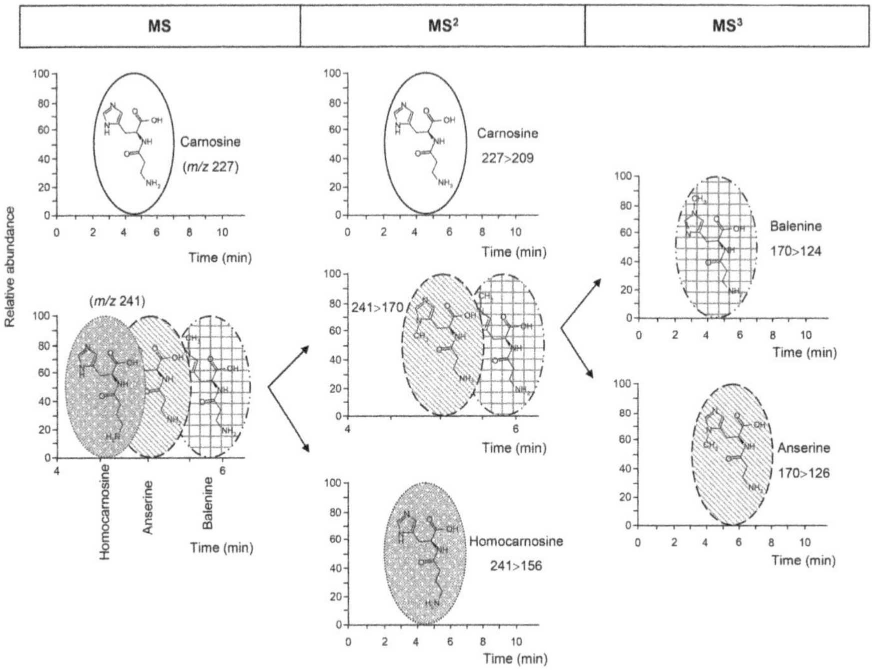

Figure 10.2 LC-MS and LC-MS^n for the analysis of carnosine, anserine, homocarnosine and balenine, in positive-ion mode. The MS experiment shows carnosine at m/z 227, and the coeluted homocarnosine, anserine and balenine at m/z 241. The MS^2 shows the selected reaction monitoring (SRM) of carnosine (227 > 209) and homocarnosine (241 > 156). Anserine and balenine coeluted. The MS^3 shows the SRM of balenine (170 > 124) and (170 > 126). The figure is adapted from Aldini *et al.* (2004). The figure shows the HPLC-MS^n strategies for the analysis of carnosine, anserine, homocarnosine and balenine by using quadrupole-ion trap. The MS experiments (on the left) show carnosine was resolved, but homocarnosine, anserine and balenine were co-eluted. The MS^2 experiments (in the middle) show homocarnosine was resolved because it has different product ion with respect to anserine and balenine, which were still coeluted. The MS^3 experiments (on the right) show balenine and anserine were resolved because they have different fragment ions from the product ion m/z 170. The figure is adapted from Aldini *et al.* (2004).

accuracy, detection limit (LOD), quantification limit (LOQ), matrix effect), and stability. The obtained results are shown in Table 10.3.

The imidazole dipeptides, carnosine, anserine and homocarnosine are commercially available standards. One exception is balenine, which is not commercially available as standard, and it has been isolated by LC semipreparative from porcine muscle (Aldini *et al.*, 2004). The stock solutions of individual standards are prepared by dissolving each compound in different aqueous solutions, such as phosphate buffer (Aldini *et al.*, 2004; Yeum *et al.*, 2010),

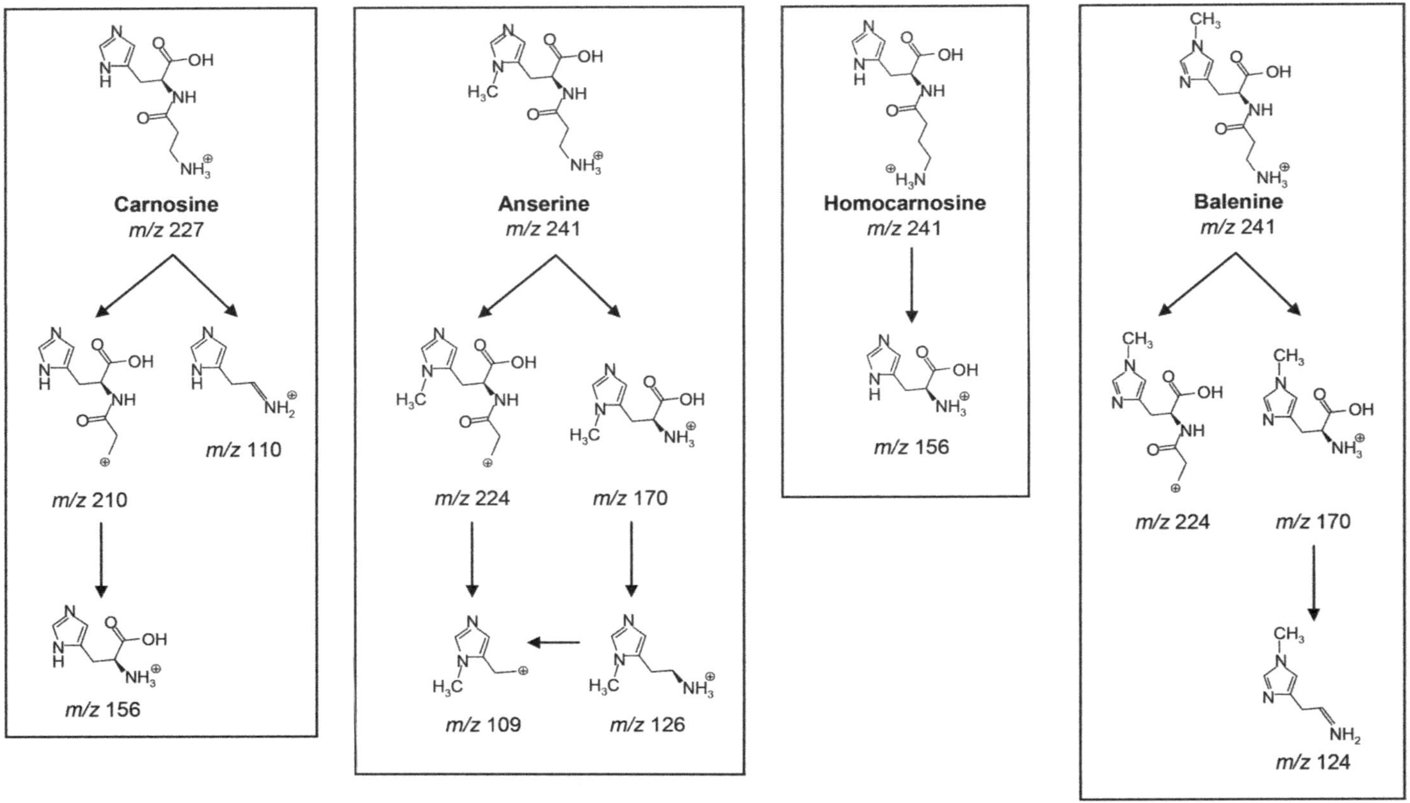

Figure 10.3 Precursor ion of carnosine, anserine, homocarnosine and balenine and their generated product ions reported in the literature by LC-MS/MS. The fragmentation pathways of carnosine and anserine is adapted from Peiretti *et al.* (2011). The figure shows the chemical structures of the precursor ion of carnosine (*m*/*z* 227), anserine (*m*/*z* 241), homocarnosine (*m*/*z* 241), and balenine (*m*/*z* 241), and their generated product ions reported in the literature. The fragmentation pathways of carnosine and anserine are adapted from Peiretti *et al.* (2011).

Table 10.3 Instrumental quality parameters of the LC-MS/MS analytical methods for the analysis of imidazole dipeptides. The instrumental quality parameters of the LC-MS/MS analytical methods for the analysis of imidazole dipeptides reported in the literature.

	Chen *et al.* (2007)	Su *et al.* (2011)	Aldini *et al.* (2004)	Jansen *et al.* (2006)	Yeum *et al.* (2010)	Peiretti *et al.* (2011)	Macià *et al.* (2012)
Imidazole dipeptide	Carnosine, anserine	Carnosine	Carnosine, anserine, homocarnosine, balenine	Homocarnosine	Carnosine, anserine	Carnosine	Carnosine, anserine
Analyzer	QqQ	QqQ	Q-ion trap	QqQ	QqQ	Ion trap–orbitrap	QqQ
Sample matrix	Standard solutions	Standard solutions	Rat lung	Cerebrospinal fluid	Rat plasma and urine	Meat samples	Chicken broth
Linearity	0–250 μM	20.50–1025 ng ml^{-1}	15–1000 nmol g^{-1} wet	0.1–5 nmol	0.5–60 μM	1–200 mg l^{-1}	7–3300 μg l^{-1}
Internal standard (IS)	No	No	H-Tyr–His-OH	Isotope homocarnosine	H-Tyr–His-OH	Phenylephrine hydrochloride	No
LOD	0.15–0.31 μM	—	5 nmol g^{-1}	0.02 μM[b]	0.2 μM	11 μg l^{-1}	0.6–1.8 μg l^{-1}
LOQ	—	0.09 μM	15 nmol g^{-1}	—	0.5 μM	—	2–6 μg l^{-1}
Precision (%RSD)	—	<12%	<14%[a]	10.9%[a]	<14%[a]	—	<14.6%
Accuracy	—	<12%[a]	<10%[a]	—	<10%[a]	—	95–105%
Matrix effect evaluation	No	Yes	Yes	No	No	Yes	Yes
Stability	—	3 freeze–thaw cycles at room temperature, 24 h	24 h, 4 C	—	24 h, 4 C	—	—

[a]Calculated as %CV.
[b]The signal-to-noise ratio was 5.

Milli-Q water (Jansen *et al.*, 2006; Chen *et al.*, 2007; Su *et al.*, 2011), Milli-Q water/methanol (50/50, v/v) (Peiretti *et al.*, 2011), Milli-Q water/acetonitrile (25/75, v/v) (Macià *et al.*, 2012), and acetonitrile/methanol (75/25, v/v) with 0.2% formic acid (Waters™, 2011; Waters™, 2013). In these studies a stock mix of the standards was prepared weekly or daily at a lower concentration. Although phosphate buffer was used to dissolve the standards of imidazole dipeptides, this solution was not the optimum for the use of MS as the detector system due to the low volatility of this buffer. This nonvolatile salt can deposit in the MS source, thus requiring more cleaning and maintenance operations of the detector equipment.

Then, the quality parameters of the reported methods were determined by spiking the aqueous solutions with the imidazole dipeptides at known concentrations (Chen *et al.*, 2007; Peiretti *et al.*, 2011; Su *et al.*, 2011) or by spiking a blank meat (Macià *et al.*, 2012) or blank biological (Aldini *et al.*, 2004; Jansen *et al.*, 2006; Yeum *et al.*, 2010) samples with these dipeptides at known concentrations. As a blank biological matrix, rat lung tissue was used since this did not contained carnosine, anserine and homocarnosine (Aldini *et al.*, 2004), and blank urine was obtained after 60 min incubation with 5% of human serum, in order to remove the dipeptides (Yeum *et al.*, 2010). When a matrix free from the studied dipeptides (blank) could not be obtained, the real sample was used as a blank, and this real standard was spiked with the standard dipeptides. Then, the real concentrations of these dipeptides were subtracted from the spiked concentration (Macià *et al.*, 2012).

Therefore, the linearity of the methods was evaluated by using real sample matrices (meat or biological samples spiked with the dipeptides) or by using aqueous standard solutions spiked with the dipeptide standards. The calibration curves, which were based on peak-area abundance, were plotted using the equation $y = a + bx$, where y is the analyte peak-area abundance and x is the analyte concentration. The concentrations of the imidazole dipeptides were calculated by interpoling their analyte peak area abundance on the calibration curve (Chen *et al.*, 2007; Peiretti *et al.*, 2011; Su *et al.*, 2011; Macià *et al.*, 2012). Each imidazole dipeptide was quantified by using its own calibration curve, with the exception of the report of Peiretti *et al.* (2011), where anserine and homocarnosine were quantified with respect to the calibration curve of carnosine.

On the other hand, in some studies, an internal standard (IS) was used to reduce the inaccuracies in the analysis. As IS were used the H-Tyr–His-OH (Aldini *et al.*, 2004; Yeum *et al.*, 2010), isotope of homocarnosine ($^{2}H_2$-L-homocarnosine) (Jansen *et al.*, 2006), and phenylephrine hydrochloride (Peiretti *et al.*, 2011). When internal calibration was used, the concentrations of the imidazole dipeptides were calculated by interpoling their (analyte peak area/IS peak area) ratio on the calibration curve.

The precision of the reported methods, which was determined by the relative standard deviations (%RSDs) of the concentrations or by the coefficient of variation (%CV), were lower than 15%. The accuracy, which was calculated from the ratio between the concentrations of the imidazole dipeptides found compared to the spiked concentration, and then this quotient multiplied by

100, was between 95 and 105% (Macià *et al.*, 2012), or also as %CV, and this was lower than 12% (Aldini *et al.*, 2004; Yeum *et al.*, 2010; Su *et al.*, 2011).

The sensitivity of the developed methods was evaluated by the LODs and LOQs. These were calculated using the signal-to-noise criterion of 3 and 10, respectively. The LODs of the reported methodologies were between 0.15 and 0.31 μM for the analysis of imidazole dipeptides in standard solutions (Chen *et al.*, 2007), between 0.6 and 11 μg l^{-1} in meat samples (Peiretti *et al.*, 2011; Macià *et al.*, 2012), and 0.2 μM in biological fluids (Jansen *et al.*, 2006; Yeum *et al.*, 2010) and 5 nmol g^{-1} in rat lung (Aldini *et al.*, 2004). On the other hand, the LOQs of the reported methodologies were 0.09 μM for the analysis of these dipeptides in standard solutions (Su *et al.*, 2011), between 2.0 and 6.0 μg l^{-1} in meat samples (Macià *et al.*, 2012), and 0.5 μM in biological fluids (Yeum *et al.*, 2010), and 15 nmol g^{-1} in rat lung (Aldini *et al.*, 2004) (see Table 10.3). Due to the sensitivity of the MS/MS detector system, these values (LODs and LOQs) were lower than those reported in the literature for the analysis of these compounds by using other detector systems, such as UV, fluorescence, and electrochemical detectors, as was expected. In addition, by comparing the LODs reported for the analysis of these dipeptides, those obtained by UPLC technology in standard solutions (Su *et al.*, 2011) and in meat samples (Macià *et al.*, 2012) were lower than the reported by HPLC technology in standard solutions (Chen *et al.*, 2007) and in meat samples (Chen *et al.*, 2007; Peiretti *et al.*, 2011), respectively. This fact is due to the higher peak efficiency (narrow peaks) when the UPLC technology is used in comparison to the conventional LC technology (HPLC) (Swartz, 2005a; Swartz, 2005b).

When in MS, the ionization technique ESI is used, it is very important to evaluate the matrix effect. This effect results from coeluting matrix components that can compete for ionization capacity, and as a consequence this competition can produce important errors in the accuracy and precision of the analytical method. Therefore, in order to reduce inaccuracies by the effect of the sample matrix, some authors have prepared the calibration curves spiking the standards of the imidazole peptides in the real sample, meat or biological samples (Aldini *et al.*, 2004; Peiretti *et al.*, 2011; Su *et al.*, 2011; Macià *et al.*, 2012). In addition, the coupling of UPLC technology with MS has been reported to be less susceptible to matrix effects due to the high peak resolution of the UPLC technology. It is reported that an efficient UPLC separation may contribute to a reduction in ion suppression. This only occurs when the suppression is produced by the coelution of two different compounds (Guillarme *et al.*, 2010).

The stability of the imidazole dipeptides in real sample matrices has also been studied for many authors (Aldini *et al.*, 2004; Yeum *et al.*, 2010; Su *et al.*, 2011). Carnosine and anserine were stables in aqueous solutions, and in the spiked biological fluids (plasma and urine) (yeum *et al.*, 2010) and tissue samples (Aldini *et al.*, 2004) when these were stored in liquid nitrogen for at least 24 h at 4 °C. Carnosine, spiked in aqueous humor, was stable after three freeze–thaw cycles and at room temperature for 24 h (Su *et al.*, 2011) (see Table 10.3).

10.4 Applications of LC-MS/MS for the Analysis of Imidazole Dipeptides

The different reported methodologies based on LC-MS/MS were applied for the determination of imidazole bioactive dipeptides in meat (Chen *et al.*, 2007; Peiretti *et al.*, 2011; Macià *et al.*, 2012) and biological samples (Aldini *et al.*, 2004; jansen *et al.*, 2006; Yeum *et al.*, 2010; Su *et al.*, 2011). Table 10.1 summarizes the applications of LC-MS/MS for the determination of these compounds in these samples.

10.4.1 Meat Samples

In the literature, there are different LC-MS/MS methodologies applied for the determination of imidazole dipeptides in meat samples, such as chicken essences (Chen *et al.*, 2007), beef, rabbit, pork, horse, chicken and turkey meats (Peiretti *et al.*, 2011), and chicken broth (Macià *et al.*, 2012). Both HPLC in RP mode (Chen *et al.*, 2007; Peiretti *et al.*, 2011) and UPLC in HILIC mode (Macià *et al.*, 2012) have been used to analyze these samples. As analyzers, triple quadrupole (QqQ) (Chen *et al.*, 2007; Macià *et al.*, 2012) and ion-trap–orbitrap (Peiretti *et al.*, 2011) have been used. The main purpose of the analysis of the imidazole dipeptides in these samples has been to assess the use of these compounds as a natural-food supplement based on their bioactive properties.

The sample pretreatment technique used for the determination of imidazole dipeptides in these meat samples is very simple. Normally this sample pretreatment includes grinding the meat to obtain a homogenized sample, weighing, a dilution or centrifugation, and a denaturalization of the proteins with trifluoroacetic acid (TFA) (Peiretti *et al.*, 2011). Previously, the meat samples can be frozen and then freeze-dried, and when it was necessary, the bones and fat can be removed (Peiretti *et al.*, 2011). On the other hand, for the analysis of imidazole dipeptides in the chicken broth, a solid-phase extraction (SPE) with a polymeric sorbent was performed to remove interferences from the sample matrix, such as salt. This interference must be removed before the chromatographic analysis when MS is used as the detector system because salt is not volatilized, and the ionization source can block and degrade the performance of MS. The presence of salt (sodium chloride) before and after the sample pretreatment technique was evaluated by adding silver nitrate. The presence of sodium chloride in the sample was verified by the presence of a white precipitate formed by the addition of the silver chloride (Macià *et al.*, 2012). In this study, the extraction recoveries (%R) for imidazole dipeptides in meat samples and chicken broth were very good, from 70 to 99%.

The determination of imidazole dipeptides in fresh meat from different animal species revealed that the highest concentration of these compounds was found in turkey meat, and the lowest in beef meat (Peiretti *et al.*, 2011). Chicken meat has also been considered by different authors as a good food source of carnosine and anserine. Chicken essences, which are a popular meat extract in Southeast Asia, have been evaluated to be used as a food supplement of carnosine and anserine. These dipeptides were detected between 0.25 and 0.53 mg g^{-1} for carnosine, and between 0.02 and 0.04 mg g^{-1} for anserine in four commercial chicken essences (Chen *et al.*, 2007). Recently, chicken broth, which is part of the Mediterranean diet and a good source of water, which helps to maintain hydration, was also analyzed in order to determine the imidazole dipeptides carnosine and anserine. For the analysis of a commercial chicken broth, the concentration of carnosine was between 6.2 and 10.7 mg l^{-1}, and anserine between 7.9 and 10.7 mg l^{-1} (Macià *et al.*, 2012).

10.4.2 Biological Samples

Imidazole dipeptides have been analyzed in biological samples for pharmacokinetic studies in order to understand and determine their distribution in biological fluids and tissues. These biological samples are rat plasma and rat urine (Yeum *et al.*, 2010), rat tissues (Aldini *et al.*, 2004), rat cerebrospinal fluid (Jansen *et al.*, 2006), and rabbit aqueous humor (Su *et al.*, 2011). The analyzers used for the analysis of these biologic samples were triple quadrupole (TQD) (Jansen *et al.*, 2006; Yeum *et al.*, 2010; Su *et al.*, 2011) and quadrupole-ion-trap (Aldini *et al.*, 2004).

Biological samples are complex matrices and they contain a number of compounds, such as salts, proteins and particulate matter, which can interfere with the analysis. Therefore, these compounds must be removed prior to LC-MS analysis by applying a suitable sample pretreatment, and in addition, to preconcentrate the analytes. The sample pretreatment used for the analysis of these dipeptides is the precipitation of the proteins in order to denaturalize them. This was achieved by the addition of perchloric acid (PCA) (Aldini *et al.*, 2004; Yeum *et al.*, 2010) or ethanol (Jansen *et al.*, 2006) into the biological sample. Then, the biological sample was centrifuged and the supernatant was diluted with the mobile phase (50/50, v/v). After precipitation of the proteins, Jansen *et al.* (2006) derivatized the studied dipeptides in order to increase their polarity, because a RP-18 as the stationary phase was used.

By using these sample preparations, carnosine, anserine, homocarnosine and balenine were determined in rat plasma and in different rat tissues, such as skeletal muscles, heart, brain, cerebellum, kidney, lung and liver, and their mapping profile or distribution was studied. The highest concentrations of carnosine and anserine were in skeletal muscles, followed by the heart, cerebellum and brain. Homocarnosine was found only in the brain and cerebellum. No imidazole dipeptides were detected in plasma, liver, kidney and lung. Carnosine and anserine were detected in all the skeletal muscles and in cerebellum, but only carnosine in brain (Aldini *et al.*, 2004). Plasma and urine from rats were also analyzed after the ingestion of pure carnosine or rich foods in imidazole dipeptides, such as beef, chicken and chicken broth. Carnosine was not detected in plasma after the intake of pure carnosine, beef, chicken or chicken broth. Nevertheless, anserine was detected in plasma samples after the ingestion of carnosine and chicken broth, and therefore it demonstrated its bioavailability. In addition, the concentration of carnosine and anserine was increased in urine samples after the ingestion of beef and chicken broth (Yeum *et al.*, 2010).

Recently, carnosine and its acetylated form (*N*-acetyl-L-carnosine) were determined in rabbit aqueous humor by UPLC-MS/MS, and the ocular pharmacokinetic research of these dipeptides was studied and compared. The ocular bioavailability of the acetylated form was higher than carnosine due to the higher hydrophobicity of acetylated form, and as a consequence, a better permeability of the acetylated form through the cornea of the treated eye was suggested. Microdialysis was used as the sample pretreatment and this allowed the continuously simultaneous monitoring of these dipeptides.

10.5 Conclusions

The different reported LC-MS/MS methodologies for the analysis of imidazole dipeptides have been reviewed. The use of the HILIC chromatographic mode for the analysis of these polar compounds has the potential of avoiding the precolumn derivatization step, increasing MS response due to the use of acetonitrile-rich mobile phases, and in addition, and high sensitivity, high selectivity, low solvent consumption and speed when this chromatographic mode is used in the UPLC technology.

The use of high-resolution MS systems is an excellent strategy for the analysis of isobaric compounds, such as anserine, homocarnosine and balenine.

Summary Points

- This chapter focuses on the use of liquid chromatography coupled to mass tandem spectrometry for the analysis of imidazole dipeptides.
- Imidazole dipeptides have been analyzed by reversed-phase chromatography with prederivatization step, to increase their retention in this stationary phase, and also by the HILIC mode without derivatization.
- The HILIC mode uses acetonitrile-rich mobile phases and as a result when this chromatographic mode is coupled to MS, the sensitivity is increased.
- Anserine, homocarnosine and balenine have the same mass nominal *m*/*z* 241 (isobaric compounds) and by means the hybrid MS systems with high-resolution MS systems, these dipeptides have been successfully resolved.
- Imidazole dipeptides have been analyzed in meat samples and biological samples by LC-MS/MS.

Key Facts of Mass Spectrometry

1. Mass spectrometry (MS) is a powerful technique that characterizes the analytes according to their mass-to-charge (*m*/*z*). In addition, this detector allows determination of the molecular weights because specific fragments of the target molecule are generated.
2. All the MS instruments have an ion source, where ions in gas phase are produced, a mass analyzer, where the ions are separated according to their *m*/*z* ratio, and an ion detector, where the resolved ions are detected and their abundance is measured.
3. The analyzer typically must be at a high vacuum, at a minimum of 10^{-4} torr, to allow the ions to reach the ion detector without collision with other gaseous molecules.
4. The electrospray ionization (ESI) is the interface between LC and MS. ESI evaporates the sample solvent and ionizes the analytes into small charged droplets.
5. The graphic representation between the ion abundance *versus* the *m*/*z* of this ion is named a mass spectrum.

Definition of Words and Terms

Accurate Mass. The mass accuracy is the measured mass value of a compound with an associated error, such as 5 ppm. Mass accuracy is typically reported to three or four decimal places.

Quadrupole. Quadrupole is an analyzer, which is based on four circular rods and different magnetic and electric fields are applied.

Detection Limit (LOD). The LOD of an analytical method is the smallest sample quantity of analyte that its instrumental response is statically different from the blank.

Hydrophilic Interaction Chromatography (HILIC). HILIC is a LC chromatographic mode used to separate analytes with high polarity. HILIC uses a stationary phase that is polar and a mobile phase that is apolar, and the analytes are resolved according to their polarity.

Liquid Chromatography (LC). LC is a powerful analytical separation technique that allows the separation of compounds with moderate polarity according to its partition between a stationary phase and a mobile phase.

Precolumn Derivatization. The derivatization modifies analytes into others with better properties by the addition of a derivatization reagent. By means of this reaction, analytes with low retention in RP can be converted to analytes with good hydrophobicity to be retained in this stationary phase. In the precolumn derivatization, the reaction is preformed before the chromatographic analysis.

Quantification Limit (LOQ). The LOQ of an analytical method is the smallest sample quantity of analyte that it can be quantified.

Reversed-Phase Chromatography (RP). RP is a LC chromatographic mode used to separate analytes with moderate polarity. RP uses a stationary phase that is apolar and a mobile phase that is polar, and the analytes are resolved according to their apolarity.

Tandem Mass Spectrometry (MS/MS). MS/MS is a powerful detector system that has more than one analyser. The second analyzer is used to fragment the ions that have been separated in the first analyzer. These generated fragments, which are specific to the molecule, are very useful for structural determinations.

Ultraperformance LC (UPLC). UPLC is an improved LC technology that has been developed in the most recent years. This system uses stationary phases whose particle size is sub-2 µm and it offers high peak efficiency, but it operates at high pressure.

List of Abbreviations

ACN	Acetonitrile
BEH	Bridged ethylene hybrid
CV	Coefficient variation
ESI	Electrospray ionization
HFBA	Hepta-fluoro-butyric-acid
HPLC	High-performance liquid chromatography
HR	High resolution
IS	Internal standard
LOD	Detection limit
LOQ	Quantification limit
MeOH	Methanol
MS	Mass spectrometry
MS/MS	Tandem mass spectrometry
ODS	Octadecylsilica
RP	Reversed-phase chromatography
RSD	Relative standard deviation
SPE	Solid-phase extraction
UPLC	Ultraperformance liquid chromatography

Acknowledgments

This chapter was supported by the company Preparados Alimenticios S.A. subsidiary of Gallina Blanca Star Group through HENUFOOD Project (CEN-20101016), which is part of the CENIT program from Spanish Minister of Economy and Competitiveness.

References

Aldini, G., Orioli, M., Carini, M. and Facino, R. M., 2004. Profiling histidine-containing dipeptides in rat tissues by liquid-chromatography/electrospray ionization tandem spectrometry. *Journal of Mass Spectrometry*. 39: 1417–1428.

Chen, Y. H., Lin, Y. P., Liou, S. E. and Chen, C. C., 2007. Mass spectrometric determination of dabsyl-chloride derivatised anserine, carnosine and taurine in commercial chicken essences. *International Journal of Food Science & Technology*. 42: 593–600.

Fekete, S., Oláh, E. and Fekete, J., 2012. Fast liquid chromatography: the domination of core-shell and very fine particles. *Journal of Chromatography A*. 1228: 57–71.

Gama, M. R., Gomes da Costa-Silva, R., Collins, C. H. and Bottoli, C. B. G., 2012. Hydrophilic interaction chromatography. *TrAC Trends in Analytical Chemistry*. 37: 48–60.

Gil-Agustí, M., Esteve-Romero, J. and Carda-Broch, S., 2008. Anserine and carnosine determination in meta samples by pure micellar liquid chromatography. *Journal of Chromatography A*. 1189: 444–450.

Guillarme, D., Schappler, J., Rudaz, S. and Veuthey, J. L., 2010. Coupling ultra-high-pressure liquid chromatography with mass spectrometry. *TrAC Trends in Analytical Chemistry*. 29: 15–27.

Guiochon, G. and Gritti, F., 2011. Shell particles, trials, tribulations and triumphs. *Journal of Chromatography A*. 1218: 1915–1938.

Grumbach, E. S., Wagrowski-Diehl, D. M., Mazzeo, J. R., Alden, B. and Iraneta, P. C., 2004. Hydrophilic interaction chromatography using silica columns for the retention of polar analytes and enhanced ESI-MS sensitivity. *LC-GC North America*. 22: 1010–1023.

Jansen, E. E. W., Gibson, K. M., Shigematsu, Y., Jaknobs, C. and Verhoeven, N. M., 2006. A novel, quantitative assay for homocarnisine in cerebrospinal fluid using stable-isotope dilution liquid chromatography-tandem mass spectrometry. *Journal of Chromatography B*. 830: 196–200.

Kovacs-Nolan, J. and Mine, J., 2010. *Bioactive proteins and peptides as functional food and neutraceuticals*. Iowa, USA, 420 pp.

Macià, A., Motilva, M. J., Romero, M. P., Labrador, A., Domínguez, A. and Peiró, L., 2012. Imporved liquid-chromatography tandem mass spectrometry method for the determination of the bioactive dipeptides, carnosine and anserine: application to analysis in chicken broth. *Talanta*. 93: 293–300.

Mazzeo, J. R., Neue, U. D., Kele, M. and Plumb, R. S., 2005. Advancing LC performance with smaller particles and higher pressure. *Analytical Chemistry*. 77: 460A–467A.

Pearson, J. D. and McCroskey, M. C., 1996. Perfluorinated acid alternatives to trifluoroacetic acid for reversed-phase high-performance liquid chromatography. *Journal of Chromatography A*. 746: 277–281.

Peiretti, P. G., Medana, C., Visentin, S., Giancotti, V., Zunino, V. and Meineri, G., 2011. Determination of carnosie, anserine, homocarnosine, pentosidine and thiobarbituric acid reactive substances contents in meat from different animals species. *Food Chemistry*. 126: 1939–1947.

Su, D., Song, Y., Geng, L., Wei, B., Xie, P., Qi, Y., Ma, C., Yang, S., Bi, K. and Chen, X., 2011. Ocular pharmacokinetic study of L-carnosine and *N*-acetyl-L-carnosine in rabbit by microdialysis coupled with UPLC-MS-MS. *Chromatographia*. 73: 1189–1195.

Swartz, M. E., 2005a. Ultra performance liquid chromatography (UPLC): an introduction. *LC-GC North America*. 23: 8–14.

Swartz, M. E., 2005b. UPLC™: an introduction and review. *Journal of Liquid Chromatography & Related Technologies*. 28: 1253–1263.

United States National Library of Medicine (NLM). U.S. Department of Health & Human Services. http://chem.sis.nlm.nih.gov/chemidplus/chemidheavy.jsp. Accessed January 2012.

Webside of the trademark Waters. Gradient separation of histidine dipeptides on Acquity UPLC BEH HILIC. http://www.waters.com/waters/library.htm. Accessed November 2011.

Webside of the trademark Waters. Analysis of Histidine dipeptides using Cortecs UPLC columns. http://www.waters.com/waters/library.htm. Accessed August 2013.

Yeum, K. J., Orioli, M., Regazzoni, L., Carini, M., Rasmussen, H., Russell, R. M. and Aldini, G., 2010. Profiling histidine dipeptides in plasma and urine after ingesting beef, chicken or chicken broth in humans. *Amino Acids*. 38: 847–858.

Section IV
Function and Effects

CHAPTER 11

Antioxidant Activity of Imidazole Dipeptides

EWA BIAZIK*[a,b] AND WIESŁAW KOPEĆ[a]

[a]Department of Animal Products Technology and Quality Management, Wrocław University of Environmental and Life Sciences, Chełmońskiego 37/41, 51-630 Wrocław, Poland; [b]Department of Quality Analysis, Wrocław University of Economics, Komandorska 118/120, 53-345 Wrocław, Poland
*E-mail: ewa.biazik@ue.wroc.pl

11.1 Introduction

Oxidation, which is the loss of an electron by any chemical substance, is caused in biological systems by reactive oxygen species, ROS, or other radicals having unpaired valence electrons. One of the most reactive radicals is the HO^* hydroperoxide radical created in biological systems as the effect of Fenton or Haber–Weiss redox reactions of iron and H_2O_2 (Kohen and Nyska, 2002). One of the molecules most susceptible to oxidation are unsaturated lipids. The process, causing creation of lipid peroxides, may have many negative effects and is associated with heart diseases, aging and mutagenic and cancerogenic processes in humans and lowers the nutritive value of food products, creating undesirable flavor and taste changes (Yen *et al.*, 2002). Autoxidation of lipids in various biological systems is accompanied by the formation of aldehydic compounds, such as hydroxylated, saturated, α,β unsaturated and polyunsaturated aldehydes. These aldehydic products of oxidation are suggested to be able to react with biomolecules that cause deleterious effects in cells and tissues (Decker *et al.*, 1992).

Food and Nutritional Components in Focus No. 8
Imidazole Dipeptides: Chemistry, Analysis, Function and Effects
Edited by Victor R Preedy

Published by the Royal Society of Chemistry, www.rsc.org

The oxidation processes are accelerated by the action of pro-oxidants or retarded by antioxidants, being reductants, *i.e.* electron donors. The oxidation processes in tissues can be prevented by natural cellular defense systems involving antioxidant enzymes (**SOD** superoxide dismutase, **GPx** glutathione peroxidase and **CAT** catalase) or low molecular weight hydrophilic and hydrophobic antioxidants exhibiting metal-chelating ability. Ions of transition metals, especially iron and copper, play important roles in peroxidation processes in different biological systems and can be inducers of oxidative changes. Ions of free transition metals, especially iron and copper, play important roles in peroxidation processes in different biological systems and can be inducers of oxidative changes. In particular, copper ions at μM concentrations induce the generation of free radicals at the surface of low-density lipoproteins (**LDL**), with consequent oxidative modifications that are of permanent importance in the pathogeneses of a number of cardiovascular and neurogenerative diseases (Bertinaria *et al.*, 2012). Oxidation processes in dispersed lipids are prevalent at the organic/water interface where hydroperoxides are decomposed into free radicals *via* transition-metal pro-oxidative action (Yuji *et al.*, 2007). An example of ion chelators occurring in animal tissues is carnosine (β-alanyl-L-histidine) and its homologs such as anserine (β-alanyl-3-methyl-L-histidine) and homocarnosine (γ-aminobutyryl-L-histidine), being the three most representative compounds of the family of the so-called imidazole dipeptides (Bellia *et al.*, 2008). The dipeptides are synthesized in the brain and muscles that are tissues susceptible to oxidants. Carnosine and its homologs play many roles in biological systems including buffering, and act as antiglycating agents. However, most research has focused on the antioxidant activity of histidine dipeptides, which are considered not only in terms of ion chelating but also radical scavenging or reactive oxygen species (**ROS**) and aldehydic product quenching.

It has been recently demonstrated that carnosine is a potential and selective quencher of α,β-unsaturated aldehydes, which are typical byproducts of membrane lipid peroxidation. This dipeptide inhibits aldehyde-induced lipid–protein and DNA–protein crosslinking in neurodegerative disorders (for instance Alzheimer's disease) and also in inflammatory diseases or in cardiovascular ischemic damage (Guiotto *et al.*, 2005). Moreover, dipeptides containing histidine exhibit the chelating properties of pro-oxidative metal ions. Dipeptides can play a role in the regulation of some microelement concentrations *via* creation of mineral–organic chelate bonds with Zn, Mn, Cu, Co in tissues or other biological systems. As a consequence of their transition-metal chelating ability, imidazole dipeptides are able to inhibit ascorbate-induced radical oxidation of lipids (Babizhayev *et al.*, 1994) and present anti-Fenton reaction properties, which may also be connected with a superoxide dismutase (**SOD**) system.

Imidazole dipetides, *i.e.* carnosine and homologs, are water-soluble substances; that is why the reactivity of these compounds towards lipid radicals seems to be low (Murase *et al.*, 1993). However, compounds from the imidazole dipeptide group are proposed to be naturally occurring scavengers of harmful reactive aldehydes emanating from the degradative oxidative pathway of endogenous compounds such as sugars, polyunsaturated fatty acids (**PUFAs**) and proteins. Carnosine and its analogs play a protective role against oxidative stress and have an antioxidant function in various tissues. L-Carnosine is also an efficient water-soluble trapping agent of oxygen-derived free radicals.

11.2 Modification of Oxidation Processes in Liposome Model Systems *via* Imidazole Dipeptides

According to Yuji *et al.* (2007), antioxidants able to scavenge free radicals are believed to be most effective in dispersion systems, especially when they are accumulated at the lipid/water interface. Hence, many researchers use carnosine and other dipeptides as agents suppressing lipid oxidation in liposome model systems (Decker *et al.*, 1992; Boldyrev *et al.*, 1993; Babizhayev *et al.*, 1994; Kensaci *et al.*, 1997; Yen *et al.*, 2002; Biazik *et al.*, 2012). Liposomes are spherical vessels composed of phospholipid bilayer membranes. These artificial structures are commonly used as a model for biological membranes. The use of liposomes in research makes it easy to establish the values of pH, temperature, composition of lipids and concentrations of various antioxidant and pro-oxidant agents. In the case of oxidation in dispersion systems the metal-related type of induction is most often used. Most research on oxidation with liposomes is conducted in ascorbate/Fe^{3+} systems. Decker *et al.* (1992) reported that carnosine in concentrations from 0.05 to 10.0 mM is capable of the inhibition of the copper-catalyzed oxidation of ascorbic acid. However, in the case of iron-ascorbate catalyzed oxidation no effect has been observed. In contrast, a concentration of 1.0–25 mM of carnosine is able to inhibit oxidation in catalyzed PC liposome model systems regardless of whether copper or iron were used. Moreover, Kensaci *et al.* (1997) reported that similar dosages of carnosine showed antioxidant activity in iron-related model systems such as liposomes made with muscle phospholipids. The decrease was dependent on incubation time, carnosine concentration and method of measurement. A significant decrease in the amounts of total 2-alkens and substances reacting with thiobarbituric acid **TBARS** (such as malonaldehyde **MDA**, hexanal, *trans*-2-undecanal) was observed when carnosine was added to previously oxidased liposomes (at concentrations of 2–10 mM). The authors reported that the antioxidant activity of carnosine is multifunctional. It includes buffering activity, prolongation of the initial step of oxidation, a decreasing effect on peroxide levels, and the ability to react with some secondary products of oxidation.

From a practical point of view, the problem of the effective concentration of dipeptides is fundamental. Biazik *et al.* (2012) studied the influence of the addition of a pure carnosine or histidine dipeptide preparation of 0.25–1.0 mM on liposome dispersions. Oxidation in phosphatidylcholine liposome model systems was ascorbic/Fe^{3+} induced. An anserine/carnosine (2 : 1) preparation was obtained from chicken breast muscle extract *via* ion-exchange chromatography and membrane filtration techniques. The antioxidant activity of histidine dipeptides was expressed as the lowering of hydroperoxides or malonaldehyde concentrations. Both carnosine and dipeptide preparations led to a 50% inhibition of PC liposome oxidation at a dosage of 0.25 mM. At a concentration of 1 mM, suppressed oxidation was observed in comparison to the control sample. The type of preparation used does not significantly influence the inhibition dynamics of oxidation processes. A 4-fold lower **TBARS** content was observed after addition of 1 mM of an anserine/carnosine preparation. At the level of 0.25 mM dipeptides, there was a 2-fold effective suppression of oxidation processes. A total inhibition of lipid peroxidation of 2.5 mM was reported regardless of whether a pure carnosine or dipeptide preparation was used. However, the effectiveness of dipeptides in liposomes has often been observed at higher concentrations, up to 10 mM (Decker *et al.*, 1992; Kensaci *et al.*, 1997). An even higher level was used by Yen *et al.* (2002), who investigated the oxidation inhibition activity of carnosine in various model systems including liposome dispersions induced by $FeCl_3$ and ascorbic acids. Carnosine in the range of 0–15 mg ml^{-1} (about 60 mM) showed an inhibition from 0 to 43.3% of **TBARS** formation.

In a study oriented more to the biological model, oxidation was induced in PC liposomes not only *via* ascorbic acid and iron ions but also in a LOOH/hemoglobin mixture (Babizhayev *et al.* 1994). The hydroxyl-radical-scavenging and inhibition of the lipid peroxidation ability of carnosine, carcinine and constituent amino acids are studied in these model systems. Both carnosine and carcinine are able to suppress phosphatidylcholine liposomal oxidation measured by lowered **TBARS** numbers. It has been shown that, chemically, dipeptide antioxidant activity is not solely due to the imidazole moieties of the molecules. Imidazole itself did not exhibit any antioxidant activity. Moreover, histamine exhibited a pro-oxidant action dependent on the concentration of catalytically active free iron ions. The antioxidant activity of the mentioned compounds may involve reduction of oxidative potential or stabilization of the imidazole radical, probably due to the peptide bond. Free β-alanine molecules and β-alanine in the mixture with histidine or histamine did not show any lipid peroxidation inhibition.

In addition, Boldyrev *et al.* (1993) showed that histidine and imidazole do not exhibit any protective activity in membrane systems against oxidative damage, while carnosine may play the role of a membrane-protecting agent. Imidazole dipeptide protects cell membrane structures *via* inhibition of lipid peroxide oxidation and production of an amount of malondialdehyde (Boldyrev *et al.*, 1993).

It can be concluded that in liposome model systems dipeptides containing an imidazole ring, especially carnosine, show a protective effect towards lipid oxidation *via* lowering the content of hydroperoxides and aldehydic compounds. Although inhibition of lipid peroxidation by imidazole dipeptides has mainly been shown in systems with oxidation catalyzed by free iron or hemoglobin, metal chelation is not the only mechanism causing this suppression. The antioxidant effect is realized *via* inhibition of membrane lipid peroxidation, the trapping action of superoxide anions, singlet oxygen and numerous products of the peroxidation processes, especially final products, *i.e.* malondialdehyde, and also *via* increasing the system's total antioxidant capacity.

11.3 Radical-Scavenging Activity of Imidazole Dipeptides

Most of the research on this activity of dipeptides has been carried out in model systems and very often artificial radicals have been used to determine so-called total antioxidant capacity. In 1993 Boldyrev reported that the behavior of carnosine in different model systems based on free-radical chain reactions shows that this dipeptide is a hydrophilic antioxidant of nonenzymatic direct action. Moreover, carnosine related compounds have been shown to be effective antioxidants that can act against many types of oxygen form radicals (Bogardus and Biossonneault, 2000).

Kohen *et al.* (1988) reported the scavenging peroxyl radical ability of carnosines and two analogs. The lipid antioxidation activity of anserine, carnosine and homocarnosine was studied in a multilamellar liposome system based on phosphatidilocholine which was induced by the water-soluble initiator 2,2′-azobis 2-amidino-propane dichloride (**AAPH**) radical. The water soluble initiator used made this system more physiological in comparison to the hydrophobic 2,2′azobis (2,3-dimethylvoleronitrile) (**AMVN**). All studied carnosine homologs presented concentration-dependent antioxidant properties. Carnosine and homocarnosine showed a similar antioxidant activity (respectively, 53% and 56% protection). A greater antioxidant activity (60% protection) was observed for anserine nitrate salt (Figure 11.1) (Kohen *et al.*, 1988).

In light lipoprotein **LDL** systems isolated directly from tissues such as human plasma Bogardus and Boissonneault investigated the ability of carnosine to suppress oxidation initiated by cooper ions Cu^{2+} or 2′,2′-azobis

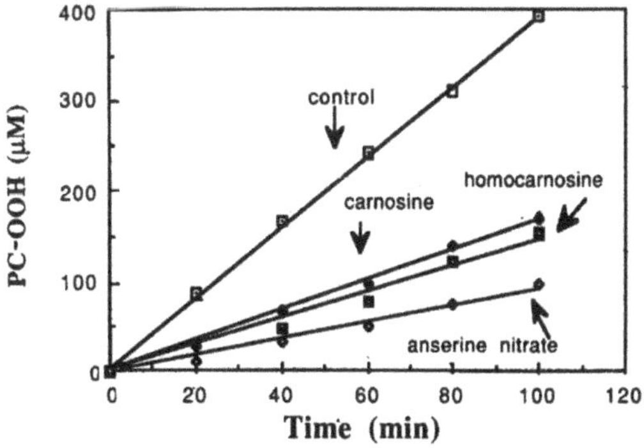

Figure 11.1 The antioxidant activity of histidine dipeptides in liposomes induced by AAPH radicals. The figure presents the antioxidant activity of carnosine, homocarnosine and anserine in liposomes induced by **AAPH** radicals. Data are from Kohen *et al.* (1988) with permission from Professor Ames.

(2-amindinopropane) hydrochloride (**AAPH**). The results showed that carnosine can act as a preventing agent against LDL oxidation in both free-radical-predicting systems (Bogardus and Boissonneault, 2000).

However, despite the research results mentioned above the dipeptide ability to radical scavenging is discussed. In the literature on carnosine, no antioxidant effect has been shown either. Gorbunov and Erin (1991) carried out comparative research into the antioxidant activity of carnosine *via* the evaluation of the electron spin resonance (**ESR**) signals of semiquinone and the 2,2-diphenyl-1-picrylhydrazyl (**DFPH**) radical of α-tocopherol. The authors reported that carnosine was not able to quench the **ESR** signals of either type of the datable radical of α-tocopherol. Carnosine was shown to be unable to interact directly with highly active free radicals and regenerate the radical of α-tocopherol that would lead to the formation of a synergistic and antiradical couple (Gorbunov and Erin, 1991). Also, Salim-Hanna *et al.* (1991) did not find any protective effect of carnosine addition (up to 17 mM) against free-radical damage of brain homogenates or red-blood-cell ghost membrane systems oxidized by the 2,2′-azobis-2-amidino-propane, but that in small concentrations inhibited the radiation-induced inactivation rates for lyzozyme and horse radish peroxidase (Salim-Hanna *et al.*, 1991).

Moreover, imidazole dipeptides can act as pro-oxidative agents. Shi *et al.* (1992) reported that carnosine, homocarnosine and anserine are able to generate alkyl and alkoxyl radicals in the reactions of Ni^{2+} with cumene peroxide or *t*-butyl hydroperoxide. This study suggests that the mentioned dipeptides, instead of presenting a protective ability, may facilitate free-radical generation in Ni^{2+}-induced system and also may play a role in Ni^{2+} mechanisms of cancerogenicity and toxicity. The free-radical-realization or scavenging activity of histidine dipetides may be related to the formation of metal–peptide chelate complexes (Shi *et al.*, 1992).

On the other hand, Wu *et al.* (2003) reported that some imidazole dipeptides and amino acids showed preventive activity against linoleic acid auto-oxidation. They used four different methods, *i.e.* inhibition of linoleic acid peroxidation, Cu^{2+} chelating properties and reducing power, and scavenging ability of α,α-diphenyl-β-picrylhydazyl (**DPPH**) radicals, for the evaluation of antioxidant activity of carnosine and anserine and some selected amino acids, *i.e.* glycine, taurine, β-alanine, 1-methylhistidine and histidine. Carnosine presented effective properties as a free-radical scavenger, copper chelator and reducing agent. β-Alanine, 1-methylhistidine and histidine were less effective as oxidation inhibitors. In addition, nonhistidine compounds did not present **DPPH** radical scavenging ability. The authors suggested that imidazole compounds were related to antioxidant properties and peptide bonds between amino acid in anserine or carnosine molecules were involved in the antioxidant activity of dipeptides (Wu *et al.*, 2003).

Also, in model studies (Biazik *et al.*, 2011) on 2,2′-azinobis-(3-ethylbenzothiazoline-6-sulfonic acid (**ABTS**$^+$) radical-scavenging ability caused by carnosine addition was tested in chicken breast muscle homogenates and myofibril model systems either immediately after sample preparation or after three days cold storage. After an increase in the dipeptide addition (0–75 mM) an increase in the total antioxidant activity of muscle tissue homogenates was observed. In myofibril suspensions a slight tendency to higher **ABTS** scavenging ability was noted (Biazik *et al.*, 2011).

11.4 Aldehyde-Quenching Activity of Imidazole Dipeptides

Carnosine is also known to react with low molecular weight aldehydic compounds (such as the advanced glycation end products, **AGEs**) and the aldehydes formed from unsaturated fatty acids during oxidation processes (such as, *trans*-2-hexanal, 4-hydroksynonenal (**HNE**) and other α,β-unsaturated aldehydes), thus protecting cell structures such as proteins, DNA, membranes from reactive species (Decker *et al.*, 1992; Bertinaria *et al.*, 2012). Hexanal and propanal are major products formed from the oxidation of omega-6 and omega-3 fatty acids (Decker *et al.*, 1992). Most studies have focused on the ability of carnosine to interact with various unsaturated aldehydes created from the breakdown of lipid peroxidations in turn formed by the action of radical oxygen species on cell membrane lipids.

Zhou and Decker (1999) studied the ability of dipeptides, such as carnosine and selected amino acids, to quench aldehydes. Carnosine was a much more reactive quencher of *trans*-2-hexanal than alanine and histidine separately; however, it showed a weak reaction with nonconjugated monounsaturated aldehydes such as *trans*-4-decanal and polyunsaturated aldehydes such as *trans*-2,3-decadienal and *trans*-2,4-hexadienal. The quenching activity of histidine and imidazole was similar (8% and 7% headspace aldehyde reduction). This suggested that the nitrogen group from an imidazole ring was the major reaction site in histidine – not the α-amino group. Dipeptides containing non-histidine amino acids were more efficient in quenching *trans*-2-hexanal than their individual components. Zhou and Decker suggested that the peptide bond increased the reactivity of the α-amino group. Moreover, all the histidine-based dipeptides showed 2.5-fold higher effectiveness than histidine. The authors also reported that the ability of other histidine-containing dipeptide (valylhistidine, leucylhistidine and isoleucylhistidine) activity to quench α,β-monounsaturated aldehydes was similar to that of carnosines (Zhou and Decker, 1999a,b).

The cited authors concluded that formation of aldehyde–peptide adducts occurs on the imidazole nitrogen, likely *via* a Michael-type reaction (Zhou and Decker, 1999a). The mechanism is based on nucleophilic addition to α,β-unsaturated carbonyl compounds and leads to formation of a single bond between reacting molecules (Mather *et al.*, 2006).

Such a mechanism has been confirmed by Aldini *et al.* (2002), who reported that during incubation of carnosine and 4-hydroxynonenal, two main intermediates with molecular weights of 364 and 382 Da were formed. The structures presented in Figure 11.2 were identified by Electrospray Ionization Mass Spectroscopy and NMR spectroscopy.

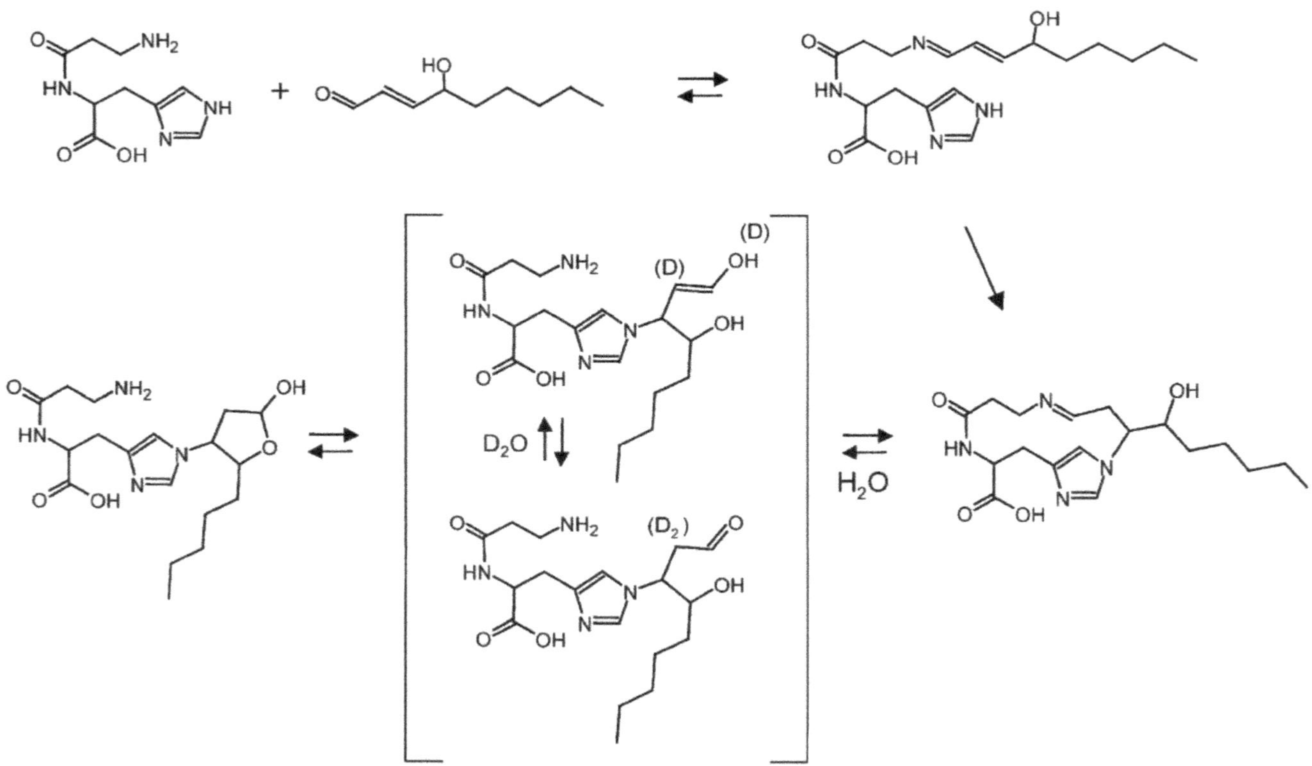

Figure 11.2 Proposed carnosine and 4-hydroxy-*trans*-2,3 nonenal mechanism reaction. The figure presents a proposed reaction mechanism for carnosine and 4-hydroxy-*trans*-2,3 nonenal. The reaction proceeds *via* a cyclization step that leads to the formation of a cyclic intermediate. This product can exist in keto–enolic equilibrium. Data are from Aldini *et al.* (2002) with permission from Elsevier.

In fact, the intermediate product of the reaction of carnosine with **HNE** is imine, which is composed of two diastereoisomers. The addition of the nitrogen atom from the histidine molecule to the double bond led to the formation of the cyclic intermediate in the Michael reaction. This form exists in equilibrium with the hemiacetal through the opened form (in keto–enolic equilibrium). The structure of the final products was confirmed by NMR spectroscopy analysis conducted directly on the incubation mixture (24 h at 37 °C) (Figure 11.3).

In the reaction of carnosine with *trans*-2,3 nonenal an imine derivate is also formed (Figure 11.3). The lack of the hydroxyl group prevented the reaction of the hemiacetalic cyclization. The imine structure is the only product which is further analyzed after imine reduction with sodium borohydrine (Aldini *et al.*, 2002).

11.5 Buffering Activity of Imidazole Dipeptides

The buffering properties of imidazole dipeptides may have nondirect influences on oxidation processes *via* regulating or changing pH levels during reactions, especially in biological systems such as tissues or meat.

In Davecy's (1960) studies on the power of carnosine and anserine to buffer in the physiological range of pH values (pH 6.5–7.5), histidine dipeptides contributed about 40% to the buffering in resting living muscle in the mentioned range of pH (Davecy, 1960). Dipetides play an important role in stabilizing pH and neutralizing its rapid drop effect in an anaerobically contracting muscle, caused by accumulation of lactic acid. In addition, carnosine has been considered a compound acting as a buffer to neutralize lactic acid produced in skeletal muscle that is undergoing anaerobic processes of glycolysis (Kohen *et al.*, 1988).

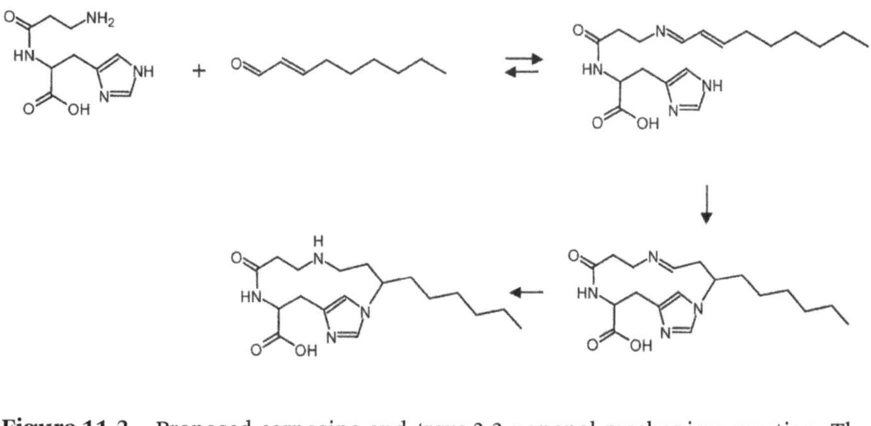

Figure 11.3 Proposed carnosine and *trans*-2,3 nonenal mechanism reaction. The figure presents a proposed mechanism reaction for carnosine and *trans*-2,3 nonenal. Imine is the final product of the reaction. Data are from Aldini *et al.* (2002) with permission from Elsevier.

Imidazole groups that exist in free L-histidine or histidine-containing dipeptides such as carnosine, anserine and ophidine and also in histidine residues of protein are characteristic of intracellular nonbicarbonate buffering capacity. Due to the similar values of the p*K* of these imidazole groups and intracellular pH (**pH$_{ic}$**), one of the nitrogen atoms of the imidazole ring may be protonated at the pH level in the physiological range. Thus, imidazole groups play the role of potent proton-buffering constituents. The regulatory process maintaining the **pH$_{ic}$** to p*K* values of imidazole groups is called "alphastat regulation". Imidazole buffer systems appear to have evolved in the muscles of vertebrates requiring high burst exercise capabilities or high anoxic tolerance for their survival (Abe, 2000).

In addition, the change of pH in natural systems can affect metal-catalyzed oxidation. Tichivangana and Morrissey (1985) reported that the pro-oxidation activity of some transition metals such as Fe^{2+}, Cu^{2+} and metmyoglobin is highly pH dependent (a relationship of increasing pH with decreasing TBARS values was observed) in heated muscle systems. The inverse pH/**TBARS** relationship was observed in muscles from fish, turkey, chicken, pork, beef and lamb. The lowest **TBARS** was observed at pH 7.0 in all cases. However, heam-catalysis is less influenced by pH level and the trend to increase **TBARS** numbers with increasing pH was observed. Also, Chen and Waimaleongora-Ek (1981) found that raw broiler meat with lower pH levels is more susceptible to oxidation. So, the buffering activity of imidazole dipeptides can also significantly influence oxidation changes, especially in more complicated biological systems, such as meat (Tichivangana and Morrissey, 1985).

11.6 Antioxidant Activity of Imidazole Dipeptides in Other Systems

11.6.1 Protein Model System

Antioxidant activity has also been studied in protein model systems. Modified forms of proteins are characterized by a higher content of carbonyl groups that can be a marker of peroxidation changes. Yen *et al.* (2002) studied the carnosine protection effect on the oxidative damage of bovine serum albumin. Protein carbonyl content was measured to determine the level of albumin oxidation in a reaction mixture with H_2O_2, $FeCl_3$ and ascorbic acid. At the level of 16.67–50 mg ml^{-1} of dipeptide the protective effect against protein oxidation increased with increasing levels of carnosine. However, in the range of 0–16.67 mg ml^{-1} dipeptide, no inhibition of albumin oxidation was observed. Also, some pseudopeptides, such as L-prolylhistamine and β-alanylhistamine (carcinie) showed effective protection ability of bovine serum albumin against peroxides generated *in situ* from phospholipids. The activity of L-carnosine was weaker, although it is able to protect protein in aqueous phase. The authors confirm that the imidazole-containing pseudopeptides may react with aldehydic oxidation products that are released during oxidative changes of unsaturated lipids (Yen *et al.*, 2002).

In myofibril model systems with added fat (25%), the relationship between **TBARS** amount and carnosine addition has also been investigated (Biazik *et al.*, 2011). Peroxidation processes were initiated naturally or by thermal treatment. In myofibril suspensions a drop in oxidation dynamics with dipeptide addition up to 50 mM was observed only for nonstored samples. After 3 days storage carnosine addition did not affect **TBARS** number. As an effect of carnosine level on **TBARS** after thermal treatment (80 °C) of myofibrils samples with added fat, a tendency to enhancive **TBARS** formation with increasing dipeptide content was observed after preparation. In contrast, after cold storage a suppression of oxidation dynamics was noted if the carnosine level was higher than 10 mM (Biazik *et al.*, 2011).

11.6.2 Tissue Model System

Also, the effect of imidazole dipeptides has been studied in more complicated systems such as meat samples or tissues homogenates. Das *et al.* (2006) investigated the influence of carnosine dosages (0%, 0.5%, 1.0% and 1.5%) on the quality of preblended buffalo meat. The authors observed an inhibition of myoglobin formation and brown color development in meat samples containing 1% and 1.5% carnosine (Das *et al.*, 2006). Also, carnosine addition inhibited lipid peroxidation of ground buffalo meat and extended shelf life to 8 days in cold storage conditions, without any undesirable changes in color or odor. A similar antioxidant effect of carnosine addition was observed by Badar (2007) in γ-irradiated beef patties and ground-beef meat. Carnosine reduced oxidative processes and metmyoglobin formation in postirradiated samples and during 15 days of storage (Badar 2007).

In homogenates with fat, no such clear dependency can be observed. In fact, Biazik *et al.* (2011) observed a decrease in **TBARS** with increased carnosine content in chicken breast meat homogenates. In samples prepared with a 25% fat addition the antioxidant effect of carnosine concentration on oxidation dynamics was more evident. However, at over 25 mM of carnosine the increase in dipeptide content did not cause further **TBARS** reduction (Biazik *et al.*, 2011).

11.7 Antioxidant Activity of Pseudopeptides

In the last 20 years there has been a trend to study carnosine derivatives resistant to carnosinases. This could enable application of substances with an imidazole moiety in food and medicine. Also, the increase of antioxidant properties in food was also a reason for such research.

In 1998 a series of pseudodipeptides were synthesized (Seguin and Babizhayev, 1998) *via* coupling between methyl-substituted histamine or histamine and a selected amino acid. Among the various structures, Seguin and Babyzhayev disclosed a number of aminoacylhistamine analogs, which were synthesized by the conjugation of histamine, 1-methylhisatamine or 3-methylhistamine with amino acids (or *N*-acetyl amino acids).

The *N*-acetylated amino acids were used in order to permit an active product sustained release after hydrolysis by acylpeptide hydrolase-type enzymes. The scope of the activity of these derivatives was demonstrated with *in vitro* tests, which were performed on the dipeptides. For instance, β-alanylhistamine was taken as an example. The dipeptides were characterized by higher activity than carnosine in a superoxide anion-scavenging assay. However, they showed slightly less activity in phosphatidylcholine liposome tests. In addition, β-alanylhistamine presented activity in cell-regeneration processes (wound-healing action). This dipeptide is able to work as an immunostimulator and chemotactic substance towards cells involved in the immune system at an early stage of healing.

A structure–activity relationship on a number of peptidomimetic analogs of L-carnosine that contain imidazole rings such as β-alanylhistamine, *N*-acetyl-β-alanylhistamine, L-glutamylhistamine, L-prolylhistamine, *p*-glutarylhistamine and L-carosine was investigated by Babizhayev *et al.* (1998). The evaluation of their antioxidant activity was based on dipeptide resistance to enzymatic degradation, free-radical scavenging ability (analyzed in different *in vitro* assays) and the deactivation *via* trapping lipid hydroperoxides.

Even minimal structure modification (for instance, the histidine decarboxilation) led to improvement in the resistance of carnosine-like dipeptides towards dipeptidases. A series of compounds where histidine was replaced by histamine were tested in PC liposome model systems, with inhibition of lipoperoxydation expressed as the percentage inhibition of malondialdehyde release (Babizhayev *et al.*, 1998) (Table 11.1).

The results obtained allowed the conclusion that the basicity of the amino acid moiety in the side chain of a dipeptide molecule is an important factor for the radical-scavenging activity, probably *via* a metal-ion-chelation action, which may play a determinant role in this ability. Moreover, the reintroduction of a carboxylic acid group in a non-α-position of L-glutamylhistamine did not enhance the antioxidant properties. Due to the presence of claw-like structures

Table 11.1 Effect of 10 mM of selected imidazole-containing peptidomimetics on lipid peroxidation in PC liposome model systems. The table presents the effects of 10 mM of selected pseudopeptides on lipid peroxidation in PC liposome model systems (ascorbate induced). The results are expressed as a percentage of MDA release inhibition. Data are from Babizhayev *et al.* (1998) with permission from Springer.

Peptidomimetic L-carnosine analogs [10 mM]	Inhibition of MDA release [%]
β-Alanylhistamine	**40**
N-Acetyl-β-alanylhistamine	**25**
L-Glutamylhistamine	**28**
L-Propylhistamine	**41**
p-Glutarynylhistamine	**25**
L-Carnosine	**58**

that favor transition-metal chelation in the pyrolidine-containing dipeptide (L-prolylhistamine), it is possible to obtain the most promising antioxidant.

Recently (*i.e.* Segunin and Babizhayev, 2000) a new series of imidazole-containing pseudodipetides, presenting antioxidant activity have been released onto the market. The following compounds that are resistant to enzymatic hydrolysis have been patented: L-glutamylhistamine, L-pyroglutamylhistamine, D-glutamylhistamine, *N*-methyl-2-amino-L-butyrylhistamine, 2-aminoisobutyrylhistamine, L-norleucylhistamine, L-2-aminooctylhistamine, L-ornitylhistamine, Boc-2-amino-L-pentylhistamine, L-phenylglicylhistamine, L-2,6-diaminopimelylhistamine, *N*-acetyl-2-aminobutyrylhistamine, *N*-phenyl-2-aminobutyrylhistamine, *N*-phenylcetyl-2-aminobutyrylhistamine, L-arginylhistamine, L-prolylhistamine and 4-hydroxy-L-propylhistamine. The increased resistance of these carnosine-like compounds to enzymatic deactivation was caused by structural features (Segunin and Babizhayev, 2000).

The results of *in vitro* experiments carried out on L-prolyhistamine and L-glutamylhistamine were reported. The free hydroxyl-radical-scavenging experiment showed the superior activity of L-prolyhistamine in comparison to pure carnosine. L-Prolyhistamine presented higher activity (62% inhibition on the free radical) than carnosine (38% inhibition). L-Glutamylhistamine was completely inactive. However, L-glutamylhistamine showed a quenching ability of the superoxide anion, which was produced by the xantine oxiadase/hypoxantine system. In this system L-prolyhistamine and carnosine were totally inactive.

11.8 Antioxidant Activity of *N*-Acyl Imidazole Dipeptide Derivatives and Adducts

Acyl derivatives, such as *N*-long-chain acyl carnosine and *N*-long-chain acyl histidine, possess antioxidant properties towards lipid peroxidation. The antioxidant activity of *N*-acyl compounds (methyl linoleate was used for acylation) has been investigated in carnosine containing multilamellar liposomes system and, alternatively, carnosine has been added after liposome preparation. Acyl-derived dipeptides are characterized by good solubility in apolar solvents due to the presence of hydrophobic fatty acid residues. These compounds can suppress the peroxidation of membrane lipids when incorporated into liposome membranes (Murase *et al.*, 1993).

Bertinaria *et al.* investigated the activity of NO-donor carnosine analogs and their des-NO derivatives which demonstrated carnosine-like properties and high stability against serum carnosinase. These nearly synthesized compounds are able to complex Cu^{2+} ions at physiological pH, displaying significant copper-ion-sequencing ability, and protect **LDL** from oxidation catalyzed by Cu^{2+} ions (Bertinaria *et al.*, 2012). All products present moderately potent **HNE** quenching properties (Table 11.2).

Also, β-cyclodextrine (**CD**) can be functionalized with carnosine molecules to synthesize compounds with antioxidant molecules, at either

Table 11.2 Lipophilicity and HNE-scavenging ability of carnosine and its NO derivatives. The table presents the lipophilicity and HNE scavenging ability of carnosine and its NO derivatives. Lipophilicity was determined *via* the shake-flask technique. **HNE** is 4-hydroxynonenal. Data are from Bertinaria *et al.* (2012) with permission from Elsevier.

No.	Compound	Lipophilicity [clog$^{D7.4}$]	HNE scavenging ability [HNE % scavenged 24 h]
1	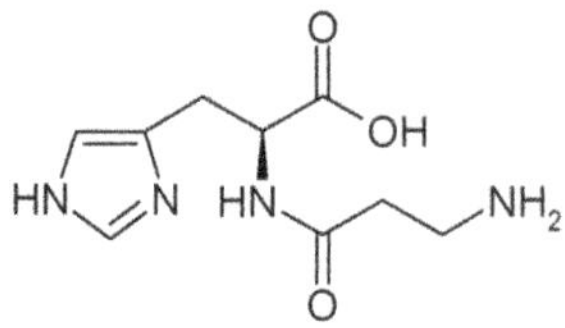	nd	88.8
2	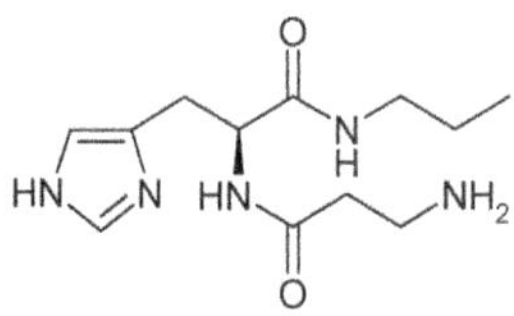	−2.72	81.0
3	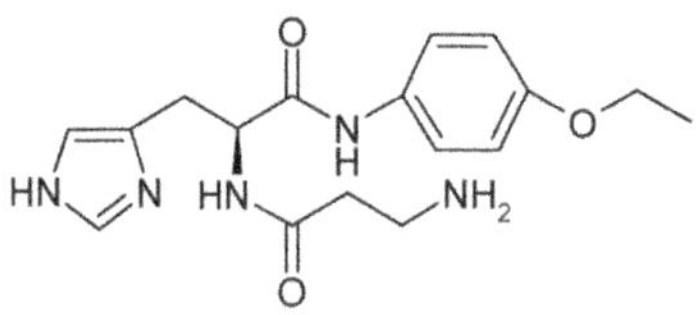	−1.15	73.3
4	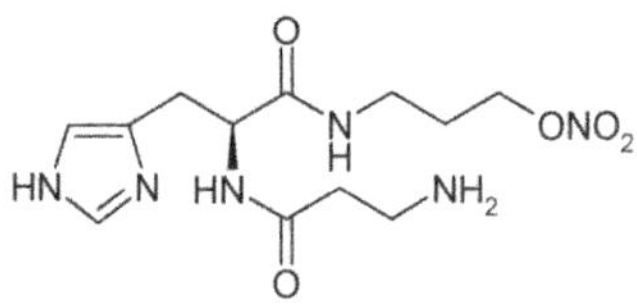	−2.43	75.4
5	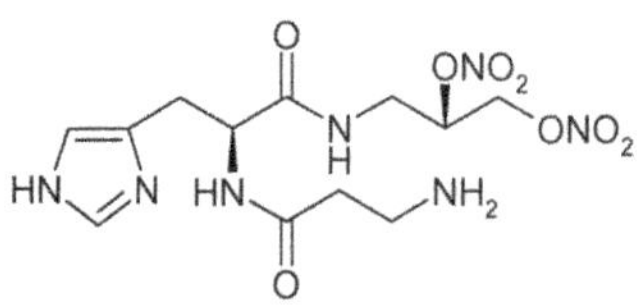	−2.19	67.2
6	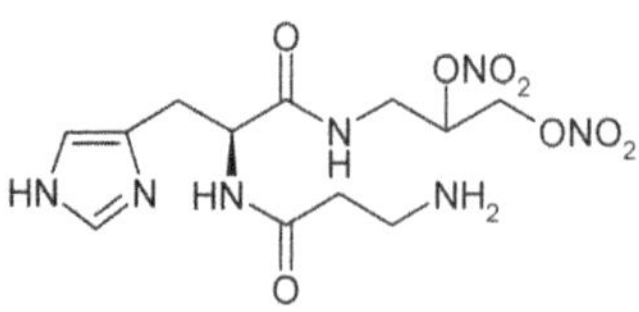	−2.19	77.6
7	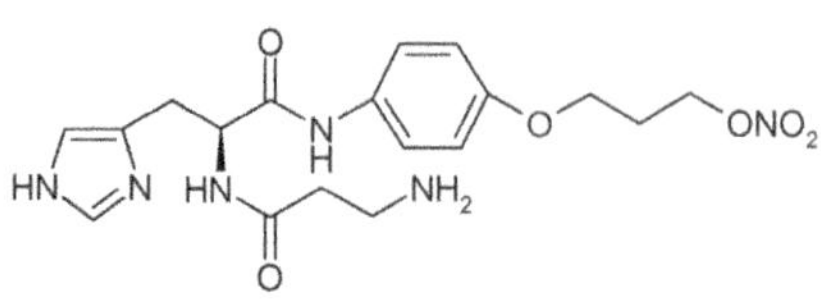	−1.01	74.3

the 3- or 6-positions. β-Cyclodextrine has to be used as a factor that increases the stability, photostability and solubility of a large number of peptide and nonpeptide drugs (Guiotto *et al.*, 2005). In 2002, La Mendola *et al.* functionalized β-cyclodextrine with carnosine and a carnosine-related compound that gave three new compounds (La Mendola *et al.*, 2002). The new derivatives were reported as exhibiting excellent scavenging activity of OH-radicals in pulse-radiolysis investigations. This effect is observed not only due to the formation of the stable imidazole-centered radical, but also because of the scavenger ability of the glucose moieties of the macrocycle. Similar quenching constant values have been reported for all three compounds (Amorini *et al.*, 2007).

The following studies on cyclodextrine derivatives were carried out by Bellia *et al.* (2008). Apart from, β-cyclodextrin and derivatives of carnosine, properties of the conjugates with anserine (**ASN**) were also studied. The increase of lipoproteins protection against oxidation when conjugates were applied was evident (Table 11.3).

Also, the pH dependency of the capacity of anserine conjugatives to lipid stabilization (expressed as **MDA** formation) in Cu^{2+}-induced human **LDL** fraction was investigated. The authors reported that mild acidification of the incubation medium had no effect either on the extent of Cu^{2+} induced **LDL** oxidation or on the inhibitory capacity of the β-cyclodextrin anserine derivatives. Similar amounts of dimalone aldehyde were generated at pH 7.40 and 5.00 (Bellia *et al.*, 2008).

In 2007, Bellia *et al.* investigated the antioxidant activity of the complexes of copper(II) ions and β-cyclodextrin (**CD**) derivatives of homocarnosine (**HC**) *via* pulse radiolysis measurements. Both the created compounds (Figure 11.4) showed a high catalytic activity towards the dismutation of the superoxide radical anion. The results demonstrated that complex formation is not detrimental to excellent scavenger activity towards hydroxyl radical exhibition by the ligands alone. These copper complexes represent very intriguing antioxidant properties against well-known toxic reactive oxygen species (Bellia *et al.*, 2007).

Table 11.3 Antioxidant effect of anserine, β-cyclodextrine and their conjugates. The table presents the antioxidative effect of selected concentrations of anserine **ASN**, β-cyclodextrine **CD** and their conjugates, expressed as percentage protection of LDL fraction (0%-protection + control LDL incubated with buffer only). Data are from Bellia *et al.* (2008) with permission from Elsevier.

Concentration [μM]	ASN	CD	CD + ASN
2	0.23	0.11	0.46
5	0.51	0.38	1.27
10	1.05	0.65	2.02
20	2.14	1.26	3.87
50	8.55	3.87	14.91
100	12.21	6.94	25.62
200	18.72	10.72	29.77

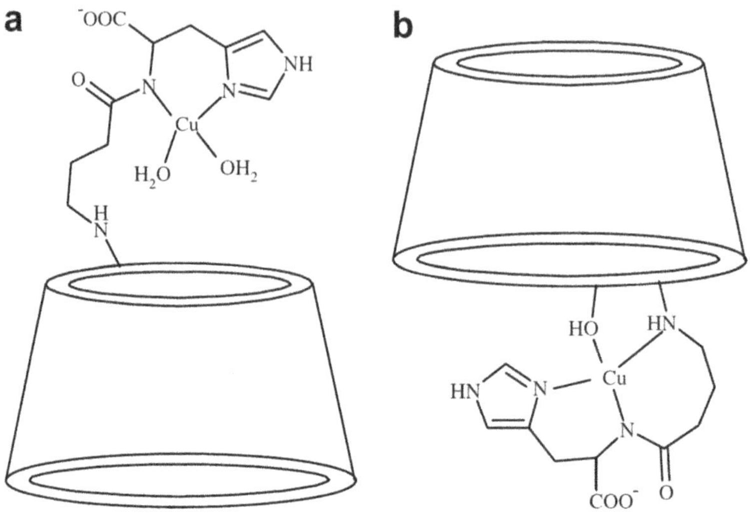

Figure 11.4 Tentative structures of the complexes of copper(II) ions and β-cyclodextrin derivatives of homocarnosine. Structure **a** shows [Cu(CDHC$_6$)H$_{-1}$] and structure **b** shows [Cu(CDHC$_3$)H$_{-1}$], where **CD** is β-cyclodextrin and **HC** is homocarnosine. Data are from Bellia *et al.* (2007) with permission from Elsevier.

In 2010, Stvolinsky *et al.* studied the activity of carnosine derivatives synthesized by *N*-acylation with (*S*) and (*R*)-trolox (Stvolinsky *et al.*, 2010). The novel compounds were characterized by resistance to serum carnosinases and exhibited antioxidant and membrane stabilizing activities in cell systems. (*S*) and (*R*)-Trolox-L-carnosines demonstrated a significantly higher ability to quench **DPPH** radicals than trolox at concentrations of 0.025–0.100 mM. At higher concentrations (about 0.3 mM) the difference was not so evident. Pure carnosine showed no activity or trace ability to quench **DPPH** at the same conditions.

In cell systems, different activities of the trolox derivatives in comparison to carnosine can be observed: *inter alia*, the inhibition activity of red blood cell oxidative hemolysis of the derivatives was lower than suppressions caused by carnosine. However, in cerebellum granule cell suspensions during induction of oxidative stress by hydrogen peroxide (H_2O_2) or *N*-methyl-D-aspartate both derivatives more efficiently lowered the accumulation of **ROS** than trolox or carnosine (Stvolinsky *et al.*, 2010).

Summary Points

- Imidazole dipeptides are a large group of compounds widely investigated by many authors since 1900.
- Compounds containing histidine and their derivatives are well-known antioxidants.

- Besides the ability to chelate pro-oxidative metal-ion antioxidant activity or trap aldehydes, which are secondary products of lipid oxidation, histidine dipeptides exhibit free-radical scavenging properties under certain conditions.
- Also, they are able to react with simple sugars and block the glycation process.
- Due to their biological activity imidazole dipeptides can be considered and applied as potential therapeutic agents against metabolic, neurodegenerative, and cardiovascular diseases.

Key Facts

Key facts about the antioxidant activity of imidazole dipeptides and their derivatives. This table lists the key facts concerning the antioxidant activity of imidazole dipeptides and their derivatives, including proposed mechanisms and dependency factors of antioxidative action.

1. Dipeptides containing histidine show antioxidant activity in various systems such as liposomes, tissues, protein and lipids systems.
2. In liposome model systems, protection ability against oxidation is observed at low dipeptide concentrations.
3. However, in some cases of metal-induced systems dipeptides present pro-oxidative activity.
4. Antioxidant action of compounds with imidazole rings is realized *via* quenching of aldehydic compounds and radical-scavenging mechanisms.
5. Nondirect influence on oxidation processes, especially in excitable animal and human tissues, may be exerted by the buffering activity of imidazole rings.
6. In terms of antioxidant activity, the concentrations and structural dependency of imidazole dipeptides are evident.
7. Structural modification is the main causes of an increase in antioxidant activity.

Definitions of Words and Terms

Lipid peroxidation. A process of oxidative degradation of lipids. Mostly, it is a radical chain reaction that leads to the formation of lipoperoxides and subsequently aldehydic compounds. However, it can also proceed *via* enzymatic reactions or photo-oxidation pathways.

Protein oxidation. Protein oxidation is connected with accumulation of a modified form of proteins during aging, oxidative stress and in some pathological conditions. It proceeds *via* radical reactions or interactions with aldehydic products of lipid oxidation.

Liposomes. Liposomes are artificial, nanosized spherical structures that can be used as models of biological membranes.

Free radicals. Free radicals are atoms, ions or molecules with unpaired valence electrons.

Antioxidant. A compound that inhibits the oxidation of other substances or molecules.

ABTS. ABTS is an abbreviation for 2,2′-azinobis-(3-ethylbenzothiazoline-6-sulfonic) acid. The ABTS test (based on the ABTS radical-scavenging activity of tested compounds) is one of several spectrophotometric analyses of antioxidant substances.

DPPH. DPPH is an abbreviation for α,α-diphenyl-β-picrylhydazyl. The DPPH test (based on the DPPH radical-scavenging activity of tested compounds) is the most common spectrophotometric method to determine antioxidant activity.

HNE. HNE is an abbreviation for 4-hydroksynonenal, one of the products of lipid peroxidation.

MDA. MDA is an abbreviation for malondialdehyde, aldehydic product of lipid peroxidation.

TBARS. TBARS is an abbreviation for substances reacting with thiobarbituric acid. The TBARS test is a method to determine oxidation changes in different systems.

List of Abbreviations

SOD	Superoxide dismutase
GPx	Glutathione peroxidase
CAT	Catalase
LDL	Low-density lipoproteins
ROS	Reactive oxygen species
PUFAs	Polyunsaturated fatty acids
TBARS	Substances reacting with thiobarbituric acid
MDA	Malondialdehyde
AAPH	2,2′-Azobis 2-amidino-propane dichloride
AMVN	2,2′-Azobis(2,3-dimethylvoleronitrile
ESR	Electron spin resonance
DFPH	2,2-Diphenyl-1-picrylhydrazyl
DPPH	α,α-Diphenyl-β-picrylhydazyl
ABTS	2,2′-Azinobis-(3-ethylbenzothiazoline-6-sulfonic acid
AGEs	Advanced glycation end products
HNE	4-Hydroxynonenal
pH_{ic}	Intracellular pH
CD	β-Cyclodextrin
ASN	Anserine
HC	Homocarnosine

References

Abe, H., 2000. Role of histidine-related compounds as intercellular proton buffering constituents in vertebrate muscle. *Biochemistry (Moscow)*. 65: 757–765.

Aldini, G., Carini, M., Beretta, G., Bradamante, S. and Facian, R. M., 2002. Carnosine is a quencher of 4-hydroxynonenal: through what mechanism of reaction. *Biochemical and Biophysical Research Communications*. 298: 699–706. Copyright 2014 Elsevier Masson SAS. All rights reserved.

Amorini, A. M., Bellia, F., Di Pietro, V., Giardina, B., La Mendole, D., Lazzarino, G., Sortino, S., Tarazzi, B., Rizzarelli, E. and Vecchio, G., 2007. Synthesis and antioxidant activity of new homocarnosine β-cyclodextrine conjugates. *European Journal of Medicinal Chemistry*. 42: 910–920.

Babizhayev, M. A., Seguin, M. C., Gueyene, J., Evstigneeva, R. P., Ageyeva, E. A. and Zheltukhina, G. A., 1994. L-Carnosine (β-alanyl-L-histidine) and carcinine (β-alanylhistamine) act as natural antioxidants with hydroxyl-radical-scavenging and lipid-peroxidase activities. *Journal of Biochemistry*. 304: 509–516.

Babizhayev, M. A., Courbebaisse, Y., Nicolay, J. F. and Semiletor, Y. A., 1998. Design and biological activity of imidazole-containing peptidomimetics with a broad-spectrum antioxidant activity. *Letters in Peptide Science*. 5: 163–169.

Badar, H. M., 2007. Antioxidative activity of carnosine in gamma irradiated ground beef and beef patties. *Food Chemistry*. 104: 665–679.

Bellia, F., La Mendola, D., Maccarrone, G., Mineo, P., Vitalini, D., Scamporrino, E., Sorino, S., Vecchio, G. and Rizzarelli, E., 2007. Copper(II) complexes with β-cyclodextrin homocarnosine conjugates and their antioxidant activity. *Inorganica Chimica Acta*. 360: 945–954. Copyright 2014 Elsevier Masson SAS. All rights reserved.

Bellia, F., Amorini, A. M., La Mendola, D., Vecchio, G., Tavazzi, B., Giardina, B., Di Pietro, V., Lazzarino, G. and Rizzarelli, E., 2008. New glycosidic derivatives of histidine-containing dipeptides with antioxidant properties and resistant to carnosinase activity. *European Journal of Medicinal Chemistry*. 43: 373–380. Copyright 2014 Elsevier Masson SAS. All rights reserved.

Bertinaria, M., Rolando, B., Giorgis, M., Montanaro, G., Marini, E., Collino, M., Benetti, E., Daniele, P. G., Fruttero, R. and Gasco, A., 2012. Carnosine analogues containing NO-donor substructures. synthesis, physico-chemical characterization and preliminary pharmacological profile. *European Journal of Medicinal Chemistry*. 54: 103–112. Copyright 2014 Elsevier Masson SAS. All rights reserved.

Biazik, E., Kopeć, W., Skiba, T. and Pudło, A., 2011. Lipids stability and total antioxidant activity in muscle tissue and myofibrils model systems depending on carnosine addition. In: *Proceedings of XX European Symposium*. W.P.S.A., Leipzig, pp. 1–13.

Biazik, E., Pudlo, A., Zaposlki, R., Kopeć, W. and Skiba, T., 2012. *Oxidation of Liposomes in the Ascorbate–Iron (III) Model System in the Presence of Histodone Dipeptides*. Przemysł chemiczny, vol. 91, pp. 684–687.

Bogardus, S. L. and Boissonneault, G. A., 2000. Carnosine inhibits in vitro low-density lipoprotein oxidation. *Nutrition Research*. 20: 967–976.

Boldyrev, A. A., Koldobski, A., Kurella, E., Maltseva, V. and Stvolinski, S., 1993. Natural histidine-containing dipeptide carnosine as a potent hydrophilic antioxidant with membrane stabilizing function. *Molecular and Chemical Neuropathology*. 19: 185–192.

Chen, T. C. and Waimaleongora-Ek, C., 1981. Effect of pH on TBA values of ground raw poultry meat. *Journal of Food Sciences*. 46: 1946–1947.

Das, A. K., Anjaneyulu, A. S. R. and Biswas, S., 2006. Effect of carnosine preblending on the quality of ground buffalo meat. *Food Chemistry*. 97: 531–538.

Davecy, C. L., 1960. The significance of carnosine and anserine in striated skeletal muscle. *Archives of Biochemistry and Biophysics*. 89: 303–308.

Decker, E. A., Crum, A. D. and Calvert, J. T., 1992. Difference in the antioxidant mechanism of carnosine in the presence of copper and iron. *Journal of Agricultural and Food Chemistry*. 40: 756–759.

Gorbunov, N. V. and Erin, A. N., 1991. Mechanism of antioxidant action of carnosine. *Bulletin of Experimental Biology and Medicine*. 111(5): 477–478.

Guiotto, A., Calderan, A., Ruzza, P. and Borin, G., 2005. Carnosine and carnosine related antioxidants: a review. *Current Medicinal Chemistry*. 12: 2293–2315.

Kensaci, G., Genot, C., Meyenier, A. and Gandemer, G., 1997. The antioxidant activity of carnosine and its consequences of the volatile profiles of liposomes during iron/ascorbate induced phospholipids oxidation. *Food Chemistry*. 60: 165–175.

Kohen, R., Yamamoto, Y., Cundy, K. C. and Ames, B. N., 1988. Antioxidant activity of carnosine, homocarnosine and anserine present in muscle and brain. *Proceedings of the National Academy of Sciences of the United States of America*. 88: 3175–3179.

Kohen, R. and Nyska, A., 2002. Invited review: oxidation of biological systems: oxidative stress phenomena, antioxidants, redox reactions, and methods for their quantification. *Toxicologic Pathology*. 30(6): 620–650.

La Mendola, D., Sortino, S., Vecchio, G. and Rizzareli, E., 2002. Synthesis of new carnosine derivatives of β-cyclodextrin and their hydroxyl radical scavenger ability. *Helvetica Chimica Acta*. 85: 1633–1643.

Mather, B. D., Viswanathan, K., Miller, K. M. and Long, T. E., 2006. Michael addition reactions in macromolecular design for emerging technologies. *Progress in Polymer Science*. 31(5): 487–531.

Murase, H., Nagao, A. and Terao, J., 1993. Antioxidant and emulsifying activity of *N*-(long-chain-acyl) histidine and *N*-(long-chain-acyl)carnosine. *Journal of Agricultural and Food Chemistry*. 41: 1601–1604.

Salim-Hanna, M., Lissi, E. and Videla, L. A., 1991. Free radical scavenging activity of carnosine. *Free Radical Research Communications*. 14: 263–270.

Seguin, M. C. and Babizhayev, M., 1998. *U.S. Pat.*, 5792784.

Seguin, M. C. and Babizhayev, M., 2000. *U.S. Pat.*, 6046340.

Shi, X., Dalal, N. S. and Kasprzak, K. S., 1992. Generation of free radicals from lipid hydroperoxides by Ni^{2+} in the presence of oligopeptides. *Archives of Biochemistry and Biophysics*. 299(1): 154–162.

Stvolinsky, S. L., Bulyginia, E. R., Federova, T. N., Meguro, K., Sato, T., Tyulina, O. V., Abe, H. and Boldyrev, A. A., 2010. Biological activity of novel synthetic derivatives of carnosine. *Cellular and Molecular Neurobiology*. 30: 395–404.

Tichivangana, J. Z. and Morrissey, P. A., 1985. The influence of pH on lipid oxidation in cooked meats from several species. *Irish Journal of Food Science and Technology*. 9: 99–106.

Wu, H. C., Shiau, C. Y., Chen, H. M. and Chiou, T. K., 2003. Antioxidant activities of carnosine, anserine, some free amino acids and their combination. *Journal of Food and Drug Analysis*. 11: 148–153.

Yen, W. J., Chang, L. W., Lee, C. P. and Duh, P. D., 2002. Inhibition of lipid peroxidation and nonlipid oxidative damage by carnosine. *Journal of the American Oil Chemists Society*. 79: 329–333.

Yuji, H., Weiss, J., Villeneuve, P., Lopez-Giraldo, L. J., Figueroa-Espinoza, M. C. and Decker, E. A., 2007. Ability of surface-active antioxidants to inhibit lipid oxidation in oil-in-water emulsions. *Journal of Agricultural and Food Chemistry*. 55: 11052–11056.

Zhou, S. and Decker, E. A., 1999a. Ability of amino acids, dipeptides, polyamines and sulfhydryls to quench hexanal, a saturated aldehydic lipid oxidation products. *Journal of Agricultural and Food Chemistry*. 47: 1932–1935.

Zhou, S. and Decker, E. A., 1999b. Ability of carnosine and other skeletal muscle components to quench unsaturated aldehydic lipid oxidation products. *Journal of Agricultural and Food Chemistry*. 47: 51–55.

CHAPTER 12

Inflammaging, Oxidative Stress and Carnosine: Role of Hormetic Vitagenes

VITTORIO CALABRESE[a], SERGIO DAVINELLI[b], MARIA LUCA[a], DAVIDE ZELLA[d], EDWARD J. CALABRESE[e], AND GIOVANNI SCAPAGNINI*[b,c]

[a]Department of Chemistry, University of Catania, Viale Andrea Doria 6, Catania 95125, Italy; [b]Department of Medicine and Health Sciences, University of Molise, Via De Sanctis s.n.c., Campobasso 86100, Italy; [c]Inter-University Consortium "Sannio Tech", Benevento, 82030, Italy; [d]Department of Biochemistry and Molecular Biology, University of Maryland-School of Medicine, 725 W. Lombard Street, Baltimore, MD 21201, USA; [e]Department Environmental Health Sciences, University of Massachusetts Amherst, Morrill I, Amherst, MA 01003, USA
*E-mail: g.scapagnini@gmail.com

12.1 Introduction

Carnosine (β-alanyl-L-histidine) is a naturally occurring endogenous dipeptide, widely expressed in many tissues, including brain and skeletal muscles at concentrations from 2 to 25 mM (Bonfanti *et al.*, 1999; Park *et al.*, 2005). Several studies have revealed that carnosine has antioxidative and anti-inflammatory activities based on its physiological and biochemical

Food and Nutritional Components in Focus No. 8
Imidazole Dipeptides: Chemistry, Analysis, Function and Effects
Edited by Victor R Preedy

Published by the Royal Society of Chemistry, www.rsc.org

functions such as free-radical scavenger, membrane-protecting properties, divalent-metal-ion chelator and immunomodulatory features (Bellia *et al.*, 2012; Boldyrev *et al.*, 1999; Quinn *et al.*, 1992). The beneficial and protective effects of carnosine (Figure 12.1) have been studied *via* both *in vitro* and *in vivo* models, but the translation of highly promising basic research into clinical applications is strongly limited. This is mainly due to its short half-life in humans and its rapid inactivation caused by serum and tissue carnosinase.

To optimize the biological activity of carnosine and its resistance to cleaving by the carnosinase enzyme, intensive research efforts have focused on the design and development of carnosine derivatives more metabolically stable *in vivo* (Bellia *et al.*, 2008). These analog products display carnosine-like properties and they can be considered capable of many anti-inflammatory effects with increased quenching efficacy against free-radical species. Indeed, carnosine attenuates many chronic inflammatory conditions and may be useful to reduce the inflammatory cascade involved in the pathogenesis of inflammation-related diseases (Kulebyakin *et al.*, 2012; Son *et al.*, 2008). Moreover, free radicals induce the formation of inflammatory mediators and one of the most studied activities of carnosine is its role in protecting cellular components from oxidative damage. Accordingly, the peculiar quenching mechanism of carnosine against reactive species or its ability to inhibit protein modifications induced by deleterious effects of oxidative and nitrosative stress, strongly support carnosine as a promising therapeutic agent for free radical-based diseases (Bellia *et al.*, 2011).

Although carnosine's mechanism of action as an anti-inflammatory or antioxidant agent remain partially obscure, recently there has been interest in identifying synergistic interactions between carnosine and other natural compounds to achieve more effective protection against oxidative damage and inflammatory processes (Davinelli *et al.*, 2013).

In this scenario, carnosine may appear particularly useful for antiaging intervention since it is able to affect redox-sensitive cytoprotective genes, called vitagenes (Calabrese *et al.*, 2006d, 2011). Successful aging and long-term health are associated with stress resistance and maintenance of homeostatic regulatory processes. The vitagene network (Figure 12.2) encompass groups of genes involved in preserving multiple components of prosurvival pathways during stressful conditions such as oxidative stress and inflammation.

Vitagenes encode for members of the heat-shock protein (Hsp) family (Table 12.1), including heme oxygenase-1 (HO-1) and Hsp72, sirtuins

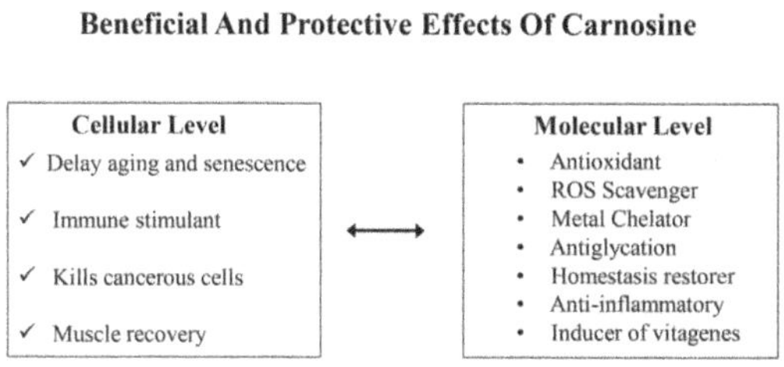

Figure 12.1 Role of carnosine in cytoprotection.

(Table 12.2) and the thioredoxin/thioredoxin reductase system (Calabrese *et al.*, 2011). Although further studies are needed to assess the *in vivo* anti-aging activity of carnosine, it produces a selective effect on the life-span and delays aging in senescence-accelerated mice and drosophila (Hipkiss, 2009a). For this reason and for its antioxidant and anti-inflammatory abilities, carnosine is beginning to emerge as an essential geroprotector with significant impact on the activation of vitagenes involved in longevity processes (Boldyrev *et al.*, 2010; Calabrese *et al.*, 2008b). In addition to its role as an antioxi-inflammaging agent, the efficacy of carnosine as an inducer of vitagenes may be described within an hormetic context (Calabrese *et al.*, 2011).

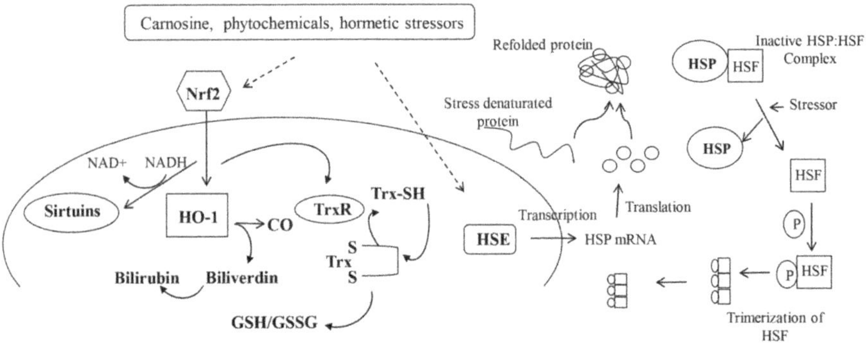

Figure 12.2 The vitagene network.

Table 12.1 Key facts of heat-shock proteins. The table reports the functions of heat-shock proteins in normal and stressful conditions and highlights their importance as neuroprotective agents. Data from: Calabrese V. *et al.*, 2010. Cellular stress responses, the hormesis paradigm, and vitagenes: novel targets for therapeutic intervention in neurodegenerative disorders. *Antioxidants and Redox Signaling.* 13: 1763–1811.

- They are so called because they were originally discovered as a cellular defence mechanism against heat-related cellular stress.
- They belong to the cellular stress response system and they are "molecular chaperones", that is proteins with a huge number of protective properties.
- Stressful conditions, such as oxidative stress, determine alterations in the folding of proteins and lead to cell death.
- Chaperones promote cell survival by sequestering damaged proteins and preventing their aggregation.
- In normal conditions, chaperones also promote the transport of macromolecules and are involved in signaling, transcription, cell division, migration, and differentiation.
- Carnosine is involved in the activation of heat-shock proteins (see text).
- Many degenerative disorders (among all, Alzheimer's disease) are characterized by the presence of misfolded proteins.
- Heat-shock proteins exert a crucial neuroprotective role and their altered expression expose neurons to the damages of oxidative stress.

The purpose of this chapter is to provide an overview of the antiaging effects of carnosine and its hormetic action in stimulating the activation of vitagenes.

12.2 Hormesis

Hormesis is a dose–response phenomenon characterized by a low-dose stimulation and a high-dose inhibition (Figure 12.3). It may be graphically represented by either an inverted U-shaped dose–response or by a J- or U-shaped

Table 12.2 Key facts of sirtuins. The table briefly resumes the molecular bases of the sirtuins' regulatory functions in order to clarify their role in the pathogenesis of degenerative disorders. Data from: Calabrese V. *et al.*, 2012. Oxidative stress, glutathione status, sirtuin and cellular stress response in type 2 diabetes. *Biochimica et Biophysica Acta (BBA) - Molecular Basis of Disease*. 1822: 729–736.

- They are also called nicotinamide adenine dinucleotide (NAD)-dependent deacetylases.
- They are a complex of proteins with regulatory functions on transcriptional silencing in the genome.
- They act through the regulation of proteins' acetylation-deacetylation pathways.
- Sirtuin-mediated protein deacetylation is affected by carnosine (see text).
- Modulating the acetylation of the respiratory chain, sirtuins protect against oxidative stress.
- They also stimulate mitocondria superoxide dismutase and isocitrate dehydrogenase, that are involved in the glutathione pathway.
- If their expression is altered, we are more vulnerable to sustained oxidative stress.
- Due to the fundamental regulatory function of sirtuins, the pathogenesis of various oxidative stress-related degenerative disorders is somehow linked to the sirtuins pathway.

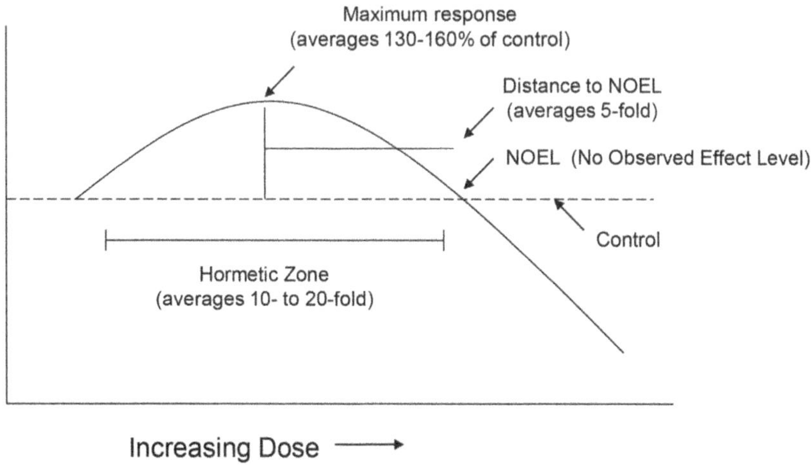

Figure 12.3 Dose–response curve depicting the qualitative features of hormesis.

dose–response. The term hormesis was first presented in the published literature in 1943 by Southam and Ehrlich who reported that low doses of extracts from the Red Cedar tree enhanced the proliferation of fungi with the overall shape of the dose–response being biphasic. However, credit for experimentally demonstrating the occurrence of hormesis belongs to Hugo Schulz (Calabrese *et al.*, 2011) who reported biphasic dose–responses in yeast following exposure to a large number of chemical disinfectant agents. The work of Schulz encouraged numerous investigators in diverse fields to assess whether such low-dose effects may be a general feature of biological systems. In fact, similar types of dose–response observations were subsequently reported by numerous researchers assessing chemicals (Calabrese V. *et al.*, 2010a) and radiation (Calabrese E. J. *et al.*, 2010a,b, 2012; Calabrese and Calabrese, 2013b,c) and they named these responses Arndt–Schulz Law, Hueppe's Rule, and other terms to describe these similar dose–response phenomena. Despite the substantial historical literature concerning hormetic dose–responses, this concept had a difficult time being incorporated into routine hazard assessment and pharmacological investigations; this was principally due to: (1) the fact that it required more rigorous evaluation in the low dose zone, (2) failure of investigators to understand its clinical significance (3) the failure to appreciate the quantitative features of the hormetic dose–response (4) the failure to understand the limitations of its agricultural and industrial applications, (5) because of the predominant interest in responses at relatively high doses during most of the 20th century as well as (6) the continuing, yet inappropriate, tendency to associate the concept of hormesis with the medical practice of homeopathy (Calabrese *et al.*, 2013a). However, since the late 1970s there has been a growing interest in hormetic-like biphasic dose–responses across the broad spectrum of biomedical sciences. This resurgence of interest resulted from a variety of factors, including the capacity to measure progressively lower doses of drugs and chemicals, the adoption of cell culture methods that has permitted more efficient testing of numerous doses and the need to re-examine the validity of linear at low-dose modeling of cancer risks due to their enormous cost implications for environmental regulations, as well the astute observations of numerous independent investigators to generalize their hormetic findings across biological systems (Calabrese *et al.*, 2007b, 2011; Mancuso *et al.*, 2007a).

The concept of hormesis is a central to the biomedical sciences because of its generalizability (Calabrese *et al.*, 2013a; Calabrese and Calabrese, 2013b,c). Hormetic dose–responses are independent of biological model, endpoint as well as chemical class and physical agent (Calabrese *et al.*, 2011). In direct head-to-head comparisons with the threshold and linear dose–response models the hormetic dose–response model was found in multiple studies with many thousands of dose–responses to make far more accurate predictions of low dose–responses (Calabrese V. *et al.*, 2010a). The hormetic dose–response has also been shown to describe the fundamental features of several dozen receptor systems, affecting a vast array of biological endpoints (Calabrese E. J. *et al.*, 2010b). The concepts of preconditioning and postconditioning are also manifestations of hormesis. When these two highly significant and general concepts are analyzed within the dose–response framework they conform to the features of the hormetic dose–response (Calabrese *et al.*, 2011). Of particular importance is that the quantitative features of the hormetic dose–response are similar across biological models and endpoints. This remarkable general feature of hormetic dose–response suggests that the hormetic dose–response may represent the first comprehensively based quantitative estimation of biological plasticity (Calabrese V. *et al.*, 2010a).

The failure to recognize and incorporate the concept of hormesis as a central biological concept during the key development period of the dose–response concept consolidation for pharmacology and toxicology during the middle decades of the 20th century had many implications. For example, the failure to recognize the generality of biphasic dose–responses led to the development of toxicology becoming a high-dose testing discipline that also only tested few doses. It became a discipline that developed flawed regulations based on an inadequate understanding of the dose–response that have inaccurately estimated the costs and the benefits of such regulations, thereby yielding an inadequate foundation for public policy decisions (Calabrese *et al.*, 2013a). Such historical failures led not only to the development of biologically unsupportable environmental regulations especially for carcinogens but also profoundly slowed the recognition and generality of adaptive responses and how they might be exploited within society. Not only was this failure highly significant in the environmental realm but it also lead to the failure of the biomedical community to adequately test chemotherapeutic agents adequately. For example, many antitumor agents that kill tumor cells at high doses may also stimulate the proliferation of these cells at lower doses, displaying the hormetic dose–response (Calabrese *et al.*, 2013a). Furthermore, since the hormetic dose–response has clearly defined quantitative features and is highly generalizable it also provides a quantitative estimate of biological performance. That is, it estimates the extent to which memory drugs can enhance learning. This concept can be readily applied to all other areas of biological performance such as bone strengthening, hair growth, decreases in anxiety, plant growth and productivity and others.

Of particular relevance to the current chapter is that the hormesis concept can be used to understand the biological limits within which pharmaceutical efforts are made to enhance the quality of aging and to affect longevity. For example, the hormesis concept argues that the biological performance can only be modestly improved in all biological systems in most situations. The term modestly is quantitatively constrained to be less than two-fold, with most maximal increases being within the 30–60% range. Such biological constraints are of vital importance to pharmaceutical companies and regulatory agencies as they develop strategies for product development and evaluation. This also suggests that attempts to extend life would also be constrained within the bounds of plasticity, which are quantitatively estimated by the hormetic dose–response. Thus, if the normal bounds of human longevity are seen to be approximately as about 100 years, the hormesis concept predicts that it may be possible to extend the human lifespan by 30–60 years at most (Calabrese *et al.*, 2013a).

The data to support these types of biological constraints are extensive. They are based on more than 20 000 dose–responses in several hormesis databases. These data sets are very general, including responses across a very broad range of biological models and endpoints to a very large number of chemical agents representing a broad spectrum of chemical classes. The range and the scope of these findings have come a long way in clarifying the limits within which biological systems operate, the overlapping features of biological control redundancies, and the constraints and flexibilities within which cells, organs and organisms operate. All of these features affect the extent to which humans can manipulate biological systems and attempt to gain improvement in function and performance. Thus, the following sections of this chapter on cellular stress responses and vitagenes all can and should be seen within an hormetic context. Even the concept of wound healing, which is essential to ensuring survival and enhancing longevity are fundamentally described within an hormetic context.

12.3 Cellular Stress Response, HSF Biology and the Vitagene Network

Cellular stress response is the ability of a cell to counteract stressful conditions. This phenomenon, which includes heat-shock response (HSR), represents an ancient and highly conserved cytoprotective mechanism (Calabrese *et al.*, 2006c; Lodi *et al.*, 2006). Production of heat-shock proteins, including protein chaperones, is essential for the folding and repair of damaged proteins, thus serving to promote cell-survival conditions that would otherwise result in apoptosis (Calabrese *et al.*, 2006b; Scapagnini *et al.*, 2006). The term "molecular chaperone" denotes a large family of ubiquitous proteins that function as part of an ancient defense system in our cells. Chaperones promote cell survival by sequestering damaged proteins and preventing their aggregation. During stressful conditions, such as elevated temperature, they prevent protein aggregation by facilitating the refolding or elimination of misfolded proteins. The stress-induced response to damaged proteins is helped by a sophisticated regulatory system, which shuts down most cellular functions and, in parallel, induces the synthesis of several chaperones and other survival-promoting proteins. Therefore, many of the chaperones are also called stress or "heat-shock" proteins in reference to the archetype of cellular stress, heat shock. Besides their role during stress, chaperones have multiple roles under normal conditions, as such they promote the transport of macromolecules (*e.g.* proteins or RNA) and participate in remodeling events involving larger protein complexes, including signaling, transcription, cell division, migration and differentiation. Cellular stress response requires the activation of prosurvival pathways that, under control of protective genes called vitagenes (Mancuso *et al.*, 2007a) produce molecules (heat-shock proteins, glutathione, bilirubin) endowed with antioxidant and antiapoptotic activities. Generally, molecular chaperones help a multitude of signaling molecules to keep their activation-competent state, and regulate various signaling processes ranging from signaling at the plasma membrane to transcription. In addition to these specific regulatory roles, recent studies have revealed that chaperones act as genetic buffers stabilizing the phenotypes of various cells and organisms. Among the cellular pathways conferring protection against oxidative stress, a key role is played by the products of vitagenes (Calabrese *et al.*, 2007b). Recent studies have shown that the HSR contributes to establishing a cytoprotective state in a wide variety of human diseases, including inflammation, cancer, aging and neurodegenerative disorders (Calabrese *et al.*, 2007a, 2008a, 2011). Given the broad cytoprotective properties of the HSR there is now strong interest in discovering and developing pharmacological agents capable of inducing the HSR (Abdul *et al.*, 2006; Calabrese *et al.*, 2008b; Mancuso *et al.*, 2007b). Molecular chaperones are known to disrupt aggregates but also to promote active aggregation when the concentration of the aggregating protein is high. Consistent with this notion, although protein aggregation is hazardous under certain circumstances, the creation of apparently less-toxic large aggregates is protective. This hypothesis is the basis of the therapeutic potential of Hsps, which prevent protein misfolding and aggregation (Calabrese *et al.*, 2008c,d).

Cellular stress response is regulated at the transcriptional, translational and post-translational levels by a family of heat-shock factors (HSFs) that are expressed and maintained in an inactive state under nonstress conditions (Calabrese and Maines 2006a; Calabrese V. *et al.*, 2008f, 2010f; Perluigi *et al.*, 2006). HSFs are essential for all organisms to survive to acute or chronic stress, and they are also important for normal development and lifespan-enhancing pathways, and the repertoire of HSF targets has thus expanded well beyond the heat-shock genes (Dinkova-Kostova *et al.*, 2013). Post-translational regulation of HSFs is emerging to integrate the metabolic state of the cell with stress biology, whereby controlling fundamental aspects of the health of the proteome and aging. At the transcriptional level the heat-shock response is mediated by *cis*-acting sequences called heat-shock elements (HSEs) present in multiple copies upstream of the Hsp genes. In contrast to a single HSF in invertebrates, multiple HSFs are expressed in plants and vertebrates. The mammalian HSF family consists of four members: HSF1, HSF2, HSF3 and HSF4, which recognize by binding the heat-shock element (HSE), composed of inverted repeats of a consensus nGAAn sequence. These distinct HSFs possess overlapping functions, display tissue-specific patterns of expression, undergoing multiple post-translational modifications, and show various interacting protein partners. In addition to thermal stress, the inducible expression of HSPs is also triggered by environmental redox changes or exposure to electrophiles that cause trimerization and DNA binding of HSF1, pointing to the importance of the cysteine redox state for the activation of this transcription factor (Dinkova-Kostova *et al.*, 2013). Thus, an intermolecular disulfide bond formation between C36 and C103 within HSF1 causes trimerization and DNA binding, whereas an intramolecular disulfide bond formation is inhibitory for the activity of the transcription factor. Hence, pharmacologic modulation of HSF-mediated gene regulation is an emerging area of research that is increasing its potential based on the current knowledge of small-molecule activators and inhibitors of HSFs, so that the impact of HSFs is further extending beyond the heat-shock response, which possesses the potential for attracting growing interest (Zhang *et al.*, 2011). HSFs, like other transcription factors, are composed of functional domains that have been well characterized for HSF1. The DNA-binding domain (DBD) belongs to the family of winged helix-turn-helix DBDs. The DBD, the signature domain of HSFs for recognition of target genes, forms a compact globular structure, except for a flexible wing or loop, located between β-strands 3 and 4, which is responsible for protein–protein interactions between adjacent subunits of the trimerized HSF, conferring high-affinity binding to DNA and mediating interactions with other factors to modulate the *trans*-activating capacity of HSFs. The trimerization of HSFs is mediated by arrays of hydrophobic heptad repeats (HR-A and HR-B) forming coils, characteristic of many Leu zippers (Akerfelt *et al.*, 2010; Fujimoto *et al.*, 2010; Westerheide *et al.*, 2012). Trimerization of HSF, facilitated by Leu zippers is abrogated by another hydrophobic repeat, HR-C located between the regulatory and *trans*-activation domains (TADs). This latter, by interacting with the HR-A/B domain prevents trimerization, maintaining HSF1 in an inactive status. The trans-activation domain in all HSFs is positioned at the extreme amino terminus, and is composed of two modules, AD1 and AD2, which are rich in hydrophobic and acidic residues that allow rapid and prolonged response to stress (Akerfelt *et al.*, 2010; Fujimoto *et al.*, 2010; Westerheide *et al.*, 2012). Although the N-terminal parts of HSF1 mediate trimer formation and acquisition of DNA-binding ability, the C-terminal parts of the protein facilitate transcriptional activation of target genes and also regulate the magnitude of HSF1 activation (Akerfelt *et al.*, 2010; Fujimoto *et al.*, 2010; Westerheide *et al.*, 2012). Although the regulatory domain of HSF has an intrinsic competence to sense heat stress, its inducibility is modulated through post-translational modifications (Akerfelt *et al.*, 2010; Fujimoto *et al.*, 2010; Westerheide *et al.*, 2012). Pharmacological inhibition of Hsp90 activity is sufficient to convert HSF1 into a DNA-binding trimeric state in cells, thus in nonstressed cells, HSF1 is maintained in an inactive monomeric state by interaction with Hsp90 (Calabrese *et al.*, 2008e; 2009b,c,d; 2011). This interaction is rapidly reduced in cells exposed to proteotoxic insults, where binding of HSF1 to the promoter occurs within seconds and in about one minute is saturated by the intervention of other positive signals such as a ribonucleoprotein complex containing eukaryotic elongation factor 1A (eEF1A) and a constitutive noncoding RNA, heat shock RNA-1 (HSR-1), which has been reported to possess a thermosensing capacity (Calabrese V. *et al.*, 2009a; 2010b,e). Consistent with the proposed model, HSR-1 undergoes a conformational change in response to heat stress facilitating, together with eEF1A, HSF1 trimerization. In this activation cycle, HSF1 undergoes extensive post-translational modifications, such as phosphorylation, sumoylation and acetylation. Under unstressed conditions HSF1 is a phosphoprotein with at least 12 Ser residues are phosphorylated (S121, S230, S292, S303, S307, S314, S319, S326, S344, S363, S419, and S444), with no detectable amount of threonine or tyrosine phosphorylation. Phosphorylation-dependent sumoylation of HSF1 on K298 requires a priming phosphorylation of S303. Phosphorylation of S303, possibly mediated by MAP kinases (MAPK), facilitates HSF1 sumoylation by the SUMO E2 conjugating enzyme Ubc9, leading to repression of HSF1 *trans*-activating capacity (Calabrese V. *et al.*, 2010a, 2011). SUMO proteins structurally resemble to ubiquitin and are transiently and covalently bound to specific lysine (K) residues of multiple cellular proteins. In many cases, sumoylation occurs on a consensus sequence consisting of the tetrapeptide ΨKxE, in which Ψ represents by a branched hydrophobic amino acid (usually a leucin, isoleucin, or valine) and K the SUMO acceptor. In HSF1, adjacent consensus sequences for sumoylation ΨKxE and phosphorylation (SP) together form an extended target motif, ΨKxESP (PDSM), mediating phosphorylation-dependent sumoylation, acting not only in the regulation of HSF biology, but also in a wide range of transcriptional regulators, where SUMO proteins are well-established repressors of transcription (Calabrese V. *et al.*, 2010a, 2011). While phosphorylation and sumoylation of HSF1 occur rapidly on heat shock, the kinetics of acetylation are delayed and coincide with the attenuation phase of the HSF1 activation cycle. Consistent with this notion, acetylation of HSF1 is regulated by the balance of acetylation by p300–CBP (CREB-binding protein) and deacetylation by the NAD^+-dependent sirtuin, SIRT1. Increased expression and activity of SIRT1 enhances and prolongs the DNA-binding activity of HSF1 at the human Hsp70 promoter, whereas downregulation of SIRT1 enhances the acetylation of HSF1 and the attenuation of DNA binding without affecting the formation of HSF1 trimers (Calabrese V. *et al.*, 2010a). Thus SIRT1 maintains HSF1 in a state that is suitable for DNA binding by counteracting acetylation. In the light of current knowledge, the attenuation phase of the HSF1 cycle is regulated by a dual mechanism: a dependency on the levels of Hsps that feedback directly by weak interactions with HSF1, and a parallel step that involves the SIRT1-dependent control of the DNA-binding activity of HSF1. Because SIRT1 has been implicated in caloric restriction and aging, the age-dependent loss of SIRT1 and impaired HSF1 activity correlate with an impairment of the heat-shock response and proteostasis in senescent cells, connecting the heat-shock response to nutrition and aging (Calabrese V. *et al.*, 2010c,d; Di Paola *et al.*, 2011).

12.4 Carnosine as an Inducer of Vitagenes in Oxi-Inflammaging

The causes of aging remains somewhat controversial but it is widely accepted that age-related changes in the organism are linked to oxidative and inflammatory stress. Recently, an integrative theory of aging has been proposed by De la Fuente *et al.* to explain that uncontrolled oxidation and a low grade of chronic systemic inflammation affect the rate of aging process and they are the basis of the loss of cellular homeostasis (De la Fuente and Miquel 2009). Antioxidant diet supplementation may provide a therapeutic strategy to restore homeostatic impairment by targeting oxidative stress and inflammation. Several lines of investigation suggest that dietary antioxidants act hormetically, displaying cytoprotective effects at low doses. Indeed, excessive antioxidant action (*i.e.*, high doses) can adversely affect key physiological processes and may have negative consequences through the generation of more reactive and harmful intermediates (Calabrese V. *et al.*, 2010b). There is evidence that suggests that carnosine has at least the potential to ameliorate a range of age-related biochemical modifications determined by oxidative stress and inflammation (Hipkiss, 2009b). A variety of cellular mechanisms are involved in the regulation of proinflammatory signaling and it is conceivable that carnosine regulates the secretion of inflammatory cytokines under inflammatory conditions. Although studies have shown that carnosine slows aging at the cellular and whole-animal levels, much work is yet to be done to establish its therapeutic value as a new agent to counteract inflammaging. However, it was proved that carnosine can regulate the production of intracellular cytokines responsible of proinflammatory mechanisms. For instance, differential induction of inflammatory cytokines produces increased levels of HO-1 and Hsp70, and these increases were abolished by pretreatment with carnosine (Calabrese *et al.*, 2005). The high amount of nitric oxide (NO) produced by inducible NO synthase (iNOS) is a typical hallmark of many pathological conditions induced by inflammatory stress. In one of our previous studies, we demonstrated that lipopolysaccharides (LPS) and interferon-γ (INF-γ) increase the expression of iNOS protein and treatment with carnosine markedly reduced iNOS expression, confirming a protective effect of this natural dipeptide (Calabrese *et al.*, 2005). The use of the D-carnosine enantiomer reduced the levels of proinflammatory cytokines, including tumor necrosis factor-α (TNF-α) and interleukin-1β (IL-1β) as well as the iNOS expression (Bellia *et al.*, 2011). In addition to TNF-α and IL-1β, it was also found that carnosine preintake dose-dependently suppressed the secretion of IL-6 and other proinflammatory cytokines in the mouse striatum, providing protection against neuroinflammation (Tsai *et al.*, 2010). Son *et al.* investigated the effect of carnosine on the intracellular degradation of the IL-8, and they suggested that the mechanism for the inhibitory effect of carnosine on the cytokine synthesis is related to its structure with a direct influence at the translation or post-translational level (Son *et al.*, 2008). An inflammatory process induces oxidative stress and there are numerous examples of protective actions of carnosine against the cellular damage caused by reactive intermediates. It is noteworthy that HSPs and redox enzymes, such as Hsp70, Hsp72, HO-1, superoxide dismutase 2 (SOD-2) are main actors in the postinflammation processes. It is generally accepted that the age-related accumulation of altered proteins or aberrant protein aggregation are a consequence of the functional decline of homeostatic process. The action of carnosine as a metal chelator may have important effects in preventing or reducing the protein aggregation during age-related diseases. Carnosine complexed to zinc(II) is able to induce the expression of important antioxidative stress proteins and enzymes, including HO-1, Hsp72 (Odashima *et al.*, 2002), Hsp70 (Odashima *et al.*, 2002; Ueda *et al.*, 2009; Nishida *et al.*, 2010). It should be pointed out that overexpression of Hsp70 or its intracellular induction by heat stress has been shown to reduce nuclear factor-kB (NF-kB) activation in astrocytes (Feinstein *et al.*, 1996). In addition, disruption of proteostasis is implicated in age-related diseases and Hsp70 system may prevent the pathological consequences of protein misfolding and aggregation (Kim *et al.*, 2013). Furthermore, carnosine has been shown to prevent β-amyloid aggregation and toxicity, and this effect is related to its potential to inhibit protein mysfolding with consequent reduced formation of advanced glycation end products (Hipkiss, 2010). Recently, it was reported that oral carnosine treatment changed the expression of Hsp27 and HO-1 and its protective effect might be mediated by the induction of Hsp27 (Pfister *et al.*, 2011). Oxidative and nitrosative are intimately involved in the aging process, and it was found that carnosine protects neurons against oxidative stress and modulates the time profile of MAPK cascade signaling (Kulebyakin *et al.*, 2012). Glutathione depletion induces HO-1 and causes extensive carbonylation of proteins and the adduction of carbonyl groups to proteins is due to the presence of excess of reactive carbonyl species (RCS). Carnosine derivatives prevented this deleterious effect sequestering cytotoxic RCS (Vistoli *et al.*, 2009). As mentioned above, oxidative and nitrosative stress are involved in several pathological conditions such as ischemia, methamphetamine neurotoxicity and neurodegenerative disorders, and in brain cells carnosine has been shown to be neuroprotective because of its ability to counteract free-radical species. Thus, carnosine has a huge potential as a therapeutic approach in clinical practice (Calabrese *et al.*, 2011). Recent observations suggest that carnosine and caloric restriction (CR) can delay aging and the onset of age-related pathologies (Hipkiss, 2009b). In particular, Hipkiss argues that carnosine may affect sirtuin-mediated protein deacetylation and ADP-ribose metabolism in different ways. First, carnosine can act as an acetyl acceptor by forming *N*-acetylcarnosine. Secondly, ADP-ribose, a product of sirtuin-mediated NAD-coupled protein deacetylation, is a potent glycating agent. Thus, carnosine could be the ultimate acceptor for ADP-ribose by generating as adduct, ADP-ribosylcarnosine (Hipkiss, 2009b). Finally, even though totally speculative, it has been postulated that CR and carnosine might regulate methylglyoxal production and consequent generation of aberrant proteins, which is a hallmark of the aged phenotype. Methylglyoxal is glycolytic byproduct and toxic glycating agent but carnosine can inhibit its reactivity. Low NAD^+ availability increases intracellular levels of methylglyoxal and carnosine and CR may be involved in the activation of redox-sensitive pathways that increase the cellular NAD^+/NADH ratio. Carnosine and CR may facilitate NAD^+ regeneration to maintain juvenile character *via* NAD^+-dependent sirtuin activity (Hipkiss, 2009b).

Summary Points

- Carnosine has antioxidative and anti-inflammatory activities.
- Carnosine derivatives are more metabolically stable.
- Carnosine may appear particularly useful for antiaging intervention since it is able to affect redox-sensitive cytoprotective genes, called vitagenes.
- Pro-survival pathways and longevity assurance processes are controlled by vitagene network.
- The concept of hormesis is central to the biomedical sciences because of its generalizability.
- Heat-shock response represents an ancient and highly conserved cytoprotective mechanism.
- Dietary antioxidant act hormetically, displaying cytoprotective effects at low doses.
- The central concept of oxidation-inflammation theory is that higher levels of oxidant and inflammatory compounds and lower antioxidant and anti-inflammatory defences affect the aging process.
- Based on its biochemical functions carnosine may modulate the oxi-inflammaging process through a vitagene network.

Definitions of Words and Terms

Vitagene. Genes involved in preserving cellular homeostasis during stressful conditions. Vitagenes encode for heat-shock proteins (Hsp) Hsp32, Hsp70, the thioredoxin and the sirtuin protein systems.
Carnosine. (β-Alanyl-L-histidine) Naturally occurring endogenous dipeptide, is widely expressed in many tissues, including brain and skeletal muscles.
Hormesis. The dose–response phenomenon characterized by a low-dose stimulation and a high-dose inhibition.
Heat-shock response. Cytoprotective mechanism against acute forms of physiological and environmental stress.
Oxi-inflammaging. Chronic oxidative stress and inflammatory stress are the basis of the age-related impairment of organism functions.

List of Abbreviations

Hsp Heat-shock protein
HO-1 Heme oxygenase-1
HSR Heat-shock response

HSF	Heat-shock transcription factor
HSE	Heat-shock element
DBD	DNA-binding domain
HR	Heptad repeats
TAD	*Trans*-Activation domain
eEF1A	Eukaryotic elongation factor 1A
HSR-1	Heat-shock RNA-1
MAPK	MAP kinases
CBP	CREB-binding protein
NO	Nitric oxide
iNOS	Inducible NO synthase
LPS	Lipopolysaccharides
INF-γ	Interferon-γ
TNF-α	Tumor necrosis factor-α
IL	Interleukin
SOD-2	Superoxide dismutase 2
NF-kB	Nuclear factor-kB
RCS	Reactive carbonyl species
CR	Caloric restriction

References

Abdul, H. M., Calabrese, V., Calvani, M. and Butterfield, D. A., 2006. Acetyl-L-carnitine-induced up-regulation of heat shock proteins protects cortical neurons against amyloid-beta peptide 1-42-mediated oxidative stress and neurotoxicity: implications for Alzheimer's disease. *Journal of Neuroscience Research*. 84: 398–408.

Akerfelt, M., Vihervaara, A., Laiho, A., Conter, A., Christians, E. S., Sistonen, L. and Henriksson, E., 2010. Heat shock transcription factor 1 localizes to sex chromatin during meiotic repression. *Journal of Biological Chemistry*. 285: 34469–34476.

Bellia, F., Amorini, A. M., La Mendola, D., Vecchio, G., Tavazzi, B., Giardina, B., Di Pietro, V., Lazzarino, G. and Rizzarelli, E., 2008. New glycosidic derivatives of histidine-containing dipeptides with antioxidant properties and resistant to carnosinase activity. *European Journal of Medicinal Chemistry*. 43: 373–380.

Bellia, F., Vecchio, G., Cuzzocrea, S., Calabrese, V. and Rizzarelli, E., 2011. Neuroprotective features of carnosine in oxidative driven diseases. *Molecular Aspects of Medicine*. 32: 258–266.

Bellia, F., Vecchio, G. and Rizzarelli, E., 2012. Carnosine derivatives: new multifunctional drug-like molecules. *Amino Acids*. 43: 153–163.

Boldyrev, A., Song, R., Lawrence, D. and Carpenter, D. O., 1999. Carnosine protects against excitotoxic cell death independently of effects on reactive oxygen species. *Neuroscience*. 94: 571–577.

Boldyrev, A. A., Stvolinsky, S. L., Fedorova, T. N. and Suslina, Z. A., 2010. Carnosine as a natural antioxidant and geroprotector: from molecular mechanisms to clinical trials. *Rejuvenation Research*. 13: 156–158.

Bonfanti, L., Peretto, P., De Marchis, S. and Fasolo, A., 1999. Carnosine-related dipeptides in the mammalian brain. *Progress in Neurobiology*. 59: 333–353.

Calabrese, E. J., Mattson, M. P. and Calabrese, V., 2010a. Resveratrol commonly displays hormesis: occurrence and biomedical significance. *Human & Experimental Toxicology*. 29: 980–1015.

Calabrese, E. J., Mattson, M. P. and Calabrese, V., 2010b. Dose response biology: the case of resveratrol. *Human & Experimental Toxicology*. 29: 1034–1037.

Calabrese, E. J., Iavicoli, I. and Calabrese, V., 2012. Hormesis: why it is important to biogerontologists. *Biogerontology*. 13: 215–235.

Calabrese, E. J., Iavicoli, I. and Calabrese, V., 2013a. Hormesis: its impact on medicine and health. *Human & Experimental Toxicology*. 32: 120–152.

Calabrese, E. J. and Calabrese, V., 2013b. Reduction of arthritic symptoms by low dose radiation therapy (LD-RT) is associated with an anti-inflammatory phenotype. *International Journal of Radiation Biology*. 89: 278–286.

Calabrese, E. J. and Calabrese, V., 2013c. Low dose radiation therapy (LD-RT) is effective in the treatment of arthritis: animal model findings. *International Journal of Radiation Biology*. 89: 287–294.

Calabrese, V., Colombrita, C., Guagliano, E., Sapienza, M., Ravagna, A., Cardile, V., Scapagnini, G., Santoro, A. M., Mangiameli, A., Butterfield, D. A., Giuffrida Stella, A. M. and Rizzarelli, E., 2005. Protective effect of carnosine during nitrosative stress in astroglial cell cultures. *Neurochemical Research*. 30: 797–807.

Calabrese, V. and Maines, M. D., 2006a. Antiaging medicine: antioxidants and aging. *Antioxidants & Redox Signaling*. 8: 362–364.

Calabrese, V., Colombrita, C., Sultana, R., Scapagnini, G., Calvani, M., Butterfield, D. A. and Stella, A. M., 2006b. Redox modulation of heat shock protein expression by acetylcarnitine in aging brain: relationship to antioxidant status and mitochondrial function. *Antioxidants & Redox Signaling*. 8: 404–416.

Calabrese, V., Butterfield, D. A., Scapagnini, G., Stella, A. M. and Maines, M. D., 2006c. Redox regulation of heat shock protein expression by signaling involving nitric oxide and carbon monoxide: relevance to brain aging, neurodegenerative disorders, and longevity. *Antioxidants & Redox Signaling*. 8: 444–477.

Calabrese, V., Giuffrida Stella, A. M., Calvani, M. and Butterfield, D. A., 2006d. Acetylcarnitine and cellular stress response: roles in nutritional redox homeostasis and regulation of longevity genes. *Journal of Nutritional Biochemistry*. 17: 73–88.

Calabrese, V., Mancuso, C., Calvani, M., Rizzarelli, E., Butterfield, D. A. and Stella, A. M., 2007a. Nitric oxide in the central nervous system: neuroprotection versus neurotoxicity. Nature Reviews. *Neuroscience*. 8: 766–775.

Calabrese, V., Guagliano, E., Sapienza, M., Panebianco, M., Calafato, S., Puleo, E., Pennisi, G., Mancuso, C., Butterfield, D. A. and Stella, A. G., 2007b. Redox regulation of cellular stress response in aging and neurodegenerative disorders: role of vitagenes. *Neurochemical Research*. 32: 757–773.

Calabrese, V., Calafato, S., Puleo, E., Cornelius, C., Sapienza, M., Morganti, P. and Mancuso, C., 2008a. Redox regulation of cellular stress response by ferulic acid ethyl ester in human dermal fibroblasts: role of vitagenes. *Clinics in Dermatology*. 26: 358–363.

Calabrese, V., Cornelius, C., Mancuso, C., Pennisi, G., Calafato, S., Bellia, F., Bates, T. E., Giuffrida Stella, A. M., Schapira, T., Dinkova Kostova, A. T. and Rizzarelli, E., 2008b. Cellular stress response: a novel target for chemoprevention and nutritional neuroprotection in aging, neurodegenerative disorders and longevity. *Neurochemical Research*. 33: 2444–2471.

Calabrese, V., Bates, T. E., Mancuso, C., Cornelius, C., Ventimiglia, B., Cambria, M. T., Di Renzo, L., De Lorenzo, A. and Dinkova-Kostova, A. T., 2008c. Curcumin and the cellular stress response in free radical-related diseases. *Molecular Nutrition & Food Research*. 52: 1062–1073.

Calabrese, V., Signorile, A., Cornelius, C., Mancuso, C., Scapagnini, G., Ventimiglia, B., Ragusa, N. and Dinkova-Kostova, A., 2008d. Practical approaches to investigate redox regulation of heat shock protein expression and intracellular glutathione redox state. *Methods in Enzymology*. 441: 83–110.

Calabrese, V., Butterfield, D. A. and Stella, A. M., 2008e. Oxidative stress response in the CNS. In: Lajtha, A., Perez-Polo, J. R. and Rossner, S. (ed.) *Handbook of Neurochemistry and Molecular Neurobiology, Development and Aging Changes in the Nervous System*. 3rd edn. p. 43.

Calabrese, V., Ientile, R., Cornelius, C., Scalia, M., Cambria, M. T., Ventimiglia, B., Pennini, G., Mancuso, C. and Butterfield, D. A., 2008f. Nutritional redox homeostasis and cellular stress response: differential role of homocysteine and acetylcarnitine. In: Surh, Y. J., Dong, Z., Cadenas, E. and Packer, L. (ed.) *Dietary Modulation of Cell Signaling Pathways*. CRC Press, New York, N.Y., USA, p. 20.

Calabrese, V., Cornelius, C., Rizzarelli, E., Owen, J. B., Dinkova-Kostova, A. T. and Butterfield, D. A., 2009a. Nitric oxide in cell survival: a janus molecule. *Antioxidants & Redox Signaling*. 11: 2717–2739.

Calabrese, V., Cornelius, C., Mancuso, C., Barone, E., Calafato, S., Bates, T., Rizzarelli, E. and Kostova, A. T., 2009b. Vitagenes, dietary antioxidants and neuroprotection in neurodegenerative diseases. *Frontiers in Bioscience, Landmark Edition*. 14: 376–397.

Calabrese, V., Cornelius, C., Dinkova-Kostova, A. T. and Calabrese, E. J., 2009c. Vitagenes, cellular stress response, and acetylcarnitine: relevance to hormesis. *BioFactors*. 35: 146–160.

Calabrese, V., Perluigi, M., Cornelius, C., Coccia, R., Di Domenico, F., Mancuso, C., Pennisi, G. and Dinkova-Kostova, A. T., 2009d. Phenolics in aging and neurodegenerative disorders. In: Fraga, C. G. (ed.) *Phenolic Compounds of Plant Origin and Health: The Biochemistry Behind Their Nutritional and Pharmacological Value*. Wiley & Sons, NY, USA, p. 25.

Calabrese, V., Cornelius, C., Dinkova-Kostova, A. T., Calabrese, E. J. and Mattson, M. P., 2010a. Cellular stress responses, the hormesis paradigm, and vitagenes: novel targets for therapeutic intervention in neurodegenerative disorders. *Antioxidants & Redox Signaling*. 13: 1763–1811.

Calabrese, V., Cornelius, C., Trovato, A., Cavallaro, M., Mancuso, C., Di Rienzo, L., Condorelli, D., De Lorenzo, A. and Calabrese, E. J., 2010b. The hormetic role of dietary antioxidants in free radical-related diseases. *Current Pharmaceutical Design*. 16: 877–883.

Calabrese, V., Cornelius, C., Stella, A. M. and Calabrese, E. J., 2010c. Cellular stress responses, mitostress and carnitine insufficiencies as critical determinants in aging and neurodegenerative disorders: role of hormesis and vitagenes. *Neurochemical Research*. 35: 1880–1915.

Calabrese, V., Cornelius, C., Maiolino, L., Luca, M., Chiaramonte, R., Toscano, M. A. and Serra, A., 2010d. Oxidative stress, redox homeostasis and cellular stress response in Ménière's disease: role of vitagenes. *Neurochemical Research*. 35: 2208–2217.

Calabrese, V., Cornelius, C., Mancuso, C., Lentile, R., Stella, A. M. and Butterfield, D. A., 2010e. Redox homeostasis and cellular stress response in aging and neurodegeneration. *Methods in Molecular Biology*. 610: 285–308.

Calabrese, V., Cornelius, C., Mancuso, C., Ientile, R., Giuffrida Stella, A. M. and Butterfield, D. A., 2010f. Redox omeostasis and cellular stress response in aging and neurodegeneration. In: Uppu, R. M., Murthy, S. N., Pryor, W. A. and Parinandi, N. L. (ed.) *Free Radical and Antioxidant Protocols*. 2nd edn, vol. 610, Humana Press, LA, USA, p. 23.

Calabrese, V., Cornelius, C., Cuzzocrea, S., Iavicoli, I., Rizzarelli, E. and Calabrese, E. J., 2011. Hormesis, cellular stress response and vitagenes as critical determinants in aging and longevity. *Molecular Aspects of Medicine*. 32: 279–304.

De la Fuente, M. and Miquel, J., 2009. An update of the oxidation-inflammation theory of aging: the involvement of the immune system in oxi-inflamm-aging. *Current Pharmaceutical Design*. 15: 3003–3026.

Davinelli, S., Di Marco, R., Bracale, R., Quattrone, A., Zella, D. and Scapagnini, G., 2013. Synergistic effect of L-carnosine and EGCG in the prevention of physiological brain aging. *Current Pharmaceutical Design*. 19: 2722–2727.

Dinkova-Kostova, A. T., Zhang, Y., Naidu, S. D., Kostov, R. V., Pheely, A. and Calabrese, V., 2013. Sulfhydryl-reactive phytochemicals as dual activators of transcription factors NRF2 and HSF1. In: Gang, D. R. (ed.) *Recent Advances in Phitochemistry, 50 Years of Phitochemistry Research*. Springer, NY, USA, vol. 43, p. 24.

Di Paola, R., Impellizzeri, D., Salinaro, A. T., Mazzon, E., Bellia, F., Cavallaro, M., Cornelius, C., Vecchio, G., Calabrese, V., Rizzarelli, E. and Cuzzocrea, S., 2011. Administration of carnosine in the treatment of acute spinal cord injury. *Biochemical Pharmacology*. 82: 1478–1489.

Feinstein, D. L., Galea, E., Aquino, D. A., Li, G. C., Xu, H. and Reis, D. J., 1996. Heat shock protein 70 suppresses astroglial-inducible nitric-oxide synthase expression by decreasing NFkappaB activation. *Journal of Biological Chemistry*. 271: 17724–17732.

Fujimoto, M., Hayashida, N., Katoh, T., Oshima, K., Shinkawa, T., Prakasam, R., Tan, K., Inouye, S., Takii, R. and Nakai, A., 2010. A novel mouse HSF3 has the potential to activate nonclassical heat-shock genes during heat shock. *Molecular Biology of the Cell*. 21: 106–116.

Hipkiss, A. R., 2009a. On the enigma of carnosine's anti-ageing actions. *Experimental Gerontology*. 44: 237–242.

Hipkiss, A. R., 2009b. Carnosine and its possible roles in nutrition and health. *Advances in Food and Nutrition Research*. 57: 87–154.

Hipkiss, A. R., 2010. Aging, proteotoxicity, mitochondria, glycation, NAD and carnosine: possible inter-relationships and resolution of the oxygen paradox. *Frontiers in Aging Neuroscience*. 2: 10.

Kim, Y. E., Hipp, M. S., Bracher, A., Hayer-Hartl, M. and Hartl, F. U., 2013. Molecular chaperone functions in protein folding and proteostasis. *Annual Review of Biochemistry*. 82: 323–355.

Kulebyakin, K., Karpova, L., Lakonsteva, E., Krasavin, M. and Boldyrev, A., 2012. Carnosine protects neurons against oxidative stress and modulates the time profile of MAPK cascade signaling. *Amino Acids*. 43: 91–96.

Lodi, R., Tonon, C., Calabrese, V. and Schapira, A. H., 2006. Friedreich's ataxia: from disease mechanisms to therapeutic interventions. *Antioxidants & Redox Signaling*. 8: 438–443.

Mancuso, C., Scapagini, G., Currò, D., Giuffrida Stella, A. M., De Marco, C., Butterfield, D. A. and Calabrese, V., 2007a. Mitochondrial dysfunction, free radical generation and cellular stress response in neurodegenerative disorders. *Frontiers in Bioscience*. 12: 1107–1123.

Mancuso, C., Bates, T. E., Butterfield, D. A., Calafato, S., Cornelius, C., De Lorenzo, A., Dinkova Kostova, A. T. and Calabrese, V., 2007b. Natural antioxidants in Alzheimer's disease. *Expert Opinion on Investigational Drugs*. 16: 1921–1931.

Nishida, T., Ohata, S., Kusumoto, C., Mochida, S., Nakada, J., Inagaki, Y., Ohta, Y. and Matsura, T., 2010. Zinc supplementation with polaprezinc protects mouse hepatocytes against acetaminophen-induced toxicity *via* induction of heat shock protein 70. *Journal of Clinical Biochemistry and Nutrition*. 46: 43–51.

Odashima, M., Otaka, M., Jin, M., Konishi, N., Sato, T., Kato, S., Matsuhashi, T., Nakamura, C. and Watanabe, S., 2002. Induction of a 72-kDa heat-shock protein in cultured rat gastric mucosal cells and rat gastric mucosa by zinc L-carnosine. *Digestive Diseases and Sciences*. 47: 2799–2804.

Park, Y. J., Volpe, S. L. and Decker, E. A., 2005. Quantitation of carnosine in humans plasma after dietary consumption of beef. *Journal of Agricultural and Food Chemistry*. 53: 4736–4739.

Perluigi, M., Joshi, G., Sultana, R., Calabrese, V., De Marco, C., Coccia, R., Cini, C. and Butterfield, D. A., 2006. In vivo protective effects of ferulic acid ethyl ester against amyloid-beta peptide 1-42-induced oxidative stress. *Journal of Neuroscience Research*. 84: 418–426.

Pfister, F., Riedl, E., Wang, Q., vom Hagen, F., Deinzer, M., Harmsen, M. C., Molema, G., Yard, B., Feng, Y. and Hammes, H. P., 2011. Oral carnosine supplementation prevents vascular damage in experimental diabetic retinopathy. *Cellular Physiology and Biochemistry*. 28: 125–136.

Quinn, P. J., Boldyrev, A. A. and Formazuyk, V. E., 1992. Carnosine: its properties, functions and potential therapeutic applications. *Molecular Aspects of Medicine*. 13: 379–444.

Scapagnini, G., Colombrita, C., Amadio, M., D'Agata, V., Arcelli, E., Sapienza, M., Quattrone, A. and Calabrese, V., 2006. Curcumin activates defensive genes and protects neurons against oxidative stress. *Antioxidants & Redox Signaling*. 8: 395–403.

Son, D. O., Satsu, H., Kiso, Y., Totsuka, M. and Shimizu, M., 2008. Inhibitory effect of carnosine on interleukin-8 production in intestinal epithelial cells through translational regulation. *Cytokine*. 42: 265–276.

Tsai, S. J., Kuo, W. W., Liu, W. H. and Yin, M. C., 2010. Antioxidative and anti-inflammatory protection from carnosine in the striatum of MPTP-treated mice. *Journal of Agricultural and Food Chemistry*. 58: 11510–11516.

Ueda, K., Ueyama, T., Oka, M., Ito, T., Tsuruo, Y. and Ichinose, M., 2009. Polaprezinc (zinc L-carnosine) is a potent inducer of anti-oxidative stress enzyme, heme oxygenase (HO)-1-a new mechanism of gastric mucosal protection. *Journal of Pharmacological Sciences*. 110: 285–294.

Vistoli, G., Orioli, M., Pedretti, A., Regazzoni, L., Canevotti, R., Negrisoli, G., Carini, M. and Aldini, G., 2009. Design, synthesis, and evaluation of carnosine derivatives as selective and efficient sequestering agents of cytotoxic reactive carbonyl species. *ChemMedChem*. 4: 967–975.

Westerheide, S. D., Raynes, R., Powell, C., Xue, B. and Uversky, V. N., 2012. HSF transcription factor family, heat shock response, and protein intrinsic disorder. *Current Protein & Peptide Science*. 13: 86–103.

Zhang, Y., Ahn, Y. H., Benjamin, I. J., Honda, T., Hicks, R. J., Calabrese, V., Cole, P. A. and Dinkova-Kostova, A. T., 2011. HSF1-dependent upregulation of Hsp70 by sulfhydryl-reactive inducers of the KEAP1/NRF2/ARE pathway. *Chemistry & Biology*. 18: 1355–1361.

CHAPTER 13

Carnosine and the Control of Blood Glucose

NOBUAKI OKUMURA*[a], MAMORU TANIDA[b], AKIRA NIIJIMA[c], NOBUO TSURUOKA[d], YOSHINOBU KISO[d], YUKO HORII[e], AND KATSUYA NAGAI[e]

[a]Institute for Protein Research, Osaka University, 3-2 Yamadaoka, Suita, Osaka 565-0871, Japan; [b]Institute for Protein Research, Osaka University, Department of Physiology II, Kanazawa Medical University, 1-1, Daigaku, Uchinada cho, Kahoku gun, Ishikawa 920-0293, Japan; [c]Niigata University, Niigata 951-8510, Japan; [d]Institute for Health Care Science, Suntory Wellness Limited, Osaka 618-8503, Japan; [e]ANBAS Corporation, Institute for Protein Research, Osaka University, 4-12-17 Toyosaki, Kita-ku, Osaka 531-0072, Japan

*E-mail: nokumura@protein.osaka-u.ac.jp

13.1 Introduction

Carnosine (β-alanyl-L-histidine) and anserine (β-alanyl-*N*-π-methyl-L-histidine) are typical imidazole dipeptides originally discovered from the skeletal muscles of vertebrates (Figure 13.1). Carnosine is synthesized from β-alanine and L-histidine by carnosine synthetase (Boldyrev and Severin, 1990), and stored at high concentrations in skeletal muscles of most vertebrates. This peptide is also detected in the brain and cardiac muscle, but is not detected

Food and Nutritional Components in Focus No. 8
Imidazole Dipeptides: Chemistry, Analysis, Function and Effects
Edited by Victor R Preedy

Published by the Royal Society of Chemistry, www.rsc.org

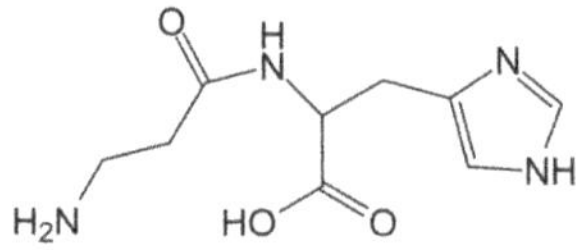

Carnosine

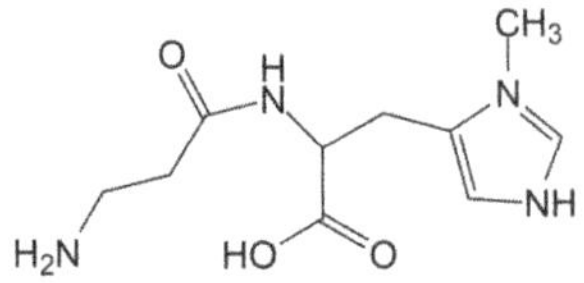

Anserine

Figure 13.1 Structures of imidazole dipeptides. Imidazole dipeptides are naturally occurring dipeptides containing histidine or its derivatives at the C-terminal side. This figure shows chemical structures of typical imidazole dipeptides; carnosine (β-alanyl-L-histidine) and anserine (β-alanyl-*N*-π-methyl-L-histidine).

in several other organs including kidney, liver, and lungs (Jackson and Lenney, 1996). Carnosine is known to possess antioxidant properties such as free-radical scavenger functions (Dahl *et al.*, 1988; Boldyrev and Severin, 1990; Hartman *et al.*, 1990). Therefore, the antioxidant effect of this compound attracted attention for the mechanism of its anti-inflammatory and wound-healing effects.

Moreover, the following findings were reported: (1) carnosine content was lower in the diaphragm of streptozotocin-diabetic rats than in that of intact rats (Buse *et al.*, 1980); (2) in a patient with a congenital deficiency of carnosinase, a carnosine-degrading enzyme, a marked decrease in the blood pressure and an increase in the blood concentration of carnosine were observed (Willi *et al.*, 1997); (3) in hypertensive rats the carnosine content of the skeletal muscle was quite low (Johnson and Hammer, 1992). These findings raised the possibilities that carnosine is implicated in the control of blood glucose and blood pressure.

With respect to these possibilities, carnosine is known to be a major component of a special chicken essence having a hypoglycemic effect. This indicates that carnosine is one of the candidate molecules by which the chicken essence normalized blood glucose levels (Yamano *et al.*, 2001). Therefore, we further investigated the effects of carnosine on the physiology and metabolism of rats, and found that carnosine dose-dependently affected blood glucose and blood pressure (Yamano *et al.*, 2001; Niijima *et al.*, 2002; Nagai *et al.*, 2003; Tanida *et al.*, 2005). In this chapter, we describe the possible physiological mechanisms that underlie these effects of carnosine, especially focusing on the autonomic nervous system and the central histaminergic system. We also mention anserine, which is another main component of the chicken essence.

13.2 Effect of Carnosine on Blood Glucose

13.2.1 Suppression of 2-Deoxy-D-Glucose-Induced Hyperglycemia by Carnosine

A metabolically inactive glucose derivative, 2-Deoxy-D-glucose (2DG), is an inhibitor of glucose utilization in mammalian cells. If 2DG is injected into the brain, mammals recognize this condition as hypoglycemic, and elevate blood glucose concentration by suppressing insulin secretion, facilitating glucagon and adrenaline secretions, and enhancing glycogenolysis *via* excitations of sympathetic nerves innervating the pancreas, adrenal glands, and liver to overcome this crisis (Frohman and Nagai, 1976; Yamamoto *et al.*, 1983; Nagai *et al.*, 1996).

Chicken extract is a health-beneficial food that shows effects on a wide variety of physiological events such as stress, fatigue and metabolism (Li *et al.*, 2012). We found that administration of a 5% water solution of a chicken essence given as sole drinking water inhibited the hyperglycemia induced by intracranial injection of 2DG (2DG-hyperglycemia) in rats (Yamano *et al.*, 2001). Since carnosine is one of the main components of the chicken essence, further studies were made on the effect of carnosine on the hyperglycemia induced by intracranial injection of 2DG in rats. Consequently, it was found that a certain amount of carnosine lowered the blood glucose levels in the rats (Figure 13.2), indicating that carnosine is involved in blood-glucose regulation (Yamano *et al.*, 2001; Nagai *et al.*, 2003).

In addition to these findings, it was reported that the carnosine content was lower in the diaphragm of streptozotocin-diabetic rats than in that of intact rats (Buse *et al.*, 1980). It was also reported that 1 mg ml^{-1} carnosine solution given as drinking water for 4 weeks increased the plasma insulin concentration, while it reduced the plasma glucagon and glucose concentrations in diabetic mice (Lee *et al.*, 2005). Moreover, transgenic db/db mice expressing human CN1 in the liver elicited elevations in blood-glucose levels and hemoglobin A1c levels (Sauerhofer *et al.*, 2007). In nontransgenic db/db mice, administration of carnosine increased serum insulin levels, while it decreased plasma glucose levels in the fasting state (Sauerhofer *et al.*, 2007). These findings indicate that carnosine is implicated in blood-glucose regulation.

13.2.2 Dose Dependence of Carnosine Action on 2DG-Hyperglycemia

In our study, suppression of 2DG-hyperglycemia by peripheral (IP) or central (intracerebroventricular, ICV) administration of carnosine was dose dependent in a bell-shaped manner in rats (Figure 13.2) (Yamano *et al.*, 2001; Nagai *et al.*, 2003). That is, the highest suppression of 2DG-hyperglycemia was observed by IP-injection of 0.5 nmol of carnosine and by ICV-injection of 0.05 nmol of carnosine, and no suppression of 2DG-hyperglycemia was observed

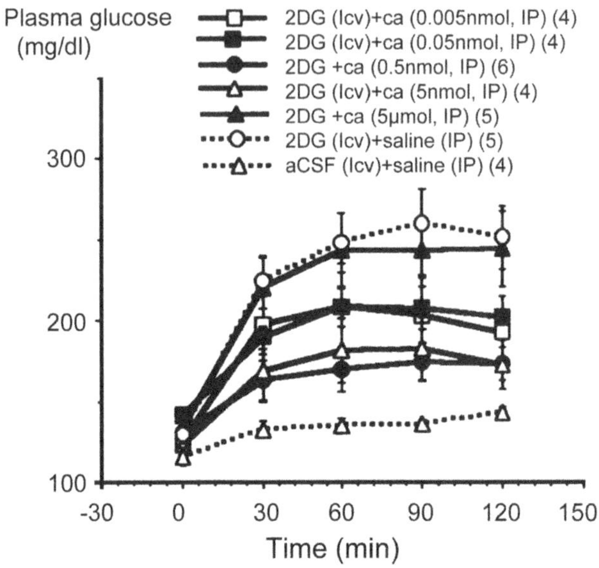

Figure 13.2 Effect of carnosine on 2DG-hyperglycemia. Carnosine was administrated in rats by IP injection, and then hyperglycemia was induced by LCV injection of 2DG. Then, blood samples were withdrawn from the heart using the heart-catheter at the indicated time points, and the blood-glucose concentrations determined. The results demonstrated that carnosine suppressed 2DG-induced hyperglycemia in a dose-dependent manner. Data are from Yamano *et al.* (2001), with permission from the publisher.

by IP-injection of a higher dose (5 μmol) of carnosine and ICV-injection of either a lower (0.005 nmol) or a higher (5 nmol) dose of carnosine (Yamano *et al.*, 2001). Moreover, when diets containing carnosine (0.1%, 0.01% and 0.001%) were given as a sole food for 7 days, it was observed that diets containing 0.01% and 0.001% carnosine but not 0.1% carnosine significantly inhibited 2DG-hyperglycemia (Nagai *et al.*, 2003). These findings indicate that carnosine may provoke different molecular events depending on dose, *e.g.* stimulation of different types of receptors for neurotransmission.

13.2.3 Effects of Carnosine on Plasma Insulin and Glucagon Levels

Previous studies have shown that 2DG-hyperglycemia is usually associated with a decrease in the blood concentration of the hypoglycemic hormone, insulin, and elevations of hyperglycemic hormones including glucagon, adrenaline, and noradrenaline (Yamamoto *et al.*, 1983; Nagai *et al.*, 1996). In the case of carnosine administration, the suppressive effect of 2DG-hyperglycemia due to IP-injection of 0.5 nmol of carnosine associated with an increase in the blood-insulin concentration and a suppression of the increase in the blood-glucagon concentration (Figure 13.3) (Yamano *et al.*, 2001). These findings could explain the mechanism by which carnosine suppresses hyperglycemic responses on 2DG-injection.

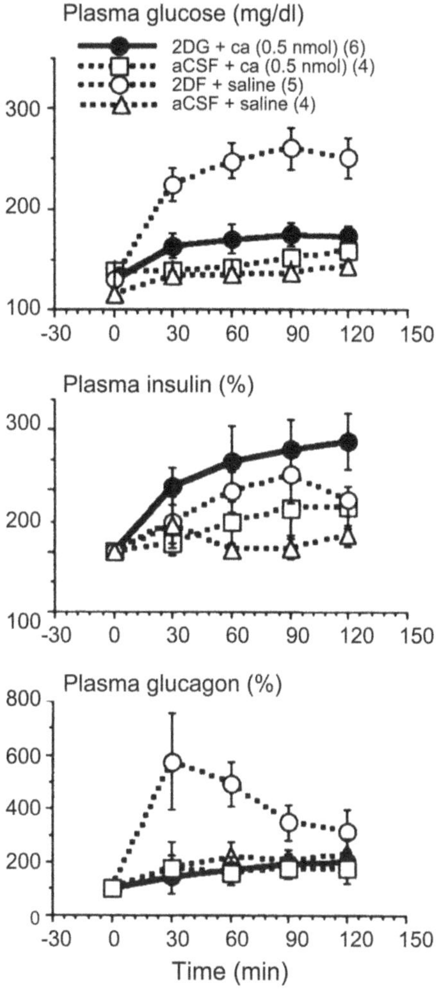

Figure 13.3 Effects of carnosine on changes in plasma concentrations of glucose, insulin and glucagon after LCV injection of 2DG. Injection (IP) of 0.5 nmol of carnosine suppressed the 2DG-induced increase in blood glucose, while it restored the increase in the plasma insulin concentration. At the same time, it suppressed the 2DG-induced increase in the plasma glucagon concentration. Data are from Yamano *et al.* (2001), with permission from the publisher.

13.2.4 Effects of Carnosine on Sympathetic and Parasympathetic Nerve Activities

In order to clarify the mechanism of the suppressive action of carnosine on 2DG-hyperglycemia, the effects of carnosine on activities of sympathetic nerves innervating the adrenal glands and liver as well as the activity of the vagal celiac nerve were examined, because an increase in the adrenal sympathetic nerve activity (ASNA) causes an elevation of the secretion of adrenaline

(Ganong, 2005). In addition, an increase in the hepatic sympathetic nerve activity (hepatic-SNA) elicits enhancement of glycogenolysis *via* activation of glycogen phosphorylase (Shimazu and Amakawa, 1968), while excitation of the pancreatic parasympathetic nerve, which is a part of the vagal celiac nerve, induces an increase in insulin secretion from the pancreatic β-cells (Woods and Porte, 1974). Consequently, it was observed that IP injection of carnosine suppressed ASNA, while it elevated vagal celiac nerve activities (Yamano *et al.*, 2001). These changes in autonomic neurotransmission by carnosine might explain the suppression of 2DG-hyperglycemia caused by carnosine, which increases insulin secretion and decreases glucagon secretion.

13.2.5 Effects of Histamine and L-Histidine on 2DG-Hyperglycemia

Since carnosine is composed of β-alanine and L-histidine, these amino acids can be liberated from carnosine when it is hydrolyzed. Histamine is synthesized from L-histidine by histidine decarboxylase, and thus carnosine can provide the substrate for an histamine synthesis. In addition, carnosine itself was suggested to be an agonist for histamine receptor (Housley *et al.*, 1988). Taken together, it is possible that histamine is involved in the mechanism of the hypoglycemic action of carnosine. Therefore, the effect of histamine on 2DG-hyperglycemia was examined in rats. When ICV injection of histamine was carried out, a bell-shaped inhibition of 2DG-hyperglycemia was observed. That is, 1–100 nmol of histamine significantly inhibited, 0.05 nmol of histamine did not affect, and 1 µmol of histamine enhanced the 2DG-hyperglycemia (Yamano *et al.*, 2001).

Furthermore, the effect of L-histidine on 2DG-hyperglycemia was examined. When IP-injection of L-histidine from 0.05 nmol to 50 nmol was carried out, no significant effect was observed on the 2DG-hyperglycemia (Nagai *et al.*, 2003). However, when ICV injection of L-histidine was performed, a bell-shaped inhibition of the 2DG-hyperglycemia was observed. That is, 0.05 nmol and 0.005 nmol, but not 5 nmol, 0.5 nmol, and 0.05 pmol of L-histidine inhibited the 2DG-hyperglycemia (Nagai *et al.*, 2003). Therefore, it was concluded that histamine and L-histidine show significant bell-shaped inhibitory actions on 2DG-hyperglycemia only when ICV injections of these agents were carried out.

13.2.6 Involvement of Histamine Receptors in Carnosine Action on 2DG-Hyperglycemia

In order to clarify the involvement of histamine in 2DG-hyperglycemia, effects of a histamine H1-receptor antagonist, diphenhydramine hydrochloride, and an inhibitor of histamine synthesis, α-fluoromethylhistidine hydrochloride (FMH) on 2DG-hyperglycemia were examined. Consequently, administration of either diphenhydramine hydrochloride or FMH significantly suppressed the 2DG-hyperglycemia (Nagai *et al.*, 2003). Moreover, a higher dose (1 μmol) of histamine enhanced the 2DG-hyperglycemia (Yamano *et al.*, 2001). These findings suggest that histamine is involved in the mechanism of 2DG-hyperglycemia.

In this regard, it was reported that histamine H3 receptor acts as an inhibitory presynaptic receptor that suppresses endogenous histamine release in central and peripheral nerves (Arrang *et al.*, 1988). Therefore, it was possible that the suppressive action of carnosine on 2DG-hyperglycemia might be elicited *via* histamine H3 receptor. Thus, the effects of ICV-injection of H3-receptor antagonist, thioperamide maleate (50 nmol/10 ml of aCSF), on the suppressive actions of carnosine and histamine on 2DG-hyperglycemia were examined. Consequently, ICV injection of thioperamide maleate eliminated both suppressive actions of carnosine (0.05 nmol, ICV) and histamine (1 nmol, ICV) on the 2DG-hyperglycemia (Yamano *et al.*, 2001). These findings suggest that the suppressive effects of carnosine and histamine on the 2DG-hyperglycemia might be elicited *via* a histamine H3 receptor. Furthermore, these facts suggest a possibility that the effect of carnosine on 2DG-hyperglycemia is due to production of L-histidine, which is in turn converted to L-histamine.

13.3 Carnosine Actions on Blood Pressure and Autonomic Regulation of Peripheral Tissues

13.3.1 Effects of Carnosine on Blood Pressure and Renal Sympathetic Nerve Activity

It has been reported that, in hypertensive rats, the carnosine content of the skeletal muscle was quite low (Johnson and Hammer, 1992), suggesting that carnosine was involved in the control of blood pressure. In our studies it was observed that a certain amount of carnosine decreased blood pressure and renal sympathetic nerve activity (RSNA) in rats (Niijima *et al.*, 2002). The carnosine action on RSNA was dose dependent (in a bell-shaped manner), that is, intravenous (IV) injection of 100 μg carnosine or ICV injection of 10 μg carnosine elevated RSNA and blood pressure, while IV injection of 1 μg carnosine or ICV-injection of 0.01 μg carnosine lowered RSNA and blood pressure in rats (Figure 13.4) (Tanida *et al.*, 2005). Since the dose dependence of carnosine effect on blood pressure and that on RSNA were in parallel, the effects of carnosine on blood pressure is likely to be mediated *via* the changes in autonomic neurotransmission. In addition, reducing responses of RSNA and BP to low-dose carnosine were eliminated by ICV injection of thioperamide maleate, and elevating responses of RSNA and BP to high dose carnosine were eliminated by ICV injection of diphenhydramine hydrochloride (Tanida *et al.*, 2005). Thus, it suggests that the central histaminergic system might be implicated in carnosine actions on the autonomic and cardiovascular system in a biphasic manner.

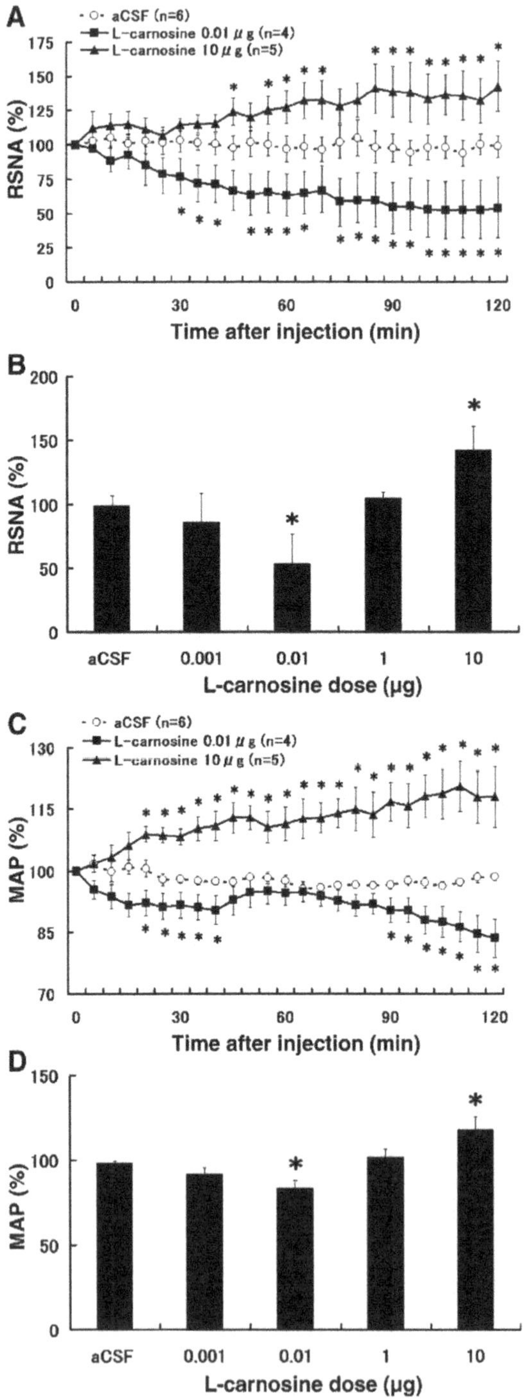
A
aCSF (n=6)
L-carnosine 0.01 μg (n=4)
L-carnosine 10 μg (n=5)
RSNA (%)
Time after injection (min)
B
RSNA (%)
aCSF
0.001
0.01
1
10
L-carnosine dose (μg)
C
aCSF (n=6)
L-carnosine 0.01 μg (n=4)
L-carnosine 10 μg (n=5)
MAP (%)
Time after injection (min)
D
MAP (%)
aCSF
0.001
0.01
1
10
L-carnosine dose (μg)

13.3.2 Effect of Carnosine on Lipid Metabolism in Adipose Tissues and Cancer

In addition to the blood glucose and blood pressure, it was also observed that carnosine dose dependently (in a bell-shaped manner) affected the brown adipose tissue sympathetic nerve activity (BAT-SNA), the white adipose tissue sympathetic nerve activity (WAT-SNA), the splenic sympathetic nerve activity (Splenic-SNA), the body temperature, lipolysis, and tumor proliferation (Tanida *et al.*, 2007a; Shen *et al.*, 2008; Horii *et al.*, 2012). That is, (1) ICV injection of 100 μg carnosine elevated BAT-SNA and the body temperature, while that of 0.01 μg carnosine lowered BAT-SNA and the body temperature in rats (about 300 g of body weight) (Tanida *et al.*, 2007a), (2) IP injection of 10 μg carnosine decreased WAT-SNA and the plasma FFA concentration, while that of 100 ng carnosine increased WAT-SNA and the plasma FFA in rats (about 300 g of body weight) (Shen *et al.*, 2008), (3) intraduodenal injection of 10 μg and 100 μg carnosine elevated, while 1 mg and 10 mg carnosine suppressed Splenic-SNA in rats (about 300 g of body weight). In addition, 1 mg ml^{-1} of carnosine solution given as the only drinking water inhibited the proliferation of human colon cancer (HCT 116) cells in subcutaneous space of athymic nude mice (Horii *et al.*, 2012).

13.4 Carnosine Synthesis, Release, and Hydrolysis

13.4.1 Carnosine Biosynthesis by Carnosine Synthetase

Carnosine is synthesized from β-alanine and L-histidine in skeletal muscles and the brain by an ATP-dependent enzymatic reaction catalyzed by carnosine synthetase. In 2010, a carnosine synthetase was identified to be ATP-grasp domain-containing protein 1 (ATPGD1) (Drozak *et al.*, 2010). This enzyme also catalyzes ligation of β-alanine with *N*-π-methyl-L-histidine, γ-aminobutyric acid, *etc.*, indicating that ATPGD1 could synthesize not only carnosine, but also anserine, homocarnosine, and several other uncharacterized dipeptides, although its catalytic efficiency of carnosine synthesis is higher than that of homocarnosine or anserine synthesis. In the skeletal muscles, carnosine synthesized by carnosine synthetase is stored at very high concentrations, partly because the skeletal muscles do not express carnosine hydrolyzing enzymes.

Figure 13.4 Effects of carnosine on renal sympathetic nerve activity (RSNA) and mean arterial pressure (MAP). Carnosine (1 or 100 μg) or saline was administrated into a rat (~350 g) by ICV injection, and then RSNA and MAP were determined. Data were expressed as means ± SE of percentages of values at 0 min. Both RSNA and MAP were decreased by a low dose of carnosine (1 μg), but increased by a high dose of carnosine (100 μg). Data are from Tanida *et al.* (2005), with permission from the publisher.

13.4.2 Carnosine Concentration in the Blood

In vertebrates, a certain amount of carnosine is detected in the blood. The plasma concentration of carnosine in rats under 12 h light and 12 h dark conditions was highest at the end of the 12 h dark period (active period for nocturnal animals like rats) and lowest at the middle to the end of the 12 h dark period (resting period) (Nagai *et al.*, 2003). This daily change suggested a possibility that physical exercise during the nocturnal dark period might increase the plasma carnosine level. Actually, when a running-wheel is attached to the rat cage, the animal showed a marked increase in their locomotive activity by turning the wheel during the 12 h dark period. In this occasion, the blood concentration of carnosine of the rat with a running-wheel increased to almost double the concentration of the control rat without the running wheel at the middle of the 12 h dark period (Nagai *et al.*, 2003). This fact suggests that physical exercise might stimulate the release of carnosine from skeletal muscles to the blood stream.

13.4.3 Carnosine Hydrolysis by Carnosine Dipeptidases

Enzyme activities that hydrolyze various dipeptides have been detected in mammalian tissue extracts since the 1940s. These enzyme activities were detected using various dipeptides as substrates, and the activities were typically referred to as names based on the substrate peptides used for enzyme assay. Carnosinase is one of these enzymes that were detected using carnosine as the substrate (Hanson and Smith, 1949). Later, it was shown that carnosinase could be classified into two types, one is Mn^{2+} independent and bestatin insensitive, while the other is Mn^{2+} dependent and bestatin-sensitive (Margolis *et al.*, 1979; Lenney, 1990). In the 2000s, two types of carnosinases, CN1 (Teufel *et al.*, 2003) and CN2 (Teufel *et al.*, 2003; Otani *et al.*, 2005), were cloned and it was found that these are M20 family metallopeptidases. Enzymatic characterization of recombinant proteins showed that CN1 is Mn^{2+} independent and bestatin insensitive, while CN2 is Mn^{2+} dependent and bestatin sensitive. Both CN1 and CN2 hydrolyze not only carnosine but also several other dipeptides with different substrate specificity. CN1 is more specific to carnosine, but also hydrolyzes homocarnosine, Ala–His, and a few other dipeptides. On the other hand, CN2 hydrolyzes carnosine and several X–His peptides including Leu–His, as well as non-X–His peptides such as Gly–Phe, while it hydrolyzes homocarnosine very weakly in the presence of Mn^{2+} (Teufel *et al.*, 2003; Otani *et al.*, 2005). Since Mn^{2+} concentrations are low in intracellular environments, the reaction condition of CN2 remains to be further investigated. In contrast to carnosine, anserine is hydrolyzed efficiently neither by CN1 nor CN2.

In fish, there is the third enzyme that hydrolyzes imidazole dipeptides. This enzyme is called anserinase, and was later suggested to be identical to *N*-acetylhistidine deacetylase. Anserinase was first cloned in 2005 (Yamada *et al.*, 2005), and found to be another member of M20 family metallopeptidases. It is a secretory enzyme, and hydrolyzes not only anserine and *N*-acetylhistidine but also various other dipeptides and *N*-acetylamino acids in the presence of Co^{2+} (Oku *et al.*, 2012).

Crystallographic analysis of CN2 in complex with bestatin suggested that its functional form is homodimer (Figure 13.5) (Unno *et al.*, 2008). Each subunit is composed of two domains, the catalytic domain and the lid or dimerization domain. The catalytic domain has a dinuclear metal-binding sites where two Zn^{2+} or two Mn^{2+} ions are coordinated, while the lid domain provides a surface for subunit–subunit interaction and also participate in catalytic reaction and substrate specificity.

13.4.4 Tissue Distribution of CN1 and CN2

In human, CN1 is expected to be a secretary protein since its cDNA has a signal sequence. It is expressed mainly in the brain and the liver, and the protein is also present in the sera. These indicate that CN1 is identical to the enzyme previously called serum carnosinase. On the other hand, murine CN1 do not have a signal peptide. In addition, it is expressed mainly in the kidney in murine, thus CN1 may have different roles in murine.

On the other hand, CN2 does not have a signal peptide in either human and murine, indicating that it is basically present in the cytosol. Tissue distribution of CN2 is more ubiquitous than that of CN1. However, in the rat brain, several neuronal populations expresses high levels of CN2 (Figure 13.6). The populations include the striata and the tuberomammillary nucleus (TMN) (Otani *et al.*, 2005). In the TMN, it is colocalized with histidine decarboxylase that converts histidine to histamine, indicating that it may provide histidine to the histamine-producing neurons (Otani *et al.*, 2008).

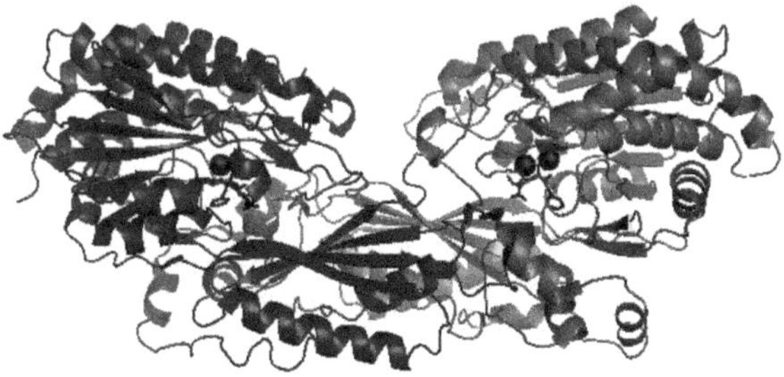

Figure 13.5 Crystal structure of CN2 in complex with a competitive inhibitor bestatin. Crystallographic studies of CN2 showed that it is a homodimer-like structure in the crystal (Unno *et al.*, 2008). Each subunit is composed of a catalytic domain and a lid domain. The catalytic domain has one active site with two metal-binding sites. Dimer formation is likely to be necessary to its enzymatic reaction. The atomic coordinates (PDB: 2ZOG) can be downloaded from Protein Data Band Japan: PDBj. The picture was drawn by Pymol 0.99.

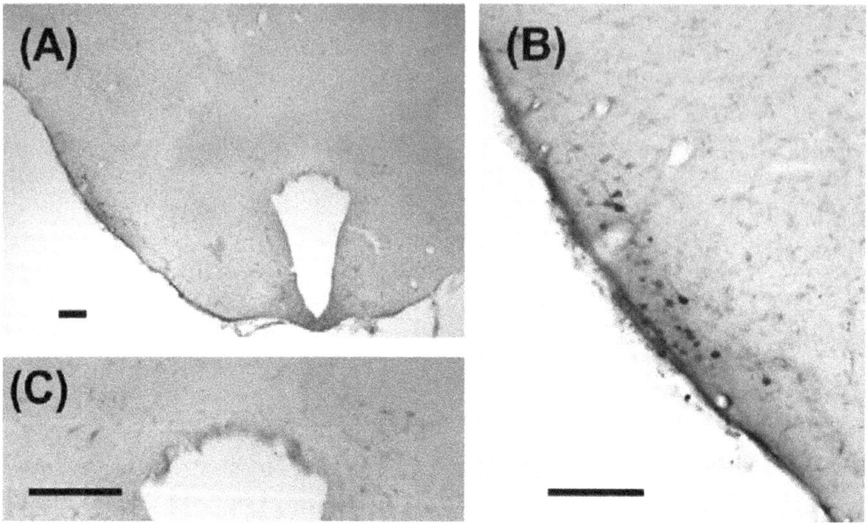

Figure 13.6 Distribution of CN2 in rat hypothalamus. Immunohistochemical localization of CN2 in the rat hypothalamus. (A) Paraformaldehyde-fixed frozen sections of a rat brain were stained with rabbit anti-CN2 IgG and the hypothalamic area of a frontal section containing the TMN was shown. (B) A higher magnification image of the brain section containing the ventral region of the TMN (VTM). (C) A higher magnification image of the dorsal TMN (DTM). Bars indicate 200 μm. Data are from Otani *et al.* (2008), with permission from the publisher.

13.5 Hypothalamic Regulation of Autonomic Nervous System and Carnosine Action

13.5.1 Involvement of Histamine Neurons in the TMN of the Hypothalamus

In mammals, histamine neurons of the central nervous system arise from the TMN, which is a pair of nuclei located in the posterior hypothalamus, and extend neurites to a wide range of regions in the entire brain. As mentioned above, histamine neurons are involved in the changes in the autonomic nerve activities, blood glucose, blood pressure and the body temperature induced by carnosine (Nagai *et al.*, 2003; Tanida *et al.*, 2005; Tanida *et al.*, 2007a). With respect to the involvement of histaminergic neurotransmission on sympathetic nerve activity, ICV injection of low-dose histamine suppressed RSNA, while high-dose histamine elevated RSNA, and a decrease in RSNA by low-dose histamine or an increase in RSNA by a high dose of histamine were abolished with ICV pretreatment of thioperamide maleate or diphenhydramine hydrochloride, respectively (Tanida *et al.*, 2007b). Therefore, it suggests that the hypothalamic histamine neurons play an important role in regulation of the autonomic nervous system in a similar manner to carnosine.

In this context, it is noteworthy that, in mouse brain, CN2 is highly expressed in the TMN (Figure 13.6) and colocalized with histidine decarboxylase (Otani *et al.*, 2008). Since histamine neurons would require a higher amount of histidine than other neurons in order to synthesize histamine, it looks like carnosine breakdown provides histidine for histamine synthesis in these neuronal populations. Therefore, it is postulated that carnosine can act as a histidine store or carrier that is synthesized outside the TMN and converted to histamine in the TMN.

13.5.2 Involvement of the SCN, the Master Circadian Clock

One of the neuronal areas with which the TMN neurons are communicating is the SCN, which has the master circadian oscillator to produce intrinsic circadian clock signals and, in addition, probably has one of the central regulators of the autonomic nervous system. Bilateral electrolytic lesions of the SCN eliminated the 2DG-hyperglycemia, indicating that SCN neurons may play an essential role in 2DG-hyperglycemia (Nagai *et al.*, 1996). Although it is not easy to determine whether the SCN is involved in the mechanism of the actions of carnosine on the 2DG-hyperglycemia, bilateral electrolytic lesions of the SCN abolished the dose-dependent effects of carnosine on RSNA and blood pressure, as well as BAT-SNA and body temperature in rats (Tanida *et al.*, 2007a). In addition, ICV injection of carnosine increased the levels of c-Fos-positive cells in the SCN (Tanida *et al.*, 2007). Therefore, it is possible that the SCN is involved in the mechanism of carnosine effects on blood glucose.

In this regard, it was shown that many dipeptide translocators existed in the hypothalamus (Yamashita *et al.*, 1997). Therefore, it is possible that carnosine released from skeletal muscles to the blood stream by physical exercise might enter the brain, transported into the histamine neurons in the TMN, hydrolyzed by CN2, the resultant L-histidine being converted to histamine by histidine decarboxylase, the synthesized histamine be released into the synaptic cleft of the synapse between the histamine neurons in the TMN and the target neurons such as SCN neuron, affect the autonomic neurotransmission of ASNA, hepatic-SNA and coeliac vagal nerve and affect glucose metabolism.

In fact, some groups (Bamshad *et al.*, 1998; Bamshad *et al.*, 1999; Sly *et al.*, 1999; Buijs *et al.*, 2001; Buijs *et al.*, 2003) presented evidence that SCN neurons have multisynaptic autonomic nerve connections to peripheral organs/tissues such as the adrenal glands, pancreas, liver, kidney, white adipose tissue, brown adipose tissue. Furthermore, the authors showed evidence that presympathetic and preparasympathetic neurons in the SCN are different (Buijs *et al.*, 2003). Moreover, evidence suggests that the circadian clock mechanism is involved in the control of autonomic neurotransmission, *e.g.*, double KO mice of Cry1 and Cry2, circadian clock genes, did not show any increase in the sympathetic and parasympathetic excitations due to olfactory stimulation with scents of grape fruit or lavender oils (Tanida *et al.*, 2007a).

Since carnosinase activity is very high in the blood, the blood concentration of carnosine might be very low. Thus, the effect of a low amount of carnosine might be elicited. Therefore, the effect of a lower amount of carnosine might be physiological ones such as lipolytic, hypoglycemic, hypotensive, and hypothermic effects. However, it is possible that in abnormal conditions such as a defect in carnosinase, the blood concentration of carnosine might dramatically increase and the effect of higher dose of carnosine might be induced. Whether these are the cases or not must be examined in the future.

13.6 Effects of Anserine on Blood Glucose and Autonomic Nervous System

The authors examined the effect of anserine on the 2DG-hyperglycemia and found that IP-injection of a certain amount of anserine (100 μg/300 g of body weight) inhibited the hyperglycemia and hyperglucagonemia induced by intracranial injection of 2DG (Kubomura *et al.*, 2010). Moreover, thioperamide, a histamine H3-receptor antagonist, inhibited the suppressive action of anserine on 2DG-hyperglycemia (Kubomura *et al.*, 2010). Since IP injection of carnosine (0.5 nmol = 0.113 μg/300 g of body weight) could inhibit the 2DG-hyperglycemia (Yamano *et al.*, 2001; Nagai *et al.*, 2003), carnosine is almost 1000 times more potent than anserine in the inhibitory action on the 2DG-hyperglycemia. The difference in the potency of the hypoglycemic actions of carnosine and anserine might be explained by the fact that carnosinases, CN1 and CN2, can hydrolyze carnosine but not anserine in rats, because the product of the hydrolysis might give L-histidine as a precursor of histamine, which might be implicated in the hypoglycemic action of carnosine and anserine.

It was also observed that intraduodenal injection of anserine (100 μg/300 g of body weight) suppressed ASNA (Kubomura *et al.*, 2010). Therefore, it is possible that a certain amount of anserine might have a hypoglycemic action *via* changes in autonomic neurotransmission. This possibility must be examined in the future. Furthermore, in our study it was found that IV injection of 1 μg/300 g of body weight of anserine suppressed RSNA and blood pressure, but IV injection of 1000 μg/300 g of body weight of anserine elevated RSNA and blood pressure (Tanida *et al.*, 2010). The suppressions of RSNA and blood pressure elicited by anserine (1 μg/300 g of body weight) was eliminated by thioperamide, a histamine H3-receptor antagonist, and the elevation of RSNA and blood pressure elicited by anserine (1000 μg/300 g of body weight) was eliminated by diphenhydramine, a histamine H1-receptor antagonist (Tanida *et al.*, 2010). These findings suggest that histamine is involved in the mechanism of anserine actions on RSNA and blood pressure, even though the potency of anserine was lower than carnosine.

13.7 Conclusions

Carnosine was shown to affect the regulation of blood glucose levels, as demonstrated by using hyperglycemia-model rats. It was also shown that carnosine affects sympathetic and parasympathetic nerve activities that are innervating various peripheral tissues, and thereby regulate insulin/glucagon secretion, blood pressure, and peripheral metabolism. One of the possible mechanisms of carnosine action is to stimulate or inhibit central histaminergic neurons through acting on histamine H1 or H3 receptors, depending on doses. Thus, the effect of carnosine on blood-glucose levels is likely to he mediated at least in part by central histaminergic neurons that send signals to various areas including autonomic centers in the brain. In addition, the carnosine action also requires the SCN, which is the location of the master circadian clock and also one of the centers of the autonomic nervous system. Therefore, the regulation of blood glucose by carnosine involves the regulation of the autonomic nervous system by the cooperative action of hypothalamic nuclei.

Summary Points

- This chapter focus on the effect of carnosine on blood-glucose regulation.
- Carnosine suppresses blood-glucose levels in hyperglycemia model (2DG-hyperglycemia) rats.
- Carnosine restores the suppression of insulin secretion after 2DG injection.
- Carnosine induces changes in sympathetic and parasympathetic nerve activities, which could affect peripheral metabolism, insulin/glucagon secretion, and blood pressure in rats. This would explain the effect of carnosine on blood glucose.
- It is possible that the effect of carnosine on the autonomic nerve activities is mediated by the histamine neurons of the central nervous system.

Key Facts

1. Glucose is required for energy production and biosynthesis of essential molecules.
2. The blood-glucose level should be kept within a certain range of optimal concentration.
3. If the blood glucose is not properly regulated and increased over the normal range for a long time, this situation is called hyperglycemia.
4. The autonomic nervous system is composed of the sympathetic and parasympathetic nervous systems, which basically act in opposite directions.
5. The sympathetic and parasympathetic neurons are innervating tissues and blood vessels to regulate blood pressure, hormone secretion, and the peripheral metabolism.
6. The blood-glucose level is regulated by hormones and the autonomic nervous system. Glucagon and adrenaline increase blood glucose, while insulin decreases it. The sympathetic nervous system increases blood glucose by direct and indirect actions.

7. The effect of carnosine on blood glucose is likely to be through the regulation of insulin and glucagon secretion by the autonomic nervous system.
8. The autonomic nervous system is under the control of several specific nuclei in the hypothalamus and the brain stem.
9. In mammalian brain, there is a population of neurons that have histamine as the neurotransmitter. The cell bodies of these neurons are located in the tuberomammillary nucleus.
10. The suprachiasmatic nucleus is a hypothalamic nucleus that produces circadian clock signals and also involved in regulation of the autonomic nervous system.

Definitions and Explanations of Key Terms

Hyperglycemia. A physical condition in which the blood-glucose level is not fully controlled and thus higher than the normal level. This is caused by a disorder of the blood-glucose regulatory mechanism.
Insulin. The hormone that decreases blood-glucose levels. Insulin secretion is stimulated when blood-glucose concentration is higher than usual. Insulin secretion is also regulated by the autonomic nervous system. Type I diabetes is caused by a lack of insulin secretion, while Type II diabetes is caused by resistance to insulin action.
Glucagon. One of the hormones that increase blood-glucose levels. Glucagon secretion is enhanced in low blood glucose conditions, and stimulates glucose output from the liver. Glucagon secretion is also regulated by the autonomic nervous system.
Autonomic nervous system. Peripheral nerves that innervate most organs and arterial walls to undergo tonic regulation of the metabolism and functions. The autonomic nervous system is composed of the sympathetic and parasympathetic nervous systems, which are activated alternatively and basically act in opposite directions. The autonomic nervous system is under the control of the central nervous system, especially several nuclei in the medulla oblongata and the hypothalamus.
Histamine neurons. Neurons that contain histamine as a neurotransmitter. In the brain, histamine acts as a neurotransmitter of a specific population of neurons, whose neuronal cell bodies are located in the TMN. They extend neurites to wide regions of the entire brain. These neurons have been implicated in a variety of functions including regulations of energy metabolism and the autonomic nervous system.
Histamine receptors. Histamine H1 receptor is expressed on the surface of postsynaptic membranes, and transmits neuronal excitation from the presynaptic cell to postsynaptic cell on histamine release. Histamine H3 receptor is an autoreceptor expressed on the presynaptic membranes, and suppresses the release of histamine from the synapses.
White adipose tissue. A peripheral tissue that acts as a lipid storage. Each cell in the tissue has one large lipid droplet in the inner space. In response

to insulin, the cells incorporate glucose, which is in turn converted to triglyceride, and stored in the lipid droplet. On the other hand, lipid breakdown is stimulated by sympathetic nerves or β-adrenergic stimulation.

Brown adipose tissue. An adipose tissue that is morphologically and functionally different from the white adipose tissue. Each cell has multiple small lipid droplets, which are more metabolically active than those of white adipose tissue. The cells express uncoupling protein 1 (UCP1), that undergo thermogenesis using metabolic energy of lipid hydrolysis in response to activation of sympathetic nerves or adrenalin.

The tuberomammillary nucleus (TMN). A neuronal population located in the posterior part of the hypothalamus. Neuronal cell bodies of histamine neurons are localized in the nucleus.

The suprachiasmatic nucleus (SCN). A neuronal population located in the anterior part of the hypothalamus. In the SCN, a circadian oscillator is located and is entrained by environmental time cues to generate clock signals to other parts of the brain. The SCN is also involved in central regulation of the autonomic nervous system.

List of Abbreviations

SCN	Suprachiasmatic nucleus
IP	Intraperitoneal
2DG	2-Deoxy-D-glucose
TMN	Tuberomammillary nucleus
ICV	Intracerebroventricular
SNA	Sympathetic nerve activity
RSNA	Renal sympathetic nerve activity
ASNA	Adrenal sympathetic nerve activity
FMH	α-Fluoromethylhistidine
aCSF	Artificial cerebrospinal fluid
BAT	Brown adipose tissue
WAT	White adipose tissue
FFA	Free fatty acid

References

Arrang, J. M., Devaux, B., Chodkiewicz, J. P. and Schwartz, J. C., 1988. H-3-receptors control histamine-release in human-brain. *Journal of Neurochemistry*. 51: 105–108.

Bamshad, M., Aoki, V. T., Adkison, M. G., Warren, W. S. and Bartness, T. J., 1998. Central nervous system origins of the sympathetic nervous system outflow to white adipose tissue. *American Journal of Physiology*. 275: R291–R299.

Bamshad, M., Song, C. K. and Bartness, T. J., 1999. CNS origins of the sympathetic nervous system outflow to brown adipose tissue. *American Journal of Physiology*. 276: R1569–R1578.

Boldyrev, A. A. and Severin, S. E., 1990. The histidine-containing dipeptides, carnosine and anserine: distribution, properties and biological significance. *Advances in Enzyme Regulation*. 30: 175–194.

Buijs, R. M., Chun, S. J., Niijima, A., Romijn, H. J. and Nagai, K., 2001. Parasympathetic and sympathetic control of the pancreas: a role for the suprachiasmatic nucleus and other hypothalamic centers that are involved in the regulation of food intake. *Journal of Comparative Neurology*. 431: 405–423.

Buijs, R. M., La Fleur, S. E., Wortel, J., Van Heyningen, C., Zuiddam, L., Mettenleiter, T. C., Kalsbeek, A., Nagai, K. and Niijima, A., 2003. The suprachiasmatic nucleus balances sympathetic and parasympathetic output to peripheral organs through separate preautonomic neurons. *Journal of Comparative Neurology*. 464: 36–48.

Buse, M. G., Weigand, D. A., Peeler, D. and Hedden, M. P., 1980. The effect of diabetes and the redox potential on amino acid content and release by isolated rat hemidiaphragms. *Metabolism*. 29: 605–616.

Dahl, T. A., Midden, W. R. and Hartman, P. E., 1988. Some prevalent biomolecules as defenses against singlet oxygen damage. *Photochemistry and Photobiology*. 47: 357–362.

Drozak, J., Veiga-Da-Cunha, M., Vertommen, D., Stroobant, V. and Van Schaftingen, E., 2010. Molecular identification of carnosine synthase as ATP-grasp domain-containing protein 1 (ATPGD1). *Journal of Biological Chemistry*. 285: 9346–9356.

Frohman, L. A. and Nagai, K., 1976. Central nervous system-mediated stimulation of glucagon-secretion in dog following 2-deoxyglucose. *Metabolism, Clinical and Experimental*. 25: 1449–1452.

Ganong, W. F., 2005. *Review of Medical Physiology*. Lange Medical Books/McGraw-Hill, New York, pp. 301–302.

Hanson, H. T. and Smith, E. L., 1949. Carnosinase; an enzyme of swine kidney. *Journal of Biological Chemistry*. 179: 789–801.

Hartman, P. E., Hartman, Z. and Ault, K. T., 1990. Scavenging of singlet molecular oxygen by imidazole compounds: high and sustained activities of carboxy terminal histidine dipeptides and exceptional activity of imidazole-4-acetic acid. *Photochemistry and Photobiology*. 51: 59–66.

Horii, Y., Shen, J., Fujisaki, Y., Yoshida, K. and Nagai, K., 2012. Effects of L-carnosine on splenic sympathetic nerve activity and tumor proliferation. *Neuroscience Letters*. 510: 1–5.

Housley, G. D., Norris, C. H. and Guth, P. S., 1988. Histamine and related substances influence neurotransmission in the semicircular canal. *Hearing Research*. 35: 87–97.

Jackson, M. C. and Lenney, J. F., 1996. The distribution of carnosine and related dipeptides in rat and human tissues. *Inflammation Research*. 45: 132–135.

Johnson, P. and Hammer, J. L., 1992. Histidine dipeptide levels in ageing and hypertensive rat skeletal and cardiac muscles. *Comparative Biochemistry and Physiology, Part B: Biochemistry & Molecular Biology*. 103: 981–984.

Kubomura, D., Matahira, Y., Nagai, K. and Niijima, A., 2010. Effect of anserine ingestion on the hyperglycemia and autonomic nerves in rats and humans. *Nutritional Neuroscience*. 13: 123–128.

Lee, Y. T., Hsu, C. C., Lin, M. H., Liu, K. S. and Yin, M. C., 2005. Histidine and carnosine delay diabetic deterioration in mice and protect human low density lipoprotein against oxidation and glycation. *European Journal of Pharmacology*. 513: 145–150.

Lenney, J. F., 1990. Separation and characterization of two carnosine-splitting cytosolic dipeptidases from hog kidney (carnosinase and non-specific dipeptidase). *Biological Chemistry Hoppe-Seyler*. 371: 433–440.

Li, Y. F., He, R. R., Tsoi, B. and Kurihara, H., 2012. Bioactivities of chicken essence. *Journal of Food Science*. 77: R105–R110.

Margolis, F. L., Grillo, M., Brown, C. E., Williams, T. H., Pitcher, R. G. and Elgar, G. J., 1979. Enzymatic and immunological evidence for two forms of carnosinase in the mouse. *Biochimica et Biophysica Acta*. 570: 311–323.

Nagai, K., Nagai, N., Shimizu, K., Chun, S., Nakagawa, H. and Niijima, A., 1996. SCN output drives the autonomic nervous system: with special reference to the autonomic function related to the regulation of glucose metabolism. *Progress in Brain Research*. 111: 253–272.

Nagai, K., Niijima, A., Yamano, T., Otani, H., Okumra, N., Tsuruoka, N., Nakai, M. and Kiso, Y., 2003. Possible role of L-carnosine in the regulation of blood glucose through controlling autonomic nerves. *Experimental Biology and Medicine*. 228: 1138–1145.

Niijima, A., Okui, T., Matsumura, Y., Yamano, T., Tsuruoka, N., Kiso, Y. and Nagai, K., 2002. Effects of L-carnosine on renal sympathetic nerve activity and DOCA-salt hypertension in rats. *Autonomic Neuroscience*. 97: 99–102.

Oku, T., Ando, S., Tsai, H. C., Yamashita, Y., Ueno, H., Shiozaki, K., Nishi, R. and Yamada, S., 2012. Purification and identification of two carnosine-cleaving enzymes, carnosine dipeptidase I and Xaa–methyl-His dipeptidase, from Japanese eel (*Anguilla japonica*). *Biochimie*. 94: 1281–1290.

Otani, H., Okumura, A., Nagai, K. and Okumura, N., 2008. Colocalization of a carnosine-splitting enzyme, tissue carnosinase (CN2)/cytosolic non-specific dipeptidase 2 (CNDP2), with histidine decarboxylase in the tuberomammillary nucleus of the hypothalamus. *Neuroscience Letters*. 445: 166–169.

Otani, H., Okumura, N., Hashida-Okumura, A. and Nagai, K., 2005. Identification and characterization of a mouse dipeptidase that hydrolyzes L-carnosine. *Journal of Biochemistry*. 137: 167–175.

Sauerhofer, S., Yuan, G., Braun, G. S., Deinzer, M., Neumaier, M., Gretz, N., Floege, J., Kriz, W., Van Der Woude, F. and Moeller, M. J., 2007. L-Carnosine, a substrate of carnosinase-1, influences glucose metabolism. *Diabetes*. 56: 2425–2432.

Shen, J., Yao, J., Tanida, M. and Nagai, K., 2008. Regulation of sympathetic nerve activity by L-carnosine in mammalian white adipose tissue. *Neuroscience Letters*. 441: 100–104.

Shimazu, T. and Amakawa, A., 1968. Regulation of glycogen metabolism in liver by autonomic nervous system. 2. neural control of glycogenolytic enzymes. *Biochimica et Biophysica Acta*. 165: 335–348.

Sly, J. D., Colvill, L., Mckinley, J. M. and Oldfield, J. B., 1999. Identification of neural projections from the forebrain to the kidney, using the virus pseudorabies. *Journal of the Autonomic Nervous System*. 77: 73–82.

Tanida, M., Gotoh, H., Taniguchi, H., Otani, H., Shen, J., Nakamura, T., Tsuruoka, N., Kiso, Y., Okumura, N. and Nagai, K., 2007a. Effects of central injection of L-carnosine on sympathetic nerve activity innervating brown adipose tissue and body temperature in rats. *Regulatory Peptides*. 144: 62–71.

Tanida, M., Kaneko, H., Shen, J. and Nagai, K., 2007b. Involvement of the histaminergic system in renal sympathetic and cardiovascular responses to leptin and ghrelin. *Neuroscience Letters*. 413: 88–92.

Tanida, M., Niijima, A., Fukuda, Y., Sawai, H., Tsuruoka, N., Shen, J., Yamada, S., Kiso, Y. and Nagai, K., 2005. Dose-dependent effects of L-carnosine on the renal sympathetic nerve and blood pressure in urethane-anesthetized rats. *American Journal of Physiology: Regulatory, Integrative and Comparative Physiology*. 288: R447–R455.

Tanida, M., Shen, J., Kubomura, D. and Nagai, K., 2010. Effects of anserine on the renal sympathetic nerve activity and blood pressure in urethane-anesthetized rats. *Physiological Research*. 59: 177–185.

Teufel, M., Saudek, V., Ledig, J. P., Bernhardt, A., Boularand, S., Carreau, A., Cairns, N. J., Carter, C., Cowley, D. J., Duverger, D., Ganzhorn, A. J., Guenet, C., Heintzelmann, B., Laucher, V., Sauvage, C. and Smirnova, T., 2003. Sequence identification and characterization of human carnosinase and a closely related non-specific dipeptidase. *Journal of Biological Chemistry*. 278: 6521–6531.

Unno, H., Yamashita, T., Ujita, S., Okumura, N., Otani, H., Okumura, A., Nagai, K. and Kusunoki, M., 2008. Structural basis for substrate recognition and hydrolysis by mouse carnosinase CN2. *Journal of Biological Chemistry*. 283: 27289–27299.

Willi, S. M., Zhang, Y., Hill, J. B., Phelan, M. C., Michaelis, R. C. and Holden, K. R., 1997. A deletion in the long arm of chromosome 18 in a child with serum carnosinase deficiency. *Pediatric Research*. 41: 210–213.

Woods, S. C. and Porte, D., 1974. Neural control of endocrine pancreas. *Physiological Reviews*. 54: 596–619.

Yamada, S., Tanaka, Y. and Ando, S., 2005. Purification and sequence identification of anserinase. *FEBS Journal*. 272: 6001–6013.

Yamamoto, H., Nagai, K. and Nakagawa, H., 1983. Role of catecholamine in time-dependent hyperglycemia due to 2-deoxyglucose, mannitol, and glucose. *Biomedical Research*. 4: 505–513.

Yamano, T., Niijima, A., Iimori, S., Tsuruoka, N., Kiso, Y. and Nagai, K., 2001. Effect of L-carnosine on the hyperglycemia caused by intracranial injection of 2-deoxy-D-glucose in rats. *Neuroscience Letters*. 313: 78–82.

Yamashita, T., Shimada, S., Guo, W., Sato, K., Kohmura, E., Hayakawa, T., Takagi, T. and Tohyama, M., 1997. Cloning and functional expression of a brain peptide/histidine transporter. *Journal of Biological Chemistry*. 272: 10205–10211.

CHAPTER 14

β-Alanine, Muscle Carnosine and Exercise

BRYAN SAUNDERS[a], GUILHERME G. ARTIOLI[a], CRAIG SALE[b], AND BRUNO GUALANO*[a]

[a]School of Physical Education and Sport, Laboratory of Applied Nutrition and Metabolism, University of São Paulo, Av. Prof. Mello de Moraes, 65, Butantã, São Paulo, SP 05508-030, Brazil; [b]Biomedical, Life and Health Sciences Research Centre, Nottingham Trent University, Nottingham, NG11 8NS, UK
*E-mail: gualano@usp.br

14.1 Introduction

It is becoming evident that carnosine and other histidine-containing dipeptides have potential therapeutic and nontherapeutic applications that are of relevance to the performance of athletic individuals and to the health of the general population (Sale *et al.*, 2013). This dipeptide has received considerable attention since it was demonstrated that the concentration of carnosine in human skeletal muscle can be increased by augmenting the dietary intake of β-alanine (Harris *et al.*, 2006), the rate-limiting factor for endogenous histidine-containing dipeptide synthesis. The potential for increased muscle carnosine to improve exercise performance and capacity becomes apparent, given that carnosine has been associated with several physiological roles including: pH regulation (Abe, 2000), enhanced calcium sensitivity (Dutka *et al.*, 2012), and protection against oxidative stress (Boldyrev *et al.*, 2003).

Food and Nutritional Components in Focus No. 8
Imidazole Dipeptides: Chemistry, Analysis, Function and Effects
Edited by Victor R Preedy

Published by the Royal Society of Chemistry, www.rsc.org

This chapter provides a comprehensive review of the literature covering the effects of increased muscle carnosine on exercise performance and capacity; physiological mechanisms and practical applications are discussed.

14.2 β-Alanine and Muscle Carnosine

Muscle carnosine content in humans ranges from 17.5 ± 4.8 mmol kg^{-1} dm in females to 21.3 ± 4.2 mmol kg^{-1} dm in males (Mannion *et al.*, 1992). The determinants of muscle carnosine content in humans include gender, fiber type, age and diet (for a review see Harris *et al.*, 2012); with dietary manipulation offering the most appropriate means to increase carnosine levels. Harris *et al.* (2006) demonstrated that dietary supplementation of free β-alanine, at an average of 5.2 g per day for 4 weeks, increased muscle carnosine concentrations by about 13 mmol kg^{-1} dm (~65%); a subsequent study showed a similar increase of about 10 mmol kg^{-1} dm (~60%) following 4 weeks of supplementation at an identical dose (Hill *et al.*, 2007). When supplementation was continued to 10 weeks, resulting in an average dose of 5.9 g per day, muscle carnosine was increased by approximately 15 mmol kg^{-1} dm (~80%). Stellingwerff *et al.* (2012) showed a linear dose–response relationship to β-alanine supplementation that is not dependent on baseline muscle carnosine, muscle type, or the daily dose of β-alanine, but is dependent on the total amount of β-alanine consumed. Supplementation with β-alanine has consistently been shown to increase muscle carnosine content (Harris *et al.*, 2006; Hill *et al.*, 2007; Derave *et al.*, 2007; Baguet *et al.*, 2009; del Favero *et al.*, 2012; Gross *et al.*, 2014), even within as little as 2 weeks (Stellingwerff *et al.*, 2012), and can therefore be considered an appropriate method to increase intracellular levels of carnosine in humans (Table 14.1).

Harris *et al.* (2006) compared the ingestion of β-alanine in free form (10, 20 and 40 mg kg^{-1} BM) with an equivalent dose (40 mg kg^{-1} BM) contained within food (in this case a chicken broth). Upon administration of the higher doses of free β-alanine (from 20 mg kg^{-1} BM), several participants began to complain of symptoms of flushing (also termed paraesthesia). Symptoms began within 20 min of administration with the free powder and were described as an unpleasant prickly sensation on the skin around the body that lasted up to one hour; this paraesthesia was evident in increasing participant number and intensity with increasing doses (Harris *et al.*, 2006). Interestingly, no participants complained of these symptoms when ingesting 40 mg kg^{-1} BM β-alanine in the chicken broth. Paraesthesia was associated with the peak concentration of plasma β-alanine; following ingestion of the chicken broth peak concentration was approximately half that of the equivalent 40 mg kg^{-1} BM β-alanine (428 ± 162 μmol L^{-1} *vs.* 833 ± 86 μmol L^{-1}) with no differences in the area under the curve (Figure 14.1). A sustained release formulation has been developed (Carnosyn SR™, Natural Alternatives International, San Marcos, California, USA), which imposes a physical restraint on the rate of release of β-alanine from the gut. This results in attenuation in the peak plasma concentration from a single dose, slower absorption kinetics and improved whole-body retention, while preventing symptoms of paraesthesia (Décombaz *et al.*, 2012).

Table 14.1 Percentage change in carnosine in studies that determined muscle carnosine content.

Author and year	Supplementation		Muscle	Change in carnosine	Method of analysis
	Dose	Duration			
Harris *et al.* (2006)	Mean: 3.2 g per day	4 weeks	*Vastus lateralis*	+42%	Muscle biopsy
	Mean: 5.2 g per day	4 weeks	*Vastus lateralis*	+64%	Muscle biopsy
Hill *et al.* (2007)	Mean: 5.2 g per day	4 weeks	*Vastus lateralis*	+59%	Muscle biopsy
	Mean: 5.9 g per day	10 weeks	*Vastus lateralis*	+80%	Muscle biopsy
Derave *et al.* (2007)	4.8 g per day	4 weeks	*Soleus*	+47%	^{1}H-MRS
	4.8 g per day	4 weeks	*Gastrocnemius*	+37%	^{1}H-MRS
Baguet *et al.* (2009)	4.8 g per day	6 weeks	*Soleus*	+39%	^{1}H-MRS
	4.8 g per day	6 weeks	*Gastrocnemius*	+23%	^{1}H-MRS
	4.8 g per day	6 weeks	*Tibialis anterior*	+27%	^{1}H-MRS
Baguet *et al.* (2010)	5 g per day	7 weeks	*Soleus*	+45%	^{1}H-MRS
	5 g per day	7 weeks	*Gastrocnemius*	+28%	^{1}H-MRS
Stellingwerff *et al.* (2012)	3.2 g per day	2 weeks	*Gastrocnemius*	+10%	^{1}H-MRS
	3.2 g per day	2 weeks	*Tibialis anterior*	+17%	^{1}H-MRS
del Favero *et al.* (2012)	3.2 g per day	12 weeks	*Gastrocnemius*	+85%	^{1}H-MRS
Gross *et al.* (2014)	3.2 g per day	5 weeks	*Vastus lateralis*	+25%	^{1}H-MRS
			vastus intermedius	+49%	^{1}H-MRS
			Gastrocnemius	+24%	^{1}H-MRS
			Tibialis anterior	+37%	^{1}H-MRS

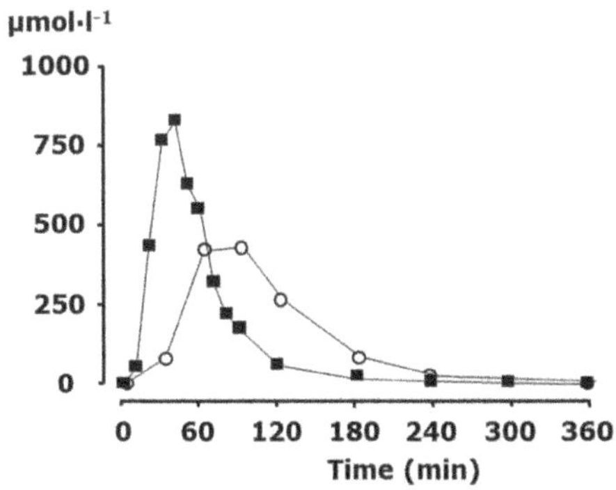

Figure 14.1 Mean plasma β-alanine concentration with time following ingestion of 40 (squares) mg kg^{-1} BM β-alanine or a chicken broth (circles) containing the equivalent of 40 mg kg^{-1} BM β-alanine in the form of anserine and carnosine. (Adapted from Harris *et al.* (2006) with permission.)

Harris *et al.* (2006), who showed that muscle carnosine concentration could be increased *via* β-alanine supplementation, paved the way for research into β-alanine and exercise since increased carnosine concentration would result in an increased buffering capacity of the muscle. Therefore, β-alanine supplementation would likely improve performance where the intensity and duration of the exercise is such that it would be limited by increasing muscle acidosis. In addition, the other purported physiological roles for carnosine (as above) also provide a rationale for an increased muscle carnosine to improve the exercise performance of differing durations and intensities to those solely limited by increasing intramuscular acidosis.

14.3 β-Alanine Supplementation and Exercise

A meta-analysis by Hobson *et al.* (2012) investigated the effects of β-alanine on exercise performance and capacity and showed that β-alanine improved exercise measures to a greater extent than placebo ($P = 0.002$). In addition, this study categorized protocols into exercise performance ($P = 0.204$) and exercise capacity ($P = 0.013$) (Figure 14.2); the lack of an effect shown in performance measures may have been due to the low number of studies employing performance protocols (Hobson *et al.*, 2012). In addition, protocols were categorized for their duration, with no effect on exercise of less than 60 s in duration ($P = 0.312$), a significant effect on exercise of 60 to 240 s ($P = 0.001$) and on exercise in excess of 240 s ($P = 0.046$). There have been several studies published since this meta-analysis that have further contributed to our understanding of the ergogenic effects of increased muscle carnosine through β-alanine supplementation (Figure 14.3).

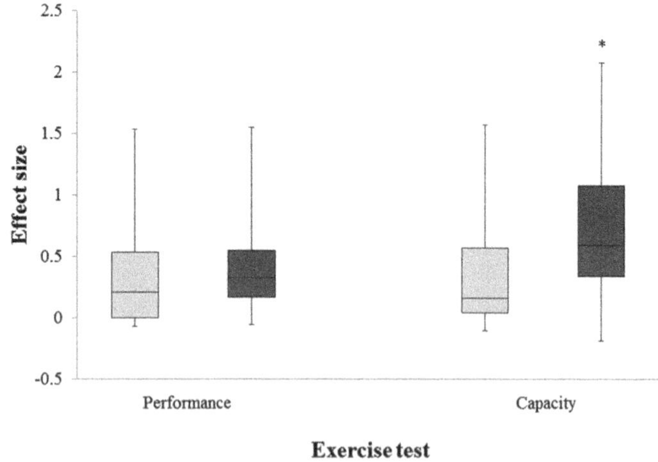

Figure 14.2 The effect size of placebo and β-alanine groups when subdivided by exercise test; exercise performance and exercise capacity. Light gray represents placebo groups and dark gray represents β-alanine groups. *Denotes significantly greater than placebo ($P = 0.013$). (Reproduced from Hobson *et al.* (2012) with permission.)

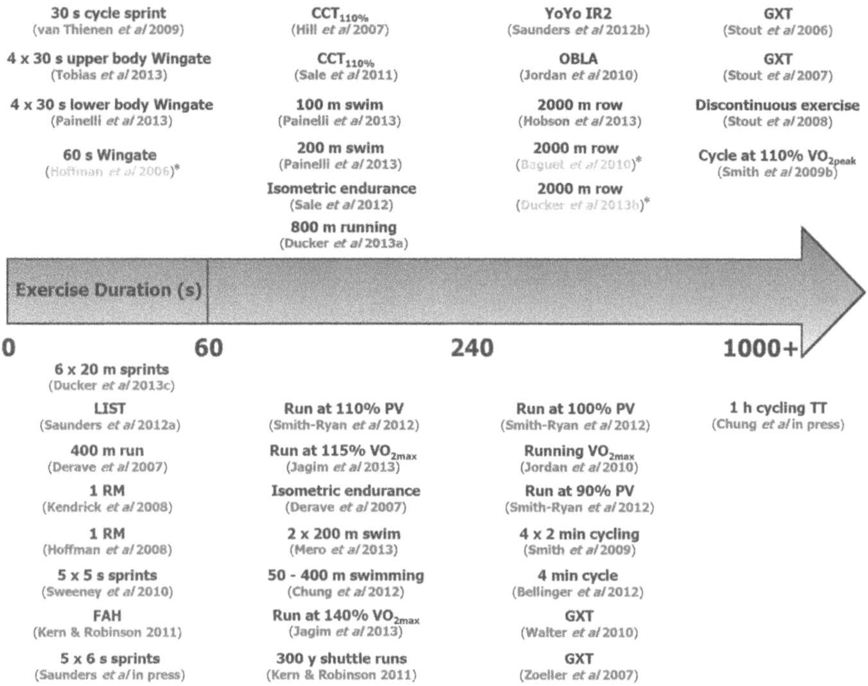

Figure 14.3 Graphic of studies to date examining the effect of β-alanine supplementation on exercise performance and capacity. Studies have been separated by the duration of the exercise test employed, grouped as those lasting less than 60 s, 60 to 240 s and over 240 s. Those below the timeline are where no effect of β-alanine supplementation was shown compared with placebo; those above the timeline either showed a trend towards a positive effect (marked with *), or showed a significant beneficial of β-alanine supplementation. (Adapted from Sale *et al.* (2013) with permission.)

14.3.1 Single-Bout High-Intensity Exercise

It is generally acknowledged that exercise of less than 60 s is insufficient in duration to be limited by muscle acidosis, and fatigue is more likely due to the decline in ATP resynthesis and an increased accumulation in ADP. The current data on β-alanine and exercise of this duration supports this notion with the majority of studies showing no beneficial effect (Derave *et al.*, 2007; Hoffman *et al.*, 2008a; Kern and Robinson, 2011). However, Hoffman *et al.* (2008b) did show a trend ($P = 0.07$) for lower rates and feelings of fatigue during a modified Wingate power test following β-alanine supplementation in football players. One possibility is that the timeframe of 60 s may just be within the lower extremity of where increased buffering capacity may have an ergogenic effect, hence the trend shown by Hoffman *et al.* (2008b).

van Thienen *et al.* (2009) did report a beneficial effect of β-alanine supplementation on a 30 s maximal isokinetic cycle sprint, although this was performed following a 110 min simulated endurance cycle race, and a 10 min time trial. Following supplementation, athletes in the β-alanine group improved their peak (+11.2%), mean (+4.9%) and final (+10.9%) power output during the isokinetic sprint. Although seemingly counter to other studies on exercise of less than 60 s duration, these findings may be explained by the preceding exercise resulting in a decrease in muscle pH prior to the start of the sprint. Indeed, increased buffering capacity is more likely to improve performance in exercise of more than 60 s in duration, as evidenced by studies using different approaches to increase buffering capacity (*i.e.*, sodium bicarbonate supplementation; Carr *et al.*, 2011).

Hill *et al.* (2007) investigated the effect of β-alanine supplementation on the $CCT_{110\%}$, a high-intensity cycling capacity test designed to last between 120 and 240 s. The $CCT_{110\%}$ is designed to induce a large accumulation of hydrogen cations (H^+) and a resultant drop in pH_i; the test has been shown to be highly reliable with a CV of 4.93% for total work done (TWD; Saunders *et al.*, 2013). Following four weeks of β-alanine supplementation, TWD was increased by 13.0% alongside a 58.8% increase in muscle carnosine; when supplementation was extended to ten weeks, carnosine was increased by 80.1% and TWD by 16.2% (Hill *et al.*, 2007). These results were confirmed by Sale *et al.* (2011), who showed similar improvements in TWD following 4 weeks of supplementation with β-alanine (14.6%). In addition, this study showed improvements in $CCT_{110\%}$ with sodium bicarbonate (+6.9%), and following cosupplementation of β-alanine and sodium bicarbonate (+18.9%), giving credence to the suggestion that the improvements were due to gains in H^+-buffering capacity.

Derave *et al.* (2007) showed no effect of β-alanine supplementation on isometric endurance of the knee extensors contracting at 45% of maximal voluntary isometric contraction (MVIC) force in trained sprinters. However, these results were in contrast to those of Sale *et al.* (2012), who showed a 13.2% increase in MIVC endurance time with β-alanine. The difference in results between these two studies may have been due to the intensity of the exercise performed; at 45% MVIC, circulation to the contracting muscles is largely occluded due to the increase in intramuscular pressure with minimal efflux of lactate and H^+ from the contracting muscle with an estimated endurance time of ~78 s (Ahlborg *et al.*, 1972). The baseline contractions times in Derave *et al.* (2007) were considerably higher than this; while Sale *et al.* (2012) showed baseline data within 3–4 s of this expected endurance time of 78 s, Derave *et al.* (2007) showed baseline holds between 170 and 200 s. Thus, the MVIC sustained by participants in the Derave *et al.* (2007) study was likely much lower than 45%, resulting in minimal intramuscular H^+ accumulation and therefore not limited by increasing acidosis.

Hobson *et al.* (2012) concluded that exercise lasting between 60 and 240 s was likely to be positively influenced by an increase in muscle buffering capacity, and further evidence has been shown by Ducker *et al.* (2013a) whose recreational club runners significantly improved 800 m track performance following β-alanine supplementation, and Painelli *et al.* (2013), whose junior-standard swimmers showed significant performance improvements of 1–2% in 100 m and 200 m freestyle swimming. Cosupplementation with sodium bicarbonate resulted in a further nonsignificant improvement in swimming performance, though the probability this difference was meaningful was 78.5% and 65.2% for the 100 and 200 m distances (Painelli *et al.*, 2013). However, not all studies using exercise protocols within this timeframe have yielded positive results; Chung *et al.* (2012) supplemented trained swimmers (50–400 m) throughout training and competition with *unclear* effects on both competition and training performance. There was no effect of β-alanine on 4 min all-out cycle performance in trained athletes (Bellinger *et al.*, 2012; Howe *et al.*, 2013). Jagim *et al.* (2013) showed no effect on two runs to exhaustion at 115 and 140% VO_{2max} lasting approximately 70 and 150 s and Smith-Ryan *et al.* (2012) similarly showed no effect of β-alanine supplementation on aerobic and anaerobic exercise capacity using three runs to exhaustion at 110, 100 and 90% of peak velocity. Although these results are in contrast to expectations, results in these studies may have been influenced by several factors including pacing strategies (Bellinger *et al.*, 2012; Howe *et al.*, 2013) and methodological shortcomings (*e.g.* nonlaboratory controlled settings; Chung *et al.*, 2012) and critical velocity and anaerobic running capacity determined upon consecutive exercise bouts conducted on the same day with only 15 min recovery (Smith-Ryan *et al.*, 2012), compared with the more appropriate and accurate method of determining this model based upon exercise bouts performed on discrete days.

Several studies have investigated the effect of β-alanine on 2000 m rowing ergometry in well-trained male rowers; this exercise is typically 6–7 min in duration with an estimated 12% of energy sources from nonoxidative glycolysis (Stellingwerff *et al.*, 2011). Baguet *et al.* (2010) recruited elite Belgian rowers who showed a 2.7 ± 4.8 s improvement in performance with β-alanine, which was close to statistical significance ($P = 0.07$); the relatively low sample size in the β-alanine group ($N = 8$) may have played a role in the lack of statistical significance. Nonetheless, individual increases in rowing speed were positively correlated to the increases in muscle carnosine. Similarly, Ducker *et al.* (2013b) showed a trend towards an improved performance in well-trained rowers with β-alanine ($P = 0.06$), which was equivalent to an improvement of 2.9 ± 4.1 s. Hobson *et al.* (2013) used magnitude-based inferences to determine small effects of β-alanine supplementation of practical importance in an applied setting, and showed a 6.4 ± 8.1 s improvement in rowing performance with β-alanine over placebo was very likely to have been due to β-alanine. Taken together, these data suggest that the optimal range within which β-alanine acts as an ergogenic aid may be extended from the 1 to 4 min suggested by Hobson *et al.* (2012) and may be more accurately 1 to 7 min. It is worth noting that factors other than exercise duration may also be important as to whether β-alanine has a positive effect on performance. Smaller muscle groups, such as upper-body muscles, may be more subjected to local acidosis during rowing than whole-body exercise such as running; this could lead to differences in the effectiveness of increased carnosine depending on the exercise modality and not only duration.

The effect of increased muscle carnosine on exercise in excess of 7 min of duration is less researched, with most studies employing graded exercise tests. Although Stout *et al.* showed improvement in the physical working capacity at the fatigue threshold during graded exercise tests lasting in excess of 10 min in men (Stout *et al.*, 2006), women (Stout *et al.*, 2007) and elderly (Stout *et al.*, 2008), this may have been due to the incremental nature of the test, with limited H^+ accumulation during the early stages. In contrast, Gross *et al.* (2014) showed no improvements with β-alanine in incremental cycling performance in active male participants. Chung *et al.* (2014) recently investigated the effect of β-alanine on 1 h time-trial performance in well-trained cyclists, though perhaps surprisingly, considering the trained nature of the athletes, both groups showed a reduction in performance following supplementation. The lack of an ergogenic effect with β-alanine in this study is in accordance with the role of carnosine as a muscle buffer, since exercise of this duration is unlikely to be influenced by increasing acidosis and more likely due to the degradation of energy stores and dehydration.

14.3.2 High-Intensity Intermittent and Repeated-Bout High-Intensity Exercise

The majority of research into β-alanine has focused on single-bout exercise, though several studies have investigated the effect on intermittent repeated bouts of exercise. There was no effect of β-alanine supplementation on repeated line drills (Hoffman *et al.*, 2008b), 5 × 5 s sprints (Sweeney *et al.*, 2010) and 3 sets of 6 × 20 m sprints (Ducker *et al.*, 2013c), or 66 repeated 15 m sprints throughout simulated games play (Saunders *et al.*, 2012a). Saunders *et al.* (2014) were the first to look at repeated sprint performance throughout simulated games play performed under hypoxic conditions using a 5 × 6 s repeated sprint test. Similarly to the other studies, repeated sprints were unaffected by β-alanine supplementation; sprint performance was similarly unaffected when β-alanine was cosupplemented with sodium bicarbonate. Taken together, these studies suggest that although these protocols are high intensity and intermittent in nature, the exercise durations of the sprints may be insufficient to be positively influenced by increased concentrations of muscle carnosine. Despite this, a beneficial effect of β-alanine on team-sport performance should not be entirely dismissed. Saunders *et al.* (2012b) showed a 34.4% improvement in team-sport-specific fitness with β-alanine, which suggests that increased muscle carnosine could be of benefit during periods of increased intensity or towards the latter stages of a match.

Although repeated sprints of short duration have shown no effect following supplementation with β-alanine, several studies employing slightly longer exercise durations have shown performance improvements. Tobias *et al.* (2013) employed 4 bouts of 30 s upper body Wingate tests separated by 3 min of recovery and β-alanine induced a significant 7% improvement in high-intensity intermittent performance; this improvement was doubled to 14% when β-alanine was cosupplemented with sodium bicarbonate. In accordance with the role of carnosine as a muscle buffer, the mean power during the first bout was unaffected with improved performance shown in the latter bouts. Painelli *et al.* (2014) showed an improved performance during 4 repeated bouts of lower-limb Wingate tests in both trained and non-trained cyclists; individuals were able to maintain their performance from the third to the fourth bout. Conversely, Mero *et al.* (2013) showed no effect on repeated 100 m swims following supplementation with β-alanine lasting ~57 s each. Surprisingly, sodium bicarbonate did have an effect though not when cosupplemented with β-alanine; participants were unblinded to the supplements, which may have contributed to the outcome and compromises the validity of the results. Despite this, these studies above suggest that repeated exercise of longer duration, which cumulatively exceed 1 min in duration, may be susceptible to improvements with β-alanine, though further research into this area is required using more sport specific exercise protocols.

14.3.3 Exercise and Training

Highly trained sprinters and rowers exhibit elevated levels of muscle carnosine (Parkhouse *et al.*, 1985), while bodybuilders have shown the highest average muscle concentrations (~43 mmol kg^{-1} dm; Tallon *et al.*, 2005), suggesting that training might increase muscle carnosine through the repeated exposure to acidosis and muscle hypoxia. Despite this, Mannion *et al.* (1994) showed no change in muscle carnosine following 16 weeks of isokinetic knee extension training and similarly, Kendrick and colleagues showed no effect of 10 weeks of strength training (Kendrick *et al.*, 2008) or 4 weeks of isokinetic training (Kendrick *et al.*, 2009) on muscle carnosine in Vietnamese sports-science students. An increase in carnosine would be expected if training increased the proportion of type II fibers, or even the cross-sectional area of these fibers, since type II fibers have approximately double the carnosine content of type I fibers (Hill *et al.*, 2007). An actual increase in carnosine with training alone would almost certainly require an increase in hepatic β-alanine, since this is the limiting factor for carnosine synthesis in the absence of dietary supplementation (Kendrick *et al.*, 2008, 2009), thus emphasizing the need to measure muscle carnosine in different fiber types in training studies.

Baguet *et al.* (2011) had 20 nonvegetarian students complete a sprint-training protocol while following a vegetarian or mixed diet for 5 weeks. There was no change in muscle carnosine in the *gastrocnemius lateralis* and *tibialis anterior* following the training period, although a nonsignificant increase in carnosine was shown in the *soleus* with the mixed diet (+11%) and a nonsignificant decrease with the vegetarian diet (−9%). This suggests that any changes in muscle carnosine were due to the dietary intake of β-alanine as opposed to the sprint training programme. However, this study would have benefited from the determination of single muscle fiber type and carnosine content. Furthermore, the timeframe in which the participants undertook training in these studies (5–16 weeks) may have been insufficient to induce long-term changes in muscle carnosine.

High-intensity interval training (HIIT) results in a large accumulation of metabolites, which may provide the mechanism for training adaptation, including the ability to buffer H^+ and delay acidosis (Weston *et al.*, 1997). Therefore, combining HIIT and β-alanine supplementation may result in additive gains, although Smith *et al.* (2009a) showed no additional effect of β-alanine to HIIT on fatigue and efficiency of electrical activity over a placebo. Similarly, Walter *et al.* (2010) showed no additional benefit of β-alanine supplementation to HIIT on VO_{2peak} during a graded exercise cycle. The training protocol used in these studies may have provided a superior stimulus, given the untrained nature of the participants. VO_{2peak} and time to exhaustion were improved during graded exercise cycles at 3 weeks in both β-alanine and placebo groups (Smith *et al.*, 2009b), although further increases from 3–6 weeks were only observed in the β-alanine group. In addition, both groups showed improvements in TWD during a 110% VO_{2peak} test over the study, highlighting the potential for β-alanine supplementation to enhance the benefits of HIIT. Gross *et al.* (2014) recently showed no enhanced adaptations to HIIT in individuals also supplemented with β-alanine, although supplementation was halted prior to the training period and resulted in a loss of metabolic effects shown with β-alanine following the supplementation period. Therefore, these results may not truly represent the potential for β-alanine to complement HIIT and further research is required.

14.4 Conclusions

There is a growing body of evidence to suggest that β-alanine supplementation can improve exercise performance and capacity, likely due to an increase in muscle buffering capacity through elevated levels of carnosine. This notion is supported by the increasing evidence base that β-alanine improves high-intensity exercise of between 1 and 4 min in duration (Hobson *et al.*, 2012), a period when anaerobic energy sources can contribute up to 60% of the total energy demand resulting in a large accumulation of H^+ (Maughan *et al.*, 1997). More recent studies suggest that this optimal timeframe may be extended to between 1 and 7 min, although exercise modality may be an additional factor affecting the efficacy of β-alanine supplementation. Importantly, the work of Painelli *et al.* (2014) showed that β-alanine supplementation can improve exercise performance to a similar extent in both trained and nontrained individuals (Figure 14.4), meaning β-alanine can be used as an ergogenic aid for high-intensity exercise regardless of training status. Although there has been a suggestion that elevated carnosine increases calcium sensitivity of the muscle (Dutka *et al.*, 2012), thereby improving muscle and exercise performance, an ergogenic effect would be expected across a much wider range of exercise durations, including very short bouts. Furthermore, no evidence of carnosine's role as an antioxidant has been demonstrated during exercise (Smith *et al.*, 2012; Smith-Ryan *et al.*, 2013), though little investigation into this area has been performed and further research is necessary.

14.5 Practical Applications

There is compelling evidence to suggest that athletes engaged in high-intensity exercise of 1 to 7 min in duration are most likely to reap the benefits of increased muscle carnosine concentration, likely due to an attenuation in muscle acidosis. These include, but are not limited to, events such as 800 m running, 100 and 200 m swimming and 2000 m rowing (Table 14.2). Furthermore, there is some suggestion that individuals involved in prolonged intermittent exercise may also benefit from β-alanine supplementation. Individuals engaged in a structured training regime should also consider supplementation as there is some evidence to suggest that supplementing with β-alanine alongside training could result in further improvements above training alone, although further evidence is required. Supplementation should begin a minimum of 2 weeks prior to an event at a high dose of 6.4 g per day; this dose has consistently been

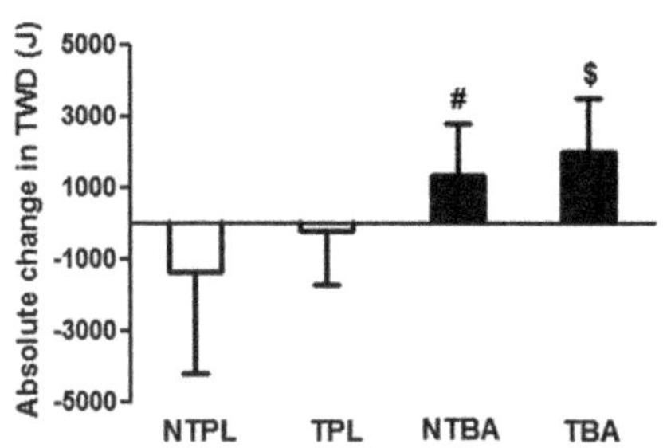

Figure 14.4 Absolute change in exercise performance for individual groups; nontrained and placebo (NTPL), trained and placebo (TPL), nontrained and β-alanine (NTBA) and trained and β-alanine (TBA). Data are mean ± 1 standard deviation. $^{\#}P = 0.008$ from NTPL. $^{\$}P = 0.037$ from TPL. (Adapted from Painelli *et al.* (2014) with permission.)

Table 14.2 The gold medal winning times in the most recent Olympics in sporting events of durations that theoretically may evoke an ergogenic benefit from β-alanine.

Sporting event		Recent olympics gold medal time	
		Men	Women
Running	400 m hurdles	47:63	52:70
	800 m	1:40:91	1:56:19
	1500 m	3:34:08	4:10:23
Swimming	100 m freestyle	47:52	53:00
	200 m freestyle	1:43:14	1:53:61
	400 m freestyle	3:40:14	4:01:45
Flat-water kayaking	500 m K1	N/A	1:51:46
	1000 m K1	3:26:46	N/A
Cycling	Track team pursuit	3:51:66	3:14:05
Rowing	2000 m 8+	5:48:75	6:10:59
	Double sculls	6:31:67	6:55:82

shown to increase muscle carnosine by ~60% following 4 weeks of supplementation. Separation of the doses by 3–4 h is required in order to minimize the incidence of paraesthesia, which is associated with the peak plasma concentration attained. Individuals should source sustained-release capsules, thereby limiting any symptoms of paraesthesia associated with β-alanine ingestion. In addition, high-level athletes may gain additive benefits from cosupplementation of β-alanine with other supplements including sodium bicarbonate. Athletes should ensure that any supplements used are screened by an accredited laboratory such as NSF (http://www.nsf.org) or HFL (http://www.hfl.co.uk) to ensure there is no contamination with steroids or stimulants.

Summary Points

- Intramuscular carnosine can play an important role during high-intensity exercise.
- The potential physiological roles of carnosine include pH regulation, calcium sensitivity and protection against oxidative stress.
- Dietary supplementation of β-alanine results in increased muscle carnosine content.
- Increased muscle carnosine results in improvements during high-intensity exercise limited by increasing muscle acidosis.
- Exercise improvements have been shown in both trained and non-trained individuals.
- Supplementation with β-alanine during a structured training regime may result in enhanced adaptations.
- β-Alanine can be considered an appropriate nutritional aid to attenuate fatigue during high-intensity exercise.

Key Facts of High-Intensity Exercise

1. Energy stores (adenosine triphosphate) are quickly depleted following the start of high-intensity exercise, resulting in dependence on alternative energy sources such as anaerobic glycolysis.
2. Anaerobic glycolysis results in an increased accumulation of hydrogen cations in the muscle, contributing to acidosis that can compromise the functionality of the working muscles.
3. Therefore, an increased ability to buffer these additional hydrogen cations may result in an improved performance or delay fatigue.

Key Facts of Increased Muscle Carnosine and Exercise

1. Intramuscular content can be increased *via* chronic supplementation with β-alanine, the rate-limiting precursor to carnosine synthesis.
2. Our understanding of the exact physiological role of carnosine remains limited though it is considered to play an important role during high-

intensity exercise due to its suggested roles as pH buffer, calcium sensitizer and antioxidant.
3. There is evidence to suggest increased carnosine concentrations can improve athletic performance likely due to expansion of the imidazole ring content of the muscle to increased buffering capacity.

Definitions of Words and Terms

β-Alanine. An amino acid that is one of the constituents of carnosine. It is found in histidine-bound forms in meats such as turkey and chicken; when ingested, this is broken down to free β-alanine and taken up into the muscle to form carnosine. It can also be ingested in nutritional supplement form to increase the carnosine concentration of muscle.
Buffering capacity. The ability of a solution to resist changes in pH when hydrogen cations are added.
Carnosine. An intracellular dipeptide consisting of histidine and β-alanine.
Ergogenic. Exercise performance or capacity enhancing. Ergogenic aids are external influences, such as supplements, than can improve exercise performance or capacity.
Exercise capacity. The maximum amount of exertion an individual can sustain at a fixed intensity.
Exercise performance. The maximum amount of work an individual can perform within a set time period or distance.
Hydrogen cation (H^+). A positively charged proton produced during exercise following the dissociation of lactate, which increases the acidity of the environment.
Muscle biopsy. An invasive procedure in which a sample of muscle is obtained directly from the selected muscle, freeze dried and stored for analysis of its contents including carnosine.
Paraesthesia. A side-effect associated with β-alanine supplementation described as an itchy sensation similar to that of pins and needles that may also result in a slight reddening of the skin. Symptoms usually occur within 20 min of ingestion and can last up to an hour.
Proton magnetic resonance spectroscopy (^{1}H-MRS). A noninvasive method by which the constitutions, including carnosine, of muscle can be determined using a magnetic resonance imaging scanner.
Sodium bicarbonate. An ergogenic aid designed to increase levels of bicarbonate in the blood, thereby increased the buffering capacity of the blood.

List of Abbreviations

ADP	Adenosine diphosphate
ATP	Adenosine triphosphate
BM	Body mass
$CCT_{110\%}$	Cycling capacity test at 110% of powermax
g per day	Grams per day

H^+	Hydrogen ions
HIIT	High intensity interval training
mg kg^{-1} BM	Milligram per kilogram of body mass
mmol kg^{-1} dm	Millimole per kilogram of dry mass
MVIC	Maximal voluntary isometric contraction
TWD	Total work done
VO_{2max}	Maximal oxygen uptake
VO_{2peak}	Peak oxygen uptake
μmol L^{-1}	Micromole per litre

Acknowledgements

Bryan Saunders has been financially supported by Fundação de Amparo à Pesquisa do Estado de Sao Paulo (FAPESP: 2011/19513-2). The authors declare no conflict of interests.

References

Abe, H., 2000. Role of histidine-related compounds as intracellular proton buffering constituents in vertebrate muscle. *Biochemistry*. 65: 757–765.

Ahlborg, B., Bergström, J., Ekelund, L., Guarnieri, G., Harris, R. C., Hultman, E. and Nordesjö, L., 1972. Muscle metabolism during isometric exercise performed at constant force. *Journal of Applied Physiology*. 33: 224–228.

Baguet, A., Reyngoudt, H., Pottier, A., Everaert, I., Callens, S., Achten, E. and Derave, W., 2009. Carnosine loading and washout in human skeletal muscles. *Journal of Applied Physiology*. 106: 837–842.

Baguet, A., Bourgois, J., Vanhee, L., Achten, E. and Derave, W., 2010. Important role of carnosine in rowing performance. *Journal of Applied Physiology*. 109: 1096–1101.

Baguet, A., Everaert, I., de Naeyer, H., Reyngoudt, H., Stegen, S., Beeckman, S., Achten, E., Vanhee, L., Volkaert, A., Petrovic, M., Taes, Y. and Derave, W., 2011. Effects of sprint training combined with vegetarian or mixed diet on muscle carnosine content and buffering capacity. *European Journal of Applied Physiology*. 111: 2571–2580.

Bellinger, P. M., Howe, S. T., Shing, C. M. and Fell, J. W., 2012. Effect of combined beta-alanine and sodium bicarbonate supplementation on cycling performance. *Medicine and Science in Sports and Exercise*. 44: 1545–1551.

Boldyrev, A. A., Bulygina, E. and Leinsoo, T., 2003. Protection of neuronal cells against reactive oxygen species by carnosine and related compounds. *Comparative Biochemistry and Physiology, Part B: Biochemistry & Molecular Biology*. 137: 81–88.

Carr, A. J., Hopkins, W. G. and Gore, C. J., 2011. Effects of acute alkalosis and acidosis on performance a meta-analysis. *Sports Medicine*. 41: 801–814.

Chung, W., Shaw, G., Anderson, M. E., Pyne, D., Saunders, P. U., Bishop, D. J. and Burke, L. M., 2012. Effect of 10 week beta-alanine supplementation on competition and training performance in elite swimmers. *Nutrients*. 4: 1441–1453.

Chung, W., Baguet, A., Bex, T., Bishop, D. J. and Derave, W., 2014. Doubling of muscle carnosine concentration does not improve laboratory 1 h cycling time trial performance. *International Journal of Sport Nutrition and Exercise Metabolism*. 24: 315–324.

Décombaz, J., Beaumont, M., Vuichoud, J., Bouisset, F. and Stellingwerff, T., 2012. Effect of slow-release β-alanine tablets on absorption kinetics and paresthesia. *Amino Acids*. 43: 67–76.

Derave, W., Ozdemir, M. S., Harris, R. C., Pottier, A., Reyngoudt, H., Koppo, K., Wise, J. A. and Achten, E., 2007. β-Alanine supplementation augments muscle carnosine content and attenuates fatigue during repeated isokinetic contraction bouts in trained sprinters. *Journal of Applied Physiology*. 103: 1736–1743.

Ducker, K. J., Dawson, B. and Wallman, K. E., 2013a. Effect of beta-alanine supplementation on 800 m running performance. *International Journal of Sport Nutrition and Exercise Metabolism*. 23: 554–561.

Ducker, K. J., Dawson, B. and Wallman, K. E., 2013b. Effect of beta-alanine supplementation on 2000 m rowing ergometer performance. *International Journal of Sport Nutrition and Exercise Metabolism*. 23: 336–343.

Ducker, K. J., Dawson, B. and Wallman, K. E., 2013c. Effect of beta-alanine and sodium bicarbonate supplementation on repeated-sprint performance. *Journal of Strength and Conditioning Research*. 27: 3450–3460.

Dutka, T. L., Lamboley, C. R., McKenna, M. J., Murphy, R. M. and Lamb, G. D., 2012. Effects of carnosine on contractile apparatus Ca^{2+} sensitivity and sarcoplasmic reticulum Ca^{2+} release in human skeletal muscle fibers. *Journal of Applied Physiology*. 112: 728–736.

del Favero, S., Roschel, H., Solis, M. Y., Hayashi, A. P., Artioli, G. G., Otaduy, M. C., Benatti, F. B., Harris, R. C., Wise, J. A., Leite, C. C., Pereira, R. M., de Sá-Pinto, A. L., Lancha-Junior, A. H. and Gualano, B., 2012. Beta-alanine (Carnosyn™) supplementation in elderly subjects (60–80 years): effects on muscle carnosine content and physical capacity. *Amino Acids*. 43: 49–56.

Gross, M., Boesch, C., Bolliger, C. S., Norman, B., Gustafsson, T., Hoppeler, H. and Vogt, M., 2014. Effects of beta-alanine supplementation and interval training on physiological determinants of severe exercise performance. *European Journal of Applied Physiology*. 114: 221–234.

Harris, R. C., Tallon, M. J., Dunnett, M., Boobis, L., Coakley, J., Kim, H. J., Fallowfield, J. L., Hill, C. A., Sale, C. and Wise, J. A., 2006. The absorption of orally supplied β-alanine and its effect on muscle carnosine synthesis in human *vastus lateralis*. *Amino Acids*. 30: 279–289.

Harris, R. C., Wise, J. A., Price, K. A., Kim, H. J., Kim, C. K. and Sale, C., 2012. Determinants of muscle carnosine content. *Amino Acids*. 43: 5–12.

Hill, C. A., Harris, R. C., Kim, H. J., Harris, B. D., Sale, C., Boobis, L. H., Kim, C. K. and Wise, J. A., 2007. Influence of β-alanine supplementation on skeletal muscle carnosine concentrations and high intensity cycling capacity. *Amino Acids*. 32: 225–233.

Hobson, R. M., Saunders, B., Ball, G., Harris, R. C. and Sale, C., 2012. Effects of β-alanine supplementation on exercise performance: a review by meta-analysis. *Amino Acids*. 43: 25–37.

Hobson, R. M., Harris, R. C., Martin, D., Smith, P., Macklin, B., Gualano, B. and Sale, C., 2013. Effect of β-alanine, with & without sodium bicarbonate, on 2000 m rowing performance. *International Journal of Sport Nutrition and Exercise Metabolism*. 23: 480–487.

Hoffman, J. R., Ratamess, N. A., Ross, R., Kang, J., Magrelli, J., Neese, K., Faigenbaum, A. D. and Wise, J. A., 2008a. β-alanine and the hormonal response to exercise. *International Journal of Sports Medicine*. 29: 952–958.

Hoffman, J. R., Ratamess, N. A., Faigenbaum, A. D., Ross, R., Kang, J., Stout, J. R. and Wise, J. A., 2008b. Short-duration β-alanine supplementation increases training volume and reduces subjective feelings of fatigue in college football players. *Nutrition Research*. 28: 31–35.

Howe, S. T., Bellinger, P. M., Driller, M. W., Shing, C. M. and Fell, J. W., 2013. The effect of β-alanine supplementation on isokinetic force and cycling performance in highly trained cyclists. *International Journal of Sport Nutrition and Exercise Metabolism*. 23: 562–570.

Jagim, A. R., Wright, G. A., Brice, A. G. and Doberstein, S. T., 2013. Effects of beta-alanine supplementation on sprint endurance. *Journal of Strength and Conditioning Research*. 27: 526–532.

Kendrick, I. P., Harris, R. C., Kim, H. J., Kim, C. K., Dang, V. H., Lam, T. Q., Bui, T. T., Smith, M. and Wise, J. A., 2008. The effects of 10 weeks of resistance training combined with β-alanine supplementation on whole body strength, force production, muscular endurance and body composition. *Amino Acids*. 34: 547–554.

Kendrick, I. P., Kim, H. J., Harris, R. C., Kim, C. K., Dang, V. H., Lam, T. Q., Bui, T. T. and Wise, J. A., 2009. The effect of 4 weeks β-alanine supplementation and isokinetic training on carnosine concentrations in type I and II human skeletal muscle fibers. *European Journal of Applied Physiology*. 106: 131–138.

Kern, B. D. and Robinson, T. L., 2011. Effects of β-alanine supplementation on performance and body composition in collegiate wrestlers and football players. *Journal of Strength and Conditioning Research*. 25: 1804–1815.

Mannion, A. F., Jakeman, P. M., Dunnet, M., Harris, R. C. and Willan, P. L. T., 1992. Carnosine and anserine concentrations in the quadriceps femoris muscle of healthy humans. *European Journal of Applied Physiology*. 64: 47–50.

Mannion, A. F., Jakeman, P. M. and Willan, P. L. T., 1994. Effects of isokinetic training of the knee extensors on high-intensity exercise performance and skeletal muscle buffering. *European Journal of Applied Physiology*. 68: 356–361.

Maughan, R. J., Gleeson, M. and Greenhaff, P. L., 1997. *Biochemistry of exercise training*. Oxford University Press.

Mero, A. A., Hirvonen, P., Saarela, J., Hulmi, J. J., Hoffman, J. R. and Stout, J. R., 2013. Effect of sodium bicarbonate and beta-alanine supplementation on maximal sprint swimming. *Journal of the International Society of Sports Nutrition*. 10: 52.

Painelli, V., Roschel, H., de Jesus, F., Sale, C., Harris, R. C., Galves, V. F., de Oliveira, N., do Carmo, C. A., Solis, M. Y., Benatti, F. B., Gualano, B., Lancha, A. H. and Artioli, G. G., 2013. Effects of β-alanine combined with sodium bicarbonate on swimming performance. *Applied Physiology, Nutrition, and Metabolism*. 38: 525–532.

Painelli, V. S., Saunders, B., Sale, C., Harris, R. C., Solis, M. Y., Roschel, H., Gualano, B., Artioli, G. G. and Lancha, Jr, A. H., 2014. Influence of training status on high-intensity intermittent performance in response to β-alanine supplementation. *Amino Acids*. 46: 1207–1215.

Parkhouse, W. S., McKenzie, D. C., Hochachka, P. W. and Ovalle, W. K., 1985. Buffering capacity of deproteinized human *vastus lateralis* muscle. *Journal of Applied Physiology*. 58: 14–17.

Sale, C., Saunders, B., Hudson, S., Wise, J. A., Harris, R. C. and Sunderland, C. D., 2011. Effect of β-alanine plus sodium bicarbonate on high-intensity cycling capacity. *Medicine & Science in Sports & Exercise*. 43: 1972–1978.

Sale, C., Hill, C. A., Ponte, J. and Harris, R. C., 2012. Beta-alanine supplementation improves isometric endurance of the knee extensor muscles. *Journal of the International Society of Sports Nutrition*. 9: 26.

Sale, C., Artioli, G. G., Gualano, B., Saunders, B., Hobson, R. M. and Harris, R. C., 2013. Carnosine: from exercise performance to health. *Amino Acids*. 44: 1477–1491.

Saunders, B., Sale, C., Harris, R. C. and Sunderland, C., 2012a. Effect of beta-alanine supplementation on repeated sprint performance during the Loughborough intermittent shuttle test. *Amino Acids*. 43: 39–47.

Saunders, B., Sunderland, C., Harris, R. C. and Sale, C., 2012b. β-Alanine supplementation improves YoYo intermittent recovery test performance. *Journal of the International Society of Sports Nutrition*. 9: 39.

Saunders, B., Sale, C., Harris, R. C., Morris, J. G. and Sunderland, C., 2013. Reliability of a high-intensity cycling capacity test. *Journal of Science and Medicine in Sport*. 16: 286–289.

Saunders, B., Sale, C., Harris, R. C. and Sunderland, C. D., 2014. Effect of sodium bicarbonate and β-alanine on repeated sprints during intermittent exercise performed in hypoxia. *International Journal of Sport Nutrition and Exercise Metabolism*. 24: 196–205.

Smith, A. E., Moon, J. R., Kendall, K. L., Graef, J. L., Lockwood, C. M. Walter, A. A., Beck, T. W., Cramer, J. T. and Stout, J. R., 2009a. The effect of β-alanine supplementation and high-intensity interval training on neuromuscular fatigue and muscle function. *European Journal of Applied Physiology*. 105: 357–363.

Smith, A. E., Walter, A. A., Graef, J. L., Kendall, K. L., Moon, J. R., Lockwood, C. M., Fukuda, D. H., Beck, T. W., Cramer, J. T. and Stout, J. R., 2009b. Effects of β-alanine supplementation and high intensity interval training on endurance performance and body composition in men; a double blind trial. *Journal of the International Society of Sports Nutrition*. 6: 5.

Smith, A. E., Stout, J. R., Kendall, K. L., Fukuda, D. H. and Cramer, J. T., 2012. Exercise-induced oxidative stress: the effects of beta-alanine supplementation in women. *Amino Acids*. 43: 77–90.

Smith-Ryan, A. E., Fukuda, D. H., Stout, J. R. and Kendall, K. L., 2012. High-velocity intermittent running: effects of beta-alanine supplementation. *Journal of Strength and Conditioning Research*. 26: 2798–2805.

Smith-Ryan, A. E., Fukuda, D. H., Stout, J. R. and Kendall, K. L., 2013. The influence of β-alanine supplementation on markers of exercise-induced oxidative stress. *Applied Physiology, Nutrition, and Metabolism*. 39: 38–46.

Stellingwerff, T., Maughan, R. J. and Burke, L. M., 2011. Nutrition for power sports: middle-distance running, track cycling, rowing, canoeing/kayaking, and swimming. *Journal of Sports Sciences*. 29: S79–S89.

Stellingwerff, T., Anwander, H., Egger, A., Buehler, T., Kries, R., Decombaz, J. and Boesch, C., 2012. Effect of two β-alanine dosing protocols on muscle carnosine synthesis and washout. *Amino Acids*. 42: 2461–2472.

Stout, J. R., Cramer, J. T., Mielke, M., O'Kroy, J., Torok, D. J. and Zoeller, R. F., 2006. Effects of twenty-eight days of beta-alanine and creatine monohydrate supplementation on the physical working capacity at neuromuscular fatigue threshold. *Journal of Strength and Conditioning Research*. 20: 928–931.

Stout, J. R., Cramer, J. T., Zoeller, R. F., Torok, D. J., Costa, P., Hoffman, J. R., Harris, R. C. and O'Kroy, J., 2007. Effects of β-alanine supplementation on the onset of neuromuscular fatigue and ventilator threshold in women. *Amino Acids*. 32: 381–386.

Stout, J. R., Graves, B. S., Smith, A. E., Hartman, M. J., Cramer, J. T., Beck, T. W. and Harris, R. C., 2008. The effect of beta-alanine supplementation on neuromuscular fatigue in elderly (55–92 years): a double-blind randomized study. *Journal of the International Society of Sports Nutrition*. 5: 21.

Sweeney, K. M., Wright, G. A., Brice, A. G. and Doberstein, S. T., 2010. The effects of β-alanine supplementation on power performance during repeated sprint activity. *Journal of Strength and Conditioning Research*. 24: 79–87.

Tallon, M. J., Harris, B. D., Boobis, L. H., Fallowfield, J. L. and Wise, J. A., 2005. The carnosine content of *vastus lateralis* is elevated in resistance-trained bodybuilders. *Journal of Strength and Conditioning Research*. 19: 725–729.

Tobias, G., Benatti, F. B., de Salles Painelli, V., Roschel, H., Gualano, B., Sale, C., Harris, R. C., Lancha Jr, A. H. and Artioli, G. G., 2013. Additive effects of β-alanine and $NaHCO_3$ on upper-body intermittent performance. *Amino Acids*. 45: 309–317.

van Thienen, R., van Proeyen, K., Vanden Eynde, B., Puype, J., Lefere, T. and Hespel, P., 2009. β-alanine improves sprint performance in endurance cycling. *Medicine and Science in Sports and Exercise*. 41: 898–903.

Walter, A. A., Smith, A. E., Kendall, K. L., Stout, J. R. and Cramer, J. T., 2010. Six weeks of high-intensity interval training with and without β-alanine supplementation for improving cardiovascular fitness in women. *Journal of Strength and Conditioning Research*. 24: 1199–1207.

Weston, A. R., Myburgh, K. H., Lindsay, F. H., Dennis, S. C., Noakes, T. D. and Hawley, J. A., 1997. Skeletal muscle buffering capacity and endurance performance after high-intensity interval training by well-trained cyclists. *European Journal of Applied Physiology*. 75: 7–13.

CHAPTER 15

Human Skeletal Muscle Carnosine and Its Function: A Focus on Homeostasis, Muscle Contractility and pH

I. EVERAERT[a] AND W. DERAVE*[a]

[a]Department of Movement and Sports Sciences, Ghent University, Watersportlaan 2, 9000 Ghent, Belgium
*E-mail: wim.derave@ugent.be

15.1 Introduction

The muscle carnosine concentration of healthy, nonspecifically trained humans has been shown to be in the range of 17–27 mmol kg^{-1} DW or 5–10 mmol L^{-1} WW. Human data concerning carnosine levels in other organs than skeletal muscles are, to our knowledge, lacking. Based on data of carnosine levels in other organs of mice, it can be suggested that over 99% of the carnosine is stored in skeletal muscle tissue. In contrast to other mammals, no methylated analogs of carnosine, *i.e.* anserine or ophidine, are present in human skeletal muscle. Between-subjects variation in muscle carnosine content is quite high, with the highest levels 3–4 times higher compared to the lowest, but within-subject muscle-carnosine levels are quite stable over the time (Baguet *et al.*, 2009). In this chapter, we will first describe the

Food and Nutritional Components in Focus No. 8
Imidazole Dipeptides: Chemistry, Analysis, Function and Effects
Edited by Victor R Preedy

Published by the Royal Society of Chemistry, www.rsc.org

determinants of human muscle carnosine and how these baseline levels can be manipulated, and subsequently discuss the current view on the physiological role of carnosine in muscle contractile behavior.

15.2 Determinants of Muscle-Carnosine Content

15.2.1 Muscle Fiber-Type Composition

An indisputable determinant of normal intramyocellular carnosine quantity is muscle fiber type composition. The observations and arguments that support this theorem are summarized below.

- By means of HPLC analysis of single muscle fibers of nonspecifically trained humans, Harris and colleagues (Harris *et al.*, 1998) were the first to describe a marked distinction between type I and II muscle fibers, with type I having on average 10.5 ± 7.6 mmol kg^{-1} DW and type II 23.3 ± 8.1 mmol kg^{-1} DW carnosine. Almost no overlap between carnosine levels of different muscle fiber types was found. Later reports (Hill *et al.*, 2007; Tallon *et al.*, 2007; Kendrick *et al.*, 2009) confirmed that type II fibers have ~30–130% more carnosine compared to type I fibers.
- It was confirmed by 3 independent labs who are using proton spectroscopy (Baguet *et al.*, 2009; Del Favero *et al.*, 2012; Gross *et al.*, 2013) that muscles with a higher proportion of fast-twitch muscle fibers (*e.g.* gastrocnemius) have indeed higher intramyocellular carnosine levels than predominantly slow muscles (*e.g.* soleus or vastus intermedius).
- When analyzing human skeletal muscle biopsies, a positive correlation was demonstrated between the fraction of type II fibers on the one hand and carnosine levels on the other ($0.43 < r < 0.71$, (Mannion *et al.*, 1995; Suzuki *et al.*, 2002; Baguet *et al.*, 2011a).
- Athletes who are typically involved in high-intensity exercises have a higher proportion of fast-twitch muscle fibers compared to endurance athletes (Costill *et al.*, 1976). In accordance, higher muscle-carnosine levels have been observed in athletes typically involved in 100–400 m track and field disciplines compared to untrained subjects or runners competing in ≥3000 m disciplines (Parkhouse *et al.*, 1985; Baguet *et al.*, 2011b).

It is possible that variations in the gene expression and activity of carnosine synthase or other related enzymes and transporters underlie the fiber-type-based difference in carnosine content. To our knowledge, no human data concerning expression patterns in different muscles are available. Mice data, however, showed that the mRNA expression of the carnosine synthase (CARNS) enzyme is almost twice as high in the more fast-twitch muscle *vastus lateralis* compared to the slower soleus (Miyaji *et al.*, 2012). In contrast, the expression of the taurine transporter that is mainly responsible for the uptake of β-alanine in mice skeletal muscle follows the opposite distribution with a 3-fold higher mRNA expression in SOL compared to EDL

(Pierno *et al.*, 2012). More research on human muscle samples is certainly needed to fully elucidate the underlying metabolic pathways for the muscle fiber-type-based variation in carnosine content.

15.2.2 Gender, Age and Androgens

If we look within an adult population, male whole-muscle (soleus and gastrocnemius, (Everaert *et al.*, 2011; Baguet *et al.*, 2012)/vastus lateralis, (Mannion *et al.*, 1992)) carnosine content is on average 21–36% higher compared to female content, except for tibialis anterior for which a 82% higher carnosine concentration was reported (Everaert *et al.*, 2011). However, these data were not confirmed in single fibers, as no difference in carnosine levels of individual type I or II muscle fibers could be observed in age-matched males and females (Tallon *et al.*, 2007). Next to gender, a subject's age has also been reported as a nonmodifiable determinant of intramyocellular carnosine. Several studies reported a negative correlation between muscle-carnosine levels and age in rodents (Johnson and Hammer, 1992; Stuerenburg and Kunze, 1999; Derave *et al.*, 2008), in patients with neuromuscular diseases (Stuerenburg and Kunze, 1999) and osteoarthritis (Tallon *et al.*, 2007), and in healthy subjects (Everaert *et al.*, 2011; Baguet *et al.*, 2012). The lower muscle-carnosine levels in older subjects compared to younger ones is mainly the result from a decrease that occurs during adulthood, even shortly after puberty, rather than a decrease from adulthood to elderly (Figure 15.1, (Baguet *et al.*, 2012)).

Several hypotheses have been proposed (and refuted) in order to explain these age- and gender-based differences in carnosine content: (1) the smaller cross-sectional area of carnosine-rich type II muscle fibers in older subjects

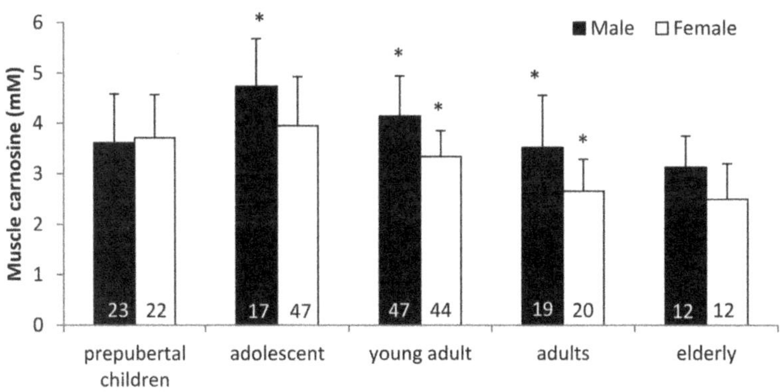

Figure 15.1 Cross-sectional comparison of muscle-carnosine levels of male and female subjects divided in different age categories. Muscle carnosine (mean ± SD) is the mean of soleus and gastrocnemius. Age categories: 8–11: prepubertal children; 15–20 (13–18): male (female) adolescent; 21–30 (19–30): male (female) young adult; 31–50: adults; 60+: elderly. $^{*}p < 0.05$ *vs.* previous age group. Number of subjects/group is indicated on the bars. Adapted from Baguet *et al.* (2012).

(Kirkendall and Garrett, 1998; Maharam *et al.*, 1999) could underlie the age-based difference in muscle carnosine. As carnosine did not differ between males and females in both type I and II muscle fibers, it was further suggested that the gender-based difference in whole-muscle carnosine (Everaert *et al.*, 2011; Baguet *et al.*, 2012) could be fully attributed to a different fiber type I:II ratio when quantifying whole-muscle carnosine (Tallon *et al.*, 2007; Harris *et al.*, 2012). (2) As female and older subjects have both lower carnosine and androgen levels compared to younger male persons, it can be suggested that, next to fiber-type composition, also the amount of circulating androgens are related to muscle carnosine concentrations. This hypothesis is supported by the fact that carnosine content is higher in adolescent compared to prepubertal boys, while no difference was observed in the same girl categories (Figure 15.1). Although there is no correlation between circulating testosterone levels and muscle carnosine in an eugonadal male population (Everaert *et al.*, 2011), more distinct changes in testosterone affected muscle carnosine concentrations in mice (see further, (Penafiel *et al.*, 2004; Everaert *et al.*, 2013a). (3) As muscle-carnosine levels of elderly subjects (Del Favero *et al.*, 2012) and females (Stegen *et al.*, 2014) can be increased to a similar extent as young and male subjects following β-alanine supplementation, it is unlikely that the capacity to transport β-alanine into the muscle or to synthesize carnosine is affected. Indeed, Baguet and coworkers (Baguet *et al.* 2011a) observed similar carnosine synthase mRNA expression in gastrocnemius of age- and diet-matched males and females.

Based on the gender- and age-based differences in muscle carnosine it was suggested that muscle carnosine content is, at least partly, regulated by testosterone. Indeed, 2 independent mice studies revealed that muscle-carnosine levels are ~40% lower 2–4 weeks after castration, which could be prevented by testosterone replacement (Penafiel *et al.*, 2004; Everaert *et al.*, 2013a). Based on mRNA expression levels it was suggested that lower intramyocellular β-alanine availability, and not a diminished carnosine synthesizing capacity, underlies this orchidectomy-induced decline in muscle carnosine (Everaert *et al.*, 2013a).

15.2.3 Diet

Carnosine is synthesized within muscle cells by combining β-alanine with L-histidine, of which the availability of the former has been shown to limit the synthesis (Harris *et al.*, 2006). So far, only the hepatic degradation of uracil has been frequently cited as an endogenous source of β-alanine. However, more research is needed to elucidate how much β-alanine can be derived from this degradation pathway and whether other endogenous pathways to synthesize β-alanine are available. β-Alanine can be exogenously derived from the degradation of carnosine and anserine that are abundantly present in meat and fish in our daily food. By using food questionnaires, it was estimated that the daily dietary intake of β-alanine is on average 0.3 g within a uniform Belgium omnivorous population (Baguet *et al.*, 2009; Everaert *et al.*, 2011; Baguet *et al.*, 2012). Although 5 weeks of vegetarianism did not affect intramuscular carnosine levels (Baguet *et al.*, 2011a), a cross-sectional comparison of chronic (≥8 years) vegetarians and age- and gender-matched controls revealed lower carnosine levels (~20%) as a result of long-term dietary β-alanine restriction (Everaert *et al.*, 2011). These results confirmed the 45% lower carnosine levels in the nonmatched female vegetarians compared to male omnivores (Harris *et al.*, 2007).

Harris *et al.* (2006) were the first to show that the enrichment of the diet with extra β-alanine (6.4 g per day) for 4 weeks enhanced the muscle-carnosine stores with 64.2%. This landmark study started the research to optimize and understand muscle-carnosine loading and the search of exercise types and modalities that are favored by enhanced muscle carnosine. Concerning carnosine loading, different modalities have been shown to further enhance net muscle-carnosine synthesis. By pooling the data of all published studies on muscle-carnosine loading, it was shown that the relative increase in muscle carnosine is dependent on the total amount of ingested β-alanine (Stellingwerff *et al.*, 2012b). The linear regression analysis revealed that approximately 230 g β-alanine is required to have a relative muscle carnosine increase of 50%. Muscle-carnosine loading is further higher when β-alanine is taken 30 min before the meal (+63.6 ± 41.6%) compared to between the meals (+40.7 ± 14.2%; (Stegen *et al.*, 2013)). Furthermore, β-alanine-induced muscle-carnosine loading is also more pronounced in trained (+77.9%) compared to untrained within-subject muscles (+42.9%; (Bex *et al.*, 2014)). Lastly, long-term β-alanine supplementation seems to enhance muscle carnosine stores in both males and females (Stegen *et al.*, 2014), trained and untrained subjects (Derave *et al.*, 2007; Baguet *et al.*, 2010a), young and old (Del Favero *et al.*, 2012) and slow and fast muscle (fibers) (Hill *et al.*, 2007; Baguet *et al.*, 2009) and seems to be independent of the use of pure *vs.* slow-release β-alanine pills (Stegen *et al.*, 2013).

Upon discontinuation of β-alanine supplementation, muscle-carnosine levels will gradually and slowly return to baseline levels in 6–20 weeks (Baguet *et al.*, 2009; Stellingwerff *et al.*, 2012a). A daily dose of 1.2 g β-alanine per day seems to be a suitable maintenance dose in order to avoid this decline in muscle carnosine following a loading phase (Figure 15.2, (Stegen *et al.*, 2014)).

It can be concluded that young, male, omnivorous subjects with a high proportion of fast-twitch muscle fibers are characterized with the highest carnosine levels within humans. These baseline carnosine levels can be easily enhanced by long-term β-alanine supplementation. The next part of this chapter will discuss the physiological role of carnosine in contracting muscle and hence, elucidate why higher muscular carnosine levels can be advantageous during exercise.

15.3 Physiological Functions of Carnosine in Contracting Muscle

A first hypothesis about the possible function of carnosine was proposed in 1938. On the basis of titration experiments with protein-free extracts of different species (rat, horse, pig, dog, *etc.*, but not human), it was concluded that both carnosine and anserine contributed strongly to the buffering capacity

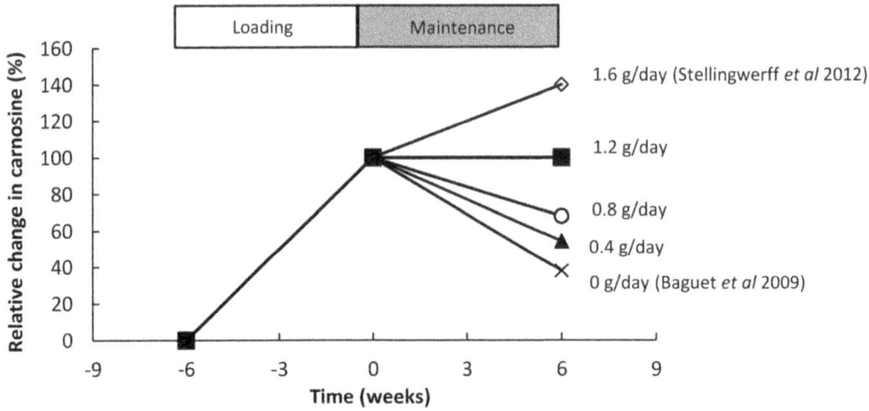

Figure 15.2 Maintenance dose following a muscle-carnosine loading protocol. Pre (−6 weeks) and post-loading (0 weeks) muscle carnosine content is set as respectively 0 and 100%. Subjects received either 1.6 g per day (Stellingwerff *et al.*, 2012a), 1.2 g per day, 0.8 g per day, 0.4 g per day or 0 g per day (Baguet *et al.*, 2009) for 6 weeks. 6–10 subjects/group. Adapted from Stegen *et al.* (2014).

of these extracts (Bate Smith, 1938). The first evidence for the potential of carnosine to attenuate peripheral fatigue was provided by Severin *et al.* in 1953 (cited in Boldyrev, 2007). The effect of carnosine on contractile activity of frog sartorius muscle was studied in isolated neuromuscular preparations, surrounded by a Ringer solution. As shown in Figure 15.3, the time to fatigue was markedly attenuated in muscles to which carnosine was added to the surrounding bath, at the stage of pronounced fatigue, compared to the control situation. Although several hypotheses were provided (*e.g.* pH buffer, enhanced contractile properties, increased energy supply, *etc.*), the mechanism for the carnosine-induced increase in the working capacity of these frog muscles was at that time unknown. Despite this very pronounced elongation of working capacity of frog muscles, similar experiments with mammalian skeletal muscles have, to our knowledge, never been reported.

The above-described ergogenic potential of carnosine is nowadays the subject of many investigations in the field of sport nutrition. The effect of muscle-carnosine loading, by supplementing participants with β-alanine for several weeks, has been verified in a wide range of different exercises and target populations. In 2012, a meta-analysis, based on 15 experimental studies, revealed that the effect size of β-alanine supplementation on performance was the largest for exercises lasting between 1 and 4 min ($p = 0.001$), but also significant for exercises longer than 4 min ($p = 0.046$), (Hobson *et al.*, 2012). Although only a minority of the studies include muscle carnosine measurements (Derave *et al.*, 2007; Hill *et al.*, 2007; Baguet *et al.*, 2010a; Del Favero *et al.*, 2012; Gross *et al.*, 2013; Chung *et al.*, 2014), it is generally accepted that the ergogenic potential of long-term β-alanine supplementation is the result of enhanced muscle-carnosine levels.

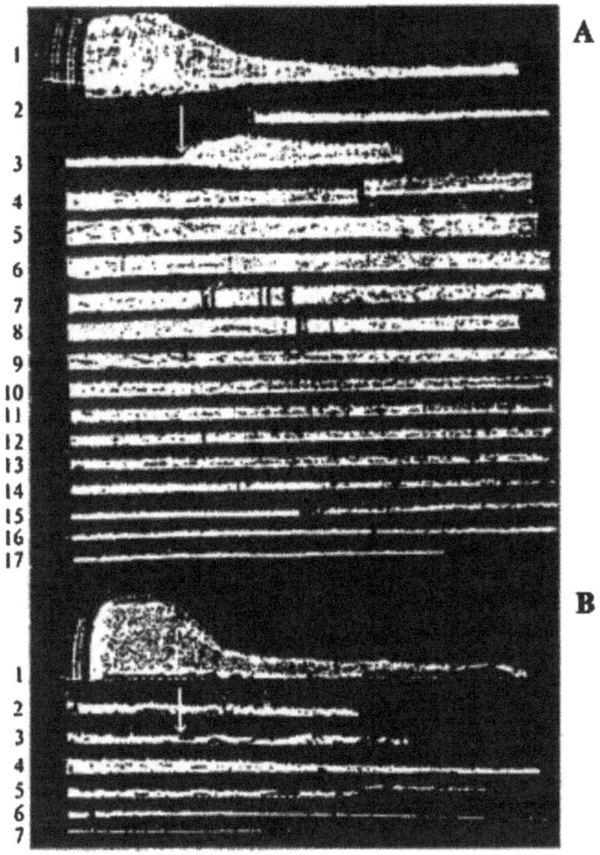

Figure 15.3 First *in vitro* report of ergogenic potential of carnosine. Working capacity of m. sartorius (frog) during *in vitro* nerve-stimulated contractions with (A, arrow line 3) and without (B) the addition of carnosine to the surrounding medium (Severin, 1953 cited in Boldyrev (2007)).

Despite the popularity of β-alanine as a supplement for athletes, the ergogenic mechanism of carnosine loading remains rather unexplored in humans. Mainly based on indirect evidences and *in vitro* rodent studies, the ergogenic potential of muscle-carnosine loading has been suggested to be the result of enhanced intracellular buffering capacity and/or improved excitation–contraction coupling during exercise. Both mechanisms will be discussed extensively in the next paragraphs.

15.3.1 Intramyocellular Buffering

The accumulation of intracellular protons during high-intensity exercises can be counteracted by different proton-buffering systems, *e.g.* inorganic phosphate, proteins, histidine-containing dipeptides, bicarbonate and phosphocreatine. Of all the amino acids, histidine is the only one with a pK_a

value (= 6.1) in the range of the physiological pH of the muscle (*i.e.* 7.0–6.0). When bounded together with β-alanine to form carnosine, the pK_a value of the imidazole ring of histidine increases to 6.83 (Bate Smith, 1938). Next to an ideal pK_a value, also a high intramuscular concentration is contributing to the buffering capacity of this dipeptide.

The contribution of carnosine to the total intracellular muscle buffering capacity is dependent on its concentration and thus on the species and muscle type under investigation. The contribution of HCDs to the total muscle-buffering capacity accounts, for example, for 23% in chicken, 25% in whale and 40% in skipjack tuna skeletal muscle (Abe, 2000). In humans, muscle contains a smaller amount of carnosine, which results in relatively lower contributions of 4.5% and 9.4% in type I and II fibers of vastus lateralis (Mannion *et al.*, 1992). Not only between but also within species there is a positive correlation ($0.34 < r < 0.69$ for humans, $r = 0.67$ for horses) between HCD content and muscle-buffering capacity (Parkhouse *et al.*, 1985; Sewel *et al.*, 1992; Baguet *et al.*, 2011a).

Despite these clear, but indirect, evidences for the role of carnosine as buffering agent, only 2 intervention studies have investigated the effect of muscle-carnosine loading on proton accumulation during high-intensity exercise. Baguet and colleagues showed that the blood pH decline following a 6 min high-intensity cycling exercise bout was attenuated following long-term β-alanine supplementation compared to control (Baguet *et al.*, 2010b). On the other hand, carnosine loading (+24%, +1.8 mmol kg^{-1} WW) following β-alanine supplementation (38 days, 3.2 g per day) was not able to attenuate the decline in vastus lateralis pH (from 7.01 to 6.76) during 90 s high-intensity cycling (Gross *et al.*, 2013). In contrast to the hypothesis, muscle-buffering capacity was also not affected by β-alanine supplementation. As β-alanine enhanced the aerobic energy contribution and decreased the O_2 deficit and muscle lactate accumulation during 90 s severe cycling, the authors hypothesized that the improved capacity for severe exercise after β-alanine supplementation could be due to mechanisms other than proton buffering (Gross *et al.*, 2013). To conclude, further research on human muscle samples during high-intensity exercise is definitely required to validate the theoretical rationale of carnosine's buffering function.

15.3.2 Calcium Release of Sarcoplasmic Reticulum

Another cause of fatigue is a reduction in the efficiency of the excitation-contraction coupling. Table 15.1 gives an overview of the studies in which the effect of enhanced muscle-carnosine levels on Ca^{2+} handling during contraction and fatigue has been studied. Different approaches to study the effect of carnosine on fatigue have been used (skinned fibers *versus* intact muscle preparations).

Data from several *in vitro* studies with both frog (Allen *et al.*, 1989), mouse (Bruton *et al.*, 2003) and rat (Lunde *et al.*, 2001) single muscle fibers showed that the myoplasmic free Ca^{2+} concentrations are reduced during repeated

Table 15.1 1Overview of *in vitro* studies investigating the effect of carnosine on muscle contractile function.[a]

	Species	Study	CARN manipulation	Calcium handling: Sensitivity	Release	Re-uptake
Skinned fiber	Frog	Lamont 1992	+ 15mM CARN	↑	ND	ND
	Mouse	Batrukova 1997	+ 30mM CARN	ND	↑	ND
	Rat	Dutka 2004	+ 4, 8, 16mM CARN	↑	=	ND
	Human	Dutka 2012	+ 8 or 16mM CARN	↑Type I & II	↑Type I, = Type II	ND
				Force Frequency	**Fatigue**	**Recovery**
Intact muscle	Mouse	Derave 2008	Oral creatine (15w, 2% food)	ND	↓SOL	↑EDL
		Everaert 2013b	Oral BA (56d, 1.2% water)	↑EDL	↓SOL ↑MaxSlope EDL ↑MinSlope SOL	=

[a]Abbreviations: BA = beta-alanine, CARN = carnosine, w = weeks, d = days, ND = not determined, SOL = soleus, EDL = extensor digitorum longus, MaxSlope = tetanic contraction velocity, MinSlope = tetanic relaxation velocity

tetanic contractions. Interestingly, restoring these Ca^{2+} levels at the stage of pronounced fatigue in single muscle fibers, by addition of caffeine or high extracellular K^+, have been shown to restore cytosolic Ca^{2+} and muscle force (reviewed by (Allen *et al.*, 2008).

A first report of an improved Ca^{2+} release by carnosine was published by Batrukova and Rubtsov (1997). A stimulating effect of both carnosine and anserine was found on the Ca^{2+} release from rabbit sarcoplasmic reticulum vesicles, with anserine being the most efficient dipeptide. Moreover, carnosine was shown to enhance the stimulating effects of caffeine and AMP on Ca^{2+} release, proving that there is no competition for the binding sites (Batrukova and Rubtsov, 1997). Similar experiments have been performed with skinned fibers of rats (Dutka and Lamb, 2004). Even though all the conditions to favor Ca^{2+} release were optimized, no Ca^{2+} release could be detected with 4–16 mM carnosine in the surrounding medium. The authors explained these contradictory results by the more physiological relevant conditions that were used: *in situ* position of sarcoplasmic reticulum under control of the DHPR/voltage sensors (*vs.* isolated Ca^{2+} release channels) and the higher Mg^{2+} concentrations (1 *vs.* 0.2 mM Mg^{2+}, which is a strong inhibitor of Ca^{2+}-release channels). The same authors repeated their experiments with human type I and II skeletal fibers and observed, only in type I fibers, a small enhancement in caffeine-induced Ca^{2+} release (Dutka *et al.*, 2012). However, it remains questionable whether these small changes are physiologically relevant as the activation of the Ca^{2+}-release channels *in vivo* are under tight control of the DHPR receptors with little contribution of the Ca^{2+}-induced Ca^{2+} release, which was used in these experiments (Dutka *et al.*, 2012).

15.3.3 Calcium Sensitivity of the Contractile Apparatus

Both *in vitro* (Westerblad and Allen, 1993) and *in vivo* (Hvid *et al.*, 2013) -induced fatigue has been shown to lower the affinity of troponin for Ca^{2+}. Although more research is required on which exercise types result in impaired Ca^{2+} sensitivity, it can be expected that restoring this reduction would have an ergogenic effect on exercise performance.

As can be observed in Table 15.1, carnosine has undoubtedly a stimulating effect on Ca^{2+} sensitivity. This has been confirmed in both frog (Lamont and Miller, 1992), rat (Dutka and Lamb, 2004) and recently in human (Dutka *et al.*, 2012) skinned muscle fiber preparations. A leftward shift of the $p\text{Ca}^{2+}$–force relationship after the addition of carnosine to the surrounding medium was observed in a concentration-dependent manner (8 and 16 mM, (Dutka and Lamb, 2004; Dutka *et al.*, 2012). This means that for a given concentration of Ca^{2+} in the sarcoplasm, the produced force was higher in the presence of carnosine. This capacity of carnosine can help the muscle to maintain higher levels of force during later stages of fatigue when Ca^{2+} release declines (Dutka and Lamb, 2004).

Acidosis of the cytoplasmic milieu occurring during high-intensity exercise has been suggested to result in lower Ca^{2+} sensitivity, as protons are able to compete with Ca^{2+} to bind troponin (Donaldson *et al.*, 1978; Fabiato and Fabiato, 1978). Therefore, it has been remarked that the effect of carnosine on Ca^{2+} sensitivity is secondary to the buffering function of carnosine (Sale *et al.*, 2013). However, as the pH was kept constant by HEPES buffer at 7.0 (Lamont and Miller, 1992) and 7.1 (Dutka and Lamb, 2004; Dutka *et al.*, 2012), there will be at least also a direct interaction between carnosine and Ca^{2+} binding at the level of troponin (Dutka *et al.*, 2012).

An improved Ca^{2+} handling as a result of enhanced carnosine levels has also been confirmed by *in vitro* contractions on intact mice skeletal muscles (Everaert *et al.*, 2013b). A leftward shift of the force–frequency relationship has been observed in EDL, but not in soleus, following long-term β-alanine supplementation, which evoked markedly higher carnosine (~160%) and anserine (~40%) concentrations (Everaert *et al.*, 2013b). Although the authors were not able to attribute these results to a better Ca^{2+} release or sensitivity, it confirms at least that the Ca^{2+} handling during contraction is also improved in intact muscle preparations that were under control of electrical stimulations.

To conclude, carnosine can clearly and dose-dependently enhance Ca^{2+} sensitivity in both fast and slow muscle fibers, but its effect on Ca^{2+} release is smaller and not present in type II fibers. More research to elucidate the modalities of the exercise types that diminish Ca^{2+} sensitivity will help to further predict which exercise performances can be favored by a muscle-carnosine loading protocol.

15.3.4 Release

Although the synthesis of carnosine occurs almost exclusively in the skeletal muscle and more than 99% of the body's carnosine is present in skeletal muscle, carnosine performs several functions in other places in the body. The question remains how carnosine arrives at these diverse places in the body. Based on indirect evidence in the literature, it can be hypothesized that skeletal muscles release carnosine into the interstitium/circulation during skeletal muscle contractions. Exercising horses and rats have indeed higher plasma carnosine levels compared to their sedentary controls (Dunnett *et al.*, 2002; Nagai *et al.*, 2003). Furthermore, the concentration of carnosine in the interstitium of *vastus lateralis* is also markedly enhanced during moderate and intense dynamic knee extensions (Norsborg *et al.*, 2003). However, it remains to be elucidated whether exercise or physical activity can, next to the interstitium, also enhance plasma carnosine content in humans and further research is certainly needed to unravel the exact mechanism and function of exercise-induced carnosine release.

15.3.5 Carnosine–Histidine–Histamine Pathway

Next to its involvement in carnosine synthesis, histidine can also be converted into histamine by the histidine decarboxylation enzyme in skeletal muscle cells. Although further research is needed, it can be hypothesized that the extra histidine needed to synthesize histamine during periods of increased demand, is mobilized through the hydrolysis of carnosine. Indeed, the acute injection of the stress-evoking compound 48/80 (5 mg kg^{-1} BW) in rats resulted in marked decreases of muscle carnosine 24 h postinjection (Greene *et al.*, 1984). Based on their experiments, the authors suggested that this decline in carnosine was the result of a higher demand of histamine (mimicked by the injection of compound 48/80). As the pH is not optimal for hydrolysis of carnosine inside the myocytes and as carnosine is possibly released during exercise into the circulation, it can be suggested that the hydrolysis of carnosine is catalyzed extracellularly rather than intracellularly. Although further research is needed on the functions and compartmentation of this proposed carnosine–histidine–histamine pathway, it is possible that carnosine, a putative nonmast cell reservoir of histidine, acts as an autocrine or paracrine regulator between skeletal muscle cells and vascular smooth muscle cells.

Summary Points

- Carnosine is a dipeptide consisting of L-histidine and β-alanine and is present in millimolar concentrations in human skeletal muscles.
- Most animals, except humans, also possess a methylated variant of carnosine, either anserine or ophidine/balenine, collectively called the histidine-containing dipeptides.
- Within-subject muscle-carnosine levels are quite stable, while a relative large between-subjects variation exists.

- Young, male subjects with a high proportion of fast-twitch muscle fibers have the highest carnosine concentration within a healthy population.
- An excess dietary intake of β-alanine (1.6–6.4 g per day) for several weeks (2–10 weeks) results in 1.3- to 2-fold higher muscle carnosine stores.
- β-Alanine is a popular nutritional supplement amongst athletes as it has been shown to improve some high-intensity exercise performances that last at least 1 min.
- As its dissociation constant is ideal (pK_a 6.83) and as the carnosine content is positively correlated to the buffering capacity, it is generally accepted that carnosine can act as a pH buffer in humans.
- *In vitro* data on both single fibers and intact muscles further suggest that carnosine can improve calcium handling during muscle contraction.
- Although more research is needed, several suggestions have been made that carnosine can be released from skeletal muscle into the interstitium and that it further acts as depot for the histidine–histamine pathway during periods of increased demands.

Key Facts

1. Carnosine was first discovered in meat extracts in 1900 by the Russian scientist Vladimir Gulevitch. Although its function has been studied intensively during the last century, its complete physiological role and metabolism in muscle is still mainly unknown.
2. Humans are the only mammals that do not have methylated analogs (*i.e.* anserine or ophidine) of carnosine in their skeletal muscles.
3. Carnosine is a dipeptide that consist of two amino acids, β-alanine and histidine, which is present in relatively high amounts in our daily food. One kilogram of chicken breast contains approximately 2 g carnosine or 800 mg β-alanine.
4. β-Alanine is a popular nutritional supplement amongst athletes (both recreational and elite athletes), as long-term supplementation (3–6 g per day, 2–10 weeks) has been shown to enhance muscle-carnosine levels and thereby improving performance during high-intensity exercises.
5. Male, young subjects with a high proportion of fast-twitch muscle fibers have the highest concentration of muscle carnosine.

Key Terms

HPLC. HPLC is the abbreviation of high-performance liquid chromatography, an analytical technique that is able to separate, identify and quantify a wide range of samples that can be dissolved in a liquid.

Proton magnetic resonance spectroscopy (H^+-MRS). H^+-MRS is an application of nuclear magnetic resonance and is used as a noninvasive method to identify and quantify metabolites that have proton(s) in their molecular structure, including carnosine.

Muscle-fiber-type composition. Skeletal muscles are composed of muscle fibers, which can be divided into more oxidative (the generation of energy, under the form of ATP, is dependent on the amount of oxygen) or more glycolytic (energy generation independent of oxygen supply) muscle fibers. Individual skeletal muscles have a mixture of various fiber types, which is called the muscle-fiber-type composition.

pH buffer. A pH buffer is a chemical compound which resists against changes in environmental pH.

Buffering capacity. The buffering capacity is the capacity of a cell or tissue to resist changes in pH. The buffering capacity can be quantified *in vitro* by the amount of acid that is needed to decrease the pH with one unit.

Excitation–contraction coupling. A term that is used to describe the transition between the electrical stimulus of a nerve and the mechanical response of the skeletal muscle.

Sarcoplasmic reticulum. This is an organelle within a muscle fiber that is responsible for the storage of calcium during rest and the release and reuptake of calcium during muscle contraction.

Calcium sensitivity. This is the sensitivity of troponin to the binding of calcium, which initiates the binding between actin and myosin and thus muscle shortening. The higher the sensitivity of troponin for the binding of calcium, the more force that will be generated for the same amount of available calcium.

Calcium release. The release of calcium from the sarcoplasmic reticulum into the myoplasm, which is initiated by a nerve-derived action potential.

List of Abbreviations

AMP	Adenosine monophosphate
BMI	Body mass index
Ca^{2+}	Calcium
CARNS	Carnosine synthase
DW	Dry weight
DHPRs	Dihydropyridine receptors
EDL	Extensor digitorum longus
HCD	Histidine-containing dipeptides
HDC	Histidine decarboxylase
HPLC	High-performance liquid chromatography
H^+-MRS	Proton-magnetic resonance spectroscopy
Mg^{2+}	Magnesium
mRNA	Messenger ribonucleic acid
POT	Proton-dependent oligopeptide transporter
SOL	Soleus
TauT	Taurine transporter
RyR	Ryanodine receptor
WW	Wet weight

Acknowledgements

This chapter describes results of studies from our lab that were financially supported by grants (to W.D.) from the Research Foundation – Flanders (FWO G.0243.11 and G.0352.13N).

References

Abe, H., 2000. Role of histidine-related compounds as intracellular proton buffering constituents in vertebrate muscle. *Biochemistry*. 65: 757–765.

Allen, D. G., Lamb, G. D. and Westerblad, H., 2008. Impaired calcium release during fatigue. *Journal of Applied Physiology*. 104: 296–305.

Allen, D. G., Lee, J. A. and Westerblad, H., 1989. Intracellular calcium and tension during fatigue in isolated single muscle fibres from *Xenopus laevis*. *Journal of Physiology*. 415: 433–458.

Baguet, A., Bourgois, J., Vanhee, L., Achten, E. and Derave, W., 2010a. Important role of muscle carnosine in rowing performance. *Journal of Applied Physiology*. 109: 1096–1101.

Baguet, A., Everaert, I., Achten, E., Thomis, M. and Derave, W., 2012. The influence of sex, age and heritability on human skeletal muscle carnosine content. *Amino Acids*. 43: 13–20.

Baguet, A., Everaert, I., De, N. H., Reyngoudt, H., Stegen, S., Beeckman, S., Achten, E., Vanhee, L., Volkaert, A., Petrovic, M., Taes, Y. and Derave, W., 2011a. Effects of sprint training combined with vegetarian or mixed diet on muscle carnosine content and buffering capacity. *European Journal of Applied Physiology*. 111: 2571–2580.

Baguet, A., Everaert, I., Hespel, P., Petrovic, M., Achten, E. and Derave, W., 2011b. A new method for non-invasive estimation of human muscle fiber type composition. *PLoS One*. 6: e21956.

Baguet, A., Koppo, K., Pottier, A. and Derave, W., 2010b. Beta-alanine supplementation reduces acidosis but not oxygen uptake response during high-intensity cycling exercise. *European Journal of Applied Physiology*. 108: 495–503.

Baguet, A., Reyngoudt, H., Pottier, A., Everaert, I., Callens, S., Achten, E. and Derave, W., 2009. Carnosine loading and washout in human skeletal muscles. *Journal of Applied Physiology*. 106: 837–842.

Bate Smith, E., 1938. The buffering of muscle in rigor: protein, phosphate and carnosine. *Journal of Physiology*. 92: 336–343.

Batrukova, M. A. and Rubtsov, A. M., 1997. Histidine-containing dipeptides as endogenous regulators of the activity of sarcoplasmic reticulum Ca-release channels. *Biochimica et Biophysica Acta*. 1324: 142–150.

Bex, T., Chung, W., Baguet, A., Stegen, S., Stautemas, J., Achten, E. and Derave, W., 2014. Muscle carnosine loading by beta-alanine supplementation is more pronounced in trained vs. untrained muscles. *Journal of Applied Physiology*. 116: 204–209.

Boldyrev, A., 2007. *Carnosine and Oxidative Stress in Cells and Tissues*. Nova Science, New York.

Bruton, J., Tavi, P., Aydin, J., Westerblad, H. and Lannergren, J., 2003. Mitochondrial and myoplasmic [Ca^{2+}] in single fibres from mouse limb muscles during repeated tetanic contractions. *Journal of Physiology*. 551: 179–190.

Chung, W., Baguet, A., Bex, T., Bishop, D. and Derave, W., 2014. Doubling of muscle carnosine concentration does not improve laboratory 1 h time trial performance. *International Journal of Sport Nutrition and Exercise Metabolism*. 24: 315–324.

Costill, D. L., Fink, W. J. and Pollock, M. L., 1976. Muscle fiber composition and enzyme activities of elite distance runners. *Medicine and Science in Sports*. 8: 96–100.

Del Favero, S., Roschel, H., Solis, M. Y., Hayashi, A. P., Artioli, G. G., Otaduy, M. C., Benatti, F. B., Harris, R. C., Wise, J. A., Leite, C. C., Pereira, R. M., de Sa-Pinto, A. L., Lancha-Junior, A. H. and Gualano, B., 2012. Beta-alanine (Carnosyn) supplementation in elderly subjects (60–80 years): effects on muscle carnosine content and physical capacity. *Amino Acids*. 43: 49–56.

Derave, W., Jones, G., Hespel, P. and Harris, R. C., 2008. Creatine supplementation augments skeletal muscle carnosine content in senescence-accelerated mice (SAMP8). *Rejuvenation Research*. 11: 641–647.

Derave, W., Ozdemir, M. S., Harris, R. C., Pottier, A., Reyngoudt, H., Koppo, K., Wise, J. A. and Achten, E., 2007. beta-Alanine supplementation augments muscle carnosine content and attenuates fatigue during repeated isokinetic contraction bouts in trained sprinters. *Journal of Applied Physiology*. 103: 1736–1743.

Donaldson, S. K., Hermansen, L. and Bolles, L., 1978. Differential, direct effects of H^+ on Ca^{2+}-activated force of skinned fibers from the soleus, cardiac and adductor magnus muscles of rabbits. *Pflügers Archiv*. 376: 55–65.

Dunnett, M., Harris, R. C., Dunnett, C. E. and Harris, P. A., 2002. Plasma carnosine concentration: diurnal variation and effects of age, exercise and muscle damage. *Equine Veterinary Journal Supplement*. 283–287.

Dutka, T. L. and Lamb, G. D., 2004. Effect of carnosine on excitation–contraction coupling in mechanically-skinned rat skeletal muscle. *Journal of Muscle Research and Cell Motility*. 25: 203–213.

Dutka, T. L., Lamboley, C. R., McKenna, M. J., Murphy, R. M. and Lamb, G. D., 2012. Effects of carnosine on contractile apparatus Ca^{2+}-sensitivity and sarcoplasmic reticulum Ca^{2+} release in human skeletal muscle fibers. *Journal of Applied Physiology*. 112: 728–736.

Everaert, I., De Naeyer, H., Taes, Y. and Derave, W., 2013a. Gene expression of carnosine-related enzymes and transporters in skeletal muscle. *European Journal of Applied Physiology*. 113: 1169–1179.

Everaert, I., Stegen, S., Vanheel, B., Taes, Y. and Derave, W., 2013b. Effect of beta-alanine and carnosine supplementation on muscle contractility in mice. *Medicine & Science in Sports & Exercise*. 45: 43–51.

Everaert, I., Mooyaart, A., Baguet, A., Zutinic, A., Baelde, H., Achten, E., Taes, Y., de, H. E. and Derave, W., 2011. Vegetarianism, female gender and increasing age, but not CNDP1 genotype, are associated with reduced muscle carnosine levels in humans. *Amino Acids*. 40: 1221–1229.

Fabiato, A. and Fabiato, F., 1978. Effects of pH on the myofilaments and the sarcoplasmic reticulum of skinned cells from cardiace and skeletal muscles. *Journal of Physiology*. 276: 233–255.

Greene, S. M., Margolis, F. L., Grillo, M. and Fisher, H., 1984. Enhanced carnosine (beta-alanyl-L-histidine) breakdown and histamine metabolism following treatment with compound 48/80. *European Journal of Pharmacology*. 99: 79–84.

Gross, M., Boesch, C., Bolliger, C. S., Norman, B., Gustafsson, T., Hoppeler, H. and Vogt, M., 2013. Effects of beta-alanine supplementation and interval training on physiological determinants of severe exercise performance. *European Journal of Applied Physiology*. 114: 221–234.

Harris, R. C., Tallon, M. J., Dunnett, M., Boobis, L., Coakley, J., Kim, H. J., Fallowfield, J. L., Hill, C. A., Sale, C. and Wise, J. A., 2006. The absorption of orally supplied beta-alanine and its effect on muscle carnosine synthesis in human vastus lateralis. *Amino Acids*. 30: 279–289.

Harris, R. C., Wise, J. A., Price, K. A., Kim, H. J., Kim, C. K. and Sale, C., 2012. Determinants of muscle carnosine content. *Amino Acids*. 43: 5–12.

Harris, R., Dunnett, M. and Greenhaff, P. L., 1998. Carnosine and taurine contents in individual fibres of human vastus lateralis muscle. *Journal of Sport Science*. 16: 639–643.

Harris, R., Jones, G., Hill, C. A., Kendrick, I. P., Boobis, L. H., Kim, C. K., Kim, H. J., Dang, V. H. and Edge Jand Wise, J. A., 2007. The carnosine content of vastus lateralis in vegetarians and omnivores. *FASEB Journal*. 21: A944.

Hill, C. A., Harris, R. C., Kim, H. J., Harris, B. D., Sale, C., Boobis, L. H., Kim, C. K. and Wise, J. A., 2007. Influence of beta-alanine supplementation on skeletal muscle carnosine concentrations and high intensity cycling capacity. *Amino Acids*. 32: 225–233.

Hobson, R. M., Saunders, B., Ball, G., Harris, R. C. and Sale, C., 2012. Effects of beta-alanine supplementation on exercise performance: a meta-analysis. *Amino Acids*. 43: 25–37.

Hvid, L. G., Gejl, K., Bech, R. D., Nygaard, T., Jensen, K., Frandsen, U. and Ortenblad, N., 2013. Transient impairments in single muscle fibre contractile function after prolonged cycling in elite endurance athletes. *Acta Physiologica*. 208: 265–273.

Johnson, P. and Hammer, J. L., 1992. Histidine dipeptide levels in ageing and hypertensive rat skeletal and cardiac muscles. *Comparative Biochemistry and Physiology, Part B: Biochemistry & Molecular Biology*. 103: 981–984.

Kendrick, I. P., Kim, H. J., Harris, R. C., Kim, C. K., Dang, V. H., Lam, T. Q., Bui, T. T. and Wise, J. A., 2009. The effect of 4 weeks beta-alanine supplementation and isokinetic training on carnosine concentrations in type I and II human skeletal muscle fibres. *European Journal of Applied Physiology*. 106: 131–138.

Kirkendall, D. and Garrett, W., 1998. The effects of aging and training on skeletal muscle. *American Journal of Sports Medicine*. 26: 598–602.

Lamont, C. and Miller, D., 1992. Calcium sensitizing action of carnosine and other endogenous imidazoles in chemically skinned striated muscle. *Journal of Physiology*. 454: 421–434.

Lunde, P. K., Dahlstedt, A. J., Bruton, J. D., Lannergren, J., Thoren, P., Sejersted, O. M. and Westerblad, H., 2001. Contraction and intracellular Ca(2+) handling in isolated skeletal muscle of rats with congestive heart failure. *Circulation Research*. 88: 1299–1305.

Maharam, L. G., Bauman, P. A., Kalman, D., Skolnik, H. and Perle, S. M., 1999. Masters athletes: factors affecting performance. *Sports Medicine*. 28: 273–285.

Mannion, A. F., Jakeman, P. M., Dunnett, M., Harris, R. C. and Willan, P. L., 1992. Carnosine and anserine concentrations in the quadriceps femoris muscle of healthy humans. *European Journal of Applied Physiology and Occupational Physiology*. 64: 47–50.

Mannion, A. F., Jakeman, P. M. and Willan, P. L., 1995. Skeletal muscle buffer value, fibre type distribution and high intensity exercise performance in man. *Experimental Physiology*. 80: 89–101.

Miyaji, T., Sato, M., Maemura, H., Takahata, Y. and Morimatsu, F., 2012. Expression profiles of carnosine synthesis-related genes in mice after ingestion of carnosine or ss-alanine. *Journal of the International Society of Sports Nutrition*. 9: 15.

Nagai, K., Niijima, A., Yamano, T., Otani, H., Okumra, N., Tsuruoka, N., Nakai, M. and Kiso, Y., 2003. Possible role of L-carnosine in the regulation of blood glucose through controlling autonomic nerves. *Experimental Biology and Medicine*. 228: 1138–1145.

Norsborg, N., Mohr, M., Pedersen, L., Nielsen, J., Langbergand, H. and Bangsbo, J., 2003. Muscle interstitial potassium kinetics during intense exhaustive exercise: effect of previous arm exercise. *American Journal of Physiology: Regulatory, Integrative and Comparative Physiology*. 285: 143–148.

Parkhouse, W. S., McKenzie, D. C., Hochachka, P. W. and Ovalle, W. K., 1985. Buffering capacity of deproteinized human vastus lateralis muscle. *Journal of Applied Physiology*. 58: 14–17.

Penafiel, R., Ruzafa, C., Monserrat, F. and Cremades, A., 2004. Gender-related differences in carnosine, anserine and lysine content of murine skeletal muscle. *Amino Acids*. 26: 53–58.

Pierno, S., Liantonio, A., Camerino, G. M., De, B. M., Cannone, M., Gramegna, G., Scaramuzzi, A., Simonetti, S., Nicchia, G. P., Basco, D., Svelto, M., Desaphy, J. F. and Camerino, D. C., 2012. Potential benefits of taurine in the prevention of skeletal muscle impairment induced by disuse in the hindlimb-unloaded rat. *Amino Acids*. 43: 431–445.

Sale, C., Artioli, G. G., Gualano, B., Saunders, B., Hobson, R. M. and Harris, R. C., 2013. Carnosine: from exercise performance to health. *Amino Acids*. 44: 1477–1491.

Sewel, D., Harris, R. C. and Marlinand Dunnett, M., 1992. Estimation of the carnosine content of different fibre type in the middle gluteal muscle of thoroughbred horses. *Journal of Physiology*. 455: 477–453.

Stegen, S., Bex, T., Vervaet, C., Vanhee, L., Achten, E. and Derave, W., 2014. The Beta-Alanine dose for maintaining moderately elevated muscle carnosine levels. *Medicine & Science in Sports & Exercise*. 46: 1426–1432.

Stegen, S., Blancquaert, L., Everaert, I., Bex, T., Taes, Y., Calders, P., Achten, E. and Derave, W., 2013. Meal and beta-alanine coingestion enhances muscle carnosine loading. *Medicine & Science in Sports & Exercise*. 45: 1478–1485.

Stellingwerff, T., Anwander, H., Egger, A., Buehler, T., Kreis, R., Decombaz, J. and Boesch, C., 2012a. Effect of two beta-alanine dosing protocols on muscle carnosine synthesis and washout. *Amino Acids*. 42: 2461–2472.

Stellingwerff, T., Decombaz, J., Harris, R. C. and Boesch, C., 2012b. Optimizing human in vivo dosing and delivery of beta-alanine supplements for muscle carnosine synthesis. *Amino Acids*. 43: 57–65.

Stuerenburg, H. and Kunze, K., 1999. Concentrations of free carnosine (a putative membrane-protective antioxidant) in human muscle biopsies and rat muscles. *Archives of Gerontology and Geriatrics*. 29: 107–113.

Suzuki, Y., Ito, O., Mukai, N., Takahashi, H. and Takamatsu, K., 2002. High level of skeletal muscle carnosine contributes to the latter half of exercise performance during 30 s maximal cycle ergometer sprinting. *Japanese Journal of Physiology*. 52: 199–205.

Tallon, M., Harris, R. C., Maffulli, N. and Tarnopolsky, M., 2007. Carnosine, taurine and enzyme activities of human skeletal muscle fibres from elderly subjects with osteoarthritis and young moderately active subjects. *Biogerontology*. 8: 129–137.

Westerblad, H. and Allen, D. G., 1993. The contribution of [Ca^{2+}]i to the slowing of relaxation in fatigued single fibres from mouse skeletal muscle. *Journal of Physiology*. 468: 729–740.

CHAPTER 16

Anserine as a Suppressor of Fatigue

YOSHINORI TAKAHASHI*[a]

[a]Central Research Institute, Maruha Nichiro Corporation, 16-2, Wadai, Tsukuba, Ibaraki 300-4295, Japan
*E-mail: yo-takahashi@maruha-nichiro.co.jp

16.1 Introduction

The imidazole dipeptides anserine and carnosine are known to exist in the skeletal muscle of migratory fish and birds and are considered to function as suppressors of fatigue. Strenuous muscular exercise promotes anaerobic glycolysis of glycogen in skeletal muscle with lactic acid production and ATP consumption, which together reduce pH. In response to low muscular pH, the constrictive efficiency is reduced by suppression of glycolytic enzyme activity, thus causing physical fatigue. Muscle cells naturally possess a buffer capacity for maintaining their pH value. Therefore, enhancement of buffer capacity can allow the cells to endure a decrease in pH and improve exercise capacity through upregulation of the anaerobic glycolytic pathway. Muscular exercise also promotes the production of reactive oxygen species, which are responsible for protein oxidation and contribute to physical fatigue. To protect against oxidative injury, muscle cells possess complex endogenous cellular defence mechanisms. Exogenous dietary antioxidants cooperatively

Food and Nutritional Components in Focus No. 8
Imidazole Dipeptides: Chemistry, Analysis, Function and Effects
Edited by Victor R Preedy

Published by the Royal Society of Chemistry, www.rsc.org

interact with endogenous antioxidants. Thus, endogenous antioxidants are candidate antifatigue substances.

Imidazole peptides have been reported to have several types of biological activity. Davey proposed that imidazole peptides in white muscle are important for stabilizing the pH in anaerobically contracting muscle (Davey, 1960). Abe and Okuma reported that imidazole peptides provided an effective buffer against proton release by anaerobic glycolysis because their pK_a is lower than the physiological pH (pH 6.5–7.5) and that anserine had the highest buffer capacity for proton ions among imidazole peptides (pK_a at 20 °C: anserine, 7.15; carnosine, 7.01) (Abe and Okuma, 1991). In addition, imidazole peptides are also known to be powerful natural antioxidants (Kohen *et al.*, 1988; MacFarlanea *et al.*, 1991; Yanai *et al.*, 2004). Similar to glutathione and vitamin C, anserine and carnosine have shown comparatively strong antioxidative activity against ClO radicals. Therefore, imidazole peptides are considered to prevent injury caused by reactive oxygen species generation during energy metabolism. These previous reports also suggested that imidazole dipeptides are effective for improvement of motor function and have antifatigue effects.

Salmon is a migratory fish whose muscle contains a large amount of anserine; the concentration of anserine is also known to increase during spawning migration (Mommsen *et al.*, 1980). Maruha Nichiro, a major fisheries company of Japan, developed a salmon muscle extract with a high content of anserine (hereafter called SEAns) and evaluated its antifatigue effects. In this chapter, the findings for SEAns with regard to improvement of exercise performance and suppression of physical and visual fatigue are described.

16.2 Effect of Single Administration of Anserine-Containing Salmon Muscle Extracts on Physical Fatigue

Anserine is present in the skeletal muscle of migratory fish such as salmon, tuna, and bonito and is considered to function as a pH regulator and antioxidant for suppressing muscle fatigue. To investigate the effect of anserine on physical fatigue, SEAns was prepared from salmon muscle as follows: minced meat of chum salmon (*Oncorhynchus keta*) caught in Iwate Prefecture, Japan was added to boiling water and centrifuged. The supernatant was then filtrated and evaporated to obtain a dried yellow powder. The effect of SEAns on the suppression of physical fatigue and improvement of exercise performance was then examined (Takahashi *et al.*, 2008).

In a placebo-controlled crossover trial, 10 healthy male volunteers were randomly assigned to orally receive 2 g of SEAns (400 mg of anserine) or a placebo. After 30 min, the subjects performed 3 sets of 4 physical fatigue-inducing tasks (press-ups, sit-ups, and back muscle strength and air-chair exercises) continuously until exhaustion, and the duration to exhaustion was measured. Blood samples were collected before administration of test foods and after 60 min of exercise.

Figure 16.1 shows the effect of SEAns on exercise persistence. Compared with the placebo group, the SEAns group tended to show improved exercise persistence in the second set and third set *versus* the first set. This finding suggests that the administration of SEAns improves exercise performance. Figure 16.2 shows changes for serum creatine phosphokinase, which is a biomarker of muscle fiber injury. Administration of SEAns significantly inhibited the elevation of creatine phosphokinase levels. In addition, the increase in white blood cells tended to be suppressed in the SEAns group. These findings suggest that SEAns decreases muscle fiber injury and increases exercise tolerance. Moreover, the serum cortisol level in the SEAns group was significantly lower than that in the placebo group (Figure 16.3). Cortisol is a glucocorticoid released from the adrenal cortex in response to stress. In general, the serum cortisol concentration is increased by psychological and physical tolerance and is used as a biomarker for stress in clinical examinations (Dickerson and Kemeny, 2004).

A single oral administration of SEAns (2 g), which contains 400 mg of anserine, is therefore effective against physical fatigue because it suppresses psychological and physical stress in addition to muscle injury caused by strenuous exercise.

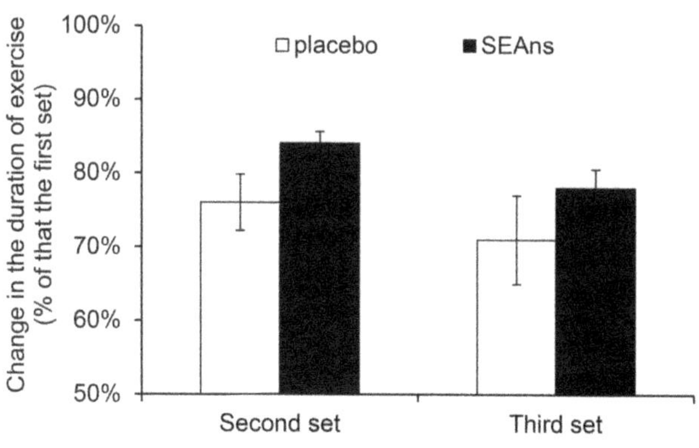

Figure 16.1 Effect of SEAns on exercise persistence. Each bar indicates changes in exercise persistence for the physical fatigue-inducing tasks (one set is consist of four tasks; press-ups, sit-ups, and back muscle strength and air-chair exercises, which were performed continuously until exhaustion, and the exercise durations were counted), compared with the exercise duration of the first set of tasks. The exercise duration for the SEAns group was longer than that for the placebo group. This finding indicates that administration of SEAns improves stamina. □: Placebo group ($n = 10$), ■: SEAns group ($n = 10$). Data are adapted from Takahashi *et al.* (2008) with a slight modification. Values are shown as the mean and SEM.

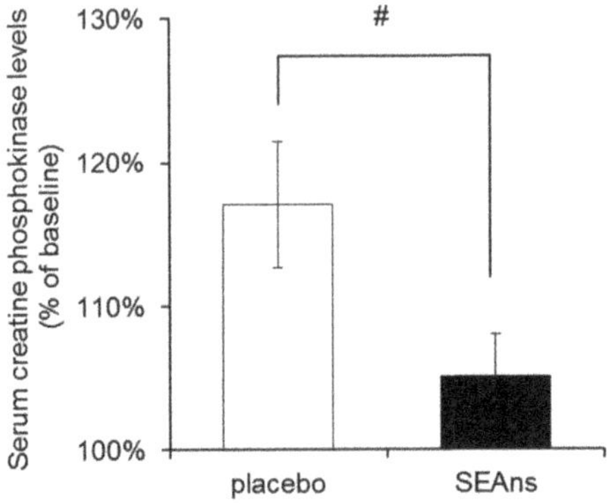

Figure 16.2 Change in serum creatine phosphokinase level. Data show the change in the serum creatine phosphokinase level after 60 min of exercise relative to the baseline. Creatine phosphokinase is a biomarker of muscle-fiber injury. Administration of SEAns significantly inhibited the elevation of creatine phosphokinase levels. This finding indicates that SEAns decreased muscle fiber injury. □: Placebo group (n = 10), ■: SEAns group (n = 10). #: Significant difference between test groups (Student's t test, p < 0.05). Data are adapted from Takahashi *et al.* (2008) with a slight modification. Values are shown as the mean and SEM.

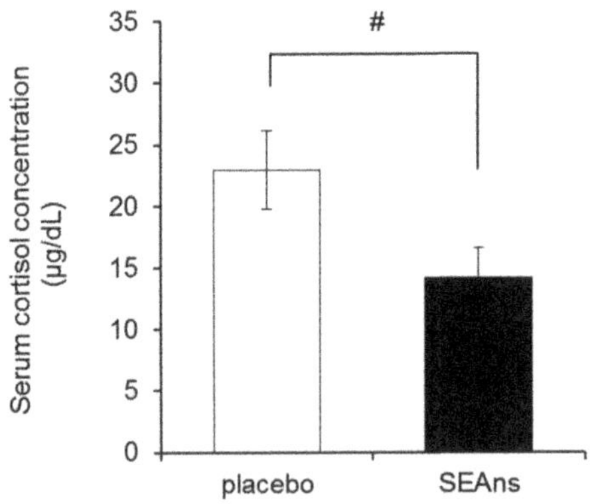

Figure 16.3 Serum cortisol concentration after 60 min of exercise tolerance. Data show the serum cortisol concentration after 60 min of exercise. Cortisol is a biomarker for stress in clinical examinations. The serum cortisol level in the SEAns group was significantly lower than that in the placebo group. This finding suggests that SEAns decreased stress and increased exercise performance. □: Placebo group (n = 10), ■: SEAns group (n = 10). #: Significant difference between test groups (Student's t test, p < 0.05). Data are adapted from Takahashi *et al.* (2008) with a slight modification. Values are shown as the mean and SEM.

16.3 Effect of Anserine-Containing Salmon Muscle Extracts on Exercise Performance and Kinetic Visual Acuity among Athletes

Harada *et al.* reported that 4 week administration of a chicken-breast extract containing imidazole dipeptides enhanced recovery from physical fatigue in mice (Harada *et al.*, 2002). Additionally, long-term intake of extracts containing imidazole dipeptides inhibited the decrease in exercise performance in a short high-intensity task over time (Sato *et al.*, 2003). The author of the present study has also investigated the effect of SEAns on the exercise performance and kinetic visual acuity of athletes (Muneda *et al.*, 2012).

The study was performed in a placebo-controlled, parallel, double-blind clinical test. Twenty male student athletes belonging to a table-tennis club were divided into 2 groups. The test group was administered an encapsulated formulation containing 0.4 g of SEAns (120 mg of anserine) per day for 8 weeks, and the placebo group was administered an encapsulated formulation that did not contain SEAns. Exercise performance was evaluated as the return accuracy against serves, as shown in Figure 16.4. Each table was divided into 4 areas and the areas were assigned points depending on the difficulty of returning a serve to the area. The subjects were paired and returned the served ball alternately. The score was counted with regard to the area in which the returned ball dropped.

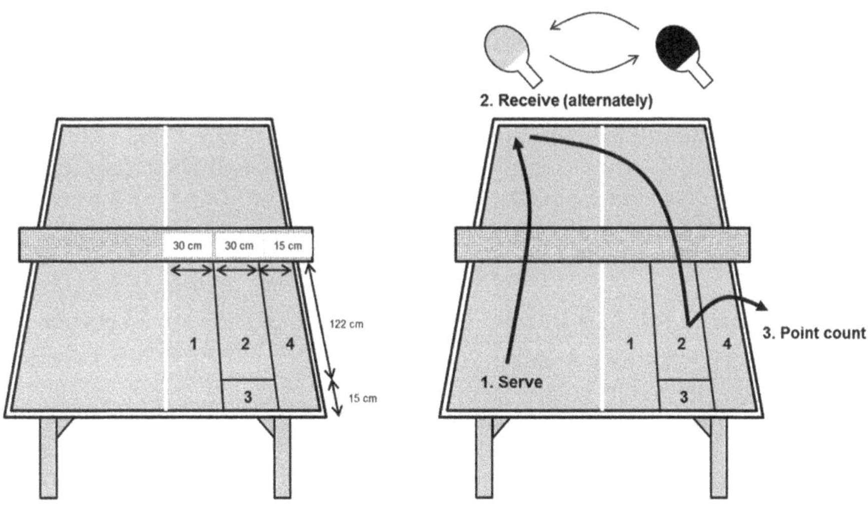

Figure 16.4 Method of exercise performance measurement: return accuracy in table tennis. The table was divided into 4 areas and assigned points depending on the difficulty of returning the serve to the area (left figure). Subjects who were student athletes belonging to a table-tennis club were paired and returned the served ball alternately. The score was counted according to the area in which the returned ball dropped. Figures are adapted from Muneda *et al.* (2012) with permission from the publisher.

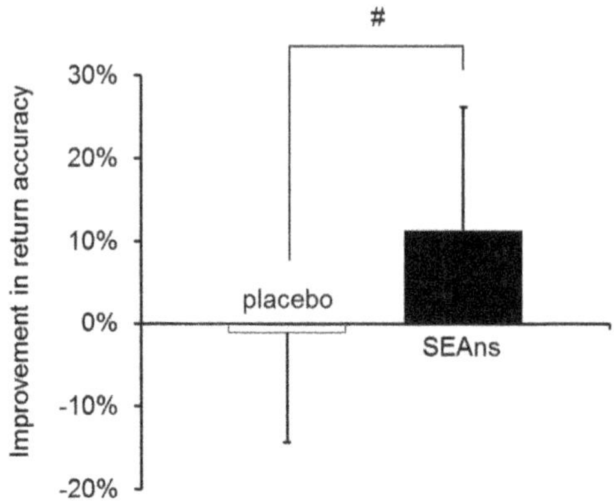

Figure 16.5 Improvement in return accuracy for table-tennis athletes by intake of SEAns for 8 weeks. Each bar indicates the improvement in return accuracy in table tennis at 8 weeks relative to week 0. Intake of SEAns significantly improved the return accuracy. This finding shows that SEAns intake for 8 weeks increases the exercise performance of table-tennis athletes. □: Placebo group ($n = 10$), ■: SEAns group ($n = 9$). #: Significant difference between test groups (Student's t test, $p < 0.05$). Data are adapted from Muneda *et al.* (2012) with permission from the publisher. Values are shown as the mean and SEM.

After 8 weeks of SEAns administration, the return accuracy against serves improved, whereas no particular pattern was observed for the placebo group, and the difference between the two groups was significant (Figure 16.5). In addition, the kinetic visual acuity of the SEAns group also improved and was significantly different from that of the placebo group (Figure 16.6). Generally, the kinetic visual acuity of high-performance athletes is higher than that of normal players (Mashimo, 1997). Therefore, intake of SEAns for 8 weeks is expected to be able to improve the exercise performance of athletes.

16.4 Effect of Anserine-Containing Salmon Muscle Extracts on Asthenopia

Recently, the use of visual display terminals (VDTs) has globally increased. Because VDT tasks impose sustained abnormal postures and expose the participant to mental stress resulting from the sequential and routine nature of the tasks, they cause neck stiffness, lumbar pain, visual fatigue, and a decline in the quality of life. Honda *et al.* reported that consecutive intake of dried skipjack soup stock, which is known to contain anserine, improved symptoms of asthenopia (Honda *et al.*, 2006). Therefore, the effects of SEAns on 39 healthy adults, who were aged in their 40 s and engaged in VDT work, were investigated in a parallel, double-blind clinical test (Muneda *et al.*, 2010). The subjects were divided into 3 groups, and each group was orally administered the encapsulated formulation with 0, 0.4, or 1.3 g of SEAns (0, 120, or 400 mg of anserine, respectively) for 8 weeks. The subjective symptoms of eyestrain and

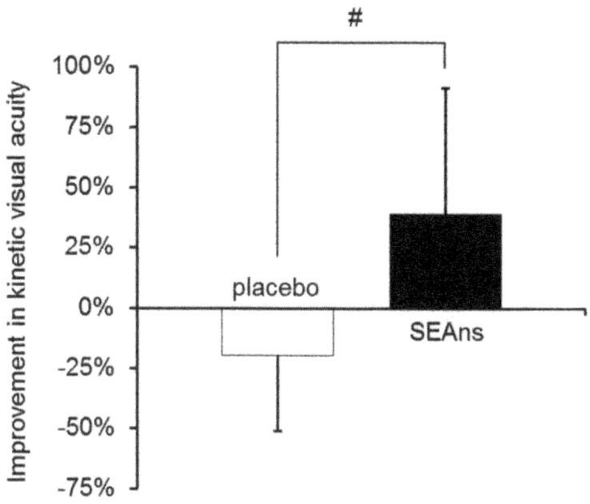

Figure 16.6 Improvement in kinetic visual acuity of table-tennis athletes by intake of SEAns for 8 weeks. Each bar indicates the improvement in kinetic visual acuity at 8 weeks relative to that at week 0. Intake of SEAns significantly improved the kinetic visual acuity. □: Placebo group ($n = 10$), ■: SEAns group ($n = 9$). #: Significant difference between test groups (Student's t test, $p < 0.05$). Data are adapted from Muneda *et al.* (2012) with permission from the publisher. Values are shown as the mean and SEM.

accommodation, which are usually used as an ophthalmologic index, were measured before and after administration to evaluate the degree of asthenopia.

As shown in Figure 16.7, the SEAns group exhibited improvement in subjective symptoms of eyestrain at 8 weeks after administration in comparison with the placebo group. Accommodation is regulated by increased lens thickness and lens surface curvature and is known to decline with aging and eyestrain. Figure 16.8 shows changes in subjective accommodation in which the lens diopter was less than 5D, that is, the focal distance was over 20 cm. Accommodation in the intake group treated with 0.4 and 1.3 g of SEAns was significantly better at 8 weeks after intake when compared to the value before ingestion. Compared to the values for the subjects in the placebo group, accommodation significantly differed with administration of 1.3 g of SEAns. It is likely that SEAns suppress the tonic accommodation of ciliary muscle caused by VDT work. Thus, daily intake of SEAns is considered to be effective for preventing asthenopia.

16.5 Conclusions

Fatigue is a common symptom in modern society, and organisms are equipped with various systems that enable recovery and protection from fatigue. Imidazole dipeptides such as anserine and carnosine have the capability to suppress physical fatigue and improve exercise performance and are known to exist in the skeletal muscle of migratory fish and birds. In salmon, the content of anserine in muscle is known to increase when the fish are swimming upstream (Mommsen *et al.*, 1980). Intake of anserine through

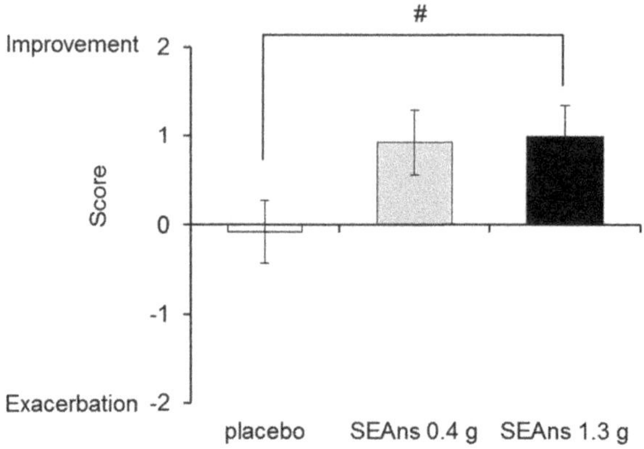

Figure 16.7 Subjective symptoms of eyestrain at 8 weeks after administration. Each bar indicates the subjective symptoms of eyestrain at 8 weeks after administration. All subjects were engaged in VDT work with asthenopia. The SEAns group exhibited improvement relative to the placebo group. □: Placebo group ($n = 14$), ■: SEAns (0.4 g) group ($n = 19$), ■: SEAns (1.3 g) group ($n = 16$). #: Significant difference between test groups (Bonferroni, $p < 0.05$). Values are shown as the mean and SEM.

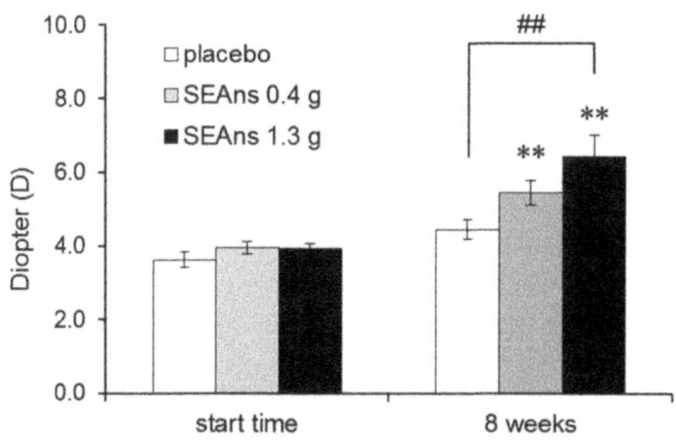

Figure 16.8 Changes in accommodation. Each bar indicates subjective accommodation in which the lens diopter is less than 5D, that is, the focal distance is >20 cm. Accommodation in the intake group treated with 0.4 and 1.3 g of SEAns was significantly improved at 8 weeks after intake relative to the baseline value. Compared to the values for the subjects in the placebo group, accommodation was significantly altered following administration of 1.3 g of SEAns. □: Placebo group ($n = 14$), ■: SEAns (0.4 g) group ($n = 19$), ■: SEAns (1.3 g) group ($n = 16$). ##: Significant difference between test groups (Dunnett, $p < 0.01$), **: significant difference from time zero (Bonferroni, $p < 0.01$). Data are from adapted from Muneda *et al.* (2010) with permission from the publisher. Values are shown as the mean and SEM.

a salmon muscle extract with high anserine levels was previously shown to suppress fatigue and improve exercise performance (Takahashi *et al.*, 2008; Muneda *et al.*, 2010; Muneda *et al.*, 2012). Other extracts containing anserine and carnosine such as chicken essence and dried skipjack soup stock have also been reported to have antifatigue effects against not only physical fatigue but also mental fatigue (Sato *et al.*, 2003; Honda *et al.*, 2006; Yamano *et al.*, 2013). Thus, intake of imidazole peptides *via* the daily diet is considered to be effective for suppression of fatigue. Humans usually absorb imidazole peptides as foods from fish and meat, and these peptides are considered to aid in recovery from fatigue.

Summary Points

- This chapter focuses on the suppressive effect of anserine against fatigue.
- Anserine is present in the skeletal muscle of migratory fish such as salmon, tuna, and bonito, and is considered to function as a pH regulator and antioxidant to suppress fatigue.
- Antifatigue effects of anserine were evaluated using salmon muscle extract containing anserine (SEAns).
- A single administration of 2 g of SEAns, which contains 400 mg of anserine, to 10 healthy subjects, suppressed elevation of the cortisol level and creatine phosphokinase activity and improved exercise endurance.
- These results show that SEAns is effective in suppression of not only muscle injury but also lassitude, thus leading to improvement in exercise performance.
- In another double-blind, placebo-controlled trial, study subjects (who belonged to a table-tennis club) were administered SEAns for 8 weeks, and the exercise performance and kinetic vision were found to improve in the SEAns group (0.4 g of SEAns containing 120 mg of anserine).
- To determine the ability of anserine to reduce asthenopia, SEAns (0, 0.4, or 1.3 g) was orally administered for 8 weeks to subjects who routinely used VDTs.
- These findings suggest that daily intake of imidazole peptides could be effective for recovery from fatigue and useful for improving the quality of life.

Key Facts

Key Facts Regarding Asthenopia

1. Asthenopia is caused by continuous concentrated and overtasked eyesight-intensive work such as driving and the use of video display terminals (VDTs).
2. Asthenopia is a series of syndromes that cause symptoms such as headache, stiff shoulders, dizziness, nausea, a sensation of pressure, and ophthalmic and frontal discomfort.

3. Long-term VDT work results in accumulation of fatigue because of an unsuitable focusing state (too near).
4. The constant focusing of eyesight causes the eyes to function unnaturally and results in asthenopia.
5. Asthenopia is classified as follows: accommodative asthenopia (ametropia and extreme shrinkage of the ciliary muscle), muscular asthenopia (imbalance of extraocular muscles), nervous asthenopia (resulting from a functional or organic neurologic disorder), neurasthenic asthenopia, and retinal asthenopia.

Definitions of Words and Terms

Accommodation. A function of the eyes that involves perspective adjustment. It is measured by sequential near point distance in diopters using an accomodometer; to do this, the eye-mark is moved from a distance of 50 mm to a more distant point, and the subject is asked to focus on the eye-mark. Accommodation is disrupted by asthenopia and presbyopia and can be used as an ophthalmologic index.
Anaerobic glycolysis. An intravital system that degrades glycogen and glucose to lactic acid to supply the requisite energy for muscle contraction. Vigorous anaerobic glycolysis occurs in mammalian muscle and is characterized by NAD-mediated oxidation without the use of oxygen.
Biomarker. Indicator of the physical state, including the health of an individual. For example, haemoglobin A1c (HbA1c) is a biomarker for diabetes.
Cortisol. A type of glucocorticoid that is an adrenal cortical hormone and controls carbohydrate, fat, and protein metabolism. The blood concentration of cortisol tends to be high in the morning, changes in the day, and is increased by stress. Depending on the amount secreted, cortisol increases the blood pressure and blood glucose level and reduces immune function.
Creatine phosphokinase. The enzyme that catalyzes the reaction in which creatine phosphate and ADP are generated from creatine and ATP. It is abbreviated as CK or CPK and is abundant in organisms that consume a large quantity of energy in a short time. It is involved in energy metabolism with muscle contraction, and its levels are increased by muscle-fiber destruction resulting from strenuous exercise.
Diopter. A unit of lens refractivity that is abbreviated as D. It is defined as the inverse of the focal distance in meters. For example, the focal distance of 0.25 m is expressed as 4D.
Kinetic visual acuity. Ability to distinguish an object that moves back and forth while maintaining a line of sight. This measure is closely related with the performance ability in ball games.
Placebo-controlled crossover trial. A method used in clinical trials. The subjects are subjected to the experiment in the active group and control group in turn. Individual differences are controlled for because the same subject experiences both conditions.

Placebo-controlled, parallel, double-blind clinical test. A high-level clinical trial method. The subjects are divided into active groups and a control group. The assistants and the subjects do not know the group assignments during the study. The study assignments are revealed after the statistical analysis.

Reactive oxygen species. These are chemically reactive molecules that contain oxygen species other than O_2; examples include the superoxide anion, hydrogen peroxide, hydroxyl radical, singlet oxygen, nitrogen monoxide, and ozone. In particular, hydroxyl radicals possess strong reactivity and participate in DNA cleavage, protein cleavage, and lipid peroxidation.

List of Abbreviations

ADP	Adenosine diphosphate
ATP	Adenosine triphosphate
DNA	Deoxyribonucleic acid
HbA1c	Haemoglobin A1c
NAD	Nicotinamide adenine dinucleotide
pH	Potential of hydrogen
pK_a	Acid dissociation constant
SEAns	Salmon muscle extract containing anserine
VDT	Visual display terminal

References

Abe, H. and Okuma, E., 1991. Effect of temperature on the buffering capacities of histidine-related compounds and fish skeletal muscle. *Nippon Suisan Gakkaishi*. 57: 2101–2107.

Davey, C. L., 1960. The significance of carnosine and anserine in striated skeletal muscle. *Archives of Biochemistry and Biophysics*. 89: 303–308.

Dickerson, S. S. and Kemeny, M. E., 2004. Acute stressors and cortisol responses: a theoretical integration and synthesis of laboratory research. *Psychological Bulletin*. 130: 355–391.

Harada, R., Urashima, K., Sato, M., Ohmori, T. and Morimatsu, F., 2002. Effect of carnosine and chicken breast extract (CBEX) on ability to recover from exhaustion in mice. *Nippon Eiyo Shokuryo Gakkaishi*. 55: 209–214.

Honda, M., Ishizaki, T. and Kuroda, M., 2006. The effect of dried skipjack soup stock on visual fatigue. *Nippon Shokuhin Kagaku Kogaku Kaishi*. 53: 443–446.

Kohen, R., Yamamoto, Y., Cundy, K. C. and Ames, B. N., 1988. Antioxidant activity of carnosine, homocarnosine, and anserine present in muscle and brain. *Proceedings of the National Academy of Sciences of the United States of America*. 85: 3175–3179.

MacFarlanea, N., McMurray1, J., O'Dowda, J. J., Dargie1, H. J. and Millera, D. J., 1991. Synergism of histidyl dipeptides as antioxidants. *Journal of Molecular and Cellular Cardiology*. 23: 1205–1207.

Mashimo, I., 1997. Kinetic visual acuity. *Japanese Journal of Physical Fitness and Sports Medicine*. 46: 321–326.

Mommsen, T. P., French, C. J. and Hochachka, P. W., 1980. Sites and patterns of protein and amino acid utilization during the spawning migration of salmon. *Canadian Journal of Zoology*. 58: 1785–1799.

Muneda, H., Kawarasaki, M., Takahashi, Y., Honda, H., Shimada, M. and Enari, H., 2010. Effects of salmon extract containing anserine on asthenopia and presbyopia. *Aminosan Kenkyu*. 4: 75–77.

Muneda, H., Mizunuma, R., Takahashi, Y., Shiina, Y. and Terasawa, N., 2012. Effects of salmon extract containing anserine on exercise performance of student athlete. *Aminosan Kenkyu*. 6: 61–63.

Sato, M., Suzuki, Y., Morimatsu, F. and Takamatsu, K., 2003. Effect of carnosine concentration in muscle and improvement of exercise performances due to long-term intake of chicken breast extract. *Tairyoku Kagaku*. 52: 255–264.

Takahashi, Y., Kawarasaki, M., Hoshino, Y., Honda, H. and Enari, H., 2008. Salmon muscle extract containing anserine as a suppressor of fatigue in humans. *Nippon Shokuhin Kagaku Kogaku Kaishi*. 55: 428–431.

Yamano, E., Tanaka, M., Ishii, A., Tsuruoka, N., Abe, K. and Watanabe, Y., 2013. Effects of chicken essence on recovery from mental fatigue in healthy males. *Medical Science Monitor*. 19: 540–547.

Yanai, N., Shiotani, S., Mizuno, M., Nabetani, H. and Nakajima, M., 2004. Characteristic of antioxidative activity of carnosine and anserine mixture isolated from chicken extract: comparison with other botanical antioxidants. *Nippon Shokuhin Kagaku Kogaku Kaishi*. 51: 238–246.

CHAPTER 17

Carnosine Promotion of Wound Healing

I. ANSURUDEEN*[a], E. FORSBERG[a], AND K. BRISMAR[a]

[a]Karolinska Institute, Department of Molecular Medicine and Surgery, Rolf Luft Research Center for Diabetes and Endocrinology, Karolinska University Hospital, Solna, L1:00, 171 76 Stockholm, Sweden
*E-mail: ishrath.ansurudeen@ki.se

17.1 Wounds

A wound is an injury to a part of tissue, especially one caused by physical trauma and characterized by tearing, cutting, piercing, or breaking of the tissue or the skin that can be superficial with limited damage to the outer layer or can be deeper, causing damage to underlying tissues or organs. Wounds can be minor or life threatening based on the site, depth and cause of the wound and can heal naturally unless it gets infected, ischemic or coexist with a chronic illness or weakened immune system. The common predicaments related to a wound include infection, necrosis (ischemia) scarring, loss of function, amputations and expensive utilization of health-care resources.

17.1.1 Type of Wounds

Wounds can be classified in many ways. Based on the healing time of the wound they can be categorized as acute or chronic wounds. Skin injury, surgical wounds and wounds that follow the implantation of prosthetic devices heal

Food and Nutritional Components in Focus No. 8
Imidazole Dipeptides: Chemistry, Analysis, Function and Effects
Edited by Victor R Preedy

Published by the Royal Society of Chemistry, www.rsc.org

normally unless infected, ischemic or in hyperglycemic environment. Acute wound care becomes necessary and expensive if infection or other complications ensue. According to the National Center for Health Statistics, 40 million inpatient surgical procedures and 31.5 million outpatient surgeries were performed in the United States in 2000 where the need for postsurgical wound care was expected to reach more than 38.0 million by 2012 (Sen *et al.*, 2009).

Chronic wounds are associated with comorbid conditions like diabetes, venous stasis, cachexia, nephropathy, peripheral arterial diseases, spinal cord injuries or obesity. Chronic wounds fail to heal in an orderly fashion and are a great threat to public health and economy. In the United States, chronic wounds affect 6.5 million patients costing an estimated 20 billion dollars annually (Singer and Clark, 1999) and in the Scandinavian countries, the costs associated with chronic wound care account for 2–4% of total health-care expenses (Gottrup *et al.*, 2001). Apart from vast economical demands, nonhealing wounds severely diminish the quality of life (Ragnarson and Apelqvist, 2000).

17.1.2 Wound Healing

Chronic nonhealing wounds and its wound care pose a significant burden to patients, and the health-care system. Knowledge of the normal healing process is essential to effectively identify the abnormality in nonhealing wounds and engineer a suitable treatment profile.

Wound healing is a complex biological process involving well-defined activation and orchestration of various intracellular and intercellular pathways (Sonnemann and Bement, 2011) (Figure 17.1). The wound-healing process can be broadly divided into three phases (1) inflammatory phase (2) proliferatory phase and (3) remodeling phase.

17.1.2.1 Inflammatory Phase

Soon after an injury happens, hemostasis sets in and a wound scaffold is formed. The damage caused to the skin, peripheral nerves, blood and lymphatic vessels lead to vasodilation (reddish color), increased cellular permeability with edema and extravasation of blood components that initiates the inflammatory process. Primarily, various cellular mediators and growth factors get accumulated at the wound area and activate blood clotting. Platelets mediate vascular constriction that prevents hemorrhage and subsequently affects perfusion around the wound area and its oxygen content further leading to changes in cellular respiration and pH changes (Diegelmann and Evans, 2004). Cytokines released by platelets and leukocytes like PDGF, ILa, ILb, IL6 and TNFa promote inflammation. Growth factors like FGF2, IGF-1, and TGFβ stimulate collagen synthesis; TGFb in addition transforms fibroblasts to myofibroblasts and has a role in angiogenesis apart from VEGFα, FGF2, and HIF1α. Other growth factors like EGF, IGF-1, and TGFa suppress re-epithelialization. Neutrophils recruited at the wound site due

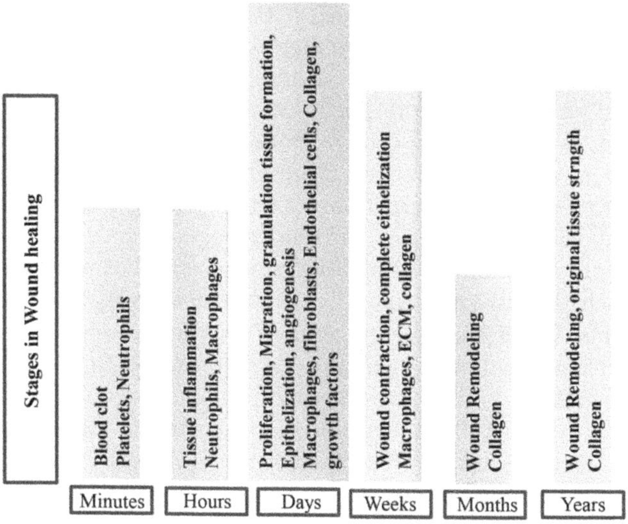

Figure 17.1 Stages of wound healing. Normal wound healing begins with hemostasis and clotting of blood that plugs the wound area. Platelets emerge from the damaged vasculature, release cytokines and other inflammatory agents to recruit inflammatory cells to the wound causing inflammation. Neutrophils initiate inflammatory response within an hour after wounding by clearing the site of pathogens and also releasing additional inflammatory cytokines. Within 24–48 h after wounding, monocytes differentiate into macrophages, and remove wound debris by phagocytosis. The macrophages also release cytokines both to stimulate angiogenesis and to recruit fibroblasts to the wound. Fibroblasts from undamaged tissue migrate to the wound, proliferate and lay down a collagen-rich matrix and form myofibroblasts. Epithelization starts with migration and proliferation of epidermal cells at the wound edge and form a new epithelial layer over the wound. Slowly the wound contracts and remodels. Fibroblasts secrete extra cellular matrix and collagen is deposited.

to byproducts of bacterial degradation and degranulation of platelets helps phagocytose local bacteria and degrade necrotic tissue in the late phase of inflammation almost 3 days after injury. The activated macrophages in the wound area release PDGF and VEGF that initiate formation of granulation tissue and studies show that the amount of inflammation determines the extent of scar formation (Kiritsy and Lynch, 1993).

17.1.2.2 Proliferatory Phase

In this phase; neovascularization, angiogenesis, re-epithelialization, and completion of tissue granulation takes place. The epidermal cells from hair follicles and keratinocytes at the wound edges ensure re-epithelialization. These cells remove clotted blood and at the same time retract intracellular tonofilaments; dissolve intercellular desmosome that connects cells and form actin filaments that enable these cells to migrate laterally. With the help of integrin receptors these cells interact with ECM proteins like fibronectin, vitronectin, *etc.* present in the stromal type 1 collagen at the margin of the wound and in the fibrin clot in wound space. Migrating epidermal cells followed by proliferating cells degrade the existing ECM by producing collagenase and activating plasmin. As re-epithelialization continues basement-membrane proteins reappear in a sequential manner from the margin inward of the wound. Soon after, the epidermal cells revert to their normal phenotype and re-establish attachments with the basement membrane. Angiogenesis begins by sprouting of new branches from endothelial cells of existing vessels. VEGF, bFGF, angiogenin, TGFb, low oxygen tension, acidic environment, *etc.* stimulate angiogenesis (Singer and Clark, 1999). In the presence of hyperglycemia angiogenesis is impaired due to destabilization of HIF1-α protein and thereby to a lack of adaptive cellular response to hypoxia (Catrina *et al.*, 2004; Bento and Pereira, 2011).

Simultaneously as angiogenesis and re-epithelialization occur, the new granulation tissue begins to fill the wound space along with macrophages, fibroblasts and blood vessels. High-density fibroblasts dominate this phase of healing and produce collagen and ECM substances. This ECM enables organized growth, movement and differentiation of cells in it. At the end of this phase contact inhibition induces fibroblast to differentiate into myofibroblast. Also, the provisional ECM is slowly replaced by a collagenous matrix and the granulation tissue in the wound space is now reinstated by an acellular scar.

17.1.2.3 Remodeling Phase

In this phase, collagen modeling and wound contraction occurs. The collagen type III produced in the proliferative phase is replaced by stronger collagen type I and becomes more organized with enhanced tensile strength. Collagen remodeling progresses with continued synthesis and catabolism of collagen at a low rate. The proteolytic enzymes that degrade collagen are produced by the macrophages, epidermal cells, endothelial cells and also the fibroblasts. The myofibroblasts that now contain contractile actin filaments cause wound contraction through multiple attachments to collagen and help reduce the scar formation. At this point the local blood flow declines and the acute wound metabolic activity slows down and ultimately stop (Kiritsy and Lynch, 1993).

17.1.3 Factors Involved in Wound Healing

Local factors and systemic factors affect wound healing. The local factors include tissue oxygenation, presence of infection, dynamic reciprocity between cells and extracellular matrix of the wound microenvironment, *etc.* (Guo and DiPietro, 2010). Systemic factors include age, stress, sex hormones, glucocorticoid steroids, hyperglycemia, hyperlipidemia, nonsteroidal anti-inflammatory drugs, chemotherapeutic drugs, obesity, alcohol consumption, smoking, nutrition, *etc.* (Guo and DiPietro, 2010).

17.1.4 Abnormal Wound Healing

Multiple physiologic and biochemical defects lead to impaired wound healing. Nonhealing wounds bypass the normal highly integrated; the above-described stages of the wound-healing process. Prolonged inflammatory phase, impaired cellular migration mediating re-epithelialization or neovascularization, abnormal cytokine secretion, decreased collagen production, enhanced levels of endopeptidases and defective immune function alter wound healing. Additionally neuropathy, ischemia, hypoxia, associated with disease states like *diabetes mellitus*, venous compression and spine disorders greatly hinder normal wound-healing processes (Singer and Clark, 1999; Botusan *et al.*, 2008).

17.2 Carnosine

Carnosine (β-alanyl-L-histidine) is a naturally occurring histidine dipeptide present in high concentration in muscle and nerve tissues. This peptide is synthesized by the enzyme carnosine synthase and metabolized by the enzyme carnosinase. Carnosinase in human is secreted by the liver and in rodents it is expressed only in the kidney. Carnosine has been implicated in numerous physiological functions, however; its biological role in mediating these effects has not been clearly elucidated. Most of its effects are attributed to its antioxidant effect, cytosolic buffering activity, role in scavenging free radicals, anti-inflammatory effects, *etc.* One interesting therapeutic effect of carnosine has been its role in wound healing.

17.2.1 Carnosine in Wound Healing

The evidence for the role of carnosine in wound healing was first published during the early 1930s. Studies by Nagai were the first to show that carnosine accelerates wound healing and mitigates pain. Carnosine was shown to stimulate granulation, regulate immune response, enhance tissue repair and thereby aid spontaneous healing (Nagai and Suda, 1988). Epithelialization of gastric and duodenal ulcers was also promoted by carnosine (Truitsina *et al.*, 1997). Carnosine was reported to enhance healing of pulmonary and corneal wounds too (Maĭchuk *et al.*, 1997; Cuzzocrea *et al.*, 2007). A radiation-induced abnormal tissue repair process could also be rescued with carnosine treatment (Tanaka *et al.*, 2005).

17.2.1.1 Animal Studies

Data from many *in vivo* studies have confirmed the enhanced wound-healing effect of carnosine. Simple animal experiments by Nagai *et al.* have characterized the various actions of carnosine that may together play a vital role in enhancing wound healing. Studies with skin-tissue homogenates from wound incisions showed that the presence of tissue carnosinase in the skin degrades carnosine to alanine and histidine. Histidine is further broken down to histamine by histidine decarboxylase present at the wound site (Figure 17.2). This early infusion of histamine accelerates the first stage of wound-healing process the inflammatory phase (Nagai *et al.*, 1986).

Similarly, carnosine enhanced tissue granulation that is an important process in the proliferative phase of wound healing (Nagai *et al.*, 1986). Using the formalin filter paper method, tissue granulation with and without carnosine treatment was studied. Rats were subcutaneously implanted with two pieces of filter paper soaked in 7% formalin in the bilateral scapular region. Nonhealing wounds were simulated by injecting animals with hydrocortisone, mitomycin, 5-fluoro uracil or bleomycin for seven consecutive days after implantation. Carnosine (8–75 mg kg^{-1} per day, I.P) treatment eliminated the inhibitory effect of hydrocortisone and other drugs and significantly increased the wet granulation tissue weight, and collagen content. Similar effects were also seen with alanine (9.8–29.5 mg kg^{-1} per day, I.P) treatment. Decrease in the tensile strength of the animal's skin induced by the administration of hydrocortisone (10 mg kg^{-1}) did not show significant differences with carnosine treatment.

These results showed the potential of carnosine in wound healing by enhancing the inflammatory phase by the release of histamine and promoting tissue granulation in the proliferative phase. The induction of collagen synthesis by the alanine released, together with enhanced tissue granulation profoundly impacts the proliferative stage of wound healing. Additionally, it was observed that the antigranulatory effects of hydrocortisone could be significantly thwarted by carnosine, suggesting an immunomodulatory role for carnosine. Other studies have further confirmed the role of carnosine as a histidine reserve for histamine synthesis in situations of stress and trauma (Fitzpatrick and Fisher, 1982).

In a mice model of bleomycin-induced lung injury, mice that were administered carnosine daily by an oral bolus of 150 mg kg^{-1} were able to prevent

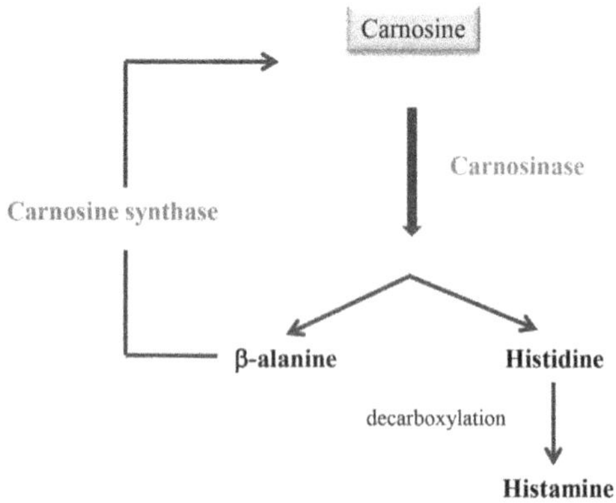

Figure 17.2 Carnosine metabolism. Carnosine is synthesized with the help of enzyme carnosine synthase and the carnosine in tissue or in circulation is metabolized into histidine and alanine by the enzyme tissue or serum carnosinase.

bleomycin-induced lung injury possibly through its direct antioxidant properties (Cuzzocrea *et al.*, 2007). Radiation-induced alterations in wound healing involve irregular proliferation of endothelial cells, decreased vessel formation, irregular capillary formation with increased permeability, reduction in granulation thickness and delayed deposition of collagen that leads to abnormal tissue repair. Interestingly, carnosine pretreatment ameliorates the deleterious effects of irradiation on wound healing. It promoted proliferation of cells and organized deposition of collagen fibers similar to the control group. The observed action of carnosine is due to its affinity for free radical and free-radical-scavenging property (Tanaka *et al.*, 2005). In adjuvant-induced arthritic rats' chronic gastric ulcers induced by thermal cauterization, polaprezinc; the chelate of zinc and carnosine significantly improved healing and the observed effect was speculated to be mediated through the increased production of IGF-1, the growth factor known to enhance epithelial wound repair (Kato *et al.*, 2001).

In a randomized study wounded rats when subjected to peptide diet or carnosine supplemented peptide diet; the group that received carnosine supplemented diet showed an increase in wound strength measured by the wound bursting pressure and accelerated wound healing when compared to controls (Roberts *et al.*, 1998). Wound bursting pressure is a measure of wound strength or dehiscence (separation of the layers of a surgical wound at the sutures) and is clinically very relevant in patients undergoing surgery with medical conditions like diabetes, cancer, immune problems, chemotherapy, or radiation therapy.

Few *in vivo* studies have studied the efficacy of carnosine in the case of chronic wounds associated with pathology like *diabetes mellitus*. Studies in streptozotocin-induced diabetic rats showed that polaprezinc (3–30 mg kg^{-1}) or its components zinc sulfate and carnosine administration orally twice for four days significantly accelerated the healing of HCl-induced gastric ulcer in spite of high blood glucose levels. The healing of the gastric lesions was greatly impaired in the diabetic animals in the absence of carnosine or polaprezinc treatment and the healing could only be improved by normalization of high blood glucose levels with daily insulin treatment (Korolkiewicz *et al.*, 2000).

In a db/db mouse model of diabetes, carnosine (100 mg kg^{-1}) significantly improved the healing of dorsal wounds when administered locally and systemically (Figure 17.3). Local and intraperitoneal injections were combined to enhance carnosine availability since microvascular dysfunction and enhanced carnosinase activity is associated with diabetes phenotype of db/db mouse model (Cho *et al.*, 2006; Riedl *et al.*, 2010; Peters *et al.*, 2011). The healing was accompanied by enhanced expression of growth factors like IGF-1, TGF-β and SDF-1 (Figure 17.4). Enhanced expression of collagen 1a and SMA1 along with better tissue granulation was also detected in animals treated with carnosine (Figure 17.4) (Ansurudeen *et al.*, 2012). IGF-1, known to promote re-epithelialization of keratinocytes, proliferation of fibroblasts and the induction of endothelial cell chemotaxis has decreased expression in human diabetic skin and diabetic mice (Grant *et al.*, 1987; Tsuboi *et al.*, 1995). Studies in mice show that IGF-1 accelerates wound healing in diabetic animal models (Bitar, 1997).

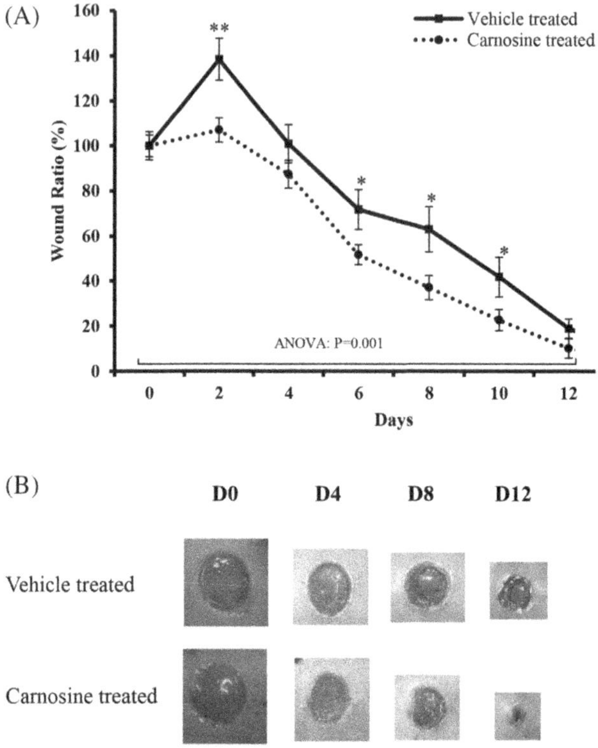

Figure 17.3 Carnosine treatment improves wound healing in the db/db, type 2 diabetes mice model (A). The rate of wound healing is accelerated in db/db mice after carnosine treatment both locally and intraperitoneally (100 mg kg^{-1}). Values are means ± SEM and * = $p < 0.05$, ANOVA: $p = 0.001$. (B) Macroscopic pictures of representative wounds of each group on day 0, 4, 8 and 12 after wounding.

Carnosine treatment induced a significant increase in IGF-1 expression in the wound area, suggesting a modulation of IGF-1 expression by carnosine enabling improved wound healing (Ansurudeen *et al.*, 2012). Similarly, TGFb, well studied for its role in cutaneous wounds in various phases of wound healing (Bhora *et al.*, 1995; Bitar and Labbad 1996), is significantly overexpressed in the wound tissue from carnosine-treated animals (Ansurudeen *et al.*, 2012). The chemokine SDF1 having an essential role in the mobilization and homing of stem/progenitor cells like endothelial progenitor cells (Ceradini *et al.*, 2004; Kucia *et al.*, 2005) is also increased after carnosine treatment in the wound tissue, suggesting improved recruitment of stem cells in the wound area. SMA1, the specific marker for myofibroblasts also exhibits increased expression in the carnosine-treated wound tissue. These myofibroblasts are specialized fibroblasts that are transiently involved in the wound-healing process and during the healing process they repair and remodel the wounded tissue, while the fibroblast from normal tissue simply maintains the matrix around them (Agarwal *et al.*, 2006). Thus, the overexpression of the growth factors and key

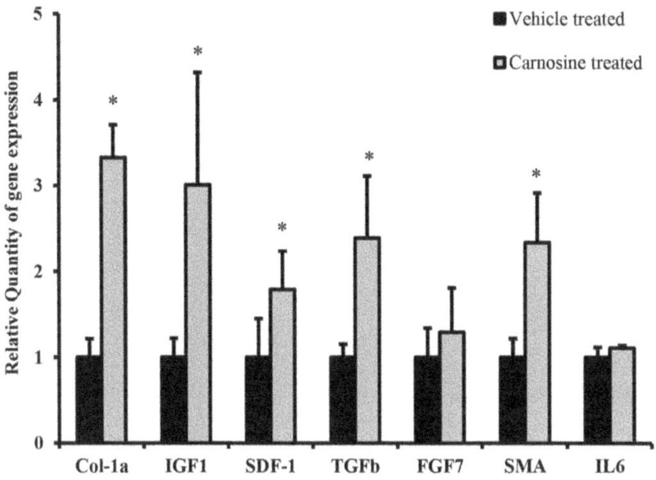

Figure 17.4 Carnosine upregulates the expression of growth factors and cytokines during wound healing. In the carnosine-treated wound tissue, the growth factors, insulin growth factor 1 (IGF-1), Transforming growth factor β (TGFβ), fibroblast growth factor (FGF7); the cytokines stromal derived cell factor 1 (SDF1), interleukin 6 (IL6) and extracellular matrix components collagen 1a (Col-1a) and smooth muscle actin (SMA) are expressed significantly more in comparison to their control counterparts except for FGF7 and IL6 proteins. Data is normalized to the expression of GAPDH in each sample and presented as the relative expression of the control group. Values are means ± SEM and * = $p < 0.05$.

proteins like SDF1 and SMA may have a role in the accelerated wound healing in the carnosine-treated animals than their control counterparts.

17.2.1.2 In Vitro Studies

Many *in vitro* studies have investigated the potential role of carnosine in wound healing. In our study we analyzed the wound-healing effect of carnosine in normal glucose conditions and high glucose conditions simulating the hyperglycemia in *diabetes mellitus* that has a negative effect on wound healing (Ansurudeen *et al.*, 2012). Using *in vitro* wound scratch assay we observed that in human dermal endothelial cells although carnosine (50 mM and 50 μM) had no significant effect on cellular migration, a lower concentration of carnosine could reverse the high glucose (30 mM) induced inhibition on cellular migration and also greatly enhance cell viability of both cells cultured in normal glucose (5 mM) and in high glucose conditions by 40% and 270%, respectively (Figures 17.5 and 17.6). Carnosine had no effect on cellular migration in human dermal fibroblasts; however, it enhanced cell viability of fibroblasts grown in high glucose concentrations (Figures 17.5 and 17.6). Similar results were observed in cells where, polaprezinc (100 μM), a chelate of zinc and carnosine enhanced proliferation and migration of the human colon cells HT29 in an *in vitro* monolayer wound assay (Mahmood *et al.*, 2007). Another study

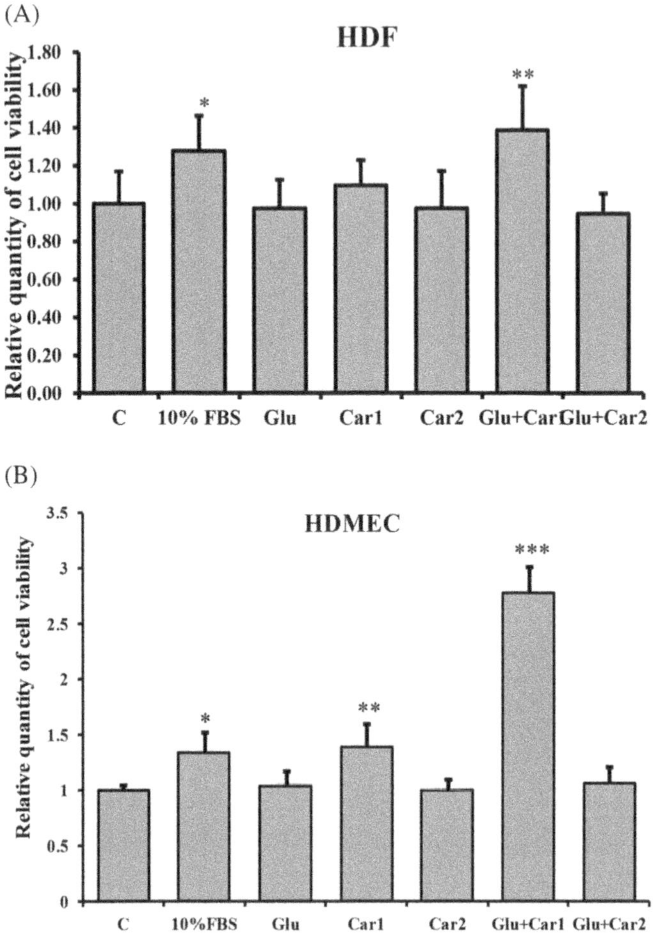

Figure 17.5 Carnosine treatment enhances cell viability in HDF and HDME cells. *In vitro* cell viability of (A). Human dermal fibroblasts (HDFs) and (B). Human dermal endothelial cells (HDMEC) without treatment (C) and after treating them with fetal bovine serum (FBS, 10%), glucose (Glu, 30 mM), carnosine (Car1–50 mM; Car2–50 μM) or a combination of glucose and two concentration of carnosine for 24 h. Values are means ± SEM and * = $p < 0.05$, ** = $p < 0.01$ *** = $p < 0.001$. The results are the mean of four–six different experiments performed in quadruplicates.

with HUVEC cells and foreskin fibroblast cells showed that polaprezinc at lower concentrations (10 nM) caused a similar increase in cellular proliferation as assessed with the BrdU incorporation assay. Increased expression of IGF-1 was involved in the proliferation of these cells. This study also showed that carnosine alone had no effect on cultured HUVEC, while zinc sulfate, a representative zinc compound, increased BrdU uptake by about 2-fold. However, the effect of zinc sulfate was weaker than that of polaprezinc. Also, the proliferative effect of polaprezinc was not seen in guinea pig gastric mucosal epithelial cells suggesting cell specific effects (Seto *et al.*, 1999).

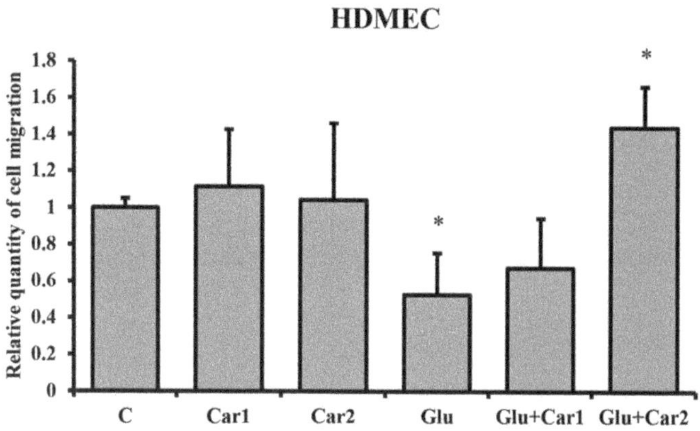

Figure 17.6 Carnosine affects cell migration more in HDMEs than in HDFs cells. *In vitro* cell migration of human dermal endothelial cells (HDMEC) without treatment (C) and after treating them with glucose (Glu, 30 mM), carnosine (Car1–50 mM; Car2–50 μM) and a combination of glucose and two concentration of carnosine for 24 h. Values are means ± SEM and * = $p < 0.05$. The results are the mean of four different experiments.

17.2.1.3 Clinical Studies

In gastric and duodenal ulcer patients subcutaneous injections of 1% carnosine solution cured ulcers in 65% of the patients and remission rate was greatly decreased. The pyloric spasms and gastric acid secretion diminished substantially, thereby alleviating pain and associated complications (Boldyrev, 2007). In Russia, clinical trials were carried out in 109 patients with corneal diseases using 5% carnosine as eye drops. Carnosine treatment significantly improved corneal erosion, trophic keratitis, postherpetic epitheliopathy, primary and secondary corneal dystrophy, and bullous keratopathy. Used in combined treatment, these eye drops accelerated healing of corneal ulcers in herpes virus and bacterial infection or dry keratoconjunctivitis (Maĭchuk *et al.*, 1997). In another group of 163 patients, lansoprazole (a proton-pump inhibitor) plus polaprezinc group showed significantly better ulcer healing ($p < 0.0001$) and significant prevention in the protrusion of the ulcer base in induced submucosal ulcers (Inaba *et al.*, 2010). A recent pilot study evaluating the effect of polaprezinc (150 mg per day for 4 weeks) in low-dose aspirin induced small bowel injury in otherwise healthy patients showed a significant decrease in inflamed lesions and ulcers (Watari *et al.*, 2013).

17.3 Conclusions

Both *in vitro* and *in vivo* studies with carnosine have highlighted its substantial role in wound healing. Of special interest is the beneficial effect of carnosine to enhance healing of wounds in mice with *diabetes mellitus* or irradiation. Existing studies show that carnosine modulates all the three phases of wound healing through multiple modes of action. Release of histamine from histidine, increased synthesis of collagen, enhanced cell viability of endothelial cells and fibroblasts, increased expression of growth factors like IGF-1, TGFb, *etc.* and scavenging the free radicals at the wound site are some of the mechanisms of action mediated by carnosine in improving wound healing. However, clinical studies with carnosine in patients with diabetic foot ulcers and patients with pressure ulcers will greatly add to the beneficial role of carnosine in wound healing. Moreover, carnosine as a wound-healing agent can be cost effective, free of side effects and be part of daily diet.

Summary Points

- Carnosine treatment significantly promotes wound healing.
- More than one mechanism may be involved in mediating the effect of carnosine in wound healing.
- Both *in vitro* and *in vivo* wound-healing studies show promising results, however, clinical studies are required to translate the observed results to humans.
- Carnosine promises to be a successful drug candidate with less or no side effects and with greater cost effectiveness.

Key Facts

1. Wound healing is a complex, tightly controlled phase of highly regulated events that involve various types of cells, cytokines and growth factors.
2. Nonhealing wounds are associated with disease states like *diabetes mellitus*, ischemia, spinal cord injuries, *etc.* and exhibit abnormal sequence of interactions in the wound-healing phases and microvascular abnormalities.
3. There exists a huge market for cost-effective and efficient wound-healing agents.
4. Carnosine an endogenous dipeptide has been documented to exhibit significant improvement of wound healing in both *in vitro* and *in vivo* studies.
5. Carnosine influences the inflammatory phase of wound healing when it is metabolized to histamine and alanine. Alanine stimulates biosynthesis of collagen and histamine stimulates the initial stage of the inflammatory process, thereby mediating the wound-healing effect of carnosine treatment.
6. Carnosine treatment increases neovascularization in the wound area and lead to greater accumulation of collagen fibers.
7. Carnosine promotes tissue granulation and enhanced expression of growth factors like IGF-1, SDF, TGFb involved in wound healing.
8. Carnosine also promotes cellular migration and cell viability in high-glucose conditions.

Key Terms

Angiogenesis. The development of new blood vessels from pre-existing vessels.

Comestibles. An edible item.

Cachexia. Disease-induced loss of body weight and physical wasting of muscle mass.

Dehiscence. Separation of the layers of a surgical wound at the sutures.

Desmosome. A site of adhesion between two adjacent cells, like in epithelial cells. It consists of a small, dense body in which the two halves are separated by an intercellular gap filled with extracellular substance.

Endogenous dipeptide. Peptides that originate from within an organism, tissue, or cell.

Granulation. It is appearance of new connective tissue with tiny blood vessels on the surface of a wound during the healing process.

Hemostasis. It is the first process in wound healing whereby the blood flow damaged blood vessels is stopped.

Hypoxia. Insufficient levels of oxygen in blood or tissue.

Inflammation. It is a process by which the body's immune system responds and protects the body from infection with foreign substances.

Ischemia. Ischemia is an insufficient supply of blood to an organ, usually due to a blocked artery.

Necrosis. An irreversible process of unprogrammed cell death.

Neovascularization. Development of new blood vessels, especially in tissues where circulation has been impaired by disease or trauma.

Nephropathy. Damage to the kidney or predominantly caused by diabetes.

Neuropathy. A functional disturbance or pathological change in the peripheral nervous system.

Tonofilaments. A proteinaceous fiber found in epithelial cells. Bundles of tonofilaments form a tonofibril, which has a supporting function.

Venous stasis. Loss of function of veins in legs to circulate blood back to the heart, which leads to blood accumulation in lower limbs.

List of Abbreviations

Ilα	Interleukin α
Ilβ	Interleukin β
IL6	Interleukin 6
TNFα	Tumor necrosis factor α
FGF2	Fibroblast growth factor 2
bFGF	Basic fibroblast growth factor
IGF-1	Insulin-like growth factor 1
TGFβ	Transforming growth factor β
TGFα	Transforming growth factor α
VEGFα	Vascular endothelial growth factor α
HIF1α	Hypoxia inducible factor α
EGF	Epidermal growth factor

PDGF	Platelet-derived growth factor
ECM	Extracellular matrix
SDF1	Stromal-derived factor 1
SMA1	Smooth muscle actin 1
HT29	Caucasian colon adenocarcinoma cell line
HUVEC	Human umbilical vein endothelial cells
BrdU	Bromodeoxyuridine

Acknowledgement

The research study was supported by the Erling-Persson foundation.

References

Agarwal, C., Britton, Z., Alaseirlis, D., Li, Y. and Wang, J., 2006. Healing and normal fibroblasts exhibit differential proliferation, collagen production, α-SMA expression, and contraction. *Annals of Biomedical Engineering*. 34(4): 653–659. DOI: 10.1007/s10439-006-9090-z.

Ansurudeen, I., Sunkari, V., Grünler, J., Peters, V., Schmitt, C., Catrina, S.-B., Brismar, K. and Forsberg, E., 2012. Carnosine enhances diabetic wound healing in the db/db mouse model of type 2 diabetes. *Amino Acids*. 43(1): 127–134. DOI: 10.1007/s00726-012-1269-z.

Bento, C. and Pereira, P., 2011. Regulation of hypoxia-inducible factor 1 and the loss of the cellular response to hypoxia in diabetes. *Diabetologia*. 54(8): 1946–1956.

Bhora, F., Dunkin, B., Batzri, S., Aly, H., Bass, B., Sidawy, A. and Harmon, J., 1995. Effect of growth factors on cell proliferation and epithelialization in human skin. *The Journal of Surgical Research*. 59: 236–244.

Bitar, M., 1997. Insulin-like growth factor-1 reverses diabetes-induced wound healing impairment in rats. *Hormone and Metabolic Research*. 29: 383–386.

Bitar, M., Labbad, Z., 1996. Transforming growth factor-beta and insulin-like growth factor-I in relation to diabetes-induced impairment of wound healing. *The Journal of Surgical Research*. 61: 113–119.

Boldyrev, A., 2007. *Carnosine and Oxidative Stress in Cells and Tissues*. Nova Science Publishers, New York.

Botusan, I., Sunkari, V., Savu, O., Catrina, A., Grünler, J., Lindberg, S., Pereira, T., Ylä-Herttuala, S., Poellinger, L., Brismar, K. and Catrina, S., 2008. Stabilization of HIF-1alpha is critical to improve wound healing in diabetic mice. *Proceedings of the National Academy of Sciences of the United States of America*. 105(49): 19426–19431.

Catrina, S., Okamoto, K., Pereira, T., Brismar, K. and Poellinger, L., 2004. Hyperglycemia regulates hypoxia-inducible factor-1alpha protein stability and function. *Diabetes*. 53(12): 3226–3232.

Ceradini, D., Kulkarni, A., Callaghan, M., Tepper, O., Bastidas, N., Kleinman, M., Capla, J., Galiano, R., Levine, J. and Gurtner, G., 2004. Progenitor cell trafficking is regulated by hypoxic gradients through HIF-1 induction of SDF-1. *Nature Medicine*. 10: 858–864.

Cho, C. H., Sung, H. K., Kim, K. T., Cheon, H. G., Oh, G. T., Hong, H. J., Yoo, O. J. and Koh, G. Y., 2006. COMP-angiopoietin-1 promotes wound healing through enhanced angiogenesis, lymphangiogenesis, and blood flow in a diabetic mouse model. *Proceedings of the National Academy of Sciences of the United States of America*. 103(13): 4946–4951.

Cuzzocrea, S., Genovese, T., Failla, M., Vecchio, G., Fruciano, M., Mazzon, E., Di Paola, R., Muià, C., La Rosa, C., Crimi, N., Rizzarelli, E. and Vancheri, C., 2007. Protective effect of orally administered carnosine on bleomycin-induced lung injury. *American Journal of Physiology: Lung Cellular and Molecular Physiology*. 292(5): L1095–L1104.

Diegelmann, R. and Evans, M., 2004. Wound healing: an overview of acute, fibrotic and delayed healing. *Frontiers in Bioscience*. 9: 283–289.

Fitzpatrick, D. and Fisher, H., 1982. Histamine synthesis, imidazole dipeptides, and wound healing. *Surgery*. 91(4): 430–434.

Gottrup, F., Holstein, P., Jørgensen, B., Lohmann, M. and Karlsmar, T., 2001. A new concept of a multidisciplinary wound healing center and a national expert function of wound healing. *Archives of Surgery*. 136(7): 765–772. DOI: 10.1001/archsurg.136.7.765.

Grant, M., Jerdan, J. and Merimee, T., 1987. Insulin-like growth factor-I modulates endothelial cell chemotaxis. *The Journal of Clinical Endocrinology and Metabolism*. 65: 370–371.

Guo, S. and DiPietro, L. A., 2010. Factors affecting wound healing. *Journal of Dental Research*. 89(3): 219–229.

Inaba, T., Ishikawa, S., Toyokawa, T., Ishikawa, H., Miyahara, K., Wato, M., Kawai, K., Okada, H. and Yamamoto, K., 2010. Basal protrusion of ulcers induced by endoscopic submucosal dissection (ESD) during treatment with proton pump inhibitors, and the suppressive effects of polaprezinc. *Hepatogastroenterology*. 57(99–100): 678–684.

Kato, S., Tanaka, A., Ogawa, Y., Kanatsu, K., Seto, K., Yoneda, T. and Takeuchi, K., 2001. Effect of polaprezinc on impaired healing of chronic gastric ulcers in adjuvant-induced arthritic rats-role of insulin-like growth factors (IGF)-1. *Medical Science Monitor*. 7: 120–25.

Kiritsy, C. P. and Lynch, S. E., 1993. Role of growth factors in cutaneous wound healing: a review. *Critical Reviews in Oral Biology & Medicine*. 4(5): 729–760. DOI: 10.1177/10454411930040050401.

Korolkiewicz, R., Fujita, A., Seto, K., Suzuki, K. and Takeuchi, K., 2000. Polaprezinc exerts a salutary effect on impaired healing of acute gastric lesions in diabetic rats. *Digestive Diseases and Sciences*. 45(6): 1200–1209.

Kucia, M., Ratajczak, J. and Ratajczak, M., 2005. Bone marrow as a source of circulating CXCR4+ tissue-committed stem cells. *Biology of the Cell*. 97: 133–146.

Mahmood, A., FitzGerald, A., Marchbank, T., Ntatsaki, E., Murray, D., Ghosh, S. and Playford, R., 2007. Zinc carnosine, a health food supplement that stabilises small bowel integrity and stimulates gut repair processes. *Gut*. 56(2): 168–175.

Maĭchuk, I., Formaziuk, V. and Sergienko, V., 1997. Development of carnosine eyedrops and assessing their efficacy in corneal diseases. *Vestnik Oftalmologii*. 113(6): 27–31.

Nagai, K. and Suda, T., 1988. Realization of spontaneous healing function by Carnosine. *Methods & Findings in Experimental & Clinical Pharmacology*. 10(8): 497–507.

Nagai, K., Suda, T., Kawasaki, K. and Mathuura, S., 1986. Action of carnosine and beta-alanine on wound healing. *Surgery*. 100: 815–821.

Peters, V., Schmitt, C., Zschocke, J., Gross, M.-L., Brismar, K. and Forsberg, E., 2011. Carnosine treatment largely prevents alterations of renal carnosine metabolism in diabetic mice. *Amino Acids*. 1–6. DOI: 10.1007/s00726-011-1046-4.

Ragnarson, T. and Apelqvist, J., 2000. Health-related quality of life in patients with diabetes mellitus and foot ulcers. *Journal of Diabetes and its Complications*. 14(5): 235–241.

Riedl, E., Koeppel, H., Pfister, F., Peters, V., Sauerhoefer, S., Sternik, P., Brinkkoetter, P., Zentgraf, H., Navis, G., Henning, R. H., Van Den Born, J., Bakker, S. J. L., Janssen, B., van der Woude, F. J. and Yard, B. A., 2010. *N*-Glycosylation of carnosinase influences protein secretion and enzyme activity. *Diabetes*. 59(8): 1984–1990.

Roberts, P., Black, K., Santamauro, J. and Zaloga, G., 1998. Dietary peptides improve wound healing following surgery. *Nutrition*. 14(3): 266–269.

Sen, C. K., Gordillo, G. M., Roy, S., Kirsner, R., Lambert, L., Hunt, T. K., Gottrup, F., Gurtner, G. C. and Longaker, M. T., 2009. Human skin wounds: a major and snowballing threat to public health and the economy. *Wound Repair and Regeneration*. 17(6): 763–771. DOI: 10.1111/j.1524-475X.2009.00543.x.

Seto, K., Yoneta, T., Suda, H. and Tamaki, H., 1999. Effect of polaprezinc (*N*-(3-aminopropionyl)-L-histidinato zinc), a novel antiulcer agent containing zinc, on cellular proliferation: role of insulin-like growth factor I. *Biochemical Pharmacology*. 58(2): 245–250.

Singer, A. J. and Clark, R. A. F., 1999. Cutaneous wound healing. *New England Journal of Medicine*. 341(10): 738–746. DOI: 10.1056/NEJM199909023411006.

Sonnemann, K. J. and Bement, W. M., 2011. Wound Repair: toward understanding and integration of single-cell and multicellular wound responses. *Annual Review of Cell and Developmental Biology*. 27(1): 237–263. DOI: 10.1146/annurev-cellbio-092910-154251.

Tanaka, R., Ramos, F., Almeida, S., Vizioli, M. and Boscolo, F., 2005. Evaluation of radioprotective effect of carnosine (beta-alanyl-L-histidine) on the wound healing in rats. *Journal of Applied Oral Science*. 13(3): 253–258.

Truitsina, I., Shabanova, M., Chikunova, B., Shavratskiĭ, V., Formaziuk, V., Sergienko, V., Stvolinskiĭ, S. and Boldyrev, A., 1997. Characterization of the anti-ulcer effectiveness of carnosine. *Patologicheskaia Fiziologiia i Eksperimental'naia Terapiia*. 4: 17–20.

Tsuboi, R., Shi, C., Sato, C., Cox, G. and Ogawa, H., 1995. Co-administration of insulin-like growth factor (IGF)-I and IGF-binding protein-1 stimulates wound healing in animal models. *J Invest Dermatol*. 104: 199–203.

Watari, I., Oka, S., Tanaka, S., Aoyama, T., Imagawa, H., Shishido, T., Yoshida, S. and Chayama, K., 2013. Effectiveness of polaprezinc for low-dose aspirin-induced small-bowel mucosal injuries as evaluated by capsule endoscopy: a pilot randomized controlled study. *BMC Gastroenterology*. 13: 108.

CHAPTER 18

Carnosine and Ischemia/ Reperfusion Injury

LATİF ABBASOĞLU*[a], SEMRA DOĞRU-ABBASOĞLU[b], AND MÜJDAT UYSAL[b]

[a]Department of Pediatric Surgery, Acibadem University Faculty of Medicine, Kozyatağı, Istanbul, Turkey; [b]Department of Biochemistry, Istanbul Faculty of Medicine, Istanbul University, Çapa, Istanbul, Turkey
*E-mail: abbasoglul@yahoo.com.tr

18.1 Ischemia/Reperfusion Injury

Cessation of blood supply causes ischemic injury and results in cell dysfunction and death. Paradoxically, restoration of blood flow following ischemia initiates a cascade of events that exaggerates the original injury. This phenomenon is known as ischemia/reperfusion (I/R) injury. I/R injury is a major determinant in many clinical conditions progressing with reduced blood flow to different tissues and organs as in hypovolemic shock, stroke, myocardial infarction, transplant patients and so forth.

During regular circulation, mitochondria consume large amounts of oxygen, generate adenosine triphosphate (ATP) and maintain a balanced production and scavenging of reactive oxygen species (ROS). Besides, mitochondria contribute to the cellular ion homeostasis including calcium (Perrelli *et al.*, 2011). At ischemic status, ATP depletion and a rise in the ATP degradation products such as adenosine, inosine and hypoxanthine take place. Decreased

Food and Nutritional Components in Focus No. 8
Imidazole Dipeptides: Chemistry, Analysis, Function and Effects
Edited by Victor R Preedy

Published by the Royal Society of Chemistry, www.rsc.org

level of ATP limits ATP-coupled ionic pumps, inactivates Na^+/K^+ ATPase and depresses active calcium efflux. Excess intracellular calcium activates calcium-dependent proteases and initiates an inflammatory process. In order to overcome energy deficiency, anaerobic metabolism dominates in the cell; thus, the increased glycolysis causes lactic acid accumulation resulting in a decrease in cell pH (Kalogeris *et al.*, 2012). Acidosis initiates a pro-oxidative state and leads to the release of oxidant iron from proteins and violation of antioxidant defense mechanisms. Xanthine oxidase, which eventuates as the conversion of xanthine dehydrogenase during hypoxia, contributes to the production of superoxide radical while metabolizing the accumulated hypoxanthine (Kalogeris *et al.*, 2012). The duration and magnitude of the ischemia are certain determinants in this event and consequent cell necrosis (Figure 18.1). Although reperfusion is mandatory to salvage the cells from death, the sudden availability of large amounts of molecular oxygen in ischemic tissues leads to

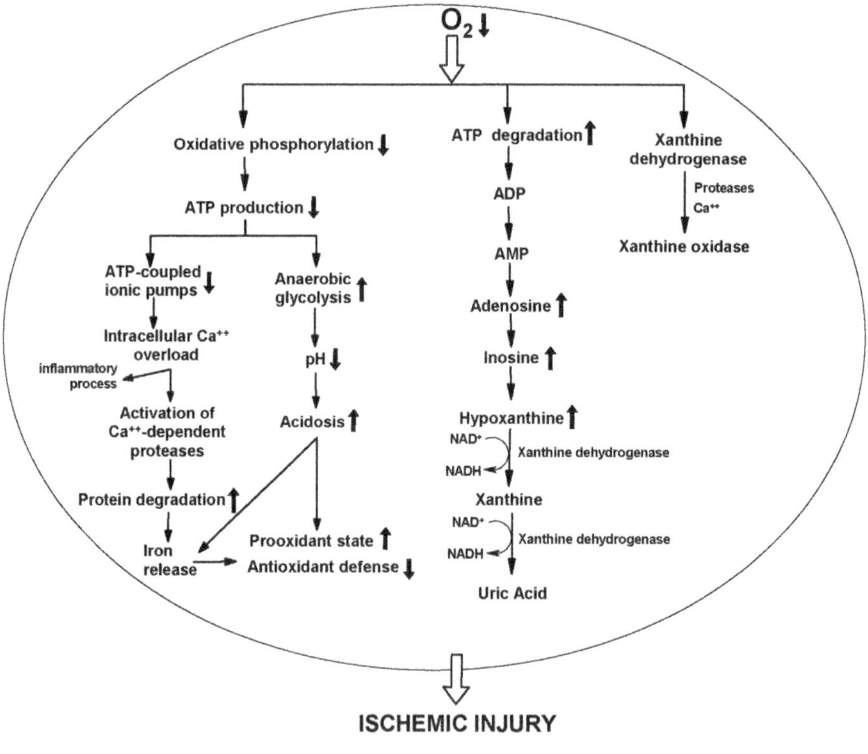

Figure 18.1 Major biochemical events causing to ischemic injury of tissues. Hypoxia promotes the interruption of the oxidative phosphorylation in mitochondria. ATP depletion causes inactivation of ATP-coupled ionic pumps (*e.g.* Na^+/K^+ ATPase), reduces active Ca^{++} influx and produces intracellular Ca^{++} overload. Activation of inflammatory processes, anaerobic glycolysis and xanthine oxidase increase the pro-oxidant state in the cell and cause the ischemic injury.

the generation of oxygen-derived free radicals. ROS oxidatively modify cellular macromolecules, such as lipids, proteins and nucleic acids, thereby paradoxically inducing cell dysfunction.

Tissue ischemia and oxidative stress activate protein kinases that converge on specific transcriptional factors regulating the expression of inflammatory genes. The resulting gene products comprise inducible nitric oxide synthase (iNOS), phospholipase A2, and cyclo-oxygenase-2, proinflammatory cytokines, prostaglandins, and adhesion molecules (Liao *et al.*, 2013). Release of proinflammatory mediators and damaged biomolecules initiates the influx of inflammatory cells such as neutrophils, which seem to be key effector cells causing tissue injury. These multifactorial mechanisms that contribute to the genesis of I/R injury ultimately initiate the pathways leading to necrotic and apoptotic cell death (Kalogeris *et al.*, 2012) (Figure 18.2).

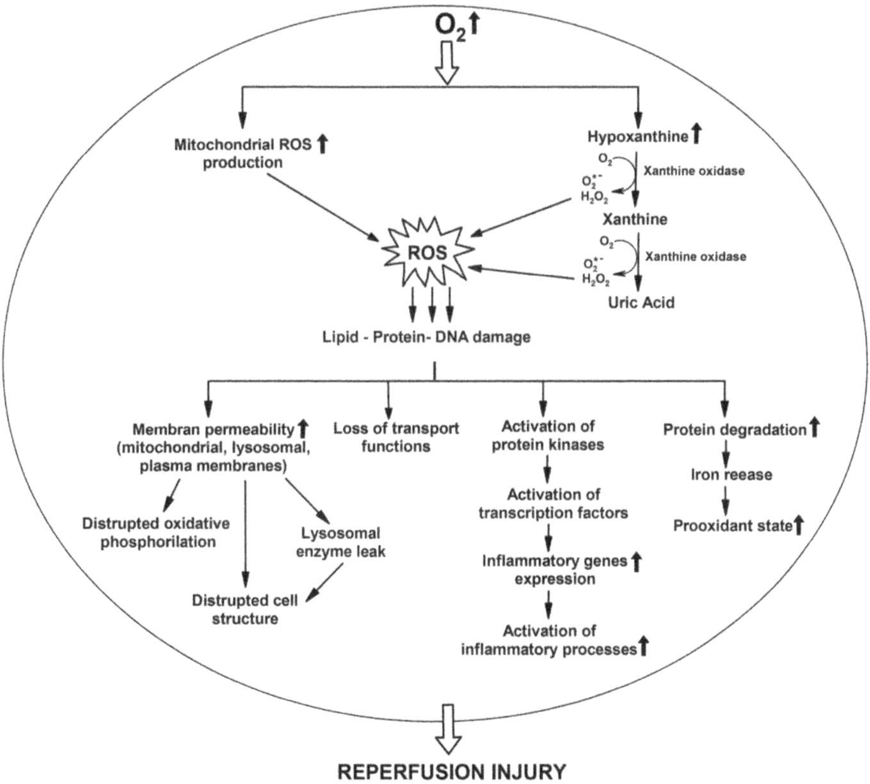

Figure 18.2 Overview of biochemical events in reperfusion injury. Reactive oxygen species (ROS) are the most important contributors to cell injury in reperfusion. The cell is equipped to overcome ROS by regulating their production and scavenging them with antioxidant enzymes and molecules. Reperfusion injury is inevitable if the defense mechanisms fail.

18.1.1 Carnosine and Ischemia/Reperfusion Injury

Carnosine (β-alanyl-L-histidine) is an endogenous water-soluble dipeptide that is found at relatively high concentrations in mammalian skeletal muscles, cardiac muscles and brain (Jackson and Lenney, 1996).

The antioxidant properties of carnosine became popular after the reports by Russian scientists at the 1980s (Dupin *et al.*, 1984; Boldyrev *et al.*, 1987; Guliaeva, 1987). Carnosine is a potent scavenger of ROS and aldehydes. It inhibits lipid peroxidation and protein oxidation. Additionally, it has numerous functions such as membrane-stabilizing activity, proton-buffering capacity, metal-ion chelating, regulation of macrophage function and antiglycating properties (Hipkiss and Brownson, 2000; Stvolinsky and Dobrota, 2000; Aldini *et al.*, 2005) (Figure 18.3).

Carnosine has a protective effect in I/R injury that is mainly based on its suppressive effect on oxygen-derived free radicals, namely superoxide anion and hydroxyl radical production in cells (Stvolinsky *et al.*, 1999). It also acts as a strong quencher of singlet oxygen that contributes to the protective effect in I/R injury (Lee *et al.*, 1999). Carnosine and related peptides stand for water-soluble equals to lipid-soluble antioxidants such as α-tocopherol in preventing oxidative damage of cell membranes (Quinn *et al.*, 1992). The histidine-containing dipeptides, carnosine and anserine also inhibit oxidation of lipids; thereby preventing cell-membrane damage (Boldyrev and Severin, 1990).

The preventive effect of carnosine is also induced by its metabolites histidine and histamine. In fact, the latter may be acting through the receptors in the central nervous system (Fujii *et al.*, 2005; Kurata *et al.*, 2006).

Carnosine is able to bind ions of bivalent metals (Cu, Co, Zn, Mn) including iron leading to the formation of chelate complexes (Severina *et al.*, 2000); herein, iron binding presents special importance accounting for its role in ROS production during I/R injury.

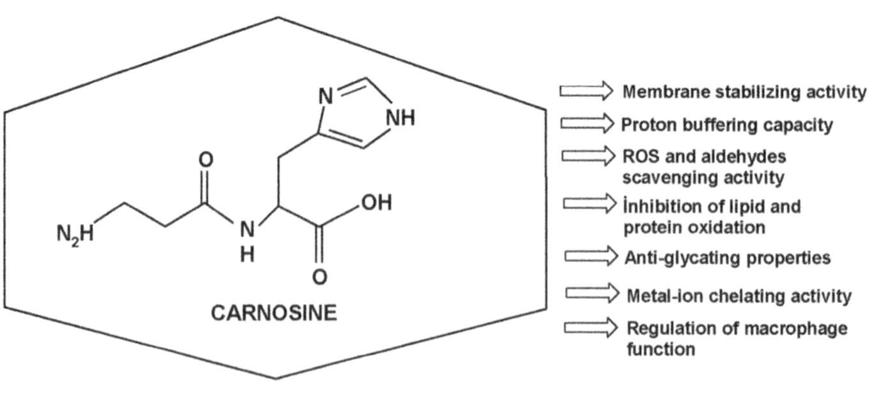

Figure 18.3 The antioxidant properties of carnosine. Mechanisms of the protective effect of CAR are briefly shown.

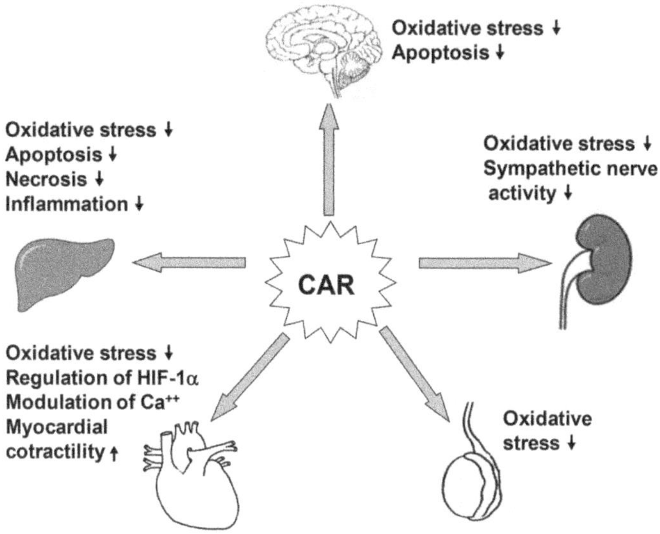

Figure 18.4 Carnosine (CAR) and ischemia/reperfusion I(I/R) injury. Mechanisms of protective effect of CAR in I/R injury in several tissues including liver, brain, kidney, heart and testes according to the published articles in the literature (HIF-1α: hypoxia inducible factor-1α).

Till now, the above-mentioned protective effects of carnosine against I/R injury have been studied in liver, kidney, heart, brain, and testes tissues (Figure 18.4). The results will be discussed in detail.

18.1.1.1 Protective Effect of Carnosine in Ischemia/Reperfusion Injury in Liver

Reduced blood supply to liver in many clinical conditions such as liver surgery, foremost liver transplantation and hemorrhagic status may lead to hepatic I/R injury. Many complex mechanisms including ATP depletion, Kuppfer cell activation and neutrophil infiltration play roles in liver I/R injury causing hepatic failure and eventually organ loss (Teoh and Farrell, 2003). ROS and nitric oxide may participate as initiating factors and modulators in this event (Kaplowitz, 2000).

The protective effect of carnosine in I/R liver injury was first investigated in a rat model (Fouad *et al.*, 2007). In their experimental model, ischemia was induced by clamping the pedicle supplying the left hepatic lobe for 60 min followed by reperfusion for 2 h. Pretreatment with a single i.p. dose of carnosine (250 mg kg^{-1}), 30 min prior to ischemia significantly improved biochemical parameters as well as light and electron microscopically examined hepatocellular changes. Fouad *et al.* demonstrated that carnosine increased the reduced nitric oxide level and decreased the caspase-3 activity in an experimental liver I/R model that is in accordance with previous studies revealing the protective effect of nitric oxide against hepatocyte damage and apoptosis during hypoxia and reoxygenation. In addition, the authors speculated that the protective effect of carnosine in I/R liver injury is mediated through the histamine receptors.

Baykara *et al.* (2009) studied the functions of carnosine and melatonin in I/R injury in the rat liver. Histopathological changes like sinusoidal dilatation, hepatocyte necrosis and neutrophil infiltration as well as biochemical parameters including alanine transaminase (ALT), aspartate transaminase (AST), glutathione reductase and myeloperoxidase were found to be ameliorated in carnosine pretreatment group (250 mg kg^{-1} i.p., 30 min before ischemia) when compared to the I/R group. A combination of carnosine and melatonin administration secured better results than carnosine alone, the investigators suggested that carnosine may be functioning synergistically with another antioxidant agent.

18.1.1.2 Protective Effect of Carnosine in Ischemia/Reperfusion Injury in Kidney

Renal ischemia is the most common cause of acute renal failure causing a functional impairment with high morbidity and mortality. However, the nature of renal tubular cell injury is still unclear; I/R initiates a complex mechanisms including the generation of ROS, rapid decrease in tissue ATP, phospholipase activation and neutrophil accumulation (Edelstein *et al.*, 1997). The first experimental study presenting the possible role of oxygen free radicals in renal ischemia was conducted by Paller *et al.* (1984).

During the last decade, some researchers have focused on the possible protective effects of carnosine in acute renal failure. Certainly, the proven data concerning the major role of I/R injury in acute renal failure inspired them.

Fujii *et al.* (2003a) created an experimental acute renal failure model by occlusion of the left renal artery and vein for 45 min followed by reperfusion, two weeks after contralateral nephrectomy. They treated the animals with carnosine before or after the ischemic period with different doses. Besides the renal function tests and histopathological examinations, norepinephrine levels in renal venous plasma were studied. Prior administration of carnosine (10 μg kg^{-1}, i.v.) markedly ameliorated the renal dysfunction after reperfusion. In addition, preischemic treatment with carnosine at a dose of 10 μg kg^{-1} significantly suppressed the increased norepinephrine levels after reperfusion. The renal sympathetic system was shown to play an important role in the development of the ischemic acute renal failure in a study from the same research center (Fujii *et al.*, 2003b). When the results of both studies are considered together, as the improving effect of carnosine in I/R-induced renal injury *via* its antioxidative activity cannot be denied, the suppressive effect on enhanced norepinephrine release may also play a role in this event.

By using a similar experimental I/R-induced renal injury model, Kurata *et al.* (2006) discussed the suppressive effect of L-carnosine on renal sympathetic nerve activity and the renal injury when given by an intracerebroventricular route. Their study design also aimed to investigate the actions of both L-carnosine and *N*-α-acetyl-L-carnosine, which is resistant to hydrolysis by carnosinase. It was shown that *N*-α-acetyl-L-carnosine has no beneficial effect on renal dysfunction, while L-carnosine or L-histidine (when given alone) ameliorates I/R-induced renal injury. In this event, they asserted that L-histidine is converted to L-histamine in the brain; thus the renoprotective effects of L-carnosine are independent from its antioxidative activity but mediated by histamine-H_3 receptors in the central nervous system. Moreover, the action of L-carnosine on the central nervous system may cause the suppression of enhanced renal sympathetic nerve activity leading to the improvement in renal I/R injury.

Another experimental study with dietary supplementation of carnosine for two weeks prior to I/R-induced renal injury also revealed the effectiveness of carnosine as a prophylactic treatment agent in the development of ischemic acute renal failure (Fujii *et al.*, 2005).

18.1.1.3 Protective Effect of Carnosine in Ischemia/Reperfusion Injury in Heart

Although early reperfusion of the myocardium during ischemia is necessary for survival, myocardial I/R-induced injury deteriorates the benefits obtained from early recirculation in acute myocardial infarction. Various complex mechanisms that take part in I/R injury and consequent ROS generation may contribute to myocardial dysfunction. Numerous agents including carnosine and related peptides targeting those mechanisms have been studied experimentally. Despite widespread interest and investigations, none of them are in routine clinical practice yet.

Lee *et al.* (1999) investigated the relative effects of carnosine and histidine on the functional recovery on isolated perfused rat hearts during reperfusion following ischemia. The study revealed that carnosine is an effective scavenger of singlet oxygen and significantly improves the functions of postischemic reperfusion injury in myocardium. They also investigated the effect of mannitol, a known hydroxyl radical scavenger, in the same experimental design. After observing the lack of protection provided by mannitol in myocardial recovery, they speculated that I/R injury in heart is mainly mediated through singlet oxygen. Carnosine was also found to increase myocardial contractility in a dose-dependent manner in isolated perfused rat hearts and this effect was explained by its modulating role on intracellular calcium (Zaloga *et al.*, 1997).

Since myocardial ischemia and infarction are due to reduced oxygen tension, and hypoxia inducible factor-1α (HIF-1α) increases hypoxia response, Bharadwaj *et al.* (2002) investigated the effect of carnosine on the expression of HIF-1α in cardiac cells. Their results suggest that carnosine may affect the regulated proteolytic degradation of HIF-1α in heart cells during hypoxia.

In recent years, ischemic preconditioning studies have been gaining popularity for many organs, particularly for heart, owing to being adapted to clinical practice. These studies are based on the theory that transient brief episodes of ischemia before a subsequent prolonged ischemia and reperfusion protects against I/R injury. The ROS have been proposed to mediate the ischemic preconditioning; therefore, Toufektsian *et al.* (2003) designed an experimental study as to whether carnosine as a ROS scavenger can block protection achieved by a single episode of ischemic preconditioning. In the end, they did not observe any blocking effect of carnosine on cardioprotection provided by ischemic preconditioning and defended that singlet oxygen production during preconditioning stimulus is unlikely to play a dominant role in triggering preconditioning *in vivo*.

18.1.1.4 Protective Effect of Carnosine in Ischemia/Reperfusion Injury in the Central Nervous System

Stroke is one of the most common causes of death and permanent disability worldwide. Reduced cerebral blood flow due to cerebrovascular diseases causes stroke, namely ischemic brain disease. The generation of ROS plays an important role in the neuronal loss after ischemic stroke and worsens antioxidant defenses causing further neuronal damage. Moreover, rapid restoration of blood flow and consequent tissue reoxygenation deepen ROS production which results in I/R injury. Therefore, antioxidant agents including carnosine have become one of the emphasized issues as a treatment strategy of ischemic brain diseases.

One of the first studies investigating the neuroprotective effect of carnosine in ischemic brain was conducted by Stvolinsky *et al.* (1999); though they did not study the reperfusion status. They observed the improving effect of carnosine with better cerebral congestion, electroencephalographic normalization as well as enzymatic activation against ROS. One year later, a study from the same research center which again only focused on the ischemic period, also revealed the beneficial effects of carnosine by showing low mortality rates in rats subjected to occlusion of the carotid arteries (Stvolinsky *et al.*, 2000). They concluded that carnosine may be a promising agent in the future treatment of stroke and brain infarction.

There are several studies researched on the protective effects of carnosine in I/R injury of brain, most being *in vivo*. One of the initial studies in this field examined the learning parameters of the pretreated rats subjected to brain ischemia induced by both sides' occlusion of common carotid arteries for 120 min period followed by reperfusion (Gallant *et al.*, 2000). Animals in the study group received carnosine with drinking water at a daily dose of 100 mg kg^{-1} for 7 months prior to experiment. Carnosine-pretreated rats showed significantly better learning results according to at least three learning parameters. This protective effect of carnosine was attributed to its own action instead of its metabolites, as histidine did not affect resistance of the subjects against oxidative stress as carnosine did.

An *in vitro* study by creating an ischemia-reperfusion model using pheochromocytoma PC12 cell cultures aimed to analyze the neuroprotective effects of carnosine and related compounds (Tabakman *et al.*, 2002). A 3–5 h hypoxia and glucose deprivation status was followed by a 18 h reoxygenation. At the end of the reoxygenation period, cell death, prostaglandin E_2 (PGE_2) and lactate dehydrogenase (LDH) release from the cells were analyzed. The authors concluded that carnosine has a neuroprotective effect against I/R injury possibly *via* its antioxidant properties.

In the field of protective effects of carnosine against I/R injury in brain, almost all studies concerned the preischemic administration of the agent. Dobrota *et al.* (2005) designed an ischemic brain model in rats by occluding both common carotid arteries as well as left arteria vertebralis for 15 min, while administering carnosine only during reperfusion period in a repeating manner for 6 days. In terms of both improved neurological symptoms and biochemical parameters, carnosine was found to have a compensatory function in antioxidant defense system in brain ischemia and this data is encouraging for its translation into clinical practice in stroke patients.

On the other hand, Pekcetin *et al.* (2009) examined the effects of carnosine treatment on apoptosis, oxidative stress and spatial memory in an animal model of cerebral ischemia that they gave the agent (250 mg kg^{-1} i.p.) 30 min prior to ischemia. They produced a cerebral ischemia model by occluding the right common carotid artery for 30 min following 24 h or 1 week of reperfusion period. As the investigators were mainly interested in the functions of hippocampus after transient cerebral ischemia, spatial learning was assessed and antioxidant enzyme activities were studied from the tissues of hippocampus and prefrontal cortex. Although pretreatment with carnosine did not prevent ischemia-induced memory impairment in subjects, significant improvement was detected in oxidative stress and ischemia-induced apoptotic neuronal cell death. Nonetheless, the authors concluded that carnosine may be beneficial as a prophylactic agent in cerebral ischemia when administered before the insult.

Experimental brain ischemia, aggravated by energy deficit induced by simultaneous administration of 3-nitropropionic acid (3-NPA) that causes selective degeneration of striatum neurons, reveals both neurochemical deviations in the brain and stronger neurological deficit (Dobrota *et al.*, 2005). In a further study design, the investigators produced a three-vessel global ischemia of brain aggravated with systemic administration of 3-NPA and evaluated the therapeutic effect of carnosine. Following the ischemic period, carnosine was injected i.p. in the very beginning of the reperfusion, and 4, 8, and 24 h after operation, then once a day for more 5 days (Dobrota *et al.*, 2010). Carnosine treatment was found to be efficient as it protected the ischemic subjects from mortality and improved neurological symptoms.

18.1.1.5 Protective Effect of Carnosine in Ischemia/Reperfusion Injury in Testes

Testicular torsion is a surgical emergency in men causing tissue hypoxia that leads to necrosis of germinal cells. The affected testis becomes permanently dysfunctional if left untreated for 4–6 h. The primary pathophysiological event in testicular torsion is ischemia of the testis followed by reperfusion upon repair. Although the exact mechanism has not been completely understood, there is increasing evidence proposing the role of ROS in the pathogenesis of testicular I/R injury. Hence, to prevent injury in testes, effects of different antioxidant agents have been studied experimentally.

Recently, we designed an experimental I/R injury model in testes to investigate the protective effects of carnosine for the first time (Abbasoğlu *et al.*, 2012). To create testicular I/R injury, the left testis was torsioned 720° for 2 h followed by 2 h of detorsion and carnosine (250 mg kg^{-1}, i.p.) was given 1 h before detorsion. Biochemical analysis including thiobarbituric acid reactive substances (TBARS), diene conjugate (DC), protein carbonyls (PC), vitamin C levels as well as histapathological evaluation were performed. Our study showed the beneficial effects of carnosine on testicular I/R injury by improving both biochemical and histological study parameters.

18.2 Conclusions

In view of all the mentioned studies, there is no doubt that carnosine has a protective effect against I/R injury in various organs including liver, kidney, heart, brain, and testes; however, all these studies were performed in experimental models. As discussed in detail in other sections, evidence of the beneficial effects of carnosine and related peptides towards human health is relatively rare in comparison with the observed improving effects in animal models. Therefore, further studies are required to prove the thesis concerning carnosine's efficacy on I/R injury and to adapt it into clinical practice.

Summary Points

- I/R injury represents tissue damage after restoration of blood flow in ischemic tissues.
- Complex mechanisms, including ROS generation in response to the sudden availability of large amounts of molecular oxygen after reperfusion in ischemic tissues, take place in I/R injury.
- Carnosine is a strong scavenger of ROS and inhibits lipid peroxidation.
- Carnosine has an improving effect in I/R injury mainly based on its ROS scavenging function.
- The preventive effect of carnosine in I/R injury may also be due to its metabolites histidine and histamine.
- Carnosine may be functioning synergistically with other antioxidants against I/R injury.
- The protective effect of carnosine in I/R injury has been shown in liver, kidney, heart, brain and testes tissues in various experimental models.
- In addition to antioxidative activity, carnosine also acts by suppressing the renal sympathetic nerve activity in I/R-induced renal injury.

- Carnosine increases myocardial contractility by modulating intracellular calcium in heart I/R injury.
- As all of the studies showing the efficacy of carnosine in I/R injury were experimental, further research is required for its clinical adaptation.

Key Facts of I/R Injury

1. Ischemia is the restriction of blood flow to the tissues, causing a deficiency of oxygen and nutrients to the cell metabolism.
2. The restoration of blood supply to the tissue after a period of ischemia is defined as "reperfusion".
3. Although reperfusion rescues cells from death by renewing reoxygenation; sudden availability of oxygen usually causes severe tissue damage, which is known as I/R injury.
4. I/R injury starts with a decrease in a high-energy phosphate supply to the mitochondrial fraction of the cells.
5. Due to energy deficiency, several events may take place in the cell such as inactivation of the ionic pumps, increase in cellular calcium and hydrogen concentrations, loss of cellular synthetic functions, and activation of membrane-degradation processes.
6. Generation of ROS that are produced during the reperfusion period plays the fundamental role in I/R injury.
7. Several factors contribute to the pathogenesis of I/R injury, including phospholipase and protease activation, inflammatory cell infiltration and release of proinflammatory mediators.
8. Consequently, I/R injury initiates the pathways leading to necrotic and apoptotic cell death.

Definitions of Words and Terms

Acute myocardial infarction (AMI). A leading cause of death worldwide that can occur when the supply of oxygen and nutrients are reduced. This event is usually caused by blocking of one of the coronary arteries due to atherosclerosis. AMI results in damage and death of the heart muscles.

Acute renal failure. Rapid loss of kidney function (filtering waste products from blood) owing to decreased renal blood flow. This event develops within 48 h and results in uremia, acidosis and fluid-electrolyte imbalance.

Apoptosis. It is known as programmed cell death and differs from necrosis by distinct morphologic and biochemical features. It is a genetically determined process of cell self-destruction and activated by several stimuli.

Hepatic failure. Loss of hepatic synthetic and detoxification functions.

Ischemia. A pathological condition due to reduced blood supply to tissues caused by arterial obstruction. Ischemia causes cell dysfunction and progresses to cell death if left untreated.

Ischemic preconditioning. An experimental technique based on the belief that after an ischemic insult; instead of rapid reperfusion, brief periods of ischemia and reperfusion attenuates I/R injury.
Reperfusion. Restoration of blood flow following ischemia.
Sympathetic nervous system (SNS). It is a part of autonomic nervous system that controls the body's internal organs and regulates homeostatic mechanisms. SNS ensures quick responses under stress conditions.
Spatial memory. Ability of a person or an animal to orient himself/itself in a tridimensional space and navigate himself/itself in a familiar environment.
Stroke. A stroke (cerebrovascular accident, brain attack) is the sudden interruption of blood flow and death of cells in a brain region due to the lack of oxygen and nutrients. Stroke may result in paralysis, speech impairment, loss of memory, reasoning ability and death.
Testicular germ cells. Specialized cells that give rise to sperms in male organisms for sexual reproduction.

List of Abbreviations

ALT	Alanine transaminase
AST	Aspartate transaminase
ATP	Adenosine triphosphate
DC	Diene conjugate
h	Hour
HIF-1α	Hypoxia inducible factor-1α
i.p.	Intraperitoneal
i.v.	Intravenous
I/R	Ischemia/reperfusion
iNOS	Inducible nitric oxide synthase
min	Minute
3-NPA	3-Nitropropionic acid
PC	Protein carbonyls
ROS	Reactive oxygen species
TBARS	Thiobarbituric acid reactive substances

References

Abbasoğlu, L., Kalaz, E. B., Soluk-Tekkeşin, M., Olgaç, V., Doğru-Abbasoğlu, S. and Uysal, M., 2012. Beneficial effects of taurine and carnosine in experimental ischemia/reperfusion injury in testis. *Pediatric Surgery International*. 28(11): 1125–1131.

Aldini, G., Facino, R. M., Beretta, G. and Carini, M., 2005. Carnosine and related dipeptides as quenchers of reactive carbonyl species: from structural studies to therapeutic perspectives. *BioFactors*. 24: 77–87.

Baykara, B., Tekmen, I., Pekcetin, C., Ulukus, C., Tuncel, P., Sagol, O., Ormen, M. and Ozogul, C., 2009. The protective effects of carnosine and melatonin in ischemia-reperfusion injury in rat liver. *Acta Histochemica*. 111: 42–51.

Bharadwaj, L. A., Davies, G. F., Xavier, I. J. and Ovsenek, N., 2002. L-Carnosine and verapamil inhibit hypoxia-induced expression of hypoxia inducible factor (HIF-1 alpha) in H9c2 cardiomyoblasts. *Pharmacological Research.* 45: 175–181.

Boldyrev, A. A., Dupin, A. M., Bunin, A. Y., Babizhaev, M. A. and Severin, S. E., 1987. The antioxidative properties of carnosine, a natural histidine containing dipeptide. *Biochemistry International.* 15(6): 1105–1113.

Boldyrev, A. A. and Severin, S. E., 1990. The histidine-containing dipeptides, carnosine and anserine: distribution, properties and biological significance. *Advances in Enzyme Regulation.* 30: 175–194.

Dobrota, D., Federova, T., Stvolinsky, S., Babusikova, E., Likavcanova, K., Drgova, A., Strapkova, A. and Boldyrev, A., 2005. Carnosine protects the brain of rats and Mongolian gerbils against ischemic injury: after-stroke-effect. *Neurochemical Research.* 30: 1283–1288.

Dobrota, D., Federova, T. N., Stepanova, M. S., Babusikova, E., Statelova, D., Tatarkova, Z., Stvolinsky, S. S. and Boldyrev, A. A., 2010. Oxidative stress induced in rat brain by a combination of 3-nitropropionic acid and global ischemia. *International Journal of Clinical and Experimental Medicine.* 3(2): 144–151.

Dupin, A. M., Boldyrev, A. A., Arkhipenko, I. V. and Kagan, V. E., 1984. Carnosine protection of Ca^{2+} transport against damage induced by lipid peroxidation. *Bulletin of Experimental Biology and Medicine.* 98(8): 186–188.

Edelstein, C. L., Ling, H. and Schrier, R. W., 1997. The nature of renal cell injury. *Kidney International.* 51(5): 1341–1351.

Fouad, A. A., El-Rehany, M. A. and Maghraby, H. K., 2007. The hepatoprotective effect of carnosine against ischemia/reperfusion liver injury in rats. *European Journal of Pharmacology.* 572: 61–68.

Fujii, T., Takaoka, M., Muraoka, T., Kurata, H., Tsuruoka, N., Ono, H., Kiso, Y., Tanaka, T. and Matsumura, Y., 2003a. Preventive effect of L-carnosine on ischemia/reperfusion-induced acute renal failure in rats. *European Journal of Pharmacology.* 474: 262–267.

Fujii, T., Kurata, H., Takaoka, M., Muraoka, T., Fujisawa, Y., Shokoji, T., Nishiyama, A., Abe, Y. and Matsumura, Y., 2003b. The role of renal sympathetic nervous system in the pathogenesis of ischemic acute renal failure. *European Journal of Pharmacology.* 481: 241–248.

Fujii, T., Takaoka, M., Tsuruoka, N., Kiso, Y., Tanaka, T. and Matsumura, Y., 2005. Dietary supplementation of L-carnosine prevents ischemia/reperfusion-induced renal injury in rats. *Biological & Pharmaceutical Bulletin.* 28(2): 361–363.

Gallant, S., Kukley, M., Stvolinsky, S., Bulygina, E. and Boldyrev, A., 2000. Effect of carnosine on rats under experimental brain ischemia. *Tohoku Journal of Experimental Medicine.* 191: 85–99.

Guliaeva, N. V., 1987. Superoxide-scavenging activity of carnosine in the presence of copper and zinc ions. *Biokhimiia.* 52(7): 1216–1220.

Hipkiss, A. R. and Brownson, C., 2000. Carnosine reacts with protein carbonyl groups: another possible role for the anti-ageing peptide? *Biogerontology.* 1(3): 217–223.

Jackson, M. C. and Lenney, J. F., 1996. The distribution of carnosine and related dipeptides in rat and human tissues. *Inflammation Research*. 45(3): 132–135.

Kalogeris, T., Baines, C. P., Krenz, M. and Korthuis, R. J., 2012. Cell biology of ischemia/reperfusion injury. In: Jeon, K. W. (ed.) *International Review of Cell and Molecular Biology*. Elsevier Academic Press, CA, USA, pp. 229–317.

Kaplowitz, N., 2000. Mechanisms of liver cell injury. *Journal of Hepatology*. 32(suppl. 1): 39–47.

Kurata, H., Fujii, T., Tsatsui, H., Katayama, T., Ohkita, M., Takaoka, M., Tsuruoka, N., Kiso, Y., Ohno, Y., Fujisawa, Y., Shokoji, T., Hishiyama, A., Abe, Y. and Matsumura, Y., 2006. Renoprotective effects of L-carnosine on ischemia/reperfusion-induced renal injury in rats. *Journal of Pharmacology and Experimental Therapeutics*. 319(2): 640–647.

Lee, J. W., Miyawaki, H., Bobst, E. W., Hester, J. D., Ashraf, M. and Bobst, A. M., 1999. Improved functional recovery of ischemic rat hearts due to singlet oxygen scavengers histidine and carnosine. *Journal of Molecular and Cellular Cardiology*. 31: 113–121.

Liao, Y. F., Zhu, W., Li, D. P. and Zhu, X., 2013. Heme oxygenase-1 and gut-ischemia/reperfusion injury. *World Journal of Gastroenterology*. 19(23): 3555–3561.

Paller, M. S., Hoidal, J. R. and Ferrisi, T. F., 1984. Oxygen free radicals in ischemic acute renal failure in the rat. *Journal of Clinical Investigation*. 74(4): 1156–1164.

Pekcetin, C., Kiray, M., Ergur, B. U., Tugyan, K., Bagriyanik, H. A., Erbil, G., Baykara, B. and Camsari, U. M., 2009. Carnosine attenuates oxidative stress and apoptosis in transient cerebral ischemia in rats. *Acta Biologica Hungarica*. 60(2): 137–148.

Perrelli, M. G., Pagliaro, P. and Penna, C., 2011. Ischemia/reperfusion injury and cardioprotective mechanisms: role of mitochondria and reactive oxygen species. *World Journal of Cardiology*. 3(6): 186–200.

Quinn, P. J., Boldyrev, A. A. and Formazuyk, V. E., 1992. Carnosine: its properties, functions and potential therapeutic applications. *Molecular Aspects of Medicine*. 13(5): 379–444.

Severina, I. S., Bussygina, O. G. and Pyatakova, N. V., 2000. Carnosine as a regulator of soluble guanylate cyclase. *Biochemistry (Mosc)*. 65(7): 783–788.

Stvolinsky, S. L., Kukley, M. L., Dobrota, D., Matejovicova, M., Tkac, I. and Boldyrev, A. A., 1999. Carnosine: an endogenous neuroprotector in the ischemic brain. *Cellular and Molecular Neurobiology*. 19: 45–56.

Stvolinsky, S. L. and Dobrota, D., 2000. Anti-ischemic activity of carnosine. *Biochemistry (Mosc)*. 65(7): 849–855.

Stvolinsky, S. L., Kukley, M., Dobrota, D., Mezesova, V. and Boldyrev, A., 2000. Carnosine protects rats under global ischemia. *Brain Research Bulletin*. 53: 445–448.

Tabakman, R., Lazarovici, P. and Kohen, R., 2002. Neuroprotective effects of carnosine and homocarnosine on pheochromocytoma PC12 cells exposed to ischemia. *Journal of Neuroscience Research*. 68: 463–469.

Teoh, N. C. and Farrell, G. C., 2003. Hepatic ischemia reperfusion injury: pathogenic mechanisms and basis for hepatoprotection. *Journal of Gastroenterology and Hepatology*. 18(8): 891–902.

Toufektsian, M. C., Morel, S., Tanguy, S., Jeunet, A., de Leiris, J. and Boucher, F. F., 2003. Involvement of reactive oxygen species in cardiac preconditioning in rats. *Antioxidants & Redox Signaling*. 5(1): 115–122.

Zaloga, G. P., Roberts, P. R., Black, K. W., Lin, M., Zapata-Sudo, G., Sudo, R. T. and Nelson, T. E., 1997. Carnosine is a novel peptide modulator of intracellular calcium and contractility in cardiac cells. *American Journal of Physiology*. 272(1 Pt 2): H462–H468.

CHAPTER 19

Antigenotoxic Effects of Carnosine

AHMET GÜNER[a], AND LOKMAN ALPSOY*[a]

[a]Department of Biology, Faculty of Art and Science, Fatih University, Istanbul 34500, Turkey
*E-mail: lalpsoy@fatih.edu.tr

19.1 Carnosine

The major cytoplasmic dipeptide carnosine is formed by bonding of histidine and β-alanine in a reaction catalyzed by the enzyme carnosine synthase that has been identified as ATP-grasp domain-containing protein 1 (Drozak *et al.*, 2013), although it is the availability of β-alanine, obtained in small amounts from hepatic synthesis and potentially in greater amounts from the diet that is limiting to synthesis (Harris *et al.*, 2012). β-Alanine, a nonproteinogenic amino acid, is produced mainly by the liver as the final metabolite of uracil and thymine degradation (Matthews and Traut, 1987; Hipkiss, 1998).

Carnosine is widely and abundantly distributed in the skeletal muscles, central nervous system, olfactory neurons and in the lens of the eye in most vertebrates, including humans throughout the body wherever there are high energy demands (Bellia *et al.*, 2011; Budzen and Rymaszewska, 2013). Hyland *et al.* (2000) indicated that carnosine can be present at concentrations as high as 20 mM in humans, around 1 mM in mice and around 5 mM in cats, dogs and sheep. Moreover, researchers note that synthesis of carnosine in muscles is dependent on the availability of the amino acid alanine in the

Food and Nutritional Components in Focus No. 8
Imidazole Dipeptides: Chemistry, Analysis, Function and Effects
Edited by Victor R Preedy

Published by the Royal Society of Chemistry, www.rsc.org

body (Budzen and Rymaszewska, 2013). Its function is to protect these vital areas from the metabolic demands of energy production and management (Baguet *et al.*, 2010; Calabrese *et al.*, 2011).

Since then, an enormous number of published scientific studies have corroborated the multiple beneficial effects of carnosine including protecting brain cells from toxic metal-ion reactions that lead to dementia, extending the life span of laboratory mice, complex, warm-blooded mammals with many of the aging features common to humans (Boldyrev *et al.*, 1999; Gallant *et al.*, 2000) and inhibiting the growth of cancer cells (Renner *et al.*, 2008). Carnosine shields us from the onslaught of oxidation, glycation, DNA damage, and other reactions that injure tissues and cripple organs (Hipkiss, 1998).

Many functions have been proposed for carnosine depicted in Figure 19.1. It is responsible for a variety of activities such as antioxidant activity in terms of protection from free radicals, buffering effects in order to help to maintain the acid–base balance, chelating effects to reduce the toxicity of metals, antiglycation effects, and protects against some neurotoxic substances (Caruso *et al.*, 2012; Aloisi *et al.*, 2013; Budzen and Rymaszewska, 2013).

Recent studies have shown that this small molecule also has membrane-protecting activity, formation of complexes with transition metals, and regulation of macrophage function. Many potential therapeutic properties have been recognized especially related to the antioxidant activity, but the therapeutic uses are strongly limited by the mechanism governing its homeostasis. This fact has been the main reason for developing the synthesis of carnosine and its derivatives with interesting potentiality, but until now there have been very few applications.

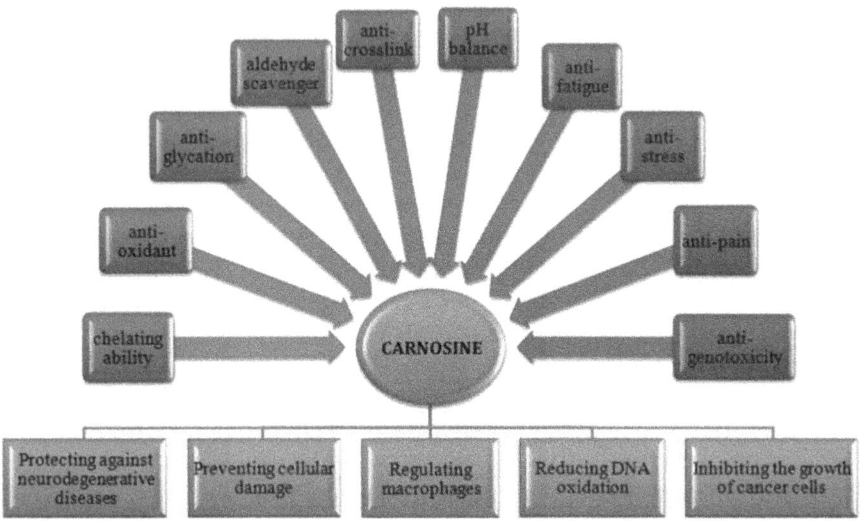

Figure 19.1 The roles of carnosine. The figure shows the functions of carnosine that play beneficial roles on diseases and human health.

In particular, it has been recently demonstrated that carnosine is a potent and selective scavenger of α,β-unsaturated aldehydes, typical byproducts of membrane lipids peroxidation (Aydin *et al.*, 2010) and considered second messengers of the oxidative stress, and inhibits aldehyde-induced protein–protein and DNA–protein crosslinking in neurodegenerative disorders such as Alzheimer's disease, in cardiovascular ischemic damage and in inflammatory diseases (Budzen and Rymaszewska, 2013). It may also regulate glycolysis, muscular contraction and oxidative phosphorylation, stimulate the immune system, bind copper, zinc and calcium (Quinn *et al.*, 1992) and purines (Neurohr and Mantsch, 1980), which suggests potential involvement in gene regulation or signal transduction (Hipkiss, 1998).

Among antioxidant mechanisms reported for carnosine are its ability to inactivate reactive oxygen species (ROS), scavenge free radicals, and chelate pro-oxidative metals (Boldyrev *et al.*, 1988; Kohen *et al.*, 1988; Yanai *et al.*, 2008). It also affects superoxide dismutase (SOD) and glutathione peroxidase (GPx) activities and glutathione (GSH) and malondialdehyde (MDA) levels (Babizhayev *et al.*, 1994; Aydin *et al.*, 2010; Tsai *et al.*, 2010). These enzymatic antioxidants have important role in reducing side effects induced by oxidative stress. Carnosine acts as a hydroxyl-radical scavenger to protect DNA cleavage and plays a role in determining the behavior of the molecule towards oxidizing and also has a different ability to protect DNA from double-strand breakage by X-ray radiation (Fu *et al.*, 2009).

It is seen that carnosine has a remarkable spectrum of health benefits. Moreover, many studies have demonstrated, both at the tissue and organelle level, that carnosine may have protective functions additional to its antioxidant and free-radical scavenging roles. Nevertheless, it has also reported that carnosine shows antigenotoxic effects.

19.2 Antigenotoxic Effects of Carnosine

Mainly toxic materials can be classified as genotoxic or nongenotoxic (Perera, 1984). The term genotoxicity was clearly defined by Perera (1984). Genotoxic agents potentially can cause a heritable change in the genetic material at the level of gene or chromosome. The genotoxic materials can react with DNA in different ways: (i) directly react with DNA, (ii) are converted to reactive substances by the cell metabolism then their metabolite can interact to DNA and (iii) increase the reactive free radicals including ROS and reactive nitrogen species (RNS), after which they cause DNA damage (Kasai *et al.*, 1986; Lindberg *et al.*, 1990; Powers *et al.*, 1995).

Polycyclic aromatic hydrocarbons, aromatic amines, nitrosamines, alkylating agents and some natural poisons such as aflatoxins exert their mutagenic effects on DNA by forming covalent adducts with bases of DNA, so they can cause DNA damage. DNA damage caused by genotoxins reveal aging, infertility and several diseases such as genetic disorders including cystic fibrosis, sickle cell anemia, Huntington disease and cancer. For instance, most of the classified carcinogens by IARC have the mutagenic impact on genetic material, so the determination of genotoxic effects of natural or artificial substances is critical to avoid the development of cancers (IARC, 1999). If the causative agents and their genotoxic impacts on DNA could be identified, many human cancers would be eliminated (Powers *et al.*, 1995; King and Robins, 2006).

Additionally, identification of genotoxic effects of substances is not only critically important in diseases but also in evaluating the safety of drugs, cosmetics and food. Varieties of natural and chemical substances are absorbed by our bodies when using drugs and cosmetics or during eating foods. Therefore, they may cause toxic and mutagenic effects on cells (Powers *et al.*, 1995).

With a lot of potent genotoxic materials, to test their mutagenic activity is crucial for the reduction of human carcinogenesis. Identification of their impacts on human have not been tested because of ethical rules, therefore we should perform these on alternative material. These tests classify three main categories: tests on micro-organisms (the most common used bacteria is *Salmonella typhimurium*), tests on DNA damage and tests on chromosome damage (King and Robins, 2006). Previous studies on the antigenotoxic effects of carnosine are mentioned below in terms of their results.

19.2.1 Protective Role of Carnosine on Mutagenicity (AMES)

If we look at the protective role of carnosine on mutagenicity, there is less information in the literature. In one study, Sato *et al.* (2008) studied chicken-breast extraction (CBEX™) containing among its primary constituents carnosine and anserine, which are histidine-containing dipeptides present in the muscle tissues of most vertebrate species. They used specific pathogen-free S1c:Wistar rats for acute (LD_{50}) and subchronic (90 days) oral toxicity studies and also used various strains of *Salmonella typhimurium* and *Escherichia coli* for mutagenicity (AMES) test. According to their results, the no observed-adverse-effect level for CBEX™ containing carnosine and anserine was considered to be 2000 kg^{-1} body weight per day for both male and female rats, the highest dose tested. Under the conditions of the mutagenicity assay, CBEX™ was determined to be nonmutagenic in *S. typhimurium* and *E. coli* strains in the presence or absence of metabolic activation. Therefore, no growth inhibition by the test substance was observed in any strain of bacteria in the presence or absence of S9 activation.

As a result, Sato *et al.* demonstrated that the use of CBEX™ in foods as a source of carnosine and anserine would not produce any adverse health effects in humans.

19.2.2 Protective Role of Carnosine on DNA Damage

Many studies reported that carnosine plays a protective role on DNA damage. We just give some examples where carnosine has positive effects on DNA damage. However, it should be determined how DNA damage can be detected and measured.

Generally, single-cell gel electrophoresis or comet assay method is used for detecting and measuring the quantification of DNA damage. DNA strand breakage causes "relaxation" or opening of circular and linear forms of DNA, which are observed as different bands on agarose gel (Kang, 2010). Carnosine could effectively inhibit acetaldehyde-mediated damage of DNA, proteins and lipids (Hipkiss, 1998).

In one study, Hsieh and coworkers focused on carnosine and whether it has a real effect on human cells or whether it is true that carnosine can prevent human cells from oxidative damage. Therefore, they investigated the effects of carnosine and anserine on DNA oxidative damage in human blood lymphocytes caused by different inducers such as Fe^{2+}, Cu^{2+} and H_2O_2 by using comet assay, and estimated their potentials to serve as a biological antioxidant. Their experiment results showed that carnosine and anserine within the concentrations 0–100 μM had a slight cytotoxicity and genotoxicity. Fe^{2+} and Cu^{2+} caused cytotoxicity and DNA damage in lymphocytes. They found that comet-like tails from damaged DNA appeared after electrophoresis. The longer ones implied severe damage of DNA. In this case they concluded that carnosine and anserine had an antioxidant effect on the DNA damage assay in lymphocytes induced by transition-metal ions and H_2O_2. This probably resulted from the chelating and scavenging effects of carnosine and anserine on free radicals that are generated from Fe^{2+}, Cu^{2+} and H_2O_2 (Hsieh *et al.*, 2002).

An investigation on the effects of carnosine, at the physiological concentration of 20 mM, on oxidative DNA damage levels *in vitro* lifespan in peripheral blood derived from human $CD4^+$ T cell clones are demonstrated that carnosine acted as a weak antioxidant, with levels of oxidative DNA damage being lower in T cells grown long term in the presence of carnosine (Hyland *et al.*, 2000). In this study under the culture conditions used (20% O_2) long-term culture with carnosine demonstrates that there was a significant increase in the lifespan of a clone derived from a healthy young subject. In the T cell clones that were cultured long term in the presence of carnosine, it was surprising to note that after one week in culture there was more DNA damage present in the supplemented clones, as compared with the controls. It is possible that an active role for carnosine in cell biochemistry and maintenance (McFarland and Holliday, 1999) caused this effect, while the cells adjusted to a carnosine-rich environment. After this time a comparison of the results between the supplemented and control clones at similar *in vitro* age revealed that carnosine protected against the accumulation of oxidative DNA damage, with less oxidative DNA damage in the carnosine-supplemented clones.

In another study, DNA damage induced by the salsolinol/ferritin system could be relatively favored in diseases in which iron concentrations are elevated (Kang, 2009). Kang reported that the protection of carnosine and related compounds against DNA damage indicates that these compounds may be explored as potential therapeutic agents for pathogenesis, which involve the oxidative damage of DNA mediated by the neurotoxin salsolinol.

As we have pointed out, these studies clearly demonstrates that carnosine has a positive effect on DNA damage because of its antioxidant and hydroxyl-radical-scavenger properties.

19.2.3 Protective Role of Carnosine on Chromosome Damage

Another role of carnosine is a protective role on chromosomal damage. It is known that DNA is organized into chromosomes, each of which contains a double-helical DNA structure carrying the genes. DNA breaks and aberrations in the chromosomes are caused by oxidative stress. Gille *et al.* (1991) studied the effect of various compounds as antioxidants, on the level of hyperoxia, exposure to nearly pure oxygen (90%), as a physiologically natural oxidative stressor. They showed the paradoxical effects of antioxidants on oxidative damage to chromosomes. Hyperoxia is thought to generate free radicals at the same intracellular sites where they normally form over time. They tested the ability of several antioxidants – including ascorbic acid, α-tocopherol, carnosine, imidazole-4-acetic acid, glutathione monoethylester, *N*-acetylcysteine and ethoxyquin – to protect the chromosomes in Chinese hamster ovary cells from oxidative damage. Some of the antioxidant tests acted instead as pro-oxidants: they increased chromosomal damage, aggravating the effects of hyperoxia. It is a well-known phenomenon that single antioxidant can sometimes exert a pro-oxidant effect in the body, which is the reason people take multiple antioxidants. In this study, only one antioxidant, carnosine, significantly reduced chromosomal damage. Cells cultured without any antioxidant exhibited 133 chromosomal aberrations per 100 cells. Carnosine reduced this level of damage by two-thirds, to only 44 chromosomal aberrations per 100 cells. Carnosine preserved 68% of cells fully intact, as compared to 46% of the control cells.

Some kinds of drugs such as cyclophosphamide and adriamycin, were used as cytostatic drugs in the treatment of cancers and connective tissue disease of the experimental animals can cause many side effects; in contrast, carnosine-administered ones were preserved against these toxic substances (Shao *et al.*, 2004; Mozdzan *et al.*, 2005). Carnosine's ability to react with deleterious aldehydes such as hydroxynonenal, MDA, acetaldehyde and methylglyoxal may also contribute to its protective functions (Kohen *et al.*, 1988).

In a recent study Naghshvar *et al.* (2012) investigated the protective effects of carnosine as a natural dipeptide in mouse bone marrow cells against genotoxicity induced by cyclophosphamide (CP). According to their study, they injected mice with solutions of carnosine at three different doses (10, 50 and 100 mg kg^{-1} bw) for five consecutive days. On the fifth day of treatment, they injected to mice with CP and killed them after 24 h. Nevertheless, they evaluated the frequency of micronuclei in polychromatic erythrocytes and the ratio of polychromatic erythrocyte/polychromatic erythrocyte + normochromatic erythrocyte [PCE/(PCE + NCE)] by May-Grunwald/Giemsa staining. Then they examined the histopathology of bone marrow in mice treated with CP and carnosine. They observed that the carnosine significantly reduced micronucleated polychromatic erythrocytes (MnPCEs) induced by CP at all three doses. Carnosine at a dose of 100 mg kg^{-1} bw reduced MnPCEs 3.76-fold and completely normalized the PCE/(PCE + NCE) ratio. Administration of carnosine inhibited bone-marrow toxicity induced by CP. It appeared that carnosine with protective activity reduced the oxidative stress and genotoxicity induced by CP in bone marrow cells of mice.

In this study they demonstrated that carnosine had potentially anticlastogenic and chemoprotective effects against genotoxicity in order to reduce the frequency of MnPCEs and increase proliferation on bone marrow cellularity resulted in the inhibition of micronuclei caused by CP.

Considering all this, carnosine can probably contribute with its antioxidative property to reduce genotoxicity induced by CP in bone marrow cells of mice. With regard to low toxicity related to consumption of carnosine, it can help defend the body against side effects induced by chemical hazardous agents (Naghshvar *et al.*, 2012).

In another study, Alpsoy *et al.* (2011) studied human lymphocytes cultures in order to investigate the effects of carnosine, a biological antioxidant, on the oxidative stress and genotoxicity by a single dose of carbon tetrachloride (CCl_4; 5 mM). They studied the antigenotoxic effects of carnosine by using a sister chromatid exchange (SCE) test system. Carnosine at 10 and 20 mM reduced SCE frequency in the human lymphocyte (Figures 19.2 and 19.3).

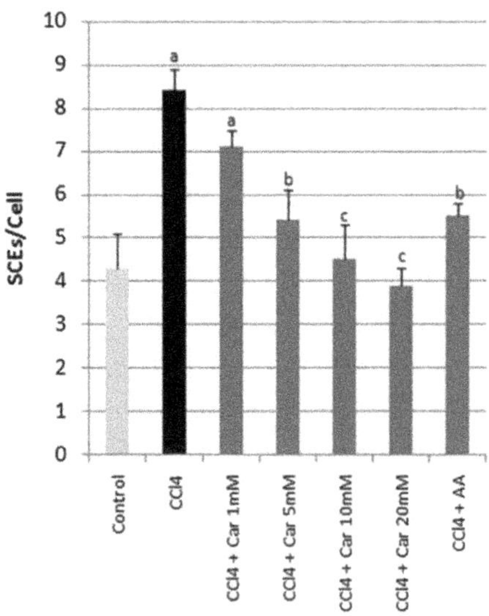

Figure 19.2 Effects of carbon tetrachloride (CCl_4) and different concentration of carnosine on the SCEs/Cell in the cultures of human peripheral lymphocytes in the study groups. The histograms show the antigenotoxic effects of carnosine that results in decreasing SCE frequency correlated with the increasing carnosine concentration. Reproduced with permission from Alpsoy *et al.* (2011). $^{a}p < 0.001$ compared with control group, $^{b}p < 0.05$ compared with CCl_4-treated group, $^{c}p < 0.001$ compared with CCl_4-treated group.

In addition, in their study CCl_4 treatment significantly depleted the level of GSH, reduced the activity of SOD and GPx and elevated the level of MDA. Carnosine treatment led to significant attenuation of CCl_4-induced oxidative stress by normalization of the activities of SOD and GPx and the level of GSH and MDA (Figure 19.5). They suggest that carnosine could provide antioxidative and antigenotoxic protection for the oxidative and genotoxic agents that cause many diseases including cancer and neurodegenerative disease.

19.3 Antigenotoxic Mechanism of Carnosine

As mentioned before, carnosine has many biological functions including metal-ion chelator, antioxidant, free-radical scavenger, antiglycating agent, and general aldehyde scavenger (Hipkiss *et al.*, 1995; Hyland *et al.*, 2000). All these properties in different portions contribute its antigenotoxic effects in the cells. Theoretically, carnosine has three broad pathways to reduce the genotoxic effects of substances. It (i) inhibits glycalization and binding advanced glycating end products (AGEs) to DNA, (ii) inhibits the production of ROS and related products (MDA), (iii) activates the antioxidant enzymes (SOD, GPx and CAT) and increases the level of antioxidant molecules (GSH) (Figure 19.4).

19.3.1 Carnosine Inhibits Glycalization

The cell metabolism produces the ROS including superoxide, hydroxyl radical and hydrogen peroxide, which oxidize nucleic acids, proteins and lipids. The oxidative damage in a cell or tissue occurs when the concentration of ROS ($O_2^{\cdot -}$, H_2O_2, and $OH^{\cdot}$) generated exceeds the antioxidant capability

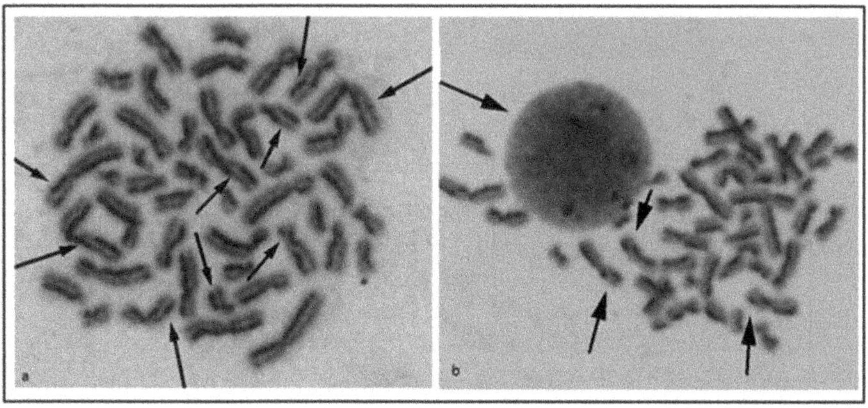

Figure 19.3 Representative examples of carbon tetrachloride (CCl_4)-treatment group (a), CCl_4 and carnosine (20 mM) treatment group (b) a cell with spontaneous sister chromatids exchanges. The image demonstrates an example of SCE assay. Reproduced with permission from Alpsoy *et al.* (2011).

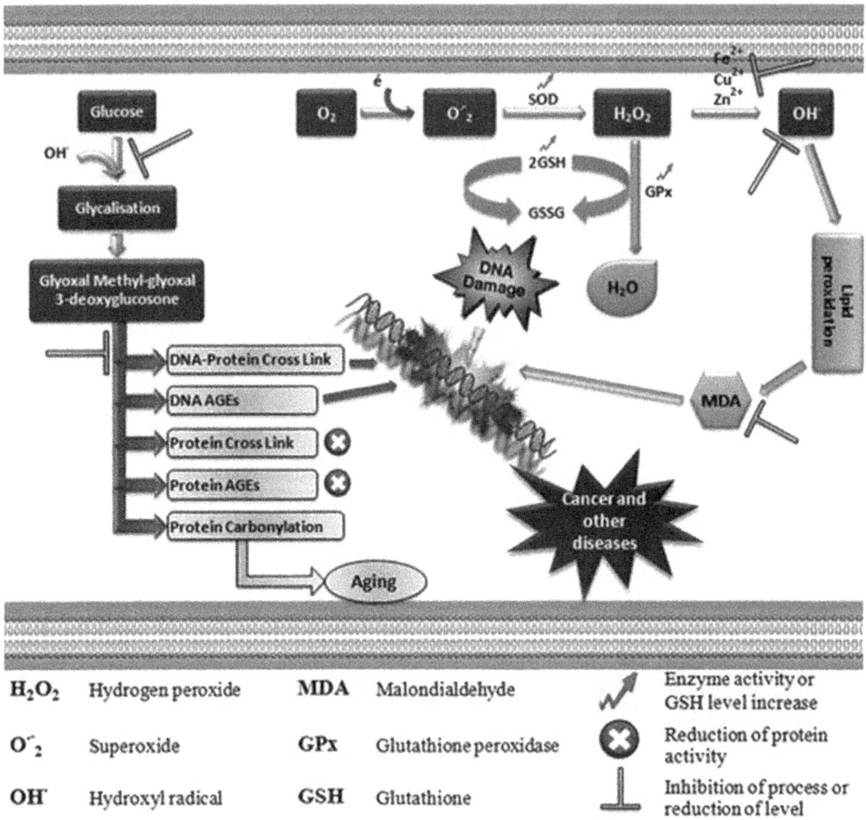

Figure 19.4 Protective mechanism of carnosine on DNA damage. This figure shows that antigenotoxic mechanism of carnosine due to inhibition of the glycalization and production of ROS and also activation of the antioxidant enzymes.

of the cell (Singh, 2006; Vandana *et al.*, 2006). During the oxidative stress the degradation of monosaccharide cause the overproduction of carbonyl species that can form stable adducts with proteins and nucleic acids. These products are known as AGEs. The AGEs induce cytotoxicity, genotoxicity, carcinogenicity and aging (Singh, 2006) (Figure 19.4).

Previous studies showed that carnosine nonenzymatically reacts with carbonyl compounds to protect macromolecules against their crosslinking actions. Carnosine reaction with proteins can termed "carnosinylation" which inhibits crosslinking of glycoxidized proteins to other macromolecules in the cells. The antiglycating, aldehyde scavenger and protective roles of carnosine result in antigenotoxic and antiaging properties (Hipkiss *et al.*, 1995; Hyland *et al.*, 2000). The formation of AGEs at the DNA level cause DNA damage and chromosome breaks, so carnosine by reducing the level of AGEs could present its antigenotoxic effects (Holliday and McFarland, 2000).

19.3.2 Carnosine has Metal-Ion-Chelator and Free-Radical-Scavenger Properties

ROS by damaging proteins cause the structural alterations in DNA and reduce DNA repair. In particular, hydroxyl radicals oxidize guanine to 8-hydroxyguanine and generate apurinic sites or strand breaks in DNA. During DNA replication 8-hydroxyguanine paired with a thymidine nucleotide can generate a changing such as G-T point mutation. Also transition metals such as Fe^{2+}, Cu^{2+} and Zn^{2+} can accelerate the level of hydroxyl radical from hydrogen peroxide (Figure 19.4). Carnosine gives an electron to decrease radical molecule damaging effects on biologic molecules. Mozdzan *et al.* (2005) in their study showed that histidine and alanine separately do not have a reducing effect on Fe^{2+} and Fe^{3+} ions, whereas carnosine has a reducing effect on these ions. In another study, it was shown that carnosine reduces the formation of free radicals and prevents lipid peroxidation in plasma of rats (Daglı *et al.*, 2005; Arzumanyan *et al.*, 2008). In addition, carnosine inhibits lipid peroxidation and formation of MDA (Guiotto *et al.*, 2007; Arzumanyan *et al.*, 2008). Alpsoy *et al.* (2011) reported that carnosine-treated groups showed a trend to a lower MDA level than a positive control group. The decrease in MDA level correlated with the increase in carnosine concentration (Figure 19.5). The most effective carnosine concentration was 20 mM on MDA level (Alpsoy *et al.*, 2011). Carnosine protects DNA against ROS and transition metals by reducing their damaging effects, so it has antigenotoxic impacts on genetic material.

19.3.3 Carnosine Activates Antioxidative Enzymes and Increases the Level of Nonenzymatic Antioxidative Compounds

The effect of ROS is balanced by the antioxidant action of nonenzymatic antioxidants such as glutathione and antioxidant enzymes including SOD, CAT and GPx. These enzymes are commonly used to eliminate ROS by cells. When the antioxidant enzymes suppress endogenous and exogenous toxic substances, ROS levels increase. Increased ROS can generate mutation in DNA and chromosomes (Valko *et al.*, 2006). Previous researches indicate that carnosine increases SOD and GPx activities and GSH (Aydin *et al.*, 2010; Alpsoy *et al.*, 2011). The effect of carnosine and/or MPTP treatment on the activity of GPX, SOD, and CAT was studied on mice. The pre-intake of carnosine dose-dependently reduced MPTP-induced GPX activity loss but only at higher concentration of carnosine significantly retained the activity of SOD and CAT (Tsai *et al.*, 2010). According to Alpsoy *et al.* decreasing SOD and GPx activities and GSH level by CCl_4 treatment has been increased by carnosine addition. This study suggested that carnosine has antigenotoxic effects on chromosomes due to its antioxidative activities (Alpsoy *et al.*, 2011).

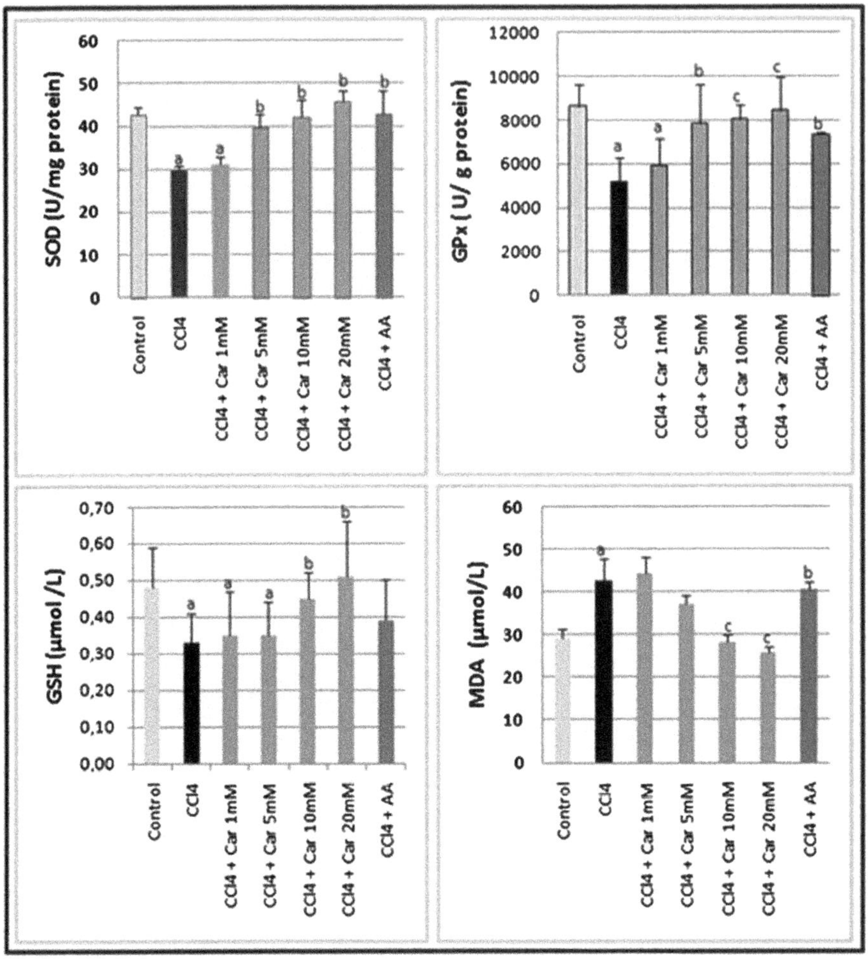

Figure 19.5 Effects of CCl_4 and different concentration of carnosine on the SOD and GPx activities and level of MDA and GSH in the cultures of human peripheral lymphocytes in the study groups. This figure depicted that activation of antioxidant enzymes and inhibition of MDA level. $^{a}p < 0.001$ compare with control group, $^{b}p < 0.05$ compare with CCl_4-treated group, $^{c}p < 0.001$ compare with CCl_4-treated group. Reproduced with permission from Alpsoy *et al.* (2011).

Summary Points

- This chapter focuses on the antigenotoxic effect of carnosine.
- Carnosine is a multifunctional dipeptide made up of a chemical combination of the amino acids β-alanine and L-histidine.
- Long-lived cells such as nerve cells (neurons) and muscle cells (myocytes) contain high levels of carnosine.
- It has an ability to inactivate reactive oxygen species, scavenges free radicals, and chelates pro-oxidative metals and also affects superoxide

dismutase and glutathione peroxidase activities and glutathione and malondialdehyde level.
- In recent studies it was shown that carnosine shields us from the onslaught of oxidation, glycation, DNA damage, and other reactions that injure tissues and cripple organs.
- However, we argue in this chapter for the antigenotoxic effects of carnosine in terms of DNA damage and chromosomal aberration and its protective mechanism.

Key Facts

1. Carnosine has antioxidative, antiaging, antigenotoxic properties against interior and exterior toxic materials.
2. As an antioxidant, carnosine potently quenches that most destructive of reactive oxygen species including hydroxyl radical, superoxide, singlet oxygen, peroxyl radicals and malondialdehyde, the product of lipid peroxidation.
3. The more reactive oxygen species and malondialdehyde levels increase, the more oxidative stress increases, causing DNA damage.
4. On the other hand, exogenic agents such as cyclophosphamide and adriamycin also cause DNA damage.
5. Antioxidant enzymes such as superoxide dismutase, glutathione peroxidase and nonenzymatic antioxidant molecules including vitamins, glutathione and carnosine protect DNA and chromosomes against oxidants, glycating agents and their reactive products due to its antioxidative and antiglycating activities.

Definitions of Words and Terms

Reactive oxygen species. ROS such as superoxide or hydroxyl radicals cause oxidative stress.

Reactive nitrogen species. RNS include nitric oxide ($NO^{\cdot}$), nitrogen peroxide ($NOO^{\cdot}$) and peroxynitrite ($ONOO^{\cdot}$) radicals, and they cause oxidative stress.

Super oxide dismutase. SOD is an enzyme that converts superoxide to hydrogen peroxide.

Glutathione peroxidase. GPx is an enzyme that converts hydrogen to water by using glutathione as a coenzyme.

Catalase. CAT is an enzyme that converts hydrogen to water.

Malondialdehyde. MDA is the reactive carbonyl end product of lipid peroxidation.

Glutathione. GSH is a tripeptide that is derived from glycine, glutamate, and cysteine. It serves as a coenzyme to glutathione peroxidase.

Cyclophosphamide. CP, a derivative of dichlorodiethylsulfide is an alkylating agent that can cause DNA and its metabolites block proliferation at several stages of the cycle.

Micronuclei. MN is a test used to identify chromosome damage in eukaryotic cells.
Sister chromatid exchange. When crossover takes place between two sister chromatids, the process is named sister chromatid exchange, it is an assay for determination of DNA breaks at the level of the chromosome.
Comet assay. Comet assay is a technique to detect DNA damage at the level of eukaryotic cell, it is also known as single-cell gel electrophoresis assay.
Genotoxicity. Genotoxicity is a property possessed by genotoxic agents and potentially can cause a heritable change in the genetic material at the level of gene or chromosome.
Advanced glycating end products. The most carbohydrates can undergo oxidative breakdown by increasing the level of hydroxyl radicals, this process produce reactive carbonyl species that can form stable adducts with proteins known as advanced glycation end products (AGEs).

List of Abbreviations

AGEs	Advanced glycating end products
CAT	Catalase
CCl_4	Carbon tetrachloride
CBEX™	Chicken breast extraction
CP	Cyclophosphamide
DNA	Deoxyribonucleic acid
GPx	Glutathione peroxidase
GSH	Glutathionc
MDA	Malondialdehyde
Mn	Micronuclei
MnPCEs	Micronucleated polychromatic erythrocytes
PAHs	Polycyclic aromatic hydrocarbons
RNS	Reactive nitrogen species
ROS	Reactive oxygen species
SCE	Sister chromatid exchange
SOD	Superoxide dismutase

References

Aloisi, A., Barca, A., Romano, A., Guerrieri, S., Storelli, C., Rinaldi, R. and Verri, T., 2013. Anti-aggregating effect of the naturally occurring dipeptide carnosine on Ab1-42 fibril formation. *PLoS One*. 8(7): e68159. doi:10.1371/journal.pone.0068159.

Alpsoy, L., Akcayoglu, G. and Sahin, H., 2011. Anti-oxidative and anti-genotoxic effects of carnosine on human lymphocyte culture. *Human & Experimental Toxicology*. 30(12): 1979–1985.

Arzumanyan, E. S., Makhro, A. V., Tyulina, O. V. and Boldyrev, A. A., 2008. Carnosine protects erythrocytes from the oxidative stress caused by homocysteic acid. *Doklady Biochemistry and Biophysics*. 418: 44–46.

Aydin, A. F., Kusku-Kiraz, Z., Dogru-Abbasoglu, S. and Uysal, M., 2010. Effect of carnosine treatment on oxidative stress in serum, apoB-containing lipoproteins fraction and erythrocytes of aged rats. *Pharmacological Reports*. 62: 733–739.

Babizhayev, M. A., Seguin, M. C., Gueyne, J., Evstigneeva, R. P., Ageyeva, E. A. and Zheltukhina, G. A., 1994. L-Carnosine (beta-alanyl-L-histidine) and carcinine (beta-alanylhistamine) act as natural antioxidants with hydroxyl-radical-scavenging and lipid-peroxidase activities. *Biochemical Journal*. 304(Pt 2): 509–516.

Baguet, A., Bourgois, J., Vanhee, L., Achten, E. and Derave, W., 2010. Important role of muscle carnosine in rowing performance. *Journal of Applied Physiology*. 109(4): 1096–1101.

Bellia, F., Vecchio, G., Cuzzocrea, S., Calabrese, V. and Rizzarelli, E., 2011. Neuroprotective features of carnosine in oxidative driven diseases. *Molecular Aspects of Medicine*. 32(4–6): 258–266.

Boldyrev, A. A., Dupin, A. M., Pindel, E. V. and Severin, S. E., 1988. Antioxidative properties of histidine-containing dipeptides from skeletal muscles of vertebrates. *Comparative Biochemistry and Physiology, Part B: Biochemistry & Molecular Biology*. 89: 245–250.

Boldyrev, A. A., Gallant, S. C. and Sukhich, G. T., 1999. Carnosine, the protective, anti-aging peptide. *Bioscience Reports*. 19(6): 581–587.

Budzen, S. and Rymaszewska, J., 2013. The biological role of carnosine and its possible applications in medicine. *Advances in Clinical and Experimental Medicine*. 22(5): 739–744.

Calabrese, V., Cornelius, C., Cuzzocrea, S., Iavicoli, I., Rizzarelli, E. and Calabrese, E. J., 2011. Hormesis, cellular stress response and vitagenes as critical determinants in aging and longevity. *Molecular Aspects of Medicine*. 32(4–6): 279–304.

Caruso, J., Charles, J., Unruh, K., Giebel, R., Learmonth, L. and Potter, W., 2012. Ergogenic effects of β-alanine and carnosine: proposed future research to quantify their efficacy. *Nutrients*. 4(7): 585–601.

Daglı, F., Yazıcı, C. and Köse, K., 2005. Investigation of antioxidant properties of carnosine in rats treated with methlglyoxal. *Journal of Health Sciences*. 14: 198–204.

Drozak, J., Chrobok, L., Poleszak, O., Jagielski, A. K. and Derlacz, R., 2013. Molecular identification of carnosine *N*-methyltransferase as chicken histamine *N*-methyltransferase-like protein (HNMT-Like). *PLoS One*. 8(5): e64805.

Fu, H., Katsumura, Y., Lin, M., Muroya, Y., Hata, K., Fujii, K., Yokoya, A. and Hatano, Y., 2009. Free radical scavenging and radioprotective effects of carnosine and anserine. *Radiation Physics and Chemistry*. 78: 1192–1197.

Gallant, S., Semyonova, M. and Yuneva, M., 2000. Carnosine as a potential anti-senescence drug. *Biochemistry (Mosc)*. 65(7): 866–868.

Gille, J. J., Pasman, P., van Berkel, C. G. and Joenje, H., 1991. Effect of antioxidants on hyperoxia-induced chromosomal breakage in Chinese hamster ovary cells: protection by carnosine. *Mutagenesis*. 6: 313–318.

Guiotto, A., Ruzza, P., Babizhayev, M. A. and Calderan, A., 2007. Malondialdehyde scavenging and aldose-derived Schiff bases' transglycation properties of synthetic histidyl-hydrazide carnosine analogs. *Bioorganic & Medicinal Chemistry*. 15: 6158–6163.

Harris, R. C., Wise, J. A., Price, K. A., Kim, H. J., Kim, C. K. and Sale, C., 2012. Determinants of muscle carnosine content. *Amino Acids*. 43: 5–12.

Hipkiss, A. R., 1998. Carnosine, a protective, anti-ageing peptide?. *International Journal of Biochemistry & Cell Biology*. 30: 863–868.

Hipkiss, A. R., Michaelis, J. and Syrris, P., 1995. Non-enzymic glycosylation of the dipeptide L-carnosine, a potential anti-protein-cross-linking agent. *FEBS Letters*. 371: 81–85.

Holliday, R. and McFarland, G. A., 2000. A role for carnosine in cellular maintenance. *Biochemistry (Moscow)*. 65(7): 843–848.

Hsieh, C. L., Ho, Y. C., Huailai, H. and Yen, G. C., 2002. Inhibitory effect of carnosine and anserine on DNA oxidative damage induced by Fe^{2+}, Cu^{2+} and H_2O_2 in lymphocytes. *Journal of Food and Drug Analysis*. 10(1): 47–54.

Hyland, P., Duggan, O., Hipkiss, A., Barnett, C. and Barnett, Y., 2000. The effects of carnosine on oxidative DNA damage levels and in vitro lifespan in human peripheral blood derived CD4_T cell clones. *Mechanisms of Ageing and Development*. 121: 203–215.

International Agency for Research on Cancer (IARC), 1999. Monographs on the evaluation of carcinogenic risks to humans, the re-evaluation of some organic chemicals, hydrazine and hydrogen peroxide. Available at: http://monographs.iarc.fr/ENG/Monographs/vol71/mono71.pdf. Accessed 11 February 2014.

Kang, J. H., 2009. Ferritin enhances salsolinol-mediated DNA strand breakage: Protection by carnosine and related compounds. *Toxicology Letters*. 188: 20–25.

Kang, J. H., 2010. Protective effects of carnosine and homocarnosine on ferritin and hydrogen peroxide-mediated DNA damage. *BMB Reports*. 43: 683–687.

Kasai, H., Crain, P., Kuchino, Y., Nishimura, S., Ootsuyama, A. and Tanooka, H., 1986. Formation of 8-hydroxyguanidine moiety in cellular DNA by agents producing oxygen radicals and evidence for its repair. *Carcinogenesis*. 7: 1849–1851.

King, R. J. B. and Robins, M. W., 2006. *Cancer Biology*. Pearson Education Limited, Harlow, England, pp. 88–111.

Kohen, R., Yamamoto, Y., Cundy, K. C. and Ames, B. N., 1988. Antioxidant activity of carnosine, homocarnosine, and anserine present in muscle and brain. *Proceedings of the National Academy of Sciences of the United States of America*. 85: 3175–3179.

Lindberg, P., Brandstrom, A., Wallmark, B., Mattson, H., Rikner, L. and Hoffman, K. J., 1990. Omeprazole: the first proton pump inhibitor. *Medicinal Research Reviews*. 10: 1–54.

Matthews, M. M. and Traut, T. W., 1987. Regulation of *N*-carbamoyl-beta-alanine amidohydrolase, the terminal enzyme in pyrimidine catabolism, by ligand-induced change in polymerization. *Journal of Biological Chemistry*. 262: 7232–7237.

McFarland, G. A. and Holliday, R., 1999. Further evidence for the rejuvenating effects of the dipeptide L-carnosine on cultured human diploid fibroblasts. *Experimental Gerontology*. 34: 35–45.

Mozdzan, M., Szemraj, J., Rysz, J. and Nowak, D., 2005. Antioxidant properties of carnosine re-evaluated with oxidizing systems involving iron and copper ions. *Basic & Clinical Pharmacology & Toxicology*. 96: 352–360.

Naghshvar, F., Abianeh, S. M., Ahmadashrafi, S. and Hosseinimehr, S. J., 2012. Chemoprotective effects of carnosine against genotoxicity induced by cyclophosphamide in mice bone marrow cells. *Cell Biochemistry and Function*. 30: 569–573.

Neurohr, K. J. and Mantsch, H. H., 1980. Complex formation of carnosine with purine nucleotides in aqueous solution. *Zeitschrift fuer Naturforschung, C: Journal of Biosciences*. 35: 557–561.

Perera, F. A., 1984. The genotoxic/epigenetic distinction: relevance to cancer policy. *Environmental Research*. 34: 175–191.

Powers, R. E., Lawton, G. P. and Modlin, I. M., 1995. Genotoxicity, carcinogenicity and acid-suppressing medications. *Pharmacology & Therapeutics*. 65: 303–317.

Quinn, P. J., Boldyrev, A. A. and Formazuyk, V. E., 1992. Carnosine: its properties, functions and potential therapeutic applications. *Molecular Aspects of Medicine*. 13: 379–444.

Renner, C., Seyffarth, A., de Arriba, S., Meixensberger, J., Gebhardt, R. and Gaunitz, F., 2008. Carnosine inhibits growth of cells isolated from human glioblastoma multiformec. *International Journal of Peptide Research and Therapeutics*. 14: 127–135.

Sato, M., Karasawa, N., Shimizu, M., Morimatsu, F. and Yamada, R., 2008. Safety evaluation of chicken breast extract containing carnosine and anserine. *Food and Chemical Toxicology*. 46: 480–489.

Shao, L., Li, Q. H. and Tan, Z., 2004. L-Carnosine reduces telomere damage and shortening rate in cultured normal fibroblasts. *Biochemical and Biophysical Research Communications*. 324: 931–936.

Singh, K. K., 2006. *Oxidative Stress, Disease and Cancer*. Imperial College Press, London, pp. 333–356.

Tsai, S. J., Kuo, W. W., Liu, W. H. and Yin, M. C., 2010. Antioxidative and anti-inflammatory protection from carnosine in the striatum of MPTP-treated mice. *Journal of Agricultural and Food Chemistry*. 58: 11510–11516.

Valko, M., Rhodes, C. J., Moncol, J., Izakovic, M. and Mazur, M., 2006. Free radicals, metals and antioxidants in oxidative stress-induced cancer. *Chemico-Biological Interactions*. 160: 1–40.

Vandana, S., Ram, S., Ilavazhagan, M., Kumar, G. D. and Banerjee, P. K., 2006. Comparative cytoprotective activity of vitamin C, E and beta-carotene against chromium induced oxidative stress in murine macrophages. *Biomedicine & Pharmacotherapy*. 60: 71–76.

Yanai, N., Shiotani, S., Hagiwara, S., Nabetani, H. and Nakajima, M., 2008. Antioxidant combination inhibits reactive oxygen species mediated damage. *Bioscience, Biotechnology, and Biochemistry*. 72: 3100–3106.

CHAPTER 20

Carnosine and Cancer

FRANK GAUNITZ*[a], HENRY OPPERMANN[a], AND ALAN HIPKISS[b]

[a]Klinik und Poliklinik für Neurochirurgie, Universitätsklinikum Leipzig AöR, Liebigstraße 20, 04103 Leipzig, Germany; [b]Aston Research Centre for Healthy Ageing (ARCHA), School of Health and Life Sciences, Aston University, Birmingham B4 7ET, UK
*E-mail: Frank.Gaunitz@medizin.uni-leipzig.de

20.1 Carnosine and Cancer – Experimental Evidence

The first experimental evidence that carnosine can inhibit the growth of cancer cells in an animal model was presented by Nagai and Suda in 1986 (Nagai and Suda, 1986). These authors subcutaneously implanted tumor cells into mice and administered carnosine by subcutaneous injection the day after implantation. This treatment was continued every second day, and tumor growth and survival was compared to animals treated with saline. In these experiments the authors demonstrated reduced tumor growth and prolonged survival of animals treated with carnosine. Ten years later Holliday and McFarland presented data from experiments that demonstrated that carnosine selectively inhibits the growth of transformed and neoplastic cells in culture (Holliday and McFarland, 1996). Renner *et al.*, again almost 10 years later, performed experiments with cells isolated from human glioblastoma in which they confirmed that growth of these tumor cells can be

Food and Nutritional Components in Focus No. 8
Imidazole Dipeptides: Chemistry, Analysis, Function and Effects
Edited by Victor R Preedy

Published by the Royal Society of Chemistry, www.rsc.org

inhibited by carnosine (Renner *et al.*, 2008). In 2010 this group also demonstrated reduced tumor cell growth in a nude mice model with subcutaneously implanted cells that expressed the human epidermal growth factor receptor 2 (Renner *et al.*, 2010b). As in the experiments of Nagai and Suda, tumor growth was significantly reduced by a daily intraperitoneal injection of carnosine, and the tumors were less pleomorphic and exhibited a reduced number of mitotic figures. Experiments with human colon cancer cells transplanted into athymic nude mice that received carnosine with the drinking water (Horii *et al.*, 2012) further confirmed the antiproliferative potential of carnosine even if it is given orally.

Recently, cell-culture experiments performed with different cells such as pheochromocytoma cells (Rybakova and Boldyrev, 2012), human colon cancer cells (Iovine *et al.*, 2011) and rat mesangial cells (Jia *et al.*, 2009) confirmed the antiproliferative effects, and the possibility to use carnosine as an anticancer agent has been discussed (Gaunitz and Hipkiss, 2012), as well as some possible mechanisms responsible for the antineoplastic effects (Hipkiss and Gaunitz, 2014).

20.2 Possible Mechanisms for Carnosine's Effects on Cancer

In the following sections we will discuss some aspects that may explain the possible mechanisms by which carnosine may influence tumor cell growth and survival with regard to its known and conceivable effects on cell signaling and metabolism.

20.2.1 Gene Regulation and Cell Signaling

Tumor cells exhibit a wide range of aberrant signaling that frequently occurs in pathways that are important for normal development and differentiation. During the last years it became obvious that carnosine may be able to influence signaling in pathways that are in a complex manner interrelated and crosstalk. Although it is beyond the scope of this chapter to discuss these pathways in detail, we will focus on known experimental data on carnosine's effects and their plausible physiological significance.

20.2.1.1 MAPK, PI3K and mTOR Signaling

In this section we will outline observations that point towards the possibility that carnosine may influence mTOR signaling (mammalian target of rapamycin signaling). Deregulation of different components of the mTOR pathway have been reported in many types of cancer (for a review see (Pópulo *et al.*, 2012)) and there are some parallels between carnosine and inhibitory effects on mTOR signaling with regard to their physiological effects in cells and organisms. The idea that carnosine may in fact be a mimic of rapamycin, which is an inhibitor of mTOR signaling, has been suggested before by Hipkiss (2011).

mTOR is a conserved serine/threonine kinase that exists in two biochemically and functionally distinct multicomponent complexes designated mTORC1 and mTORC2 of which only mTORC1 is sensitive to rapamycin. mTORC1 is a downstream regulator of tyrosine kinase receptors in the signaling pathways of phosphatidylinositol-3-kinase (PI3K) and of mitogen-activated protein kinases (MAPKs), whereas mTORC2 is located upstream of the serine–threonine protein kinase AKT (commonly referred to as protein kinase B or PKB) (Figure 20.1). The PI3K and the MAPK pathway play important roles in normal physiological regulation and are also fundamental in the metabolism, growth and survival of several human cancers. Activated PI3K acts *via* phosphorylation of phosphatidylinositol-4,5-phosphate ($PI(4,5)P_2$) to produce phosphoinositol-3,4,5-phosphate ($PI(3,4,5)P_3$), which leads to the activation of the serine/threonine-specific protein kinase AKT and finally to

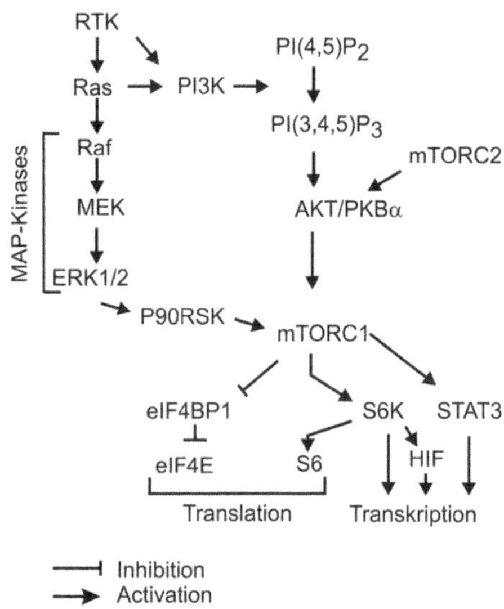

Figure 20.1 Signal transduction pathways upstream and downstream from the mammalian target of rapamycin complex 1. This figure illustrates two major pathways leading to the activation of the mammalian target of rapamycin complex 1 (mTORC1). One pathway signals *via* mitogen-activated kinases (MAP-kinases: Raf, MEK and ERK) and the p90 ribosomal S6 kinase (P90RSK) on mTORC1. The other pathway is the phosphatidylinositol 3-kinase (PI3K) pathway. RTK: receptor tyrosine kinase; Ras (rat sarcoma G-protein); $PI(4,5)P_2$: phosphatidylinositol-4,5-phosphate; $PI(3,4,5)P_3$: phosphoinositol-3,4,5-phosphate; AKT/PKBα: Ak strain transforming/protein kinase Bα; eIF4BP1: eukaryotic initiation factor 4E binding protein 1; eIF4E: eukaryotic initiation factor 4E; S6K: ribosomal protein S6 kinases; S6: ribosomal protein S6; STAT3: signal transducer and mediator of transcription 3; HIF: hypoxia-inducible factor.

the activation of mTORC1. The main components of the MAPK pathway in humans are Raf, MEK and ERK. ERKs are known to activate many transcription factors and can also lead to the activation of mTORC1 *via* p90 ribosomal S6 kinase (P90RSK). Activation of mTORC1 has a great number of effects since it activates several specific transcription factors that can promote growth and proliferation by promoting protein synthesis, lipid biogenesis and metabolism. Active mTORC1 does also inhibit lysosomal biogenesis, reduces autophagy and enhances mitochondrial biogenesis.

Among the targets that can be phosphorylated by mTORC1 are eukaryotic initiation factor 4E binding protein 1 (eIF4BP1) and the ribosomal protein S6 kinases 1 and 2 (S6K1/2) that are of interest with regard to tumor biology. Upon phosphorylation of eIF4BP1 by mTORC1 this protein releases the cap-binding protein eIF4E (eukaryotic initiation factor 4E) thereby initiating cap-dependent translation. Since cancer cells are dependent on a continuous high rate of protein synthesis for cell growth and proliferation the activation of eIF4E *via* mTORC1 and eIF4BP1 is considered to support this need. The first identified target of S6 kinases was the ribosomal protein S6, which is a component of the 40S ribosomal subunit. In addition to influencing translation and protein synthesis it has to be noted that protein 6 kinases have a number of other targets, and control fundamental cellular processes such as transcription, glucose homoeostasis, insulin sensitivity, adipocyte metabolism, body mass and energy balance, tissue and organ size, learning, memory and aging.

The idea that carnosine may be a mimic of rapamycin comes from a number of observations demonstrating that both compounds exert comparable effects on cells and organisms. For example, carnosine and mTOR signaling have comparable effects on lifespan, and there are parallels with regard to effects on tumor growth, glycolytic activity, proteolysis and β-amyloid accumulation in mice (reviewed in (Hipkiss, 2011)).

Although, there is yet no confirmation that carnosine may directly influence mTORs there is experimental evidence showing that carnosine affects mTOR signaling: Son *et al.* reported a suppression of IL-8 production in Caco-2 cells under the influence of carnosine at the post-transcriptional level (Son *et al.*, 2008). The reduced production and secretion of IL-8 in carnosine-treated cells was accompanied by a significant reduction of the hydrogen peroxide stimulated phosphorylation of eIF4E, AKT, ERK1/2 and p38 MAP kinase. Since Son *et al.* found a suppression of IL-8 secretion not only in cells stimulated with hydrogen peroxide (2 mM) but also in cells stimulated with TNF-α, it was thought unlikely that carnosine's action on secretion may simply be interpreted by its effect on free-radical species.

20.2.1.2 HIF Signaling

Hypoxia-inducible factors (HIFs) are activated under conditions of oxygen deprivation and are in turn responsible for the activation of different genes that enable the cell to cope with the reduced supply of oxygen and nutrients. The main cellular components of HIF signaling are the constitutively expressed subunits HIF-1α and HIF-1β. Under normoxic conditions HIF-1α is continuously hydroxylated by oxygen-dependent prolyl hydroxylases (PHD) which promotes the interaction between Von Hippel–Lindau (VHL) protein and HIF1-α, which is followed by ubiquitinylation and degradation of HIF1-α (for a more detailed discussion of the pathway see (Cavadas *et al.*, 2013)). Under hypoxic conditions degradation is halted and HIF-1α translocates to the nucleus where it dimerizes with HIF-1β. Together with the transcriptional coactivator p300/CBP this complex binds to hypoxia-responsive elements (HREs) on different target genes and induces their expression.

Among the more than 200 genes known to be affected are vascular endothelial growth factors (VEGFs), erythropoietin, several genes encoding glycolytic enzymes and the glucose transporter-1 (GLUT1) (for a review see (Semenza, 2012)). In general, HIF signaling up-regulates anaerobic glycolysis, the conversion of pyruvate to lactate and inhibits pyruvate dehydrogenase (PDH). HIF overexpression is seen in many types of cancer and there is a high degree of similarity between HIF target genes and those upregulated in cancerous tissues. Since carnosine reduces glycolytic ATP production (Renner *et al.*, 2010a), which is stimulated by HIF signaling, the analysis of a possible interaction between carnosine and the HIF pathway appears to be rational.

20.2.1.3 STAT3 Signaling

The signal transducer and activator of transcription 3 (STAT3) regulates physiological functions in normal and malignant tissues, such as proliferation, differentiation, survival, migration, invasion and angiogenesis. Since constitutive activation has been observed in a wide range of human cancers, STAT3 inhibitors are considered as potential therapeutics for the treatment of solid tumors. STAT3 is a cytoplasmic transcription factor that is phosphorylated upon stimulation of Janus kinase (JAK). Phosphorylation finally results in dimerization, nuclear translocation and binding to DNA of different target genes. As Yokogami *et al.* have demonstrated, phosphorylation, which promotes transcriptional activity of STAT3, can be inhibited by rapamycin's action on mTORC1 (Yokogami *et al.*, 2000). Therefore, an influence of carnosine on mTOR signaling may also affect STAT3 signaling.

20.3 Cell Cycle

Experimental evidence that carnosine can promote cell-cycle arrest in tumor cells has been demonstrated by several groups: Iovine *et al.* (2011) observed cell-cycle arrest in G1 phase in HCT116 cells in the presence of 50 and 100 mM carnosine and Jia *et al.* (2009) reported arrest between G1 and S phase in high glucose-induced rat mesangial cells at a carnosine concentration of 20 mM. Although Rybakowa and Boldyrev also observed an inhibition of cell-cycle progression in PC-12 pheochromocytoma cells at a concentration of 50 mM carnosine, they found an accumulation of cells in S- and G2/-phase (Rybakova and Boldyrev, 2012). In the experiments of Iovine *et al.* (2011) G1 arrest was associated with ERK1/2 phosphorylation.

20.4 Apoptosis

It is generally accepted that proapoptotic stimuli have antineoplastic effects, and tumor cell growth is frequently accompanied by effective mechanisms by which tumor cells prevent the induction of apoptosis (for a review see Fulda (2013)). However, at least in nontumor cells carnosine exhibits antiapoptotic effects instead of inducing apoptosis (Zhang *et al.*, 2011). In addition, neither the experiments of Renner *et al.* (2008) and Jia *et al.* (2009) (aside from late apoptosis in only a small number of cells) nor those of Iovine *et al.* (2011) point towards an induction of apoptosis in tumor cell culture experiments. Up to now, there is only one report in which carnosine prevented tumor cell growth *via* apoptosis (Zhang *et al.*, 2005), and it cannot be ruled out that pro- or antiapoptotic effects are dependent on which signaling pathways are active in the cells studied.

20.5 Metastasis

There is no information on whether carnosine inhibits metastasis in animal tumor models (Nagai and Suda, 1986; Renner *et al.*, 2010b; Horii *et al.*, 2012), but Chuang and Hu demonstrated that in cell-culture experiments carnosine inhibited both cell migration and invasion of the highly invasive hepatocarcinoma cell line SK-Hep-1 (Chuang and Hu, 2008). These authors also found that the inhibitory effects were associated with a decreased expression of matrix metalloproteinase 9 (MMP-9), a protein involved in tumor invasion.

20.6 Glycolysis

Tumor cells are highly dependent on glycolysis even in the presence of sufficient oxygen. This phenomenon, which is known as the Warburg effect, has been critically reviewed in recent years and was greatly resuscitated in the field of cancer research (for a review see Bensinger and Christofk (2012)). Although, many theories about the mechanisms driving the Warburg effect have been critically re-evaluated in recent years, it is still without doubt that most anabolic processes required for accelerated tumor growth are accomplished by increased glycolysis, which is observed in many tumors. Because of the great importance of glycolysis for tumor cell growth, glucose deprivation or the inhibition of glycolytic activity is considered as a strategy towards new therapies. Since the antiproliferative effect of carnosine is accompanied by a reduced production of ATP in cells derived from glioblastoma, that did not exhibit any contribution of oxidative phosphorylation to ATP production (Renner *et al.*, 2010a), the effect of carnosine on the metabolic flux through glycolysis may be regarded as an important aspect of its antitumor activity.

20.6.1 Regulation of Glycolysis

The control of glycolysis is mediated by complex signaling mechanisms, involving different signal transduction pathways, regulators of enzyme activity and feedback loops, by which metabolites in turn control and regulate enzymatic activities or even growth factor receptors and their associated pathways. Therefore, a lot of possible influences of carnosine on glycolytic flux are imaginable and we will focus on possible mechanisms for which there is experimental evidence.

20.6.1.1 Transcription and Signaling

As discussed above, different pathways appear to be affected by carnosine that in turn can affect the regulation of glycolysis. Among the targets of HIF signaling are the glucose transporters 1 and 3 (GLUT1/GLUT3), hexokinase 2, glucose-6-phosphate isomerase, phosphofructokinase 1, aldolase, triose-phosphate isomerase, glyceraldehyde-3-phosphate dehydrogenase, phosphoglycerate kinase, phosphoglycerate mutase, enolase, pyruvate kinase and 6-phospho-2-kinase/fructose-2,6 bisphosphatase (Denko, 2008). Since HIF can be regulated by mTORC1 independently of hypoxia, an influence of carnosine on mTORC1 may cause expressional alterations of the aforementioned genes. STAT3 signaling, on the other hand, was shown to suppress key gluconeogenic enzymes (Wang *et al.*, 2012), which may indicate that suppression of STAT3 activity may result in upregulation of enzymes that catalyze reactions in the opposite direction of glycolytic degradation of glucose to pyruvate and lactate. However, experiments with three cell lines derived from glioblastoma did not show any unequivocal influence on mRNA expression of the aforementioned enzymes involved in glycolysis under the influence of carnosine aside from a strongly enhanced mRNA expression of pyruvate dehydrogenase kinase 4 (PDK4) and hexokinase 4 (Letzien *et al.*, 2014).

20.6.1.2 Enzymes

Some enzymes have been shown to be directly activated by carnosine such as purified (or at least cell-free) glycogen phosphorylase (Severin *et al.*, 1990). This observation points towards the possibility that carnosine can affect enzymatic activity at a post-translational level, independent from indirect effects on other modulatory factors such as reactive oxygen species or methylglyoxal-induced protein glycation. However, nothing is known of whether such influences may also occur *in vivo* and may involve enzymes important for glycolysis.

20.6.1.3 *Possible Interactions of Carnosine with Glycolytic Intermediates*

Carnosine has been shown to react nonenzymatically with glucose and fructose, and presumably with their phosphorylated derivatives and also with galactose, deoxyribose and the triose dihydroxyacetone (Hipkiss, 2009). To our knowledge, there have been no reports that these carnosine adducts are present in tissue, but it has been shown that carnosine is able to form stable carnosine–aldehyde conjugates with different unsaturated aldehydes generated during peroxidation of lipids (Liu *et al.*, 2003). Therefore, it cannot be ruled out that carnosine may react with glycolytic intermediates, thus preventing the flux of glucose through glycolysis.

20.7 Carnosine and Induction of PDK4 mRNA Expression

Just recently, carnosine was shown to induce the expression of mRNA encoding the pyruvate dehydrogenase kinase 4 (PDK4) in glioblastoma tumor cell lines (Letzien *et al.*, 2014). Since the regulation of the expression of this enzyme has been studied for many years, this new observation opens a window for the further investigation of possible molecular targets of carnosine (for a detailed review on PDK4 mRNA regulation see Jeong *et al.* (2012)).

Dependent on the cell type and tissue studied, the expression of the human PDK4 gene (hPDK4) is known to be regulated by different signal transducers, transcription factors and epigenetic mechanisms. In general, its expression is low under a carbohydrate-rich diet but strongly increased by a high-fat diet, in starvation, during and after different forms of exercise or by hibernation and in diabetes.

Among the factors known to interact with specific elements within the ~850 bp long 5′-upstream region of the human PDK4 gene are the activated glucocorticoid receptor (GR) and the forkhead box proteins FOXO1 and FOXO3a (Figure 20.2). Furthermore, there are two retinoic acid response elements (RAREs) for the activated retinoid/retinoic receptors (RXRα/RARα), a binding site for the transcription factor Sp1, and a binding site for p300/CBP. In addition, a candidate monomeric ERR (estrogen-related receptor) response element was identified. Experiments with different human and animal models also revealed that the peroxisome proliferator-activated receptors (PPARs) can modulate transcription of PDK4, with PPAR-γ functioning as an inhibitor and PPAR-δ and PPAR-α being positive regulators. At this point, it is interesting to note that PPARs can be regulated by mTORC1 that activates PPARγ but inhibits PPARα (see Figure 20.2). The influence of PPARs on expression is supposed to be at least partially caused by interaction of PPARs with RXRα/RARα and ERRα in a cell and species specific manner, and in some models it requires the presence of the peroxisome proliferator-activated receptor γ coactivator 1α (PGC-1α), which may also directly cooperate with ERRα or

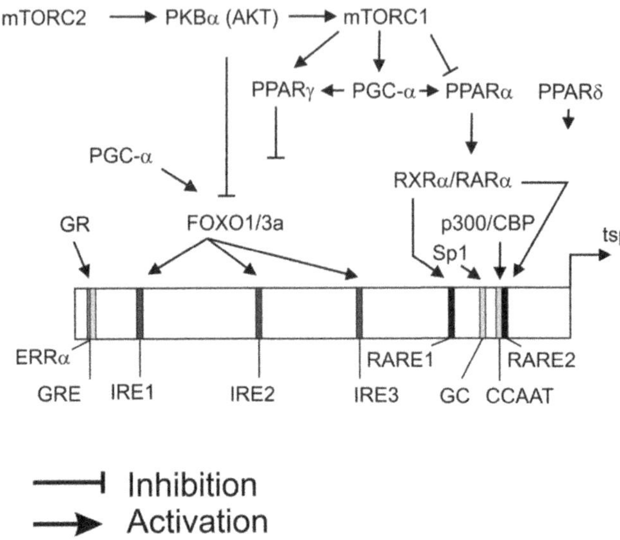

Figure 20.2 Regulation of the human pyruvate dehydrogenase kinase 4 gene. This figure shows an 850-bp long region 5′-upstream from the transcriptional start point (tsp) of the human pyruvate dehydrogenase kinase (hPDK4) gene with transcription factor binding sites and the factors interacting with them. In addition, upstream regulators are depicted that are known to influence transcription of hPDK4. IRE1/2/3: insulin responsive elements 1/2 and 3; FOXO1 and FOXO3a: forkhead transcription factors 1 and 3a; GRE: glucocorticoid receptor element; GR: activated glucocorticoid receptor; ERRα: estrogen-related receptor α; RARE1/2: retinoic acid response element 1 and 2; RXRα/RARα: retinoid X receptor/retinoic acid receptor; GC: GC-box; SP1: specificity protein 1; CCAAT: CCAAT-box; p300/CBP: Adenovirus early region 1A binding protein 300/cAMP-response element-binding protein (CREB)-binding protein; mTORC1/2: mammalian target of rapamycin 1/2; PKBα (AKT): protein kinase Bα (Ak strain transforming); PPARα, γ, δ: peroxisome proliferator activated receptor α, γ, δ; PGC-1α: peroxisome proliferator-activated receptor γ coactivator 1α.

FOXOs. PGC-1α on the other hand, can also induce expression of ERRα. However, the role of PGC-1α is difficult to estimate since it is supposed to interact with different mediators aside from PPARγ, including PPARα and FOXOs.

Since the aforementioned factors are regulated by different pathways it will be important to investigate whether carnosine influences phosphorylation or dephosphorylation of major components of these pathways or their expression. One of the pathways most likely to be influenced appears to be the phosphoinositol-3-kinase pathway (PI3K) acting *via* protein kinase B (PKB/AKT) on FOXOs, since it has been demonstrated that a constitutively active PKBα completely represses PDK4 promoter activity stimulated by glucocorticoids through the inactivation of these factors. At this point it should also be noted that the mammalian FOXO proteins are orthologs of the *Caenorhabditis elegans* protein DAF-16 that is involved in the regulation of aging, thereby pointing towards the possibility that the known effects of carnosine on aging may be linked to an influence on FOXO activity. It can also not be ruled out that carnosine acts upstream of PI3K *via* the MAP kinase pathway (see Figure 20.1) since it has been described that Her2/neu (human epidermal growth factor receptor 2/neuroblastoma) activity can suppress PDK4 expression in an ERK-dependent manner. Since ERK signals *via* P90RSK also on the mTORC1, which in the activated state inhibits PPARα and activates PGC-1α and PPARγ, the observation of enhanced PDK4 mRNA expression may be mediated *via* inactivation of mTORC1 followed by loss of PPARγ inhibiting function, and at the same time by activation of PPARα, which is in agreement with the discussion about a possible role of carnosine as a mimic of rapamycin (compare with Figure 20.1).

With regard to possible epigenetic mechanisms involved in the regulation of PDK4 mRNA expression, it is important to note that the hPDK4 promoter exhibits a marked hypomethylation in healthy sedentary men and women after acute exercise. In addition, PDK4 gene expression was shown to be regulated by triochostatin A (TSA), an inhibitor of histone deacetylase, what may be linked to the observation that FOXOs and activated RXRα/RARα interact cooperatively with the acetyltransferase p300/CBP, whereas unliganded RAR recruits a histone deacetylase complex.

20.8 Carnosine as a Drug for Cancer Treatment

Whether carnosine may efficiently be used for therapeutic purposes is a question that has to be addressed from different points of view, including its bioavailability depending on the route of administration, and its uptake by the tumor cells to be targeted. In addition, the mode of administration and the dosage of carnosine have to be taken into account as well as its degradation by serum carnosinase in the blood or by tissue carnosinase in the tumor cells. In addition, possible side effects have to be considered.

20.8.1 Bioavailability and Mode of Administration

To be considered as a potential antineoplastic drug for the treatment of cancer patients, the central question is how carnosine must be administered, especially with regard to serum carnosinase activity, which may degrade carnosine to β-alanine and L-histidine if it is passaged in the blood. In the case of brain tumors that have been surgically removed, one may consider to apply a depot of carnosine into the excavation cavity from which carnosine is constantly released in order to prevent recurrence of the tumor, which is usually the case in glioblastoma patients. Another possibility might be an administration *via* the nasal cavity that will also bypass the blood–brain barrier *via* the olfactory and trigeminal nerves, and may be enhanced by use of nanoparticles.

20.8.2 Transport and Delivery

Considering an oral administration of carnosine, the first event to be considered is intestinal transport. Upon oral administration carnosine is absorbed through the intestine (Bauchart *et al.*, 2007), and at least a part of it is not degraded by absorption (Gardner *et al.*, 1991). Since Gardner *et al.* measured intact urinary carnosine over a period of 5 h after ingestion of 4 g carnosine in all of their 6 subjects tested (Gardner *et al.*, 1991), it has to be assumed that orally ingested carnosine may also in humans reach the tissues, provided it can leave the blood stream. We do not know which of the four described transporters that can transport carnosine are responsible for the final transport through endothelia and final delivery of carnosine into target cells, but the animal experiments performed by several groups indicate that this transport occurs in animals. A further indication for successful delivery are experiments that demonstrate that carnosine has beneficial effects in behavior and/or cognitive function in patients treated for Gulf War Illness (Baraniuk *et al.*, 2013), schizophrenia (Chengappa *et al.*, 2012) and for autistic disorders (Chez *et al.*, 2002) and Parkinson's disease (Boldyrev *et al.*, 2008).

20.8.3 Carnosinase Activity and Plasma Levels of Carnosine

The main obstacle to the therapeutic use of carnosine is its rapid degradation by carnosinase 1 (CN1) that is secreted from the brain (Jackson *et al.*, 1991). Although, as discussed above, serum carnosine levels are enhanced after ingestion of carnosine, and physiological responses in patients have been described, strategies to reach higher effective concentrations in patients should be beneficial. Aside from the development of carnosine derivatives that are not degraded by CN1, specific inhibition of CN1 has been considered (Vistoli *et al.*, 2006). However, it is uncertain how patients would respond to inhibition of serum carnosinase since little or no carnosinase activity was reported to be associated with neurological disease and mental defects in two children (Perry *et al.*, 1968).

20.8.4 Dosage and Side Effects

Although carnosine is a naturally occurring dipeptide and therefore regarded as a safe drug, a responsible estimation of possible consequences for patients receiving high dosages of carnosine needs to be made. Given the high activity of CN1 in plasma, it is tempting to speculate whether evolution "invented" serum carnosinase to prevent unwanted side effects such as impairment of glycolytic flux in erythrocytes. However, several animals such as mice do not have a serum carnosinase activity (Jackson *et al.*, 1991), and the possible counter argument that carnosinase activity may be related to a carnivorous diet seems to be unjustified since dogs also do not have a serum carnosinase activity (Jackson *et al.*, 1991).

As stated above, there is at least one report on neurological disease and mental defects in children with impaired carnosinase activity (Perry *et al.*, 1968). However, Gardner *et al.*, who performed experiments in order to determine the intestinal uptake of carnosine in participants who ingested a single dose of 4 g of carnosine, reported no adverse effects, aside from a mild and transient digital paresthesia that some subjects experienced within the first hour. Therefore, at least an oral administration of carnosine in these dosages appears to be safe.

20.8.5 Derivatives of Carnosine and Related Compounds for Cancer Treatment

Since rapid hydrolysis of carnosine in plasma may limit the therapeutic use of carnosine, as discussed above, different strategies are under investigation to stabilize carnosine. Among the different derivatives that have been reported to have a higher stability towards degradation and that are at least able to maintain carnosine's ability to scavenge reactive oxygen species are (*S*)-Trolox™ (Stvolinsky *et al.*, 2010), and derivatives with saccharides such as β-cyclodextrin and trehalose (for recent reviews see (Vistoli *et al.*, 2012)). Recently, Bellia *et al.* also reported the design of a promising biotin-derivative of carnosine with an exceptionally high stability towards carnosinase (Bellia *et al.*, 2013). However, it is currently not known which derivatives can be successfully delivered *in vivo*.

20.9 Conclusions

More than 20 years after the first and widely unrecognized observation that carnosine has an antiproliferative potential toward tumor cells, experiments with cells from glioblastoma initiated a revival of research on carnosine's antineoplastic effects. New data on carnosine's potential to interfere with glycolytic flux and its ability to modulate pathways critical for tumor growth, survival and metastasis, accompanied by the fact that it appears to be a safe drug, raise the hope that it might pave the way towards new therapeutic strategies for the treatment of cancer. The development of potent new derivatives that are not degraded by serum carnosinase and new routes of administration that are under development might additionally contribute to this hope.

Summary Points

- This chapter focuses on carnosine's effects on tumor cells.
- Carnosine inhibits the growth of tumor cells in cell culture experiments and in animal models.
- The antineoplastic effect of carnosine appears to be related to its effect on signaling pathways around the mammalian target of rapamycin complex.
- Carnosine affects progression through the cell cycle.

- The influence of carnosine on apoptosis in tumor cells appears to be of minor importance for its antineoplastic effects.
- Carnosine may influence metastasis.
- Carnosine influences glycolytic flux in tumor cells.
- The influence of carnosine on glycolytic flux may be dependent on its influence on signal transduction pathways, glycolytic enzymes or glycolytic intermediates.
- Carnosine induces the mRNA expression of pyruvate dehydrogenase kinase 4 that is known to be regulated by signal transduction pathways that are important for tumor cell survival.
- Carnosine can be considered as a drug for cancer treatment.
- Serum carnosinase activity and possible side effects have to be taken into account considering the use of carnosine as a therapeutic drug.
- Derivatives of carnosine may be possible alternatives for cancer therapy circumventing problems with side effects or premature destruction of carnosine by serum carnosinase.

Key Facts of Signal Transduction Pathways

1. Signal transduction pathways are induced by extracellular signaling molecules that react with receptors on the surface of a cell or with intracellular receptors.
2. Interaction of a signaling molecule with its receptor induces the propagation of the signal by a second messenger.
3. The second messenger activates or inactivates other intracellular molecules that may result in a cascade of factors being activated or inactivated.
4. The propagation of the extracellular signal through intracellular molecules frequently occurs by phosphorylation or dephosphorylation of these molecules.
5. The most-downstream targets of signaling cascades are transcription factors.
6. Transcription factors influence the transcription of genes that have binding sites for the specific transcription factor mostly but not exclusively in their promoters.
7. Altered transcription results in physiological changes in the cell.

Key Facts of Pyruvate Dehydrogenase Kinase

1. Glucose that is taken up by a cell can be used for the production of energy in the form of ATP.
2. ATP production through the glycolytic pathway results in the formation of pyruvate
3. Pyruvate can be reduced to lactate especially under hypoxic conditions or enter the citric acid cycle under normoxic conditions.

4. In order to produce more intracellular ATP by use of the citric acid cycle and oxidative phosphorylation, pyruvate has to be converted to acetyl-CoA.
5. Conversion of pyruvate to acetyl-CoA is catalyzed by the enzyme pyruvate dehydrogenase complex.
6. The pyruvate dehydrogenase complex can be inhibited by phosphorylation catalyzed by one of four pyruvate dehydrogenase kinases designated PDK1 to PDK4.

Key Facts of the Warburg Effect

1. The Warburg effect describes the best-studied metabolic phenotype of cancer cells, which is aerobic glycolysis.
2. Aerobic glycolysis of cancer cells is characterized by an increased metabolism of glucose to lactate in the presence of sufficient oxygen.
3. The molecular mechanisms driving the Warburg effect are highly heterogeneous and mitochondrial defects are only a rare phenomenon in cancer cells.
4. Most anabolic processes required for accelerated tumor growth are accomplished by increased glycolysis.
5. Because of the great importance of glycolysis for tumor cell growth, glucose deprivation or the inhibition of glycolytic activity is considered as a strategy for new therapies.

Definition of Words and Terms

Apoptosis. A form of programmed cell death, in which a highly controlled "suicide" program becomes activated, followed by the activation of caspases, which are proteolytic enzymes that kill the cell.

Autophagy. Autophagy describes a process by which cells destroy worn-out organelles or unnecessary cellular components by their own lysosomes.

Cell cycle. The cell cycle is a series of physiological events called phases that finally leads to the duplication of a cell. A typical cell cycle consists of the phases *G1*, *S* (in which DNA is replicated), *G2* and *M* (the phase in which the cell is divided). A special resting phase called *G0* is a phase into which cells enter from G1. The entry from one phase into the other is highly controlled in normal cells and frequently aberrantly regulated in tumor cells.

DAF-16. DAF-16 is the *Caenorhabditis elegans* ortholog of the mammalian Forkhead box proteins. It is critical for diverse biological processes such as longevity and stress resistance.

Estrogen-related receptors (ERRs). ERRs are transcription factors that bind to the same DNA sequences as the estrogen receptor (and other elements) without the need to be activated by estrogen.

Forkhead box, class O proteins (FOXOs). FOXOs are transcriptional regulators that constitute a subfamily within a superfamily of transcriptional regulators that have a conserved DNA binding domain termed forkhead box. FOXOs are involved in various signaling pathways and control a wide range of physiological processes such as cellular differentiation, metabolism, stress response and cell-cycle control.

Glioblastoma (multiforme). Glioblastoma multiforme is classified by the World Health Organization (WHO) as a grade IV tumor of the central nervous system. It is the most malignant neoplasm. Due to its invasive nature it cannot be completely resected and despite recent progress in radio/chemotherapy less than half of patients survive more than one year after diagnosis.

Glucocorticoids. Glucocorticoids are a class of steroid hormones that bind to intracellular glucocorticoid receptors that after binding of the hormone homodimerize at DNA elements of different genes, thereby regulating their transcription. In experiments, the most frequently used steroid for the activation of the glucocorticoid receptor is the synthetic compound dexamethasone.

Glucose transporters (GLUTs). The GLUTs comprise a family of transporters that dependent on the subtype of transporter may, in addition to glucose, also transport other carbohydrates across cell membranes.

Hypoxia-inducible factors (HIFs). HIFs are transcription factors and the central components of a signal transduction pathway that is activated under hypoxic conditions and inhibited under normoxic conditions. Under low oxygen conditions HIFs enter the nucleus where they bind to HIF response elements in target genes (HREs) thereby affecting their transcription.

Histone-acetylases (HATs)/deacetylases (HDACs). HATs and HDACs constitute classes of enzymes that can modify histones by linking acetyl moieties to these proteins (HATs) or removing them (HDAC). In general, acetylation of the histones associated with a gene can be a prerequisite for its activation whereas deacetylation generally leads to transcriptional repression.

Mitogen-activated protein kinases (MAP kinases). MAP kinases are members of signal transduction pathways that phosphorylate downstream targets that can be other MAP kinases or different effectors. The phosphorylation modifies the activity of the substrates, finally resulting in a number of physiological processes in the cell, including apoptosis, cell survival, proliferation and differentiation.

Matrix metalloproteinases (MMPs). MMPs are enzymes that, dependent on the specific type, can breakdown different components of the extracellular matrix. They are required for several physiological processes such as differentiation, regeneration or tissue remodeling. However, aberrant expression of certain MMPs may enhance the metastatic potential of tumor cells.

Mammalian target of rapamycin (mTOR). mTOR is a protein kinase that controls different physiological processes such as translation, glucose metabolism, proliferation and cell survival. It is targeted by other signal transduction pathways and named after its specific inhibitor rapamycin.

It has to be noted that strong inhibition by rapamycin is limited to mTOR complex 1, whereas the second complex can only be inhibited in certain cell types and under prolonged exposure.

Metastasis. Metastasis is the process by which cancer cells spread from their tissue of origin to other locations in the body. Metastasis may involve detachment of single cells from the tumor mass, migration, degradation of extracellular matrix proteins, escape into the blood stream and extravasation out of the circulation to a new tissue.

p38 mitogen-activated protein kinases. p38 mitogen-activated protein kinases are a special class of mitogen-activated protein kinases. They respond to external stimuli, such as stress induced by heat, osmotic changes or ultraviolet irradiation and they are involved in processes such as cell differentiation, apoptosis and autophagy.

Peroxisome proliferator activated receptors (PPARs). PPARs are nuclear receptor proteins that upon activation by different ligands, and dependent on the subtype, affect transcription of their target genes. PPARs play an important role in lipid and glucose homeostasis and are considered to be lipid and insulin sensors.

Peroxisome proliferator γ coactivator 1α (PGC-1α). PGC-1α is a regulatory protein that, in association with other factors, regulates the transcription of genes. It was first identified as a coactivator of PPARγ but also interacts with other PPARs and transcription factors. PGC-1α is a highly versatile coactivator that influences different physiological processes dependent on the tissue. It can control mitochondrial biogenesis, metabolism, and antioxidant defense and it controls the expression of key gluconeogenesis pathway enzymes in the liver.

Phosphoinositide 3-kinase (PI3K). PI3K is a lipid kinase that produces the key signaling lipid phosphatidyl-inositol-3,4,5-triphosphate (PI(3,4,5)P3), which affects the activity of a multitude of downstream targets including AKT and mTOR. In most cancer cells (PI(3,4,5)P3) is constitutively elevated. This contributes to the aberrant regulation of physiological processes that are hallmarks of cancer, such as cell cycle, survival, metabolism, motility and genomic instability.

Protein kinase B (PKB). PKB, which is also known as AKT, is a downstream effector in the phosphoinositide 3-kinase signaling pathway. It can be phosphorylated by mTORC2 and is an upstream effector of mTORC1. In addition, there is also a feedback loop by which mTORC1 signals back on PKB. Among others, PKB regulates cell survival, metabolism, proliferation and angiogenesis.

Rapamycin. Rapamycin is a bacterial antifungal that is named after the Easter Island (which is known to the natives as Rapa Nui), since the bacterial species that produces it was identified in a soil sample from this island. It is a specific inhibitor of the mTOR complex 1 and supposed to be useful for the treatment of different kinds of cancer.

SP1. SP1 is a transcription factor that binds to GC-rich elements in the promoters of many genes. It is supposed to act with other transcription factors, and the interaction of SP1 bound to one side on a DNA strand may, by interaction with an SP1 bound to a different place on the DNA, contribute to scaffolding of DNA strands into a higher structure, required for proper interactions of transcription factors on certain promoters.

Signal transducers and activators of transcription (STATs). The STATs comprise a family of transcription factors that are activated by the so-called Janus kinases that are themselves activated by external ligands such as cytokines. Aside from their role in immunological processes, STATs are especially known for their role in the regulation of growth and proliferation.

Trichostatin A (TSA). TSA is an antibiotic that inhibits certain classes of histone deacetylases. This inhibition can cause higher levels of histone acetylation, which in turn may lead to the constant activation of genes that are transcriptionally inhibited by a low level of acetylated histones, because this can block access of other transcription factors to their corresponding binding sites in a promoter.

Vascular endothelial growth factors (VEGFs). The VEGFs comprise a family of signaling proteins that upon activation stimulate vasculogenesis and angiogenesis. The high activity of these factors can promote the growth of solid tumors, since it stimulates the formation of new blood vessels that in turn results in a better supply of nutrients and oxygen for the tumor.

List of Abbreviations

AKT	Ak strain transforming
ATP	Adenosine triphosphate
Bp	Base pair
CN1	Carnosinase 1
DAF-16	(Abnormal) dauer formation protein 16
eIF4E	Eukaryotic initiation factor 4E
eIF4BP1	Eukaryotic initiation factor 4E binding protein 1
FOXO1/3	Forkhead box protein O1/3
ERK1/2	Extracellular-signal regulated kinase
ERR	Estrogen-related receptor
GLUT	Glucose transporter
GR	Glucocorticoid receptor
GRE	Glucocorticoid response element
HER2/neu	Human epidermal growth factor receptor 2/ neuroblastoma
HIF	Hypoxia inducible factor
hPDK4	Human pyruvate dehydrogenase kinase gene
HRE	Hypoxia-responsive element
IL-8	Interleukin 8
IRE	Insulin response element
JAK	Janus kinase
MAPK	Mitogen-activated protein kinase

MEK	Mitogen-activated protein kinase
MMP9	Matrix metallopeptidase 9
mTOR/mTORC	Mammalian target of rapamycin/mTOR complex
P90RSK	p90 ribosomal S6 kinase
PDH	Pyruvate dehydrogenase
PDK4	Pyruvate dehydrogenase kinase 4
PGC-1α	Peroxisome proliferator-activated receptor γ coactivator 1α
PHD	Prolyl hydroxylase
(PI(4,5)P_2)	Phosphatidylinositol-4,5-diphosphate
(PI(3,4,5)P3)	Phosphatidylinositol-3,4,5-triphosphosphate
PI3K	Phosphatidylinositol 3-kinase (phosphoinositide 3-kinase)
PKB	Protein kinase B
PPAR	Peroxisome proliferator activated receptor
Raf	Rapidly accelerated fibrosarcoma
RARE	Retinoic acid response element
RAR	Retinoic acid receptor
Ras	Rat sarcoma
rpS6	Ribosomal protein S6
RTK	Receptor tyrosine kinase
ROS	Reactive oxygen species
RXR	Retinoid X receptor
S6K1/2	Ribosomal protein S6 kinases 1 and 2
SP1	Specificity protein 1
STAT3	Signal transducer and activator of transcription 3
TNF-α	Tumor necrosis factor α
TSA	Trichostatin A
VEGF	Vascular endothelial growth factor
VHL	von Hippel–Lindau protein

References

Baraniuk, J. N., El-Amin, S., Corey, R., Rayhan, R. and Timbol, C., 2013. Carnosine treatment for Gulf War Illness: a randomized controlled trial. *Global Journal of Health Science*. 5: 69–81.

Bauchart, C., Savary-Auzeloux, I., Patureau Mirand, P., Thomas, E., Morzel, M. and Remond, D., 2007. Carnosine concentration of ingested meat affects carnosine net release into the portal vein of minipigs. *Journal of Nutrition*. 137: 589–593.

Bellia, F., Oliveri, V., Rizzarelli, E. and Vecchio, G., 2013. New derivative of carnosine for nanoparticle assemblies. *European Journal of Medicinal Chemistry*. 70: 225–232.

Bensinger, S. and Christofk, H. R., 2012. New aspects of the Warburg effect in cancer cell biology. *Seminars in Cell and Developmental Biology*. 23: 352–361.

Boldyrev, A., Fedorova, T., Stepanova, M., Dobrotvorskaya, I., Kozlova, E., Boldanova, N., Bagyeva, G., Ivanova-Smolenskaya, I. and Illarioshkin, S., 2008. Carnosine [corrected] increases efficiency of DOPA therapy of Parkinson's disease: a pilot study. *Rejuvenation Research*. 11: 821–827.

Cavadas, M. A. S., Nguyen, L. K. and Cheong, A., 2013. Hypoxia-inducible factor (HIF) network: insights from mathematical models. *Cell Communication and Signaling*. 11: 11–42.

Chengappa, K. N., Turkin, S. R., DeSanti, S., Bowie, C. R., Brar, J. S., Schlicht, P. J., Murphy, S. L., Hetrick, M. L., Bilder, R. and Fleet, D., 2012. A preliminary, randomized, double-blind, placebo-controlled trial of l-carnosine to improve cognition in schizophrenia. *Schizophrenia Research*. 142: 145–152.

Chez, M. G., Buchanan, C. P., Aimonovitch, M. C., Becker, M., Schaefer, K., Black, C. and Komen, J., 2002. Double-blind, placebo-controlled study of L-carnosine supplementation in children with autistic spectrum disorders. *Journal of Child Neurology*. 17: 833–837.

Chuang, C. H. and Hu, M. L., 2008. L-Carnosine inhibits metastasis of SK-Hep-1 cells by inhibition of matrix metaoproteinase-9 expression and induction of an antimetastatic gene, nm23-H1. *Nutrition and Cancer*. 60: 526–533.

Denko, N. C., 2008. Hypoxia, HIF1 and glucose metabolism in the solid tumor. *Nature Reviews Cancer*. 8: 705–713.

Fulda, S., 2013. Regulation of apoptosis pathways in cancer stem cells. *Cancer Letters*. 338: 168–173.

Gardner, M. L. G., Illingworth, K. M., Kelleher, J. and Wood, D., 1991. Intestinal-absorption of the intact peptide carnosine in man, and comparison with intestinal permeability to lactulose. *Journal of Physiology*. 439: 411–422.

Gaunitz, F. and Hipkiss, A., 2012. Carnosine and cancer: a perspective. *Amino Acids*. 43: 135–142.

Hipkiss, A. R., 2009. Carnosine and its possible roles in nutrition and health. *Advances in Food and Nutrition Research*. 57: 87–154.

Hipkiss, A. R., 2011. Energy metabolism, proteotoxic stress and age-related dysfunction – protection by carnosine. *Molecular Aspects of Medicine*. 32: 267–278.

Hipkiss, A. R. and Gaunitz, F., 2014. Inhibition of tumor cell growth by carnosine: some possible mechanisms. *Amino Acids*. 46: 327–373.

Holliday, R. and McFarland, G. A., 1996. Inhibition of the growth of transformed and neoplastic cells by the dipeptide carnosine. *British Journal of Cancer*. 73: 966–971.

Horii, Y., Shen, J., Fujisaki, Y., Yoshida, K. and Nagai, K., 2012. Effects of L-carnosine on splenic sympathetic nerve activity and tumor proliferation. *Neuroscience Letters*. 510: 1–5.

Iovine, B., Iannella, M. L., Nocella, F., Pricolo, M. R., Baldi, M. R. and Bevilacqua, M. A., 2011. Carnosine inhibits KRas-mediated HCT-116 proliferation by affecting ATP and ROS production. *Cancer Letters*. 315: 122–128.

Jackson, M. C., Kucera, C. M. and Lenney, J. F., 1991. Purification and properties of human serum carnosinase. *Clinica Chimica Acta*. 196: 193–205.

Jeong, J. Y., Jeoung, N. H., Park, K. G. and Lee, I. K., 2012. Transcriptional regulation of pyruvate dehydrogenase kinase. *Diabetes & Metabolism Journal*. 36: 328–335.

Jia, H. J., Qi, X. D., Fang, S. H., Jin, Y. H., Han, X. Y., Wang, Y., Wang, A. and Zhou, H., 2009. Carnosine inhibits high glucose-induced mesangial cell proliferation through mediating cell cycle progression. *Regulatory Peptides*. 154: 69–76.

Letzien, U., Oppermann, H., Meixensberger, J. and Gaunitz, F., 2014. The antineoplastic effect of carnosine is accompanied by induction of PDK4 and can be mimicked by L-histidine. *Amino Acids*. 46: 1009–1019.

Liu, Y., Xu, G. and Sayre, L. M., 2003. Carnosine inhibits (*E*)-4-hydroxy-2-nonenal-induced protein cross-linking: structural characterization of carnosine–HNE adducts 1. *Chemical Research in Toxicology*. 16: 1589–1597.

Nagai, K. and Suda, T., 1986. Antineoplastic effects of carnosine and beta-alanine – physiological considerations of its antineoplastic effects. *Journal of the Physiological Society of Japan*. 48: 741–747.

Perry, T. L., Hansen, S. and Love, D. L., 1968. Serum-carnosinase deficiency in carnosinaemia. *Lancet*. 291: 1229–1230.

Pópulo, H., Lopes, J. M. and Soares, P., 2012. The mTOR signalling pathway in human cancer. *International Journal of Molecular Sciences*. 13: 1886–1918.

Renner, C., Asperger, A., Seyffarth, A., Meixensberger, J., Gebhardt, R. and Gaunitz, F., 2010a. Carnosine inhibits ATP production in cells from malignant glioma. *Neurological Research*. 32: 101–105.

Renner, C., Seyffarth, A., de Arriba, S., Meixensberger, J., Gebhardt, R. and Gaunitz, F., 2008. Carnosine inhibits growth of cells isolated from human glioblastoma multiforme. *International Journal of Peptide Research and Therapeutics*. 14: 127–135.

Renner, C., Zemitzsch, N., Fuchs, B., Geiger, K. D., Hermes, M., Hengstler, J., Gebhardt, R., Meixensberger, J. and Gaunitz, F., 2010b. Carnosine retards tumor growth in vivo in an NIH3T3-HER2/neu mouse model. *Molecular Cancer*. 9: 2.

Rybakova, Y. S. and Boldyrev, A. A., 2012. Effect of carnosine and related compounds on proliferation of cultured rat pheochromocytoma PC-12 cells. *Bulletin of Experimental Biology and Medicine*. 154: 136–140.

Semenza, G. L., 2012. Regulation of metabolism by hypoxia-inducible factor 1. *Cold Spring Harbor Symposia on Quantitative Biology*. 76: 347–353.

Severin, S. E., Skolysheva, L. K., Shur, S. A. and Vulfson, P. L., 1990. The pH-dependent conformational transition in glycogen phosphorylase b. The effect of carnosine and anserine on its activity. *Biochemistry International*. 20: 227–238.

Son, D. O., Satsu, H., Kiso, Y., Totsuka, M. and Shimizu, M., 2008. Inhibitory effect of carnosine on interleukin-8 production in intestinal epithelial cells through translational regulation. *Cytokine+*. 42: 265–276.

Stvolinsky, S., Antipin, M., Meguro, K., Sato, T., Abe, H. and Boldyrev, A., 2010. Effect of carnosine and its trolox-modified derivatives on life span of drosophila melanogaster. *Rejuvenation Research*. 13: 453–457.

Vistoli, G., Pedretti, A., Cattaneo, M., Aldini, G. and Testa, B., 2006. Homology modeling of human serum carnosinase, a potential medicinal target, and MD simulations of its allosteric activation by citrate. *Journal of Medical Chemistry*. 49: 3269–3277.

Vistoli, G., Carini, M. and Aldini, G., 2012. Transforming dietary peptides in promising lead compounds: the case of bioavailable carnosine analogs. *Amino Acids*. 43: 111–126.

Wang, B., Hsu, S.-H., Frankel, W., Ghoshal, K. and Jacob, S. T., 2012. Stat3-mediated activation of microRNA-23a suppresses gluconeogenesis in hepatocellular carcinoma by down-regulating glucose-6-phosphatase and peroxisome proliferator-activated receptor gamma, coactivator 1 alpha. *Hepatology*. 56: 186–197.

Yokogami, K., Wakisaka, S., Avruch, J. and Reeves, S. A., 2000. Serine phosphorylation and maximal activation of STAT3 during CNTF signaling is mediated by the rapamycin target mTOR. *Current Biology*. 10: 47–50.

Zhang, J. H., Lu, Q., Shi, W. J., Wu, Z. Z. and Wang, L. S., 2005. The induction apoptosis of HL-60 cells by low molecular weight compounds of taurine, ornithine and carnosine from new born calf liver. *Chinese Journal of Applied Physiology*. 21: 200–205.

Zhang, Y., Song, L., Cheng, X., Yang, Y., Luan, B., Jia, L., Xu, F. and Zhang, Z., 2011. Carnosine pretreatment protects against hypoxia-ischemia brain damage in the neonatal rat model. *European Journal of Pharmacology*. 667: 202–207.

CHAPTER 21

L-Carnosine and Human Colon Cancer

BARBARA IOVINE[a], MARIANGELA GAROFALO[a], MARIA OREFICE[a], AND MARIA ASSUNTA BEVILACQUA*[a]

[a]Department of Molecular Medicine and Medical Biotechnology, University of Naples Federico II, Via Pansini, 5, 80131 Napoli, Italy
*E-mail: mabevil@unina.it

21.1 Colon Cancer

Colon cancer is the most common type of malignant tumor arising from the inner wall of the large intestine and is the result of cancer cells that form in the lining of the colon and rectum. Colorectal cancer (CRC) often begins as a benign growth, known as a polyp. Most polyps will remain benign, but some have the potential to turn into cancer over the long term and only early removal prevents this transformation. Currently, CRC represents one of the major threats to human health and it accounts for 15% of all tumors, with the highest incidence in the United States and the lowest in Africa and the Far East. In Europe and other western countries it is the second leading cause of death in men (after lung cancer) and in women (after breast cancer) (Haggar and Boushey, 2009).

Food and Nutritional Components in Focus No. 8
Imidazole Dipeptides: Chemistry, Analysis, Function and Effects
Edited by Victor R Preedy

Published by the Royal Society of Chemistry, www.rsc.org

21.1.1 A Comprehensive Summary of Colorectal Cancers

CRC occurs as sporadic forms (85%) or as hereditary familial forms (15%). Several epidemiological studies have shown that at least 15% of the hereditary types of CRC have a well-defined hereditary autosomal dominant pattern or, more rarely, recessive pattern. Of the hereditary forms, 1 to 3% consist of the polyposis syndromes such as familial adenomatous polyposis (FAP) and hereditary nonpolyposis colorectal cancer (HNPCC) or Lynch syndrome. HNPCC is an autosomal dominant form, manifests generally under the age of forty years does not have a specific phenotype and the malignancy develops in the absence of adenomatosis (Jasperson, 2012). Another 10% of the hereditary CRCs are characterized by other familial syndromes and 1% by the presence of chronic inflammatory bowel disease (IBD), such as ulcerative colitis and Crohn's disease (Figure 21.1). All CRCs are classified according to stage. Staging determines how far a tumor has spread beyond its original site and is not related to the size of the tumor. The majority of adenocarcinomas progress through several stages, from the normal colonic mucosa an adenomatous polyp or adenoma can arise. This tumor mass, which protrudes into the lumen of the intestine, can grow and become dysplastic and then transform into carcinoma "*in situ*" and subsequently an invasive carcinoma. The malignancy is characterized by autonomous growth and widespread invasion of surrounding tissues and can metastasise, mainly to the lymphnodes, lung and liver. CRC can remain asymptomatic for years before being diagnosed and the symptoms can vary greatly according to the site of the tumor. It is important to know the stage of a cancer in order to determine prognosis and to plan the best treatment.

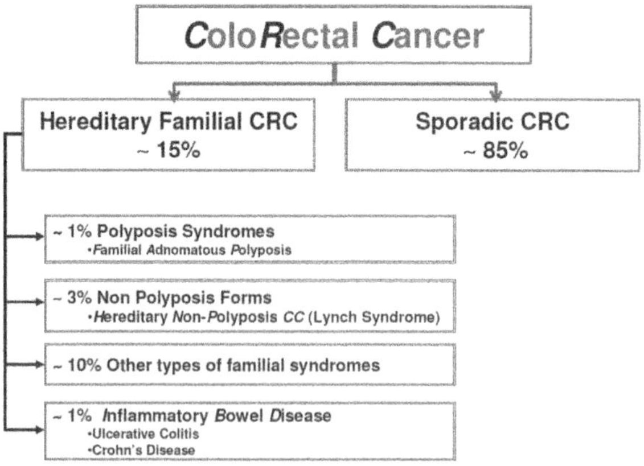

Figure 21.1 Distribution and types of colon cancer. CRC may occur as sporadic forms (85%) or as hereditary familial forms (15%). Hereditary forms are classified by type and incidence.

21.1.2 Colorectal Carcinoma Genetics

In recent years, the understanding of colorectal carcinogenesis has greatly improved thanks to research on the genetic basis of the disease. Two genetic pathways have been defined, 80% of cases of sporadic CRC are characterized by the chromosomal instability (CIN) pathway, in the form of aneuploidy involving amplifications and deletions of chromosomal regions (Derks *et al.*, 2008). The CIN pathway involves an accumulation of defects in long segments of DNA coding sequences that result in the loss of tumor-suppressor genes and activation of oncogenes. It is through this pathway that cancers of the classical adenoma–carcinoma sequence in FAP and the majority of sporadic CRCs develop. The remaining 15–20% of cases of sporadic CRC, in addition to almost all of the hereditary forms (HNPCC) are characterized by the microsatellite instability (MSI). This pathway involves the loss of DNA microsatellite mismatch repair (MMR) protein function resulting in multiple defects in repetitive noncoding regions of DNA (microsatellites) (Figure 21.2). This form of genetic instability is associated with inactivation of genes involved in the repair of DNA damage that in the sporadic form and in HNPCC are due to mutations of MLH1 gene (MutL Homolog 1 gene) (Mori *et al.*, 2004). A third mechanism that has been proposed is the CpG island methylator phenotype (CIMP) that is associated with a prevalent tendency for DNA methylation and describes tumors with high degrees of methylation in CpG islands of promoters. The components of this pathway are not fully understood and there appears to be some overlap with MMR deficiency and MSI (Hughes *et al.*, 2012). Colon cancer is triggered by a series of point mutations and genetic alterations that progressively cause normal cells

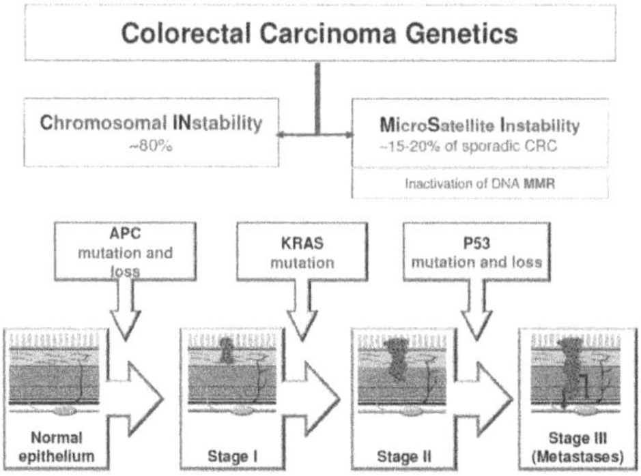

Figure 21.2 The genetics of colorectal cancer. Two genetic pathways have been defined: chromosomal instability (CIN) (80%) and microsatellite instability (MSI) (15–20%); the latter is associated with inactivation of DNA microsatellite mismatch repair (MMR). The pathological progression of adenoma to carcinoma depends on reproducible genetic alterations: the loss of function of APC allows the development of Stage I adenoma; then, mutation in KRAS transforms cells and leads to Stage II adenoma; the loss of function of the p53 gene causes the progression of the adenoma to Stage III carcinoma and subsequent metastases.

to transform into adenomas. The pathological progression of adenoma to carcinoma depends on reproducible genetic alterations such as the adenomatous polyposis coli (APC) inactivation, KRAS oncogene mutation, and p53 mutation (Figure 21.2). According to the model proposed by Vogelstein the early molecular event that characterizes the majority of CRCs arise as a consequence of APC gene mutations that lead to aberrant Wnt pathway activation, cellular proliferation and polyp formation (Vogelstein *et al.*, 1988; Neufeld, 2009). Mutations in the APC gene, a tumor-suppressor gene that controls tumor initiation, have been identified in 60% of sporadic tumors and in 80–90% of patients with FAP (von Allmen, 2006). Furthermore, inactivation of the APC gene has been found to result in activation of the Wnt signal transduction pathway that regulates crucial cellular processes including cell-fate determination, motility, and cell polarity, as well as general cell-maintenance processes such as homeostasis and cell-cycle regulation. Indeed, APC inhibits WNT protein, responsible for the expression of the c-Myc gene, a transcription factor that in turn induces the expression of many genes required for the transition from G1 to S phase of the cell cycle (Bilir *et al.*, 2013). Activation of Myc increases cell growth and can lead to development of a small benign tumor, called an adenoma class I. In addition, a series of studies supports that 30% of human CRC have a mutation in the KRAS oncogene rendering it constitutively active (Prior *et al.*, 2012). KRAS (Kirsten rat sarcoma viral oncogene) is a gene encoding a protein that plays a key role in the intracellular signal transduction pathway of the receptor for the epidermal growth factor (EGFR). EGFR is a tyrosine kinase receptor able to initiate the signal transduction cascades promoting cell division, migration, angiogenesis, and inhibiting apoptosis. There are multiple pathways downstream of the KRAS oncogene making it a focal point for a network of signals; in fact components of the family of RAS–RAF–MAPKs have a key role in many cellular programs including maintenance of cell proliferation, differentiation and survival (Raman *et al.*, 2007). By contributing to the stages of initiation, progression and metastasis typical of tumor pathologies and by triggering the mechanisms of resistance to apoptosis, cell proliferation, angiogenesis, cell migration and tissue invasion overexpression and/or uncontrolled activation of the components of this pathway play a part in oncogenic transformation by promoting proliferative diseases such as cancers (Rajagopalan *et al.*, 2002). Genes that belong to the family of RAS (H, K, N) are frequently mutated in human cancer and are located on chromosome 12 and such mutations all have the same biological significance in determining the constitutive activation of a protein, altering its GTPase activity (Edkins *et al.*, 2006; Er *et al.*, 2014). The loss of this enzyme activity and the continued translation of the signal that occurs through the use of GTP, no longer transformed into GDP, would be able to transform cells and to induce them to form adenomas of class II, the largest class after class I (Gryfe *et al.*, 1997). A late molecular event in the process of colon carcinogenesis is the loss of function of the p53 gene, located on chromosome 17. p53 is a protein that regulates the cell cycle and blocks progression when DNA is aberrant, directing the cell to apoptosis. The loss of both copies of p53 indicates the progression of the adenoma to a class III carcinoma and from that stage to distant metastases (Figure 21.2).

21.2 What are the Main Causes of Colon Cancer?

There are many risk factors for colorectal cancer, some of them are beyond our control, and include: an inflammatory bowel disease, the presence of colon polyps or long-standing ulcerative colitis, age over 50, also hereditary factors for example a family history of colorectal, breast or ovarian cancer. Additionally there are risk factors that we can control for example maintenance of a healthy weight and prevention of obesity by controlling the diet and taking adequate exercise, also avoiding smoking and the consumption of alcohol (Figure 21.3).

21.2.1 Diet and Colorectal Cancer

It is well known that diet is one of the main risk factors for the development of colon cancer. A diet high in calories, high in animal fat and low in fiber is believed to be associated with an increase in intestinal tumors predisposing people to colorectal cancer (Chao *et al*,. 2005). In countries with high colorectal cancer rates, the fat intake is much higher than in countries with low cancer rates. Diets rich in vegetables and high-fiber foods such as whole-grain breads and cereals contain less of the fat that produces these carcinogenic effects and may actually counter the effects of the carcinogens themselves, both of which help reduce the risk of cancer. The small intestine is the primary site for the absorption and transportation of nutrients; more accurately the intestinal epithelial cells (IEC) perform these functions. These cells form a monolayer covering the internal intestinal surface and function both as a barrier and as the site of absorption, besides their role in signal recognition and transduction. IEC have many transporters and receptors to recognize food-derived nutrients; in addition, they form between them tight junctions with gaps and pores through which small compounds can pass. All these functions are controlled by hormones and cytokines, in fact the IEC, besides

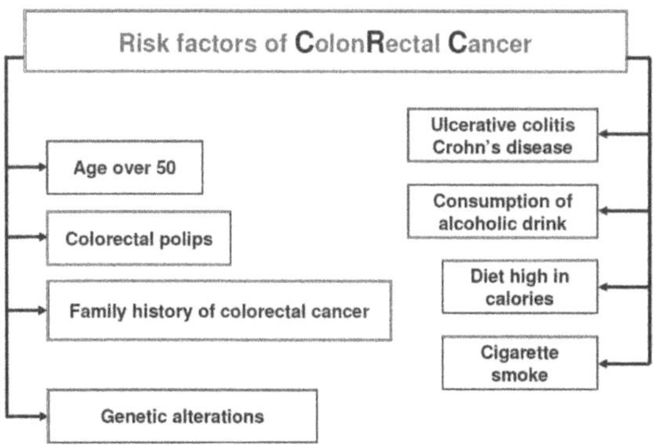

Figure 21.3 Main risk factors for colorectal cancer. Risks factors of CRC include hereditary factors or genetic alteration. Behavioral factors as high-calorie diet, alcohol, and smoking contribute to the susceptibility of CRC.

being involved in the metabolism of food substances, are also involved in the intestinal immune system (Shimizu, 2010).

21.2.2 Functional Relationship between Inflammation and Colon Cancer

Recently, many studies highlight that chronic inflammation such as gastritis and esophagitis increase the risk of colon cancer. Inflammation facilitates the growth and progression of normal cells to malignancy by production of proinflammatory cytokines, reactive oxygen species (ROS) and nitrogen derivates (Ohshima *et al.*, 2005). Inflammation may contribute to tumor development in three different ways:

(1) inducing mediators that promote DNA synthesis and cell proliferation: this promotes genomic instability and facilitates the action of genotoxic agents (aflatoxins, *N*-nitroso derivatives) that cause permanent damage to the DNA;
(2) inducing the formation of hydrogen peroxide (H_2O_2) that in turn can be converted into ROS, that can damage DNA by oxidation of the nitrogenous bases that compose it;
(3) activating nuclear factor kB (NF-kB) a key protein in the induction of inflammation in response to bacterial and viral infections.

NF-kB plays a key role in the immunoresponse to inflammation, in fact it is activated by proinflammatory cytokines such as tumor necrosis factor (TNF)-α and interleukin 6, a cytokine involved in tumorigenesis of colorectal cancer. Its regulation has been linked to initiation and progression of cancer by controlling the expression of genes involved in the cell cycle. Crohn's disease and ulcerative colitis are chronic IBD recognized to be precancerous conditions for CRC (Itzkowitz and Yio, 2004). These diseases, through repeated cycles of inflammation and regeneration of the mucosa, transform IEC into malignant cells. It is known that cell proliferation alone does not cause cancer, but continued proliferation in an environment rich in inflammatory cells, growth factors, and agents that promote DNA damage certainly potentiates and/or promotes the neoplastic risk. The most significant evidence for the role of inflammation in the progression to neoplastic colonic adenoma–carcinoma is provided by the reduction in the percentage of CRC diagnoses in regular users of aspirin and other nonsteroidal anti-inflammatory drugs (NSAID). Study of the link between cancer and inflammation is beginning to have a significant impact on the prevention and treatment of cancer through the administration of drugs, not only to slow or stop the growth of precancerous cells but also to make them regress. It has been demonstrated that several mitogenic activities can be ascribed to inflammatory molecules including the promotion of angiogenesis, stimulation of the production of free radicals that act on the DNA and promotion of metastases. From this it can be deduced that it is possible to control the tumor by blocking these inflammatory pathways. Indeed,

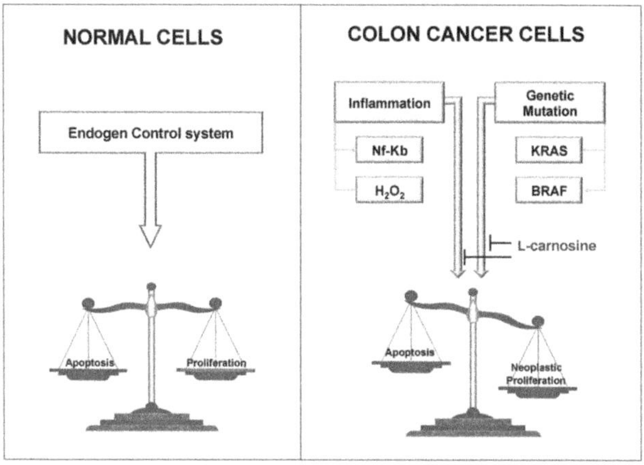

Figure 21.4 L-Carnosine involvement in the balance between cell death and proliferation. Cell growth is regulated by a dynamic balance between cell death and cell proliferation. This dynamic equilibrium is controlled by proteins regulating cell cycle, DNA damage repair and apoptotic process. Inflammation and genetic mutations interfere with the physiological mechanisms of cell growth and can trigger neoplastic transformation. In colon cancer cells L-carnosine prevents cell proliferation and inflammation, thus inhibiting tumor progression.

some anti-inflammatory-like drugs have been shown in clinical trials to prevent human colon cancer at several sites. Therefore, downregulation of the production of intestinal proinflammatory cytokines by nutrients could represent a promising therapeutic approach. On this basis, the dipeptide L-carnosine plays an increasingly interesting role in colon cancer, indeed, in recent years, experimental data demonstrates that this dipeptide can downregulate the inflammatory response of IEC. In addition, it is able to suppress the tumorigenesis of colon cancer by inhibiting proliferation of the cells (Figure 21.4) (Son *et al.*, 2004; Iovine *et al.*, 2012).

21.3 The Dipeptide L-Carnosine in Human Colon Cancer

L-Carnosine (β-alanyl-L-histidine) is a naturally occurring histidine dipeptide, normally found in brain and muscle, known to possess numerous functions including antioxidant activity.

21.3.1 L-Carnosine Inhibits the Inflammatory Response of Cancer Intestinal Epithelial Cells

An interesting feature of L-carnosine is its inhibitory effect on the inflammatory response of the human intestinal epithelial cancer cell line, an extensively used model of the intestinal barrier. Son *et al.*, demonstrated that L-carnosine is able to inhibit H_2O_2 or TNF-α-induced interleukin-8 secretion in the Caco-2 cell line, by suppressing its translation. Overexpression of IL-8 is induced in IEC by several stimuli such as bacteria and oxidative stress and plays a pivotal role in the inflammatory diseases of the intestine. The authors reported that L-carnosine does not change IL-8 mRNA levels but exerts its effect on IL-8 by decreasing the eukaryotic translation initiation factor 4E (eIF4E) phosphorylation (Son *et al.*, 2008). In eukaryotic cells translation is divided into three phases: initiation, elongation and termination; the most important step is the first one that occurs by the recruitment of ribosomes to the mRNA 5′-cap structure (m7GpppN). This mechanism involves the eIF4F complex, composed of eIF4E, eIF4A, and eIF4G. eIF4E is the protein that specifically interacts with the mRNA 5′-cap structure and regulates cap-dependent protein translation initiation. eIF4A is a protein that exhibits ATPase and RNA helicase activity. eIF4G is a scaffolding protein that binds to both eIF4E and eIF4A. eIF4E protein represents one of the main regulatory initiation factors, because it is present in a limited molar amount in cells, its activation is regulated by abundance, phosphorylation state and binding with other proteins. Phosphorylated eIF4E has been reported to have a higher binding affinity for the cap and to form a more stable eIF4F complex (Figure 21.5A). It has been demonstrated that L-carnosine not only strongly inhibits H_2O_2-induced eIF4E phosphorylation but also significantly inhibits the phosphorylation of Akt (a serine/threonine-specific protein kinase). Phosphorylation of Akt is the key event regulating eukaryotic translation initiation factor 4E-binding protein 1(4E-BP1) phosphorylation in the mTOR pathway, known to favor both proliferation and cellular survival. Indeed, hypo-phosphorylated 4E-BP1 sequesters eIF4E and the interaction between 4E-BP1 and eIF4E inhibits the complex assembly and represses translation (Figure 21.5A). Consequently, the suppression of Akt phosphorylation by L-carnosine could inhibit 4E-BP1 phosphorylation and lead to the suppression of eIF4F complex formation (Azar *et al.*, 2013). The phosphorylation of this protein may be critical in development of gastric and CRCs. eIF-4E is principally phosphorylated by ERK1/2, whose phosphorylation is in turn inhibited by L-carnosine treatment. The activation of eIF4E phosphorylation by MAPK signaling could represent the L-carnosine target and could underlie the inhibition of IL-8 secretion during inflammation. A particular aspect of this study is the more marked inhibitory effect on IL-8 secretion in colon cancer cells pretreated with L-carnosine when compared to cells treated simultaneous with L-carnosine and H_2O_2. This molecular mechanism could be related to an increase in L-carnosine uptake during pretreatment. These data suggest that L-carnosine could be a positive regulator of the immune response in carcinoma colon cells because it is able to inhibit cytokine secretion, thus preventing dysfunction of the immune response that could result in tumorigenesis.

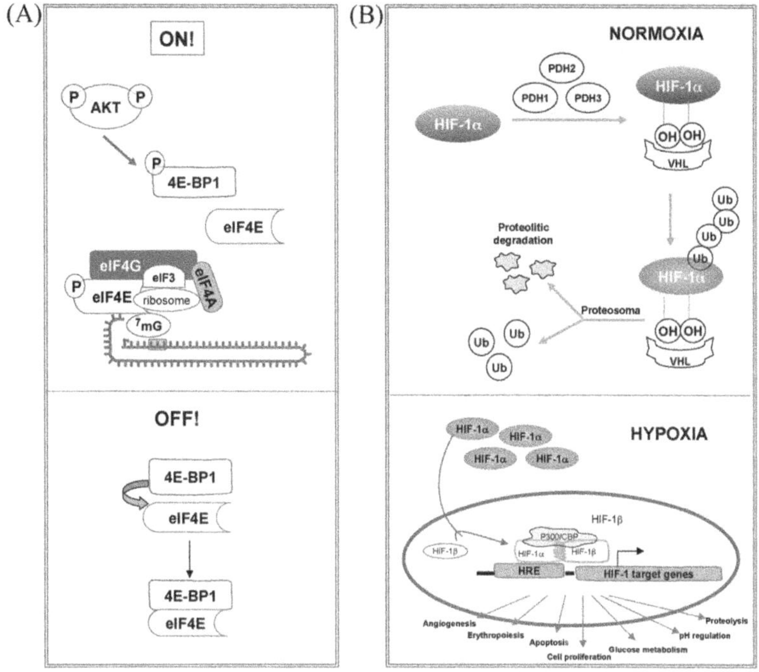

Figure 21.5 (A) Cap-dependent protein translation; (B) HIF-1α signaling. (A) The initiation phase of cap-dependent protein translation occurs by recruitment of ribosomes to mRNA 5′-cap structure (m7GpppN). This mechanism involves the eIF4F complex, composed of eIF4E, eIF4A, and eIF4G. (ON!) Akt-mediated phosphorylation of 4E-BP1 results in its dissociation from eIF4E and activation of mRNA translation; (OFF!) hypophosphorylated 4E-BP1 sequesters eIF4E. Interaction of 4E-BP1 with eIF4E inhibits complex assembly and represses translation. (B) Under normoxia conditions, HIF-1α is hydroxylated by the PHD2 and binds to the E3 ubiquitin ligase VHL complex, which promotes its ubiquitination and degradation. During hypoxia, HIF-1α moves into the nucleus and forms a complex with HIF-1β. HIF1 complex (HIF-1α/HIF-1β) binds to the hypoxia responsive elements (HRE), and activates the transcription of genes involved in several cell mechanisms.

21.3.2 L-Carnosine Induces Cell-Cycle Arrest in Colon Cancer Cells

Another aspect of the effect of L-carnosine on colon cancer concerns its capacity to induce cell-cycle arrest in HCT116, a cell line with a mutation in codon 13 of the RAS proto-oncogene, by targeting essential molecules for cancer metabolism. L-Carnosine treatment decreases the intracellular concentration of ATP and ROS levels. It has been demonstrated that, in colon cancer cells, generation of ROS is essential for KRAS-induced cell proliferation and tumorigenesis through regulation of the ERK–MAPK signaling pathway (Figure 21.6). In the HCT116 cell line L-carnosine reduced the level of ERK1/2 phosphorylation in a concentration-dependent manner. This result was consistent with a decrease in cell growth and correlates to an arrest in the G1 phase of the cell cycle. Indeed, under these conditions the protein level of the cyclin-dependent kinase inhibitor 1 (p21) increased and cyclin A protein levels decreased. The role of these factors as cell-cycle regulators has been well described. p21 is part of the cyclin-dependent kinase inhibitor (CKI) family and binds to cyclin/cyclin-dependent kinase (cyclin–CDK) complexes, regulating G1-S phase transition. In particular the interaction between cyclin A and CDK2 activates the enzymatic activity of CDK and is required for passage into S phase. The addition of 50–100 mM L-carnosine increases the cell number in the G1 phase by 73–87%, respectively, compared to 58% of control cells, without inducing significant levels of apoptosis. A reduction in ERK1/2 phosphorylation and cell proliferation is also reported in SW480 cells, an additional colon-rectal cancer cell line with a mutation in codon

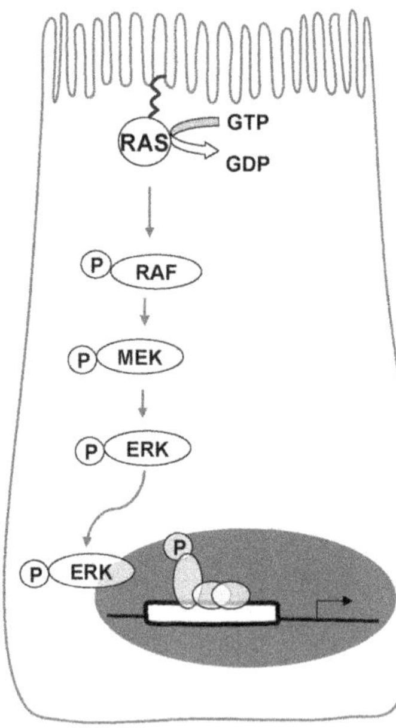

Figure 21.6 Ras/Raf/Mek/Erk transduction pathway. Extracellular signals, such as mitogens, inflammatory and growth factors, activate Ras/Raf/MEK/ERK transduction pathway. The activation of Ras, a membrane-anchored GTPase, directly induces phosphorylation of the three serine/threonine kinases Raf, MEK and ERKs. Activated ERKs moves into the nucleus and phosphorylates transcription factors involved in cellular functions, including cell proliferation, cell-cycle differentiation, and apoptosis.

12 of the RAS proto-oncogene. Accordingly, the antiproliferative effect of L-carnosine is not cell-line specific and could be mediated to KRAS signaling (Iovine *et al.*, 2012). The proto-oncogene KRAS is mutated in a wide array of human cancer and is important in tumor progression and in failure of response to antitumoral drugs. In HCT-116 cells the oncogenic KRAS contributes to transformation by inducing hypoxia inducible factor 1 (HIF-1) activation. Indeed, the activation of HIF-1 downstream of RAS signaling is responsible for the expression of several HIF-1-induced genes, which are involved in cell growth and survival. HIF-1 is a heterodimeric transcription factor consisting of an O_2-regulated HIF-1α and a constitutively expressed HIF-1β subunit. During hypoxia HIF-1α expression is rapidly induced, however, under normoxic conditions this subunit is degraded following hydroxylation by the prolyl hydroxylases (PHDs) (Fong and Takeda, 2008). The hydroxylated Pro residues promote the recruitment of HIF-1α by von Hippel–Lindau tumor suppressor protein (VHL), a recognition module of the E3-ubiquitin ligase complex responsible for its ubiquitination and subsequent degradation (Aragonés *et al.*, 2009). The HIF-1 complex (HIF1α/HIF1β) in the nucleus binds to the hypoxia-responsive elements (HRE), activating the transcription of genes involved in angiogenesis, iron metabolism, glycolysis, cell proliferation and cell survival (Figure 21.5B). This transcription factor, in fact, responds to the oxygen deficit by activating the transcription of genes encoding enzymes of the glycolytic pathway and activating pyruvate dehydrogenase kinase 1 (PDK1), which phosphorylates the pyruvate dehydrogenase preventing the conversion of pyruvate to acetyl CoA and consequently the utilization of pyruvate by the mitochondria (Semenza, 2013). Under hypoxic conditions the cells adapt to the oxygen deficit through activation of a definite gene program dependent on HIF-1α. This condition induces a "metabolic switch" that reduces the use of the tricarboxylic acid cycle and increases the activity of anaerobic glycolysis as the main source of ATP and is known as the Warburg effect. In HCT116 cells HIF-1α and oncogenic KRAS signaling could modulate cancer metabolism promoting ATP production by glycolysis and increasing ROS generation by mitochondrial metabolism, respectively. These observations support the theory of the involvement of L-carnosine also in the degradation signaling pathway of HIF-1α and are in agreement with the idea that in HCT116 cells HIF-1α and HIF-2α regulate cancer metabolism by overlapping with oncogenic KRAS (Chun *et al.*, 2010; Weinberg *et al.*, 2010; Asperger *et al.*, 2011; Iovine *et al.*, 2014).

In conclusion, on the basis of these considerations, a model is proposed to explain the antiproliferative effect of L-carnosine on colon cancer cells. In Figure 21.7 we introduce the possible targets of L-carnosine by hypothesizing that this dipeptide acts on cancer metabolism. This model also considers the possibility that HIF-1 signaling is one of the L-carnosine targets (Iovine *et al.*, 2014).

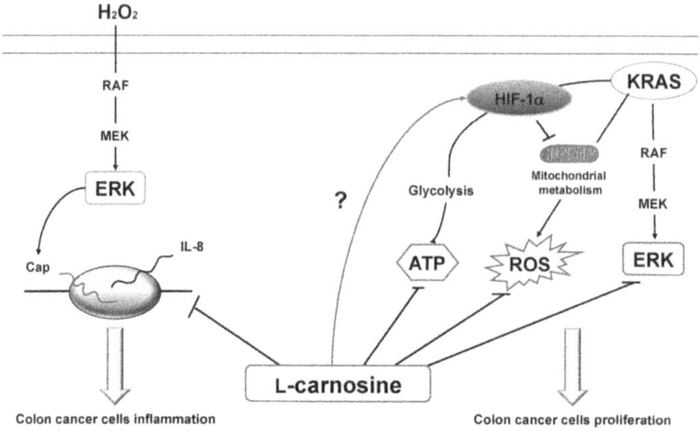

Figure 21.7 Summary of the molecular targets of L-carnosine in colon cancer cells. The oncogenic KRAS contributes to colon cancer cell proliferation by inducing hypoxia inducible factor 1 (HIF1) activation, increasing ROS generation by mitochondrial metabolism, and activating the pathways of MAP-family kinase. Furthermore, H_2O_2 activates MAPK, including ERK, which phosphorylates eIF4E. This event regulates cap-dependent protein translational initiation of interleukin-8 (IL-8), causing colon cancer cell inflammation. L-Carnosine decreases the intracellular concentration of ATP and ROS levels, reduces ERK1/2 phosphorylation, and arrests the cell cycle. In addition, during hydrogen-peroxide-induced inflammation L-carnosine affects ERK1/2 pathway by suppressing IL-8 translation.

21.4 Conclusions

The reported data suggest that the positive affect of L-carnosine on colon cancer cells is the result of a complex cellular response to several properties of this dipeptide. Although much is known about the properties of L-carnosine, its physiological functions remain unknown.

21.4.1 Critical Molecular Pathways Involved in the L-Carnosine Effect on Colon Cancer

An interesting feature of L-carnosine is its antioxidant activity. It has been reported that L-carnosine could exercise its antioxidant activity inhibiting the radical production by binding copper ions, interacting with lipid peroxidation products and inhibiting AGE (advanced glycation end product) formation, which are end products of the nonenzymatic glycation of proteins. AGEs bind the receptors for AGEs (RAGE) on the cellular membrane, inducing oxidative stress and increasing ROS production. Focusing the attention on the prevention of colon cancer by modulation of chronic inflammatory processes and cell proliferation, we cannot rule out that the effect of L-carnosine on colon cancer can be attributed mainly to its antioxidant action. Indeed, the common feature shared by inflammation and cell proliferation is ROS production. ROS is a principal player in the activation of the inflammatory pathway in colon cancer and also represents a good target to inhibit tumor progression. Indeed, under inflammatory conditions cells respond to tissue injury by inducing and activating oxidant-generating enzymes, which produce high concentrations of several free oxygen radicals. The ability of some antioxidants to suppress tumorigenesis has been reported, it has been hypothesized that a possible antitumorigenic effect of antioxidants is related to proteolysis of HIF-1α. As reported above, the HIF pathway represents a good candidate target for L-carnosine in colon cancer (Figure 21.7). Another critical molecular signaling pathway that may elucidate the L-carnosine effect on colon cancer is the MAPK pathway, in particularly ERK1/2 phosphorylation (Figure 21.6). L-Carnosine activity may have a pivotal role in the inactivation of ERK1/2. Indeed, the hypophosphorylation of ERK1/2 could explain the inhibition of IL-8 mRNA translation in H_2O_2-induced inflammation, supported by the finding that eIF-4E is not phosphorylated in CaCo-2 cells. On the other hand, it is noted that, among their several functions, these kinases could mediate the activation and regulation of HIF-1α. A major characteristic of L-carnosine that makes this dipeptide a good candidate for new therapeutic strategies is its bioavailability. Indeed, cellular uptake of drug-like molecules is the major obstacle in the development of an effective cancer therapy. New anticancer strategies aim to deliver and facilitate the uptake and intracellular accumulation of drugs into cancer cells while reducing the nonspecific uptake into normal cells. From this point L-carnosine could represent a good adjuvant candidate for the treatment of colon cancer because it is thought to be absorbed primarily *via* intestinal epithelium cells. Indeed, this dipeptide was transported across the apical membrane of CaCo-2 cells by the proton-coupled peptide transporter (PePT1), a molecule mainly expressed in the small intestine, and its uptake was dependent on lowering the apical pH value (Son *et al.*, 2004). The level of uptake of L-carnosine seems to be important in fact a competition experiment performed with a PepT1 substrate demonstrated that L-carnosine inhibits IL-8 secretion only at high intracellular concentrations. Preincubation with L-carnosine for 3 h before treatment with H_2O_2 increases its uptake and appears to be necessary in the inhibition of IL-8 secretion under inflammatory conditions (Son *et al.*, 2008). In addition, it should be remembered that L-carnosine can be endogenously synthesized and its tissue concentration can be increased by supplementing with a normal diet of meat, poultry and fish.

Summary Points

- This chapter focuses on the effects of L-carnosine in human colon cancer.
- Colon cancer is a common malignancy and a leading cause of cancer death worldwide.
- Some risk factors for getting colorectal cancer are beyond our control, while other risk factors can be controlled.

- Diet and inflammatory diseases are the main risk factors for colon cancer development.
- In recent years it has been demonstrated that L-carnosine has a protective effect against oxidative stress in intestinal epithelial cells.
- L-Carnosine inhibits interleukin-8 (IL-8) secretion that plays a pivotal role in inflammatory diseases within the intestine.
- L-Carnosine also inhibits the proliferation of human colon cancer cells.
- The antioxidant activity of L-carnosine might be the major feature by which this dipeptide performs its function.
- We argue that L-carnosine could represent a good adjuvant candidate for the treatment of colon cancer. L-Carnosine, indeed, is primarily absorbed by intestinal epithelial cells, where it plays a key role as an antioxidant molecule.

Key Facts

Key Features of Signal Transduction Mediated by the Ras/Raf/MEK/ERK Pathway

1. The primary function of the Ras/Raf/MEK/ERK transduction pathway is to couple signals from cell surface receptors to transcription factors which regulate expression of genes involved in the cell growth, differentiation and survival pathway.
2. The specific components of this pathway vary greatly between different stimuli and cell types and consequently it can transmit signals that result in the prevention or induction of apoptosis or cell-cycle progression.
3. The overexpression and/or uncontrolled activation of the components of this pathway are frequently implicated in proliferative diseases, such as cancers, and contribute to the stages of initiation, progression, and metastasis.
4. Additional components of this pathway are discovered daily and there are numerous kinases and transcription factors that can be activated or inactivated by protein phosphorylation.
5. The activation of Ras directly induces the sequential phosphorylation of the three serine/threonine kinases Raf, MEK and ERKs.
6. The genes of the family RAS are involved in tumorigenesis of CRC. 30% of colon tumors have a mutation in KRAS.
7. Extracellular signal-regulated protein kinases 1 and 2 (ERK1/2), which are 44 and 42 kDa serine/threonine kinases. ERKs activation requires both threonine and tyrosine residue phosphorylation to be fully activated.
8. Activated ERKs moves into the nucleus and phosphorylates transcription factors regulating gene expression involved in cellular proliferation, differentiation, and survival.
9. Inhibitors of Ras, Raf, MEK and some downstream targets have been developed and many of them are currently in clinical trials.

Definitions of Words and Terms

Adenomatous polyposis coli (APC). A tumor-suppressor gene located on the long arm of chromosome 5. Mutations in this gene are responsible for colorectal cancer. The inactivation of the APC gene leads to an increase in cell proliferation and contributes to the formation of colonic polyps.

Cap structure. Structure of 5′end of all eukaryotic cellular mRNAs, which consists of a 7-methylguanosine linked to the 5′nucleoside of the mRNA chain through a 5′,5′-triphosphate bridge (m^7GpppN; where N is any nucleotide and m is a methyl group).

Cytokines. A group of molecules involved in the transmission of intracellular signals during immune responses. They are produced by several cell types, particularly lymphocytes and activated macrophages, but also endothelial cells, epithelial cells and connective tissue cells.

DNA mismatch repair (MMR). Repair system responsible for the integrity of the genome. This system is an evolutionarily conserved process that corrects mismatches generated during DNA replication.

Eukaryotic translation initiation factor 4E (eIF4E). This 24 kDa polypeptide specifically interacts with the 5′-cap structure (m^7GpppN; where N is any nucleotide) of mRNAs. The eIF4E-binding protein 1 (4E-BP1) belongs to a family of translation repressor proteins, interacts and regulates eIF4E activity by preventing its assembly into the eIF4F complex. Interaction of this protein with eIF4E inhibits complex assembly and represses the translation. This protein is phosphorylated in response to various signals, resulting in its dissociation from eIF4E and activation of mRNA translation.

Inflammatory bowel disease (IBD). Inflammatory bowel disease describes a group of chronic gastrointestinal disorders characterized by the presence of inflammatory cells in the lining of the digestive tract.

Intestinal epithelial cells (IEC). Highly specialized cells lining the small intestine and involved in the process of absorption of water, glucose molecules, and mineral ions. The epithelial cells have huge numbers of finger-like projections on their surface, called microvilli, which contain the enzymes that break down disaccharides into monosaccharides.

Kirsten rat sarcoma viral oncogene (KRAS). Oncogene with an intrinsic GTPase enzymatic activity. Its activation occurs by hormones or proteins binding their associated receptors, representing a central component of the system of signal transduction downstream of EGFR (epidermal growth factor receptor).

Nuclear factor kappa-light-chain-enhancer of activated B cells (NF-KB). Transcriptional factor that plays a critical role in regulation of genes involved in a wide range of biological processes including innate and adaptive immunity, inflammation and stress responses.

Reactive oxygen species (ROS). Molecules that have an unpaired electron in the outermost orbital, which confers high instability and, therefore, high reactivity with other molecules. Reactive oxygen species are divided into two main categories: free radicals, including superoxide and hydroxyl radical, and nonradical molecules, including hydrogen peroxide (H_2O_2).

Uptake of drug. This is the absorption system of a substance, food material or mineral, by a tissue for permanent or temporary retention.

Wnt signal transduction pathways. The name Wnt results from a fusion of the name of the Drosophila segment polarity gene *wingless* and the name of the vertebrate homolog, *integrated or int-1*. Wnt proteins bind to specific receptors and trigger a multistep process activating β-catenin; activated protein moves into the nucleus where it promotes cell growth and differentiation. The activation of β-catenin is known as the canonical Wnt signaling pathway.

Warburg effect. Theory postulated by Otto Heinrich Warburg describing the observation that cancer cells prefer glycolysis to produce energy even when enough oxygen is present to properly respire. The Warburg effect may be a consequence of damage to the mitochondria in cancer, or an adaptation to low-oxygen environments within tumors.

List of Abbreviations

FAP	Familial adenomatous polyposis
HNPCC	Hereditary nonpolyposis colorectal cancer
IBD	Inflammatory bowel disease
CIN	Chromosomal instability
MSI	Microsatellite instability
MMR	Microsatellite mismatch repair
MLH1	MutL homolog 1
CIMP	CpG island methylator phenotype
APC	Adenomatous polyposis coli
KRAS	Kirsten rat sarcoma viral oncogene homolog
EGFR	Epidermal growth factor receptor
IEC	Intestinal epithelial cells
TNF-α	Tumor necrosis factor-α
ERK1/2	Extracellular signal-regulated protein kinases 1 and 2
HIF-1	Hypoxia-inducible factor-1
MAPK	Mitogen-activated protein kinase
VHL	von Hippel–Lindau tumor suppressor protein
PHD	Prolyl hydroxylase domain
HRE	Hypoxia responsive element
PDK1	Pyruvate dehydrogenase kinase 1
NSAID	Nonsteroidal anti-inflammatory drugs
eIF4E	Eukaryotic translation initiation factor 4E
AKT	Serine/threonine-specific protein kinase
4E-BP	Eukaryotic translation initiation factor 4E-binding protein 1
mTOR	Mammalian target of rapamycin
PePT1	Proton-coupled peptide transporter 1

References

von Allmen, D., 2006. Intestinal polyposis syndromes: progress in understanding and treatment. *Current Opinion in Pediatrics*. 18: 316–320.

Aragonés, J., Fraisl, P., Baes, M. and Carmeliet, P., 2009. Oxygen sensors at the crossroad of metabolism. *Cell Metabolism*. 9: 11–22. Review.

Asperger, A., Renner, C., Menzel, M., Gebhardt, R., Meixensberger, J. and Gaunitz, F., 2011. Identification of factors involved in the anti-tumor activity of carnosine on glioblastomas using a proteomics approach. *Cancer Investigation*. 29: 272–281.

Azar, R., Lasfargues, C., Bousquet, C. and Pyronnet, S., 2013. Contribution of HIF-1α in 4E-BP1 gene expression. *Molecular Cancer Research*. 11: 54–61.

Bilir, B., Kucuk, O. and Moreno, C. S., 2013. Wnt signaling blockage inhibits cell proliferation and migration, and induces apoptosis in triple-negative breast cancer cells. *Journal of Translational Medicine*. 11: 280.

Chao, A., Thun, M. J., Connell, C. J., McCullough, M. L., Jacobs, E. J., Flanders, W. D., Rodriguez, C., Sinha, R. and Calle, E. E., 2005. Meat consumption and risk of colorectal cancer. *Journal American Medical Association*. 293: 172–182.

Chun, S. Y., Johnson, C., Washburn, J. G., Cruz-Correa, M. R., Dang, D. T. and Dang, L. H., 2010. Oncogenic KRAS modulates mitochondrial metabolism in human colon cancer cells by inducing HIF-1α and HIF-2α target genes. *Molecular Cancer*. 13; 9: 293.

Derks, S., Postman, C., Carvalho, B., van den Bosch, S. M., Moerkerk, P., Herman, J. G., Weijenberg, M. P., de Brűine, A. P., Meijer, G. A. and van Engeland, M., 2008. Integrated analysis of chromosomal, microsatellite and epigenetic instability in colorectal cancer identifies specific associations between promoter methylation of pivotal tumour suppressor and DNA repair genes and specific chromosomal alterations. *Carcinogenesis*. 29: 434–439.

Edkins, S., O'Meara, S., Parker, A., Stevens, C., Reis, M., Jones, S., Greenman, C., Davies, H., Dalgliesh, G., Forbes, S., Hunter, C., Smith, R., Stephens, P., Goldstraw, P., Nicholson, A., Chan, T. L., Velculescu, V. E., Yuen, S. T., Leung, S. Y., Stratton, M. R. and Futreal, P. A., 2006. Recurrent KRAS codon 146 mutations in human colorectal cancer. *Cancer Biology & Therapy*. 5: 928–932.

Er, T. K., Chen, C. C., Bujanda, L. and Herreros-Villanueva, M., 2014. Clinical relevance of KRAS mutations in codon 13: where are we? *Cancer Letter*. 343: 1–5.

Fong, G. H. and Takeda, K., 2008. Role and regulation of prolyl hydroxylase domain proteins. *Cell Death & Differentiation*. 15: 635–641.

Gryfe, R., Swallow, C., Bapat, B., Redston, M., Gallinger, S. and Couture, J., 1997. Molecular biology of colorectal cancer. *Current Problems in Cancer*. 21: 233–300.

Haggar, F. A. and Boushey, R. P., 2009. Colorectal cancer epidemiology: incidence, mortality, survival, and risk factors. *Clinics in Colon and Rectal Surgery*. 22: 191–197.

Hughes, L. A., Khalid-de Bakker, C. A., Smits, K. M., van den Brandt, P. A., Jonkers, D., Ahuja, N., Herman, J. G., Weijenberg, M. P. and van Engeland, M., 2012. The CpG island methylator phenotype in colorectal cancer: progress and problems. *Biochimica et Biophysica Acta*. 1825: 77–85.

Iovine, B., Iannella, M. L., Nocella, F., Pricolo, M. R. and Bevilacqua, M. A., 2012. Carnosine inhibits KRAS-mediated HCT116 proliferation by affecting ATP and ROS production. *Cancer Letter*. 315: 122–128.

Iovine, B., Oliviero, G., Garofalo, M., Orefice, M., Nocella, F., Borbone, N., Piccialli, V., Centore, R., Mazzone, M., Piccialli, G. and Bevilacqua, M. A., 2014. The Anti-proliferative effect of L-carnosine correlates with a decreased expression of hypoxia inducible factor 1 alpha in human colon cancer cells. *PLoS One*. 9(5): e96755.

Itzkowitz, S. H. and Yio, X., 2004. Inflammation and cancer IV. Colorectal cancer in inflammatory bowel disease: the role of inflammation. *American Journal of Physiology: Gastrointestinal and Liver Physiology*. 287: G7–G17. Review.

Jasperson, K. W., 2012. Genetic testing by cancer site: colon (polyposis syndromes). *Cancer Journal*. 18: 328–333.

Mori, Y., Yin, J., Sato, F., Sterian, A., Simms, L., Selaru, F., Schulmann, K., Xu, Y., Olaru, A., Wang, S., Deacu, E., Abraham, J. M., Young, J., Leggett, B. A. and Meltzer, S. J., 2004. Identification of genes uniquely involved in frequent microsatellite instability colon carcinogenesis by expression profiling combined with epigenetic scanning. *Cancer Research*. 64: 2434–2438.

Neufeld, K. L., 2009. Nuclear APC. *Advances in Experimental Medicine and Biology*. 656: 13–29.

Ohshima, H., Tazawa, H., Sylla, B. S. and Sawa, T., 2005. Prevention of human cancer by modulation of chronic inflammatory processes. *Mutation Research*. 591: 110–122.

Prior, I. A., Lewis, P. D. and Mattos, C., 2012. A comprehensive survey of RAS mutations in cancer. *Cancer Research*. 72: 2457–2467.

Rajagopalan, H., Bardelli, A., Lengauer, C., Klinzer, K. W., Vogelstein, B. and Velculescu, V. E., 2002. Tumorigenesis: RAF/RAS oncogenes and mismatch-repair status. *Nature*. 418: 934.

Raman, M., Chen, W. and Cobb, M. H., 2007. Differential regulation and properties of MAPKs. *Oncogene*. 26: 3100–3112.

Semenza, G. L., 2013. HIF-1 mediates metabolic responses to intratumoral hypoxia and oncogenic mutations. *Journal of Clinical Investigation*. 123: 3664–3671.

Shimizu, M., 2010. Interaction between food substances and the intestinal epithelium. *Bioscience Biotechnology and Biochemistry*. 74: 232–241.

Son, D. O., Satsu, H., Kiso, Y. and Shimizu, M., 2004. Characterization of carnosine uptake and its physiological function in human intestinal epithelial Caco-2 cells. *BioFactors*. 21: 395–398.

Son, D. O., Satsu, H., Kiso, Y., Totsuka, M. and Shimizu, M., 2008. Inhibitory effect of carnosine on interleukin-8 production in intestinal epithelial cells through translational regulation. *Cytokine*. 42: 265–276.

Vogelstein, B., Fearon, E. R., Hamilton, S. R., Kern, S. E., Preisinger, A. C., Leppert, M., Nakamura, Y., White, R., Smits, A. M. and Bos, J. L., 1988. Genetic alterations during colorectal-tumor development. *New England Journal of Medicine*. 319: 525–532.

Weinberg, F., Hamanaka, R., Wheaton, W. W., Weinberg, S., Joseph, J., Lopez, M., Kalyanaraman, B., Mutlu, G. M., Budinger, G. R. and Chandel, N. S., 2010. Mitochondrial metabolism and ROS generation are essential for Kras-mediated tumorigenicity. *Proceedings of the National Academy of Sciences of the United States of America*. 11; 107: 8788–8793.

CHAPTER 22

Carnosine-Induced Neuroprotection

S. M. FAYAZ[a], ARSHAD MAJID[b], AND G. K. RAJANIKANT*[a]

[a]School of Biotechnology, Coordinator, DBT – Centre for Bioinformatics, National Institute of Technology Calicut, Calicut-673601, India; [b]Sheffield Institute for Translational Neuroscience (SITraN), University of Sheffield, 385A Glossop Road, Sheffield S10 2HQ, UK

*E-mail: rajanikant@nitc.ac.in

22.1 Introduction

Carnosine was discovered by Russian chemist W. S. Gulewitsch in 1900 from Liebig's meat extract and was subsequently identified as β-alanyl-histidine in 1918. It is found exclusively in tissues like brain, cardiac muscle, kidney, stomach, olfactory bulbs and in large amounts in skeletal muscle. A high rate of carnosine synthesis occurs in glial cells, and is subsequently delivered to neurons. Carnosine is endogenously synthesized by carnosine synthetase from β-alanine and L-histidine. It is hydrolyzed into its constituent amino acids by carnosinase.

Carnosine is an important antioxidant, metal-ion chelator, aldehyde-scavenger, antiglycating and anticrosslinking agent. It is responsible for a variety of beneficial activities related to the detoxification of the body from free-radical species and the byproducts of membrane lipid peroxidation. It also has membrane-protecting activity, proton-buffering capacity, formation of complexes with transition metals, and regulation of macrophage

Food and Nutritional Components in Focus No. 8
Imidazole Dipeptides: Chemistry, Analysis, Function and Effects
Edited by Victor R Preedy

Published by the Royal Society of Chemistry, www.rsc.org

function. Carnosine acts as a natural scavenger of dangerous reactive aldehydes from degradative oxidative pathway of endogenous molecules such as sugars, polyunsaturated fatty acids (PUFAs) and proteins. In particular, it is a potent and selective scavenger of α,β-unsaturated aldehydes that are typical byproducts of membrane lipid peroxidation and are also considered as secondary messengers of oxidative stress. Carnosine inhibits aldehyde-induced protein–protein and DNA–protein crosslinking in neurodegenerative disorders.

Carnosine also operates as an intracellular pH buffer. The amino acid constituents are accountable for its buffering capacity. The imidazole side chain of histidine has a pK_a of approximately 6.0, and overall, the amino acid has a pK_a of 6.5. Histidine is a relatively weak buffer over the physiological intramuscular pH range. This means that, at physiologically relevant pH values, relatively small shifts in pH will change its average charge. However, when bound to other amino acids, the pK_a increases nearer to 6.8–7.0. In particular, when bound to β-alanine, the pK_a value is 6.83, making it a very efficient buffer.

Carnosine is a pleiotropic molecule and confers neuroprotection against experimental models in a wide array of neurological disorders by virtue of its diverse physiological properties. In addition, its neuroprotective capability is backed by its ability to cross the blood–brain barrier and reach the site of action. The mechanisms through which carnosine induces neuroprotection, and the various proteins that are bound and influenced by carnosine are presented in Figure 22.1 and Table 22.1.

In this chapter, we have discussed the properties of carnosine and how these varied properties may make it highly beneficial in a number of neurological disorders.

22.2 Pleiotropic Properties of Carnosine

22.2.1 Carnosine as an Antioxidant

Oxidative stress is caused by an imbalance between the production of reactive oxygen species (ROS) and a biological system's ability to readily detoxify the reactive intermediates or easily repair the resulting damage. The cellular processes involved in the production of metabolic energy, like Adenosine triphosphate (ATP), uses oxygen. During these processes, ROS are produced as byproducts. ROS include peroxides like hydrogen peroxide (H_2O_2) and oxygen free radicals such as the hydroxyl radical ($^{\bullet}OH$) and the superoxide anion (O^{2-}). Thus, all forms of life maintain a reducing environment within their cells. This reducing environment is preserved by natural antioxidant defense mechanisms present inside the cells in the form of antioxidant enzymes, like glutathione peroxidase, catalase, and superoxide dismutase, as well as antioxidant compounds like carnosine. However, disturbances in this normal redox state in the form of severe oxidative stress can cause damage to major cellular components, including proteins, lipids, and DNA, thereby eventually

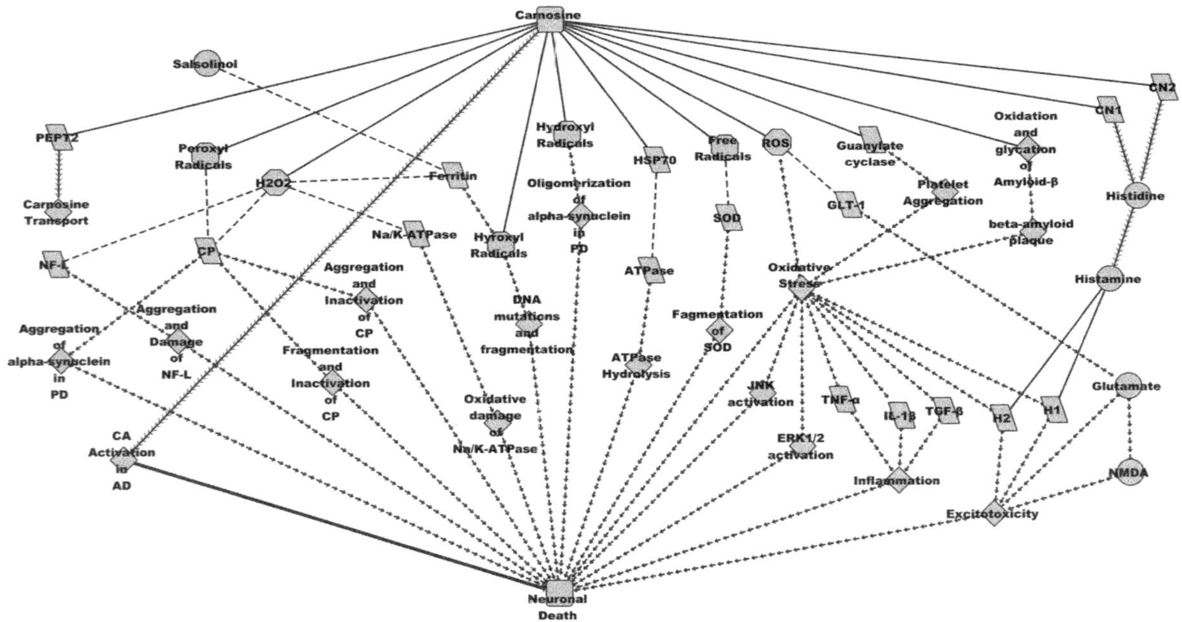

Dash Line – Interaction, when carnosine is absent
Solid Line – Interaction, when carnosine is present
Solid Line with blunt end – Inhibition, when carnosine is present
Separate Arrow – Lead to/Result, when carnosine is absent
Continuous Arrow – Lead to/Result, when carnosine is present

Figure 22.1 Mechanisms through which carnosine induces neuroprotection. Neuronal death is seen in most of the neurological disorders like cerebral ischemia, Alzheimer's disease and Parkinson's disease. The mechanisms that are involved in this neuronal death are shown. The fate of these mechanisms changes when carnosine is present. Therefore, the overall mechanism of action of carnosine in neuroprotection is presented.

Table 22.1 Proteins that are influenced by carnosine. Carnosine induces neuroprotection by interacting with various proteins and influencing them either directly, by binding to them, or indirectly, by influencing other proteins. These proteins are involved in pathological mechanisms related to various neurological disorders.

Sl no.	Proteins	Interaction type
1	Na/K-ATPase	Indirect
2	HSP70	Direct
3	ATPase	Indirect
4	Ceruloplasmin (CP)	Indirect
5	Carnosinase CN1	Direct
6	Carnosinase CN2	Direct
7	Histamine receptor H1	Indirect
8	Histamine receptor H2	Indirect
9	GLT-1	Indirect
10	Neurofilament-L (NF-L)	Indirect
11	Guanylatecyclise	Direct
12	Transforming growth factor-β (TGF-β)	Indirect
13	Interleukin (IL)-1β	Indirect
14	Tumor necrosis factor (TNF)-α	Indirect
15	Superoxide dismutase (SOD)	Indirect
16	PEPT2	Direct
17	Ferritin	Indirect
18	STAT3	Indirect
19	Amyloid-β	Direct

leading to cell death through apoptosis or necrosis. Since oxidative stress is a major underlying deleterious process behind cell death in ischemia, neurodegenerative diseases, and abnormal cell division in brain tumors, antioxidants are being investigated as potential neurotherapeutic agents against these diseases.

Carnosine is a well-established antioxidant. It is a hydroxyl and superoxide radical scavenger and a strong quencher of singlet molecular oxygen (Pavlov *et al.*, 1993). Apart from ROS, carnosine has been demonstrated to scavenge α,β-unsaturated aldehydes, formed from peroxidation of cell membrane fatty acids, during oxidative stress. When L-carnosine was administered with epigallocatechin gallate (EGCG), it had a profound effect on the activation of two stress-responsive pathways, namely HO-1 and Hsp72 (the inducible form of Hsp70), which play an important role in cytoprotection against oxidative stress-induced cell damage. Other studies on age-related oxidative stress in liver, heart and brain tissues of aged rats illustrated that the combination of carnosine with vitamin E increased its antioxidant efficiency. This indicates that the carnosine based combined treatment may be beneficial in the prevention of oxidative stress-related neurological disorders.

Carnosine exhibited neuroprotection against nitric oxide (NO) induced cell death in an *in vitro* model of primary rat astroglial cell cultures treated with lipopolysaccharide (LPS) and interferon-γ (INF-γ) (Nicoletti *et al.*, 2007). In primary neuronal cell cultures and *ex vivo* brain homogenates, carnosine exhibited robust antiexcitotoxic, antioxidant, and mitochondria protecting activity (Bae *et al.,* 2013). Carnosine and its derivative, *N*-acetyl carnosine could protect antioxidant enzymes during oxidative stress. Both carnosine and *N*-acetyl carnosine were shown to significantly attenuate the LPS-induced NO synthesis and the expression of inducible nitric oxide synthase (iNOS). A direct interaction of NO with carnosine and *N*-acetyl carnosine was also demonstrated (Nicoletti *et al.*, 2007).

22.2.2 Carnosine as Antiglycating Agent

Glycation impairs the functioning of biomolecules through the conversion of simple sugars such as glucose, fructose, and galactose to advanced glycation end products (AGEs). Glycation is implicated in the onset of various neurological disorders. The role of glycation, AGEs and the receptor for AGEs (RAGE) in CNS disorders has been reviewed elsewhere (Münch *et al.*, 2012).

AGEs are thought to be an important factor in many age-related chronic diseases such as cardiovascular diseases, Alzheimer's disease (AD), cancer and peripheral neuropathy. This range of diseases is the result of the very basic level at which glycations interfere with molecular and cellular functioning throughout the body. Enhanced oxidative stress is caused when AGEs induce the secretion of a variety of cytokines by binding to macrophage, endothelial, and mesangial cells. Carnosine could inhibit AGE formation both *in vitro* and *in vivo*. It can also protect the antioxidant enzymes, namely superoxide dismutase and catalase, from nonenzymatic glycation and protein crosslinking (Hipkiss, 2007). Thus, carnosine opposes glycation, which is a common event seen in neurological disorders.

22.2.3 Carnosine as Metal-Ion Chelator

Metals such as iron, copper, chromium, vanadium and cobalt are capable of redox cycling, in which, a single electron may be accepted or donated by the metal. This action catalyzes reactions that produce ROS, thereby leading to oxidative stress. Zinc and copper are endogenous transition metals that are synaptically released during neuronal activity. Under normal conditions, these metals probably function to modulate neuronal excitability. However, they can also be neurotoxic and may even contribute to the neuropathology associated with a variety of neurological disorders. Carnosine modulates the effects of zinc and copper on their neuronal excitability and neurotoxic effects as well. Carnosine could rescue neurons from zinc- and copper-mediated neurotoxicity (Horning *et al.*, 2000). Thus, the action of metals in neuronal death is inhibited by the natural metal-chelation property of carnosine.

22.2.4 Carnosine and Altered Proteins

Accumulation of altered proteins is seen in neurological disorders such as AD and Parkinson's disease (PD). Altered protein accumulation may be caused by decreased intracellular elimination of aberrant polypeptides by proteasomal activity and autophagy. It should be noted that, compared to normal gene products, error-proteins or altered proteins are more readily glycated and undergo oxidative damage by ROS. Carnosine may improve cellular ability to deal with altered proteins. There is much evidence, mostly obtained from model systems, suggesting that carnosine could suppress protein modifications mediated by ROS. It could stimulate intracellular proteolysis in cultured human fibroblasts (Hipkiss *et al.*, 1998). Further, it could also stimulate neutral, nonlysosomal proteolysis in cell-free extracts of rat brain.

All these properties craft carnosine as a potent neuroprotective agent. The effect of carnosine in cerebral ischemia, AD, PD, brain tumor and other neurological disorders has been discussed below.

22.3 Role of Carnosine in Neurological Disorders

22.3.1 Cerebral Ischemia

Cerebral stroke continues to be a major cause of death and the leading cause of long-term disability worldwide. Severe cerebral ischemic insult leads to metabolic stress, ionic perturbations, and a complex cascade of biochemical and molecular events ultimately causing the loss of cellular integrity and tissue destruction. Mechanisms of cell damage include excitotoxicity, oxidative stress, calcium overload and inflammation. Ischemia in brain, if prolonged and severe enough, ultimately results in neuronal cell death by apoptosis and necrosis.

Carnosine and homocarnosine were found to induce neuroprotection against ROS in an ischemic model comprising PC12 cell cultures exposed to oxygen–glucose deprivation (OGD) followed by reoxygenation. In our previous studies, we have shown that carnosine has antioxidant and antiapoptotic roles in the animal models of focal cerebral ischemia (Rajanikant *et al.*, 2007; Min *et al.*, 2008; Bae *et al.*, 2013). Carnosine conferred cerebral protection in both the permanent and transient cerebral ischemia models through the attenuation of oxidative stress and apoptosis. In rats with transient cerebral ischemia, induced by right unilateral common carotid arteries occlusion (rUCCAO), carnosine elicited a significant decrease in lipid peroxidation and increase in antioxidant enzyme activities. In a mouse model of permanent focal cerebral ischemia induced by the middle cerebral artery occlusion (pMCAO), carnosine was able to improve neurological function (Min *et al.*, 2008). Carnosine and its analogs namely homocarnosine and anserine, could exert neuroprotection in the brains of neonatal rats against free-radical damage during hypoxia–ischemia insults. They were also able to reduce the activity of brain tyrosine hydroxylase, an enzyme activated by free radicals (Boldyrev *et al.*, 1997). Carnosine could inhibit the mRNA expression of apoptosis-inducing factor (AIF) and caspase-3 that are involved in ischemic cell death. It also improved the spatial learning and memory abilities of rats. Further, increase in superoxide dismutase (SOD) activity and decrease in the malondialdehyde (MDA) level were also observed.

Carnosine had a therapeutic effect in the brain of rats and Mongolian gerbils against global ischemic injury. It showed a pronounced protective effect on neurological symptoms and animal mortality. It also protected the brain against membrane disordering, which is seen during ischemia (Stvolinsky *et al.*, 2000). During ischemia, matrix metalloproteinases (MMPs) have been shown to be involved in the degradation of extracellular matrix, disrupting the blood–brain barrier and allowing inflammatory cells to enter the brain. Carnosine treatment was found to decrease MMP-9 activity and protein levels in the brain after ischemia (Rajanikant *et al.*, 2007). The possible mechanism of its MMP inhibition may be due to the chelation of calcium or catalytic zinc required for MMP activity. In addition, it might also be due to the reduction of ROS levels by carnosine, because ROS has been shown to be involved in MMP activation.

Carnosine is hydrolyzed into histidine, which is a precursor of histamine. The conversion of histidine to histamine is catalyzed by histidine decarboxylase (HDC) enzyme. However, it was reported that carnosine exerts neuroprotection against oxidative stress and necrotic cell death through the histidine/histamine pathway in other disease and organ models, the same does not seem to apply during ischemia. The mechanistic pathways that mediate the neuroprotective effect of carnosine in ischemic brain injury were elucidated using *in vitro* primary astrocytes and cortical neurons, and an *in vivo* model of ischemic stroke (Bae and Majid, 2013). It was demonstrated that the protective effects of carnosine is only partially mediated by its activity at the histamine type 1 receptor on astrocytes, and the histidine/histamine pathway did not play a critical role in carnosine induced neuroprotection. In another study, carnosine ameliorated white-matter lesion and cognitive impairment after rUCCAO in both wild-type mice and HDC knockout mice, which are lacking in endogenous histamine. Further, it significantly attenuated the increase in ROS generation after rUCCAO. In contrast, administration of histidine did not show the same effect. Therefore, the neuroprotective effect of carnosine may not be dependent on the histaminergic pathway during ischemic cell death.

However, other studies have shown that histamine receptors and histaminergic pathway are involved in the protection of carnosine against NMDA-induced neurotoxicity. Carnosine protected brain microvascular endothelial cells against rotenone-induced oxidative stress injury *via* histamine H(1) and H(2) receptors. After rotenone exposure, carnosine decreased the number of apoptotic cells that was reversed either by histamine H(1) antagonists pyrilamine and diphenhydramine, or H(2) antagonists cimetidine and zolatidine. Similar results were observed in another study where carnosine protected HDC knockout mice with permanent cerebral ischemia, by reducing glutamate excitotoxicity. It decreased the glutamate levels and preserved the expression of glutamate transporter-1 (GLT-1) but not the glutamate/aspartate transporter in astrocytes exposed to ischemia *in vivo* and *in vitro*. Thus, these studies revealed that the neuroprotection of carnosine against ischemia may also be attributed to its antiexcitotoxic effects.

Carnosine could inhibit neuronal cell apoptosis through signal transducer and activator of transcription 3 (STAT3) signaling pathway. Investigation on the antiapoptotic mechanisms of carnosine in the focal cerebral ischemia using rat models of pMCAO revealed that L-carnosine-treated rats showed an upregulation in the expression of STAT3 mRNA levels and higher protein expression of pSTAT3 (Wang *et al.*, 2013).

Ideal therapy for ischemic stroke should have high efficacy, a long therapeutic time window, and acceptable toxicity. Carnosine was found to have all these characteristics (Bae *et al.*, 2013). Intravenous treatment of carnosine in rat models of pMCAO exhibited robust cerebroprotection in a dose-dependent manner. It did not exhibit any evidence of adverse effects or toxicity in safety and tolerability assessments. Moreover, histological evaluation of organs, complete blood count, coagulation tests, and serum chemistry did not reveal any abnormalities. Thus, carnosine was suggested to be an effective neuroprotectant against ischemic insults (Figure 22.2).

22.3.2 Alzheimer's Disease

It has been postulated that β-amyloid (Aβ) deposits in neuronal cells are a fundamental cause of Alzheimer's disease. While AGEs are considered a part of the normal aging process, accelerated production is implicated in the formation of β-amyloid plaques and neurofibrillary tangles. Amyloid proteins were produced as side products during the reactions progressing to AGEs that lead to glycation and accumulation of Aβ proteins, which are usually found in the cerebrospinal fluid (CSF) of AD patients. The aberrant accumulation of Aβ is directly involved in the acceleration of neuronal injuries during AD and its plasma levels were found to strictly correlate to the severity of the disease. Carnosine, through its antiglycating property, showed potential in counteracting neurodegenerative effects arising from altered protein accumulation and toxicity (Kawahara *et al.*, 2011). Carnosine supplementation in 3xTg-AD mice promoted a strong reduction in the hippocampal intraneuronal accumulation of Aβ and completely rescued AD and aging-related mitochondrial dysfunctions. It also rescued cognitive deficits caused by feeding a high fat diet in a transgenic mouse model of AD.

In vitro studies on the effect of carnosine against a truncated form of the Aβ peptide (Aβ25–35) on rat brain vascular endothelial cells showed that it could completely prevent the cellular damage caused by this peptide (Preston *et al.*, 1998). The Aβ(25–35) is a highly toxic peculiar peptide, present in senile plaques and degenerating hippocampal neurons of AD brains. Carnosine was also effectively engaged in the protection of differentiated rat PC12 cells against Aβ42-induced neurotoxicity. It could attenuate the aggregation of

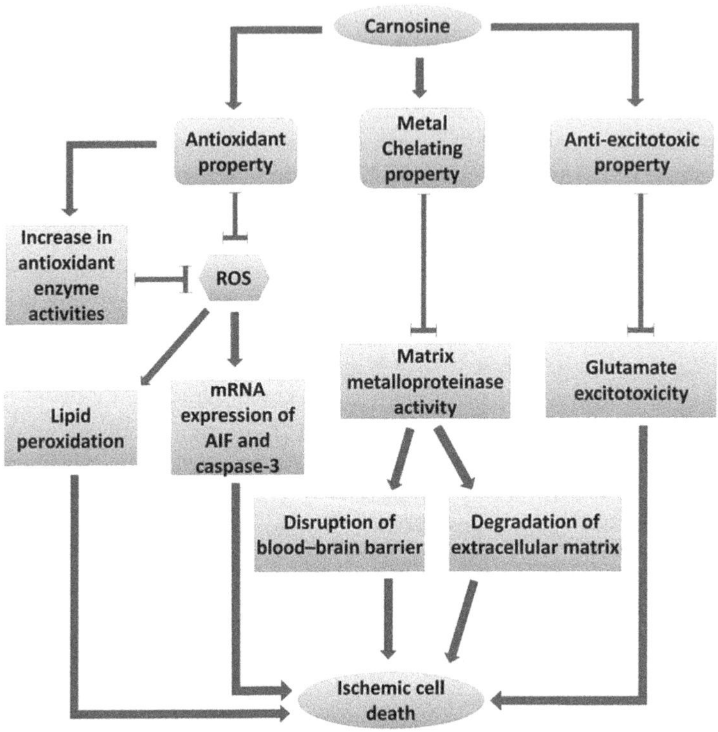

Figure 22.2 Carnosine-induced neuroprotection in cerebral ischemia. During cerebral ischemia, ROS induces cell death through lipid peroxidation and expression of apoptotic proteins. Matrix metalloproteinases induces disruption of blood–brain barrier and degradation of extracellular matrix that leads to cell death. Glutamate excitotoxicity also leads to cell death. Carnosine induces neuroprotection in cerebral ischemia through its antioxidant property that inhibits ROS. It also leads to an increase in antioxidant enzymes that in turn inhibit ROS. Carnosine inhibits matrix metalloproteinase activity through its metal-chelating property. It also inhibits glutamate excitotoxicity through its antiexcitotoxic property.

amyloidogenic peptide fragments of Aβ1–42 that are involved in AD injury. The fluorescence assays confirmed that Aβ1–42 polymerization of amyloid aggregates was quantitatively reduced in the presence of carnosine in a dose-dependent manner (Alsoisi *et al.*, 2013).

In silico studies supported by experimental results proved that carnosine is a very good ligand of Aβ1–42 (Alsoisi *et al.*, 2013). Through docking studies, the predicted binding mode for carnosine and Aβ1–42 fibril displayed the close interactions between carnosine and the residues D23 (L-aspartate) and K28 (L-lysine) of Aβ1–42. This was in agreement with the antiaggregating properties of carnosine observed *in vitro*, being the D23 and K28 residues directly involved in the Aβ1–42 self-association process by forming an intermolecular salt bridge between two adjacent Aβ1–42 monomers. In this context, the binding of carnosine to the Aβ1–42 monomer could prevent the intermolecular salt-bridge formation, thus inhibiting the fibril aggregation process. Molecular dynamics (MD) simulations indicated that carnosine interacts transiently with monomeric Aβ42 through salt bridges and van der Waals contacts. NMR experiments on the nonaggregative fragment Aβ12–28 were in agreement with MD simulations (Attanasio *et al.*, 2013). Thus, the experimental and simulation results suggest that carnosine might block the pathway toward toxic aggregates by perturbing the hydrogen-bond network near residues with key roles in fibrillogenesis.

Oxidative stress is a significant factor in the development and pathology of AD. Mitochondria-driven overproduction of ROS and imbalanced homeostasis of endogenous metals like zinc (Zn^{2+}) are important cofactors in the development and progression of AD. In Alzheimer's patients, peroxidation of lipids and proteins was observed, which were subsequently accumulated in the lysosomes of brain cells. Due to the presence of histidine residue, carnosine has been proposed as a chelator for divalent cations like Cu^{2+} and Zn^{2+}. For instance, Cu^{2+} and Zn^{2+} are released during normal synaptic activity. However, in the presence of a mildly acidic environment, which is a characteristic of AD, they are reduced to their ionic forms and become toxic to the nervous system. In this case, carnosine could buffer copper and zinc toxicity in the brain and rescue the neuronal cells. Loss of olfactory function and oxidative damage to olfactory tissue are early symptoms of AD. The olfactory lobe is normally enriched in carnosine and zinc. In the olfactory bulb, carnosine prevented the inhibitory effects of copper on synaptic transmission, through chelation, thus acting as a neuromodulator (Trombley *et al.*, 1998). Therefore, carnosine exerts neuroprotection against AD through its antiglycating, antiaggregating, antioxidant and metal-ion-chelation properties (Figure 22.3).

Investigation on the genes involved in carnosine metabolism, in the context of AD, revealed that the activity of brain-specific carnosinase was altered in fluids from patients with AD dementia (Balion *et al.*, 2007). Therefore, carnosinase has been proposed as a novel biomarker in the cerebrospinal fluid for staging early AD.

22.3.3 Parkinson's Disease (PD)

PD is characterized by the accumulation of α-synuclein into inclusions called Lewy bodies in neurons. When α-synuclein was incubated with both Cu,Zn-SOD and H_2O_2, then the amount of oligomerization of A53T mutant, which is associated with rare familial forms of PD, increased relative to that of the wild-type protein. This oligomerization of α-synuclein is mediated by the generation of hydroxyl radical through metal-catalyzed reaction. Carnosine was able to effectively inhibit oligomerization of the wild-type and mutant proteins, through its free-radical scavenging and copper-chelating properties (Kang and Kim, 2003). Aggregation of α-synuclein was induced when it

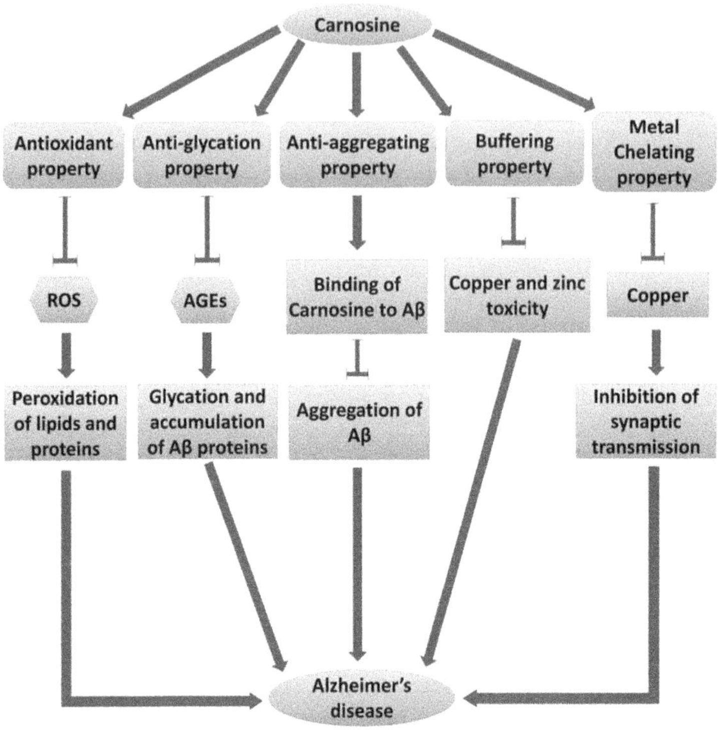

Figure 22.3 Carnosine-induced neuroprotection in Alzheimer's disease. ROS, AGEs, aggregation of Aβ and copper and zinc toxicity leads to Alzheimer's disease. Carnosine induces neuroprotection in Alzheimer's disease through its antioxidant, antiglycation, antiaggregating, metal-chelating and buffering properties.

was incubated with ceruloplasmin (CP) and H_2O_2. This resulted in the formation of free radicals and the release of copper ions. Antioxidant molecules carnosine, homocarnosine and anserine significantly inhibited the CP/H_2O_2 system-induced α-synuclein aggregation (Kim *et al.*, 2002) (Figure 22.4).

Dysfunction and reduced mitochondrial activity was seen in PD. This may be related to oxidative stress and neuronal death. Carnosine could provide antioxidant and anti-inflammatory protection for striatum against the development of PD. Mice treated with 1-methyl-4-phenyl-1,2,3,6-tetrahydropyridine (MPTP) showed significant depletion in striatal glutathione content, reduction in the activity of glutathione peroxidase (GPX), superoxide dismutase (SOD) and catalase, increase in the levels of ROS, as well as enhanced iNOS activity in the striatum. The preintake of carnosine in these mice significantly attenuated these effects.

Proteins like ferritin undergo modifications in PD. When ferritin was incubated with a catechol neurotoxin, salsolinol, protein aggregation increased in a time-dependent manner. Carnosine and its derivative, *N*-acetyl-carnosine,

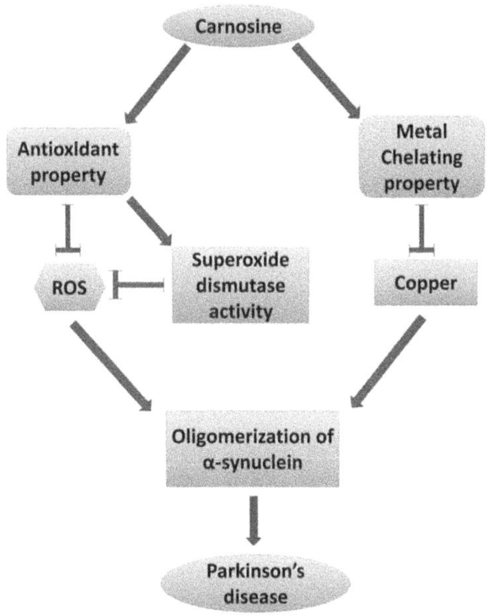

Figure 22.4 Carnosine-induced neuroprotection in Parkinson's disease. ROS and metals like copper leads to oligomerization of α-synuclein in Parkinson's disease. In this case, carnosine induces neuroprotection through its antioxidant and metal-chelating properties.

could protect neuronal cells against salsolinol-mediated ferritin modifications through their free-radical-scavenging activity (Kang, 2010). In a pilot study in humans, addition of carnosine as a food additive to the basic protocol of PD treatment resulted in the significant improvement of neurological symptoms along with increase in Cu/Zn-SOD and decrease in blood plasma protein carbonyls and lipid hydroperoxides. The combination of carnosine with basic therapy was suggested to be a useful way to increase efficiency of PD treatment.

22.3.4 Brain Tumor

Cancer shares similar pathophysiological characteristics with other neurological disorders. Oxidative stress and ROS are considered as the important factors in the development of cancer. Apart from antioxidant and antiglycating properties, carnosine also has an antiproliferative property. But, the exact physiological and biochemical mechanisms responsible for its antineoplastic activity are not known (Figure 22.5).

Glioblastomas (GBMs) are the most common and lethal primary tumors of the CNS with high level of recurrence despite aggressive therapy. Altered levels of carnosinase enzyme were identified in the plasma specimens of GBM patients. Carnosinase has been reported for its role in the regulation

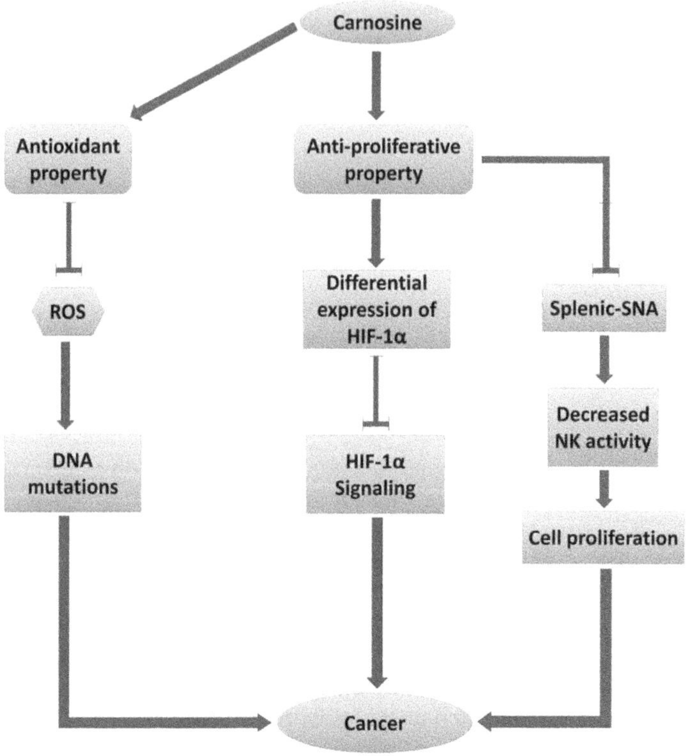

Figure 22.5 Carnosine-induced neuroprotection in cancer. ROS, HIF-1α signaling and splenic-SNA leads to cancer. Here, carnosine induces neuroprotection through its antioxidant and antiproliferative properties.

of carnosine levels. Carnosine treated GBM cells showed differentially expressed proteins such as BCL2-associated athanogene 2 and the von Hippel–Lindau binding protein 1, linking the action of carnosine to protein folding and HIF-1α signaling in GBM. Influence of carnosine on hypoxia inducible factor-1α (HIF-1α) signaling may be attributed to its antitumor activity since upregulated hypoxic response activity is a frequent characteristic of tumor cells.

Intraduodenal (ID) injection of L-carnosine significantly suppressed splenic sympathetic nerve activity (splenic-SNA) in rats. Reduction in splenic-SNA increases natural killer (NK) activity of splenic cells, which in turn elevates tumor immunity. It was suggested that L-carnosine suppresses splenic-SNA and inhibits cancer-cell proliferation, probably by elevating NK activity (Horii *et al.*, 2012).

Pheochromocytoma is a neuroendocrine tumor of the medulla of the adrenal glands. The effect of carnosine and related compounds on proliferation and cell-cycle progression of cultured rat pheochromocytoma PC-12 cells showed that carnosine treatment led to the deceleration of cell proliferation, reduction of G0/G1 peak, and accumulation of S- and G2/M-phase cells. Further, methylation of carnosine molecule at the 1N position of the imidazole ring potentiated its effect on cell proliferation and acetylation (Rybakova and Boldyrev, 2012).

Carnosine attenuated the proliferation of human HCT116 colon cancer cells by decreasing ATP and ROS concentration and arresting cell cycle in G1 phase. These effects were related to decreased ERK1/2 phosphorylation and increased p21waf1 protein. Extracellular signal-regulated kinase (ERK) could be activated by ROS and its signaling regulates many functions including cell survival and proliferation in a broad range of human tumors. However, the role of carnosine in ERK1/2 phosphorylation in brain tumor cells need to be investigated.

Similar investigation is needed to discover the influence of carnosine on matrix metalloproteinases (MMPs) in brain tumors. MMPs, especially MMP-9, play a critical role in tumor metastasis. Active MMP-9 expression was associated with primary GBM subtype. L-Carnosine inhibited metastasis of hepatocellular carcinoma SK-Hep-1 cells through the inhibition of MMP-9 expression. However, the role of carnosine on MMP-9 expression in brain tumor cells is not known.

In gliomas, glycolysis was identified as a crucial process for ATP production. Carnosine inhibited the growth of human malignant gliomacells through the inhibition of vital anaerobic glycolysis (Renner *et al.*, 2010).

22.3.5 Other Neurological Disorders

Carnosine had a protective role in other neurological disorders like schizophrenia (Chengappa *et al.*, 2012), epilepsy (Jin *et al.*, 2005), spinal cord injury (SCI) (Di Paola *et al.*, 2011), autistic spectrum disorders (ASD) (Chez *et al.*, 2002), prion diseases (Kawahara *et al.*, 2011) and nociceptive pain (Ohsawa *et al.*, 2012).

Glutamate excitotoxicity with implication of free radicals is the main pathophysiological condition in schizophrenia. L-Carnosine could improve cognition in schizophrenia patients (Chengappa *et al.*, 2012). It was hypothesized that L-carnosine, which is colocalized and released with glutamate, would improve executive dysfunction, a cognitive domain associated with glutamate. It was suggested that L-carnosine can be considered as an adjunctive treatment to improve executive dysfunction in persons with schizophrenia.

Investigation on the efficacy of carnosine in an animal model of SCI suggested that carnosine treatment reduces the development of inflammation and tissue injury associated with spinal-cord trauma (Di Paola *et al.*, 2011).

Carnosine was effective in models of prion diseases. These are progressive neurodegenerative diseases that are associated with the conversion of normal cellular prion protein (PrP(C)) to abnormal pathogenic prion protein (PrP(SC)) by conformational changes. Prion protein is a metal-binding protein, suggested to be involved in metal homeostasis. It was reported that the prion protein, PrP106–126 exhibited the formation of β-sheet structures and enhanced neurotoxicity during the aging process. Carnosine significantly inhibited both the β-sheet formation and the neurotoxicity of PrP106–126 (Kawahara *et al.*, 2011).

Carnosine has antinociceptive potency. Nociception is the neural processes of encoding and processing noxious stimuli. It is the afferent activity produced in the peripheral and central nervous systems by stimuli that have the potential to damage tissue. Carnosine had antinociceptive effects in a mouse model of inflammation-induced nociceptive pain (Ohsawa *et al.*, 2012).

Reduced levels of antioxidants such as carnosine were found in ASD. Greater oxidative stress, greater free-radical production, impaired energetics and cholinergics, and higher excitotoxic markers are seen in autism. Double-blind, placebo-controlled trials of carnosine showed significant improvement in autistic behavior. Another similar study of L-carnosine supplementation in children with ASD showed enhanced neurological functions (Chez *et al.*, 2002).

Carnosine, through its anticonvulsant effects, plays an important role in the control of epilepsy, which is a common and diverse set of chronic neurological disorders characterized by seizures. Carnosine had protective effects on amygdaloid-kindled seizures in rats (Jin *et al.*, 2005). Amygdala kindling is an epilepsy model involving long-term network plasticity in the nervous system. Therefore, carnosine was suggested to be an endogenous anticonvulsant factor in the brain and it could be used as a potential therapy for epilepsy in the future.

22.4 Conclusions

Carnosine is an antioxidant, metal-ion chelator, aldehyde-scavenger, antiglycating and anticrosslinking agent. It is also an intracellular pH buffer. These properties reciprocate the effects caused due to oxidative stress, metal ions and reactive aldehydes during neurological disorders. It is amazing to find all these properties in a single molecule and thus, carnosine is considered as an important neuroprotectant in most of the neurological disorders like cerebral ischemia, AD, PD, brain tumor, schizophrenia, epilepsy, SCI, ASD, prion diseases and nociceptive pain. Carnosine-based combined treatment, with molecules like vitamin E, may be beneficial in the prevention of oxidative-stress-related neurological disorders. Carnosine and its derivatives could protect antioxidant enzymes from oxidative stress and AGEs. Carnosine improves cellular ability to remove altered proteins by stimulating intracellular proteolysis.

This detailed insight into the properties of carnosine reflects its probable mechanism of action in various neurological disorders. Carnosine imparts neuroprotection against oxidative stress, ROS and glutamate excitotoxicity and also decreases the effect of MMPs. It interacts with amyloid proteins and reduces the accumulation of Aβ. It reduces oligomerization of α-synuclein. It could inhibit cancer-cell proliferation by arresting cell-cycle progression and also by inhibiting anaerobic glycolysis. Therefore, administration of carnosine might induce neuroprotection and exert a great impact on the treatment of neurological disorders. Further, additional research needs to be carried out to see whether carnosine is effective in other neurological disorders that share similar pathophysiological states with the diseases discussed in this chapter.

Summary Points

- This chapter focuses on carnosine and its neuroprotection ability.
- The composition of carnosine is β-alanyl-histidine.
- A high rate of carnosine synthesis occurs in glial cells.
- Carnosine is an important antioxidant, metal-ion chelator, aldehyde-scavenger, antiglycating and anticrosslinking agent. It also operates as an intracellular pH buffer.
- Carnosine has membrane-protecting activity, proton-buffering capacity, formation of complexes with transition metals, and regulation of macrophage function.
- Carnosine is an important neuroprotectant in most of the neurological disorders like cerebral ischemia, Alzheimer's disease, Parkinson's disease, brain tumor, schizophrenia, epilepsy, spinal cord injury, autistic spectrum disorder, prion diseases and nociceptive pain.
- This chapter gives a detailed insight into the properties of carnosine and its probable mechanism of neuroprotection in various neurological disorders.

Key Facts of Oxidative Stress

1. Oxidative stress is an important pathological state seen in most of the neurological disorders.
2. Oxidative stress leads to the formation of free radicals.
3. Free radicals cause damage to the cellular macromolecules like proteins and DNA.
4. Free radicals lead to the modification of proteins and these modified proteins undergo aggregation and accumulation in the cells thereby leading to cell death.
5. DNA damage at a higher extent may lead to DNA fragmentation and cell death.
6. Free radicals may cause DNA damage at a lower extent in the form of DNA mutations.
7. Mutations in DNA may lead to abnormal cell division and cancer.
8. Carnosine protects the cells from oxidative-stress-induced cell death and cancer.

Definitions of Words and Terms

Endogenously synthesized. Synthesized within an organism.

Rate-limiting precursor. The progression of the reaction depends on the availability of this precursor.

Metal-ion chelator. Binds to the metal ions and acts as a metal-chelating ligand thereby minimizing the effects of free metal ions.

Neurodegenerative disorders. These are diseases that affect the brain and cause the death of brain cells or neurons.

Neuroprotection. The mechanisms and strategies used to protect brain against acute neuronal injury (*e.g.* stroke or nervous system injury/trauma) or chronic degeneration processes (*e.g.* Parkinson's, Alzheimer's, multiple sclerosis) is known as neuroprotection.

Pleiotropic molecule. A molecule that has more than one property and also multiple effects.

Reactive oxygen species (ROS). These are molecules or ions that contain oxygen atom. They are formed as byproducts during oxidative reactions. They are very reactive and react with cellular macromolecules.

Antioxidants. Molecules that inhibit the oxidation of other molecules, thereby preventing cellular damage.

Glycation. Enzyme-controlled addition of sugars, such as fructose or glucose, to biomolecules like proteins or lipids is termed glycosylation, whereas addition of sugars to protein or lipid molecules without the interference of an enzyme is known as glycation.

Advanced glycation end products (AGEs). Due to glycation, simple sugars such as glucose, fructose, and galactose are converted to advanced glycation end products.

Oxygen-glucose deprivation (OGD). Reduction in the levels of both oxygen and glucose.

Excitotoxicity. Due to excessive concentration of neurotransmitters like glutamate, neuronal cells are damaged and killed by a pathological process known as excitotoxicity.

List of Abbreviations

AD	Alzheimer's disease
AGEs	Advanced glycation end products
AIF	Apoptosis-inducing factor
ASD	Autistic spectrum disorders
ATP	Adenosine triphosphate
Aβ	β-Amyloid
CNS	Central nervous system
CP	Ceruloplasmin
CSF	Cerebrospinal fluid
EGCG	Epigallocatechin gallate
ERK	Extracellular signal-regulated kinase

GBMs	Glioblastomas
GLT-1	Glutamate transporter-1
GPX	Glutathione peroxidase
H_2O_2	Hydrogen peroxide
HDC	Histidine decarboxylase
HIF-1α	Hypoxia inducible factor-1 α
IL	Interleukin
INF	Interferon
iNOS	Inducible nitric oxide synthase
LDH	Lactate dehydrogenase
LPS	Lipopolysaccharide
MAPK	Mitogen-activated protein kinases
MD	Molecular dynamics
MDA	Malondialdehyde
MMPs	Matrix metalloproteinases
MPTP	1-Methyl-4-phenyl-1,2,3,6-tetrahydropyridine
NF-L	Neurofilament-L
NO	Nitric oxide
OGD	Oxygen–glucose deprivation
PD	Parkinson's disease
pMCAO	Permanent ischemia induced by the middle cerebral artery occlusion
PUFAs	Polyunsaturated fatty acids
RAGE	Receptor for AGEs
ROS	Reactive oxygen species
rUCCAO	Right unilateral common carotid arteries occlusion
SAMP	Senescence-accelerated mice
SCI	Spinal cord injury
SNA	Sympathetic nerve activity
SOD	Superoxide dismutase
STAT3	Signal transducer and activator of transcription 3
TGF-β	Transforming growth factor-β
TNF	Tumor necrosis factor

References

Alsoisi, A., Barca, A., Romano, A., Guerrieri, S., Storelli, C., Rinaldi, R. and Verri, T., 2013. Anti-aggregating effect of the naturally occurring dipeptide carnosine on aβ1–42 fibril formation. *PLoS One*. 8: e68159.

Attanasio, F., Convertino, M., Magno, A., Caflisch, A., Corazza, A., Haridas, H., Esposito, G., Cataldo, S., Pignataro, B., Milardi, D. and Rizzarelli, E., 2013. Carnosine inhibits Aβ(42) aggregation by perturbing the H-bond network in and around the central hydrophobic cluster. *ChemBioChem*. 14: 583–592.

Bae, O. N. and Majid, A., 2013. Role of histidine/histamine in carnosine-induced neuroprotection during ischemic brain damage. *Brain Research*. 1527: 246–254.

Bae, O. N., Serfozo, K., Baek, S. H., Lee, K. Y., Dorrance, A., Rumbeiha, W., Fitzgerald, S. D., Farooq, M. U., Naravelta, B., Bhatt, A. and Majid, A., 2013. Safety and efficacy evaluation of carnosine, an endogenous neuroprotective agent for ischemic stroke. *Stroke*. 44: 205–212.

Balion, C. M., Benson, C., Raina, P. S., Papaioannou, A., Patterson, C. and Ismaila, A. S., 2007. Brain type carnosinase in dementia: a pilot study. *BMC Neurology*. 7: 38.

Boldyrev, A. A., Stvolinsky, S. L., Tyulina, O. V., Koshelev, V. B., Hori, N. and Carpenter, D. O., 1997. Biochemical and physiological evidence that carnosine is an endogenous neuroprotector against free radicals. *Cellular and Molecular Neurobiology*. 17: 259–271.

Chengappa, K. N., Turkin, S. R., DeSanti, S., Bowie, C. R., Brar, J. S., Schlicht, P. J., Murphy, S. L., Hetrick, M. L., Bilder, R. and Fleet, D., 2012. A preliminary, randomized, double-blind, placebo-controlled trial of L-carnosine to improve cognition in schizophrenia. *Schizophrenia Research*. 142: 145–152.

Chez, M. G., Buchanan, C. P., Aimonovitch, M. C., Becker, M., Schaefer, K., Black, C. and Komen, J., 2002. Double-blind, placebo-controlled study of L-carnosine supplementation in children with autistic spectrum disorders. *Journal of Child Neurology*. 17: 833–837.

Di Paola, R., Impellizzeri, D., Salinaro, A. T., Mazzon, E., Bellia, F., Cavallaro, M., Cornelius, C., Vecchio, G., Calabrese, V., Rizzarelli, E. and Cuzzocrea, S., 2011. Administration of carnosine in the treatment of acute spinal cord injury. *Biochemical Pharmacology*. 82: 1478–1489.

Hipkiss, A. R., Preston, J. E., Himsworth, D. T., Worthington, V. C., Keown, M., Michaelis, J., Lawrence, J., Mateen, A., Allende, L., Eagles, P. A. and Abbott, N. J., 1998. Pluripotent protective effects of carnosine, a naturally occurring dipeptide. *Annals of the New York Academy of Sciences*. 854: 37–53.

Hipkiss, A. R., 2007. Could carnosine or related structures suppress Alzheimer's disease?. *Journal of Alzheimer's Disease*. 11: 229–240.

Horii, Y., Shen, J., Fujisaki, Y., Yoshida, K. and Nagai, K., 2012. Effects of L-carnosine on splenic sympathetic nerve activity and tumor proliferation. *Neuroscience Letters*. 510: 1–5.

Horning, M. S., Blakemore, L. J. and Trombley, P. Q., 2000. Endogenous mechanisms of neuroprotection: role of zinc, copper, and carnosine. *Brain Research*. 852: 56–61.

Jin, C. L., Yang, L. X., Wu, X. H., Li, Q., Ding, M. P., Fan, Y. Y., Zhang, W. P., Luo, J. H. and Chen, Z., 2005. Effects of carnosine on amygdaloid-kindled seizures in Sprague-Dawley rats. *Neuroscience*. 135: 939–947.

Kang, J. H. and Kim, K. S., 2003. Enhanced oligomerization of the alpha-synuclein mutant by the Cu,Zn-superoxide dismutase and hydrogen peroxide system. *Molecules and Cells*. 15: 87–93.

Kang, J. H., 2010. Salsolinol, a catechol neurotoxin, induces modification of ferritin: protection by histidine dipeptide. *Environmental Toxicology and Pharmacology*. 29: 246–251.

Kawahara, M., Koyama, H., Nagata, T. and Sadakane, Y., 2011. Zinc, copper, and carnosine attenuate neurotoxicity of prion fragment PrP106-126. *Metallomics*. 3: 726–734.

Kim, K. S., Choi, S. Y., Kwon, H. Y., Won, M. H., Kang, T. C. and Kang, J. H., 2002. The ceruloplasmin and hydrogen peroxide system induces alpha-synuclein aggregation in vitro. *Biochimie*. 84: 625–631.

Min, J., Senut, M. C., Rajanikant, K., Greenberg, E., Bandagi, R., Zemke, D., Mousa, A., Kassab, M., Farooq, M. U., Gupta, R. and Majid, A., 2008. Differential neuroprotective effects of carnosine, anserine, and *N*-acetyl carnosine against permanent focal ischemia. *Journal of Neuroscience Research*. 86: 2984–2991.

Münch, G., Westcott, B., Menini, T. and Gugliucci, A., 2012. Advanced glycation endproducts and their pathogenic roles in neurological disorders. *Amino Acids*. 42: 1221–1236.

Nicoletti, V. G., Santoro, A. M., Grasso, G., Vagliasindi, L. I., Giuffrida, M. L., Cuppari, C., Purrello, V. S., Stella, A. M. and Rizzarelli, E., 2007. Carnosine interaction with nitric oxide and astroglial cell protection. *Journal of Neuroscience Research*. 85: 2239–2245.

Ohsawa, M., Mutoh, J., Asato, M., Yamamoto, S., Ono, H., Hisa, H. and Kamei, J., 2012. Carnosine has antinociceptive properties in the inflammation-induced nociceptive response in mice. *European Journal of Pharmacology*. 682: 56–61.

Pavlov, A. R., Revina, A. A., Dupin, A. M., Boldyrev, A. A. and Yaropolov, A. I., 1993. The mechanism of interaction of carnosine with superoxide radicals in water solutions. *Biochimica et Biophysica Acta*. 1157: 304–312.

Preston, J. E., Hipkiss, A. R., Himsworth, D. T., Romero, I. A. and Abbott, J. N., 1998. Toxic effects of beta-amyloid(25–35) on immortalised rat brain endothelial cell: protection by carnosine, homocarnosine and beta-alanine. *Neuroscience Letters*. 242: 105–108.

Rajanikant, G. K., Zemke, D., Senut, M. C., Frenkel, M. B., Chen, A. F., Gupta, R. and Majid, A., 2007. Carnosine is neuroprotective against permanent focal cerebral ischemia in mice. *Stroke*. 38: 3023–3031.

Renner, C., Asperger, A., Seyffarth, A., Meixensberger, J., Gebhardt, R. and Gaunitz, F., 2010. Carnosine inhibits ATP production in cells from malignant glioma. *Neurological Research*. 32: 101–105.

Rybakova, Y. S. and Boldyrev, A. A., 2012. Effect of carnosine and related compounds on proliferation of cultured rat pheochromocytoma PC-12 cells. *Bulletin of Experimental Biology and Medicine*. 154: 136–140.

Stvolinsky, S., Kukley, M., Dobrota, D., Mezesova, V. and Boldyrev, A., 2000. Carnosine protects rats under global ischemia. *Brain Research Bulletin*. 53: 445–448.

Trombley, P. Q., Horning, M. S. and Blakemore, L. J., 1998. Carnosine modulates zinc and copper effects on amino acid receptors and synaptic transmission. *NeuroReport*. 9: 3503–3507.

Wang, J. P., Yang, Z. T., Liu, C., He, Y. H. and Zhao, S. S., 2013. L-Carnosine inhibits neuronal cell apoptosis through signal transducer and activator of transcription 3 signaling pathway after acute focal cerebral ischemia. *Brain Research*. 1507: 125–133.

CHAPTER 23

Carnosine and Vascular Dementia

JING MA[a,b] AND ZHONG CHEN*[a]

[a]Department of Pharmacology, Key Laboratory of Medical Neurobiology of the Ministry of Health of China, Zhejiang Province Key Laboratory of Neurobiology, College of Pharmaceutical Sciences, School of Medicine, Zhejiang University, Hangzhou, Zhejiang 310058, P. R. China; [b]Department of Pharmacy, Xinhua Hospital, Shanghai Jiaotong University, School of Medicine, Shanghai 200092, China
*E-mail: chenzhong@zju.edu.cn

23.1 Carnosine

Carnosine (β-alanyl-L-histidine) (Figure 23.1), a natural dipeptide, is synthesized from the constituent amino acids β-alanine and histidine under the hydrolysis of adenosine triphosphate (ATP) and catalysis by carnosine synthetase (Figure 23.2). Carnosine was first isolated from Liebig's meat and found to be widely distributed in animal and human tissues. Carnosine are usually abundant in excitable tissues, such as muscle and nerves. In the brain, carnosine has been found exclusively in glial and ependymal cells, but not in mature neurons (De Marchis *et al.*, 1997). Recent studies have demonstrated that carnosine is not present in all glial cells, but in mature astrocytes and oligodendrocytes (Sakai *et al.*, 1988). With cultures highly enriched in oligodendrocyte progenitor cells tracer in corporation studies, demonstrated that carnosine is rapidly synthesized by oligodendrocytes and then

Food and Nutritional Components in Focus No. 8
Imidazole Dipeptides: Chemistry, Analysis, Function and Effects
Edited by Victor R Preedy

Published by the Royal Society of Chemistry, www.rsc.org

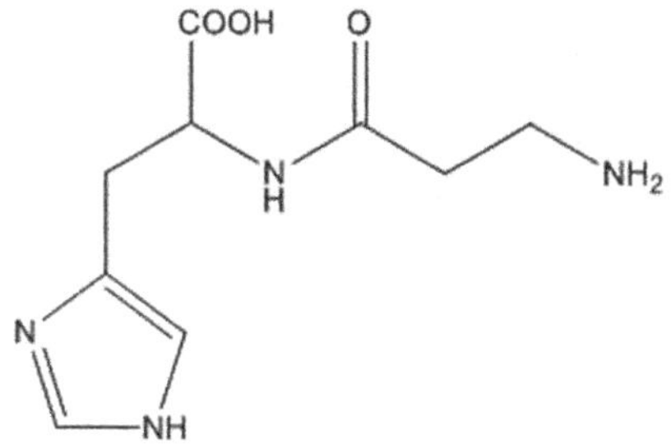

Figure 23.1 Structure of carnosine.

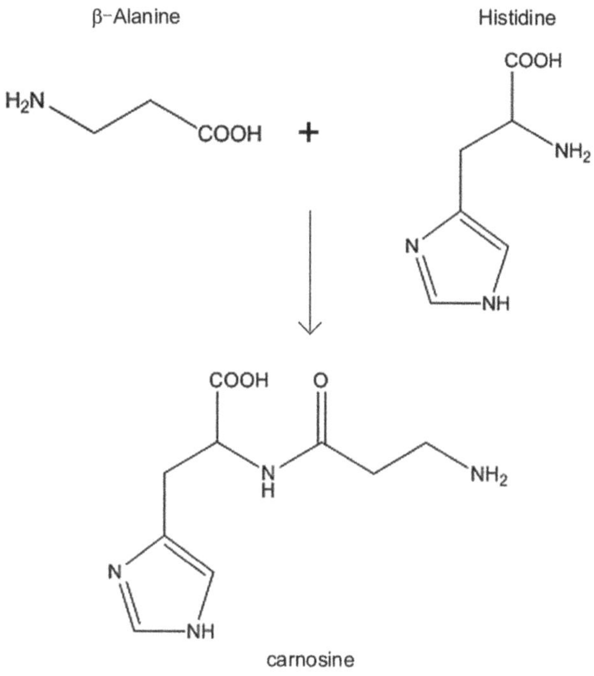

Figure 23.2 Synthesis of carnosine from the constituent amino acids β-alanine and histidine.

released into medium. Some reports addressed that carnosine is actively synthesized by the C-6 glioma cell line. Moreover, carnosine can easily enter the brain from the periphery. The release of carnosine by glial cells has been ascribed to several mechanisms such as membrane depolarization, swelling of type 1 and type 2 astrocytes due to high K^+, and reversal of membrane transport (Bonfanti *et al.*, 1999). In general, carnosine is found throughout the brain at the low concentration, with the exception of the olfactory bulb where high amounts of carnosine are detected. For this reason, carnosine has been largely considered as a neurotransmitter/modulator in olfaction.

The level of carnosine declines under the stress, which suggests that carnosine is involved in the modulation of the central nervous system function.

23.2 Carnosine in Nervous System Disease

Over the last fifty years, hundreds of studies have reported the function and the effect of carnosine (Shen *et al.*, 2010). Carnosine is found to protect against several nervous system diseases, such as Alzheimer's disease (AD), Parkinson's disease (PD) and cerebral ischemia. Many beneficial roles are attributed to its ion-chelating, free-radical scavenging and protein glycosylation inhibition action. In AD patients, the plasma level of carnosine is lower than that in age-matched controls (Fonteh *et al.*, 2007). Fu *et al.* (2008) found that Aβ 42 causes a concentration-dependent reduction of viability in PC12 cells, while pretreatment with carnosine ameliorates this impairment. Some studies have revealed the protective effects of carnosine on rat brain endothelial cell against the damage from amyloid-β (Aβ) peptide fragment Aβ (25–35) and postulate that the protection may be due to their antiglycating and/or antioxidation activities. Carnosine also showed a disassembling effect on α-crystalline amyloid fibrils. Carnosine supplementation in triple-transgenic AD mice (3xTg-AD) promoted a strong reduction in the hippocampal intraneuronal accumulation of Aβ and completely rescues AD and aging-related mitochondrial dysfunctions (Sensi *et al.*, 2008). These results suggest that carnosine can be a potential therapeutic agent for AD.

Some studies have demonstrated the efficiency of carnosine when it is used as a food additive to increase the efficiency of basic therapy of PD patients, so carnosine may be a reasonable supplement to modify the protocol of PD treatment (Boldyrev *et al.*, 2008). The α-synuclein is a major component of Lewy bodies that are found in the brain of patients with PD. The oligomerization of α-synuclein is mediated by the generation of the hydroxyl radical through the metal-catalyzed reaction (Tsigelny *et al.*, 2008). Carnosine inhibit the a-synuclein oligomerization by inhibiting the generation of reactive oxygen specie (ROS) and protecting mitochondria (Kang and Kim, 2003).

The effects of carnosine on cerebral ischemia have been extensively studied both *in vitro* and *in vivo*. Carnosine has been reported to alleviate memory impairment and decrease infarct volume after hypoxia–ischemia in neonatal rats (Zhang *et al.*, 2011). Meanwhile, carnosine protects rats and mice against hypoxia or ischemia-induced brain damage by antioxidation and attenuation of apoptosis (Pekcetin *et al.*, 2009). *In vitro*, carnosine reduces ischemic cell death after oxygen-glucose deprivation (OGD), and this effect is abolished by histamine receptor type 1 antagonist (Bae and Majid, 2013). Carnosine also showed the protective effect on neuronal and PC12 cell following OGD (Shen *et al.*, 2007). The protective effect in cerebral ischemia is probably through the regulation of matrix metalloprotease activity, the recovery of glutamate transporter which alleviates glutamate excytotoxicity, and its conversion to histidine and then histamine, which offers a beneficial effect against neuronal damage (Shen *et al.*, 2010).

23.3 Hypothetical Functions of Carnosine in the Nervous System

Carnosine has been assigned to many other putative actions beyond working on a specific disease, such as free-radical scavenging, protein glycosylation inhibition anti-inflammation, and mobile organic pH buffering. Carnosine has been studied with the aim to test its effect on oxidative and nitrosative stress *in vitro* and *in vivo*. In cultured astrocytes and oligodendrocyte cells, carnosine was found to reduce inducible nitric oxide synthase (iNOS) expression, poly-ADP-ribose polymerase 1 (PARP-1) and PARP-2 activation induced by lipopolysaccharide (LPS) and interferon-γ (IFN-γ) (Spina-Purrello *et al.*, 2010). Carnosine has an antioxidative effect on cultured human lymphocyte (Alpsoy *et al.*, 2011). *In vivo* carnosine alleviates oxidative stress and tissue damage induced by D-galactose in rat liver (Kalaz *et al.*, 2013). Carnosine also shows an antioxidative action in liver, heart and brain tissues of aged rats (Coban *et al.*, 2013).

Carnosine also acts as a competitive inhibitor of nonenzymic glycosylation of proteins (Hipkiss, 2009). In the process of protein glycosylation, subsequent oxidations and dehydrations lead to the formation of a wide range of products referred to as advanced glycation end products (AGEs). There is evidence suggesting that in tumor cells carnosine somehow inhibits ATP synthesis during glycolysis. Treatment of carnosine inhibits the growth of yeast *Saccharomyces cerevisiae* when cultured fermentatively on glucose (Cartwright *et al.*, 2012).

Other reports showed that carnosine reduces glutamate excitotoxicity by reversing the decrease of glutamate transporter-1 (GLT-1) expression in astrocytes induced after ischemia (Shen *et al.*, 2010) (Figure 23.3). Carnosine decreases the glutamate levels and preserves the expression of GLT-1 but not the glutamate/aspartate transporter in astrocytes exposed to ischemia *in vivo* and *in vitro*. This action is probably due to its inhibition of the dissipation of $\Delta\Psi_m$ and generation of mitochondrial ROS induced by OGD, which leads to oxidative damage of GLT-1 on the astrocyte membrane and then increase of the extracellular glutamate level.

Moreover, carnosine can reduce the levels of proinflammatory cytokines TNF-α, IL-1β and nitrotyrosine, which provides antioxidative and anti-inflammatory protection for the striatum against the development of Parkinson's disease. Some studies suggest that carnosine has a beneficial effect in the reduction of the cellular apoptosis by recovering the Bcl-2 levels and inhibiting the increase of Bax.

The above evidences suggest that it is worth studying the therapeutic value of carnosine in various CNS diseases.

23.4 Vascular Dementia

23.4.1 The Background of Vascular Dementia

Vascular dementia (VaD) is a term currently used to define any type of dementia resulting from cerebral blood vessel disease (Micieli, 2006). VaD does not represent a single entity, but quite a large group of conditions

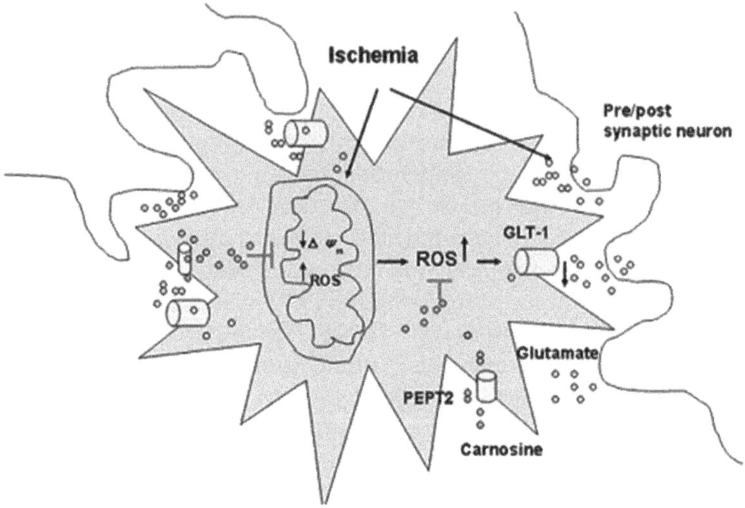

Figure 23.3 Schematic showing hypothesized mechanisms underlying the carnosine action on cerebral ischemic excitotoxicity. Ischemia first induces mitochondrial ROS generation and dissipation of the $\Delta\Psi_m$ in cortical astrocytes, which leads to oxidative damage of GLT-1 on the astrocyte membrane and then an increase of extracellular glutamate level. Carnosine suppresses the mitochondrial ROS generation and $\Delta\Psi_m$ dissipation and therefore alleviates the oxidative damage of GLT-1 and prevents glutamatergic excitotoxicity in ischemic brain. Adapted from (Shen *et al.*, 2010) and reproduced with permission from Free Radic Biol Med/Elsevier Inc.

characterized by different pathophysiological correlates and various clinical findings. From a clinical point of view, classification of VaD is based on the main diagnostic point: history of stroke and/or focal vascular neurological deficits, cognitive deficits (memory, strategic abilities, executive functions), neuroimaging showing neurovascular focal or diffuse lesion and temporal association between stroke and the onset of dementia.

VaD is recognized as the second most prevalent type of dementia (Kalaria *et al.*, 2008). Prevalence estimates of VaD in developing countries range from 0.6% to 2.1% in those aged over 65 years. A third of 4.5 million Chinese patients with dementia are predicted to have VaD and up to 30% of Chinese people in urban areas develop poststroke cognitive impairment or delayed dementia after stroke (Zhang *et al.*, 2005). In European countries, it has a prevalence ranging from 22% to 45%, while the incidence is equal to 27%.

Its pathological changes are the development of ischemic white matter (WM) lesion, glial activation and cognitive impairment. Cognitive signs of VaD are represented by memory deficit, executive function disturbances, slow processing of information and behavioral and mood changes. It is assumed that risk factors for VaD should be the same as for stroke. Old age represents the first of a series of demographic factors including race/ethnic group (mainly Asian), sex (male), low education. Among the risk factors for VaD, hypertension has a major role: several studies suggest that an elevated blood pressure measured in midlife increases the risk of dementia or accelerates age-related cognitive decline (Micieli, 2006).

The symptoms of vascular dementia may begin suddenly, for example after a stroke. Some symptoms may be similar to those of other types of dementia such as AD. People with vascular dementia may particularly experience, such as problems with speed of thinking, concentration and communication, depression and anxiety accompanying the dementia, symptoms of stroke, memory problems and seizures.

There are different types of vascular dementia including stroke-related dementia, subcortical vascular dementia (small vessel disease or Binswanger's disease), mixed dementia (vascular dementia and Alzheimer's disease). The difference between these types depends on what has caused the damage in the brain, and which part of the brain has been damaged. Subcortical ischemic vascular dementia (SIVD) induced by chronic hypoperfusion as a result of small-artery disease is a common cause for vascular dementia.

Recently, the therapy of VaD has received extensive attention. Most clinical trials pertaining to VaD have produced disappointing results and there are no known compounds that have regulatory approval for prevention and treatment of this disease (Baskys and Hou, 2007). Pharmaceuticals that prevent strokes or treat Alzheimer's dementia offer limited benefits for VaD. Acetylcholinesterase inhibitors may improve cognition but have significant side effects, such as an increase in patient mortality (Roman *et al.*, 2010). Pharmacological treatment of hypertension, if started early, reduces VaD incidence and slows progression but there could require more time to develop the correlation between VaD and hypertension. Taken together, despite increasing investigations in experimental and clinical neuroscience, there are no effective pharmacological compounds approved for the treatment of vascular dementia.

23.4.2 The Cellular Pathology of VaD

VaD arises as a consequence of ischemic insults such as hypoperfusion that trigger neurodegeneration by depriving nerve cells of oxygen and glucose (Kalaria, 2003). Oxygen and glucose deprivation results in depletion of nerve cell energy supplies, leading to membrane depolarization, followed by an excessive release of glutamate and overactivation of the *N*-methyl-D-aspartate (NMDA) receptor/channel complex (Baskys and Hou, 2007). Overactivation of neuronal NMDA receptors allows influx of toxic levels of Ca^{2+} into nerve cells (Coyle and Puttfarcken, 1993), which enables activation of various intracellular calcium-dependent enzymes. VaD also leads the glia activation and myelin degeneration. Activated microglia and astrocyte are the major sources of proinflammatory cytokines, including TNF-α, IL-1β, and IL-6, which may result in the degeneration of myelin and apoptosis of oligodendrocytes (Benarroch, 2009). The primary consequence of oligodendrocyte death is axonal demyelination, resulting in delay or blocking of conduction of the action potential in the axon.

23.4.3 The Animal Model of VaD

There are many animal models to study the VaD as occlusion of four major cephalic vessels and occlusion of two major cephalic vessels, but there are some defects in these animal models as rats were often used for VaD that cannot be applicable to transgenic or knockout mice and with high mortality rates. Later, Shibata *et al.* developed a mouse model induced by the stenosis of the bilateral common carotid arteries (BCCAs) using microcoils, which also results in WM rarefaction and delayed memory impairment, but the degree of stenosis is controlled by adjusting the internal diameter of the microcoils, which seems a little complicated (Shibata *et al.*, 2004). Recently, Yoshizaki *et al.* (2008) have reported a novel right unilateral common carotid arteries (rUCCAO) model, which is more simple and convenient with low mortality rate and can induce impairment of WM and nonspatial memory in C57BL/6 mice. Ma *et al.* (2012) further found that after rUCCAO, mice also showed a significant damage in learning and memory as evaluated by passive avoidance task and Morris water maze, which reflect the fear memory and spatial working memory, respectively. The WM was associated with the behavioral effects pertaining to emotionality, spatial working memory and motivated exploration (Miu *et al.*, 2006). Taken together, it is likely that chronic cerebral ischemia induced by rUCCAO impairs both the spatial and nonspatial working memory, as well as fear memory, which better simulates the clinical cognitive impairment due to WM damage in VaD.

23.5 Carnosine in VaD

23.5.1 The Effect of Carnosine on VaD

Carnosine has been reported to have the protective effect on various experimental VaD investigations. A long-term brain ischemia induced by both sides' occlusion of common carotid arteries (BCCAO) has been used to study the effect of dietary carnosine on the behavioral and biochemical characteristics in rats (Gallant *et al.*, 2000). It was found that BCCAO results in 55% mortality of rats and those who survived are characterized by partial suppression of orientation in T-maze. In the group of rats treated with carnosine, mortality after ischemic attack decreases (from 55% to 17%) and most of the learning parameters are kept at the preischemic level. Carnosine improves the animal orientation and learning in "open field" and "T-maze" tests, and this effect is accompanied with an increase in glutamate binding to NMDA receptors in brain synaptosomes. The cerebral blood flow decreases after occlusion of the common carotid arteries, while administration of carnosine reverses the decrease during the early stage after ischemia (Kuklei and Gannushkina, 1997).

Effective reduction of mortality by carnosine is also observed in another model of ischemia in rats induced by occlusion of four major cephalic vessels including the spinal and common carotid arteries (Stvolinsky *et al.*, 1999).

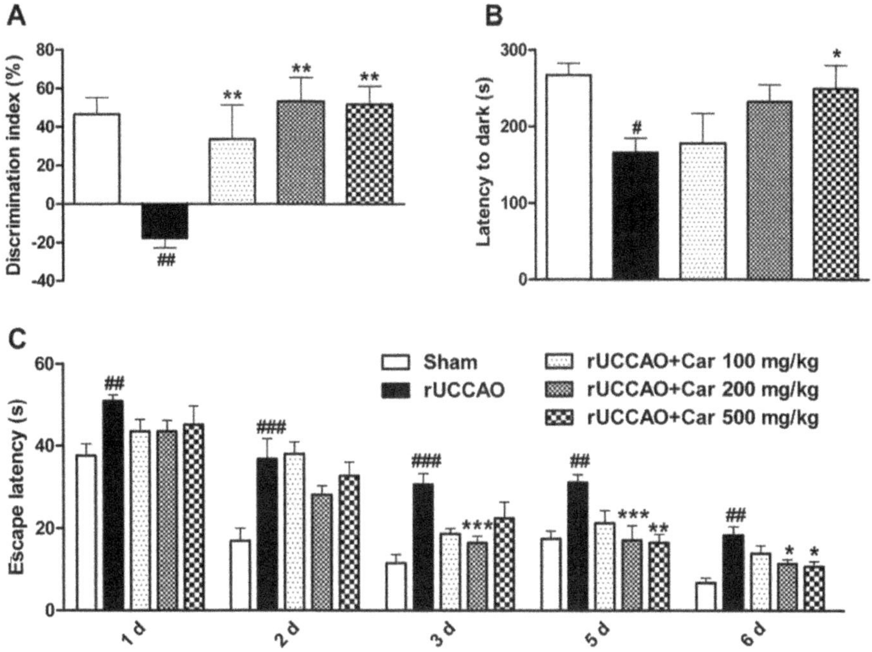

Figure 23.4 Effect of carnosine on learning and memory in object recognition test (A), passive avoidance task (B) and Morris water maze test (C) after rUCCAO. n = 8–14. #P < 0.05, ##P < 0.01, *vs.* sham group; *P < 0.05, **P < 0.01, *vs.* rUCCAO group. Adapted from (Ma *et al.*, 2012) and reproduced with permission from CNS Neurosci Ther/Blackwell Publishing Ltd.

The levels of lactate increase in the ischemic injury zone by 4–5-fold during 45 min observation after ischemia induced by occlusion of four major cephalic vessels, whereas in carnosine-treated animals, the level of lactate increase by only 2-fold and the effect is preserved for long period (up to 90 min and more). Meanwhile, carnosine recovers the Na, K-ATPase activity, inhibition of which alters ion transport, results in the membrane depolarization and loss of electrical conductivity of neurons, probably through its free-radical scavenging action.

Ma *et al.* (2012) have found that carnosine improves the learning and memory and delays the progression of WM rarefaction after VaD induced by rUCCAO. After rUCCAO, mice decrease discriminative ability in the object recognition test, while carnosine significantly elevates the discrimination index. In a passive avoidance task, the escape latency decreases after rUCCAO, which is completely reversed by carnosine. In Morris water maze training trials, rUCCAO-treated mice shows prolonged escape latency in acquisition phase and reversal phase (Figure 23.4), while carnosine markedly shortens the escape latency during both the acquisition and reversal phase. Carnosine also delayed the progress of the decline of fiber density and reversed the

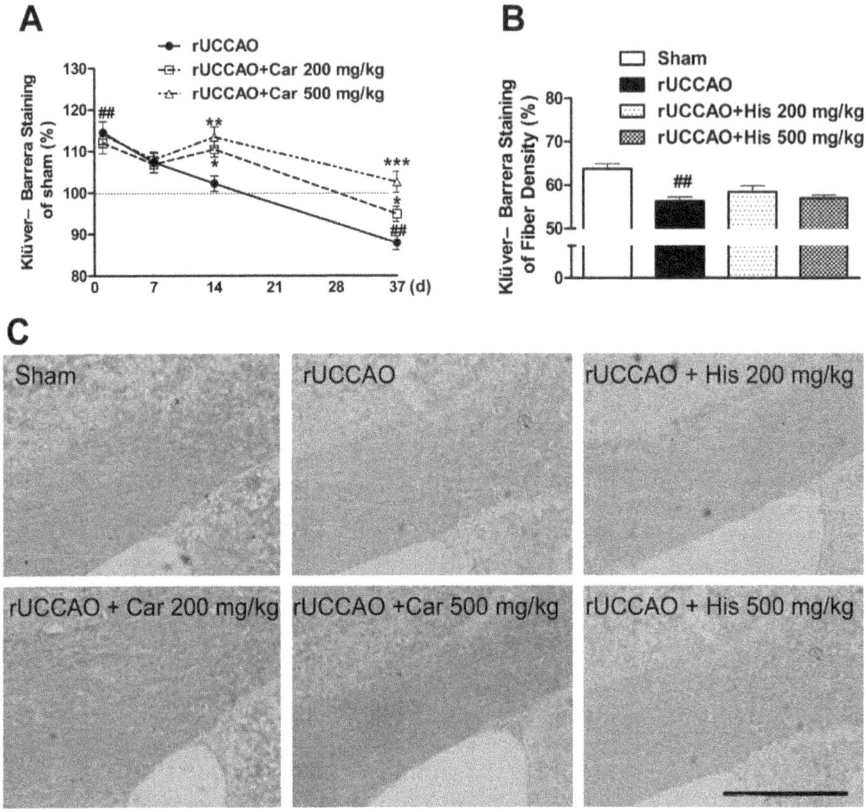

Figure 23.5 Effect of carnosine on fiber density in corpus callosum is independent of the histaminergic pathway. The fiber density in corpus callosum in WT mice was analyzed by Klüver–Barrera staining after carnosine (A) and histidine (B) treatment in rUCCAO or sham groups. The representative photomicrographs of fiber density in corpus callosum 37 days after rUCCAO are shown in C. Scale bar, 50 μm. $n = 8$. ##$P < 0.01$, *vs.* sham group; *$P < 0.05$, **$P < 0.01$, ***$P < 0.001$, *vs.* rUCCAO group. Adapted from (Ma *et al.*, 2012) and reproduced with permission from CNS Neurosci Ther/Blackwell Publishing Ltd.

decrease of myelin basic protein expression (Figure 23.5). These results suggest carnosine may have potent clinical value to treat VaD.

23.5.2 Mechanisms of Carnosine in VaD

The protective effect of carnosine in the brain is often attributed to their beneficial actions such as free-radical scavenging, ion-chelating and protein glycosylation inhibition. However, the conversion to histidine and then histamine from carnosine has also been suggested to contribute to their protection after ischemia. Carnosine serves as a nonmast-cell reservoir for histamine in the brain (Jin *et al.*, 2005). Histamine plays an important role in the alleviation of spatial memory deficits induced by MK-801 or scopolamine, and protect against cerebral ischemia induced by acute blockade of global or focal cerebral perfusion. It has been reported that carnosine ameliorates acute renal failure induced by ischemia/reperfusion in rats and NMDA induced excitotoxic injury in differentiated PC12 cells through its conversion to histidine and histamine (Kurata *et al.*, 2006). Hence, it is proposed that carnosine may protect against VaD through the histaminergic pathway. However, Ma *et al.* (2012) found that carnosine attenuates the cognitive impairment and WM injury after rUCCAO in both HDC-KO and WT mice almost to the same extent, which indicates that carnosine has the same protective effect against VaD whether or not mice lack histamine. Meanwhile, histidine, which will be transformed to histamine, does not show the protective action in WT mice after rUCCAO, although histamine is considered to participate in normal learning and memory, indicated by the impaired learning and memory in object recognition test and Morris water maze in HDC-KO mice without rUCCAO. These data revealed that the protective action of carnosine on rUCCAO may not involve the carnosine–histidine–histamine metabolic pathway.

Ma *et al.* also found that glia activation and myelin degeneration are the major changes of WM other than rarefaction (Ma *et al.*, 2012). They found that carnosine markedly inhibit the activation of astrocyte and microglia as attenuating the elevation of GFAP and Iba-1 fluorescent intensities (Figure 23.6). Meanwhile, the MBP expression in the corpus callosum is largely reversed by carnosine. Carnosine has also been reported to significantly inhibit LPS-induced inflammatory mediators in cultured BV2 microglia. Activated microglia and astrocyte are the major sources of proinflammatory cytokines, including TNF-α, IL-1β, and IL-6, which may result in the degeneration of myelin and apoptosis of oligodendrocyte. So, the protection of carnosine on VaD may be related to the inhibition of the glia activation and then myelin degeneration in the corpus callosum.

Meanwhile, it has been found that carnosine attenuates the increase of ROS to the level in the sham group after rUCCAO (Figure 23.7). ROS often participates in the pathogenesis of WM injury, and it initiates lesion formation by inducing activation of astrocyte and microglia and myelin phagocytosis (van der Goes *et al.*, 1998). Carnosine is considered to be a free-radical scavenger (Hipkiss, 2009) to improve mitochondrial function by reducing the ROS generation, which recovers the glutamate transporter expression in astrocytes, and then ameliorates the neurological dysfunction after permanent focal cerebral ischemia (Shen *et al.*, 2007). Some studies have found that the administration of carnosine before ischemia prevents the lipid components of the brain membranes from peroxidation during ischemia. This effect is more pronounced in animals with high sensitivity to ischemia. The neurological symptoms are less expressed in animals whose membrane lipids are protected from peroxidation.

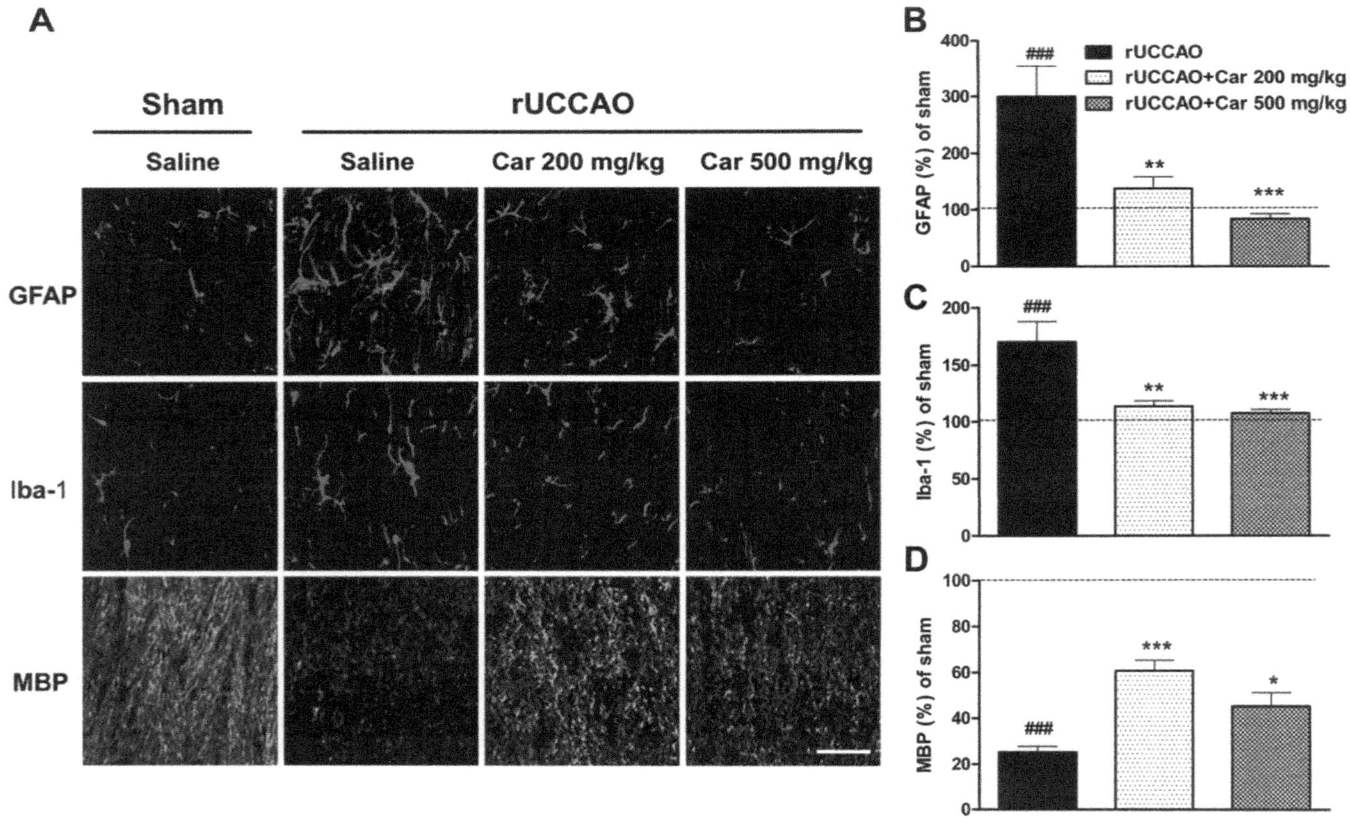

Figure 23.6 Effect of carnosine on glia activation and MBP at 37 days after rUCCAO. The astrocyte and microglia activation in corpus callosum was analyzed by immunohistochemically staining of GFAP (A and B) and Iba-1 (A and C), respectively. MBP immunohistochemical results are shown in (A and D). Scale bar, 50 μm. $n = 8$. ###$P < 0.001$, *vs.* sham; *$P < 0.05$, **$P < 0.01$, ***$P < 0.001$, *vs.* rUCCAO group. Adapted from (Ma *et al.*, 2012) and reproduced with permission from CNS Neurosci Ther/Blackwell Publishing Ltd.

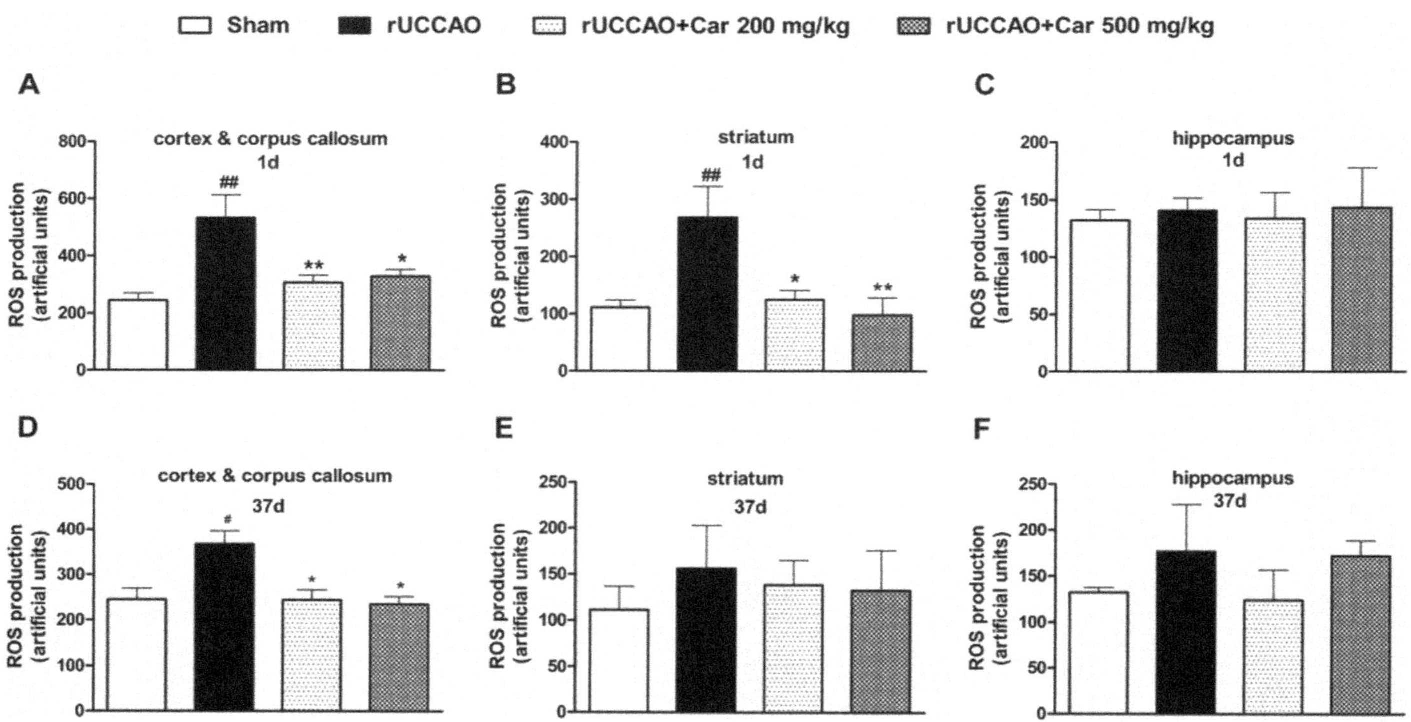

Figure 23.7 Attenuation of ROS production by carnosine after rUCCAO. At days 1 (A–C) and 35 (D–F) after rUCCAO, the ROS production in the cortex and corpus callosum (A and D), striatum (B and E), and hippocampus (C and F) was measured after carnosine (200, 500 mg kg^{-1}) treatment. $n = 6$. #$P < 0.05$, ##$P < 0.01$, *vs.* the sham group; *$P < 0.05$, **$P < 0.01$, *vs.* the rUCCAO group. Adapted from (Ma *et al.*, 2012) and reproduced with permission from CNS Neurosci Ther/Blackwell Publishing Ltd.

Therefore, these results suggest that the protection of carnosine against WM damage after VaD may be through inhibiting ROS generation and then reducing glia activation and myelin degeneration.

23.6 Conclusions

Since carnosine has exhibited protective effects in multiple experimental VaD models, it is imperative to investigate its therapeutic value in clinics. On the other hand, more substantial evidence for its molecular and pathophysiological mechanism will benefit the formation of a solid scientific basis for its clinical application. The current use of carnosine to prevent aging and complications of diabetes such as nerve damage, eye disorders, and kidney problems, increases the possibility of its clinical application in VaD regarding drug safety.

Summary Points

- Vascular dementia (VaD) is a common condition recognized as the second most prevalent type of dementia.
- Carnosine is a natural dipeptide that is expressed in the central nervous system, and can easily enter the brain from the periphery.
- Carnosine can converse to histidine and then histamine, and serves as a nonmast-cell reservoir for histamine.
- Carnosine was shown to improve the animal orientation and learning in "open field" and "T-maze" tests, and this effect was accompanied with an increase in glutamate binding to NMDA receptors in brain synaptosomes.
- The protective action of carnosine on right unilateral common carotid arteries may not involve the carnosine–histidine–histamine metabolic pathway.
- Carnosine has a protective effect in VaD perhaps due to an inhibition of ROS generation, glia activation, and myelin degeneration.
- Carnosine has also been assigned many other putative roles, such as an anti-inflammatory agent, free-radical scavenger, and mobile organic pH buffer.

Key Facts

Key Facts of Vascular Dementia

Vascular dementia (VD) is recognized as the second most prevalent type of dementia. A third of 4.5 million Chinese patients with dementia are predicated to have VD. VD might be more common among the Chinese. Subcortical VD caused by small-artery disease, associated with hypertensive disease, seems to be a common (73%) cause of VD.

Key Facts of Alzheimer's Dementia

Alzheimer's dementia (AD) is the most common subtype of age-related dementia. Total population projections suggest that 3.1 million people in China could have AD. AD is a progressive neurocognitive disorder without preventive intervention.

Definitions of Words and Terms

Dementia. Dementia is a syndrome characterized by progressive deterioration of cognitive function, most commonly of memory, but other domains such as language, praxis, visual perception and most notably executive function are often affected. Most of the causes of this syndrome are progressive, but not invariably so. As cognitive function worsens, there is increasing interference with the patients' daily activities leading to loss of independence and eventually, for some, the need for nursing home care.

Reactive oxygen species (ROS). ROS, including superoxide (O_2^-), hydroxyl radical ($^{\cdot}OH$) and H_2O_2, are constantly generated in aerobic organisms. The endogenous sources usually are oxidative phosphorylation, P450 metabolism, peroxisomes and inflammatory cell activation. ROS can randomly react with lipids, proteins and nucleic acids causing oxidative stress and damage in these macromolecules, leading to pathogenesis of age-related and chronic diseases, which include cancer, cardiovascular disease, diabetes, chronic inflammation, and neurodegenerative disorders.

Oligodendrocytes. Oligodendrocytes are the myelin-producing cells in the central nervous system (CNS) and are critical for function and survival of the axons. Oligodendrocytes originate from progenitor cells. After migration, oligodendrocyte progenitor cells reach their destination and begin to differentiate; they upregulate expression of myelin proteins, increase the number and complexity of their processes, and begin to form membrane wraps around the axons. Each oligodendrocyte can produce up to 30–50 myelin segments in different axons. The oligodendrocytes are also critical for maintenance of axon structure and survival and may exert a neuroprotective function, as they provide trophic support for the neuronal cell bodies by producing several trophic factors.

Microglia. Microglial are the brain-specific tissue macrophages, to initiate bystander injury by producing and releasing multiple cytoactive factors. Microglia functional activation can reduce proinflammatory mediators involved in repair and reconstruction of the damage brain, and increase the release of anti-inflammatory cytokines. Microglial are identified by the expression of ionized calcium-binding adapter-1 (Iba-1).

Astrocyte. Astrocytes are a very heterogeneous group that comprises several cell types with many putative, but not always well-proven, functions. Astrocytes are identified by the expression of glial fibrillary acidic protein (GFAP), a specific component of glial intermediate filament. Astrocyte perform many functions, including biochemical support of endothelial cells that form the blood–brain barrier, provision of nutrients to the nervous tissue, maintenance of extracellular ion balance, and a role in the repair and scarring process of the brain and spinal cord following traumatic injuries.

Myelin. Myelin is a principal component of white matter that is generated postnatally in vertebrates by oligodendrocytes through complex developmental steps that culminate in wrapping a spiral insulating sheath around axons. This insulation greatly enhances the rapid propagation of action potentials from neuronal cell bodies to axon terminals and its destruction or malformation results in many neurological disorders. The production of the myelin sheath is called myelination, and successful myelin repair in the injury nervous system relies on the regenerative capacity of progenitor cells.

Myelin basic protein (MBP). MBP is a major constituent of the myelin sheath of oligodendrocytes that is important in the process of myelination of nerves in the CNS, and the deficiency of expression hints at the demyelination. MBP is a very basic protein that is normally associated with membranes and maintains the correct structure of myelin, interacting with the lipids in the myelin membrane. Some studies have shown a role for antibodies against MBP in the pathogenesis of CNS diseases.

Glial fibrillary acidic protein (GFAP). GFAP is a protein that is expressed by numerous cell types of the CNS including astrocytes and ependymal cells and expressed in glomeruli and peritubular fibroblasts. GFAP is involved in many important CNS processes, including cell communication and the functioning of the blood–brain barrier. GFAP is shown to be important in repair after CNS injury. GFAP is often used to be a marker of astrocyte.

Histidine decarboxylase (HDC). HDC is the enzyme that catalyzes the reaction that produces histamine from histidine. This enzyme is a pyridoxal phosphate-dependent decarboxylase and is highly specific for its histidine substrate. HDC-knock-out mice lack histamine.

Neurotransmitter. Neurotransmitters are the brain chemicals that communicate information throughout our brain and body. They relay signals between neurons and can also affect mood, sleep, concentration, weight, and can cause adverse symptoms when they are out of balance. There are two kinds of neurotransmitters, inhibitory and excitatory. Inhibitory neurotransmitters can calm the brain including GABA and serotonin. Excitatory neurotransmitters are not necessarily exciting, including glutamine and norepinephrine.

γ-Aminobutyric acid (GABA). GABA is an inhibitory neurotransmitter in the mammalian CNS. It plays a role in the regulation of neuronal excitability throughout the nervous system. GABA acts at inhibitory synapses in the brain by binding to specific transmembrane receptors in the plasma membrane of both pre- and postsynaptic neuronal processes. When GABA is out of range, it is likely that an excitatory neurotransmitter is firing too often in the brain. GABA will be sent out to attempt to balance this stimulating overfiring.

Glutamate transporter-1 (GLT-1). GLT-1 also called excitatory amino-acid transporters 2 (EAAT2), is glutamate transporters, which belongs to the family of neurotransmitter transporters. GLT-1 serves to terminate the excitatory signal by removal (uptake) of glutamate from the neuronal synapse into neuroglia and neurons. GLT-1 is a membrane-bound secondary transporter that superficially resembles ion channels. GLT-1 plays an important role in regulating concentrations of glutamate in the extracellular space by transporting it along with other ions across cellular membranes.

Apoptosis. Apoptosis is the process of programmed cell death that is an ATP-dependent physiological process. Apoptosis is important for embryogenesis, tissue homeostasis and pathological conditions, such as infections and autoimmunity. The major morphological changes of the apoptosis cell are cell shrinkage, chromatin aggregating, nuclear condensation and the disassembly of the cell into membrane blebs, called apoptosis bodies, without the loss of membrane integrity. Caspases (cysteinyl-aspartate-specific proteinases) are the main effectors of the apoptosis pathway.

Excitotoxicity. Excitotoxicity is the pathological process by which nerve cells are damaged and killed by excessive stimulation by neurotransmitters such as glutamate and similar substances. Excitotoxicity may be involved in spinal cord injury, stroke, traumatic brain injury and neurodegenerative disease of the CNS. Glutamate is the principal excitatory neurotransmitter in the CNS, but it is also a potent neurotoxin that can kill nerve cells.

List of Abbreviations

AD	Alzheimer's disease
ATP	Adenosine triphosphate
Aβ	Amyloid-β
AGEs	Advanced glycation end products
BCCAs	Bilateral common carotid arteries
CNS	Central nervous system
GABA	γ-Aminobutyric acid
GFAP	Glial fibrillary acidic protein
GLT-1	Glutamate transporter-1
HDC-KO	Histidine decarboxylase knockout
Iba1	Ionized calcium-binding adapter-1
IL-1β	Interleukin-1β
IL-6	Interleukin-6
LPS	Lipopolysaccharide
MBP	Myelin basic protein
MCAO	Middle cerebral artery occlusion
NMDA	*N*-Methyl-D-aspartate
PARP	Poly-ADP-ribose polymerase 1

PD	Parkinson's disease
ROS	Reactive oxygen specie
rUCCAO	Right unilateral common carotid arteries
SIVD	Subcortical ischemic vascular dementia
TNF	Tumor necrosis factor
VaD	Vascular dementia
WT	Wild type
WM	White matter

References

Alpsoy, L., Akcayoglu, G. and Sahin, H., 2011. Anti-oxidative and anti-genotoxic effects of carnosine on human lymphocyte culture. *Human & Experimental Toxicology*. 30: 1979–1985.

Bae, O. N. and Majid, A., 2013. Role of histidine/histamine in carnosine-induced neuroprotection during ischemic brain damage. *Brain Research*. 1527: 246–254.

Baskys, A. and Hou, A. C., 2007. Vascular dementia: pharmacological treatment approaches and perspectives. *Clinical Interventions in Aging*. 2: 327–335.

Benarroch, E. E., 2009. Oligodendrocytes: susceptibility to injury and involvement in neurologic disease. *Neurology*. 72: 1779–1785.

Boldyrev, A., Fedorova, T., Stepanova, M., Dobrotvorskaya, I., Kozlova, E., Boldanova, N., Bagyeva, G., Ivanova-Smolenskaya, I. and Illarioshkin, S., 2008. Carnosine increases efficiency of DOPA therapy of Parkinson's disease: a pilot study. *Rejuvenation Research*. 11: 821–827.

Bonfanti, L., Peretto, P., De Marchis, S. and Fasolo, A., 1999. Carnosine-related dipeptides in the mammalian brain. *Progress in Neurobiology*. 59: 333–353.

Cartwright, S. P., Bill, R. M. and Hipkiss, A. R., 2012. L-Carnosine affects the growth of *Saccharomyces cerevisiae* in a metabolism-dependent manner. *PLoS One*. 7: e45006.

Coban, J., Bingul, I., Yesil-Mizrak, K., Dogru-Abbasoglu, S., Oztezcan, S. and Uysal, M., 2013. Effects of carnosine plus vitamin E and betaine treatments on oxidative stress in some tissues of aged rats. *Current Aging Science*. 6: 199–205.

Coyle, J. T. and Puttfarcken, P., 1993. Oxidative stress, glutamate, and neurodegenerative disorders. *Science*. 262: 689–695.

De Marchis, S., Melcangi, R. C., Modena, C., Cavaretta, I., Peretto, P., Agresti, C. and Fasolo, A., 1997. Identification of the glial cell types containing carnosine-related peptides in the rat brain. *Neuroscience Letters*. 237: 37–40.

Fonteh, A. N., Harrington, R. J., Tsai, A., Liao, P. and Harrington, M. G., 2007. Free amino acid and dipeptide changes in the body fluids from Alzheimer's disease subjects. *Amino Acids*. 32: 213–224.

Fu, Q., Dai, H., Hu, W., Fan, Y., Shen, Y., Zhang, W. and Chen, Z., 2008. Carnosine protects against Abeta42-induced neurotoxicity in differentiated rat PC12 cells. *Cellular and Molecular Neurobiology*. 28: 307–316.

Gallant, S., Kukley, M., Stvolinsky, S., Bulygina, E. and Boldyrev, A., 2000. Effect of carnosine on rats under experimental brain ischemia. *Tohoku Journal of Experimental Medicine*. 191: 85–99.

van der Goes, A., Brouwer, J., Hoekstra, K., Roos, D., van den Berg, T. K. and Dijkstra, C. D., 1998. Reactive oxygen species are required for the phagocytosis of myelin by macrophages. *Journal of Neuroimmunology*. 92: 67–75.

Hipkiss, A. R., 2009. Carnosine and its possible roles in nutrition and health. *Advances in Food and Nutrition Research*. 57: 87–154.

Jin, C. L., Yang, L. X., Wu, X. H., Li, Q., Ding, M. P., Fan, Y. Y., Zhang, W. P., Luo, J. H. and Chen, Z., 2005. Effects of carnosine on amygdaloid-kindled seizures in Sprague-Dawley rats. *Neuroscience*. 135: 939–947.

Kalaria, R. N., 2003. Comparison between Alzheimer's disease and vascular dementia: implications for treatment. *Neurological Research*. 25: 661–664.

Kalaria, R. N., Maestre, G. E., Arizaga, R., Friedland, R. P., Galasko, D., Hall, K., Luchsinger, J. A., Ogunniyi, A., Perry, E. K., Potocnik, F., Prince, M., Stewart, R., Wimo, A., Zhang, Z. X. and Antuono, P., 2008. Alzheimer's disease and vascular dementia in developing countries: prevalence, management, and risk factors. *Lancet Neurology*. 7: 812–826.

Kalaz, E. B., Coban, J., Aydin, A. F., Dogan-Ekici, I., Dogru-Abbasoglu, S., Oztezcan, S. and Uysal, M., 2013. Carnosine and taurine treatments decreased oxidative stress and tissue damage induced by D-galactose in rat liver. *Journal of Physiology and Biochemistry*. 70: 15–25.

Kang, J. H. and Kim, K. S., 2003. Enhanced oligomerization of the alpha-synuclein mutant by the Cu,Zn-superoxide dismutase and hydrogen peroxide system. *Molecules and Cells*. 15: 87–93.

Kuklei, M. L. and Gannushkina, I. V., 1997. Effect of carnosine on cerebral blood flow and amplitude of EEG theta-rhythm in cerebral ischemia in rats with varied motor activities. *Doklady Akademii Nauk*. 352: 416–419.

Kurata, H., Fujii, T., Tsutsui, H., Katayama, T., Ohkita, M., Takaoka, M., Tsuruoka, N., Kiso, Y., Ohno, Y., Fujisawa, Y., Shokoji, T., Nishiyama, A., Abe, Y. and Matsumura, Y., 2006. Renoprotective effects of L-carnosine on ischemia/reperfusion-induced renal injury in rats. *Journal of Pharmacology and Experimental Therapeutics*. 319: 640–647.

Ma, J., Xiong, J. Y., Hou, W. W., Yan, H. J., Sun, Y., Huang, S. W., Jin, L., Wang, Y., Hu, W. W. and Chen, Z., 2012. Protective effect of carnosine on subcortical ischemic vascular dementia in mice. *CNS Neuroscience & Therapeutics*. 18: 745–753.

Micieli, G., 2006. Vascular dementia. *Neurological Sciences*. 27(suppl. 1): S37–S39.

Miu, A. C., Heilman, R. M., Pasca, S. P., Stefan, C. A., Spanu, F., Vasiu, R., Olteanu, A. I. and Miclea, M., 2006. Behavioral effects of corpus callosum transection and environmental enrichment in adult rats. *Behavioural Brain Research*. 172: 135–144.

Pekcetin, C., Kiray, M., Ergur, B. U., Tugyan, K., Bagriyanik, H. A., Erbil, G., Baykara, B. and Camsari, U. M., 2009. Carnosine attenuates oxidative stress and apoptosis in transient cerebral ischemia in rats. *Acta Biologica Hungarica*. 60: 137–148.

Roman, G. C., Salloway, S., Black, S. E., Royall, D. R., Decarli, C., Weiner, M. W., Moline, M., Kumar, D., Schindler, R. and Posner, H., 2010. Randomized, placebo-controlled, clinical trial of donepezil in vascular dementia: differential effects by hippocampal size. *Stroke*. 41: 1213–1221.

Sakai, M., Kani, K., Karasawa, N., Yoshida, M. and Nagatsu, I., 1988. Carnosine-like immunoreactivity in the olfactory bulb of the rat: an electron microscopic study. *Brain Research*. 438: 335–338.

Sensi, S. L., Rapposelli, I. G., Frazzini, V. and Mascetra, N., 2008. Altered oxidant-mediated intraneuronal zinc mobilization in a triple transgenic mouse model of Alzheimer's disease. *Experimental Gerontology*. 43: 488–492.

Shen, Y., He, P., Fan, Y. Y., Zhang, J. X., Yan, H. J., Hu, W. W., Ohtsu, H. and Chen, Z., 2010. Carnosine protects against permanent cerebral ischemia in histidine decarboxylase knockout mice by reducing glutamate excitotoxicity. *Free Radical Biology & Medicine*. 48: 727–735.

Shen, Y., Hu, W. W., Fan, Y. Y., Dai, H. B., Fu, Q. L., Wei, E. Q., Luo, J. H. and Chen, Z., 2007. Carnosine protects against NMDA-induced neurotoxicity in differentiated rat PC12 cells through carnosine–histidine–histamine pathway and H(1)/H(3) receptors. *Biochemical Pharmacology*. 73: 709–717.

Shibata, M., Ohtani, R., Ihara, M. and Tomimoto, H., 2004. White matter lesions and glial activation in a novel mouse model of chronic cerebral hypoperfusion. *Stroke*. 35: 2598–2603.

Spina-Purrello, V., Giliberto, S., Barresi, V., Nicoletti, V. G., Giuffrida Stella, A. M. and Rizzarelli, E., 2010. Modulation of PARP-1 and PARP-2 expression by L-carnosine and trehalose after LPS and INFgamma-induced oxidative stress. *Neurochemical Research*. 35: 2144–2153.

Stvolinsky, S. L., Kukley, M. L., Dobrota, D., Matejovicova, M., Tkac, I. and Boldyrev, A. A., 1999. Carnosine: an endogenous neuroprotector in the ischemic brain. *Cellular and Molecular Neurobiology*. 19: 45–56.

Tsigelny, I. F., Sharikov, Y., Miller, M. A. and Masliah, E., 2008. Mechanism of alpha-synuclein oligomerization and membrane interaction: theoretical approach to unstructured proteins studies. *Nanomedicine*. 4: 350–357.

Yoshizaki, K., Adachi, K., Kataoka, S., Watanabe, A., Tabira, T., Takahashi, K. and Wakita, H., 2008. Chronic cerebral hypoperfusion induced by right unilateral common carotid artery occlusion causes delayed white matter lesions and cognitive impairment in adult mice. *Experimental Neurology*. 210: 585–591.

Zhang, X., Song, L., Cheng, X., Yang, Y., Luan, B., Jia, L., Xu, F. and Zhang, Z., 2011. Carnosine pretreatment protects against hypoxia-ischemia brain damage in the neonatal rat model. *European Journal of Pharmacology*. 667: 202–207.

Zhang, Z. X., Zahner, G. E., Roman, G. C., Liu, J., Hong, Z., Qu, Q. M., Liu, X. H., Zhang, X. J., Zhou, B., Wu, C. B., Tang, M. N., Hong, X. and Li, H., 2005. Dementia subtypes in China: prevalence in Beijing, Xian, Shanghai, and Chengdu. *Archives of Neurology*. 62: 447–453.

CHAPTER 24

Carnosine and Alzheimer's Disease-Related Fibril Formation

ALESSANDRA ALOISI[b] AND ROSS RINALDI*[a]

[a]Department of Mathematics and Physics "E. De Giorgi", Università del Salento, Lecce, Italy; [b]Istituto Nanoscienze-CNR, Euromediterranean Center for Nanomaterial Modelling and Technology (ECMT), via Arnesano 73100, Lecce, Italy
*E-mail: ross.rinaldi@unisalento.it

24.1 Carnosine and Alzheimer's Disease-Related Fibril Formation

The biochemistry of amyloid proteins represents a key and stimulating area of research for its contribution to the understanding of protein folding dynamics and assembly, also in the pathogenic mechanisms of human disorders. Several proteins go through cycles of unfolding and refolding due to a multiplicity of factors including transport across a membrane, cellular secretion, or exposure to stress conditions (*e.g.* changes in pH, temperature) (Ecroyd and Carver, 2008). Also, many globular and natively disordered proteins can organize into amyloid fibrils. Evidence has shown that the morphological, histochemical, and cytotoxic behavior of the aggregates resulting from nondisease-related proteins are analogous to those of disease-associated amyloid proteins, suggest that amyloid fibril-forming propensity is a generic

Food and Nutritional Components in Focus No. 8
Imidazole Dipeptides: Chemistry, Analysis, Function and Effects
Edited by Victor R Preedy

Published by the Royal Society of Chemistry, www.rsc.org

property of all polypeptides (Naeem *et al.*, 2011; Zhang *et al.*, 2010). In summary, these structures are associated with many disorders (Westermark *et al.*, 2007) as well as with ordinary cellular functions (Fowler *et al.*, 2007; Maji *et al.*, 2009), and often arise throughout protein denaturation (Astbury and Dickinson, 1935; Calamai *et al.*, 2005).

24.2 Amyloid Proteins and AD

On November 3, 1906, Alois Alzheimer gave a lecture to the Meeting of the Psychiatrists of South West Germany, describing the clinical features of the case of Auguste D., a 51 year-old German housewife, who had showed a quick decline into dementia, and who was shown, postmortem, to have, within the gray matter, the pathological features – plaques (miliary foci) and tangles (fibrils) – that nowadays are seen as the hallmark of Alzheimer's disease (AD) (Hardy and Selkoe, 2002; Hardy, 2006). These assemblies, predominantly composed of filamentous aggregates – fibrils of the 39- to 42-residue-long amyloid-β (Aβ) peptide, are formed from the proteolytic cleavage of the amyloid precursor protein (APP) by β- and γ-secretases (Kang *et al.*, 1987; Thinakaran and Koo, 2008).

The major species in Aβ assembly are the Aβ40 and Aβ42 peptides, with Aβ42 being predominant in neuritic plaques of AD patients and displaying a higher *in vitro* tendency to aggregate and organize into fibrils under specific conditions (Burdick *et al.*, 1992; Roher *et al.*, 1993; Iwatsubo *et al.*, 1996; Ahmed *et al.*, 2010).

Many proteins enter the so-called amyloid state, in which they form extended fibers, with spines consisting of many-stranded β-sheets. A substantial body of structural information about amyloid fibril structures, shadowy until 15 years ago, now exists. By combining structural constraints from solid-state nuclear magnetic resonance (NMR) measurements (Tycko and Wickner, 2013) and other analytical techniques – including X-ray fiber diffraction, cryoelectron microscopy, scanning transmission electron microscopy, and atomic force microscopy – the atomic resolution structures of three amyloid polymorphs formed by an 11-residue peptide have been reported. In particular, details of the packing interactions by which the constituent β-strands are assembled hierarchically into protofilaments, filaments, and mature fibrils have been given (Fitzpatrick *et al.*, 2013).

24.2.1 Cross-β Structure of Aβ Amyloid Fibrils

Essentially, amyloid fibrils can be described as fibrillar polypeptide aggregates with a cross-β structure. A protein's amyloid state occurs when a segment exposes its backbone amide N–H groups and C=O groups, allowing them to associate into hydrogen bonds with other protein chains. The β-sheet plane and the backbone hydrogen bonds that connect the β-strands are oriented parallel to the main fibril axis. Thus, the β-strands run perpendicular ("cross") to this direction. It is certain that the cross-β structure of Aβ fibrils has been identified from numerous research groups, resulted formed by discrete sequence segments at the peptide center or C-terminus. The local concentration of the exposed segment must be sufficiently copious to overcome the entropy that opposes establishment of ordered fibers. Thus, amyloid fibers can be obtained in the laboratory from a protein at high concentration by denaturing it slightly with a destabilizing solvent, or a change in pH, heating, or surface denaturation from agitation. Once the nucleus is shaped within a slow nucleation phase, a more rapid growth phase follows: single molecules can join the emergent fibril one at a time, as they open up to expose the proper segment and bond at the ends of fibrils. (See for a full review: Fändrich *et al.*, 2011; Eisenberg and Jucker, 2012.)

24.3 Screening for Negative Modulators and Carnosine Activity

The outcome that a number of pathologies, including Alzheimer's disease, are associated with amyloid fibrils has stimulated research on fibril inhibition. Constraining Aβ aggregation at the level of the monomer is a smart therapeutic approach: it represents the earliest opportunity to interfere with the aggregation cascade (Sievers *et al.*, 2011; Lemkul and Bevan, 2012). Extensive efforts are ongoing to screen small inhibitory compounds that are able to counteract the fibrillogenesis of disease-related amyloid proteins. Evaluation of these modulators could be helpful in the development of therapeutics, of course providing likewise that the tested active molecules do not interfere with physiological processes. To this aim, several compounds have been assayed to prevent or reduce the aggregation of Aβ, including small molecules such as glycosaminoglycan (GAG) mimetic, dye compounds, polyphenols, metal-ion chelators, self-recognition elements (SREs) antibodies, osmolytes, peptidic β-sheet breakers, chaperone proteins and many more. An extensive review has been prepared on this topic (Hamley, 2012). In the same way, different methodological approaches have been developed, among these, basically, the most promising one is to favor the structure stabilization of the soluble form of a protein, diminishing the rate at which it is likely to undergo conversion to the amyloid state. An alternative approach is to screen for small molecules that may disrupt fibril and oligomer formation. A third strategy uses the self-associating property of amyloid fibers to poison the growth of amyloid fibers with short peptide segments. The fourth approach is a variation of the third: to inhibit fiber growth by the structure-based design of small peptides targeted to block the ends of fibrils (Eisenberg and Jucker, 2012).

At the present time, molecules screening for drug therapies is still in progress (Table 24.1), thus it is difficult to propose specific recommendations for the prevention of AD some strategies only exist for delay the development of symptoms.

Table 24.1 Examples of chemical inhibitors of amyloid β aggregation. Compounds for amyloid-based therapeutics with the potential to demonstrate disease-modifying activity. Reprinted from Nie *et al.* (2011), with permission from the Nature Publishing Group.

Chemical compound	Structure	Current trial state
Bapineuzamab	Monoclonal antibody	Phase III
Tramiprosate		Failed in phase III
PTB2	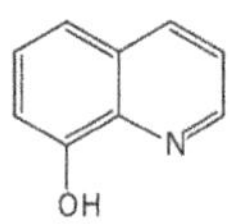	Phase IIa
Scyllo-inositol	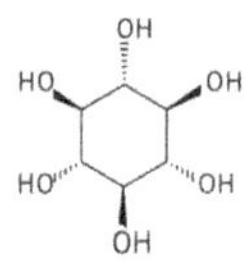	Phase II
PPI1019	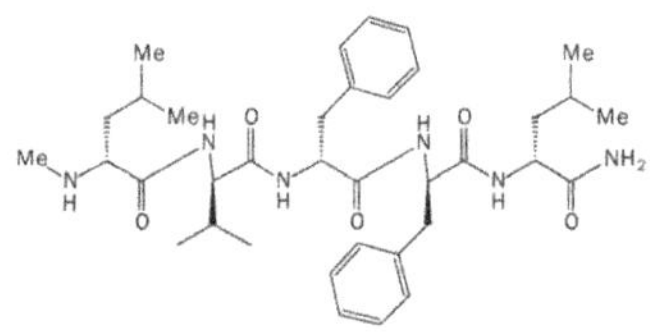	Completed phase II
RS0406	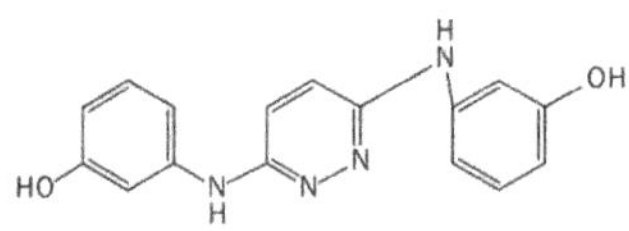	Phase II
SP-233	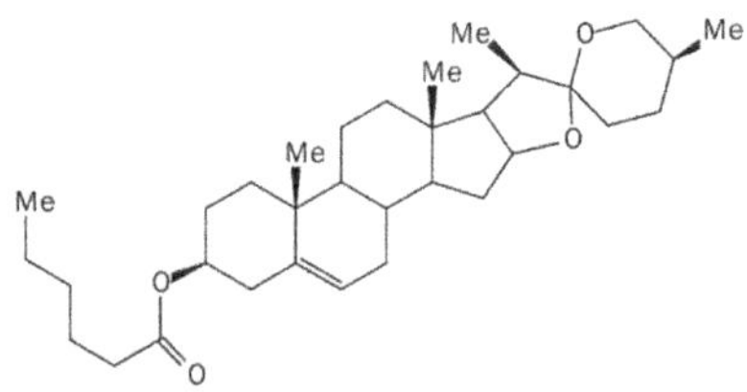	Phase I
EGCG	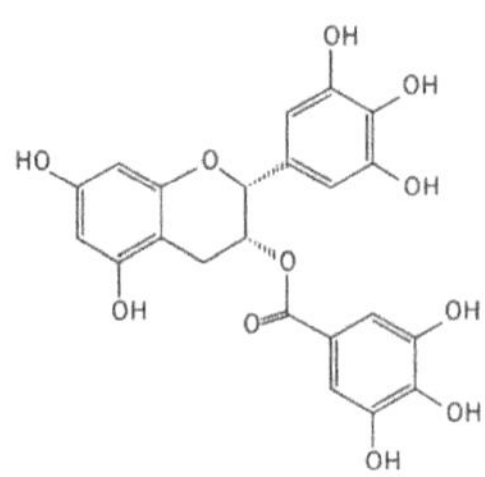	Phase II

Xebryl-1	No structural details are available	Phase I
SEN606	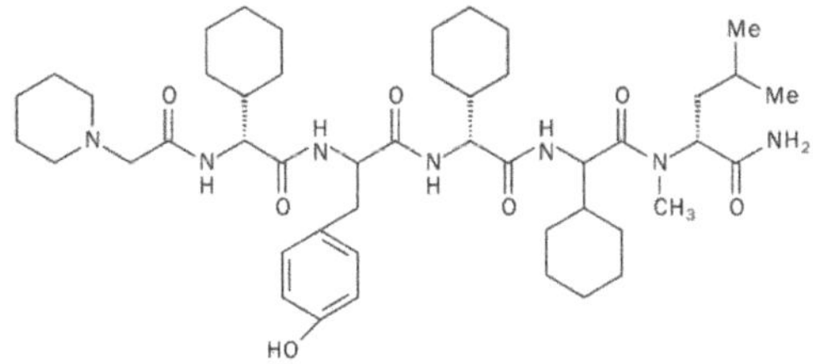	Preclinical

Moreover, the recent failure of the small molecule tramiprosate, a GAG mimetic, in phase III clinical trials, highlights the necessity for new inhibitors with other mechanisms of modulation of Aβ aggregation. To this aim, the potential therapeutic role of nutraceuticals has been considered during recent years, even if epidemiological analysis of the relationships between nutrient intake and cognitive failure is complex, and it is highly unsure that a single element plays a major role.

In this context, the naturally occurring carnosine is becoming a clinically accepted nutritional supplement with application over a considerable range of chronic diseases. It has been suggested as a promising therapeutic agent as it may be involved in defeating age-related dysfunction in either energy metabolism or proteostasis (Hipkiss, 2009, 2011). In particular, its role as a neuroprotective agent was shown from animal studies (Bellia *et al.*, 2011). Furthermore, the idea that carnosine might be used to control Alzheimer's disease (Hipkiss, 2007a,b) is supported by its ability to suppress the toxic effects of amyloid β towards cultured cells (Preston *et al.*, 1998; Fu *et al.*, 2008). Nonetheless, owing to the occurrence of carnosinase enzyme action, carnosine effectiveness is reported to be reduced in clinical treatments (Bellia *et al.*, 2012). Conjugation strategy with vitamin E derivatives (Stvolinsky *et al.*, 2010) or small sugars, (Lanza *et al.*, 2011; Grasso *et al.*, 2011) represents the first attempts to stabilize carnosine against carnosinase degradation. Despite the consistent effects of carnosine on human and animals widely shown in the literature, the biomolecular pathways originating these effects are still unclear.

Zinc-ion modulation, upregulation of heat-shock protein expression, and/or amplified proteolysis of the abnormal polypeptide could be at the basis of the mechanism. Aggregation of Aβ may be promoted by metal dyshomeostasis, in fact unusually high concentration of metal ion (such as Cu^{2+}, Zn^{2+}) are found to colocalize with senile plaques in AD brains. Treatment with metal-ion chelators can reduce the deposition of Aβ in brains (Hegde *et al.*, 2009) – carnosine is a metal-ion chelator (Trombley *et al.*, 2000), feasibly its function in metal homeostasis may represent a significant topic to be further examined at the molecular level. Additionally, carnosine reacts with and inactivates aldehydes and ketones, decreasing protein glycosylation and the rate of formation of advanced glycation end products (AGEs) (Münch *et al.*,

2003) considered not only as a triggering factor in amyloidosis but also a player in later stages of the aggregation process. Other results suggest that histidine dipeptides like carnosine have antioxidant effects on brain proteins (such as Neurofilament L) under pathophysiological conditions leading to degenerative damage, such as those caused by neurodegenerative disorders (Kim and Kang, 2007).

On the whole, the inhibitory activity against fibrillogenesis could represent the keystone of a new path to be explored. Notably, carnosine has been found to disaggregate glycated α-crystallin protein (Seidler *et al.*, 2004) and to prevent the native protein from forming amyloid fibrils (Attanasio *et al.*, 2009). Recently, carnosine has been shown to suppress the gathering of β-amyloid in the central nervous system of transgenic mice model for Alzheimer's disease (Corona *et al.*, 2011) and reduce the *in vitro* fibrillogenesis of β-amyloid peptide Aβ42 (Aloisi *et al.*, 2013; Attanasio *et al.*, 2013). This ability could be of particular effect considering, as mentioned above, that direct interference with Aβ aggregation is a major target in AD therapeutics (Hamley, 2012).

24.3.1 Carnosine and β-Amyloid Peptide: Interaction Dynamics

Given the consistent body of evidence supporting the neuroprotective activity of carnosine, independent studies have recently indicated that millimolar concentrations of this dipeptide are able to effectively hinder Aβ aggregation in tube tests. By combining NMR data and molecular dynamics (MD) simulations of heptapeptide segments of Aβ, a significant decrease of the number of interpeptide H-bonds in the region 18–24 was observed (Attanasio *et al.*, 2013). It is known that the salt bridge between the side chains of D23 and K28 is involved in Aβ toxicity, as they are crucial in the intermolecular interactions between two adjacent Aβ42 monomers during the elongation of protofibrils, agreeing with the model of growth by single monomer addition (Lührs *et al.*, 2005). In view of this, agents able to drive these residues towards nonsalt-bridge-forming conformations are supposed to have a therapeutic role in AD (Mithu *et al.*, 2011). Docking analysis clearly shows arrangement of carnosine in such a region (Aloisi *et al.*, 2013) supporting the self-aggregation reduction evidence (Figure 24.1). Specifically, the extensive and comparative molecular docking studies performed on numerous natural or synthetic recognized inhibitors of the Aβ1–42 fibril formation, enclose carnosine together with the top-ranked compounds. In particular, the *in silico* screening of Aβ1–42 ligands predicted this dipeptide as the best one among the natural histidine-containing tested dipeptides, and as a very good ligand when matched with other compounds capable of inhibiting the aggregation of the Aβ1–42 amyloid peptide (Table 24.2). Interactions between the imidazole ring of the L-histidine of carnosine and the D23 residue of Ab1–42, and between the β-alanine of the dipeptide and the K28 residue of Aβ1–42 are evidenced, supporting the potential interfering role of carnosine on fibrillogenesis, and its ability to disturb the growing fibril by impeding direct

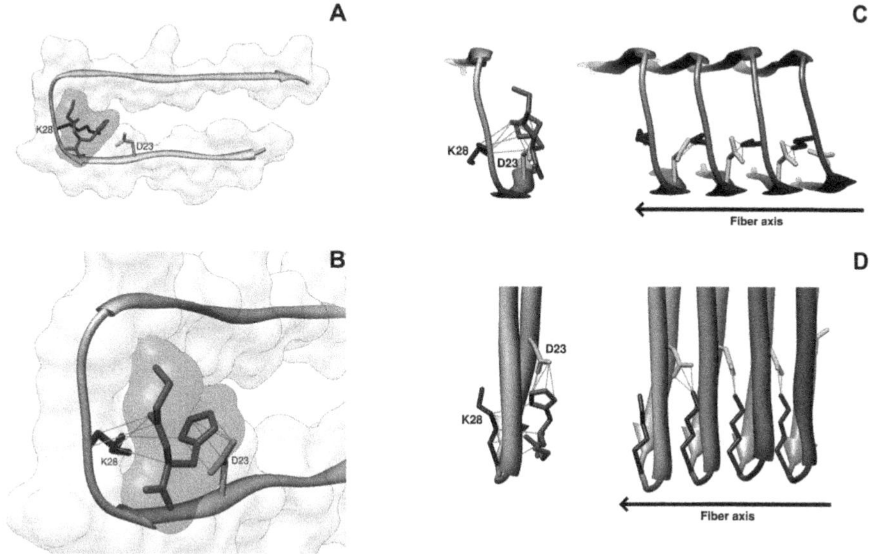

Figure 24.1 Cartoon model of carnosine–Aβ42 peptide fibril interaction complex. Interaction between carnosine and the fibril monomer of Aβ42 peptide at the level of the coiled region, dipeptide and fibril representation (A). Binding contacts, between (1) the imidazole ring of L-histidine in carnosine and Aβ42 L-aspartic residue (D23) and (2) β-alanine end and Aβ42 L-lysine (K28) (B). Aβ42 self-association model inhibited by carnosine (C); a 90° right-handed rotation view (D). The direct binding of D23 of a monomer with the K28 of the adjacent monomer along the oligomer axis is prevented by carnosine. Reprinted from Aloisi *et al.* (2013), licensed under http://creativecommons.org/licenses/by/4.0/.

Table 24.2 Carnosine and carnosine-like dipeptides: ligand efficiency for the β-fibril model of Aβ42. Indices of ligand efficiency obtained from molecular docking analysis of the β-fibril model of Aβ42 (PDB:2BEG) used as the receptor of carnosine and carnosine-like dipeptides. Ligand efficiency indices are indicated as BE/MW (molecular-weight-based efficiency index), BE/NHA (number of heavy atoms-based efficiency index) BE/NoC (number of carbons-based efficiency index) and BE/W (Wiener index-based efficiency index). Autodock Vina and Open Babel Package and UCSF Chimera package, the software. Modified from Aloisi *et al.* (2013), licensed under http://creativecommons.org/licenses/by/4.0/.

			Ligand efficiency index			
	Carnosine and carnosine-like dipeptides	PubChe	BE/MW 10^{-2} kcal mol^{-1} kDa^{-1}	BE/NHA 10^{-1} kcal mol^{-1} NHA^{-1}	BE/NoC 10^{-1} kcal mol^{-1} NoC^{-1}	BE/W 10^{-2} kcal mol^{-1} W^{-1}
1	Carnosine	439224	−1.64	−2.33	−4.13	−0.78
2	L-Anserine	112072	−1.61	−2.28	−3.88	−0.70
3	Homocarnosine	102433	−1.59	−2.25	−3.82	−0.66
4	Balenine	308054	−1.44	−2.04	−3.46	−0.58

binding of the D23 of one Aβ1–42 monomer and the K28 of the contiguous monomer (Aloisi *et al.*, 2013).

Circular dichroism (CD) spectroscopy performed in the presence of the dipeptide has considered the effect on the conformational behavior of Aβ42, confirming that the presence of carnosine does not significantly modify the conformational features of Aβ42 but, reasonably, interferes with its self-assembling route, resulting in a lessening in the total amount of fibrils formed at the end of the process.

A comparison concerning the distribution of the amount of *inter*-peptide backbone hydrogen bonds in the MD simulations with and without carnosine also shows that the dipeptide reduces the aggregation tendency of the octameric Aβ18–24 system significantly, but only marginally for octameric Aβ14–20 and Aβ16–22. The number of *intra*-peptide backbone hydrogen bonds increased significantly only for the Aβ18–24 heptapeptide, which is consistent with the decrease in *inter*-peptide hydrogen bonds (Attanasio *et al.*, 2013). The evidence of reduced Aβ1–42 aggregation has been quantitatively confirmed by ThT fluorescence assay, showing a reduction of the Aβ1–42 polymerization process in the presence of increasing concentrations of carnosine. Correspondingly, by AFM imaging of dehydrated samples, it has been revealed that, in the presence of carnosine, extended Aβ1–42 fibrils items are drastically underrepresented, while few deposited aggregates predominantly include shorter fibrils and small globular formations (Figure 24.2) (Aloisi *et al.*, 2013).

On a parallel route, a positive correlation between the concentration and inhibitory effect of carnosine against lysozyme fibril formation has been also demonstrated (Wu *et al.*, 2013). Actually, hen egg white lysozyme (HEWL), often used as a nice model system to study *in vitro* phenomena associated with fibril formation, has been examined in terms of the thermal denaturation profiles in the presence of various concentrations of carnosine, showing carnosine aptitude to increase the thermal stability of HEWL (Figure 24.3) (Wu *et al.*, 2013).

In this scenario, the relatively new concept of pharmacological chaperones in the treatment of certain chronic disabling diseases may hold the biological and therapeutic activities of the family of imidazole-containing peptidomimetics carcinine (β-alanylhistamine), *N*-acetylcarnosine (*N*-acetyl-β-alanylhistidine) and carnosine (β-alanyl-L-histidine) that have been described in a recent review (Babizhayev and Yegorov, 2010). In fact, cells retain a complete set of functionally competent proteins normally and in the face of injury or environmental stress with the use of various mechanisms, including systems of proteins called molecular chaperones. The aforementioned are molecules with established aptitude to be effective in preventing misfolding of different disease-causing proteins, and, basically, reducing the severity of several neurodegenerative disorders (like age-related macular degeneration). Chaperones, including the *heat-shock proteins* (Hsps), facilitate refolding or degradation of misfolded polypeptides, deter protein aggregation and take part in formation of aggresome,

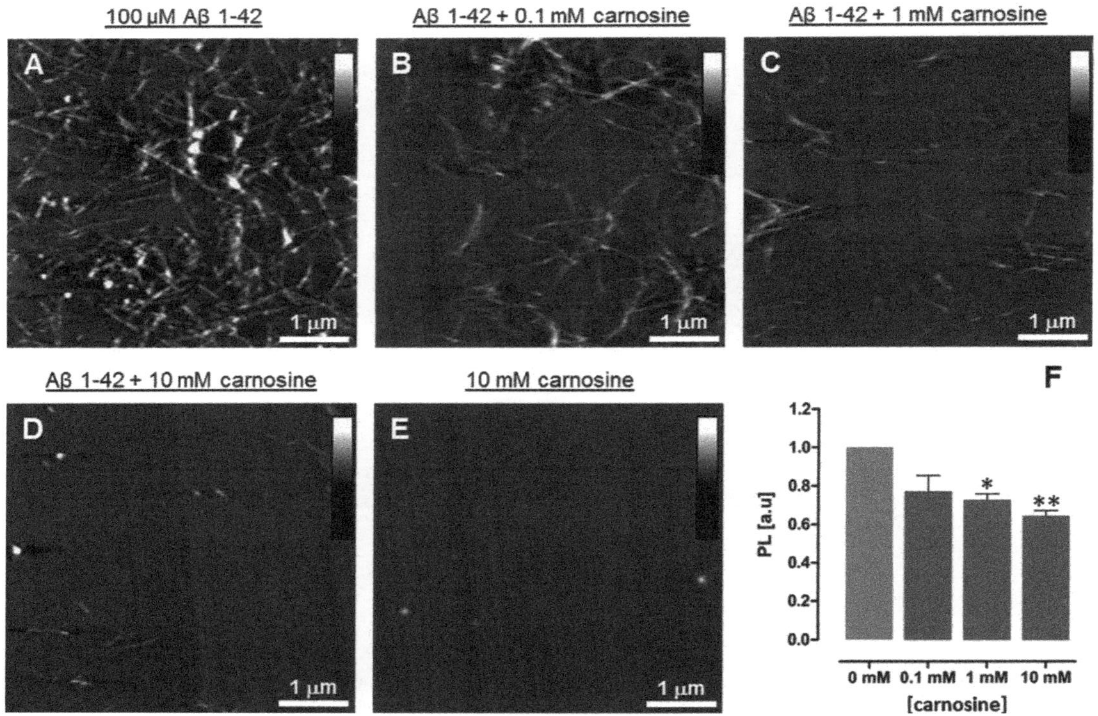

Figure 24.2 Effects of carnosine on Aβ42 fibrillogenesis: qualitative and quantitative evaluation. Analysis of the deposited amyloid aggregates as assessed by atomic force microscopy (AFM) and thioflavin T (ThT) fluorescence assays. AFM representative pictures of Aβ42 samples (control; (A)), Aβ42 samples incubated with 0.1 mM (B), 1 mM (C) and 10 mM (D) carnosine, and 10 mM carnosine alone (E). Height mode imaging with bar color code: 0.0 nm, total black, 30 nm, total white. In (F), data relative to the quantitative effects of increasing concentrations of carnosine on Aβ42 fibrillogenesis, represented as ThT photoluminescence (PL) values at 480 nm (means ± S.E.M., $n = 3$) in solutions of Aβ1–42 (100 mM) incubated for 30 min in the absence (control, 0 mM carnosine) and presence of 0.1, 1 and 10 mM carnosine The stoichiometric interaction between ThT and amyloid fibrils provides accurate quantification as a function of amyloid fibril length and number. (***p*, 0.01; **p*, 0.05; one-way ANOVA analysis of variance of the means; Bonferroni *post hoc* test). Slightly modified from Aloisi *et al.* (2013), licensed under http://creativecommons.org/licenses/by/4.0/.

a centrosome-associated body to which small cytoplasmic aggregates are transported. Interestingly enough, there is proof that carnosine, either as the free dipeptide or in complex with zinc ions, can induce expression of heat-shock proteins. Currently, while these findings recently reviewed (Hipkiss *et al.*, 2013), strengthening the suggestion that carnosine might aid to uphold proteostasis, further and deeper study is required to meritoriously confirm this hypothesis and to recommend this naturally occurring dipeptide as promising new therapeutic strategy against protein misfolding processes.

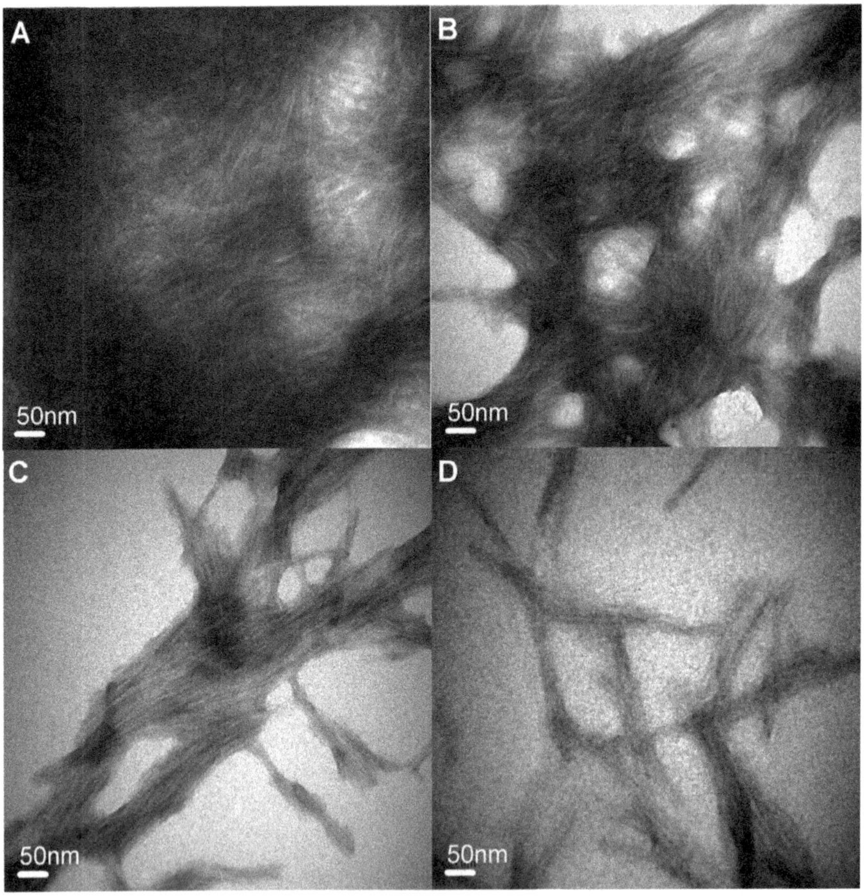

Figure 24.3 Transmission electron micrographs of HEWL samples incubated with various concentrations of carnosine. Negatively stained electron micrographs of hen egg-white lysozyme (HEWL) alone (A), HEWL coincubated with 20 mM carnosine (B), HEWL coincubated with 40 mM carnosine (C), and HEWL coincubated with 50 mM carnosine (D). The scale bar represents 50 nm. Reprinted from Wu *et al.* (2013), licensed under http://creativecommons.org/licenses/by/4.0/.

24.4 Critical Points in Perspective

In the physical chemistry of amyloid fibril other questions remain, in which, thinking and methodology can be expected to provide complementary contributions. It is commonly expected that nucleation arises from a pool of unfolded – aggregation-prone regions – exposing peptides. On the other hand, important issues, concerning the mechanisms of self-association, exist. In particular, about how broad is the Aβ42 primary aggregation pathway and how broad is secondary nucleation mechanism among amyloid forming proteins and peptides (Härd, 2014). Primary pathways generate new aggregates at a monomer-concentration-dependent rate, while it is independent of the concentration of existing fibrils. Conversely, secondary pathways generate new aggregates at a rate dependent on the concentration of existing fibrils (Figure 24.4). The latter class include diverse processes that may be: (1) monomer independent (with a rate depending only upon the concentration of existing fibril) and (2) monomer dependent (with a rate dependent on both the concentration of monomers and that of existing fibrils). Once a small but critical amount of amyloid fibrils has formed, the toxic oligomers are predominantly generated from the monomeric peptide in a secondary nucleation reaction, catalyzed by the larger fibrillar species. The identification of secondary nucleation underlines the significance of elucidating the detailed structures of amyloid fibrils and their surfaces. The origins of their surface–catalytic activity should be studied by means of molecular simulations (Cohen *et al.*, 2013).

This assumption suggests that a key approach for suppressing the production of neurotoxic Aβ42 oligomers could be to focus on the secondary in addition to the primary nucleation pathway. Once fibrils reach a critical concentration, the only perturbation of the primary nucleation pathway ceases to be fully effective in reducing the whole propagation of oligomers, as most new

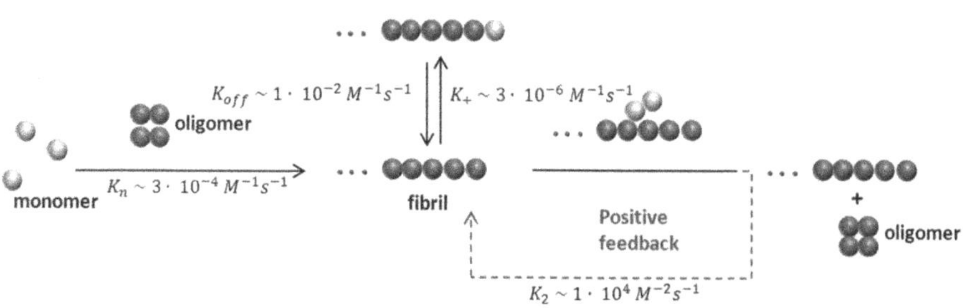

Figure 24.4 Schematic representation of amyloid fibril formation, primary and secondary nucleation pathways. Once a critical concentration of amyloid fibrils has formed, secondary nucleation becomes the prevailing mechanism generating new aggregates. The surfaces of existing fibrils catalyze the nucleation of new aggregates from the monomeric state, and further proliferation becomes exponential due to positive feedback. Slightly modified from Cohen *et al.* (2013), permission is granted.

aggregates are created *via* secondary mechanisms (Cohen *et al.*, 2013). This could be significant, because it may inspire effectively innovative and accurate approaches for quantitatively evaluate active molecules in kinetics experiments. In this perspective, based on the recent encouraging results concerning the "antiaggregating effect of the naturally occurring dipeptide carnosine on Aβ1–42 fibril formation", the aforementioned topic may be the next challenge to be performed in order to develop a novel effective therapeutic strategy in the context of AD as well of other severe neurodegenerative pathologies.

Summary Points

- This chapter focused on the impact of carnosine on amyloid-β peptide (Aβ) dynamics in the context of Alzheimer's disease (AD).
- AD is a degenerative brain disease that impairs mainly the memory and cognitive function in the elderly.
- Protein crosslinking and proteolytic dysfunction contribute to AD.
- Extracellular β-amyloid deposition is one of the pathological events that are thought to cause neuronal damage.
- Amyloid β monomers tend to aggregate and polymerize, forming oligomers, which in turn, contribute to protofibrils, and the protofibrils to fibrils.
- Reduction in the generation of the amyloidogenic arrangement of proteins, increase in the clearance degree of misfolded or aggregated proteins, recovery of the native state stability in amyloidogenic proteins, and direct inhibition of the self-assembly process have been proposed as therapeutic strategies.
- Currently, the action of small molecules able to interfere with the structural organization of soluble or aggregating Aβ peptide during fibrillogenesis is under investigation.
- Carnosine has been shown to affect fibril elongation *in vitro*.
- *In silico* simulations support the evidence that carnosine binds to the Aβ1–42 peptide.
- Several studies have revealed the protective effect of carnosine against amyloid cytotoxicity in cell culture and in animal models.
- The role of carnosine as homeostatic agent in the context of Aβ1–42 amyloid fibril formation is a theme that needs further investigation, with respect to pathophysiological potential arising from carnosine natural existence in the central nervous system (CNS).

Key Facts

Key Features of Amyloid Fibrils

Listed are the key requirements for fibrils definition, in particular from two different points of view: those of the pathologists and biophysicists community.

1. Amyloid formation is a function of the concentration of the amyloid forming peptide.
2. Fewer than 25 amyloid-forming proteins have been recognized and related with severe diseases, comprising amyloid-β peptide (Aβ) with Alzheimer's disease (AD).
3. Pathologists define the fibers as structures, usually extracellular, found *in vivo* and disease associated.
4. The fibers could be identified by Congo Red dye binding and green birefringence between crossed polarizers.
5. Disease-related proteins form analogous fibers *in vitro*, hence, biophysicists, differently from pathologists, define fibers as those structures which exhibit the cross-β diffraction pattern.

Key Features of Alzheimer's Disease (AD)

Listed are the key facts of Alzheimer's disease, briefly outlining definition, principal etiopathogenesis and therapeutic strategies.

1. AD is the most common type of dementia, a term describing a variety of diseases and conditions that develop when nerve cells in the brain die or no longer function normally.
2. Among the brain alterations thought to contribute to the development of AD are the extracellular accumulation of the protein β-amyloid and the intracellular accumulation of the protein tau.
3. The Aβ42 peptide self-assembles to form oligomers, protofibrils and fibrils evolving in the deposition of amyloid plaques.
4. It is now thought that oligomers formed in the initial self-associating process of Aβ are critical as toxic agents.
5. Oxidative stress involves the production of free radicals (especially hydroxyl radicals) in the presence of metal ions, which can influence metabolism and also promote Aβ aggregation.
6. Numerous attempts have been made to slow disease progression, including active and passive immunization, hormones, antioxidants, anti-inflammatory drugs or supposed mitochondrial protectors, however, no effective treatment has emerged from these efforts so far.
7. Drugs approved by the FDA are inhibitors of acetylcholine esterase, able to increase the levels of the neurotransmitter acetylcholine (depleted in AD).
8. The impact of these drugs on disease manifestations is modest and transient.
9. The field, currently, is guessing the results of several *disease-modifying* approaches, including Aβ deposition interference by anti-Aβ aggregation agents, or selective Aβ 42-lowering agents, as well as tau deposition interference and methods for reduction of inflammation and oxidative damage.

Key Features of Protein Folding into Functional Protein

Listed are the key facts of protein folding including the information necessary in order to reach the functional form, the possibility to misfolding and the consequence of this process.

1. All the information required for a protein to fold into its appropriate three-dimensional (3D) structure is enclosed within its amino acid sequence.
2. This occurs through an intermediate form, known as secondary structures, the most common of which are the rod-like α-helix and the plate-like β-pleated sheet.
3. The result of 3D correct folding is the protein functional form.
4. Aminoacids interact locally, thus limiting the space that the protein has to explore and driving the protein to follow a funnel-like energy setting that allows it to fold quickly.
5. The failure of a protein to fold correctly can have serious consequences, frequently evolving into a severe degenerative disease (like AD).

Key Features of Chaperones in Proteostasis

Listed are the key facts of chaperones' role within proteostatic network and the potentiality of modulation of their function by small active molecules.

1. Chaperones consist of a functionally related group of proteins that assists protein folding in the cell under physiological and stress conditions.
2. They do not determine the tertiary structure, but help proteins to find it more efficiently.
3. By binding to their targets, chaperones act as "collectors" of damaged non-native proteins preventing unspecific aggregation.
4. The search for therapeutics to treat neurodegeneration is focusing on the identification of small-molecule proteostasis regulators.
5. These small molecules could act by enhancing the expression or work of molecular chaperones.
6. This mechanism may act by increasing the rates of folding of assisted proteins, as well as, by enhancing degradation/clearance or by controlling protein translation rates.

Definitions of Words and Terms

Key Terms

Amyloid. Originally meaning starch-like, the term amyloid refers to protein deposits resembling those first observed for starch. It is now specifically associated with proteins and peptides adopting fibrils based on the cross-β structure, in which the peptide backbone is orthogonal to the fibril axis. β-Amyloid is a protein fragment derived from proteolytic processing of the amyloid precursor protein (APP).

Atomic force microscopy (AFM). A technique for imaging the topography of surfaces at high resolution. The core of an AFM instrument is a probe fixed near to the end of a flexible microcantilever arm. This sharp probe is moved toward the area of interest and allowed to cross it by raster scanning and simultaneously monitoring the deflection of the cantilever arm as it meets the sample topographic features.

Fibrils. Oligomers associated into typical β-sheets structure.

Glycosaminoglycan (GAGs). Large unbranched polysaccharides constructed of repeating units of alternating uronic acids and amino sugars.

Hen egg-white lysozyme (HEWL). A 129-residue monomeric protein with a molecular weight of 14.3 kD, it has been widely chosen as a model protein in research relating to the subjects of protein folding, unfolding, and aggregation.

Heat-shock proteins (HSPs). Stress- or heat-shock proteins are the most abundant and conserved proteins in both prokaryotes and eukaryotes. Their expression is induced as a reaction to a wide range of physiological inputs and environmental insults.

Molecular dynamics. Computer simulation of the physical movements of atoms and molecules, used to bring biomolecular structures alive. The method gives insights into the natural behavior of biomolecules in solution, on different timescales, exploring which conformations of a complex are thermally accessible. This technique is used for exploring conformational space, for instance, in ligand-docking applications, a method that predicts the predominant binding mode of one molecule to a second, used as a key tool in structural molecular biology and computer-assisted drug design.

Proteostasis. Combination of two words: *protein* and *homeostasis*, this term refers to the process that regulates proteins within the cell in order to maintain the health of the cellular proteome system. In detail, protein folding homeostasis screens and ensures folding, assembly and targeting of synthesized proteins, repair of damaged proteins, and finally clearance.

Model. Structural representation based on the homology between molecules, general side-chain geometry or chemical bond lengths, obtained from a database.

Neuritic plaques. Also termed senile plaques, are spherical bodies that, typically, contain a central dense core of amyloid surrounded by degenerating neuronal processes.

Oligomers. Aggregated β-strand structures not organized into β-sheets (as recently shown by NMR spectroscopy by Ahmed M. and colleagues in a work focused on "Structural conversion of neurotoxic amyloid-β(1–42) oligomers to fibrils").

Thioflavin T. A benzothiazole salt, widely used as dye to visualize and quantify the presence of misfolded protein aggregates, both *in vitro* and *in vivo*, (*e.g.*, plaques composed of amyloid b found in the brains of Alzheimer's disease patients).

List of Abbreviations

AGEs	Advanced glycation end products
Aβ	Amyloid-β
AD	Alzheimer's disease
AFM	Atomic force microscopy
APP	Amyloid precursor protein
FDA	Food and Drug Administration
GAG	Glycosaminoglycan
HEWL	Hen egg-white lysozyme
Hsps	Heat-shock proteins
NMR	Nuclear magnetic resonance
TEM	Transmission electron microscopy
ThT	Thioflavin T
3D	Three-dimensional

References

Aloisi, A., Barca, A., Romano, A., Guerrieri, S., Storelli, C., Rinaldi, R. and Verri, T., 2013. Antiaggregating effect of the naturally occurring dipeptide carnosine on abeta1–42 fibril formation. *PLoS One*. 8: e68159.

Ahmed, M., Davis, J., Aucoin, D., Sato, T., Ahuja, S., Aimoto, S., Elliott, J. I., Van Nostrand, W. E. and Smith, S. O., 2010. Structural conversion of neurotoxic amyloid-β1–42 oligomers to fibrils. *Nature Structural & Molecular Biology*. 17: 561–567.

Astbury, W. T. and Dickinson, S., 1935. The X-ray interpretation of denaturation and the structure of the seed globulins. *Biochemical Journal*. 29: 2351–2360.

Attanasio, F., Cataldo, S., Fisichella, S., Nicoletti, S., Nicoletti, V. G., Pignataro, B., Savarino, A. and Rizzarelli, E., 2009. Protective effects of L- and D-carnosine on alpha-crystallin amyloid fibril formation: implications for cataract disease. *Biochemistry*. 48: 6522–6531.

Attanasio, F., Convertino, M., Magno, A., Caflisch, A., Corazza, A., Haridas, H., Esposito, G., Cataldo, S., Pignataro, B., Milardi, D. and Rizzarelli, E., 2013. Carnosine inhibits A beta 42 aggregation by Perturbing the H-bond network in and around the central hydrophobic cluster. *ChemBioChem*. 14: 583–592.

Babizhayev, M. A. and Yegorov, Y. E., 2010. Advanced drug delivery of *N*-acetylcarnosine (*N*-acetyl-beta-alanyl-L-histidine), carcinine (beta-alanylhistamine) and L-carnosine (beta-alanyl-L-histidine) in targeting peptide compounds as pharmacological chaperones for use in tissue engineering, human disease management and therapy: from in vitro to the clinic. *Recent Patents on Drug Delivery & Formulation*. 4: 198–230.

Bellia, F., Vecchio, G., Cuzzocrea, S., Calabrese, V. and Rizzarelli, E., 2011. Neuroprotective features of carnosine in oxidative driven diseases. *Molecular Aspects of Medicine*. 32: 258–266.

Bellia, F., Vecchio, G. and Rizzarelli, E., 2012. Carnosine derivatives: new multifunctional drug-like molecules. *Amino Acids*. 43: 153–163.

Burdick, D., Soreghan, B., Kwon, M., Kosmoski, J., Knauer, M., Henschen, A., Yates, J., Cotman, C. and Glabe, C., 1992. Assembly and aggregation properties of synthetic Alzheimer's A4/β amyloid peptide analogs. *Journal of Biological Chemistry*. 267: 546–554.

Calamai, M., Chiti, F. and Dobson, C. M., 2005. Amyloid fibril formation can proceed from different conformations of a partially unfolded protein. *Biophysical Journal*. 89: 4201–4210.

Corona, C., Frazzini, V., Silvestri, E., Lattanzio, R., La Sorda, R., Piantelli, M., Canzoniero, L. M. T., Ciavardelli, D., Rizzarelli, E. and Sensi, S. L., 2011. Effects of dietary supplementation of carnosine on mitochondrial dysfunction, amyloid pathology, and cognitive deficits in 3xTg-AD Mice. *PLoS One*. 6: e17971.

Cohen, S. I., Linse, S., Luheshi, L. M., Hellstrand, E., White, D. A., Rajah, L., Otzen, D. E., Vendruscolo, M., Dobson, C. M. and Knowles, T. P. J., 2013. Proliferation of amyloid-β42 aggregates occurs through a secondary nucleation mechanism. *Proceedings of the National Academy of Sciences of the United States of America*. 110: 9758–9763.

Ecroyd, H. and Carver, J. A., 2008. Unraveling the mysteries of protein folding and misfolding. *IUBMB Life*. 60: 769–774.

Eisenberg, D. and Jucker, M., 2012. The amyloid state of proteins in human diseases. *Cell*. 148: 1188–1203.

Fitzpatrick, A. W. P., Debelouchina, G. T., Bayro, M. J., Clare, D. K., Caporini, M. A., Bajaj, V. S., Jaroniec, C. P., Wang, L., Ladizhansky, V., Müller, S. A., MacPhee, C. E., Waudby, C. A., Mott, H. R., De Simone, A., Knowles, T. P. J., Saibil, H. R., Vendruscolo, M., Orlova, E. V., Griffin, R. G. and Dobson, C. M., 2013. Atomic structure and hierarchical assembly of a cross-β amyloid fibril. *Proceedings of the National Academy of Sciences of the United States of America*. 110: 5468–5473.

Fändrich, M., Schmidt, M. and Grigorief, N., 2011. Recent progress in understanding Alzheimer's β-amyloid structures. *Trends in Biochemical Sciences*. 36: 338–345.

Fowler, D. M., Koulov, A. V., Balch, W. E. and Kelly, J. W., 2007. Functional amyloid – from bacteria to humans. *Trends in Biochemical Sciences*. 32: 217–224.

Fu, Q., Dai, H., Hu, W., Fan, Y., Shen, Y., Zhang, W. and Chen, Z., 2008. Carnosine protects against Abeta 42-induced neurotoxicity in differentiated rat PC12 cells. *Cellular and Molecular Neurobiology*. 28: 307–316.

Grasso, G. I., Arena, G., Bellia, F., Maccarrone, G., Parrinello, M., Pietropaolo, A., Vecchio, G. and Rizzarelli, E., 2011. Intramolecular weak interactions in the thermodynamic stereoselectivity of copper(II) complexes with carnosine–trehalose conjugates. *Chemistry – A European Journal*. 17: 9448–9455.

Hamley, I. W., 2012. The amyloid beta peptide: a chemist's perspective. Role in Alzheimer's and fibrillization. *Chemical Reviews*. 112: 5147–5192.

Härd, T., 2014. Amyloid fibrils: formation, polymorphism, and inhibition. *Journal of Physical Chemistry Letters*. 5: 607–614.

Hardy, J. and Selkoe, D. J., 2002. The amyloid hypothesis of Alzheimer's disease: progress and problems on the road to therapeutics. *Science*. 297: 353–356.

Hardy, J., 2006. A hundred years of essay Alzheimer's disease research. *Neuron*. 52: 3–13.

Hegde, M. L., Bharathi, P., Suram, A., Venugopal, C., Jagannathan, R., Poddar, P., Srinivas, P., Sambamurti, K., Rao, K. J., Scancar, J., Messori, L., Zecca, L. and Zatta, P., 2009. Challenges associated with metal chelation therapy in Alzheimer's disease. *Journal of Alzheimer's Disease*. 17: 457–468.

Hipkiss, A. R., 2007a. Dietary restriction, glycolysis, hormesis and ageing. *Biogerontology*. 8: 221–224.

Hipkiss, A. R., 2007b. Could carnosine or related structures suppress Alzheimer's disease? *Journal of Alzheimer's Disease*. 11: 229–240.

Hipkiss, A. R., 2009. Carnosine and its possible roles in nutrition and health. *Advances in Food and Nutrition Research*. 57: 87–154.

Hipkiss, A. R., 2011. Energy metabolism, proteotoxic stress and age-related dysfunction – protection by carnosine. *Molecular Aspects of Medicine*. 32: 267–278.

Iwatsubo, T., Saido, T. C., Mann, D. M., Lee, V. M. and Trojanowski, J. Q., 1996. Full-length amyloid-β (1–42(43)) and amino-terminally modified and truncated amyloid-β 42(43) deposit in diffuse plaques. *American Journal of Pathology*. 149: 1823–1830.

Kang, J., Lemaire, H. G., Unterbeck, A., Salbaum, J. M., Masters, C. L., Grzeschik, K. H., Multhaup, G., Beyreuther, K. and Hill, B. M., 1987. The precursor of Alzheimer's disease amyloid A4 protein resembles a cell-surface receptor. *Nature*. 325: 733–736.

Kim, N. H. and Kang, J. H., 2007. Protective effects of histidine dipeptides on the modification of neurofilament-L by the cytochrome c/hydrogen peroxide system. *Journal of Biochemistry and Molecular Biology*. 40: 125–129.

Lanza, V., Bellia, F., D'Agata, R., Grasso, G., Rizzarelli, E. and Vecchio, G., 2011. New glycoside derivatives of carnosine and analogs resistant to carnosinase hydrolysis: synthesis and characterization of their copper(II) complexes. *Journal of Inorganic Biochemistry*. 105: 181–188.

Lemkul, J. A. and Bevan, D. R., 2012. The role of molecular simulations in the development of inhibitors of amyloid β-peptide aggregation for the treatment of Alzheimer's disease. *ACS Chemical Neuroscience*. 3: 845–856.

Lührs, T., Ritter, C., Adrian, M., Riek-Loher, D., Bohrmann, B., Döbeli, H., Schubert, D. and Riek, R., 2005. 3D structure of Alzheimer's amyloid-beta(1–42) fibrils. *Proceedings of the National Academy of Sciences of the United States of America*. 102: 17342–17347.

Maji, S. K., Perrin, M. H., Sawaya, M. R., Jessberger, S., Vadodaria, K., Rissman, R. A., Singru, P. S., Nilsson, K. P. R., Simon, R., Schubert, D., Eisenberg,

D., Rivier, J., Sawchenko, P., Vale, W. and Riek, R., 2009. Functional amyloids as natural storage of peptide hormones in pituitary secretory granules. *Science*. 325: 328–332.

Mithu, V. S., Sarkar, B., Bhowmik, D., Chandrakesan, M., Maiti, S. and Madhu, P. K., 2011. Zn^{++} binding disrupts the Asp^{23}–Lys^{28} salt bridge without altering the hairpin-shaped cross-β structure of $A\beta_{42}$ amyloid aggregates. *Biophysical Journal*. 101: 2825–2832.

Münch, G., Kuhla, B., Lüth, H.-J., Arendt, T. and Robinson, S. R., 2003. Anti-AGEing defences against Alzheimer's disease. *Biochemical Society Transactions*. 31: 1397–1399.

Naeem, A., Khan, T., Muzaffar, M., Ahmad, S. and Saleemuddin, M., 2011. A partially folded state of ovalbumin at low ph tends to aggregate. *Cell Biochemistry and Biophysics*. 59: 29–38.

Nie, Q., Du, X. and Geng, X., 2011. Small molecule inhibitors of amyloid β peptide aggregation as a potential therapeutic strategy for Alzheimer's disease. *Acta Pharmacologica Sinica*. 32: 545–551.

Preston, J. E., Hipkiss, A. R., Himsworth, D. T., Romero, I. A. and Abbott, J. N., 1998. Toxic effects of beta-amyloid(25–35) on immortalised rat brain endothelial cell: protection by carnosine, homocarnosine and beta-alanine. *Neuroscience Letters*. 242: 105–108.

Roher, A. E., Lowenson, J. D., Clarke, S., Woods, A. S., Cotter, R. J., Gowing, E. and Ball, M. J., 1993. β-Amyloid-(1–42) is a major component of cerebrovascular amyloid deposits—implications for the pathology of Alzheimer-disease. *Proceedings of the National Academy of Sciences of the United States of America*. 90: 10836–10840.

Seidler, N. W., Yeargans, G. S. and Morgan, T. G., 2004. Carnosine disaggregates glycated alpha-crystallin: an in vitro study. *Archives of Biochemistry and Biophysics*. 427: 110–115.

Sievers, S. A., Karanicolas, J., Chang, H. V., Zhao, A., Jiang, L., Zirafi, O., Stevens, J. T., Münch, J., Baker, D. and Eisenberg, D., 2011. Structure-based design of non-natural amino-acid inhibitors of amyloid fibril formation. *Nature*. 475: 96–100.

Stvolinsky, S. L., Bulygina, E. R., Fedorova, T. N., Meguro, K., Sato, T., Tyulina, O. V., Abe, H. and Boldyrev, A., 2010. Biological activity of novel synthetic derivatives of carnosine. *Cellular and Molecular Neurobiology*. 30: 395–404.

Thinakaran, G. and Koo, E. H., 2008. Amyloid precursor protein trafficking, processing, and function. *Journal of Biological Chemistry*. 283: 29615–29619.

Trombley, P. Q., Horning, M. S. and Blakemore, L. J., 2000. Interactions between carnosine and zinc and copper: implications for neuromodulation and neuroprotection. *Biochemistry (Moscow)*. 65: 807–816.

Tycko, R. and Wickner, R. B., 2013. Molecular structures of amyloid and prion fibrils: consensus versus controversy. *Accounts of Chemical Research*. 46: 1487–1496.

Westermark, P., Benson, M. D., Buxbaum, J. N., Cohen, A. S., Frangione, B., Ikeda, S., Masters, C. L., Merlini, G., Saraiva, M. J. and Sipe, J. D., 2007. A primer of amyloid nomenclature. *Amyloid*. 14: 179–183.

Wu, J. W., Liu, K.-N., How, S.-C., Chen, W.-A., Lai, C.-M., Liu, H.-S., Hu, C.-J. and Wang, S. S.-S., 2013. Carnosine's effect on amyloid fibril formation and induced cytotoxicity of lysozyme. *PLoS One*. 8: e81982.

Zhang, C., Jackson, A. P., Zhang, Z. R., Han, Y., Yu, S., He, R. Q. and Perrett, S., 2010. Amyloid-like aggregates of the yeast prion protein ure2 enter vertebrate cells by specific endocytotic pathways and induce apoptosis. *PLoS One*. 5: e12529.

CHAPTER 25

The Function of Carnosine and Its Homologs on Behavior

MITSUHIRO FURUSE*[a]

[a]Laboratory of Regulation in Metabolism and Behavior, Faculty of Agriculture, Kyushu University, Fukuoka 812-8581, Japan
*E-mail: furuse@brs.kyushu-u.ac.jp

25.1 The Presence of Carnosine in Animals

The imidazole dipeptide carnosine (β-alanyl-L-histidine) and its derivative anserine (β-alanyl-1-methyl-L-histidine) are present in high levels in the muscle and brain of animals, while they are negligible in plants (Tamaki *et al.*, 1976; Kohen *et al.*, 1988; O'Dowd *et al.*, 1988; Biffo *et al.*, 1990). Carnosine is synthesized from β-alanine (β-Ala) and L-histidine (L-His) by carnosine synthase, and anserine is mainly formed from carnosine by methylation (Boldyrev and Severin, 1990). As it is produced *via* a different pathway, Horinishi *et al.* (1978) indicated that the purified enzyme from chicken meat could use the methylated histidine to generate anserine by direct synthesis. Both carnosine and anserine are hydrolyzed to constitutive amino acids by carnosinase (Pegova *et al.*, 2000).

In terms of how it affects the brain, in rats carnosine may directly influence brain function by passing through the blood–brain barrier (Yamano *et al.*, 2001) and increasing the brain carnosine level itself (Jin *et al.*, 2005). These dipeptides have antioxidant functions (Kohen *et al.*, 1988) and buffering capacities (Abe, 2000). Therefore, supplementation of carnosine would

Food and Nutritional Components in Focus No. 8
Imidazole Dipeptides: Chemistry, Analysis, Function and Effects
Edited by Victor R Preedy

Published by the Royal Society of Chemistry, www.rsc.org

benefit the brain. In fact, in rats, long-term supplementation of carnosine has been shown to have a protective effect against experimental brain ischemia (Gallant *et al.*, 2000). When injected intraperitoneally, carnosine was found to prevent convulsions *via* brain histaminergic neurons in mice (Zhu *et al.*, 2007) and rats (Jin *et al.*, 2005). In the central nervous system (CNS), localization of carnosine-related dipeptides exhibits a complex pattern, involving both neuronal and glial cells. The glial cells, which include a subset of mature astrocytes and oligodendrocytes, are widely distributed throughout the whole brain and spinal cord. On the other hand, the neuronal cells are restricted to a particular type of neuron (the olfactory receptor neurons), and to a restricted population of putative migrating neurons and neuroblasts (Marchis *et al.*, 2000). In addition, carnosine has been characterized as a putative neurotransmitter in the olfactory receptor neurons (Bonfanti *et al.*, 1999) and has been tentatively implicated in neuron–glial cell interactions (Marchis *et al.*, 2000).

However, the central mechanisms and functions of carnosine and its homologs in the exchange of amino acid residues have not been well understood. The level of anserine in the brain is particularly high (approximately 50 μmol/100 g) in day-old chicks (Fischer *et al.*, 1977). Thus, it can be predicted that carnosine-related dipeptides play a significant role in the chick brain. Furthermore, the organization of the human genome is closer to that of the chicken than it is to that of the mouse (Burt *et al.*, 1999). Therefore, in order to demonstrate the facts mentioned below, our study used young chicks as a model for humans.

25.2 The Behavioral Effects of the Central Administration of Carnosine and Its Constituents

It has previously been reported that carnosine functions differently from its constituents in terms of the ability of the ryanodine receptor in skeletal muscle to release calcium ions (Batrukova and Rubtsov, 1997), as well as in relation to vasodilatory effects (Ririe *et al.*, 2000) and antioxidant activity (Kohen *et al.*, 1988). Other reports have shown that carnosine functions as a reservoir of L-His (Greene *et al.*, 1984) and that it functions in the same way as L-His in attenuating 2-deoxy-D-glucose-induced hyperglycemia (Yamano *et al.*, 2001). Some reports have indicated that glial cells may be involved in the control of behavior. Willoughby *et al.* (2003) reported that cerebral cortex injections of fluorocitrate induced seizures when given at a dose that selectively inhibited astroglial metabolism, and in this way, astroglial activity may affect locomotion. The basomedial hypothalamus and the area postrema react to changes in circulating glucose levels by altering feeding behavior and by secreting both pituitary and nonpituitary hormones.

Tomonaga *et al.* (2004) investigated the effect on food intake of the intracerebroventricular (i.c.v.) injection of both carnosine and anserine. Figure 25.1

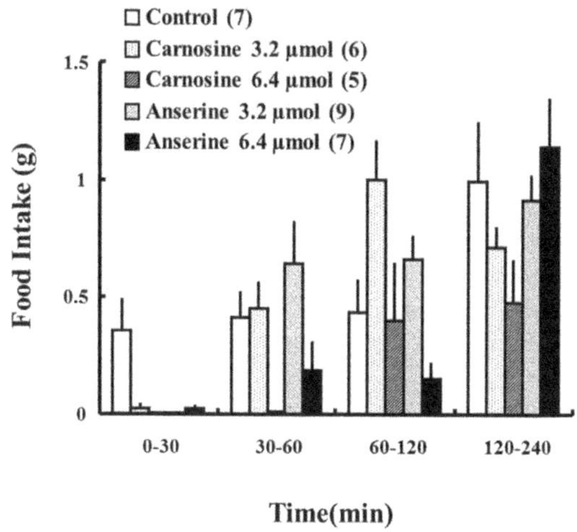

Figure 25.1 Effect of intracerebroventricular injection of carnosine and anserine on food intake in chicks. Under *ad libitum* feeding conditions, chicks were injected i.c.v. with saline (control), carnosine (3.2 and 6.4 μmol) or anserine (3.2 and 6.4 μmol). Figures in parentheses indicate the number of chicks in each group. Values are expressed as means ± S.E.M. Data from Tomonaga *et al.* (2004) reproduced with permission from Elsevier as the authors' right (MF is coauthor).

shows the effect on the food intake of chicks when they are given i.c.v. injections of carnosine and its derivative anserine. Both carnosine and anserine clearly inhibited food intake. In addition, the effects were comparable between the two dipeptides. This implies that substitution of the methyl group in anserine was not implicated in its appetite-inhibiting function. Furthermore, the results led to the hypothesis that either carnosine itself or its constituents are involved in the action obtained since L-His, a constituent of carnosine, has previously been shown to suppress food intake in rats (Sheiner *et al.*, 1985; Yoshimatsu *et al.*, 2002). Thus, a comparison was made between the efficacy of carnosine on food intake and that of its constituents (Tomonaga *et al.*, 2004). L-His alone tended to be stronger than β-Ala alone in inhibiting food intake during the first 30 min (Figure 25.2). On the other hand, carnosine suppressed feeding for 2 h postinjection (Figure 25.2). Therefore, the inhibitory effect of carnosine on feeding was stronger than that of L-His. The effect of jointly injecting β-Ala and L-His was to significantly suppress food intake in a way that was similar to the effect of carnosine itself (Figure 25.2). However, Tomonaga *et al.* (2004) reported that behaviors differed between the two treatments, at least immediately postinjection. Some of the birds treated with β-Ala plus L-His did not move for a while after injection, though in contrast, some of the carnosine-treated birds vocalized loudly and seemed hyperactive. Therefore, at least for a short time after its injection, the effect of carnosine may be somewhat different from that of its constituents.

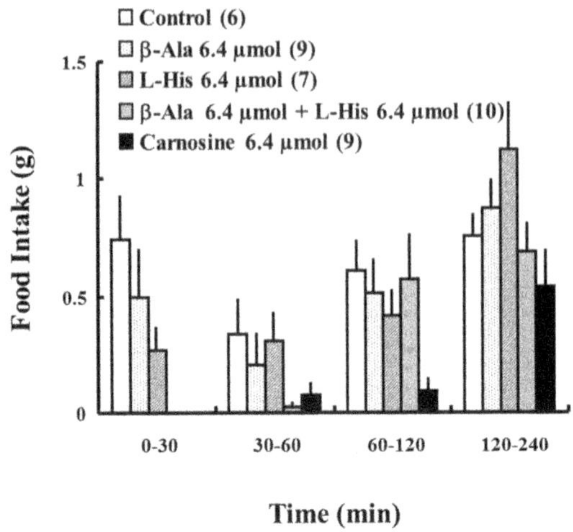

Figure 25.2 Effect of intracerebroventricular injection of carnosine and its constituents on food intake in chicks. Under *ad libitum* feeding conditions, chicks were injected i.c.v. with saline (control), β-Ala (6.4 μmol), L-His (6.4 μmol), β-Ala (6.4 μmol) plus L-His (6.4 μmol) or carnosine (6.4 μmol). Figures in parentheses indicate the number of chicks in each group. β-Ala and L-His indicate β-alanine and L-histidine, respectively. Values are expressed as means ± S.E.M. Data from Tomonaga *et al.* (2004) reproduced with permission from Elsevier as the authors' right (MF is coauthor).

Tomonaga *et al.* (2004) investigated behavioral changes caused by i.c.v. carnosine. They found that carnosine treatment induced hyperactivity and increased vocalization in a dose-dependent manner (Figure 25.3). Hyperactivity and increased vocalization are characteristic behaviors of chicks administered corticotropin releasing factor (CRF) intracerebroventricularly (Zhang *et al.*, 2001), and it was confirmed that the plasma corticosterone level increased in chicks injected with CRF (Furuse *et al.*, 1997) as it did in a dose-dependent manner in carnosine-treated chicks (Tomonaga *et al.*, 2004). These findings may therefore indicate that central carnosine action is related in some degree to the stimulation of the hypothalamic pituitary adrenal axis. On the other hand, Makara and Haller (2001) have suggested that glucocorticoids themselves may have locomotor effects, and additionally, in birds, glucocorticoids may increase locomotion by means of rapid nongenomic mechanisms (Breuner *et al.*, 1998). It is thus unclear whether central carnosine is directly related solely to CRF or not. It has previously been reported that carnosine speeded up the metabolism of serum noradrenaline and cortisol in rats given electric shocks (Nagai *et al.*, 1990), and carnosine has also been confirmed to have a hypotensive effect (Ririe *et al.*, 2000). Thus, the above findings seem to contradict these previous ones. However, Tomonaga *et al.* (2004) used a high dose, and this makes it difficult to speculate on whether carnosine has a direct physiological role; the contradictory results may instead point to a pharmacological effect.

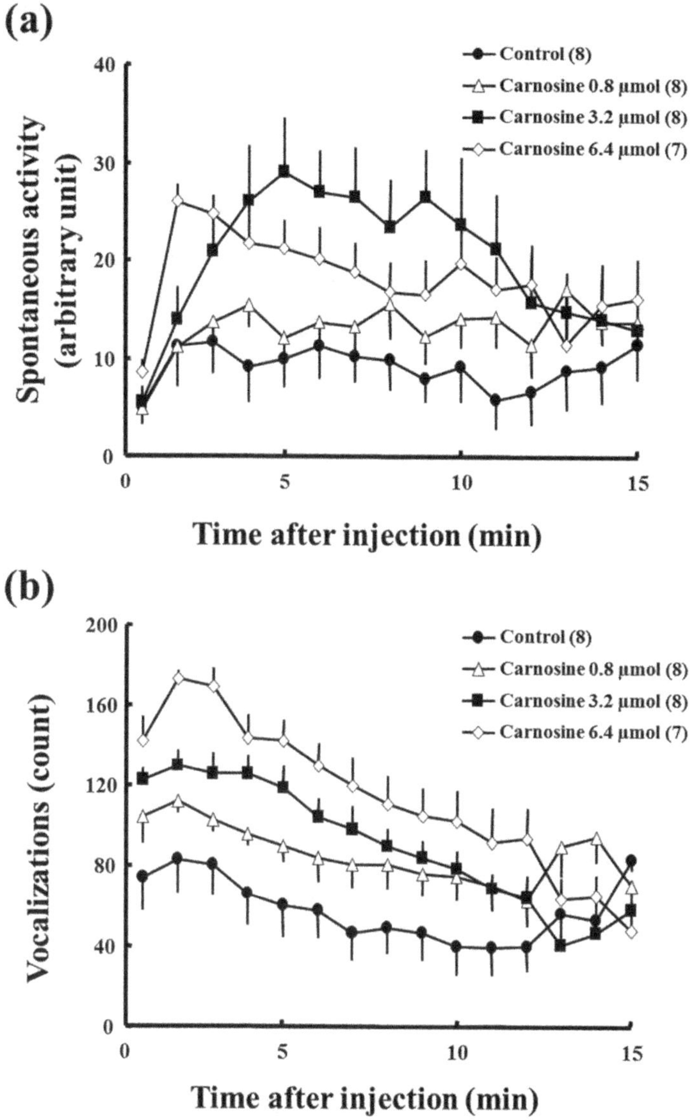

Figure 25.3 Effect of intracerebroventricular injection of carnosine on (a) spontaneous activity and (b) vocalization in chicks. Chicks were injected i.c.v. with saline (control) or carnosine (0.8, 3.2 and 6.4 μmol). Spontaneous activity was automatically determined utilizing infrared beam sensors, and the number of distress vocalizations, which are shrill and intense calls, was simultaneously recorded and counted. Figures in parentheses indicate the number of chicks in each group. Values are expressed as means ± S.E.M. Data from Tomonaga *et al.* (2004) reproduced with permission from Elsevier as the authors' right (MF is coauthor).

In contrast to carnosine's central effect on behavior, administrating β-Ala and L-His jointly and dose-dependently induced hypoactivity (Figure 25.4). It has previously been reported that i.c.v. injection of β-Ala (9.7 μmol) induced hypoactivity in rats (Mena Gomez *et al.*, 1978), and that β-Ala activated the γ-aminobutyric acid (GABA)-A receptor and the glycine receptor *in vitro* (Wu *et al.*, 1993). GABA-A receptor activation is known to induce sleep (Gottesmann, 2002). However, it was also previously reported that L-His induced hypoactivity in rats (Dutra-Filho *et al.*, 1989). Following this, Tomonaga *et al.* (2004) conducted an experiment to reveal whether the hypoactivity that was observed was due to β-Ala, L-His or an additive effect of β-Ala and L-His. They found that the hypoactivity induced by joint administration of β-Ala and L-His was mainly due to the β-Ala (Figure 25.5). A higher dose of L-His may influence some behaviors of chicks more strongly than β-Ala due to an appetite-inhibiting effect of L-His, confirmed by Tomonaga *et al.* (2004), that was stronger than that of β-Ala, as mentioned above. In terms of the hypoactive effect, however, β-Ala seemed to exert a stronger effect than L-His. In conclusion, carnosine and its constituents may affect behavior through different mechanisms. The acute degradation of carnosine – for example, as a result of abrupt oxidative stress – may increase the levels of β-Ala and L-His in the brain, and it may be the effect of the constituents themselves that influences brain function. The regulation of brain function and/or behavior in the chicken may be determined by the particular forms of carnosine and anserine that are involved. Therefore, carnosine-related dipeptides in glial cells may have physiological significance in the CNS.

25.3 Nitric Oxide is Involved in Carnosine-Induced Hyperactivity

Nitric oxide (NO) is a gaseous mediator that transmits signals between cells, or from one part of a cell to another. NO synthase (NOS) catalyzes the conversion of L-arginine to NO. The NO that is generated is soluble in both water and lipid, and diffuses freely within and between cells. NOS can be found throughout the brain, including in neurons and glial cells. Enzymatic activity seems to vary with brain region (Forstermann *et al.*, 1990; Wiesinger, 2001).

As mentioned above, i.c.v. injection of carnosine induced hyperactivity in chicks, and it has been hypothesized that this hyperactivity might be one of the roles of carnosine in neuron–glial interactions. However, Alaghband-Zadeh *et al.* (2001) suggested that carnosine generates NO *via* NOS in the liver *in vitro*. Thus, carnosine-induced hyperactivity may be linked to the generation of NO *via* NOS in the brain.

Tomonaga *et al.* (2005a) demonstrated that a nonselective NOS inhibitor, N^G-nitro-L-arginine methyl ester HCl (L-NAME), could attenuate carnosine-induced hyperactivity (Figure 25.6) and they also showed that the effect was dose dependent (Figure 25.7). It was hypothesized that the attenuation of carnosine-induced hyperactivity by L-NAME might be due to the specific

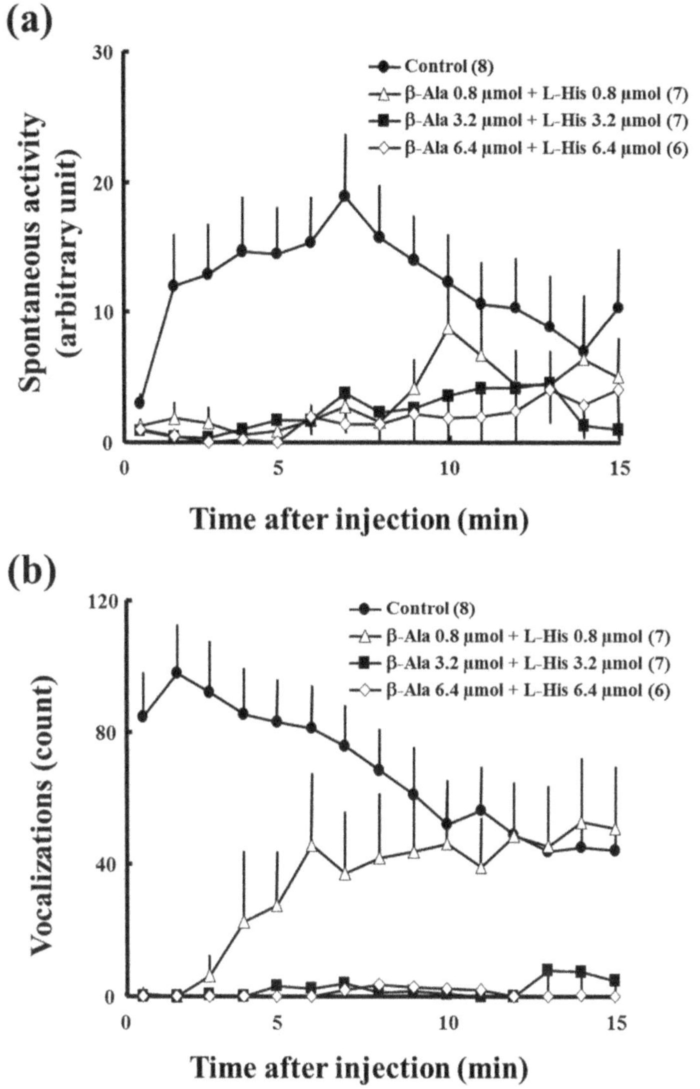

Figure 25.4 Effect of intracerebroventricular injection of constituents of carnosine on (a) spontaneous activity and (b) vocalization in chicks. Chicks were injected i.c.v. with saline (control), β-Ala (0.8 µmol) plus L-His (0.8 µmol), β-Ala (3.2 µmol) plus L-His (3.2 µmol), or β-Ala (6.4 µmol) plus L-His (6.4 µmol). Spontaneous activity and distress vocalizations were analyzed as described in the legend to Figure 25.3. Figures in parentheses indicate the number of chicks in each group. β-Ala and L-His indicate β-alanine and L-histidine, respectively. Values are expressed as means ± S.E.M. Data from Tomonaga *et al.* (2004) reproduced with permission from Elsevier as the authors' right (MF is coauthor).

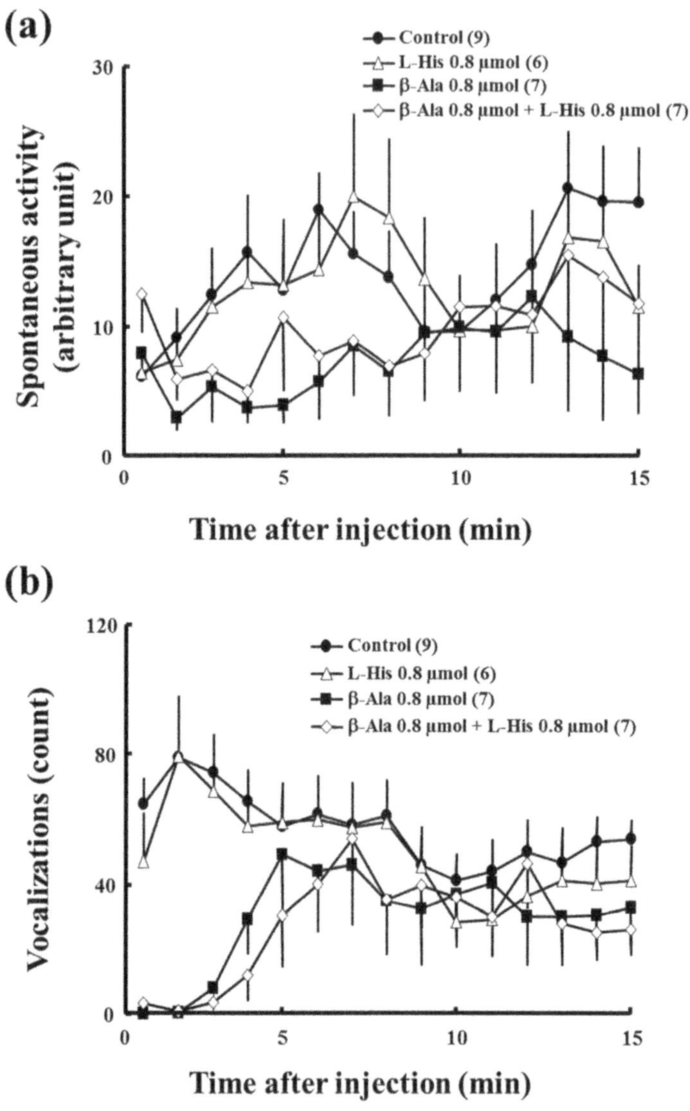

Figure 25.5 Effect of intracerebroventricular injection of constituents of carnosine on (a) spontaneous activity and (b) vocalization in chicks. Chicks were injected i.c.v. with saline (control), L-His (0.8 µmol), β-Ala (0.8 µmol), or β-Ala (0.8 µmol) plus L-His (0.8 µmol). Spontaneous activity and distress vocalizations were analyzed as described in the legend to Figure 25.3. Figures in parentheses indicate the number of chicks in each group. β-Ala and L-His indicate β-alanine and L-histidine, respectively. Values are expressed as means ± S.E.M. Data from Tomonaga *et al.* (2004) reproduced with permission from Elsevier as the authors' right (MF is coauthor).

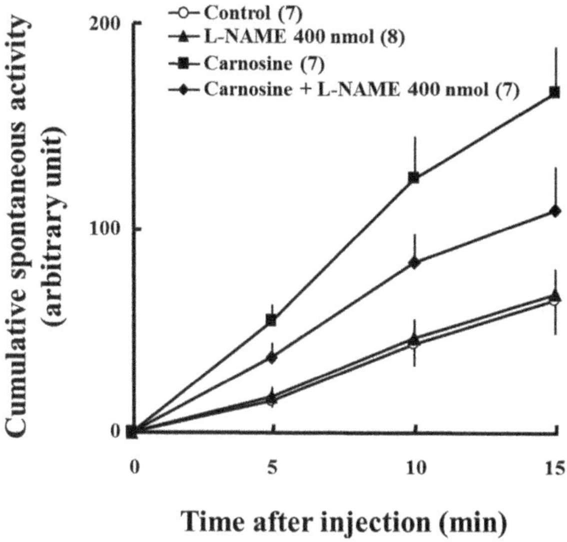

Figure 25.6 Effect of N^G-nitro-L-arginine methyl ester HCl on carnosine-induced spontaneous activity for 15 min in chicks. Chicks were injected i.c.v. with saline (control), L-NAME (400 nmol), carnosine (3.2 μmol) or carnosine (3.2 μmol) plus L-NAME (400 nmol). Spontaneous activity and distress vocalizations were analyzed as described in the legend to Figure 25.3. Figures in parentheses indicate the number of chicks in each group. L-NAME indicates N^G-nitro-L-arginine methyl estel HCl. Values represent cumulative amount of spontaneous activity per 5 min. Values are expressed as means ± S.E.M. Data from Tomonaga *et al.* (2005a) reproduced with permission from Elsevier as the authors' right (MF is coauthor).

inhibition of NOS activity in the brain because the inactive isomer of the NOS inhibitor, N^G-nitro-D-arginine methyl ester HCl (D-NAME), was not involved in this attenuation. Therefore, carnosine-induced hyperactivity may be linked to the generation of NO *via* NOS in the brain.

NOS isoforms are constitutively expressed forms of NOS (cNOS), such as neuronal cNOS (ncNOS), endothelium cNOS (ecNOS) or inducible NOS (iNOS). The cNOSs produce picomolar–nanomolar amounts of NO for short periods in response to receptor stimulation or as a result of shear stress. Unlike the ecNOS and ncNOS isoforms, iNOS is expressed as a result of exposure to diverse stimuli such as inflammatory cytokines and lipopolysaccharide (LPS), and it generates significantly larger and more sustained amounts of NO in comparison with cNOS (Abramson *et al.*, 2001). For example, 3 h after the intraperitoneal injection of the iNOS inducer LPS in mice, no iNOS protein in the lung was detected, whereas the iNOS protein appeared 6 h after it was injected (Lim *et al.*, 2004). This demonstrates that, in contrast to cNOS, NO derived from iNOS does not appear to be generated immediately after its stimulation. On the other hand, carnosine-induced hyperactivity

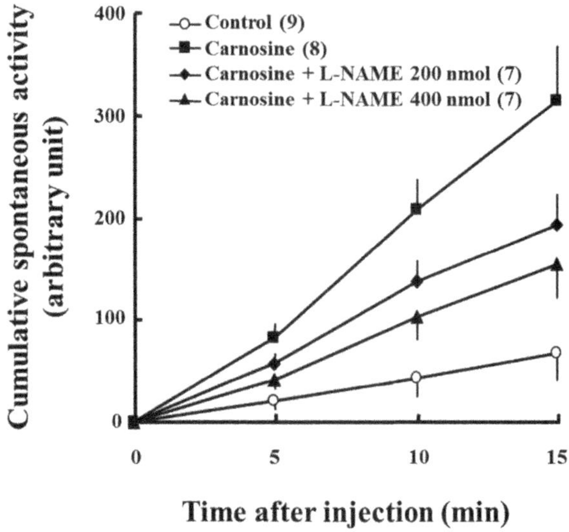

Figure 25.7 Dose-dependent effects of N^G-nitro-L-arginine methyl estel HCl on carnosine-induced spontaneous activity for 15 min in chicks. Chicks were injected i.c.v. with saline (control), carnosine (3.2 μmol), carnosine (3.2 μmol) plus L-NAME (200 nmol) or carnosine (3.2 μmol) plus L-NAME (400 nmol). Spontaneous activity and distress vocalizations were analyzed as described in the legend to Figure 25.3. Figures in parentheses indicate the number of chicks in each group. L-NAME indicates N^G-nitro-L-arginine methyl estel HCl. Values represent the cumulative amount of spontaneous activity per 5 min. Values are expressed as means ± S.E.M. Data from Tomonaga *et al.* (2005a) reproduced with permission from Elsevier as the authors' right (MF is coauthor).

was an acute effect because it could be observed within 5 min, and these acute effects were suppressed by L-NAME. Therefore, carnosine-induced hyperactivity may be linked to cNOS rather than to iNOS. To clarify whether this is the case or not, Tomonaga *et al.* (2005a) investigated the effect of a selective iNOS inhibitor, L-N^6-(1-iminoethyl) lysine HCl (L-NIL), on carnosine-induced hyperactivity. The dose of L-NIL in the experiment was the same as the amount of L-NAME that is used to inhibit carnosine-induced hyperactivity. L-NIL was not found to affect carnosine-induced hyperactivity. Even though L-NAME has been widely used as a nonselective NOS inhibitor, it tends to inhibit cNOS selectively rather than iNOS (Bryk and Wolff, 1999). Furthermore, L-NIL inhibited iNOS more strongly than L-NAME did (Chlopicki *et al.*, 1999). This finding indicates that carnosine-induced hyperactivity may be linked to the stimulation of cNOS as opposed to iNOS.

Tomonaga *et al.* (2007) reported that in rats, in both the hippocampus and the hypothalamus, carnosine, anserine and citrulline levels significantly increased in a time-dependent manner after the animals were administered chicken breast extract. The increase in brain citrulline, an NO coproduct, could be an index of brain NO generation *via* NOS (Igarashi *et al.*, 2000). Whatever the case, it was suggested that in chicks, central carnosine might regulate brain function and/or behavior through the generation of NO *via* cNOS.

25.4 The Effect of Replacing the Amino Acid Residue of Carnosine

25.4.1 The Central Administration of Dipeptides in the Form of β-Alanyl Branched Chain Amino Acids Induces Hyperactivity

β-Ala is the only naturally occurring β-amino acid found in its free state in the brain. It is a component of carnosine and anserine. β-Ala is synthesized in the liver as the final metabolite following the degradation of uracil and thymine (Matthews and Traut, 1987). It has been reported that β-Ala decreased the toxic effects of β-amyloid in rat brain endothelial cells (Preston *et al.*, 1998) and that, also in rats, it reduced bacterial lipopolysaccharide-induced hepatotoxicity (Kim and Kim, 2002). Although β-Ala is a nonproteinaceous amino acid, it is particularly rich in the brain and breast muscle of chickens, where it is found both as a component of carnosine or anserine and as a free amino acid (Tomonaga *et al.*, 2005b), and it is thus reasonable to speculate that β-Ala may have other functions as well. Tsuneyoshi *et al.* (2007) examined its role as a neurotransmitter and investigated the effect of dipeptides that have β-Ala at the amino terminus, replacing the imidazole group of carnosine, L-His, by another amino group. In general, novel dipeptides have been discovered through the process of isolation and purification, but Tsuneyoshi *et al.* (2007) designed and manufactured dipeptides related to carnosine and screened these dipeptides, looking in particular at their effects on behavior.

Carnosine and β-analyl-L-leucine (β-Ala-L-Leu) significantly enhanced both spontaneous activity as a whole and distress vocalizations, though the effect of β-alanyl-glycine (β-Ala-Gly) was no different to that of the control. Carnosine and β-Ala-L-Leu increased plasma corticosterone concentration significantly, and β-Ala-Gly did so to a lesser extent. Although the detailed mechanism is not known, β-Ala-L-Leu and carnosine may have a similar function in the CNS. The effect of other dipeptides (β-alanyl-L-alanine, β-alanyl-β-alanine and β-alanyl-L-lysine) has also been investigated, but no effect was detected (unpublished data). Branched chain amino acids (BCAAs) – namely, L-leucine, L-isoleucine and L-valine – have a similar structure, and Tsuneyoshi *et al.* (2007) thought that the effects of β-alanyl dipeptides having BCAAs at the carboxyl terminus might have a similar function. They therefore investigated the effect of β-Ala-L-Leu, β-analyl-L-isoleucine (β-Ala-L-Ile) and β-analyl-L-valine (β-Ala-L-Val). The i.c.v. injection of β-Ala-L-Ile was found to have the same effect on hyperactivity, distress vocalizations and plasma corticosterone concentration as observed with carnosine and β-Ala-L-Leu. On the other hand, the effect of β-Ala-L-Val was somewhat weaker than the effects of β-Ala-L-Leu and β-Ala-L-Ile. Although the difference in efficacy between the three β-alanyl-BCAAs could not be completely determined, it is thought that the low molecular weight of L-valine may be associated with its weak activity. It was concluded that one of the central functions in β-alanyl-BCAAs is the induction of hyperactivity. Tsuneyoshi *et al.* (2007) proposed naming β-Ala-L-Leu as Excitin-1, β-Ala-L-Ile as Excitin-2 and β-Ala-L-Val as Excitin-3 as a result of their demonstrated functions in generating excitement. Later (Tsuneyoshi *et al.*, 2009), it was confirmed that the oral administration of Excitin-1 in rats led to hyperactivity in the open field.

25.4.2 The Effect of the Substitution of Amino Terminal β-Ala on Carnosine-Induced Sedative Effects Under an Acute Stressful Condition in Chicks

L-His has been found to accelerate lipolysis in the white adipose tissue through activation of the sympathetic nerve (Yoshimatsu *et al.*, 2002), and to suppress fat accumulation (Kasaoka *et al.*, 2004). L-His induces both pro-oxidant (Tansini *et al.*, 2004) and antioxidant (Obata and Yamanaka, 2000) effects, depending upon its concentration. L-His-related dipeptide might have a novel role, given these positive and negative oxidant effects. Tsuneyoshi *et al.* (2008a) investigated the function of dipeptides that have L-His at the carboxyl terminus. First, they investigated L-seryl-L-histidine (L-Ser-L-His), L-valyl-L-histidine (L-Val-L-His) and L-isoleucyl-L-histidine (L-Ile-L-His). Then they examined L-alanyl-L-histidine (L-Ala-L-His), glycyl-L-histidine (Gly-L-His) and L-methionyl-L-histidine (L-Met-L-His). The i.c.v. injection of L-Ser-L-His and L-Ile-L-His led to decreased spontaneous activity and distress vocalizations, and that of L-Val-L-His decreased spontaneous activity alone.

The precise mechanisms as to why L-Ser-L-His, L-Ile-L-His and L-Val-L-His had sedative effects were unclear. Asechi *et al.* (2006) demonstrated that the i.c.v. injection of L-serine resulted in sedative and hypnotic effects in chicks, but these effects did not decrease the plasma corticosterone concentrations. The effect of L-Ser-L-His was similar to that of L-serine as reported by Asechi *et al.* (2006). These findings suggest that it may be the terminal amino acid residue of L-serine that affects the latter's function. Asechi *et al.* (2006) also reported that in a similar way, i.c.v. injection of glycine also induced sedative and hypnotic effects in chicks. However, Gly-L-His had no effect. Thus, it is the combination of amino acid and carboxyl terminal L-His, rather than the amino terminal amino acids, that is important in producing sedative effects.

25.4.3 The Reverse Structure of Carnosine Induced Sedative and Hypnotic Effects Under Acute Stress Conditions

It has been suggested that β-Ala can activate the glycine and GABA-A receptors (Wu *et al.*, 1993), and that it can act as an inhibitory transmitter in certain CNS regions (Fykse and Fonnum, 1996). In fact, Tomonaga *et al.* (2004) demonstrated that the i.c.v. injection of β-Ala induced hypoactivity in the form of sleep-like behavior in the chick brain. However, the role or precise mechanism of β-Ala in the brain has not yet been clarified. The effect of carnosine-related dipeptides in which the carboxyl terminal L-His was replaced with other amino acids was mentioned above. β-Alanyl-BCAAs have been found to increase hyperactivity in chicks in the same way as has been observed with carnosine. Accordingly, Tsuneyoshi *et al.* (2008b) investigated carboxyl β-Ala related dipeptides in which terminal amino acids were replaced with other amino acids such as glycyl-β-alanine (Gly-β-Ala), L-valyl-β-alanine (L-Val-β-Ala), L-histidyl-β-alanine (L-His-β-Ala) and L-seryl-β-alanine (L-Ser-β-Ala). Of the dipeptides having carboxyl terminus β-Ala, L-His-β-Ala alone had a sedative and hypnotic effect (Figure 25.8). This implies that even if the dipeptide has β-Ala, it is the combination of amino acids at the amino terminal that has an effect on function. Although L-His-β-Ala induced sedative and hypnotic effects in chicks, carnosine was found to induce hyperactivity. L-His-β-Ala has the reverse structure to carnosine, and its effect of i.c.v. injection was the opposite of that of carnosine. Following the combination of the two peptides, the resultant behaviors were between the behavior of each peptide (Figure 25.9). Tsuneyoshi *et al.* (2008b) gave reverse carnosine (L-His-β-Ala) the name rev-carnosine. As mentioned above, Excitin-1 (β-Ala-L-Leu) and Excitin-2 (β-Ala-L-Ile) induced similar hyperactivity to that observed with carnosine, and Excitin-3 (β-Ala-L-Val) also had a tendency to increase distress vocalizations (Tsuneyoshi *et al.*, 2007). However, L-Val-β-Ala, which has the reverse structure to β-Ala-L-Val, had no significant effects. The i.c.v. injection of β-Ala, glycine, L-serine or L-Ser-L-His induced sedative and hypnotic effects in the brain of chicks (Tomonaga *et al.*, 2004; Asechi *et al.*, 2006; Tsuneyoshi *et al.*, 2008a). In contrast, Gly-β-Ala and L-Ser-β-Ala had no significant effects (Tsuneyoshi *et al.*,(2008b). In addition, β-Ala-Gly and Gly-β-Ala also had no significant effects (Tsuneyoshi *et al.*, 2007; Tsuneyoshi *et al.*, 2008a). Therefore, it can be concluded that constitutive amino acids and the construction of dipeptides have a significant influence on the effects of dipeptides in the chick brain.

Summary Points

- This chapter focuses on the central function of carnosine and its homologs regarding behavior in young chicks.
- Carnosine inhibited food intake in a dose-dependent fashion when injected intracerebroventricularly.
- The constituents of carnosine, β-alanine (β-Ala) and L-histidine (L-His) also inhibited food intake.
- The injection of carnosine induced hyperactivity, whereas β-Ala plus His induced hypoactivity.
- Carnosine-induced hyperactivity may be linked to nitric oxide generation *via* constitutive nitric oxide synthase.
- The replacement of amino acid residues of carnosine with other amino acids resulted in hyperactivity or hypoactivity being induced by different processes.
- Mental motivation in humans may be modified by carnosine and its homologs.

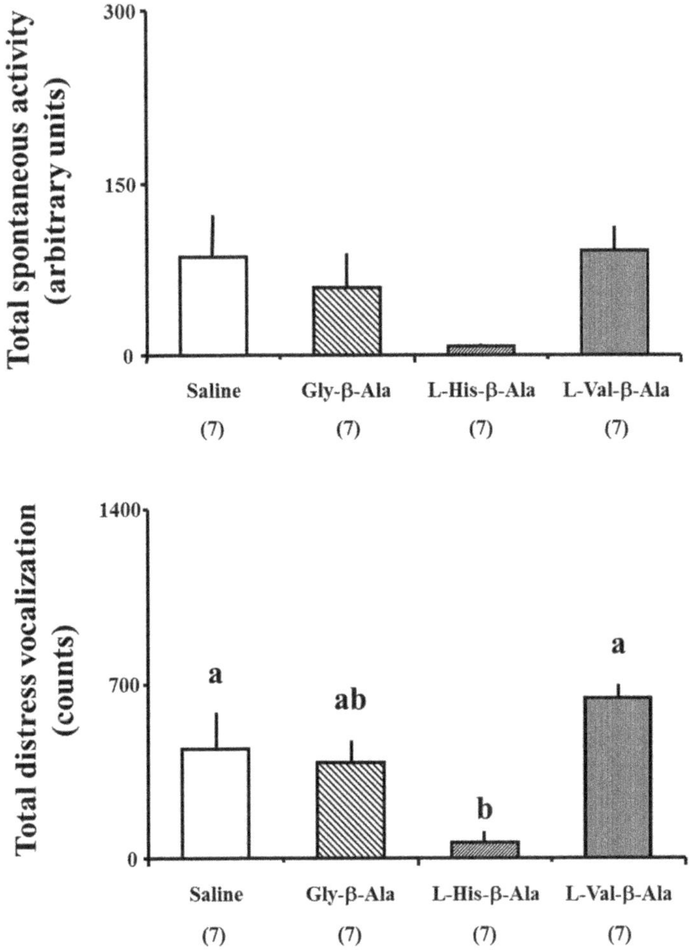

Figure 25.8 Effect of i.c.v. injection of Gly-β-Ala, L-His-β-Ala and L-Val-β-Ala on total amount of spontaneous activity (upper panel) and total number of distress vocalizations (lower panel) in chicks during the 10 min post-injection period. Chicks were divided into four groups, and given i.c.v. injections of saline as a control and 2.8 μmol of Gly-β-Ala, L-Val-β-Ala and L-His-β-Ala. Spontaneous activity and distress vocalizations were analyzed as described in the legend to Figure 25.3. Figures in parentheses indicate the number of chicks in each group. Gly-β-Ala, L-His-β-Ala and L-Val-β-Ala indicate glycyl-β-alanine, l-histidyl-β-alanine and L-valyl-β-alanine, respectively. Values are expressed as means ± S.E.M. There was a significant difference between "a" and "b" ($p < 0.05$), and there was not a significant difference between "a" and "ab" or "b" and "ab". Values are expressed as means ± S.E.M. Data from Tsuneyoshi *et al.* (2008b) reproduced with permission from Elsevier as the authors' right (MF is coauthor).

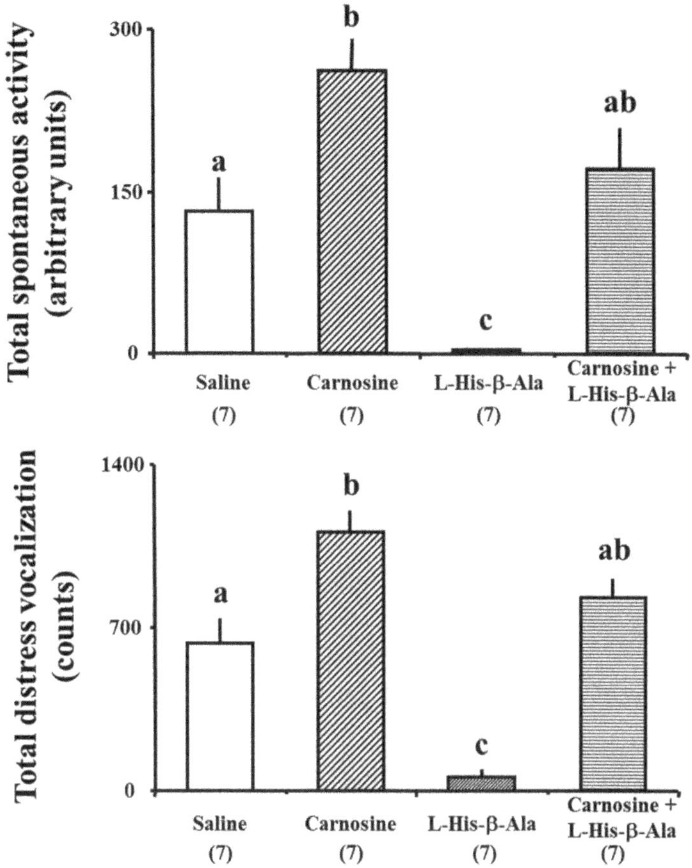

Figure 25.9 Effect of i.c.v. injection of carnosine, L-His-β-Ala and carnosine plus L-His-β-Ala on total amount of spontaneous activity (upper panel) and total number of distress vocalizations (lower panel) in chicks for 10 min postinjection. Chicks were given i.c.v. injections of saline, carnosine (2.8 μmol), L-His-β-Ala (2.8 μmol) or carnosine (2.8 μmol) plus L-His-β-Ala (2.8 μmol). Spontaneous activity and distress vocalizations were analyzed as described in the legend to Figure 25.3. Figures in parentheses indicate the number of chicks in each group. L-His-β-Ala indicates L-histidyl-β-alanine. Results are expressed as means ± S.E.M. There was a significant difference between "a" and "b", "a" and "c", or "ab" and "c" ($p < 0.05$). There was not a significant difference between "a" and "ab" or "b" and "ab". Data from Tsuneyoshi *et al.* (2008b) reproduced with permission from Elsevier as the authors' right (MF is coauthor).

Key Facts

Key facts of the chick model for behavioral studies. This table lists the merits of the chick model for screening drugs and nutrients in a behavioral regulation investigation.

1. The organization of the human genome is closer to that of the chicken than it is to that of the mouse.
2. Many types of chicken are available, *e.g.* obese (broiler), those that are sensitive to stress (layer), and so on.
3. The chick-separation stress paradigm has traditionally been used as a biobehavioral test, and is recognized as a valid anxiolytic screening method.
4. Chicks feel comfortable in crowds, but they become stressed when isolated. Such social separation stress increases spontaneous activity and vocalization in chicks.
5. Various drugs that produce anxiolytic activity in rodent models of anxiety have also had anxiolytic effects in this chick-separation stress model.
6. Furthermore, this chick-separation stress model has the practical advantage that chicks are relatively inexpensive to purchase and maintain, and require only small quantities of drugs in the screening process.

Definitions of Words and Terms

β-Alanine. β-Alanine is a naturally occurring β amino acid that is not used in the biosynthesis of any major proteins or enzymes, and that is formed by the degradation of dihydrouracil and carnosine *in vivo*.

Anserine. Anserine (β-alanyl-1-methyl-L-histidine), an antioxidant, is mainly formed from carnosine by a process of methylation. *Via* a different pathway, methylated histidine is available to generate anserine through direct synthesis.

Branched-chain amino acids. These are amino acids having aliphatic side chains with a branch such as leucine, isoleucine or valine.

Carnosine. Carnosine (β-alanyl-L-histidine), synthesized from β-alanine and L-histidine by carnosine synthase, is present, together with anserine, in high levels in the muscle and brain of mammalian and avian species.

Corticotropin-releasing factor (Corticotropin-releasing hormone). This peptide hormone and neurotransmitter is involved in the stress response, and stimulates the pituitary synthesis of adrenocorticotropic hormone.

Central nervous system. This is the part of the nervous system consisting of the brain and the spinal cord.

Excitin. The central functions of excitin are related to the generation of excitement. β-Ala-L-Leu is named Excitin-1, β-Ala-L-Ile is called Excitin-2 and β-Ala-L-Val is Excitin-3.

L-Histidine. This is one of the proteinogenic α-amino acids with an imidazole functional group.

Hyperactivity. Many different mental diseases and medical conditions have the symptom of hyperactivity – the state of being more active than is usual or desirable.

Hypoactivity. Hypoactivity, an inhibition of behavioral or locomotor activity, is a characteristic effect of sedative agents and many centrally acting anaesthetics.

Hypothalamic pituitary adrenal axis. When animals are stressed, a hormonal response in the body termed the hypothalamus–pituitary–adrenal axis is triggered. The hypothalamus secretes corticotrophin-releasing factor, which in turn binds to specific receptors on the pituitary cells, which produce adrenocorticotropic hormone. Adrenocorticotropic hormone is transported to its target, the adrenal gland, which stimulates the secretion of glucocorticoids.

Nitric oxide. Nitric oxide is a gaseous mediator that transmits signals between cells or from one part of a cell to another.

Nitric oxide synthase. Nitric oxide synthase acts as a catalyst in the conversion of L-arginine to nitric oxide. There are several forms of nitric oxide synthase, all of which operate in different parts of the body.

Rev-carnosine. Rev-carnosine (L-histidyl-β-alanine) has not only the reverse structure of carnosine, but also the reverse central function – to induce sedation.

List of Abbreviations

β-Ala	β-Alanine
β-Ala-Gly	β-Alanyl-glycine
L-Ala-L-His	L-Alanyl-L-histidine
β-Ala-L-Ile	β-Analyl-L-isoleucine
β-Ala-L-Leu	β-Analyl-L-leucine
β-Ala-L-Val	β-Analyl-L-valine
BCAAs	Branched-chain amino acids
CNS	Central nervous system
CRF	Corticotropin releasing factor
Gly-L-His	Glycyl-L-histidine
L-His	L-Histidine
L-His-β-Ala	L-Histidyl-β-alanine
i.c.v.	Intracerebroventricular
L-Ile-L-His	L-Isoleucyl-L-histidine
GABA	γ-Aminobutyric acid
Gly-β-Ala	Glycyl-β-alanine
LPS	Lipopolysaccharide
L-Met-L-His	L-Methionyl-L-histidine
L-NAME	N^G-Nitro-L-arginine methyl ester HCl

D-NAME	N^G-Nitro-D-arginine methyl ester HCl
L-NIL	L-N^6-(1-Iminoethyl) lysine HCl
NO	Nitric oxide
NOS	NO synthase
cNOS	Constitutively expressed NOS
ecNOS	Endothelium cNOS
iNOS	Inducible NOS
ncNOS	Neuronal cNOS
L-Ser-L-His	L-Seryl-L-histidine
L-Ser-β-Ala	L-Seryl-β-alanine
L-Val-β-Ala	L-Valyl-β-alanine
L-Val-L-His	L-Valyl-L-histidine

References

Abe, H., 2000. Role of histidine-related compounds as intracellular proton buffering constituents in vertebrate muscle. *Biochemistry (Mosc)*. 65: 757–765.

Abramson, S. B., Amin, A. R., Clancy, R. M. and Attur, M., 2001. The role of nitric oxide in tissue destruction. Best Practice & Research. *Clinical Rheumatology*. 15: 831–845.

Alaghband-Zadeh, J., Mehdizadeh, S., Khan, N. S., O'Farrell, A., Bitensky, L. and Chayen, J., 2001. The natural substrate for nitric oxide synthase activity. *Cell Biochemistry and Function*. 19: 277–280.

Asechi, M., Tomonaga, S., Tachibana, T., Han, L., Hayamizu, K., Denbow, D. M. and Furuse, M., 2006. Intracerebroventricular injection L-serine analogs and derivatives induces sedative and hypnotic effects under an acute stressful condition in neonatal chicks. *Behavioural Brain Research*. 170: 71–77.

Batrukova, M. A. and Rubtsov, A. M., 1997. Histidine-containing dipeptides as endogenous regulators of the activity of sarcoplasmic reticulum Ca-release channels. *Biochimica et Biophysica Acta*. 1324: 142–150.

Biffo, S., Grillo, M. and Margolis, F. L., 1990. Cellular localization of carnosine-like and anserine-like immunoreactivities in rodent and avian central nervous system. *Neuroscience*. 35: 637–651.

Boldyrev, A. A. and Severin, S. E., 1990. The histidine-containing dipeptides, carnosine and anserine: distribution, properties and biological significance. *Advances in Enzyme Regulation*. 30: 175–194.

Bonfanti, L., Peretto, P., De Marchis, S. and Fasolo, A., 1999. Carnosine-related dipeptides in the mammalian brain. *Progress in Neurobiology*. 59: 333–353.

Breuner, C. W., Greenberg, A. L. and Wingfield, J. C., 1998. Noninvasive corticosterone treatment rapidly increases activity in Gambel's white-crowned sparrows (*Zonotrichia leucophrys gambelii*). *General and Comparative Endocrinology*. 111: 386–394.

Bryk, R. and Wolff, D. J., 1999. Pharmacological modulation of nitric oxide synthesis by mechanism-based inactivators and related inhibitors. *Pharmacology & Therapeutics*. 84: 157–178.

Burt, D. W., Bruley, C., Dunn, I. C., Jones, C. T., Ramage, A., Law, A. S., Morrice, D. R., Paton, I. R., Smith, J., Windsor, D., Sazanov, A., Fries, R. and Waddington, D., 1999. The dynamics of chromosome evolution in birds and mammals. *Nature*. 402: 411–413.

Chlopicki, S., Olszanecki, R., Jakubowski, A., Lomnicka, M. and Gryglewski, R. J., 1999. L-N^6-(1-Iminoethyl)-lysine (L-NIL) but not *S*-methylisothiourea sulphate (SMT) displays selectivity towards NOS-2. *Polish Journal of Pharmacology*. 51: 443–447.

Dutra-Filho, C. S., Wannmacher, C. M., Pires, R. F., Gus, G., Kalil, A. M. and Wajner, M., 1989. Reduced locomotor activity of rats made histidinemic by injection of histidine. *Journal of Nutrition*. 119: 1223–1227.

Fisher, D. E., Amend, J. F. and Strumeyer, D. H., 1977. Anserine and carnosine in chicks (*Gallus gallus*) rat pups (*Rattus rattus*) and ducklings (*Anas platyrhynchos*): comparative ontogenetic observations. *Comparative Biochemistry and Physiology, Part B: Biochemistry & Molecular Biology*. 56: 367–370.

Forstermann, U., Gorsky, L. D., Pollock, J. S., Schmidt, H. H., Heller, M. and Murad, F., 1990. Regional distribution of EDRF/NO-synthesizing enzyme(s) in rat brain. *Biochemical and Biophysical Research Communications*. 168: 727–732.

Furuse, M., Matsumoto, M., Saito, N., Sugahara, K. and Hasegawa, S., 1997. The central corticotropin-releasing factor and glucagon-like peptide-1 in food intake of the neonatal chick. *European Journal of Pharmacology*. 339: 211–214.

Fykse, E. M. and Fonnum, F., 1996. Amino acid neurotransmission: dynamics of vesicular uptake. *Neurochemical Research*. 21: 1053–1060.

Gallant, S., Kukley, M., Stvolinsky, S., Bulygina, E. and Boldyrev, A., 2000. Effect of carnosine on rats under experimental brain ischemia. *Tohoku Journal of Experimental Medicine*. 191: 85–99.

Gottesmann, C., 2002. GABA mechanisms and sleep. *Neuroscience*. 111: 231–239.

Greene, S. M., Margolis, F. L., Grillo, M. and Fisher, H., 1984. Enhanced carnosine (β-alanyl-L-histidine) breakdown and histamine metabolism following treatment with compound 48/80. *European Journal of Pharmacology*. 99: 79–84.

Horinishi, H., Grillo, M. and Margolis, F. L., 1978. Purification and characterization of carnosine synthetase from mouse olfactory bulbs. *Journal of Neurochemistry*. 31: 909–919.

Igarashi, K., Sugiyama, Y., Kasuya, F., Inoue, H., Matoba, R. and Castagnoli, N., 2000. Analysis of citrulline in rat brain tissue after perfusion with haloperidol by liquid chromatography-mass spectrometry. *Journal of Chromatography B: Biomedical Sciences and Applications*. 746: 33–40.

Jin, C. L., Yang, L. X., Wu, X. H., Li, Q., Ding, M. P., Fan, Y. Y., Zhang, W. P., Luo, J. H. and Chen, Z., 2005. Effects of carnosine on amygdaloid-kindled seizures in Sprague-Dawley rats. *Neuroscience*. 135: 939–947.

Kasaoka, S., Tsuboyama-Kasaoka, N., Kawahara, Y., Inoue, S., Tsuji, M., Ezaki, O., Kato, H., Tsuchiya, T., Okuda, H. and Nakajima, S., 2004. Histidine supplementation suppresses food intake and fat accumulation in rats. *Nutrition*. 20: 991–996.

Kim, S. K. and Kim, Y. C., 2002. Attenuation of bacterial lipopolysaccharide-induced hepatotoxicity by betaine or taurine in rats. *Food and Chemical Toxicology*. 40: 545–549.

Kohen, R., Yamamoto, Y., Cundy, K. C. and Ames, B. N., 1988. Antioxidant activity of carnosine, homocarnosine, and anserine present in muscle and brain. *Proceedings of the National Academy of Sciences of the United States of America*. 85: 3175–3179.

Lim, S., Kang, K. W., Park, S. Y., Kim, S. I., Choi, Y. S., Kim, N. D., Lee, K. U., Lee, H. K. and Pak, Y. K., 2004. Inhibition of lipopolysaccharide-induced inducible nitric oxide synthase expression by a novel compound, mercaptopyrazine, through suppression of nuclear factor-kappaB binding to DNA. *Biochemical Pharmacology*. 68: 719–728.

Makara, G. B. and Haller, J., 2001. Non-genomic effects of glucocorticoids in the neural system. Evidence, mechanisms and implications. *Progress in Neurobiology*. 65: 367–390.

Marchis, S. D., Modena, C., Peretto, P., Migheli, A., Margolis, F. L. and Fasolo, A., 2000. Carnosine-related dipeptides in neurons and glia. *Biochemistry (Mosc)*. 65: 824–833.

Matthews, M. M. and Traut, T. W., 1987. Regulation of *N*-carbamoyl-β-alanine amidohydrolase, the terminal enzyme in pyrimidine catabolism, by ligand-induced change in polymerization. *Journal of Biological Chemistry*. 262: 7232–7237.

Mena Gomez, M. A., Carlsson, A. and Garcia de Yebenes, J., 1978. The effect of β-alanine on motor behaviour, body temperature and cerebral monoamine metabolism in rat. *Journal of Neural Transmission*. 43: 1–9.

Nagai, K., Suda, T., Kawasaki, K. and Yamaguchi, Y., 1990. Acceleration of metabolism of stress-related substances by L-carnosine (in Japanese). *Nippon Seirigaku Zasshi*. 52: 221–228.

Obata, T. and Yamanaka, Y., 2000. Protective effect of histidine on potassium chloride depolarization enhances 1-metyl-4-phenylpyridinium ion-induced hydroxyl radical generation in the rat striatum. *Life Sciences*. 68: 689–697.

O'Dowd, J. J., Robins, D. J. and Miller, D. J., 1988. Detection, characterisation, and quantification of carnosine and other histidyl derivatives in cardiac and skeletal muscle. *Biochimica et Biophysica Acta*. 967: 241–249.

Pegova, A., Abe, H. and Boldyrev, A., 2000. Hydrolysis of carnosine and related compounds by mammalian carnosinases. *Comparative Biochemistry and Physiology, Part B: Biochemistry & Molecular Biology*. 127: 443–446.

Preston, J. E., Hipkiss, A. R., Himsworth, D. T. J., Romero, I. A. and Abbott, J. N., 1998. Toxic effects of β-amyloid (25–35) on immortalised rat brain endothelial cell: protection by carnosine, homocarnosine and β-alanine. *Neuroscience Letters*. 242: 105–108.

Ririe, D. G., Roberts, P. R., Shouse, M. N. and Zaloga, G. P., 2000. Vasodilatory actions of the dietary peptide carnosine. *Nutrition*. 16: 168–172.

Sheiner, J. B., Morris, P. and Anderson, G. H., 1985. Food intake suppression by histidine. *Pharmacology Biochemistry and Behavior*. 23: 721–726.

Tamaki, N., Iizumi, H., Masumitu, N., Kubota, A. and Hama, T., 1976. Species specificity on the contents of anserine and carnosine (in Japanese). *Yakugaku Zasshi*. 96: 1481–1486.

Tansini, C. M., Durigon, K., Testa, C. G., Béllo-Klein, A., Wajner, M., Wannmacher, C. M. D., Wyse, A. T. S. and Dutra-Filho, C. S., 2004. Effects of histidine and imidazolelactic acid on various parameters of the oxidative stress in cerebral cortex of youn rats. *International Journal of Developmental Neuroscience*. 22: 67–72.

Tomonaga, S., Tachibana, T., Takagi, T., Saito, E.-S., Zhang, R., Denbow, D. M. and Furuse, M., 2004. Effect of central administration of carnosine and its constituents on behaviors in chicks. *Brain Research Bulletin*. 63: 75–82.

Tomonaga, S., Tachibana, T., Takahashi, H., Sato, M., Denbow, D. M. and Furuse, M., 2005a. Nitric oxide involves in carnosine-induced hyperactivity in chicks. *European Journal of Pharmacology*. 524: 84–88.

Tomonaga, S., Kaji, Y., Tachibana, T., Denbow, D. M. and Furuse, M., 2005b. Oral administration of β-alanine modifies carnosine concentrations in the muscles and brains of chickens. *Animal Science Journal*. 76: 249–254.

Tomonaga, S., Hayakawa, T., Yamane, H., Maemura, H., Sato, M., Takahata, Y., Morimatsu, F. and Furuse, M., 2007. Oral administration of chicken breast extract increases brain carnosine and anserine concentrations in rats. *Nutritional Neuroscience*. 10: 181–186.

Tsuneyoshi, Y., Tomonaga, S., Asechi, M., Morishita, K., Denbow, D. M. and Furuse, M., 2007. Central administration of dipeptides, β-alanyl-BCAAs, induces hyperactivity in chicks. *BMC Neuroscience*. 8: 37.

Tsuneyoshi, Y., Tomonaga, S., Yamane, H., Morishita, K., Denbow, D. M. and Furuse, M., 2008a. Central administration of L-Ser-L-His and L-Ile-L-His induced sedative effects under an acute stressful condition in chicks. *Letters in Drug Design & Discovery*. 5: 65–68.

Tsuneyoshi, Y., Yamane, H., Tomonaga, S., Morishita, K., Denbow, D. M. and Furuse, M., 2008b. Reverse structure of carnosine-induced sedative and hypnotic effects in the chick under acute stress. *Life Sciences*. 82: 1065–1069.

Tsuneyoshi, Y., Sato, M., Yamane, H., Tomonaga, S., Morishita, K., Denbow, D. M. and Furuse, M., 2009. Oral administration of excitin-1 (β-alanyl-L-leucine) alters behavior and brain monoamine and amino acid concentrations in rats. *Nutritional Neuroscience*. 12: 175–182.

Wiesinger, H., 2001. Arginine metabolism and the synthesis of nitric oxide in the nervous system. *Progress in Neurobiology*. 64: 365–391.

Willoughby, J. O., Mackenzie, L., Broberg, M., Thoren, A. E., Medvedev, A., Sims, N. R. and Nilsson, M., 2003. Fluorocitrate-mediated astroglial dysfunction causes seizures. *Journal of Neuroscience Research*. 74: 160–166.

Wu, F. S., Gibbs, T. T. and Farb, D. H., 1993. Dual activation of $GABA_A$ and glycine receptors by β-alanine: inverse modulation by progesterone and 5α-pregnan-3α-ol-20-one. *European Journal of Pharmacology*. 246: 239–246.

Yamano, T., Niijima, A., Iimori, S., Tsuruoka, N., Kiso, Y. and Nagai, K., 2001. Effect of L-carnosine on the hyperglycemia caused by intracranial injection of 2-deoxy-D-glucose in rats. *Neuroscience Letters*. 313: 78–82.

Yoshimatsu, H., Chiba, S., Tajima, D., Akehi, Y. and Sakata, T., 2002. Histidine suppresses food intake through its conversion into neuronal histamine. *Experimental Biology and Medicine*. 227: 63–68.

Zhang, R., Nakanishi, T., Ohgushi, A., Ando, R., Yoshimatsu, T., Denbow, D. M. and Furuse, M., 2001. Interaction of corticotropin-releasing factor and glucagon-like peptide-1 on behaviors in chicks. *European Journal of Pharmacology*. 430: 73–78.

Zhu, Y. Y., Zhu-Ge, Z. B., Wu, D. C., Wang, S., Liu, L. Y., Ohtsu, H. and Chen, Z., 2007. Carnosine inhibits pentylenetetrazol-induced seizures by histaminergic mechanisms in histidine decarboxylase knock-out mice. *Neuroscience Letters*. 416: 211–216.

CHAPTER 26

Carnosine and Kidney Diseases

KATARZYNA KILIŚ-PSTRUSIŃSKA*[a]

[a]Department of Paediatric Nephrology, Wroclaw Medical University, Ul. Borowska 213, 50-556 Wrocław, Poland
*E-mail: katarzyna.kilis-pstrusinska@umed.wroc.pl

26.1 Introduction

Carnosine (β-alanyl-L-histidine) is a dipeptide present in different human tissues *i.e.* in kidneys. It is synthesized from L-histidine and β-alanine by an ATP-dependent carnosine synthase enzyme (Quinn *et al.*, 1992). Carnosine is hydrolyzed by the enzyme carnosinase to β-alanine and L-histidine (Lenney *et al.*, 1982). Carnosinase (aminoacyl histidine dipeptidase) is a member of M20/M28 metalloproteases, which appears in human organisms in the form of two isoenzymes: serum carnosinase (EC 3.4.13.20) with a molecular mass of 167 kDa and tissue carnosinase (EC 3.4.13.18), currently called cytosolic nonspecific dipeptidase, with a molecular mass of 90 kDa. These isoenzymes are encoded respectively by *CNDP1* and *CNDP2* genes, located in chromosome 18 (18q22.3) (Teufel *et al.*, 2003).

Carnosine and carnosine metabolites may be subject to further changes. Histamine – a mediator of allergic and inflammatory reactions – is the product of histidine decarboxylation. On the other hand, β-alanine constitutes an essential element of coenzyme A that may stimulate collagen synthesis in tissues. Since carnosine is immediately metabolized in serum, its concentration in this fluid is very low (Lenney *et al.*, 1982; Peters *et al.*, 2011). An inherited deficiency of the serum carnosinase leads to the development of

Food and Nutritional Components in Focus No. 8
Imidazole Dipeptides: Chemistry, Analysis, Function and Effects
Edited by Victor R Preedy

Published by the Royal Society of Chemistry, www.rsc.org

autosomal recessive inheritance disease, which manifests itself in mental retardation and is associated with carnosinuria (Perry *et al.*, 1968). Higher concentrations of carnosine were reported in skeletal muscles, heart and brain cells. The level of carnosine depended on, *inter alia*, its tissue activity (Teufel *et al.*, 2003; Boldyrev *et al.*, 2013).

Carnosine inhibits the development of advanced glycation end products (AGEs), oxygen free radicals, autocrine signaling of angiotensine II and fibronectine synthesis (Guiotto *et al.*, 2005; Boldyrev *et al.*, 2013). It maintains the alkaline–acid balance in tissues, decreases toxicity of metal ions and regulates the activity of reticular calcium canals in, *inter alia*, cardiac cells. Carnosine reportedly inhibits aging and has a beneficial effect on the cardiovascular system (Hipkiss, 2011). Experimental studies showed that carnosine reduces the level of proinflammatory and profibrotic cytokines, such as transforming growth factor-β type 1 (TGF-β1) and tumor necrosis factor-α (TNF-α) (Cuzzocrea *et al.*, 2007).

The above proves that carnosine is engaged in metabolic pathways, which may take part in the pathophysiology of renal failure, whereas biological properties of carnosine suggest its renal protective effect. However, the studies on carnosine, as well as reports regarding carnosinase have rarely appeared in the academic literature.

26.2 Biological Activity of Carnosine in Relation to Kidneys

Studies have demonstrated that carnosine possesses antioxidant properties. Carnosine was found to be a potent "scavenger" of singlet oxygen because it deactivates hydroxyl and peroxide radicals (Quinn *et al.*, 1992; Guiotto *et al.*, 2005). Due to its hydrophilic features, carnosine supplies cytosolic fraction of the antioxidative barrier and represents a water-soluble counterpart to lipophilic antioxidants such as α-tocopherol (Gariballa and Sinclair, 2000). Experiments on animals indicated that administration of both carnosine and α-tocopherol resulted in higher concentrations of these two compounds in tissues than after the administration of α-tocopherol alone. This may suggest that carnosine and α-tocopherol form a synergistic couple (Chan *et al.*, 1994). The antioxidant properties of α-tocopherol have been repeatedly examined in various renal diseases. α-Tocopherol reportedly decreases oxidative stress in hemodialyzed patients and protects against secondary cardiovascular diseases (Coombes and Fassett, 2012). Additional administration of carnosine would probably escalate this beneficial effect. Experimental studies demonstrated that carnosine lowers the concentration of malonyldialdehyde (MDA), which is an indicator of a lipid peroxidation, depending on the applied dose (Hipkiss *et al.*, 1998). It should be emphasized that oxygen-metabolism disorders are advances in chronic kidney disease (CKD).

Studies demonstrated that carnosine has a buffering effect, which is significant in tissues where abnormalities of alkaline–acid balance easily occur.

Nitrogen atoms of carnosine within the imidazole ring of histidine may be subject to protonation when acidification of the intracellular environment takes place (Gariballa and Sinclair, 2000). Carnosine weakens lipids peroxidation at the weakly alkaline pH. Since it binds heavy metals (zinc, cobalt, copper), carnosine inhibits many enzymatic reactions (Boldyrev *et al.*, 2013). In this way it regulates metal concentration in kidneys and lowers the toxicity of these compounds.

Carnosine is reported to have the capacity to modulate immune response in human neutrophils, where it increases the production of interleukin-1β (IL-1β) and suppresses apoptosis (Tan and Candlish, 1998).

Carnosine plays an important role in regulating carbonyl stress as it binds reactive aldehydes, which are toxic for cells, because they adhere to proteins and lipoproteins (Alhamdani *et al.*, 2007; Hipkiss, 2011). Carnosine may also react with carbonyl groups of already oxidative-modified proteins. It facilitates the ubiquitination process and stimulates degradation of proteasomes. In this way carnosine deactivates carbonylated proteins, which accumulate in cells, leading to cell aging and death.

Carnosine added to human fibroblast culture reportedly prolongs the lives of cells. It was reported that carnosine delays aging symptoms and increases the number of programmed cell divisions (McFarland and Holliday, 1999). This effect may result from the fact that carnosine was found to reduce telomeric DNA sequences (Shao *et al.*, 2004). On the other hand, Son *et al.* reported that carnosine inhibits the translation of IL-8 mRNA by inhibiting the initiation factor, eIF4E, in stimulated Caco2 cells (Son *et al.*, 2008). These results indicate that the synthesis of this proinflammatory cytokine decreases. It was also reported that carnosine may block other regulatory proteins such as ERK1/2 and p38MAP kinases. In studies conducted on *Caenorhabditis elegans*, a model organism in genetics, it was reported that among persons with eIF4E factor mutation, which initiates mRNA translation, aging is delayed, stress resistance is increased and lifespan is prolonged (Pan *et al.*, 2007). According to Hipkiss, the beneficial effect of carnosine on survival of human fibroblasts may take place by analogical mechanisms (Hipkiss, 2011). Table 26.1

26.3 Carnosine and Kidney Diseases

Carnosine, thanks to its biological activity presented above, may block pathological processes in kidneys and inhibit the development of renal diseases. Table 26.2

26.3.1 Hypertension

Carnosine may act as a mediator in regulating hypertension, in which – as we know – kidneys play a dominant role.

According to Hou *et al.*, carnosine possesses angiotensin-converting enzyme (ACE) inhibitor properties (Hou *et al.*, 2003). Willi *et al.* reported a

Table 26.1 Summary table of selected studies reporting renoprotective properties of carnosine. TGF-β1 – transforming growth factor-β 1; PD – peritoneal dialysis.

Effect of carnosine	Material/methods	Authors
Reduction in proteinuria and renal vascular permeability	Experimental study, rats	Peters *et al.* (2012)
Inhibition of matrix accumulation in human mesangial cells by interfering with TGF-β1 production and signalling	Human mesangial cells culture	Köppel *et al.* (2011)
Control of blood pressure by autonomic nerves in rats	Experimental study, rats	Nagai *et al.* (2012)
Regulation of blood glucose through controlling autonomic nerves	Experimental study, rats	Nagai *et al.* (2003)
Prevention of ischemia/reperfusion-induced renal injury	Experimental study, rats	Fujii *et al.* (2003a, 2003b), Kurata *et al.* (2006)
Prevention of apoptosis of glomerular cells and podocyte loss in diabetic rats	Experimental study, rats	Riedl *et al.* (2011).
Inhibition of apoptosis in mesangial cells	Experimental study, rats	Liu *et al.* (2003)
Antioxidant effect on cisplatin-induced renal oxidative stress in rats	Experimental study, rats	Noori and Mahboob (2010)
Prevention of gentamicin-induced nephrotoxicity (antioxidant action, protein molecule protection, activation of immune system, preservation of membrane fluidity, and cytosolic buffering)	Experimental study, rats	Soliman *et al.* (2007)
Decreased formation of advanced glycation end-products in PD-fluid	Incubation PD-fluid with human serum albumin or collagen	Alhamdani *et al.* (2007)

Table 26.2 Kidney diseases in which carnosine may play renoprotective role (hypotheses or results of experimental studies).

Kidney disease
Hypertension
Acute kidney injury: ischemia/reperfusion-induced acute renal failure, drug-induced nephrotoxicity
Diabetic nephropathy
Renal failure treated with peritoneal dialysis

case of a child with hypotension and increased serum carnosine due to the inherited carnosinase deficiency (Willi *et al.*, 1997).

Niijima *et al.* have documented that L-carnosine administered *via* the diet inhibited the elevation of blood pressure in deoxycorticosterone acetate salt hypertensive rats (Niijima *et al.*, 2002). Tanida *et al.* noticed that in urethane-anesthetized rats intravenous injection of L-carnosine suppressed renal sympathetic nerve activity, one of the important pathogenetic factors of hypertension (Tanida *et al.*, 2005). To identify the pathomechanism of L-carnosine, the authors examined the effects of intravenous or intralateral cerebral ventricular injection of various doses of L-carnosine on renal sympathetic nerve activity and hypertension in rats. Lower doses of the dipeptide suppressed nerve activity and lowered blood pressure, whereas higher doses elevated them. Furthermore, the antagonists of histaminergic receptors (H1 and H3), given peripherally or centrally, influenced L-carnosine-induced effects. Furthermore, bilateral lesions of the hypothalamic suprachiasmatic nucleus eliminated both effects on renal sympathetic nerve activity and blood pressure induced by the lower (1 μg) and higher (100 μg) doses of peripheral L-carnosine. These observations suggest that L-carnosine takes part in control of blood pressure and the suprachiasmatic nucleus and histaminergic nerve are involved in the L-carnosine-induced effect. The antioxidant activity of carnosine may also be responsible for the effect of this dipeptide on blood pressure (Nagai *et al.*, 2012).

26.3.2 Acute Kidney Injury

Japanese scientists examined the renoprotective effects of L-carnosine on ischemia/reperfusion-induced acute renal failure in rats (Fujii *et al.*, 2003a,b; Kurata *et al.*, 2006). Molecular mechanisms that underlie this type of renal failure are not entirely understood. There are, however, some casual factors, *e.g.* ATP depletion, reactive oxygen species, phospholipase activation and vasoactive peptides, which contribute to this renal damage. Activity of renal sympathetic nervous system is increased, which enhances noradrenaline release from nerve endings. This finding is substantiated by decreased symptoms of this acute renal failure in rats by means of a surgical and pharmacological blockade of renal sympathetic nerves and, consequently, by means of suppressing the increased noradrenaline level in renal venous vessels (Fujii *et al.*, 2003a). A similar effect was observed after intravenous application of L-carnosine (Fujii *et al.*, 2003b). According to scientists, the nephroprotective effect of carnosine in ischemia/reperfusion-induced acute renal failure in rats is strictly connected with the inhibiting effect of dipeptide on the renal sympathetic nervous system; however, the influence of antioxidative properties of carnosine should also not be discounted. Studies demonstrated the beneficial effects of carnosine on acute renal failure when it was administered intraperitoneally (Kuarata *et al.*, 2006). The authors observed improvement of renal function and reduction of histopathological changes, such as tubular necrosis, proteinaceous casts in tubulli and medullary congestion, in rats that were given carnosine for 2 weeks before the experimental renal ischemia. According to the authors, histamine and receptor H3 activation in the central nervous system may take part in the protective effect of carnosine on ischemic renal failure. Nevertheless, the above hypothesis requires further research.

Other authors analyzed the effect of carnosine on gentamicin-induced nephrotoxicity in rats (Soliman *et al.*, 2007). After administering antibiotics, in biopsied kidneys in glomeruli dilated capillaries with a focus of necrosis were found as well as some fragmented podocytes; however, in the proximal tubules various degrees of necrosis. Animals that were administered carnosine in combination with gentamicin reportedly reconstructed the appropriate structure of podocytes and proximal tubular cells and significantly ameliorated renal activity (serum creatinine decreased up to 72% and urea to 33%). Carnosine weakens after-drug nephrotoxicity that, according to Soliman *et al.*, is attributed to its many properties: carnosine preserves membrane fluidity, has cystolic buffering properties, ameliorates the toxic decrease of the body antioxidants such as superoxide dismutase, catalase and glutathione, and scavenges reactive oxygen species, which prevents lipid peroxidation of cell membranes and regaining those already oxidized.

In experimental studies the authors also evaluated the effect of carnosine pretreatment on cisplatin-induced nephrotoxicity. Cisplatin is an antitumor drug used in the treatment of various cancers. The therapeutic effect depends on the drug doses; however, high-dose therapy with cisplatin is limited by its cumulative nephrotoxicity The study of Noori and Mahboob showed that administrating cisplatin to rats caused a reduction in glomerular filtration rate (GFR) (Noori and Mahboob, 2010). The impairment in kidney function coexisted with increased MDA contents in the kidney tissue and histological abnormalities: glomeruli congestion, marked acute necrosis in proximal and distal tubules, decrease in the height of epithelial cells, shedding of atypical cytoplasm and loss of brush boarder. Marked congestion in vessels was also seen. Pretreatment with carnosine significantly prevented congestion in glomeruli and vessels and other alterations in kidney tissue. The mechanism of cisplatin-induced nephrotoxicity probably is associated with increased lipid peroxidation in renal tissue. Pretreatment of carnosine significantly reduced the lipid peroxides and restored endogenous antioxidant enzymes.

26.3.3 Diabetic Nephropathy

Carnosine in diabetes and its complications is still subject to many studies. Diabetic nephropathy (DN) is a major microvascular complication of diabetes. The pathomechanism of DN is not entirely understood. Clinical studies demonstrate that hyperglycemia plays a significant role. Morphologically, DN is characterized by excessive extracellular matrix (ECM) formation in the mesangium and thickening of glomerular and tubular basement membranes, ultimately progressing to glomerulosclerosis and tubulo-interstitial fibrosis. A key molecule in DN pathogenesis is the profibrotic cytokine TGF-β1.

Janssen *et al.* reported that in cultured human podocytes and mesangial cells (MC) carnosine inhibited the increased production of fibronectin and collagen type VI in podocytes and the increased production of TGF-β2 in mesangial cells induced by glucose (Janssen *et al.*, 2005). Similar observations were made by Köppel *et al.* (2011). In their study, human mesangial cells were cultured in high-glucose medium alone or in high-glucose medium to which L-carnosine was added. The authors observed that under high-glucose conditions increased collagen VI and fibronectin were inhibited by L-carnosine. Additionally, increased TGF-β production, due to hyperglycaemia, was normalised thanks to L-carnosine. The dipeptide can modulate matrix accumulation in two ways. First, inhibition of TGF-β production might result in an overall inhibition of matrix accumulation. Secondly, L-carnosine inhibits TGF-β-induced matrix accumulation *via* inhibition of the ALK5 pathway. These observations suggest that carnosine supplementation or carnosinase inhibition might therefore be of therapeutic relevance in diabetic patients with an increased risk for developing DN.

It was reported that carnosine in cultured rat mesangial cells in high glucose concentrations can inhibit mesangial cell proliferation by modulating cell cycle progress, *i.e.* increasing the cell population in G1 and reducing that in S-phase (Jia *et al.*, 2009).

Furthermore, carnosine could reduce the activity of p-ERK and p-p38 kinases, whose phosphorylation is activated by high glucose concentrations. These kinases play an important role in signal transduction cascades for cell growth and other important functions. Activation of kinase p38 in the pathogenesis of cell proliferation and accumulation of ECM in ischemic, toxic and inflammatory renal damage and in diabetic nephropathy was reported (Liu *et al.*, 2003). Oxidative free radicals and AGEs are also significant in mesangial cell proliferation, which take place under the influence of high glucose concentrations. Methylglyoxal is included in the toxic aldehydes which take part in protein glycation (Liu *et al.*, 2003). Methylglyoxal is suggested to be responsible for the majority of organ complications in diabetes. According to Hipkiss, carnosine, by its influence on the TOR signaling pathway, inhibits glycosis processes and reduces the secretion of methylglyoxal (Hipkiss, 2011). In this context protective influence of carnosine on kidney tissue may be multidimensional: by inhibiting cell cycle in an antioxidative and antiglycosis manner.

Other authors indicate that supplementation with carnosine may protect kidneys from glomerular apoptosis in diabetes. In the experimental model of streptozotocin-induced diabetes, it was reported that administration of this dipeptide reverses changes triggered by hyperglycemia (Riedl *et al.*, 2011). The study demonstrated a decreased number of apoptotic cells of renal glomeruli and the reduction of lost podocytes percentage, despite the increased level in kidneys in rats AGEs and modified proteins of hexosamine pattern (GlcNA). Therefore, regardless of biochemical abnormalities, the protective properties of carnosine in relation to kidneys may result from its ability to inhibit proapoptotic signals.

The above observations suggest the possibility of future application of carnosine in the treatment of diabetic nephropathy.

26.3.4 Others

Alhamdani *et al.* studied the effect of carnosine against the formation of AGEs in peritoneal fluid (1.5% solution of dextrose, Dianeal PD-2, Baxter, after being heat sterilized), which was incubated with human serum albumin, type IV collagen, carnosine, homocarnosine and anserine (Alhamdani *et al.*, 2007). The authors reported that carnosine and related peptides can inhibit modification of these two proteins by glucose degradation products, which are formed *in vitro* in peritoneal fluid upon its sterilization.

26.4 Carnosinase and Kidney

Carnosinase, which disposes carnosine, may play a significant role in renal dysfunctions. Genes encoding the secreted serum carnosinase and cytosolic nonspecific dipeptidase (respectively, *CNDP1 i CNDP2*) are located in 18q22.3 chromosome (Teufel *et al.*, 2003), whose connection with diabetic renal disease has been shown in numerous studies (Janssen *et al.*, 2005; Peters *et al.*, 2011). Table 26.3

CNDP1 gene consists of 12 exons. In 2005 Janssen *et al.* examined polymorphisms of gene *CNDP1* in diabetic patients of Caucasian and Arab origin (from Qatar) and reported that the most significant association was found with marker D18S880 (OR 4.77; 95% CI 1.01-22.5), a trinucleotide repeat in exon 2 (Janssen *et al.*, 2005). As this polymorphism lies in the 5′ coding part of the transcript, the number of trinucleotide repeats is directly related to the number of leucine (L) residues in the leader peptide of the carnosinase precursor in: 5, 6 or 7 leucines. Furthermore, it was reported that the shortest allelic form of D18S880 variation (5) results in a weaker predisposition to renal damage in diabetes and is connected with a reduced activity of

Table 26.3 Human studies concerning polymorphisms in *CNDP1* gene in kidney diseases. The *CNDP1* gene encodes a secreted serum carnosinase that degrades carnosine.

Kidney disease	Authors
Diabetic nephropathy	Janssen *et al.* (2005) Wanic *et al.* (2008) McDonough *et al.* (2009) Alkhalaf *et al.* (2010) Ahluwalia *et al.* (2011)
Chronic kidney disease in the course of tubulointerstitial nephritis	Kiliś-Pstrusińska *et al.* (2010)
Chronic kidney disease in the course of glomerulonephritis	Kiliś-Pstrusińska *et al.* (2010)

carnosinase in serum than longer allelic forms 6 and 7. Adversely, the higher number of leucines repeats in the leader peptide would demonstrate the development of diabetic nephropathy in type 1 and 2 diabetes.

The results of other studies are different. Wanic *et al.* excluded polymorphisms of genes *CNDP1* and *CNDP2* as a result of nephropathy in diabetes 2 (Wanic *et al.*, 2008). Similar results were obtained by Alkhalaf *et al.* In the study of a large cohort of white European patients with type 1 diabetes, the prevalence of homozygosity for 5L–5L repeats in the *CNDP1* gene was not significantly different between patients with DN and patients without DN at study entry (Alkhalaf *et al.*, 2010). Patients with DN and 5L–5L were at increased risk of progression to ESRD compared with those with other allelic variations, but this increase in risk became apparent only after 8 years of follow-up. Probably other unidentified risk alleles may be involved in the deterioration of renal function in diabetic patients. McDonough *et al.* have shown that other variants in the region of chromosome 18 contribute to the risk of DN. These other variants may mask or modify the effect of 5L–5L (McDonough *et al.*, 2009).

Another potential explanation for the inverse association between *CNDP1* and ESRD could lie in environmental factors that alter over time, such as glycaemic control.

Ahluwalia *et al.* genotyped nine single-nucleotide polymorphisms (SNPs) and one trinucleotide repeat polymorphism (D18S880, five to seven leucine repeats) in *CNDP1* and *CNDP2* in a case–control set-up including 4888 unrelated type 2 diabetic patients (with and without nephropathy) from Sweden (Scania Diabetes Registry) (Ahluwalia *et al.*, 2011). They found that two SNPs, rs2346061 in *CNDP1* and rs7577 in *CNDP2* were associated with an increased risk of DN, but only the first influenced estimated GFR. The authors did not confirm the association between D18S880 marker and DN. There were no significant differences in D18S880 genotype frequencies between DN and type 2 diabetic patients. However, they found that the haplotype C-C-G (including alleles from rs7577, rs2346061 and rs7244370) was associated with a threefold increased risk of DN by altering kidney function. These observations suggest that other modifying SNPs are needed to increase the risk of progression towards kidney dysfunction.

Quantitative dependency between carnosine and carnosinase is not simple (Peters *et al.*, 2011). The intensity of carnosine effect cannot be explained only in reference to the activity of carnosinase. Human serum carnosinase also leads to degradation of other histidine peptides, *i.e.* anserine and homocarnosine but carnosine remains a main substrate of human serum carnosinase (Lenney *et al.*, 1982; Teufel *et al.*, 2003). Peters *et al.* reported that in healthy people, homocarnosine degradation, compared to carnosine, is almost 50 times slower, while anserine degradation almost 200 times slower (Peters *et al.*, 2011). The authors also reported that the carnosinase activity in serum is differentiated interindividually (*e.g.* carnosinase activity in relation to carnosine is 0.9–5.9 μmoL mL^{-1} h^{-1}) and does not correlate with the histidine-dipeptides concentration. Studies demonstrated that homocarnosine and anserine can lead to nonenzymatic inhibition of carnosine. The above observations suggest that the concentration of serum carnosine as well as the results of its systemic effect do not depend only on activation of serum carnosinase but also on the concentration of other dipeptides and their transport in blood. This regulation may also happen by carnosine passing to tissues and being eliminated by kidneys. According to Peters *et al.*, due to the low concentration of carnosine in serum and to the fact that it is not correlated with the activity of serum carnosinase, it is very probable that local carnosine metabolism plays an important role in the development of diabetic nephropathy (DN). The above conclusion is supported by the study of Janssen *et al.*, who reported elevated expression of *CNDP1* gene in podocytes and glomeruli in patients with diabetic nephropathy (Janssen *et al.*, 2005).

Another study investigated the renal carnosine metabolism in db/db mice (mice model of diabetic nephropathy) (Peters *et al.*, 2012). In the kidneys of mice with diabetic nephropathy, the authors demonstrated increased carnosinase activity and ten-fold decreased anserine concentration compared to controls, but the concentration of carnosine remained almost unaltered. According to Riedl *et al.* *N*-glycosylation plays a significant role in secretion and carnosinase activity in diabetes (Riedl *et al.*, 2011). The increased activity of carnosinase in sick mice may be explained by hyperglycemia triggered by poor glucose control. Peters *et al.* reported that the treatment of db/db mice with egzogenic carnosine normalized carnosinase activity and concentrations of anserine in kidneys; furthermore, this treatment lowered vascular permeability, reduced proteinuria and glycaemia (Peters *et al.*, 2012). The authors reported two mechanisms of reducing carnosinase activity despite supplying substrate for the enzyme: lowering glucose concentration in blood, which reduced nonenzymatic carnosine glycosylation and increased anserine concentration as an egzogenic carnosine product.

The importance of polymorphism of *CNDP1* gene and the activity of carnosine was also examined in the development of chronic kidney disease (CKD) on nondiabetic etiology. It was demonstrated that in families with one child suffering from CKD and two biological healthy parents there were no significant differences in the transmission of alleles of the D18S880 microsatellite polymorphism in *CNDP1* gene from heterozygous parents to their offspring with CKD caused by tubulointerstitial nephritis (Kiliś-Pstrusińska *et al.*, 2010). This suggests that this polymorphism not increases risk for development of CKD caused by chronic tubulointerstitial nephritis. It was also reported that there is a connection of *CNDP1* polymorphism with the development of chronic kidney disease in the course of glomerulopathy. In contrast to people with diabetic nephropathy, the shortest allelic form of the D18S880 variant (5) predisposed more to renal failure.

The study reported also significantly higher carnosinase activity in the serum of sick people compared to those observed in their parents without renal failures, which may suggest pathophysiological role of this enzyme in the development of various nephropathies, not only of diabetic etiology. However, these results should be interpreted cautiously. It was reported that many factors may influence the carnosinase activity (Peters *et al.*, 2011; Riedl *et al.*, 2011). Higher serum carnosinase and, presumably, the lower concentration of its natural substrate – carnosine, may manifest its weakened antioxidative defense and the cytoprotective effect in chronic kidney disease. Metabolism disorders of histidine-contained peptides and/or lack of appropriate adaptive reactions of damaged kidneys should be taken into consideration. Numerous studies have not proven that there is, reported by Janssen *et al.*, statistically significant correlation between genotype of variant D18S880 of *CNDP1* gene and the activity of carnosinase in serum (Kiliś-Pstrusińska *et al.*, 2010; Chakkera *et al.*, 2011). It was reported that carnosinase activity is lower in genotype 6–6 carriers in comparison with genotype 5–6 and 6–7 carriers (Kiliś-Pstrusińska *et al.*, 2010). One of the potent carnosinase activators is reportedly cadmium (Lenney *et al.*, 1982; Teufel *et al.*, 2003). In advanced stages of CKD cadmium accumulates in, *e.g.* kidneys and bones. Elevated cadmium concentration was reported in the blood of patients treated with renal replacement therapy (Vanholder *et al.*, 2002). Cadmium has a pro-oxidative effect that, *e.g.* lowers the concentration of antioxidants in blood. It also alters enzyme activity, which depends on metal ions. The relationship between increasing activity of carnosinase in serum of patients with chronic kidney disease and the accumulation of this trace element cannot be excluded.

26.5 Summary

Since carnosine in serum appears only in very low concentrations and is barely detected, it poses limitations on research, but the fact that carnosine contributes to many metabolic pathways inspires workers to further research. In particular, renal diseases require further studies due to the renal-protective properties of carnosine. Many authors reported that carnosine may be useful in the treatment of diabetic nephropathy, hypertension or in preventing toxic or ischemic renal failure. Carnosine was regarded as a rapamycin and macrolide mimic, presenting immunosuppressive and antiproliferative properties, with a well-grounded position in nephrology and transplantology (Hipkiss, 2011). Carnosine, similarly to rapamycin, regulates phosphorylation of the factor initiating protein binding TOR/FRAP pathway. All these facts inspire more hope for further therapeutic studies of carnosine. However, further research on carnosine effects in kidney and the mechanisms regulating tissue concentration of carnosine is essential.

Summary Points

- This chapter focus on the role of carnosine in kidneys and kidney diseases.
- Carnosine is a dipeptide involved in many metabolic pathways in kidneys due to its multifunctional properties. Several experimental

studies have indicated that carnosine has the propensity to function as a natural angiotensin converting enzyme inhibitor, as a pH buffer, as a scavenger for reactive oxygen species and as an inhibitor of advanced glycation end-product formation and excessive extracellular accumulation.

- Carnosine might be protective for hypertension, diabetic nephropathy, acute kidney injury and other kidney diseases.
- Serum secreted carnosinase, the enzyme actively degrading carnosine into L-histidine and β-alanine, influences the carnosine effect in kidney. A polymorphism in the serum carnosinase gene (*CNDP1*) may be associated with development and progression of kidney disease. However, further studies are needed, concerning also other candidate genes and their interactions with *CNDP1* or environmental factors.
- A better understanding of carnosine effects in kidney and the mechanisms regulating tissue concentration are required in order to use carnosine as a therapeutic or renal protective molecule.

Key Facts

Key Facts of Renoprotection

1. Chronic kidney disease (CKD) can lead to total loss of kidney function and patients enter end-stage renal failure (ESRF).
2. CKD is caused by a wide range of diseases: diabetes mellitus, hypertension, glomerulopathies, infections and physical obstruction of urinary tract.
3. Patients with ESRF must be treated by dialysis or kidney transplantation and their quality of life is decreased.
4. Therapies to slow CKD progression (*i.e.* renoprotection) are not fully effective.
5. One of the promising therapeutic and renoprotective molecule is carnosine.

Key Facts of Importance of Carnosine in Kidney Diseases

1. Carnosine is an endogenously synthesized dipeptide. It is degraded by the enzyme carnosinase.
2. Carnosine is engaged in many metabolic pathways in kidney.
3. Experimental studies showed positive effect of carnosine for kidney, among others anti-ischemic activity, antioxidant and antiapoptotic effect, lowering the blood pressure. It is a protective factor regarding diabetic nephropathy.
4. Extrapolation of these findings to humans remains a major challenge.
5. Carnosine supplementation or carnosinase inhibition could be of therapeutic or renoprotective relevance in patients with kidney diseases or diabetes.

Definitions of Words and Terms

Carnosine. It is a naturally occurring dipeptide involved in many metabolic pathways in the human body.

CNDP1. *CNDP1* is carnosinase 1 gene located on chromosome 18q22.3. *CNDP1* encodes secreted serum carnosinase, enzyme that hydrolyzes carnosine to β-alanine and L-histidine.

Kidney. One of most important organs of human homeostasis. Kidneys perform a wide range of vital functions in the body, such as: removing wastes from the blood, regulation of the body's fluids, electrolytes and acid–base balance, blood-pressure control and producing hormones (*i.e.* vitamin D, erythropoietin).

Diabetic nephropathy (DN). A major microvascular complication of diabetes mellitus. Pathophysiologic changes in DN lead to deterioration of renal functions.

Hypertension. Abnormally high arterial blood pressure. It is the most important and modifiable risk factors for cardiovascular diseases.

Chronic kidney disease (CKD). A progressive loss of kidney functions. The last stage of CKD is named end-stage renal failure. CKD is caused by a wide range of diseases, *i.e.* diabetes mellitus, hypertension and glomerulopathies. Regardless of the primary underlying disease, chronically injured kidneys are histomorphologically characterized by tubulointerstitial fibrosis.

Transforming growth factor β 1 (TGF-β1). A multipotential cytokine that regulates cell growth, its differentiation, processes of apoptosis, extracellular matrix production as well inducing fibrosis in a variety of tissues. TGF-β1 is the key player in renal fibrogenesis.

Nephrotoxicity. An effect of exposure the body to varies substances (medications, chemicals) that causes damage to kidneys. Cisplatin- or aminoglycosides-induced nephrotoxicity can by a cause of acute kidney injury.

Renoprotection. All factors or strategies that protect the kidneys from damage or slowing the progression of kidney disease to renal failure. One of the promising renoprotective molecules might be a dipeptide carnosine.

List of Abbreviations

ACE	Angiotensin converting enzyme
AGEs	Advanced glycation end products
ATP	Adenosine triphosphate
CKD	Chronic kidney disease
CNDP1	Carnosine peptidase 1 gene
DM	Diabetes mellitus
DN	Diabetic nephropathy
ESRD	End-stage renal disease
GFR	Glomerular filtration rate
IL	Interleukin
TGF-β1	Transforming growth factor β 1

References

Ahluwalia, T. S., Lindholm, E. and Groop, L. C., 2011. Common variants in CNDP1 and CNDP2, and risk of nephropathy in type 2 diabetes. *Diabetologia*. 54: 2295–2302.

Alhamdani, M. S., Al-Azzawie, H. F. and Abbas, F. K., 2007. Decreased formation of advanced glycation end-products in peritoneal fluid by carnosine and related peptides. *Peritoneal Dialysis International*. 27: 86–89.

Alkhalaf, A., Bakker, S. J., Bilo, H. J., Gans, R. O., Navis, G. J., Postmus, D., Forsblom, C., Groop, P. H., Vionnet, N., Hadjadj, S., Marre, M., Parving, H. H., Rossing, P. and Tarnow, L., 2010. A polymorphism in the gene encoding carnosinase (CNDP1) as a predictor of mortality and progression from nephropathy to end-stage renal disease in type 1 diabetes mellitus. *Diabetologia*. 53: 2562–2568.

Boldyrev, A. A., Aldini, G. and Dereve, W., 2013. Physiology and pathophysiology of carnosine. *Physiological Reviews*. 93: 1803–1845.

Chan, W. K., Decker, E. A., Chow, C. K. and Boissonneault, G. A., 1994. Effect of dietary carnosine on plasma and tissue antioxidant concentrations and on lipid oxidation in rat skeletal muscle. *Lipids*. 29: 461–466.

Coombes, J. S. and Fassett, R. G., 2012. Antioxidant therapy in hemodialysis patients: a systematic review. *Kidney International*. 81: 233–246.

Cuzzocrea, S., Genovese, T., Failla, M., Vecchio, G., Fruciano, M., Mazzon, E., Di Paola, R., Muia, C., La Rosa, C., Crimi, N., Rizzarelli, E. and Vancheri, C., 2007. Protective effect of orally administered carnosine on bleomycin-induced lung injury. *American Journal of Physiology: Lung Cellular and Molecular Physiology*. 292: 1095–1104.

Chakkera, H. A., Hanson, R. L., Kobes, S., Millis, M. P., Nelson, R. G., Knowler, W. C. and Distefano, J. K., 2011. Association of variants in the carnosine peptidase 1 gene (CNDP1) with diabetic nephropathy in American Indians. *Molecular Genetics and Metabolism*. 103: 185–190.

Fujii, T., Kurata, H., Takaoka, M., Muraoka, T., Fujisawa, Y., Shokoji, T., Nishiyama, A., Abe, Y. and Matsumura, Y., 2003a. The role of renal sympathetic nervous system in the pathogenesis of ischemic acute renal failure. *European Journal of Pharmacology*. 481: 241–248.

Fujii, T., Takaoka, M., Muraoka, T., Kurata, H., Tsuruoka, N., Ono, H., Kiso, Y., Tanaka, T. and Matsumura, Y., 2003b. Preventive effect of L-carnosine on ischemia/reperfusion-induced acute renal failure in rats. *European Journal of Pharmacology*. 474: 261–267.

Gariballa, S. E. and Sinclair, A. J., 2000. Carnosine: physiological properties and therapeutic potential. *Age Ageing*. 29: 207–210.

Guiotto, A., Calderan, A., Ruzza, P. and Borin, G., 2005. Carnosine and carnosine-related antioxidants: a review. *Current Medicinal Chemistry*. 12: 2293–2315.

Hipkiss, A. R., Worthington, V. C., Himsworth, D. T. and Herwig, W., 1998. Protective effects of carnosine against protein modification mediated by malondialdehyde and hypochlorite. *Biochimica et Biophysica Acta*. 1380: 46–54.

Hipkiss, A. R., 2011. Energy metabolism, proteotoxic stress and age-related dysfunction – protection by carnosine. *Molecular Aspects of Medicine*. 32: 267–278.

Hou, W. C., Chen, H. J. and Lin, Y. H., 2003. Antioxidant peptides with angiotensin converting enzyme inhibitory activities and applications for angiotensin converting enzyme purification. *Journal of Agricultural and Food Chemistry*. 51: 1706–1709.

Janssen, B., Hohenadel, D., Brinkkoetter, P., Peters, V., Rind, N., Fischer, C., Rychlik, I., Cerna, M., Romzova, M., de Heer, E., Baelde, H., Bakker, S. J., Zirie, M., Rondeau, E., Mathieson, P., Saleem, M. A., Meyer, J., Köppel, H., Sauerhoefer, S., Bartram, C. R., Nawroth, P., Hammes, H. P., Yard, B. A., Zschocke, J. and van der Woude, F. J., 2005. Carnosine as a protective factor in diabetic nephropathy: association with a leucine repeat of the carnosinase gene CNDP1. *Diabetes*. 54: 2320–2327.

Jia, H., Qi, X., Fang, S., Jin, Y., Han, X., Wang, Y., Wang, A. and Zhou, H., 2009. Carnosine inhibits high glucose-induced mesangial cell proliferation through mediating cell cycle progression. *Regulatory Peptides*. 154: 69–76.

Kiliś-Pstrusińska, K., Zwolińska, D. and Grzeszczak, W. 2010. Study Group. Is carnosinase 1 gene (CNDP1) polymorphism associated with chronic kidney disease progression in children and young adults? Results of a family-based study. *Archives of Medical Research*. 41: 356–362.

Köppel, H., Riedl, E., Braunagel, M., Sauerhoefer, S., Ehnert, S., Godoy, P., Sternik, P., Dooley, S. and Yard, B. A., 2011. L-Carnosine inhibits high-glucose-mediated matrix accumulation in human mesangial cells by interfering with TGF-β production and signalling. *Nephrology Dialysis Transplantation*. 26: 3852–3858.

Kurata, H., Fujii, T., Tsutsui, H., Katayama, T., Ohkita, M., Takaoka, M., Tsuruoka, N., Kiso, Y., Ohno, Y., Fujisawa, Y., Shokoji, T., Nishiyama, A., Abe, Y. and Matsumura, Y., 2006. Renoprotective effects of L-carnosine on ischemia/reperfusion-induced renal injury in rats. *Journal of Pharmacology and Experimental. Therapeutics*. 319: 640–647.

Lenney, J. F., George, R. P., Weiss, A. M., Kucera, C. M., Chan, P. W. and Rinzler, G. S., 1982. Human serum carnosinase: characterization, distinction from cellular carnosinase, and activation by cadmium. *Clinica Chimica Acta*. 123: 221–231.

Liu, B. F., Miyata, S., Hirota, Y., Higo, S., Miyazaki, H., Fukunaga, M., Hamada, Y., Ueyama, S., Muramoto, O., Uriuhara, A. and Kasuga, M., 2003. Methylglyoxal induces apoptosis through activation of p38 mitogen-activated protein kinase in rat mesangial cells. *Kidney International*. 63: 947–957.

McDonough, C. W., Hicks, P. J., Lu, L., Langefeld, C. D., Freedman, B. I. and Bowden, D. W., 2009. The influence of carnosinase gene polymorphisms on diabetic nephropathy risk in African-Americans. *Human Genetics*. 126: 265–275.

McFarland, G. A. and Holliday, R., 1999. Further evidence for the rejuvenating effects of the dipeptide L-carnosine on cultured human diploid fibroblasts. *Experimental Gerontology*. 34: 35–45.

Nagai, K., Niijima, A., Yamano, T., Otani, H., Okumra, N., Tsuruoka, N., Nakai, M. and Kiso, Y., 2003. Possible role of L-carnosine in the regulation of blood glucose through controlling autonomic nerves. *Experimental Biology and Medicine (Maywood, N. J.)*. 228: 1138–1145.

Nagai, K., Tanida, M., Niijima, A., Tsuruoka, N., Kiso, Y., Horii, Y., Shen, J. and Okumura, N., 2012. Role of L-carnosine in the control of blood glucose, blood pressure, thermogenesis, and lipolysis by autonomic nerves in rats: involvement of the circadian clock and histamine. *Amino Acids*. 43: 97–109.

Niijima, A., Okui, T., Matsumura, Y., Yamano, T., Tsuruoka, N., Kiso, Y. and Nagai, K., 2002. Effects of L-carnosine on renal sympathetic nerve activity and DOCA-salt hypertension in rats. *Autonomic Neuroscience. Basic & Clinical*. 97: 99–102.

Noori, S. and Mahboob, T., 2010. Antioxidant effect of carnosine pretreatment on cisplatin-induced renal oxidative stress in rats. *Indian Journal of Clinical Biochemistry*. 25: 86–91.

Pan, K. Z., Palter, J. E., Rogers, A. N., Olsen, A., Chen, D., Lithgow, G. J. and Kapahi, P., 2007. Inhibition of mRNA translation extends lifespan in Caenorhabditis elegans. *Aging Cell*. 6: 111–119.

Perry, T. L., Hansen, S. and Love, D. L., 1968. Serum-carnosinase deficiency in carnosinaemia. *Lancet*. 1: 1229–1230.

Peters, V., Jansen, E. E., Jakobs, C., Riedl, E., Janssen, B., Yard, B. A., Wedel, J., Hoffmann, G. F., Zschocke, J., Gotthardt, D., Fischer, C. and Köppel, H., 2011. Anserine inhibits carnosine degradation but in human serum carnosinase (CN1) is not correlated with histidine dipeptide concentration. *Clinica Chimica Acta*. 412: 263–267.

Peters, V., Schmitt, C. P., Zschocke, J., Gross, M. L., Brismar, K. and Forsberg, E., 2012. Carnosine treatment largely prevents alterations of renal carnosine metabolism in diabetic mice. *Amino Acids*. 42: 2411–2416.

Quinn, P. J., Boldyrev, A. A. and Formazuyk, V. E., 1992. Carnosine: its properties, functions and potential therapeutic applications. *Molecular Aspects of Medicine*. 13: 379–444.

Riedl, E., Pfister, F., Braunagel, M., Brinkkötter, P., Sternik, P., Deinzer, M., Bakker, S. J., Henning, R. H., van den Born, J., Krämer, B. K., Navis, G., Hammes, H. P., Yard, B. and Koeppel, H., 2011. Carnosine prevents apoptosis of glomerular cells and podocyte loss in STZ diabetic rats. *Cellular Physiology and Biochemistry*. 28: 279–288.

Shao, L., Li, Q. H. and Tan, Z., 2004. L-Carnosine reduces telomere damage and shortening rate in cultured normal fibroblasts. *Biochemical and Biophysical Research Communication*. 324: 931–936.

Soliman, K. M., Abdul-Hamid, M. and Othman, A. I., 2007. Effect of carnosine on gentamicin-induced nephrotoxicity. *Medical Science Monitor*. 13: 73–83.

Son, D. O., Satsu, H., Kiso, Y., Totsuka, M. and Shimizu, M., 2008. Inhibitory effect of carnosine on interleukin-8 production in intestinal epithelial cells through translational regulation. *Cytokine*. 42: 265–276.

Tan, K. M. and Candlish, J. K., 1998. Carnosine and anserine as modulators of neutrophil function. *Clinical and Laboratory Haematology*. 20: 239–244.

Tanida, M., Niijima, A., Fukuda, Y., Sawai, H., Tsuruoka, N., Shen, J., Yamada, S., Kiso, Y. and Nagai, K., 2005. Dose-dependent effects of L-carnosine on the renal sympathetic nerve and blood pressure in urethane-anesthetized rats. *American Journal of Physiology. Regulatory, Integrative and Comparative Physiology*. 288: 447–455.

Teufel, M., Saudek, V., Ledig, J. P., Bernhardt, A., Boularand, S., Carreau, A., Cairns, N. J., Carter, C., Cowley, D. J., Duverger, D., Ganzhorn, A. J., Guenet, C., Heintzelmann, B., Laucher, V., Sauvage, C. and Smirnova, T., 2003. Sequence identification and characterization of human carnosinase and a closely related non-specific dipeptidase. *Journal of Biological Chemistry*. 278: 6521–6531.

Vanholder, R., Cornelis, R., Dhondt, A. and Lameire, N., 2002. The role of trace elements in uraemic toxicity. *Nephrology Dialysis Transplantation*. 17(suppl. 2): 2–8.

Wanic, K., Placha, G., Dunn, J., Smiles, A., Warram, J. H. and Krolewski, A. S., 2008. Exclusion of polymorphisms in carnosinase genes (CNDP1 and CNDP2) as a cause of diabetic nephropathy in type 1 diabetes: results of large case-control and follow-up studies. *Diabetes*. 57: 2547–2551.

Willi, S. M., Zhang, Y., Hill, J. B., Phelan, M. C., Michaelis, R. C. and Holden, K. R., 1997. A deletion in the long arm of chromosome 18 in a child with serum carnosinase deficiency. *Pediatric Research*. 41: 210–213.

CHAPTER 27

Carnosine Protection Against Liver Injury

MÜJDAT UYSAL*[a], NECLA KOÇAK-TOKER[a], AND SEMRA DOĞRU-ABBASOĞLU[a]

[a]Department of Biochemistry, Istanbul Medical Faculty, Istanbul University, Çapa, Istanbul, Turkey
*E-mail: m.mujdatuysal@hotmail.com

27.1 Introduction

The liver is a miraculous organ conducting a wide spectrum of functions. The digestion and metabolism of carbohydrates, lipids, proteins and amino acids, storage of glycogen, lipids, metals and vitamins, synthesis of hormones, excretion of bile acids and bilirubin, detoxification of xenobiotics, bilirubin and ammonia, and accomplishment of innate and systemic immune regulation can be counted among its many functions. It is comprised of parenchymal, sinusoidal and perisinusoidal cells. Most of its functions are actualized in the parenchymal hepatocytes that form over 70% of the total liver mass. The sinusoidal cells are made up of Kupffer cells as hepatic macrophages and fenestrated endothelial cells, whereas the hepatic stellate cells (HSC) that store vitamin A and lipids are the cells of perisinusoidal areas (Hernandez-Gea and Friedman, 2011).

In simple terms, liver injury can be classified as acute or chronic. The persistence of the responsible insult may be the determinant in this rough classification. The liver has the distinguished capacity of regenerating its

Food and Nutritional Components in Focus No. 8
Imidazole Dipeptides: Chemistry, Analysis, Function and Effects
Edited by Victor R Preedy

Published by the Royal Society of Chemistry, www.rsc.org

acutely damaged cells. However, the presence of a persistent threat that overwhelms this capacity, is usually followed by hepatocyte death. Cell death, mostly by apoptosis and/or necrosis is induced by exposure to intense reactive oxygen species (ROS) production, ATP depletion, xenobiotics, elevated levels of bile acids, and cytokines such as tumor necrosis factor-α (TNF-α), interleukin1 (IL-1) and interferon-γ, and prevalent injury (Jaeschke, 2011).

ROS are generated by normal hepatocyte metabolism to assist protection against various pathogens. The cell has a strong defense system consisting of antioxidant enzymes, free-radical scavengers and damage-repairing enzymes, collectively known as the antioxidant system. There exists a delicate balance between the prooxidants and the antioxidants to maintain the health of the cell. When this balance is disrupted in favor of the pro-oxidant species, oxidative stress is staged. In the presence of the oxidative stress cellular biomolecules like proteins, lipids and DNA are damaged by ROS resulting in function loss (Muriel, 2009; Jaeschke, 2011) (Figure 27.1).

Oxidative stress appears to be the common determinant in liver diseases such as those from toxic or viral origins. Kupffer cells and neutrophils along with the mitochondria and cytochrome P450 system of hepatocytes can be counted among the hepatic sources of ROS. ROS-induced lipid peroxidation products such as malondialdehyde (MDA) and 4-hydroxynonenal (4-HNE), together with other reactive species cause damage to DNA and its repair systems. Thus, oxidative stress brings the inevitable end of the hepatocyte that occurs either by apoptosis or necrosis (Malhi *et al.*, 2010).

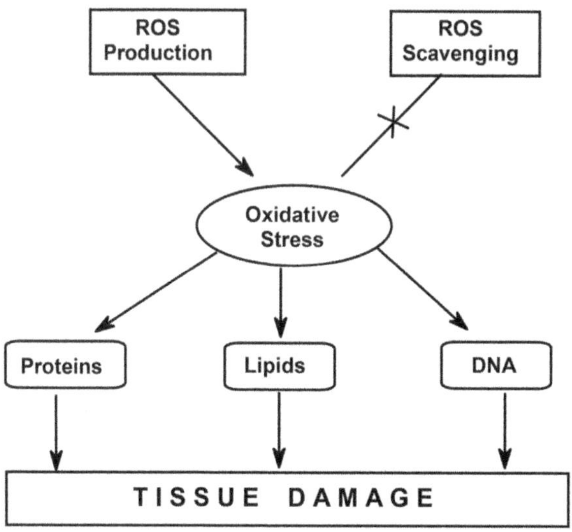

Figure 27.1 Oxidative stress. When the balance between reactive oxygen species (ROS) production and ROS scavenging is disrupted in favor of production, oxidative stress occurs, leading to the loss of function of biomolecules and concomitantly to tissue damage.

Mitochondria dysfunction is the hallmark of hepatocyte death. When mitochondrial injury that can be caused by various factors, increases mitochondrial membrane permeability, known as mitochondrial permeability transition (MPT); membrane potential loss, swelling and outer membrane rupture occur. MPT resulting in ATP depletion pursues necrosis, whereas the release of cytochrome c triggers apoptosis *via* caspase activation when ATP is conserved (Malhi *et al.*, 2010).

Necrosis eventuates from deprivation of the hepatocyte ATP, swelling of the mitochondria and rupture of the plasma membrane. The resultant leakage of cellular contents contributes to inflammation. Apoptosis is a morphologically defined cell-death programme that is responsible for removing damaged cells and leaked cellular contents, as cellular defense. Blebbing of the membrane, cell shrinkage, chromatin condensation and nuclear fragmentation result in the appearance of membranous bodies called apoptotic bodies. Intense apoptosis promotes the activation of Kupffer cells that engulf the apoptotic bodies. Usually, Kupffer cells located near the parenchymal and nonparenchymal cells of the liver become activated to respond to an acute insult that may cause infection, so as to protect the healthy state of the liver. In the case of being exposed to conditions that may lead to severe injury, activation of the Kupffer cells contributes to amplification of the injury. Upon engulfment of apoptotic bodies, Kupffer cells express death ligands like TNF-α, tumor necrosis factor-related apoptosis inducing ligand and Fas ligand. The release of these death ligands induces death receptor-mediated apoptosis that not only aggravate hepatocyte apoptosis, but also contribute to inflammation and fibrosis. Meanwhile, HSCs are activated by the engulfment of apoptotic bodies and transform into myofibroblast-like cells that secrete the fibrogenic cytokine, transforming growth factor-β1 (TGF-β1) and collagen type I, to promote inflammation and fibrosis. The prominent indicator of chronic injury is progressive fibrosis that may result in cirrhosis or hepatocellular carcinoma (Malhi *et al.*, 2010) (Figure 27.2).

A balance between the synthesis and breakdown of extracellular matrix (ECM) exists for proper functioning of ECM. Increased synthesis and decreased degradation of ECM proteins lead to fibrosis. Matrix metalloproteinases (MMPs) that are a family of endopeptidases active in the breakdown of ECM proteins, play an important role in the regulation of ECM. Tissue inhibitors of metalloproteinases (TIMPs) that are naturally present, prevent degradation of ECM by MMPs to maintain homeostasis. On the initiation of fibrosis, although the expressions of most of the MMPs are held constant, activated HSCs secrete MMP-2 to destroy the normal liver architecture. This act, in turn allows the activation and proliferation of HSCs to increase. The elevated secretion of TIMPs by the activated HSCs also contributes to the progression of fibrosis. On the other hand, when apoptosis of activated HSCs is promoted, TIMP inhibition is removed as a result. Thus, proteolysis of fibrosis becomes possible by high MMP-2 collagenase activity for reconstructing of normal liver architecture (Hernandez-Gea and Friedman, 2011).

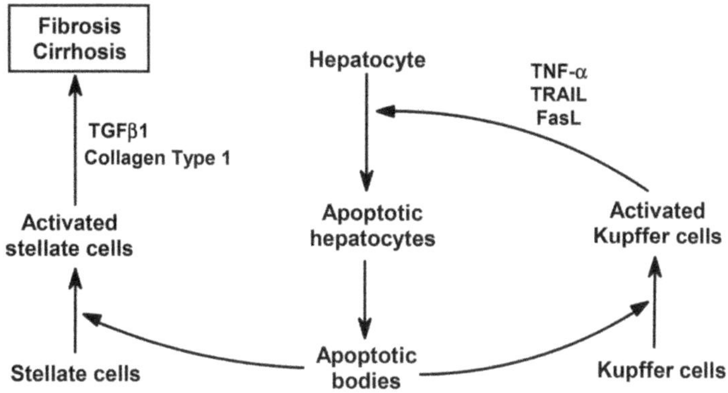

Figure 27.2 Hepatic injury and fibrosis. Apoptosis of hepatocytes results in apoptotic bodies that promote the activation of Kupffer cells. Activated Kupffer cells secrete death ligands like tumor necrosis factor α (TNF-α), tumor necrosis factor-related apoptosis inducing ligand (TRAIL) and Fas ligand (FasL). These ligands contribute to the amplification of hepatocyte apoptosis. Upon engulfing apoptotic bodies stellate cells are activated and secrete transforming growth factor (TGB)-β1 and collagen type 1, leading to the development of fibrosis and cirrhosis.

Carnosine (β-alanyl-L-histidine) is an endogenous dipeptide possessing various functions like membrane stabilizing, buffering, metal chelating especially with zinc and copper ions and preventing inflammation. It is known to scavenge ROS as an antioxidant. It inhibits lipid peroxidation and reacts with the harmful aldehydes like 4-HNE, MDA and acetaldehyde. It also prevents protein oxidation and formation of advanced glycation end products (AGE). These functions enable carnosine to be used as an efficient antioxidant and cell protector. Therefore, it has been employed in many oxidative stress-induced pathologies like atherosclerosis, diabetes, aging and ischemia–reperfusion (Hipkiss, 2009; Boldyrev *et al.*, 2013). Indeed, its actions not only as antioxidative but also as anti-inflammatory, antiapoptotic, antifibrotic and antiglycating may prove helpful in the treatment of liver diseases of various origins in rodents. L-Carnosine has a long half-life in rodents that lack serum carnosinase, and carnosine administration has been found to elevate their plasma and tissue carnosine levels (Liu *et al.*, 2008). However, L-carnosine has a limited use in humans due to its breakdown to L-histidine and β-alanine by serum and tissue carnosinases. To overcome this unfortunate issue, carnosinase-resistant enantiomer D-carnosine and L-carnosine derivatives like *N*-acetyl carnosine or trolox-L-carnosine are being tried instead of L-carnosine (Hipkiss, 2009; Bellia *et al.*, 2012).

27.2 Liver Injuries and Carnosine

There are a limited number of studies regarding the effect of carnosine on liver injury as a result of various insults. Most are experimental models, only a few are concerned with humans.

27.2.1 Ischemia–Reperfusion-Induced Liver Injury and Carnosine

In the liver, when blood flow and thus oxygen supply is reduced, ischemia develops concomitantly. As the blood supply is restored, inflammation and oxidative damage occur due to induced oxidative stress. The inflammatory factors being led to the damaged area by the returning blood and the reintroduction of oxygen causing ROS production worsen the injury. The use of the antioxidants has been proposed to be beneficial for protecting the liver against ischemia–reperfusion injury (Muriel, 2009). There are two studies about the role of carnosine in hepatic ischemia–reperfusion.

Fouad *et al.* (2007), using light and electron microscopy found that carnosine (250 mg kg^{-1}; i.p.) corrected ischemia–reperfusion-induced hepatic histopathological damage when applied to rats prior to ischemia generation. Serum transaminase activities and hepatic MDA levels and caspase-3 activity were reduced, while hepatic nitrite levels and catalase and glutathione peroxidase (GPx) activities were elevated. These results support the ameliorating effect of carnosine in ischemia–reperfusion injury.

When carnosine (250 mg kg^{-1}; i.p.) is applied after ischemia generation and before reperfusion, liver histopathology has also been found to improve (Baykara *et al.*, 2009). Serum transaminase and hepatic myeloperoxidase activities were decreased and hepatic glutathione (GSH) content was increased. When carnosine was administered together with melatonin, the amelioration of ischemia–reperfusion-induced damage was more pronounced than that produced by the use of both compounds alone (Baykara *et al.*, 2009).

27.2.2 Acetaminophen-Induced Liver Injury and Carnosine

Acetaminophen (paracetamol; *N*-acetyl-*p*-aminophenol) is a drug with analgesic and antipyretic effects. Although its use at therapeutic doses is safe, the overdose is known to cause liver injury through its metabolic activation by cytochrome P450 2E1 (CYP2E1). *N*-Acetyl-*p*-benzoquinonemine, which is the reactive metabolite produced, covalently binds to cellular GSH and proteins in the liver. The depletion of reduced GSH stimulates hepatic oxidative stress that produces mitochondrial dysfunction that in turn amplifies the generation of ROS and peroxynitrite. Together with the release of proinflammatory cytokines like TNF-α and IL-6, oxidative stress affects liver functions leading to liver failure (Malhi *et al.*, 2010; Rashid *et al.*, 2013).

One study is present in the literature reporting the role of carnosine on acetaminophen-induced liver injury in mice (Yan *et al.*, 2009). Pretreatment with carnosine or histidine (0.5, 1, 2 g L^{-1} in drinking water, for 4 weeks) followed by acetaminophen application (350 mg kg^{-1}; i.p.) ameliorated the antioxidant system by increasing reduced GSH and α-tocopherol, retaining hepatic GPx, catalase and superoxide dismutase (SOD) activities. A high dose of carnosine was found to raise the ascorbic acid levels. Apparently, carnosine is helpful in sparing or regenerating GSH, α-tocopherol and ascorbic acid. Reductions in hepatic IL-6 and TNF-α levels, lipid peroxidation and CYP2E1 activity evidently support antioxidant actions of carnosine. Interestingly, carnosine is remarked to act effectively at the transcriptional level by suppressing the mRNA expressions of TNF-α and monocyte chemoattractant protein and enhancing GPx mRNA expression. Histidine also exhibited protective actions similar to carnosine.

27.2.3 Thioacetamide-Induced Liver Injury and Carnosine

The fungicide thioacetamide (TAA) is used in several experimental models to propagate liver injury. Its biotransformation is mediated mainly by CYP2E1. TAA is converted to thioacetamide sulfoxide that undergoes further conversion to thioacetamide sulfdioxide. During its bioactivation, ROS like superoxide anion and hydrogen peroxide, are also produced. Covalent binding of TAA metabolites and ROS to hepatocellular molecules generate oxidative stress that triggers the steps that proceed to liver injury. While a single dose of TAA administration has been shown to cause acute liver injury characterized by centrilobular necrosis, chronic applications result in hepatic cirrhosis (Aydın *et al.*, 2010b).

Prevention or attenuation of oxidative stress in the treatment of liver injury caused by TAA has been attempted by using various antioxidants. However, there are few studies about the use of carnosine in literature. The study by Mehmetçik *et al.* (2008) is the only study implicating the effective antioxidant role of carnosine in acute TAA-induced liver necrosis. Carnosine (250 mg kg^{-1}; i.p.) efficiently prevented TAA-induced necrosis and decreased oxidative stress parameters like hepatic MDA and diene conjugate (DC) levels in rats, when coadministered with TAA (500 mg kg^{-1}). It did not improve the TAA-induced depression in the antioxidant system, except for the increase in vitamin E levels.

Hepatic fibrosis produced by chronic application of TAA in rats has been implicated by elevated hydroxyproline and lipid peroxide levels, while the antioxidant system was depressed (Aydın *et al.*, 2010b). Carnosine supplementation (2 g L^{-1} in drinking water, for three months) together with TAA application was not found to be helpful for the regression of fibrosis, although it could depress the hepatic lipid peroxidation without any change in hepatic GSH levels, SOD and GPx activities and their protein expressions.

27.2.4 Cadmium-Induced Liver Injury and Carnosine

Cadmium is a very toxic heavy metal that is extensively used in industry. Electroplating, production of industrial paints and nickel–cadmium batteries can be counted among various sources of cadmium. Cigarette smoking is another important source of cadmium. Exposure to cadmium present in contaminated water, food, and polluted air, especially near industrial areas, may be severely hazardous.

Acute cadmium exposure is one of the important experimental models of heavy-metal toxicity. Cadmium is known to have a prominent role in the activation of inflammatory cells that inevitably leads to the generation of ROS and subsequently to lipid peroxidation. It has also a depleting effect on cellular antioxidants. Acutely exposed cadmium accumulates in the liver and causes hepatotoxicity (Rashid *et al.*, 2013).

There is only one reported study that evaluated the hepatoprotective effect of carnosine in acute cadmium toxicity (Fouad *et al.*, 2009). Carnosine (10 mg kg^{-1} per day; i.p.) that was given to mice during cadmium exposure (6.5 mg kg^{-1}; i.p.), depressed hepatic lipid peroxidation and restored the hepatic antioxidant system by increasing GSH levels, and SOD and catalase activities. The improvement of liver injury was detected by light and electron microscopy. Carnosine lowered the hepatic cadmium ion concentration, acting as a metal chelator. Its effect on increasing levels of GSH, which is also an important cadmium chelator, helped in decreasing hepatic cadmium levels. Attenuation of myeloperoxidase and caspase-3 activities exhibited the anti-inflammatory and antiapoptotic actions of carnosine. Carnosine is explicitly proved to be a hopeful remedy in ameliorating the hepatotoxicity of acute cadmium exposure.

27.2.5 Sepsis-Induced Liver Injury and Carnosine

An infective stimulus eliciting an exaggerated immune response may generate sepsis that is also called systemic inflammatory response syndrome. Fighting the infection is exacerbated leading to the damage of body organs. Immense inflammation and coagulation coupled with the suppression of fibrinolysis and production of cytokines and chemokines amplify the inflammatory response. This brings about tissue hypoxia and ischemia proceeding to the ultimate end as organ failure and death (Jaeschke, 2011).

Sepsis in the liver is manifested by hepatic dysfunction. Endotoxins are removed by Kupffer cells, which also produce the mediators like TNF-α, IL-1, superoxide radical and nitric oxide. The prooxidant chaos created in the liver may lead to apoptosis and necrosis. In the only study investigating the effect of carnosine on sepsis, septic shock in rats was induced by laparotomy and cecal ligation (Sahin *et al.*, 2013). Carnosine (250 mg kg^{-1}; i.p.) was applied when the first signs of septic shock emerged, which were determined using blood pressure measurements. It not only returned serum transaminase activities back to normal; but also lowered both liver and serum MDA levels. The antioxidant action of carnosine along with its efficient improvement of the liver histopathology is overtly described by the study.

27.2.6 Stress-Induced Liver Injury and Carnosine

Stress is a condition encountered in everyday life, affecting the adaptation of an organism to the environment and impairing the quality of life. Exposure to stress conditions may lead to an organ's dysfunction by causing the secretion of neuroendocrine hormones and inhibiting the immune system. Stress activates the hypothalamic–pituitary–adrenal (HPA) axis, resulting in the enhanced release of glucocorticoids (GCs). The liver, having a high metabolic capacity becomes the primary target of elevated GCs in stress conditions. Prolonged exposure to high concentrations of GCs causes increased ROS production and depressed antioxidant system in tissues including liver (Kalaz *et al.*, 2012).

Carnosine is reported to ameliorate the detrimental effects that emerge in several stress conditions. It relieves the hyperactivated HPA axis and decreases GCs levels. There are few studies about the hepatic stress and carnosine in the literature. In the study of Tsoi *et al.* (2011), the disturbed hepatic glucose metabolism was improved by carnosine (150 mg kg^{-1} and 300 mg kg^{-1} per day, for 7 days) in restraint-stressed mice, while GC levels decreased. In the other study (Kalaz *et al.*, 2012), carnosine (250 mg kg^{-1}; i.p. for 21 days) efficaciously amended the disrupted pro-oxidant–antioxidant balance in the brains and livers of rats exposed to chronic (cold and immobilization) stress. Although its effects were more significant in the brain, it was capable of depressing hepatic oxidative stress. Carnosine treatment elevated hepatic ascorbic acid levels, but it was not influential on the other elements of the antioxidant system.

27.2.7 Age-Induced Liver Injury and Carnosine

Getting old is a frightening episode of the drama called life. Weakening of the body functions and losing strength are unpleasant for most people. Aging also affects the liver in which apparent changes are observed. The reductions in blood supply and number of cells concomitantly reduce the metabolic and detoxifying functions of the liver and make it vulnerable to injury. Mitochondrial dysfunction, ROS production and oxidative stress are well known to contribute to liver aging. The triggered apoptosis that is explicit during aging, may attribute to the enhancement of the process thus leading to diseases related to increased age. Thus supplementation of antioxidative compounds has been proposed to ameliorate the changes in the liver caused by aging.

When carnosine (250 mg kg^{-1}; i.p., for one month) was supplied to aged rats, the hepatic oxidative stress was suppressed, as indicated by reduced MDA, DC and protein carbonyl (PC) levels. Hepatic GSH and vitamin E content were elevated but the antioxidant enzyme activities remained unchanged. Carnosine was observed to be selectively effective on liver but not on heart and brain (Aydın *et al.*, 2010a).

The efficiency of carnosine increases when used in combination with lipophilic antioxidants like vitamin E or melatonin (Baykara *et al.*, 2009; Çoban *et al.*, 2013). Indeed supplementation of carnosine (250 mg kg^{-1}; i.p.) in combination with vitamin E (200 mg kg^{-1}) was influential on depressing the oxidative stress in the liver as well as the heart and the brain of aged rats (Çoban *et al.*, 2013).

D-Galactose overload is an experimental model that mimics the normal aging process in experimental animals. While accumulated galactitol leads to osmotic and oxidative stresses, the reaction of galactose with amine groups of proteins produces AGEs. In galactose-treated rats (300 mg kg^{-1}; s.c for two months) hepatic oxidative stress was induced, as indicated by increased MDA and PC levels and decreased GSH content and antioxidant enzyme activities (Kalaz *et al.*, 2014). Elevated hepatic Bax expression was the indicator of induced apoptosis, while decreased Ki-67 expression showed reduced proliferation. Carnosine supplementation (250 mg kg^{-1}; i.p. for two months) that was simultaneously applied with galactose, decreased serum transaminase activities and improved the disrupted liver histopathology. As an effective antioxidant it was able to depress the oxidative stress and increase the GSH content and SOD and GPx activities without changing the mRNA expressions. Carnosine-induced decrease in Bax expression and increase in Ki-67 expression exhibited the antiapoptotic and proliferative actions of the compound, respectively.

27.2.8 Alcoholic Liver Injury and Carnosine

Although certain health benefits are known when ethanol, under the general name of alcohol, is taken in moderate doses, excessive consumption may produce different types of alcoholic liver disease. Generation of fatty liver, alcoholic hepatitis and cirrhosis are types of alcoholic liver disease that appear succeeding as heavy drinking continues.

Ethanol metabolism, which takes place mainly in the liver, is based on oxidations. The three hepatic enzyme systems that catalyze the oxidation of ethanol to acetaldehyde, are cytosolic alcohol dehydrogenase, microsomal ethanol oxidizing system of the smooth endoplasmic reticulum and catalase in peroxisomes. Acetaldehyde, in return, is oxidized to acetate by aldehyde dehydrogenase. These oxidation reactions, which enhance NADH production and thus change the ratio of NAD^+/NADH, cause the inhibition of gluconeogenesis and fatty acid oxidation leading to the generation of fatty liver. Aldehyde oxidase and xanthine oxidase can also oxidize acetaldehyde to acetate. CYP2E1, the main enzyme of the microsomal ethanol oxidizing system, is especially upregulated in chronic ethanol drinking. ROS are produced by CYP2E1, and the free-radical generation accompanies the oxidation of NADPH to $NADP^+$ in chronic alcohol consumption. In addition, activated hepatic macrophages generate proinflammatory cytokines such as TNF-α and IL-6. These factors induce inflammation and fibrosis through the increased production of ROS. On the other hand, acetaldehyde may form antigenic adducts when it is bound to cellular proteins, to promote inflammation (Muriel, 2009) (Figure 27.3).

The hepatic prooxidant–antioxidant balance during ethanol metabolism was investigated by many researchers and several substances are used to recover the balance that is disturbed in favor of pro-oxidation, and the consequent liver injury. Few studies are present that investigate the role of carnosine in the oxidative stress caused by ethanol.

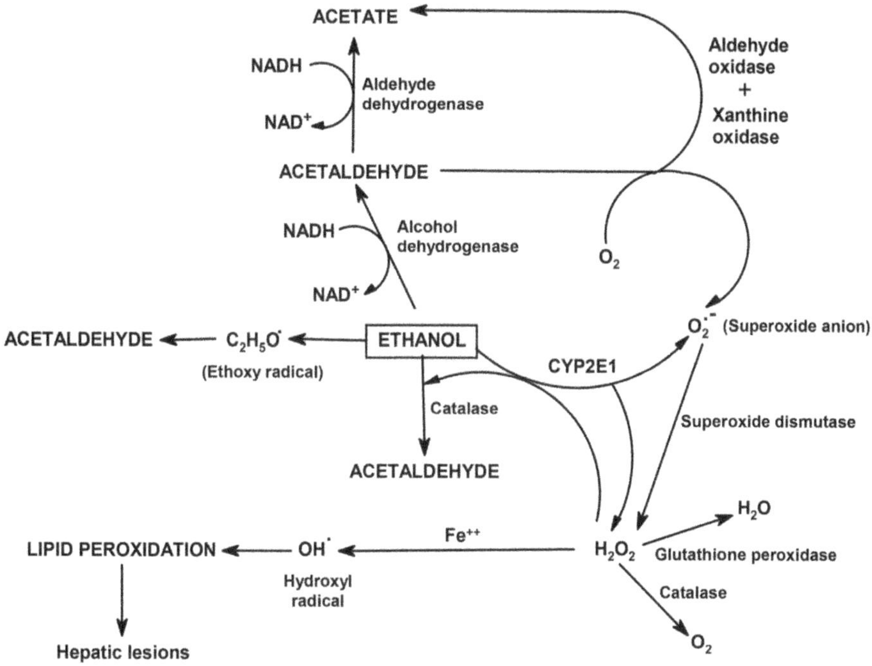

Figure 27.3 Reactive oxygen species production during hepatic ethanol metabolism. Ethanol is oxidized to acetaldehyde by alcohol dehydrogenase. Acetaldehyde can be oxidized to acetate by aldehyde dehydrogenase or by xanthine oxidase and aldehyde oxidase that produce superoxide anion. Superoxide anion can also be generated by cytochrome P450 2E1 (CYP2E1) of the microsomal ethanol oxidizing system. Superoxide anion is dismutated by superoxide dismutase to produce hydrogen peroxide (H_2O_2). Glutathione peroxidase converts H_2O_2 to water (H_2O) by oxidizing glutathione or it is converted to H_2O and O_2 by catalase. H_2O_2 may produce a hydroxyl radical ($OH^{\cdot}$) with the help of cytosolic iron (Fe^{++}). $OH^{\cdot}$ can promote lipid peroxidation.

In the study of Artun *et al.* (2010) carnosine pretreatment (2 g L^{-1} in drinking water, for 4 weeks) was followed by binge ethanol administration (5 g kg^{-1} as three doses in every 12 h) to rats. Carnosine was shown to decrease the elevated plasma transaminase activities and hepatic lipid peroxidation. It was unable to recover the depressed antioxidant system and hepatic steatosis. However, hepatic GSH levels were increased due to carnosine treatment. In another study of binge ethanol administration (Baykara *et al.*, 2014), carnosine (100 mg kg^{-1}; oral) was applied one hour before every ethanol treatment (5 g kg^{-1}; oral in three doses) in rats. Carnosine also lowered plasma transaminase activities and hepatic lipid peroxidation and myeloperoxidase activity and increased GSH content. Binge ethanol administration-induced apoptosis was suppressed as demonstrated by the declines in Bax and caspase-3 expressions and the increase in antiapoptotic Bcl-2 expression using immunohistochemical evaluations. Carnosine treatment also ameliorated the alcohol-induced morphological changes in hepatocytes as observed using electron microscopy.

There is one study regarding the role of carnosine in chronic ethanol treatment in mice (Liu *et al.*, 2008). The chronic ethanol application of four weeks was followed by carnosine usage in three different doses (0.5, 1, 2 g L^{-1} in drinking water for 3 weeks). Carnosine post-treatment proved to abate alcoholic hepatotoxicity in all doses. Similar to the findings of the two above-mentioned studies, plasma transaminase levels and hepatic lipid peroxidation were lower and hepatic GSH levels were increased with carnosine. The antioxidant and anti-inflammatory actions of carnosine were clearly manifested in this study. Elevated hepatic activities of GPx and catalase and mRNA expression of catalase and decreased CYP2E1 activity are reflections of its protective properties. The depressions of mRNA expressions and hepatic levels of IL-6 and TNF-α and the decreased C-reactive protein levels by carnosine prohibited the generation of inflammation due to chronic ethanol metabolism. Meanwhile, when histidine was used with similar doses in this study, antioxidative and anti-inflammatory effects like carnosine's were also exhibited.

27.2.9 Nonalcoholic Fatty Liver Disease and Carnosine

One of the common chronic liver diseases is nonalcoholic fatty liver disease (NAFLD). Although there is no abuse of alcohol, a range of liver injury from simple steatosis to nonalcoholic steatohepatitis (NASH), advanced fibrosis and cirrhosis may be present. A metabolic syndrome with the involvements of insulin resistance, obesity, dyslipidemia, hypertension and diabetes may promote the development of NAFLD by stimulation of hepatocyte fat accumulation. The subsequent generation of oxidative stress followed by lipid peroxidation and inflammation are also accepted to play the leading roles in the manifestation of hepatic injury related to NAFLD (Muriel, 2009). In the treatment of NAFLD, apart from lifestyle interventions, various drug therapies have been proposed to be beneficial. Insulin sensitizers, anti-inflammatory agents, antioxidants, lipid-lowering drugs and compounds with antiapoptotic action can be counted among these therapies. There are few reports on the effects of carnosine in NAFLD.

A high-fat diet (HFD) is known to induce obesity, insulin resistance, dyslipidemia and hepatic steatosis, together with the production of ROS and lipid peroxidation. In the study of Mong *et al.* (2011), carnosine (1 g L^{-1} in drinking water) was given to rats fed on HFD (containing 60% of calories as fat) for eight weeks. Carnosine effectively reduced the body weights, improved insulin resistance and dyslipidemia. Acting as an antilipogenic agent, it diminished the activities and mRNA expressions of some lipogenic enzymes like malic enzyme, fatty acid synthase and 3-hydroxy-3-methylglutaryl coenzyme A reductase in the liver and adipose tissue. Sterol regulatory element binding proteins (SREBRs) are transcription factors responsible for fatty acid and cholesterol synthesis. SREBR-1c modulates the mRNA expressions of enzymes that catalyze fatty acid synthesis, and SREBR-2 modulates the mRNA expressions of those that catalyze cholesterol synthesis. Carnosine was capable of suppressing expressions of both transcription factors in liver and adipose tissue, thus attenuating hepatic steatosis through an antisteatotic action.

Another animal model of insulin resistance is feeding rats with diet that contains high fructose (HFr). HFr diet is observed to cause disrupted glucose tolerance, hypertriglyceridemia and fatty liver besides insulin resistance. Since oxidative stress is brought forward as a mechanism that plays a role in high fructose model, usage of several antioxidants became an issue in various researches. In the study of Giriş *et al.* (2014) rats fed with HFr containing 60% fructose, were given carnosine (2 g L^{-1} in drinking water) alone or with α-tocopherol (200 mg kg^{-1}; i.m) for 8 weeks. Insulin resistance and hypertriglyceridemia were not affected by carnosine alone, although hepatic steatosis and lipid peroxidation were diminished. Carnosine in combination with α-tocopherol, on the other hand, was found to be effective in decreasing inflammation, insulin resistance, hepatic steatosis and lipid peroxidation and elevating GPx activity and expression. The results of this recent study also supports the increased efficiency of carnosine when used in combination with α-tocopherol.

There is a poignant relationship between diabetes and NAFLD. Although it is obscure whether diabetes foretells the development of NAFLD or *vice versa*; both play an eminent role for the progression of one another. The effectiveness of carnosine on liver has also been reported in streptozotocin-induced diabetic mice (Lee *et al.*, 2005). Carnosine treatment (0.5, 1 g L^{-1} in drinking water, for 4 weeks) following the induction of diabetes, improved the diabetic complications by enhancing hepatic antioxidant activity and restoring insulin secretion. It suppressed lipid peroxidation and increased catalase and GPx activities. Reduction in the levels of the proinflammatory cytokines of IL-6 and TNF-α also supported the anti-inflammatory action of carnosine. In both studies separate supplementations of histidine had similar effectiveness to carnosine (Lee *et al.*, 2005; Mong *et al.*, 2011).

27.3 Liver Disease and Polaprezinc

Polaprezinc contains L-carnosine chelated with zinc ions. The use of polaprezinc in different experimental models is quite recent (Figure 27.4).

Feeding experimental animals with methionine- and choline-deficient (MCD) diet exhibits a good model for studying NASH. In the study of Sugino *et al.* (2008), polaprezinc (2.2 g kg^{-1}, for 10 weeks) was given to mice fed on MCD diet, which displayed indications of steatohepatitis with mild fibrosis. Although polaprezinc supplementation was not able to alter fatty liver development, it promoted fibrolysis. Along with the reduction in lipid peroxidation, HSC activation and mRNA expressions of the proinflammatory (TNF-α) and fibrogenic (TGF-β) cytokines and of procollagen α1 were suppressed. While enhancements of mRNA expressions of MMP-2 and TIMP-1 and TIMP-2 were apparent in fibrosis due to MCD diet, polaprezinc was capable of depressing

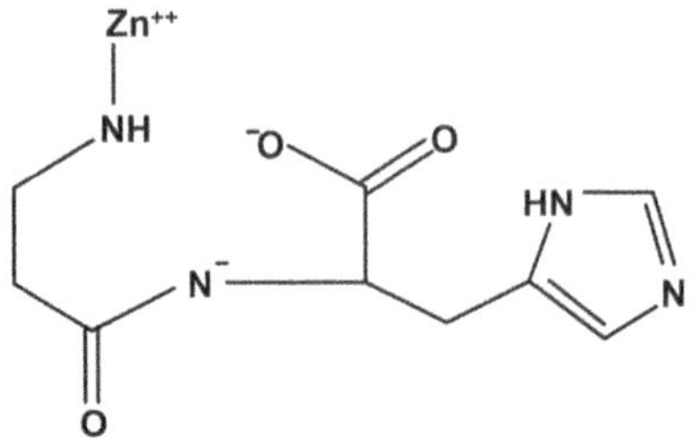

Figure 27.4 Polaprezinc. Zinc *N*-(3-aminopropionyl)histidine.

the enhancements of TIMP-1 and TIMP-2 mRNA expressions. Thus, by the inhibition of fibrogenesis, the anti-NASH effect of polaprezinc was marked. Although the authors favored zinc as the effective element of polaprezinc, the contribution of carnosine in this antifibrotic action has to be emphasized.

Indeed, in the study of Kono *et al.* (2012) TAA-induced liver fibrosis was generated in rats. In agreement with Sugino *et al.* (2008), at the end of a 10-week period, mild fibrosis was distinctly seen. A second 10-week period, together with polaprezinc, L-carnosine and zinc supplementations revealed striking results. Polaprezinc appeared efficacious in preventing the progression of fibrosis as indicated by reduced hepatic hydroxyproline and TGF-β levels, and inhibited HSC activation. It was capable of reducing both MMP-2 activity and TIMP-1 levels to induce fibrolysis. Also, polaprezinc restored decreased hepatic zinc levels. When zinc or L-carnosine, in amounts equal to those present in polaprezinc were supplemented individually, no amelioration in the ongoing fibrosis and oxidative stress was observed. Actually, the inefficiency of L-carnosine in inhibiting the development of fibrosis caused by TAA, as observed by Aydın *et al.* (2010b), supports this finding, although it was given together with TAA. The competence of polaprezinc in the regression of fibrosis as an antifibrolytic and antioxidative agent has been attributed to the combined effects of zinc and L-carnosine.

There are some clinical trials of polaprezinc that are associated with patients with hepatitis C virus (HCV)-related chronic liver disease. Chronic hepatitis C is one of the major causes of liver cirrhosis and hepatocellular carcinoma. Oxidative stress is proposed to be an underlying mechanism of HCV infection and progression of the developing liver disease. Along with inflammation caused by the viral infection, the core protein of the virus is also known to induce oxidative stress as indicated by elevated generation of ROS and decreased GSH levels (Muriel, 2009; Malhi *et al.*, 2010). Polaprezinc was given to patients either alone or together with the antiviral therapy consisting of interferon-α, or interferon-α and ribavirin. With polaprezinc alone for a long-term usage (7 years), the pathology of the liver was improved and incidence of hepatic carcinoma was reduced (Matsumura *et al.*, 2012). Similarly, when polaprezinc was given together with interferon-α for 20 weeks, the therapeutic potential of interferon-α was more pronounced (Takagi *et al.*, 2001).

Summary Points

- Carnosine is a dipeptide consisting of L-histidine and β-alanine.
- Acute and chronic liver diseases are among important health problems.
- Carnosine is employed in many animal models exhibiting liver injury.
- Carnosine has a protective effect in liver injury induced by ischemia–reperfusion, acetaminophen, thioacetamide, cadmium and endotoxins as well as alcoholic and nonalcoholic fatty liver disease.
- Carnosine exhibits antioxidative, anti-inflammatory, antiapoptotic, antifibrotic and antiglycating actions in the amelioration of liver injury.

Key Facts of Liver Injury

1. Liver injury can be classified as acute or chronic due to the persistence of insult.
2. When persistence of threat overwhelms regenerating capacity of liver, hepatocyte death follows.
3. Increase of mitochondrial membrane permeability cause mitochondrial dysfunction leading to hepatocyte death occurring by apoptosis and/or necrosis.
4. When extracellular matrix protein synthesis increase and protein degradation decrease, fibrosis initiates.
5. Activated hepatic stellate cells (HSC) secrete metalloproteinase 2 to destroy normal liver architecture.
6. Proliferation of HSC increase and tissue inhibitors of metalloproteinases are secreted, all leading to progression of fibrosis.

Definition of Words and Terms

Bcl-2. It is encoded by Bcl-2 gene and is important as an antiapoptotic protein.

Bax. Apoptosis regulator protein that is a member of Bcl-2 gene family. It promotes apoptosis by binding to Bcl-2 protein.

Ki-67. It is a nuclear protein that is associated with cellular proliferation.

Reactive oxygen species (ROS). They are chemically reactive molecules, mostly being free radicals in nature. They contain an oxygen atom and form during normal oxygen metabolism. Although they are helpful in cellular reactions, elevated levels cause damage in the cell.

Oxidative stress. The cumulative damage caused by cellular oxidants when the balance between oxidant production and antioxidant defense is disturbed in favor of oxidants.

Lipid peroxidation. The oxidation chain reactions of lipids containing carbon–carbon double bonds. It is triggered by reactive oxygen species and leads to degradation of the molecules.

Advanced glycation end products (AGE). These molecules form when carbohydrate residues bind with proteins by the reaction called glycation. They may be formed in the cells or in food by heating. The presence of these products leads to premature aging and they play a role in age-related diseases.
Cytokine. They are a group of molecules taking part in hormonal regulation and cell signaling. They are mostly protein or glikoprotein in nature. They are mostly referred to as agents responsible for immune reactions. Those that start an inflammatory response are grouped as proinflammatory and those that reduce inflammation are grouped as anti-inflammatory.
Chemokine. A family of small cytokines that are chemotactic.
High-fat diet (HFD). It is a type of experimental diet in which 35–60% of total calories are obtained from fat.
Methionine- and choline-deficient diet (MCD). It is a type of experimental diet that is known to cause impairment of mitochondrial β-oxidation and induction of CYP2E1 expression.

List of Abbreviations

AGE Advanced glycation end products
CYP2E1 Cytochrome P450 2E1
DC Diene conjugate
ECM Extracellular matrix
GCs Glucocorticoids
GSH Glutathione
GPx Glutathione peroxidase
HSC Hepatic stellate cells
HCV Hepatitis C virus
HFD High-fat diet
4-HNE 4-Hydroxynonenal
HPA Hypothalamic–pituitary–adrenal
IL Interleukin
MDA Malondialdehyde
MMPs Matrix metalloproteinases
MCD Methionine and choline deficient
MPT Mitochondrial permeability transition
NAFLD Nonalcoholic fatty liver disease
NASH Nonalcoholic steatohepatitis
PC Protein carbonyl
ROS Reactive oxygen species
SREBR Sterol regulatory element binding proteins
SOD Superoxide dismutase
TAA Thioacetamide
TIMPs Tissue inhibitors of metalloproteinases
TGF-β Transforming growth factor-β
TNF-α Tumor necrosis factor-α

References

Artun, B. C., Küskü-Kiraz, Z., Güllüoğlu, M., Çevikbaş, U., Koçak-Toker, N. and Uysal, M., 2010. The effect of carnosine pretreatment on oxidative stress and hepatotoxicity in binge ethanol administered rats. *Human & Experimental Toxicology*. 329: 659–665.

Aydın, A. F., Küçükgergin, C., Özdemirler-Erata, G., Koçak-Toker, N. and Uysal, M., 2010a. The effect of carnosine treatment on prooxidant–antioxidant balance in liver, heart and brain tissues of male aged rats. *Biogerontology*. 11: 103–109.

Aydın, A. F., Küskü-Kiraz, Z., Doğru-Abbasoğlu, S., Güllüoğlu, M., Uysal, M. and Koçak-Toker, N., 2010b. Effect of carnosine against thioacetamide-induced liver cirrhosis in rat. *Peptides*. 31: 67–71.

Baykara, B., Cilaker-Mıcılı, S., Tugyan, K., Tekmen, I., Bagriyanik, H. A., Sonmez, U., Sonmez, A., Oktay, G., Yener, N. and Ozbal, S., 2014. The protective effects of carnosine in alcohol-induced hepatic injury in rats. *Toxicology and Industrial Health*. 30: 25–32.

Baykara, B., Tekmen, I., Pekcetin, C., Ulukus, C., Tuncel, P., Sagol, O., Ormen, M. and Ozogul, C., 2009. The protective effects of carnosine and melatonin in ischemia–reperfusion injury in the rat liver. *Acta Histochemica*. 111: 42–51.

Bellia, F., Vecchio, G. and Rizzarelli, E., 2012. Carnosine derivatives: new multifunctional drug-like molecules. *Amino Acids*. 43: 153–163.

Boldyrev, A. A., Aldini, G. and Derave, W., 2013. Physiology and pathophysiology of carnosine. *Physiological Reviews*. 93: 1803–1845.

Çoban, J., Bingül, I., Yeşil-Mızrak, K., Doğru-Abbasoğlu, S., Öztezcan, S. and Uysal, M., 2013. Effects of carnosine plus vitamin E and betaine treatments on oxidative stress in some tissues of aged rats. *Current Aging Science*. 6: 199–205.

Fouad, A. A., El-Rehany, M. A. A. and Maghraby, H. K., 2007. The hepatoprotective effect of carnosine against ischemia/reperfusion liver injury in rats. *European Journal of Pharmacology*. 572: 61–68.

Fouad, A. A., Qureshi, H. A., Yacoubi, M. T. and Al-Melhim, W. N., 2009. Protective role of carnosine in mice with cadmium-induced acute hepatotoxicity. *Food and Chemical Toxicology*. 47: 2863–2870.

Giriş, M., Doğru-Abbasoğlu, S., Kumral, A., Olgaç, V., Koçak-Toker, N. and Uysal, M., 2014. Effect of carnosine alone or combined with α-tocopherol on hepatic steatosis and oxidative stress in fructose-induced insulin-resistant rats. *Journal of Physiology and Biochemistry*. 70: 385–395.

Hernandez-Gea, V. and Friedman, S. L., 2011. Pathogenesis of liver fibrosis. *Annual Review of Pathology: Mechanisms of Disease*. 6: 425–456.

Hipkiss, A. R., 2009. Carnosine and its possible roles in nutrition and health. *Advances in Food and Nutrition Research*. 57: 87–154.

Jaeschke, H., 2011. Reactive oxygen and mechanisms of inflammatory liver injury: present concepts. *Journal of Gastroenterology and Hepatology*. 26(suppl. 1): 173–179.

Kalaz, E. B., Çoban, J., Aydın, A. F., Doğan-Ekici, I., Doğru-Abbasoğlu, S., Öztezcan, S. and Uysal, M., 2014. Carnosine and taurine treatments decreased oxidative stress and tissue damage induced by D-galactose in rat liver. *Journal of Physiology and Biochemistry*. 70: 15–25.

Kalaz, E. B., Evran, B., Develi-İş, S., Vural, P., Doğru-Abbasoğlu, S. and Uysal, M., 2012. Effect of carnosine on prooxidant–antioxidant balance in several tissues of rats exposed to chronic cold plus immobilization stress. *Journal of Pharmacological Sciences*. 120: 98–104.

Kono, T., Asama, T., Chisato, N., Ebisawa, Y., Okayama, T., Imai, K., Karasaki, H., Furukawa, H. and Yoneda, M., 2012. Polaprezinc prevents ongoing thioacetamide-induced liver fibrosis in rats. *Life Sciences*. 90: 122–130.

Lee, Y., Hsu, C., Lin, M. and Yin, M., 2005. Histidine and carnosine delay diabetic deterioration in mice and protect human low density lipoprotein against oxidation and glycation. *European Journal of Pharmacology*. 513: 145–150.

Liu, W., Liu, T. and Yin, M., 2008. Beneficial effects of histidine and carnosine on ethanol-induced chronic liver injury. *Food and Chemical Toxicology*. 46: 1503–1509.

Malhi, H., Guicciardi, M. E. and Gores, G. J., 2010. Hepatocyte death: a clear and present danger. *Physiological Reviews*. 90: 1165–1194.

Matsumura, H., Nirei, K., Nakamura, H., Arakawa, Y., Higuchi, T., Hayashi, J., Yamagami, H., Matsuoka, S., Ogawa, M., Nakajima, N., Tanaka, N. and Moriyama, M., 2012. Zinc supplementation therapy improves the outcome of patients with chronic hepatitis C. *Journal of Clinical Biochemistry and Nutrition*. 51: 178–184.

Mehmetçik, G., Özdemirler, G., Koçak-Toker, N., Çevikbaş, U. and Uysal, M., 2008. Role of carnosine in preventing in thioacetamide-induced liver injury in the rat. *Peptides*. 29: 425–429.

Mong, M. C., Chao, C. Y. and Yin, M. C., 2011. Histidine and carnosine alleviated hepatic steatosis in mice consumed high saturated fat diet. *European Journal of Pharmacology*. 653: 82–88.

Muriel, P., 2009. Role of free radicals in liver diseases. *Hepatology International*. 3: 526–536.

Rashid, K., Sinha, K. and Sil, P. C., 2013. An update on oxidative stress-mediated organ pathophysiology. *Food and Chemical Toxicology*. 62: 584–600.

Sahin, S., Oter, S., Burukoglu, D. and Sutken, E., 2013. The effects of carnosine in an experimental rat model of septic shock. *Medical Science Monitor Basic Research*. 19: 54–61.

Sugino, H., Kumagai, N., Watanabe, S., Toda, K., Takeuchi, O., Tsunematsu, S., Morinaga, S. and Tsuchimoto, K., 2008. Polaprezinc attenuates liver fibrosis in a mouse model of non-alcoholic steatohepatitis. *Journal of Gastroenterology and Hepatology*. 23: 1909–1916.

Takagi, H., Nagamine, T., Abe, T., Takayama, H., Sato, K., Otsuka, T., Kakizaki, S., Hashimoto, Y., Matsumoto, T., Kojima, A., Takezawa, J., Suzuki, K., Sato, S. and Mori, M., 2001. Zinc supplementation enhances the response to interferon therapy in patients with chronic hepatitis C. *Journal of Viral Hepatitis*. 8: 367–371.

Tsoi, B., He, R.-R., Yang, D.-H., Li, Y.-F., Li, X.-D., Li, W.-X., Abe, K. and Kurihara, H., 2011. Carnosine ameliorates stress-induced glucose metabolism disorder in restrained mice. *Journal of Pharmacological Sciences*. 117: 223–229.

Yan, S. L., Wu, S. T., Yin, M. C., Chen, H. T. and Chen, H. C., 2009. Protective effects from carnosine and histidine on acetaminophen-induced liver injury. *Journal of Food Science*. 74: H259–H265.

CHAPTER 28

The Blood–Retinal Barrier and Carnosine

YOSHIYUKI KUBO[a], SHIN-ICHI AKANUMA[a], AND KEN-ICHI HOSOYA*[a]

[a]Department of Pharmaceutics, Graduate School of Medicine and Pharmaceutical Sciences, University of Toyama, 2630 Sugitani, Toyama, 930-0194, Japan
*E-mail: hosoyak@pha.u-toyama.ac.jp

28.1 Introduction

The retina is a neural tissue that controls the visionary sense, and it is known that severe retinal diseases, such as diabetic retinopathy, macular degeneration and glaucoma, cause blindness. In the retina of vertebrates, L-carnosine (β-alanyl-L-histidine) has been detected by high-performance liquid chromatography (Margolis and Grillo, 1984), and the involvement of L-carnosine has been suggested in the functional modulation of γ-aminobutyric acid (GABA) and glutamate receptors expressed by photoreceptors and bipolar cells in the retina since immunohistochemical analysis showed the colocalization of L-carnosine and glutamate in retinal tissue (Panzanelli *et al.*, 1997). In addition, the protective effect of L-carnosine against retinal vascular damage was suggested in a recent report on model rats with diabetic retinopathy (Pfister *et al.*, 2011), and the study of Alzheimer's disease and brain ischemia supports the neuroprotective effects of L-carnosine, such as its antioxidant,

Food and Nutritional Components in Focus No. 8
Imidazole Dipeptides: Chemistry, Analysis, Function and Effects
Edited by Victor R Preedy

Published by the Royal Society of Chemistry, www.rsc.org

antiglycation and metal ion-chelating actions. In a study using mouse model of retinal ischemia, the intraperitoneal administration of L-carnosine was shown to have an inhibitory effect on the loss of retinal ganglion cells by affecting the expression of hypoxia-inducing factor-1 α (HIF-1α), glial fibrillary acidic protein (GFAP), dynamin-related protein-1 (Drp-1), Bcl-2-associated X protein (Bax) and B-cell lymphoma 2 (Bcl-2) (Ji *et al.*, 2013). This evidence supports the physiological importance of L-carnosine in the retina, and it is thought that clarification of the details of the physiological system for the supply of L-carnosine will provide us with beneficial information for the treatment of retinal diseases and maintenance of the healthy retina. In this chapter, we shall focus on the system supplying L-carnosine to the retina from the circulating blood.

28.2 Blood–Retinal Barrier (BRB) and Membrane Transporters

In the retina, the blood–retinal barrier (BRB) separates the circulating blood and the neural tissue, and contributes to the protection of the neural retina by preventing nonspecific paracellular transport of low-weight molecules, such as toxic and metabolized waste compounds, to the retina from blood. The BRB is formed by two different types of cell, retinal capillary endothelial cells and retinal pigment epithelial (RPE) cells, that from the inner and outer BRB, respectively, and the tight junction formed by these cells contribute to the restriction of nonspecific transport at the BRB (Figure 28.1). Furthermore, the contribution of a carrier-mediated process has been shown to be associated with the blood-to-retina transport of essential nutrients, suggesting the involvement of membrane transporters expressed at the plasma membrane of retinal capillary endothelial cells and RPE cells (Stewart and Tuor, 1994; Cunha-Vaz, 2004; Hosoya *et al.*, 2011) (Figure 28.1). In particular, two-thirds of the retina is nourished *via* the inner BRB formed by retinal capillary endothelial cells. The establishment of an *in vitro* model cell line of the inner BRB, the conditionally immortalized rat retinal capillary endothelial cell line (TR-iBRB2 cells), has improved our understanding of different transport mechanisms operating at the inner BRB (Hosoya *et al.*, 2001), and recent advances in transporter research have revealed the expression of nutrient transporters, such as glucose transporter (GLUT1/SLC2A1), equilibrative nucleoside transporter 2 (ENT2/SLC29A2), Na^+-dependent multivitamin transporter (SMVT/SLC5A6) and glycine transporter (GlyT1/SLC6A9), at the inner BRB (Kumagai *et al.*, 1996; Nagase *et al.*, 2006; Okamoto *et al.*, 2009; Ohkura *et al.*, 2010). In addition to essential nutrients, a carrier-mediated influx transport system for cationic drugs, such as verapamil, pyrilamine and propranolol, has recently been identified at the inner BRB, supporting the important role of the BRB in the systemic delivery of drugs used to treat retinal diseases (Kubo *et al.*, 2013).

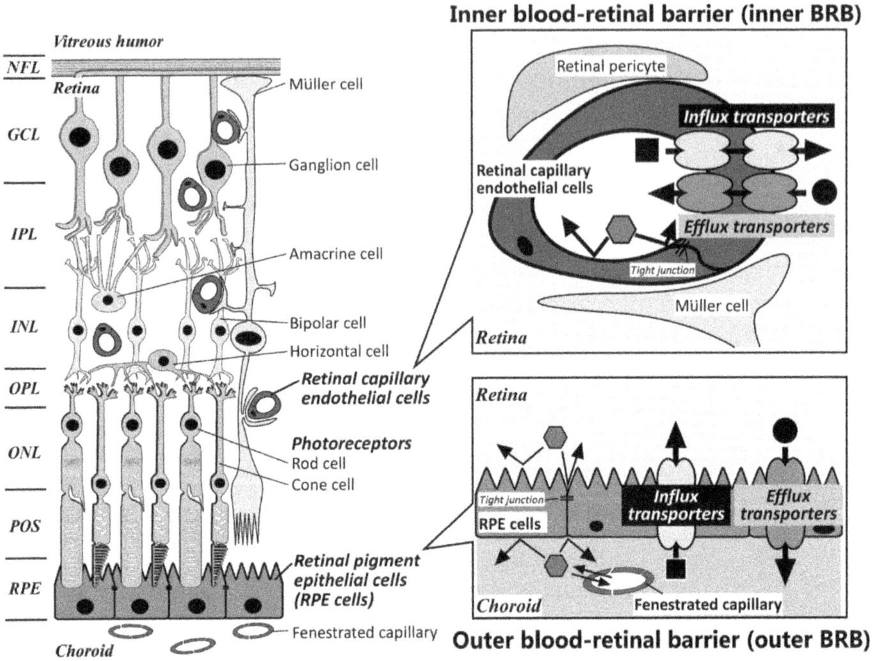

Figure 28.1 Structure of the blood–retinal barrier (BRB). The neural retina and the circulating blood are separated by the BRB, that consists of two different barrier structures, the inner and outer BRB. These barrier structures are formed by retinal capillary endothelial cells and retinal pigment epithelial (RPE) cells, respectively. At the inner and outer BRB, tight junction formed by the cells restricts the paracellular transport, and membrane transporters contribute to the nutrient supply to the retina. GCL, ganglion cell layer; IPL, inner plexiform layer; INL, inner nuclear layer; OPL, outer plexiform layer; ONL, outer nuclear layer; POS, photoreceptor outer segment; RPE, retinal pigment epithelium.

28.3 Carrier-Mediated L-Carnosine Transport in the Different Tissues

L-Carnosine is a bioactive dipeptide of β-alanine and L-histidine, and its transport system has been studied in several organs and tissues. A study using rabbit renal brush-border membrane vesicles (BBMVs) has demonstrated Na^+-independent L-carnosine transport, the activity of which is inhibited in the presence of di- and tripeptides, suggesting carrier-mediated transport of L-carnosine at the apical membrane of renal proximal tubular epithelial cells (Ganapathy and Leibach, 1982). A study using primary-cultured conjunctival epithelial cells of rabbits suggested the contribution of an H^+-driven oligopeptide transporter to the carrier-mediated transport of L-carnosine in the conjunctiva (Basu *et al.*, 1998). In a study with rabbit intestinal BBMV, the H^+-dependent and Na^+-independent carrier-mediated transport of penicillin-G, a representative β-lactam antibiotic, was inhibited in the presence of L-carnosine, suggesting the involvement of H^+-driven oligopeptide transporters in the intestinal absorption of L-carnosine (Poschet *et al.*, 1996). In the field of membrane-transporter research, a study with cultured cells expressing oligopeptide transporter (PEPT1/SLC15A1) showed that L-carnosine is recognized and transported by PEPT1 (Nakanishi *et al.*, 2000), and a study with Caco-2, an *in vitro* model cell line of human intestinal epithelial cell, also supports the involvement of PEPT1 in the intestinal absorption of L-carnosine (Son *et al.*, 2004). In the brain, a study with primary-cultured choroid plexus epithelial cells showed the possible contribution of PEPT2 (SLC15A2), a high-affinity oligopeptide transporter, to L-carnosine transport at the blood–cerebrospinal fluid barrier (BCSFB), and the expression of PEPT2 in guinea pig cardiomyocytes has been reported (Shu *et al.*, 2002; Lin and King, 2007). Therefore, these research findings support the contribution of oligopeptide transporters belonging to the SLC15A family to L-carnosine transport in mammals.

28.4 Oligopeptide Transporters (SLC15A) in the Retina

Carrier-mediated transport of L-carnosine in the kidney, small intestine and brain suggests the possible involvement of members of the SLC15A family, such as PEPT1, PEPT2 and peptide/histidine transporters (PHT1/SLC15A4 and PHT2/SLC15A3), in the supply of L-carnosine across the BRB (Daniel and Kottra, 2004). In an expression study of the SLC15A family, the transcripts of PEPT2, PHT1 and PHT2 are expressed in the human and bovine retinas in spite of the undetectable expression of PEPT1. Furthermore, the expression of PHT1 mRNA was shown in an *in vitro* model cell line of human RPE (ARPE-19) and freshly isolated bovine and human RPE cells while mRNA expression of PEPT2 and PHT2 was undetectable (Ocheltree *et al.*, 2003). Recently, we investigated the expression of SLC15A family members in TR-iBRB2 cells and primary-cultured rat RPE cells, and our results suggest that rat PHT1 is expressed at the inner and outer BRB in spite of the undetectable expression of rat PEPT1, rat PEPT2 and rat PHT2 at the BRB (Figure 28.2), supporting a previous report about the SLC15A family in human and bovine subjects. Although PHT1 is reported to transport dipeptides, including L-carnosine, the *in vitro* uptake study of [^{3}H]glycylsarcosine ([^{3}H]Gly-Sar), a representative dipeptide, by ARPE-19 cells suggested that carrier-mediated dipeptide transport plays only a minor role at the outer BRB (Ocheltree *et al.*, 2003). On the other hand, in an *in vivo* study involving bolus injection of [^{3}H]Gly-Sar into rabbits, the influx transport of [^{3}H]Gly-Sar across the BRB exhibited a time-dependent increase for 30 min at least, and was significantly inhibited in the presence of L-carnosine, captopril and glycylproline, suggesting the possible contribution of a carrier-mediated process at the BRB (Atluri *et al.*, 2004). These findings strongly suggest that the

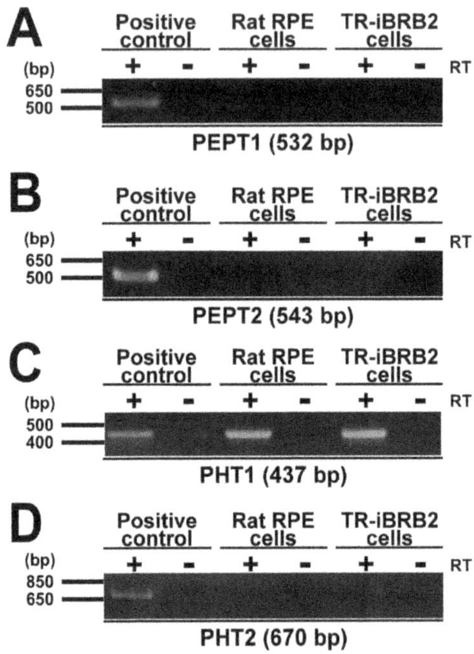

Figure 28.2 Expression of oligopeptide transporters at the inner and outer BRB. RT-PCR analysis was performed to investigate the expression of PEPT1/SLC15A1 (A; 532 bp), PEPT2/SLC15A2 (B; 543 bp), PHT1/SLC15A4 (C; 437 bp) and PHT2/SLC15A3 (D; 670 bp) in the inner and outer BRB. TR-iBRB2 cells and primary-cultured rat RPE cells were used as an *in vitro* model of the inner and outer BRB, respectively. Specific primers for PEPT1, PEPT2, PHT1 and PHT2 were designed with reference to the sequences registered as NM_005073, NM_001145998, NM_145648 and NM_016582 in National Center for Biotechnology Information (NCBI).

contribution of a carrier-mediated process to the L-carnosine transport at the BRB is controversial.

28.5 Minor Contribution of Dipeptide Transport at the Inner and Outer BRB

Recently, we have investigated the controversial contribution of a carrier-mediated process to the blood-to-retina transport of dipeptides (Usui *et al.*, 2013). The results obtained in an *in vivo* analysis with rats suggested that the transport of oligopeptide is minor at the BRB. In a study involving carotid artery injection, the retinal uptake index (RUI) of [^{3}H]Gly-Sar was estimated to be 21.3%, similar to that of [^{3}H]D-mannitol (11.6%), a nonpermeable paracellular marker compound, and no significant effect was observed in the presence of unlabeled Gly-Sar (Figure 28.3A). The minor transport of

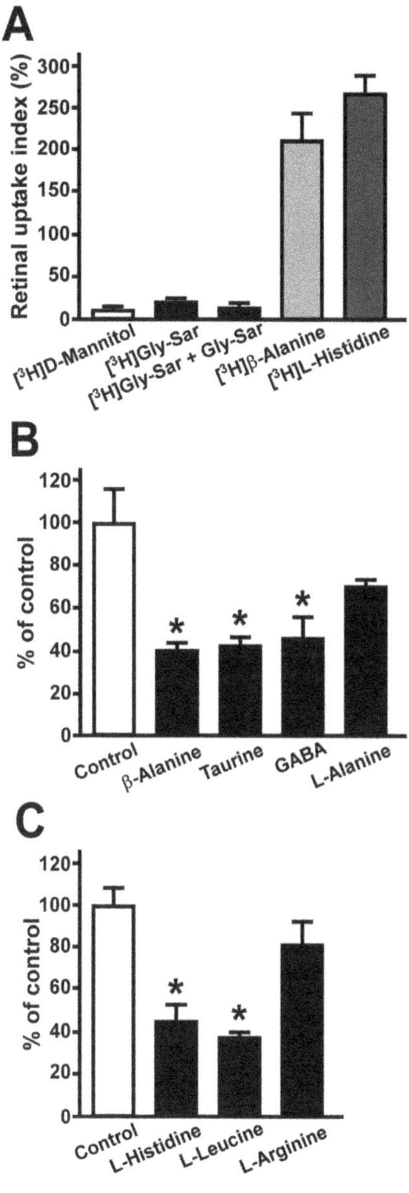

Figure 28.3 Retinal uptake index (RUI) of [^{3}H]D-mannitol, [^{3}H]Gly-Sar, [^{3}H]β-alanine and [^{3}H]L-histidine. (A) The RUI values were estimated from data obtained from the carotid injection analysis of rats, in which [^{3}H]D-mannitol, [^{3}H]Gly-Sar, [^{3}H]β-alanine or [^{3}H]L-histidine were injected into rats with a diffusible internal reference compound, [^{14}C]*n*-butanol. Unlabeled Gly-Sar, at a concentration of 100 mM, was coinjected to investigate its inhibitory effect on the RUI value of [^{3}H]Gly-Sar. The inhibitory effects for [^{3}H]β-alanine (B) and [^{3}H]L-histidine (C) were examined in the presence of 10 mM compounds, and were expressed as a percentage of the control. Each value represents the means ± S.E.M. (n = 3–7). *p < 0.01, significantly different from the control. The figure was prepared by reference to Nagase *et al.* (2006) and Usui *et al.* (2013). Gly-Sar, glycylsarcosine; GABA, γ-aminobutyric acid.

dipeptide was also supported by a cellular uptake study with TR-iBRB2 cells and primary-cultured RPE cells, which are *in vitro* models of the inner and outer BRB, respectively, and no concentration- or temperature-dependent uptake of [^{3}H]Gly-Sar was exhibited by these cells, suggesting the minor contribution of carrier-mediated dipeptide transport at the inner and outer BRB. TR-iBRB2 cells and primary-cultured RPE cells exhibited initial uptake rates of 0.0374 μL per min per mg per protein and 0.0385 μL per min per mg per protein (Table 28.1), and these values were much lower than that of dopamine (0.684 μL per min per mg per protein) which undergoes passive diffusion transport. These findings support the minor contribution of a carrier-mediated process to the transport of dipeptides, suggesting that the carrier-mediated transport of L-carnosine is minor at the BRB.

28.6 Is L-Carnosine Synthesized from β-alanine and L-histidine in the Retina?

Our *in vivo* and *in vitro* analyses suggested that dipeptides, such as Gly-Sar and L-carnosine, are not directly transported into the retina across the BRB (Usui *et al.*, 2013), and it has been proposed that L-carnosine is synthesized from β-alanine and L-histidine to express its physiological or pharmacological effects in the retina. In the transporter research field, carrier-mediated transport of β-alanine and L-histidine is known, and it has been reported that β-alanine is recognized and transported by betaine/GABA transporter (BGT-1/SLC6A12), GABA transporters (GAT-1/SLC6A1, GAT-2/SLC6A13 and GAT-3/SLC6A11), system A (SLC38A1-4) and taurine transporter (TAUT/SLC6A6) (Smith *et al.*, 1992; Borden, 1996; Matskevitch *et al.*, 1999; Sugawara *et al.*, 2000), while L-histidine is transported by system y^+L (y^+LAT2/SLC7A6, y^+LAT1/SLC7A7), cationic amino acid transporters CATs (SLC7A1-4) and L (leucine-referring)-type amino acid transporters (LAT1/SLC7A5, LAT2/SLC7A8) (Kanai *et al.*, 1998; Pineda *et al.*, 1999; Segawa *et al.*, 1999). These findings strongly supported the possible involvement of influx transport systems for β-alanine and L-histidine in the retinal function of L-carnosine in spite of the unclear expression of carnosine synthase in the retina.

28.7 Influx Transport of β-alanine and L-Histidine across the BRB

The involvement of carrier-mediated transport has been suggested in the influx transport of various amino acids. An *in vivo* study in rats demonstrated greater retinal permeability of L-serine, L-proline, L-leucine, L-arginine, glycine and L-cystine compared with that of a nonpermeable paracellular marker, and cellular and molecular biological investigations have revealed that such carrier-mediated amino acid transport is carried out by LAT1, CAT1, GlyT1, amino acid transporter A2 (ATA2/SLC38A2) and cystine/glutamate exchanger

Table 28.1 Kinetic parameters for the uptake of [^{3}H]Gly-Sar, [^{3}H]β-alanine and [^{3}H]L-histidine. The values were estimated from the data obtained in the uptake study of [^{3}H]Gly-Sar (0.1 μCi, 18 nM), [^{3}H]β-alanine (0.1 μCi, 10 nM) and [^{3}H]L-histidine (0.1 μCi, 10.4 nM) by TR-iBRB2 cells and primary-cultured retinal pigment epithelial (RPE) cells. N.D., not determined; Gly-Sar, glycylsarcosine. The table was prepared by reference to Usui *et al.* (2013).

Substrates	Cells	Initial uptake rate (μL per min per mg per protein)	K_m (μM)	V_{max} (nmol per min per mg per protein)	K_d (μL per min per mg per protein)	Temperature-dependence	Ion dependence
[^{3}H]Gly-Sar	TR-iBRB2 cells	0.0374	N.D.	N.D.	N.D.	Independent	N.D.
	RPE cells	0.0385	N.D.	N.D.	N.D.	Independent	N.D.
[^{3}H]β-Alanine	TR-iBRB2 cells	10.5	124	3.49	N.D.	Dependent	Na^{+}- and Cl^{-}-dependent
[^{3}H]L-Histidine	TR-iBRB2 cells	38.5	86.3	4.74	3.14	Dependent	Na^{+}-independent

(xCT/SLC7A11) expressed at the BRB (Tomi *et al.*, 2002; Tomi *et al.*, 2005; Tomi *et al.*, 2009; Okamoto *et al.*, 2009; Yoneyama *et al.*, 2010; Tachikawa *et al.*, 2011). In our study (Usui *et al.*, 2013), [^{3}H]β-alanine had an 18-fold greater RUI value (212%) than that of a nonpermeable paracellular marker, and its RUI value was decreased in the presence of taurine and GABA (a reduction of more than 53%) (Figures 28.3A and B). Regarding L-histidine, the RUI values of [^{3}H]L-histidine (268%) was shown to be 23-fold greater than that of a nonpermeable paracellular marker, and this value was reduced in the presence of L-leucine (a reduction of more than 61%) (Figures 28.3A and C). These findings suggest the involvement of taurine- and GABA-sensitive, and the L-leucine-sensitive carrier-mediated transport processes in the influx transport of β-alanine and L-histidine, respectively, across the BRB.

28.8 Carrier-Mediated Influx Transport of β-Alanine at the Inner BRB

In an *in vitro* uptake study with TR-iBRB2 cells, the transport of β-alanine at the inner BRB was investigated (Usui *et al.*, 2013). The results obtained supported the involvement of a carrier-mediated process in the influx transport of β-alanine at the inner BRB since TR-iBRB2 cells showed that the uptake of [^{3}H]β-alanine took place in a temperature-, concentration-, Na^+- and Cl^--dependent manner (Table 28.1). The kinetic parameters (K_m and V_{max}) were estimated to be 124 μM and 3.49 nmol per min per mg per protein, respectively (Table 28.1). The [^{3}H]β-alanine uptake was inhibited in the presence of taurine, β-guanidinopropionic acid (GPA), GABA, guanidinoacetic acid (GAA) and nipecotic acid (a reduction of more than 34%) in spite of the weak effects observed in the presence of betaine and L-alanine (Figure 28.4A). In the transporter research field, it is known that taurine and GPA are substrates of TAUT and GABA transporters (GAT-1, GAT-2 and GAT-3). In addition, it is known that GAA and nipecotic acid are substrates of GAT-2 and GAT-3, while betaine and L-alanine are substrates of BGT-1 and system A, suggesting the possible contribution of TAUT, GAT-2 and GAT-3 to the influx transport of β-alanine at the inner BRB (Smith *et al.*, 1992; Borden, 1996; Matskevitch *et al.*, 1999; Sugawara *et al.*, 2000).

28.9 Carrier-Mediated Influx Transport of L-Histidine across the Inner BRB

Our *in vivo* analysis revealed the contribution of a carrier-mediated transport process to the influx transport of L-histidine at the BRB (Usui *et al.*, 2013). The *in vitro* uptake study supports the carrier-mediated transport of L-histidine at the inner BRB, and TR-iBRB2 cells exhibited temperature- and concentration-dependent uptake of [^{3}H]L-histidine (Table 28.1). The analysis of kinetic parameters showed that the K_m, V_{max}, and K_d were 86.3 μM, 4.74 nmol per min per mg per protein and 3.14 μL per min per mg per protein,

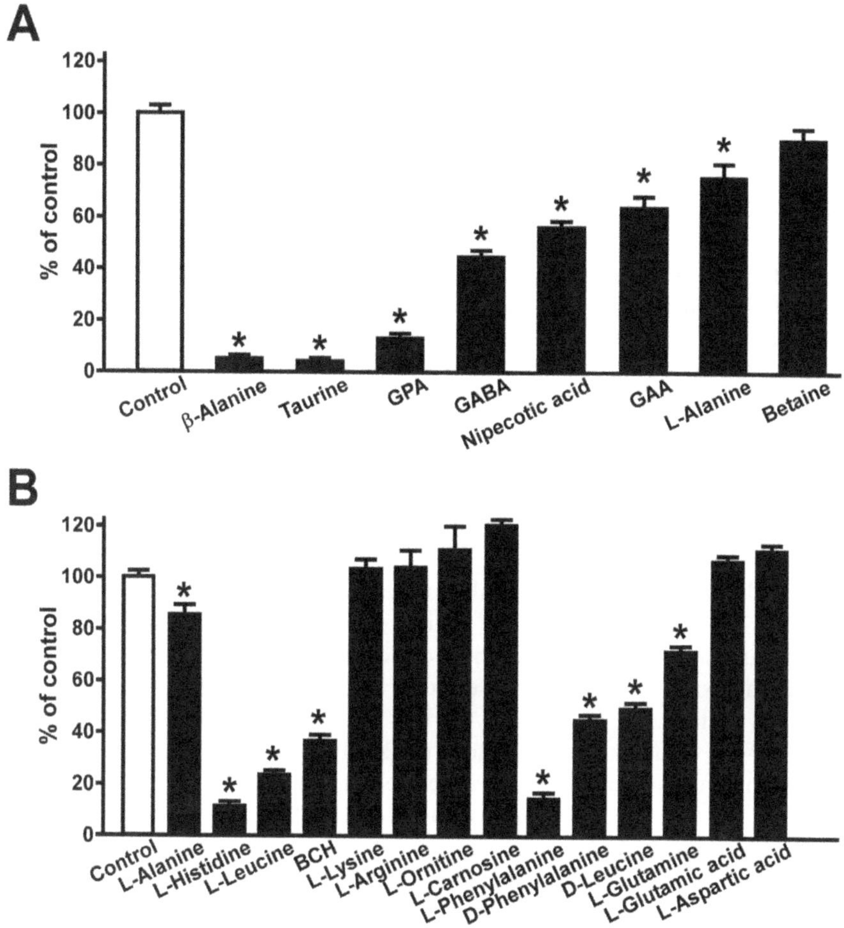

Figure 28.4 Inhibitory effects on the uptake of [^{3}H]β-alanine (A) and [^{3}H]L-histidine (B) by TR-iBRB2 cells. [^{3}H]β-Alanine and [^{3}H]L-histidine uptake by TR-iBRB2 cells was performed in the absence (control) or presence of 1 mM compounds at 37 °C. Each value represents the means ± S.E.M. ($n = 3$–9). $^*p < 0.01$, significantly different from the control. BCH, 2-aminobicyclo-2(2,2,1)-heptane-2-caboxylic acid; GABA, γ-aminobutyric acid; GPA, β-guanidinopropionic acid; GAA, guanidinoacetic acid. The figure was prepared by reference to Usui *et al.* (2013).

respectively (Table 28.1), and this suggests the involvement of a nonsaturable process (K_d) in the transport of L-histidine at the inner BRB. The L-histidine transport clearance (V_{max}/K_m) was estimated to be 56.7 μL per min per mg per protein, which is approximately 18-fold greater than the K_d values. In addition, the contribution of the carrier-mediated process to L-histidine transport was estimated to be 89% at the physiological plasma concentration of L-histidine (approximately 100 μM), supporting a major contribution of the carrier-mediated process to the influx transport of L-histidine at the inner BRB. In the *in vitro* inhibition study, the uptake of [^{3}H]L-histidine by TR-iBRB2 cells was inhibited in the presence of L-leucine and 2-aminobicyclo-(2,2,1)-heptane-2-carboxylic acid (BCH) (a reduction of more than 62%) in spite of no significant effect being observed in the presence of L-arginine, L-lysine and L-ornithine (Figure 28.4B). It is known that L-arginine, L-lysine and L-ornithine are substrates of CATs and y$^+$LATs (Kanai *et al.*, 1998; Pineda *et al.*, 1999; Segawa *et al.*, 1999; Tomi *et al.*, 2009), and the results obtained suggest the minor involvement of CATs and y$^+$LATs in the influx transport of L-histidine at the inner BRB although CAT1 is expressed at the inner BRB. On the other hand, L-leucine and BCH are known to inhibit the function of LATs (Kanai *et al.*, 1998; Pineda *et al.*, 1999; Segawa *et al.*, 1999; Tomi *et al.*, 2009), supporting the possible involvement of LATs in the influx transport of L-histidine across the inner BRB.

28.10 TAUT and LAT1 are Involved in Retinal L-Carnosine Synthesis

The cellular uptake study of β-alanine showed the possible involvement of TAUT, GAT-2 and GAT-3 in the influx transport of β-alanine at the inner BRB. In our study, the concentration-dependent inhibition of [^{3}H]β-alanine uptake was examined in the presence of taurine and GABA, respectively. It is known that the K_m value of GATs has been reported to be less than 12 μM for GABA, and that the K_m values of TAUT are reported to be 43 μM and 1.5 mM for taurine and GABA, respectively. The comparison of IC_{50} and K_m values suggested the major involvement of TAUT in the influx transport of β-alanine across the inner BRB since the IC_{50} values of GABA and taurine in [^{3}H]β-alanine uptake have been estimated to 612 μM and 21.8 μM, respectively (Clark *et al.*, 1992; Smith *et al.*, 1992; Liu *et al.*, 1993; Borden, 1996; Tomi *et al.*, 2008; Usui *et al.*, 2013) (Figure 28.5). The expression of TAUT at the inner BRB has previously been reported (Tomi *et al.*, 2008), and the involvement of TAUT is also supported by the Na^+- and Cl^--dependence shown in a functional study of TAUT. Regarding the influx transport of L-histidine at the inner BRB, the possible involvement of LATs has been shown (Usui *et al.*, 2013). It is known that LAT1 and LAT2 have different sensitivities to several amino acids such as D-phenylalanine, D-leucine, L-alanine and L-glutamine (Kanai *et al.*, 1998; Segawa *et al.*, 1999), and a study of the differential inhibitory effect with these amino acids suggests the major involvement of LAT1 since [^{3}H]L-histidine uptake by TR-iBRB2 cells was significantly inhibited in the presence of D-phenylalanine and D-leucine (a reduction of more than 50%) in spite of the weaker inhibitory effect observed in the presence of L-alanine and L-glutamine (a reduction of more than 13%) (Figures 28.4B and 28.5). The expression of LAT1 has previously been reported at the inner BRB to support the involvement of LAT1 in the influx transport of L-histidine across the inner BRB (Tomi *et al.*, 2005).

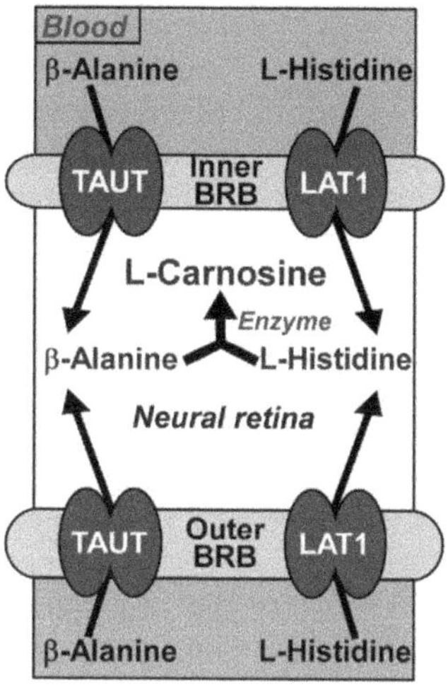

Figure 28.5 Involvement of LAT1 and TAUT in the retinal supply of L-carnosine. In the retina, L-carnosine is suggested to be synthesized from L-histidine and β-alanine transported by LAT1 and TAUT expressed at the inner and outer BRB.

28.11 β-Alanine and L-Histidine Transport at the Outer BRB

The *in vitro* uptake study with TR-iBRB2 cells suggests the involvement of LAT1 and TAUT in the influx transport of L-histidine and β-alanine at the inner BRB, respectively. A study with ARPE-19 cells, the cultured cell line of human retinal pigment epithelial cells, reported that the contribution of carrier-mediated dipeptide transport is minor at the outer BRB (Ocheltree *et al.*, 2003), suggesting the possible significance of L-histidine and β-alanine transport at the outer BRB. Previously, reverse transcription polymerase chain reaction (RT-PCR) and Western blot analyses of ARPE-19 clearly showed the expression of LAT1 and TAUT at the outer BRB (El-Sherbeny *et al.*, 2004; Yamamoto *et al.*, 2010), and these reports suggest that LAT1 and TAUT are involved in the influx transport of L-histidine and β-alanine at the outer BRB, respectively, contributing to the supply of L-carnosine to the retina.

28.12 Concluding Remarks

Although several studies have described the transport of L-carnosine, the mechanism by which L-carnosine is supplied to the retina is unclear. Our *in vivo* and *in vitro* studies suggest the minor contribution of carrier-mediated oligopeptide transport processes, including members of the SLC15A family, to the supply of L-carnosine to the neural retina from the circulating blood. Therefore, it was hypothesized that L-carnosine is enzymatically synthesized from β-alanine and L-histidine in the retina. The *in vivo* studies suggest the carrier-mediated influx transport of β-alanine and L-histidine at the BRB, and the *in vitro* studies suggest the involvement of taurine transporter (TAUT) and L-type amino acid transporter (LAT1) in the influx transport of β-alanine and L-histidine at the inner BRB, respectively (Figure 28.5). In addition to the expression of TAUT and LAT1 at the inner BRB, their expression at the outer BRB has also been reported, and these findings strongly support the hypothesis that the retinal supply of L-carnosine involves the carrier-mediated transport of β-alanine and L-histidine across the inner and outer BRB. In addition to the retinal biosynthesis and physiological effects of L-carnosine, pharmacological activities of β-alanine and L-histidine have been suggested in the retina, and β-alanine has been reported to function as an agonist of GABA-C receptors expressed in the retina, and the enzymatic conversion of L-histidine into L-histamine has been reported to be involved in flush sensitivity regulation under dark-adapted conditions, supporting the physiological significance of these transport systems in retinal function. The findings described in this chapter suggest the importance of carrier-mediated β-alanine and L-histidine transport across the BRB in the molecular function of L-carnosine in the retina.

Summary Points

- This chapter focuses on the supply of L-carnosine across the BRB.
- *In vivo* and *in vitro* studies of Gly-Sar transport suggests that the carrier-mediated transport of dipeptide, including L-carnosine, is minor at the BRB.
- Carrier-mediated transport processes are involved in the influx transport of β-alanine and L-histidine across the BRB.
- Taurine transporter (TAUT) is involved in the influx transport of β-alanine at the inner BRB
- L-Type amino acid transporter 1 (LAT1) is involved in the influx transport of L-histidine at the inner BRB
- L-Carnosine is not directly transported into the retina, and is suggested to be synthesized from β-alanine and L-histidine in the retina.
- The expression of LAT1 and TAUT in RPE cells supports their involvement in the influx transport of β-alanine and L-histidine across the outer BRB.
- TAUT and LAT1 are involved in the retinal biosynthesis of L-carnosine from β-alanine and L-histidine.

Key Facts

Key Features of Dipeptide Transport at the Inner and Outer BRB

1. TR-iBRB2 cells and primary-cultured RPE cells derived from rat were used as *in vitro* models of the inner and outer BRB (Hosoya *et al.*, 2001).
2. Gly-Sar is a representative dipeptide and [^{3}H]Gly-Sar (Moravek Biochemicals, USA) was used as a test compound to investigate dipeptide transport.

3. Carrier-mediated transport of dipeptides, including L-carnosine, was known to include a temperature- and concentration-dependent function (Ganapathy and Leibach, 1982; Poschet *et al.*, 1996; Basu *et al.*, 1998; Nakanishi *et al.*, 2000; Daniel and Kottra, 2004; Son *et al.*, 2004).
4. Retinal uptake index (RUI) analysis was performed as an *in vivo* method to investigate the blood-to-retina transport of dipeptides.
5. [^{3}H]D-Mannitol was used as a nonpermeable paracellular marker, and its RUI value was 11.6% (Nagase *et al.*, 2006).

Key Features of β-Alanine and L-Histidine Transport at the BRB

1. RUI analysis was performed to investigate the blood-to-retina transport of L-histidine and β-alanine across the BRB.
2. L-[ring-2,5-^{3}H]Histidine ([^{3}H]L-histidine, American Radiolabeled Chemicals, USA) and β-[3-^{3}H]alanine ([^{3}H]β-alanine, Moravek Biochemicals, USA) were used as a test compound to investigate the transport of L-histidine and β-alanine at the BRB, respectively.

Key Features of L-histidine Transport at the Inner BRB

1. The influx transport of L-histidine at the inner BRB was examined in a cellular uptake study with TR-iBRB2 cells, *in vitro* model of the inner BRB.
2. Membrane transporters, such as y$^+$LATs, CATs and LATs, recognize L-histidine as a substrate (Kanai *et al.*, 1998; Pineda *et al.*, 1999; Segawa *et al.*, 1999).
3. L-Leucine and BCH were used as inhibitors of LATs, and L-arginine, L-lysine and L-ornithine were used as inhibitors of CATs and y$^+$LATs (Kanai *et al.*, 1998; Pineda *et al.*, 1999; Segawa *et al.*, 1999; Tomi *et al.*, 2009).

Key Facts about β-Alanine Transport at the Inner BRB

1. The influx transport of β-alanine at the inner BRB was examined by a cellular uptake study with TR-iBRB2 cells, *in vitro* model of the inner BRB.
2. Membrane transporters, such as BGT-1, GATs, system A and TAUT, recognize β-alanine as a substrate (Smith *et al.*, 1992; Borden, 1996; Matskevitch *et al.*, 1999; Sugawara *et al.*, 2000).
3. Taurine and GPA were used as inhibitors of TAUT, GAT-1, GAT-2 and GAT-3 (Smith *et al.*, 1992; Borden, 1996).
4. GAA and nipecotic acid were used as inhibitors of GAT-2 and GAT-3 transporters (Borden, 1996), and betaine and L-alanine were used as inhibitors of BGT-1 and system A (Matskevitch *et al.*, 1999; Sugawara *et al.*, 2000).

Key Facts about Transporter Involved in the Supply of L-Carnosine to the Retina

1. The K_m values of TAUT are reported to be 43 μM and 1.5 mM for taurine and GABA, and the K_m values of GAT-1, 2 and 3 are reported to be less than 12 μM for GABA (Clark *et al.*, 1992; Smith *et al.*, 1992; Liu *et al.*, 1993; Borden, 1996; Tomi *et al.*, 2008).
2. D-Phenylalanine and D-leucine were used as the inhibitors of LAT1, and L-alanine and L-glutamine were used as inhibitors of LAT2 (Kanai *et al.*, 1998; Segawa *et al.*, 1999).
3. The IC_{50} values of taurine and GABA for [^{3}H]β-alanine uptake by TR-iBRB2 cells were estimated to be 21.8 μM and 612 μM, respectively (Usui *et al.*, 2013).
4. LAT1 and TAUT are reported to be expressed by TR-iBRB2 cells, an *in vitro* model of the inner BRB (Tomi *et al.*, 2005; Tomi *et al.*, 2008).
5. LAT1 and TAUT are shown to be expressed by ARPE-19 cells, an *in vitro* model of the outer BRB (El-Sherbeny *et al.*, 2004; Yamamoto *et al.*, 2010).

Definition of Words and Terms

β-Alanine. β-Alanine is one of the β-amino acids, that is involved in the biosynthesis of L-carnosine. The plasma concentration of β-alanine is estimated to be approximately 20 μM.

Blood–retinal barrier. The neural retina and the circulating blood are separated by a two-barrier structure, the inner and outer BRB, to maintain an optimal environment for retinal function (Stewart and Tuor, 1994; Cunha-Vaz, 2004; Hosoya *et al.*, 2011).

L-Histidine. L-Histidine is one of the essential amino acids, and is involved in the biosynthesis of L-carnosine. The plasma concentration of L-histidine is estimated to be approximately 100 μM.

Inner BRB. Barrier structure formed by the retinal capillary endothelial cells, the tight junction of which limits nonspecific paracellular solute transport (Hosoya *et al.*, 2011).

L-Type amino acid transporter 1 (LAT1). This membrane transporter belongs to the SLC7A family, and is one of the molecules responsible for system L-amino acid transport. LAT1 forms a heterodimer with a 4F2 heavy chain to recognize large neutral amino acids, such as L-leucine, L-isoleucine, L-valine, L-phenylalanine and L-histidine as substrates (Kanai *et al.*, 1998; Tomi *et al.*, 2005).

Membrane transporter. Membrane transporters can be roughly classified into two groups, ATP-binding cassette (ABC) and solute carrier (SLC) transporters. In particular, SLC transporters include influx transporters for glucose, amino acids, nucleosides, vitamins, peptides, cationic drugs, and anionic drugs (Hosoya *et al.*, 2011; Kubo *et al.*, 2013).

Oligopeptide transporter. The SLC15A family includes H^+-driven oligopeptide transporters that recognize di- and tripeptides as substrates (Nakanishi *et al.*, 2000; Daniel and Kottra, 2004).

Outer BRB. Barrier structure formed by retinal pigment epithelial (RPE) cells, the tight junction of which limits nonspecific paracellular solute transport (Hosoya *et al.*, 2011).

Retinal uptake index (RUI). The RUI value can be estimated from data obtained by the carotid artery injection method, and used as an indicator for the retinal permeability of endogenous and exogenous compounds.

Taurine transporter (TAUT). This membrane transporter belongs to the SLC6A family, and recognizes taurine and β-alanine as substrates, suggesting its involvement in cellular osmoregulation (Smith *et al.*, 1992).

TR-iBRB2 cells. A conditionally immortalized retinal capillary endothelial cell line derived from transgenic rats harboring temperature-sensitive simian virus 40 large T-antigen gene is used as an *in vitro* model of the inner BRB (Hosoya *et al.*, 2001).

List of Abbreviations

BRB	Blood–retinal barrier
Bax	Bcl-2-associated X protein
BBMV	Brush-border membrane vesicle
BCH	2-Aminobicyclo-(2,2,1)-heptane-2-carboxylic acid
Bcl-2	B-Cell lymphoma 2
BCSFB	Blood-cerebrospinal fluid barrier
BGT-1	Betaine/GABA transporter
CATs	Cationic amino acid transporters
Drp-1	Dynamin-related protein-1
GAA	Guanidinoacetic acid
GAT	GABA transporter
GFAP	Glial fibrillary acidic protein
Gly-Sar	Glycylsarcosine
GPA	β-Guanidinopropionic acid
HIF-1α	Hypoxia-inducing factor-1 α
K_d	Nonsaturable uptake rate in Michaelis–Menten kinetics
K_m	Michaelis constant in Michaelis–Menten kinetics
LATs	L-Type amino acid transporters
RPE cells	Retinal pigment epithelial cell
RT-PCR	Reverse transcription polymerase chain reaction
RUI	Retinal uptake index
TAUT	Taurine transporter
V_{max}	Maximal uptake rate in Michaelis–Menten kinetics
y^+LATs	y^+L-Type amino acid transporters

References

Atluri, H., Anand, B. S., Patel, J. and Mitra, A. K., 2004. Mechanism of a model dipeptide transport across blood–ocular barriers following systemic administration. *Experimental Eye Research*. 78: 815–822.

Basu, S. K., Haworth, I. S., Bolger, M. B. and Lee, V. H., 1998. Proton-driven dipeptide uptake in primary cultured rabbit conjunctival epithelial cells. *Investigative Ophthalmology & Visual Science*. 39: 2365–2373.

Borden, L. A., 1996. GABA transporter heterogeneity: pharmacology and cellular localization. *Neurochemistry International*. 29: 335–356.

Clark, J. A., Deutch, A. Y., Gallipoli, P. Z. and Amara, S. G., 1992. Functional expression and CNS distribution of a beta-alanine-sensitive neuronal GABA transporter. *Neuron*. 9: 337–348.

Cunha-Vaz, J. G., 2004. The blood–retinal barriers system. Basic concepts and clinical evaluation. *Experimental Eye Research*. 78: 715–721.

Daniel, H. and Kottra, G., 2004. The proton oligopeptide cotransporter family SLC15 in physiology and pharmacology. *Pfluegers Archiv*. 447: 610–618.

El-Sherbeny, A., Naggar, H., Miyauchi, S., Ola, M. S., Maddox, D. M., Martin, P. M., Ganapathy, V. and Smith, S. B., 2004. Osmoregulation of taurine transporter function and expression in retinal pigment epithelial, ganglion, and müller cells. *Investigative Ophthalmology & Visual Science*. 45: 694–701.

Ganapathy, V. and Leibach, F. H., 1982. Peptide transport in rabbit kidney. Studies with L-carnosine. *Biochimica et Biophysica Acta*. 691: 362–366.

Hosoya, K., Tomi, M., Ohtsuki, S., Takanaga, H., Ueda, M., Yanai, N., Obinata, M. and Terasaki, T., 2001. Conditionally immortalized retinal capillary endothelial cell lines (TR-iBRB) expressing differentiated endothelial cell functions derived from a transgenic rat. *Experimental Eye Research*. 72: 163–172.

Hosoya, K., Tomi, M. and Tachikawa, M., 2011. Strategies for therapy of retinal diseases using systemic drug delivery: relevance of transporters at the blood–retinal barrier. *Expert Opinion on Drug Delivery*. 8: 1571–1587.

Ji, Y. S., Park, J. W., Heo, H., Park, J. S. and Park, S. W., 2013. The Neuroprotective Effect of Carnosine (β-Alanyl-L-Histidine) on Retinal Ganglion Cell Following Ischemia-Reperfusion Injury. *Current Eye Research*. 39: 634–641.

Kanai, Y., Segawa, H., Miyamoto, K., Uchino, H., Takeda, E. and Endou, H., 1998. Expression cloning and characterization of a transporter for large neutral amino acids activated by the heavy chain of 4F2 antigen (CD98). *Journal of Biological Chemistry*. 273: 23629–23632.

Kubo, Y., Shimizu, Y., Kusagawa, Y., Akanuma, S. and Hosoya, K., 2013. Propranolol transport across the inner blood–retinal barrier: potential involvement of a novel organic cation transporter. *Journal of Pharmaceutical Sciences*. 102: 3332–3342.

Kumagai, A. K., Vinores, S. A. and Pardridge, W. M., 1996. Pathological upregulation of inner blood–retinal barrier Glut1 glucose transporter expression in diabetes mellitus. *Brain Research*. 706: 313–317.

Lin, H. and King, N., 2007. Demonstration of functional dipeptide transport with expression of PEPT2 in guinea pig cardiomyocytes. *Pfluegers Archiv*. 453: 915–922.

Liu, Q. R., López-Corcuera, B., Mandiyan, S., Nelson, H. and Nelson, N., 1993. Molecular characterization of four pharmacologically distinct gamma-aminobutyric acid transporters in mouse brain. *Journal of Biological Chemistry*. 268: 2106–2112.

Margolis, F. L. and Grillo, M., 1984. Carnosine, homocarnosine and anserine in vertebrate retinas. *Neurochemistry International*. 6: 207–209.

Matskevitch, I., Wagner, C. A., Stegen, C., Bröer, S., Noll, B., Risler, T., Kwon, H. M., Handler, J. S., Waldegger, S., Busch, A. E. and Lang, F., 1999. Functional characterization of the Betaine/gamma-aminobutyric acid transporter BGT-1 expressed in Xenopus oocytes. *Journal of Biological Chemistry*. 274: 16709–16716.

Nagase, K., Tomi, M., Tachikawa, M. and Hosoya, K., 2006. Functional and molecular characterization of adenosine transport at the rat inner blood–retinal barrier. *Biochimica et Biophysica Acta*. 1758: 13–19.

Nakanishi, T., Tamai, I., Takaki, A. and Tsuji, A., 2000. Cancer cell-targeted drug delivery utilizing oligopeptide transport activity. *International Journal of Cancer*. 88: 274–280.

Ocheltree, S. M., Keep, R. F., Shen, H., Yang, D., Hughes, B. A. and Smith, D. E., 2003. Preliminary investigation into the expression of proton-coupled oligopeptide transporters in neural retina and retinal pigment epithelium (RPE): lack of functional activity in RPE plasma membranes. *Pharmaceutical Research*. 20: 1364–1372.

Ohkura, Y., Akanuma, S., Tachikawa, M. and Hosoya, K., 2010. Blood-to-retina transport of biotin via Na^+-dependent multivitamin transporter (SMVT) at the inner blood–retinal barrier. *Experimental Eye Research*. 91: 387–392.

Okamoto, M., Akanuma, S., Tachikawa, M. and Hosoya, K., 2009. Characteristics of glycine transport across the inner blood–retinal barrier. *Neurochemistry International*. 55: 789–795.

Panzanelli, P., Cantino, D. and Sassòe-Pognetto, M., 1997. Co-localization of carnosine and glutamate in photoreceptors and bipolar cells of the frog retina. *Brain Research*. 758: 143–152.

Pfister, F., Riedl, E., Wang, Q., vom Hagen, F., Deinzer, M., Harmsen, M. C., Molema, G., Yard, B., Feng, Y. and Hammes, H. P., 2011. Oral carnosine supplementation prevents vascular damage in experimental diabetic retinopathy. *Cellular Physiology and Biochemistry*. 28: 125–136.

Pineda, M., Fernández, E., Torrents, D., Estévez, R., López, C., Camps, M., Lloberas, J., Zorzano, A. and Palacín, M., 1999. Identification of a membrane protein, LAT-2, that Co-expresses with 4F2 heavy chain, an L-type amino acid transport activity with broad specificity for small and large zwitterionic amino acids. *Journal of Biological Chemistry*. 274: 19738–19744.

Poschet, J. F., Hammond, S. M. and Fairclough, P. D., 1996. Characterisation of penicillin-G uptake in rabbit small-intestinal brush-border membrane vesicles. *Biochimica et Biophysica Acta*. 1278: 233–240.

Segawa, H., Fukasawa, Y., Miyamoto, K., Takeda, E., Endou, H. and Kanai, Y., 1999. Identification and functional characterization of a Na^+-independent neutral amino acid transporter with broad substrate selectivity. *Journal of Biological Chemistry*. 274: 19745–19751.

Shu, C., Shen, H., Teuscher, N. S., Lorenzi, P. J., Keep, R. F. and Smith, D. E., 2002. Role of PEPT2 in peptide/mimetic trafficking at the blood-cerebrospinal fluid barrier: studies in rat choroid plexus epithelial cells in primary culture. *Journal of Pharmacology and Experimental Therapeutics*. 301: 820–829.

Smith, K. E., Borden, L. A., Wang, C. H., Hartig, P. R., Branchek, T. A. and Weinshank, R. L., 1992. Cloning and expression of a high affinity taurine transporter from rat brain. *Molecular Pharmacology*. 42: 563–569.

Son, D. O., Satsu, H., Kiso, Y. and Shimizu, M., 2004. Characterization of carnosine uptake and its physiological function in human intestinal epithelial Caco-2 cells. *BioFactors*. 21: 395–398.

Stewart, P. A. and Tuor, U. I., 1994. Blood-eye barriers in the rat: correlation of ultrastructure with function. *Journal of Comparative Neurology*. 340: 566–576.

Sugawara, M., Nakanishi, T., Fei, Y. J., Martindale, R. G., Ganapathy, M. E., Leibach, F. H. and Ganapathy, V., 2000. Structure and function of ATA3, a new subtype of amino acid transport system A, primarily expressed in the liver and skeletal muscle. *Biochimica et Biophysica Acta*. 1509: 7–13.

Tachikawa, M., Okamoto, M., Hirose, S., Yoneyama, D., Akanuma, S., Terasaki, T. and Hosoya, K., 2011. Inner blood–retinal barrier mediates l-isomer-predominant transport of serine. *Journal of Pharmaceutical Sciences*. 100: 3892–3903.

Tomi, M., Hosoya, K., Takanaga, H., Ohtsuki, S. and Terasaki, T., 2002. Induction of xCT gene expression and L-cystine transport activity by diethyl maleate at the inner blood–retinal barrier. *Investigative Ophthalmology & Visual Science*. 43: 774–779.

Tomi, M., Mori, M., Tachikawa, M., Katayama, K., Terasaki, T. and Hosoya, K., 2005. L-Type amino acid transporter 1-mediated L-leucine transport at the inner blood–retinal barrier. *Investigative Ophthalmology & Visual Science*. 46: 2522–2530.

Tomi, M., Tajima, A., Tachikawa, M. and Hosoya, K., 2008. Function of taurine transporter (Slc6a6/TauT) as a GABA transporting protein and its relevance to GABA transport in rat retinal capillary endothelial cells. *Biochimica et Biophysica Acta*. 1778: 2138–2142.

Tomi, M., Kitade, N., Hirose, S., Yokota, N., Akanuma, S., Tachikawa, M. and Hosoya, K., 2009. Cationic amino acid transporter 1-mediated L-arginine transport at the inner blood–retinal barrier. *Journal of Neurochemistry*. 111: 716–725.

Usui, T., Kubo, Y., Akanuma, S. and Hosoya, K., 2013. β-Alanine and L-histidine transport across the inner blood–retinal barrier: potential involvement in L-carnosine supply. *Experimental Eye Research*. 113: 135–142.

Yamamoto, A., Akanuma, S., Tachikawa, M. and Hosoya, K., 2010. Involvement of LAT1 and LAT2 in the high- and low-affinity transport of L-leucine in human retinal pigment epithelial cells (ARPE-19 cells). *Journal of Pharmaceutical Sciences*. 99: 2475–2482.

Yoneyama, D., Shinozaki, Y., Lu, W. L., Tomi, M., Tachikawa, M. and Hosoya, K., 2010. Involvement of system A in the retina-to-blood transport of L-proline across the inner blood–retinal barrier. *Experimental Eye Research*. 90: 507–513.

CHAPTER 29

L-Carnosine and Zinc in Gastric Protection

HYUK SOON CHOI[a], EUN SUN KIM[a], BORA KEUM[a], HOON JAI CHUN*[a], AND MI-KYUNG SUNG[b]

[a]Division of Gastroenterology and Hepatology, Department of Internal Medicine, Institute of Digestive Disease and Nutrition, Korea University College of Medicine, Inchon-ro 73, Seongbuk-gu, Seoul, 136-705, Korea; [b]Department of Food and Nutrition, Sookmyung Women's University, 100 Cheongpa-ro 47-gil, Yongsan-gu, Seoul, 140-742, Korea
*E-mail: drchunhj@chol.com

29.1 Introduction

Zinc L-carnosine (or polaprezinc) is a chelated compound that consists of L-carnosine and zinc (Choi *et al.*, 2013). Zinc is an essential mineral and a component of various important enzymes required for different biological systems. These zinc-dependent enzymes play crucial roles in cell proliferation during tissue repair through DNA and RNA polymerases. Therefore, zinc is useful in the process of wound healing in cases of skin, connective tissue, and gastric ulcer (Watanabe *et al.*, 1998). L-Carnosine is well known to accelerate gastric ulcer healing and to increase tissue granulation in rats. Zinc L-carnosine was designed to merge the beneficial effects of L-carnosine and zinc for treating gastric mucosal injury. L-Carnosine acts as a catalyst

Food and Nutritional Components in Focus No. 8
Imidazole Dipeptides: Chemistry, Analysis, Function and Effects
Edited by Victor R Preedy

Published by the Royal Society of Chemistry, www.rsc.org

maximizing the therapeutic effect of zinc L-carnosine during gastric mucosa injury.

Recent studies have shown that L-carnosine is not only involved in gastric mucosal protection but also in the protection of other organs, especially in inflammatory colon disease and disorders of the oral mucosa and skin (Sakae *et al.*, 2013; Itagaki *et al.*, 2014). Other studies showed that zinc L-carnosine is beneficial in the treatment of idiopathic taste disorder, aspirin-induced small-bowel injury, ulcerative colitis, liver injury such as hepatitis C, and liver fibrosis. Basic studies on zinc L-carnosine have also showed sufficient evidence of its therapeutic effects on other organs (Table 29.1). This chapter focuses on the efficacy of L-carnosine and zinc in gastric mucosal protection.

29.2 L-Carnosine and Zinc

Zinc L-carnosine (or polaprezinc) is a chelated complex that consists of L-carnosine and zinc (Figure 29.1). *N*-β-Alanyl-L-histidine (L-carnosine) is present in the muscles of vertebrates, in the form of a dipeptide, which is also present in consumed meats. With regard to pharmacological action, L-carnosine is a biomolecule involved in immunomodulating and promoting tissue-restoration action. In particular, the antiulcer effect of L-carnosine has been reported in ethanol- and cortisone-induced gastric ulcers. Carnosine is also associated with cardiac muscle, eye lens, and nervous tissues, including the brain, especially the olfactory lobe. The protective activity of carnosine has been reported in wound healing, diabetes, osteoporosis, loss of vision, and immune function. Carnosine possesses many biochemical homeostatic functions such as pH buffering, antioxidant effect, antiglycation capability, metal-ion chelating, and carbonyl scavenging (Hipkiss, 2009). Carnosine has attracted attention in studies when it was determined as an antiaging agent. These studies focused on delaying senescence of human fibroblasts and the protective functions of cardiac ischemia *via* antioxidant activity (Shao *et al.*, 2004). Carnosine is a suitable chelator of metal ions, so compounds with calcium, copper, and zinc ions can be used in conjunction with carnosine. Zinc levels in the neuronal tissue, especially the olfactory lobe, can be controlled with carnosine and zinc.

Zinc is an essential trace mineral that plays important roles in many important body functions and physiological processes, including the growth and reproduction of the cells (Opoka *et al.*, 2010). Zinc is also associated with cell proliferation and gene transcription, with the activation of DNA and RNA. One study revealed zinc deficiency in ulcerated gastric mucosa and suggested that the zinc supplements can help heal gastric ulcers. Zinc is not produced or accumulated in the body, but needs to be obtained through diet. It is found in meat, eggs, shellfish, cheese, legumes, and tofu. Zinc is a potent antioxidant that is involved in protection of cellular membranes and macromolecules and helps reduce the proportion of destructive free radicals. Polaprezinc is an antiulcer medication developed for the treatment of gastric

Table 29.1 Clinical trials of zinc L-carnosine according to disease. Recent studies have shown that L-carnosine is not only involved in gastric mucosal protection but also in the protection of other organs, especially in inflammatory colon disease and disorders of the oral mucosa and skin. Other studies showed that zinc L-carnosine is beneficial in the treatment of idiopathic taste disorder, aspirin-induced small-bowel injury, ulcerative colitis, liver injury such as hepatitis C, and liver fibrosis. Basic studies on zinc L-carnosine have also shown sufficient evidence of its therapeutic effects on other organ.

Related disease	Intervention	Conclusion	References
Aspirin-induced small-bowel injury	Clinical trial, aspirin-induced erosion and ulcers, polaprezinc (PZ) *vs.* control	Significant decrease in erosion and ulcers in the PZ group	Ikue Watari *et al.* (2013)
Ulcerative colitis	Clinical trial, PZ enema *vs.* control enema	PZ enema: significant score (Matt' endoscopic/Mayo score)	Munenori Itagaki *et al.* (2014)
Pressure ulcer	Clinical trial, PZ orally 150 mg per day and L-carnosine orally 116 mg per day	PZ orally: significantly greater (PUSH score)	Kensaku Sakae *et al.* (2013)
Idiopathic taste disorder	Clinical trial, PZ groups (17 mg/34 mg/68 mg) *vs.* placebo group	68 mg PZ group, significant improvement	Masafumi Sakagami *et al.* (2009)
Hepatitis C	Clinical trial, HCC incidence, PZ groups *vs.* control group, 7 years	HCC incidence: decreased in PZ group, 7 years	Hiroshi Matsumura *et al.* (2012)
Oral mucositis	Clinical trial, head and neck cancer, radiation-induced oral mucositis, PZ group *vs.* azulene oral rinse group	Mucositis, pain, xerostomia, taste disturbance: markedly lower in PZ group	Tomoko Watanabe *et al.* (2010)

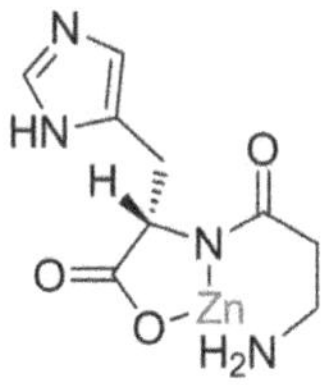

Figure 29.1 Structural formula of zinc L-carnosine (Matsukura and Tanaka, 2000). Zinc L-carnosine is a chelated complex that consists of L-carnosine and zinc.

ulcers in Japan. The gastroprotective ability of zinc L-carnosine is usually attributed to its zinc content (Ko and Leung, 2010).

The bioavailability of zinc is higher in the presence of L-carnosine than in other compounds. These studies showed that L-carnosine is a good carrier of zinc into living organs, especially the stomach (Matsukura and Tanaka, 2000).

29.3 Mechanism Underlying Gastric Mucosal Damage

The incidence of gastric mucosal injury has been increasing rapidly because of increased frequency of nonsteroidal anti-inflammatory drugs (NSAIDs) and antithrombotic agent prescriptions, especially to elderly patients. The continuous gastric mucosal injury caused by these medications leads to gastric ulcers.

A gastric ulcer involves a distinct breach in the gastric wall that extends through the muscularis mucosa into deeper tissue layers. A gastric ulcer has diverse etiologies, but it is mainly caused by an imbalance in the defense and offensive mechanism.

Gastric acid plays an important role in the injury of gastric mucosal lesions. The dictum of Karl Schwarz, "no acid, no ulcer" has been quoted for decades, with regard to the pathogenesis of gastric ulcer (Gustafson and Welling, 2010). Histamine-H2 receptor antagonists like cimetidine and ranitidine have been developed to reduce the amount of gastric acid; these drugs are used to treat gastroduodenal ulcers. After development of various proton-pump inhibitors, the incidence of gastric ulcers was expected to decrease considerably. However, the incidence of gastric ulcer and gastroduodenal bleeding has not decreased.

Helicobacter pylori was first identified by Marshall in 1982 and was recognized as a risk factor for gastroduodenal ulcers, gastritis, gastric cancer, and mucosa-associated lymphoid tissue lymphoma (Choi *et al.*, 2012). The concept of "no *Helicobacter pylori*, no ulcer" has also been suggested after *H. pylori* was proven to be the virulent factor involved in gastric mucosal damage. *Helicobacter pylori* eradication therapy is prescribed to infected patients; gastroduodenal ulcers were expected to decrease with this therapy. The majority of gastric ulcers do resolve with the eradication therapy, but gastric ulcers due to NSIADs, aspirin, and various antiplatelet agents persist, causing the remainder of the ulcer.

NSAIDs are considered the second leading cause of gastroduodenal ulcer development (Gustafson and Welling, 2010). NSAIDs inhibit prostaglandin biosynthesis and act on the target organ for eliciting therapeutic effects. However, the long-term administration of these drugs causes gastrointestinal complications such as bleeding, gastric ulcer, and severe gastritis. Selective cyclo-oxygenase-2 (COX-2)-specific drugs that maintain the anti-inflammatory action of NSAIDs and retain their gastroprotective properties *via* the COX-1 pathway have been used for protecting the gastric mucosa, but these alternative drugs have limited suitability and efficacy in managing complications (Sinha *et al.*, 2013). Numerous studies on minimizing NSAID-induced adverse gastrointestinal effects have suggested novel strategies to maintain the anti-inflammatory functions of NSAIDs as well as effective gastrointestinal mucosa defense, especially for the prevention and cure of gastropathy, including gastric ulcers.

Both NSAIDs and *Helicobacter pylori* infection are independent risk factors for gastric mucosal injury that have synergistic effects on gastric ulcer and ulcer bleeding. In a meta-analysis, adverse risk for the incidence of peptic ulcer was significantly higher in patients positive for *H. pylori* infection and also taking NSAIDs, compared with control group negative for *H. pylori* not taking NSAIDs (Huang *et al.*, 2002; Yuan *et al.*, 2006) (Figure 29.2).

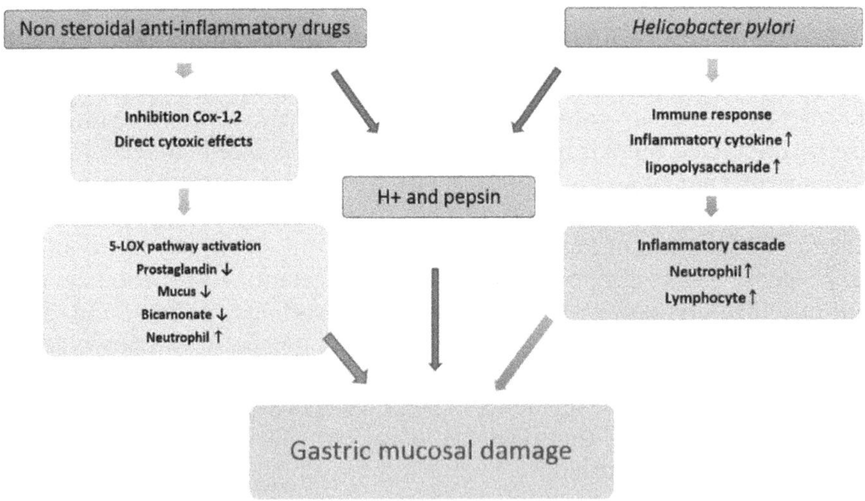

Figure 29.2 Pathogenesis of nonsteroidal anti-inflammatory drugs and *Helicobacter pylori*-induced gastric mucosal damage (COX: cyclo-oxygenase, LOX: lipoxygenase, modified and updated) (Yuan *et al.*, 2006) The major cause factors of gastric mucosal damage are *Helicobacter pylori* and non steroidal anti-inflammatory drugs. Both NSAIDs and *Helicobacter pylori* infection are independent risk factors for gastric mucosal injury that have synergistic effects on gastric ulcer and ulcer bleeding. The use of this figure was permitted by the Nature Publishing Group.

29.4 Defense and Repair Mechanisms in the Gastric Mucosa

The gastric mucosa is continuously exposed to numerous injurious substances. The gastric mucosa must resist damage from the offensive environment of the lumen, such as acid, *H. pylori*, and NSAIDS. Various defense mechanisms come into action to protect the stomach mucosa from injury: the gastric epithelium secretes a mucosal barrier composed of water, bicarbonate, mucin glycoproteins, and trefoil factor family (TFF) peptides (Palileo and Kaunitz, 2011). This mucus-bicarbonate-phospholipid barrier plays a crucial role in gastric mucosal defense (Figure 29.3).

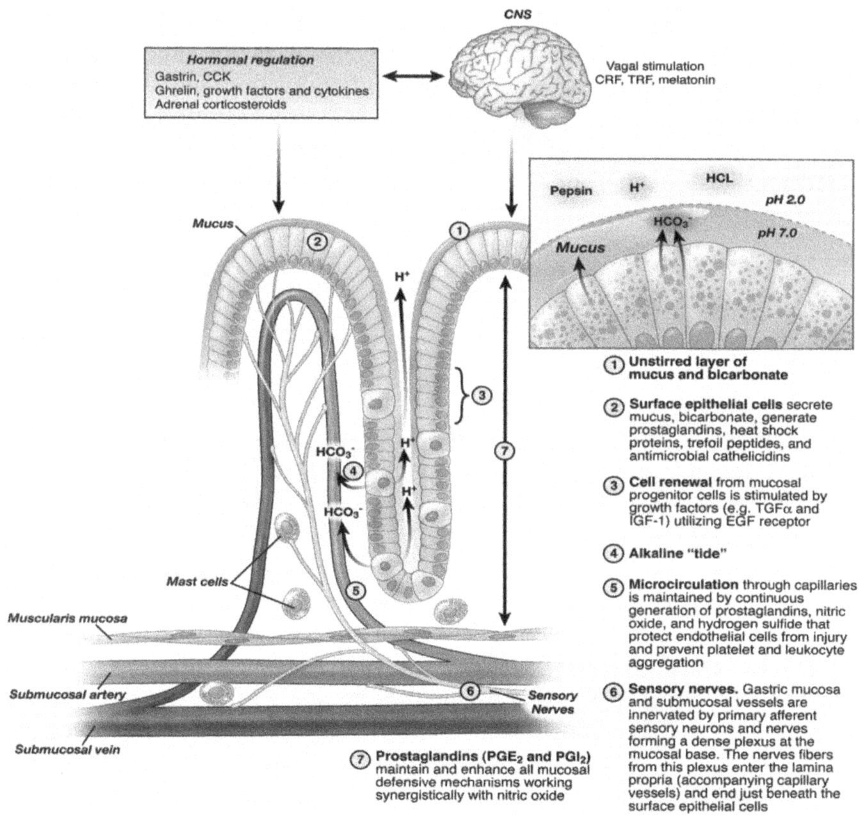

Figure 29.3 Diagrammatic representation of gastric mucosal defense (Laine *et al.*, 2008). Various defense mechanisms come into action to protect the stomach mucosa from injury: the gastric epithelium secretes a mucosal barrier composed of water, bicarbonate, mucin glycoproteins, and trefoil factor family (TFF) peptides. This mucus-bicarbonate-phospholipid barrier plays a crucial role in gastric mucosal defense. The use of this figure was permitted from W.B./Saunders Co. (Copyright Clearance Center).

29.4.1 Bicarbonate Secretion

Gastric epithelial cells secrete bicarbonate, a primary regulator of the acid–base balance in the stomach. This bicarbonate, mucosal gel, and surfactant phospholipid together create a barrier that is an essential defense mechanism. This barrier prevents penetration of pepsin and the subsequent proteolytic breakdown of the epithelium (Laine *et al.*, 2008).

29.4.2 Prostaglandin E Synthase

Ulcerogenic agents Prostaglandin E synthase (PGES) catalyze the conversion of PGH_2 to PGE_2, and PGE_2 increases bicarbonate secretion. PGE_2 plays a key role in the gastric defense mechanism. Membrane-bound PGES-1 level is significantly increased in gastric ulcers and plays an important role in ulcer healing (Palileo and Kaunitz, 2011). Rebamipide, a gastric mucoprotectant, acts as a prostaglandin inducer. Prostaglandins inhibit acid secretion and stimulate mucus, phospholipid, and bicarbonate secretion. COX-1 and COX-2 are crucial enzymes in the biosynthesis of prostaglandins. As mentioned above, NSAIDs suppress COX enzymes and this leads to damage of the mucosa of the stomach.

29.4.3 Mucin Glycoproteins

Mucins play an important role in gastric protection and healing. Mucin genes (MUC) produce mucin glycoproteins and secret mucus gel that is secreted by the gastric epithelial cell. The gel-forming mucin units polymerize into large mucin multimers important for gel formation (Laine *et al.*, 2008). *MUC-1*, *MUC-2*, *MUC-5AC*, *MUC-5B*, and *MUC-6* genes have been known to be related to gastric mucosal defense. The genetic polymorphisms in the *MUC-1* gene are related to an increased prevalence of gastritis and gastric adenocarcinoma. *MUC-1* deficiency leads to a significant increase in *H. pylori* colonization and also increased the severity of atrophic gastritis with the loss of parietal cells (Nayeb-*Hashemi* and Kaunitz, 2009). MUC-2 gene is also normally expressed in gastric adenocarcinoma, but NSAID-induced MUC-2 expression may be part of an adaptive response to gastric damage (Wallace *et al.*, 2011). MUC-5AC is present in the epithelial cells of the antrum, cardia, and fundus in the stomach, and MUC-6 is expressed in the mucosal neck cells of the antral glands in the cardia and antrum and of the fundus (Allen and Flemstrom, 2005). The proportion of multimeric mucin in the adherent gastric mucus layer decreases with *H. pylori* infection; the mucus gel also loses its stability in this case. Mucus gel is secreted together with the TFF peptides. Trefoil family peptides (TFPs) play an essential role in epithelial defense by facilitating mucosal restoration.

29.4.4 Growth Factors

The healing of gastric ulcers requires the re-formation of epithelial structures as well as the underlying connective tissue, vessels, and muscle layers. Some growth factors are involved in this process. Growth factors play important roles in cell functions of tissue healing, such as cell migration and proliferation. Transforming growth factor-a (TGF-a) and epidermal growth factor (EGF), and TFPs are mostly related within the reconstruction of the epithelial structures. Basic fibroblast growth factor (bFGF), platelet-derived growth factor (PDGF), vascular endothelial growth factor (VEGF), and TGF-b play a crucial role in the reconstruction of connective tissue with vessels and smooth muscle cells (Milani and Calabro, 2001). These growth factors are involved in the acceleration of ulcer healing (Table 29.2).

Table 29.2 Summary of the cellular effects of growth factors involved in ulcer healing (Milani and Calabro, 2001). Several growth factors play important roles in cell functions of gastric-ulcer healing, such as cell migration and proliferation. Transforming growth factor-a (TGF-a) and epidermal growth factor (EGF), and TFPs are mostly related within the reconstruction of the epithelial structures. Basic fibroblast growth factor (bFGF), platelet-derived growth factor (PDGF), vascular endothelial growth factor (VEGF), and TGF-b play a crucial role in the reconstruction of connective tissue with vessels and smooth muscle cells. These growth factors are involved in the acceleration of ulcer healing. This table was permitted from John/Wiley & Sons, Inc (Copyright Clearance Center).

Repair (restitution and healing) Protection		
EGF/TGF-α	Cell proliferation and migration, angiogenesis	Inhibition of gastric secretion, synthesis of mucus and bicarbonate, maintenance of gastric blood flow
TFFs	Epithelial cell migration	Mucus oligomerization
HGF	Cell proliferation, migration, and morphogenesis	
bFGF	Cell proliferation and migration, angiogenesis	
PDGF	Cell proliferation and migration, cell and extracellular matrix contraction, angiogenesis	
TGF-β1	Inhibition of cell proliferation, modulation of cell motility and adhesion, extracellular matrix synthesis and degradation, angiogenesis	
VEGF	Endothelial cell viability, proliferation, and chemotaxis, vascular permeability	

29.5 Mechanisms Underlying Zinc L-Carnosine Action

29.5.1 Antiulcer Effect

Previous studies have shown the antiulcer effect of zinc L-carnosine in stomach ulcers. Zinc L-carnosine not only reduced gastric injury lesion but also accelerated the healing process of the gastric mucosal injury lesion. In the ethanol-induced gastro-injury rat model, zinc L-carnosine treatment decreased the ulcer index of the rat stomach and showed a significant ulcer-healing effect similar to that elicited by the gastric mucoprotective agent rebamipide (Choi *et al.*, 2013). Zinc L-carnosine also showed significant ulcer healing effect compared with the placebo treatment group in an aspirin-induced gastroduodenal injury animal model (Naito *et al.*, 2001; Baan *et al.*, 2011). A study involving acetic acid-induced rat model reported that zinc L-carnosine treatment group showed a significant antiulcer effect and healing action than the control treatment group (Ito *et al.*, 1990).

29.5.2 Anti-Inflammatory Action

The main action of zinc L-carnosine in gastric ulcer is its anti-inflammatory action. In the ethanol-induced mucosal injury rat model, the expression of interleukin-1β (IL-1β), interleukin-8 (IL-8), interleukin-6 (IL-6), and tumor necrosis factor (TNF-α) significantly decreased in a dose-dependence manner in the zinc L-carnosine treatment group as compared with the control group (Choi *et al.*, 2013). This result shows that polaprezinc treatment affects gastric mucosal healing by inhibiting the overexpression of proinflammatory mediators. An experimental study using gastric epithelial cells showed a decrease in proinflammatory cytokine-induced nuclear factor-κB activation and IL-8 level after zinc L-carnosine treatment (Shimada *et al.*, 1999). Zinc L-carnosine reduced the expression of the TNF-α, IL-1β, macrophage inflammatory protein-2, and cytokine-induced neutrophil chemoattractant-2α in the ulcerated tissues of acetic acid-induced gastric injury rat model (Ko and Leung, 2010). In the aspirin-induced gastric mucosal injury rat model, the gastric concentration of TNF-α downregulated in a dose-dependent manner after treatment with zinc L-carnosine (Naito *et al.*, 2001).

29.5.3 Antioxidant Action

Antioxidants action plays a crucial role in gastric ulcer healing and repair. In the ethanol-induced mucosal injury rat model, the protein expression of antioxidant factors, such as superoxide dismutase-1 (SOD-1), superoxide dismutase-2 (SOD-2), heme oxygenase-1 (HO-1), peroxiredoxin-1 (PRX-1), peroxiredoxin-V (PRX-V) was increased in the polaprezinc treatment group as compared with that in the control group (Choi *et al.*, 2013). One study suggested that zinc L-carnosine protects rat gastric mucosa from damage caused by ethanol through glutathione peroxidase (GSH-px) and an SOD-like factor (Ito *et al.*, 1992). The levels of HO-1 mRNA, antioxidants, and cytoprotectants, in the zinc L-carnosine-treated rats increased in a dose-dependent manner as compared with those in the control group (Ueda *et al.*, 2009). In the aspirin-induced rat gastric mucosal injury model, the increase in thiobarbituric acid (TBA)-reactive substance and tissue-associated myeloperoxidase (MPO) in the gastric mucosa was significantly inhibited by zinc L-carnosine in a dose-dependent manner (Naito *et al.*, 2001). Zinc-carnosine reduces gastric mucosal damage and inhibits the upregulation of lipid peroxide in the gastric mucosa induced by ischemia-reperfusion or environmental shock (Yoshikawa *et al.*, 1991). H_2O_2-induced ^{51}Cr release means quantified cytotoxicity that can be prevented by the addition of zinc carnosine. These studies highlight the antioxidant property of polaprezinc.

29.5.4 Growth Factors

A large number of studies have shown the role of growth factors in the process of gastric ulcer healing and repair. One study reported that as compared to the control group, the zinc L-carnosine-treated group showed an increase in the expression of growth factors in the rat gastric mucosa. Several growth factors, such as vascular endothelial growth factor (vEGF), nerve growth factor (NGF), and PDGF-B were significantly overexpressed in the zinc L-carnosine treatment group, as evidenced by western blot analyses (Choi *et al.*, 2013). Zinc L-carnosine-induced overexpression of bFGF and vEGF in the ulcerated tissues of the acetic acid-induced gastric injury rat model (Ko and Leung, 2010). Insulin-like growth factor-I (IGF-I) mRNA expression in the fibroblast increased after zinc L-carnosine treatment in rabbit gastric epithelial cells (Watanabe *et al.*, 1998).

29.5.5 Other Potential Cytoprotective Factors

Endogenous cytoprotective factors such as heat-shock proteins, recently have received attention as important mediators in gastric mucosal protection. Heat-shock protein 72 (HSP72), is a cytoprotectant present in the cells; HSP72 was found in rat gastric mucosal cells, observed by western blot analysis after zinc L-carnosine treatment (Odashima *et al.*, 2002). Another study revealed that HSP72 played an important role in the mucosal protective action, especially when HSP72 expression was decreased by acid-suppressing agents such as histamine H2-receptor antagonists or proton-pump inhibitors (Wada *et al.*, 2006). Another study using ethanol-induced gastric mucosal injury rat model showed that expressions of several cytoprotectants such as HSP90, HSP70, HSP60, HSP27, HSP10 increased in the zinc L-carnosine administration group compared to the expressions in the control group (Choi *et al.*, 2013) (Table 29.3).

Table 29.3 Mechanism underlying L-carnosine and zinc action in the stomach. Zinc L-carnosine protects the mucosa against aggressive injury factors and has been found to prevent gastric mucosal injury in diverse experimental models through anti-inflammatory action, antioxidant action, growth factors, and cytoprotective factors such heat-shock proteins.

Mechanism	References
Inflammation factor	
↓Tumor necrosis factor, ↓interleukin-1β, ↓macrophage inflammatory protein-2, ↓cytokine-induced neutrophil chemoattractant-2α, ↓interleukin-1β, ↓interleukin-8, ↓interleukin-6, ↓nuclear factor-κB	Ko and Leung (2010), Naito *et al.* (2001), Choi *et al.* (2013), Shimada *et al.* (1999)
Antioxidant factor	
↑Superoxide dismutase-1, ↑superoxide dismutase-2, ↑heme oxygenase-1, ↑peroxiredoxin-1, ↑peroxiredoxin-V, ↑glutathione peroxidase, ↑superoxide dismutase, ↓thiobarbituric acid-reactive substance, ↓tissue-associated myeloperoxidase, ↓H_2O_2-induced ^{51}Cr release	Choi *et al.* (2013), Ueda *et al.* (2009), Ito *et al.* (1992), Naito *et al.* (2001), Yoshikawa *et al.* (1991), Hiraishi *et al.* (1999)
Growth factor	
↑Basic fibroblast growth factor, ↑vascular endothelial growth factor, ↑insulin-like growth factor-I, ↑nerve growth factor, ↑platelet-derived growth factor B	Ko and Leung (2010), Watanabe *et al.* (1998), Choi *et al.* (2013)
Cytoprotective factor	
↑Heat-shock protein 90, ↑heat-shock protein 70, ↑heat-shock protein 60, ↑heat-shock protein 27, ↑heat-shock protein 10, heat-shock protein 72	Choi *et al.* (2013), Odashima *et al.* (2002), Wada *et al.* (2006)
Prostaglandin	
Prostaglandin: independent effect	Arakawa *et al.* (1990)

29.6 Clinical Evidence of, and Future Research of Zinc L-Carnosine in Gastric Mucosal Protection

Zinc L-carnosine is found to prevent gastric mucosal injury in diverse experimental models through its anti-inflammatory and antioxidant action and the promotion of growth and cytoprotective factors. However, its effect is not associated with prostaglandin's cytoprotective action (Arakawa *et al.*, 1990).

One of the important etiological factors of gastric ulcer is *H. pylori*. Several studies showed that zinc L-carnosine is effective in treating *H. pylori* infection. Triple therapy for *H. pylori* infection treatment with zinc L-carnosine showed significant eradication effect (Kashimura *et al.*, 1999). One study suggested that zinc L-carnosine may inhibit *H. pylori*-induced polymorphonuclear leukocyte-mediated gastric inflammation by reducing leukocyte CD11b/CD18 integrin expression and the production of proinflammatory cytokine IL-8 in gastric epithelial cells (Handa *et al.*, 2002).

Table 29.4 Gastroprotective and ulcer-healing actions of zinc L-carnosine in experimental gastric models. Several experimental studies showed that zinc L-carnosine has mucosal protective and ulcer healing actions in stomach.

Model and mechanism	References
Helicobacter pylori	
Anti-inflammatory effect, inhibitory effect for *H. pylori*-associated gastritis, ↑cure rate of *H. pylori*	Handa *et al.* (2002), Suzuki *et al.* (2001), Ishihara *et al.* (2002), Kashimura *et al.* (1999)
Nonsteroidal anti-inflammatory drugs	
↑Gastroscopy score, ↓tumor necrosis factor, ↓thiobarbituric acid-reactive substance, ↓tissue-associated myeloperoxidase ↓apoptosis	Baan *et al.* (2011), Naito *et al.* (2001), Fuji *et al.* (2000)
Ethanol	
↓Ulcer index, ↓interleukin-1β, ↓interleukin-8, ↓interleukin-6, ↑nerve growth factor, ↑platelet-derived growth factor B, ↑vascular endothelial growth factor, ↑superoxide dismutase-1, ↑superoxide dismutase-2, ↑heme oxygenase-1, ↑peroxiredoxin-1, ↑peroxiredoxin-V, ↑heat-shock protein 90, ↑heat-shock protein 70, ↑heat-shock protein 60, ↑heat-shock protein 27, ↑heat-shock protein 10, ↑glutathione peroxidase, ↑superoxide dismutase	Choi *et al.* (2013), Ito *et al.* (1992)
Acetic acid	
↓Ulcer size and depth, ↑basic fibroblast growth factor, ↑vascular endothelial growth factor, ↓tumor necrosis factor, ↓interleukin-1β, ↓macrophage inflammatory protein-2, ↓cytokine-induced neutrophil chemoattractant-2α	Ito *et al.* (1990), Ko and Leung (2010)

A study involving Mongolian gerbils orally inoculated with *H. pylori* showed that gastric mucosal damage in the inoculated gerbils was significantly inhibited by therapy with zinc L-carnosine (Suzuki *et al.*, 2001). Another study showed that zinc L-carnosine attenuated the development of *H. pylori*-induced gastritis through its eliminating action against monochloramine in the *H. pylori* Mongolian gerbils model (Ishihara *et al.*, 2002). The reason the mucoprotectant zinc L-carnosine is effective against *H. pylori* is unknown. Some studies suggest that zinc ion is responsible for eliminating the *H. pylori* (Kashimura *et al.*, 1999). Zinc plays a role in attenuating urease activity of *H. pylori* and inhibits its growth by preventing its adhesion to the gastric mucin (Suzuki *et al.*, 2001) (Table 29.4).

A number of clinical studies have shown the healing effect of zinc L-carnosine in patients with gastric mucosal damage. Zinc L-carnosine is also effective in NSAID-induced gastric injury, ethanol-induced gastropathy, and acetic acid-induced gastric damage. In conclusion, zinc L-carnosine is a valuable mucosal protective agent with multiple targets that can be used in gastric mucosal damage healing. In addition, zinc L-carnosine has been studied for treating idiopathic taste disorder, aspirin-induced small-bowel injury, ulcerative colitis, liver injury such as hepatitis C, and liver fibrosis. Because the general principles of tissue-damage recovery are similar in various tissue, one can expect that zinc L-carnosine will provide protective and healing effect on a variety of tissues. Further studies for clinical relevance of zinc L-carnosine should be proven by both basic research and clinical trials.

Summary Points

- This chapter focuses on the effect of zinc L-carnosine in gastric protection.
- Zinc is an essential trace mineral that plays important roles in many important body functions and physiological processes, including the growth and reproduction of the cells. Zinc is also associated with cell proliferation and gene transcription, with the activation of DNA and RNA.
- L-Carnosine is present in the muscles of vertebrates, in the form of a dipeptide, which is also present in consumed meats. Carnosine is also associated with cardiac muscle, eye lens, and nervous tissues, including the brain, especially the olfactory lobe. The protective activity of carnosine has been reported in wound healing, diabetes, osteoporosis, loss of vision, and immune function. Carnosine possesses many biochemical homeostatic functions such as pH buffering, antioxidant effect, antiglycation capability, metal-ion chelating, and carbonyl scavenging.
- Zinc L-carnosine is a chelated compound that consists of L-carnosine and zinc.
- A gastric ulcer has diverse etiologies, but it is mainly caused by an imbalance in the defense and offensive mechanisms.
- Various defense mechanisms come into action to protect the stomach mucosa from injury: the gastric epithelium secretes a mucosal barrier composed of water, bicarbonate, mucin glycoproteins, and trefoil factor family (TFF) peptides. This mucus-bicarbonate-phospholipid barrier plays a crucial role in gastric mucosal defense.
- Zinc L-carnosine protects the mucosa against aggressive injury factors and has been found to prevent gastric mucosal injury in diverse experimental models through anti-inflammatory action, antioxidant action, growth factors, and cytoprotective factors such as heat-shock proteins.
- Zinc L-carnosine is effective in treating *Helicobacter pylori* infection, NSAID-induced gastric injury, ethanol-induced gastropathy, and acetic acid-gastric damage.

- Studies on the therapeutic efficacy on zinc L-carnosine have shown that zinc L-carnosine relives gastric mucosal injury and gastric ulcer
- Zinc L-carnosine has been used for treating idiopathic taste disorder, aspirin-induced small-bowel injury, ulcerative colitis, liver injury such as hepatitis C, and liver fibrosis.

Key Facts

Figures 29.2 and 29.3. Key features of gastric mucosa damage and ulcer.

1. Gastric ulcers involve a distinct breach in the gastric wall that extends through the muscularis mucosa into deeper tissue layers.
2. A gastric ulcer is mainly caused by an imbalance in the defense and offensive mechanism.
3. The major cause factors of gastric ulcer are *Helicobacter pylori*, nonsteroidal anti-inflammatory drugs, and antithrombotic agents.
4. Proton-pump inhibitor and histamine-2 blocker are the drugs for inhibition of acid production in stomach.
5. Zinc L-carnosine is used for strengthening the defense system of gastric mucosa.

Figure 29.4, Tables 29.3 and 29.4. Key features of the defense mechanism of zinc L-carnosine in stomach.

1. Zinc L-carnosine protects the mucosa against aggressive injury factors and has been found to prevent gastric mucosal injury through anti-inflammatory action, antioxidant action, growth factors, and cytoprotective factors such as heat-shock proteins.
2. Zinc L-carnosine is effective in treating *Helicobacter pylori* infection, NSAID-induced gastric injury, ethanol-induced gastropathy, and acetic acid-gastric damage.
3. Studies on the therapeutic efficacy on zinc L-carnosine have shown that zinc L-carnosine relives gastric mucosal injury and gastric ulcers.

Definitions of Words and Terms

Gastric ulcer. A distinct breach in the gastric wall that extends through the muscularis mucosa into deeper tissue layers.

Helicobacter pylori. A helix-shaped, flagellated, Gram-negative bacterium and is first identified by Marshall in 1982 and was recognized as a risk factor for gastroduodenal ulcers, gastritis, gastric cancer, and mucosa-associated lymphoid tissue lymphoma.

Nonsteroidal anti-inflammatory drugs (NSAIDs). These drugs is used mainly for relieving pain, reducing fever and blocking blood-clotting function of platelets. NSAIDs block cyclooxygenase (COX) enzymes and

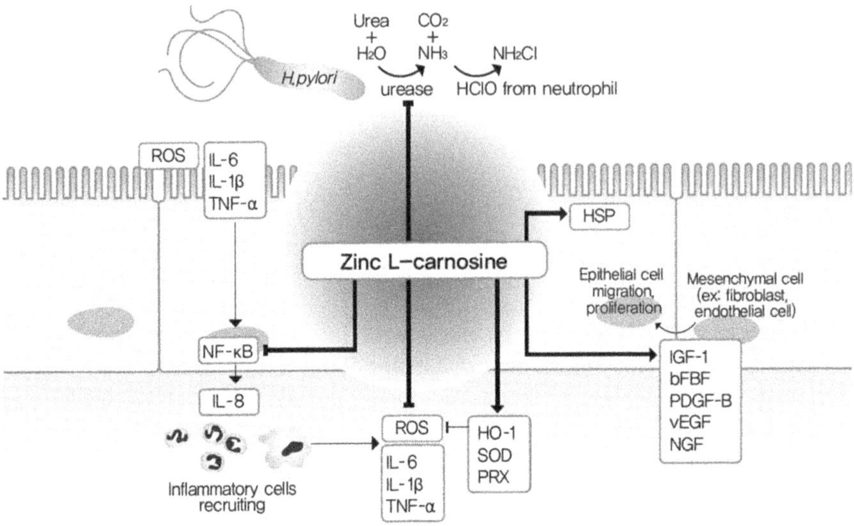

Figure 29.4 Proposed mechanisms of mucosal protective, ulcer healing actions of zinc L-carnosine. 1. Anti-*Helicobacter pylori* effect: attenuated urease activity of *Helicobacter pylori*, and inhibited adhesion of *Helicobacter pylori* to the gastric mucosa. 2. Inhibition of inflammatory cytokines IL-1β, IL-8, and IL-6 and TNF-α in gastric epithelial cells. 3. Inhibition of reactive oxygen species by the increased production of antioxidants (HO-1, SOD, and PRX). 4. Production of endogenous cytoprotectants: heat-shock proteins. 5. Promotion of tissue repair through increased the production of growth factors (bFBF, vEGF, IGF-1, NGF, and PDGF-B) (TNF-α: Tumor necrosis factor, IL-1β: Interleukin-1β, IL-8: Interleukin-8, IL-6: interleukin-6, NF-κB: Nuclear factor-kappa B, SOD: superoxide dismutase, HO-1: heme oxygenase-1, PRX: peroxiredoxin, bFBF: basic fibroblast growth factor, vEGF: vascular endothelial growth factor, IGF-1: Insulin-like growth factor-I, NGF: nerve growth factor, PDGF-B: platelet-derived growth factor B).

decrease prostaglandins through the body. NSAIDs reduce inflammation such as swelling and arthritis, but it can cause gastrointestinal mucosal damage and ulcer bleeding.

Proton-pump inhibitor (PPI). These drugs inhibit the proton pump (H^+/K^+ ATPase) in the acid-secreting parietal cells of the stomach, and are used for most potent gastric acid suppressing agents in the present clinical field. Proton-pump inhibitors are used for treatment of gastroesophageal reflux disease, gastric-ulcer disease, and other conditions that cause hypersecretion of acid such as gastrinomas, Zollinger–Ellison syndrome.

Histamine-H2 receptor antagonists (H2RA). These medications reduce the acid made by the stomach. This drug selectively blocks the histamine receptor on parietal cells in the stomach. H2RA is used in the treatment of peptic ulcer disease, gastroesophageal reflux disease and gastric acid related disease.

L-Carnosine. L-Carnosine is present in the muscles of vertebrates, in the form of a dipeptide, which is also present in consumed meats. Carnosine is also associated with cardiac muscle, eye lens, and nervous tissues, including the brain, especially the olfactory lobe. The protective activity of carnosine has been reported in wound healing, diabetes, osteoporosis, loss of vision, and immune function.

Zinc L-carnosine. A chelated complex that consists of L-carnosine and zinc, it is also called polaprezinc. Zinc L-carnosine is a mucosal protective agent with multiple targets that is mainly used for gastric mucosal damage healing.

Heat-shock proteins (HSPs). The proteins induced by heat shock, first discovered in 1962, are regarded as a crucial factor of cellular defense mechanism. Several groups of the HSPs families maintain cellular hemeostasis by stabilizing unstable conformer of another proteins. HSPs can be produced by exposure to diverse stressful environmental condition. HSPs are grouped into several major families according to their biological activities and apparent molecular weight: HSP100, HSP90, HSP70, HSP60, HSP40 and small HSPs including HSP27 and HSP10. HSP groups are regarded as endogenous cytoprotective factors in several pathophysiological conditions.

References

Allen, A. and Flemstrom, G., 2005. *American Journal of Physiology: Cell Physiology*. 288: C1–C19.

Arakawa, T., Satoh, H., Nakamura, A., Nebiki, H., Fukuda, T., Sakuma, H., Nakamura, H., Ishikawa, M., Seiki, M. and Kobayashi, K., 1990. *Digestive Diseases and Sciences*. 35: 559–566.

Baan, M., Sherding, R. G. and Johnson, S. E., 2011. *Journal of Veterinary Internal Medicine/American College of Veterinary Internal Medicine*. 25: 39–46.

Choi, H. S., Chun, H. J., Park, S. H., Keum, B., Seo, Y. S., Kim, Y. S., Jeen, Y. T., Um, S. H., Lee, H. S., Kim, C. D. and Ryu, H. S., 2012. *World Journal of Gastroenterology*. 18: 2377–2382.

Choi, H. S., Lim, J. Y., Chun, H. J., Lee, M., Kim, E. S., Keum, B., Seo, Y. S., Jeen, Y. T., Um, S. H., Lee, H. S., Kim, C. D., Ryu, H. S. and Sul, D., 2013. *Life Sciences*. 93: 69–77.

Fuji, Y., Matsura, T., Kai, M., Kawasaki, H. and Yamada, K., 2000. *Japanese Journal of Pharmacology*. 84: 63–70.

Gustafson, J. and Welling, D., 2010. *Journal of the American College of Surgeons*. 210: 110–116.

Handa, O., Yoshida, N., Tanaka, Y., Ueda, M., Ishikawa, T., Takagi, T., Matsumoto, N., Naito, Y. and Yoshikawa, T., 2002. *Canadian Journal of Gastroenterology*. 16: 785–789.

Hipkiss, A. R., 2009. *Advances in Food and Nutrition Research*. 57: 87–154.

Hiraishi, H., Sasai, T., Oinuma, T., Shimada, T., Sugaya, H. and Terano, A., 1999. *Alimentary Pharmacology and Therapeutics Symposium Series*. 13: 261–269.

Huang, J. Q., Sridhar, S. and Hunt, R. H., 2002. *Lancet*. 359: 14–22.
Ishihara, R., Iishi, H., Sakai, N., Yano, H., Uedo, N., Narahara, H., Iseki, K., Mikuni, T., Ishiguro, S. and Tatsuta, M., 2002. *Helicobacter*. 7: 384–389.
Itagaki, M., Saruta, M., Saijo, H., Mitobe, J., Arihiro, S., Matsuoka, M., Kato, T., Ikegami, M. and Tajiri, H., 2014. *Scandinavian Journal of Gastroenterology*. 49: 164–172.
Ito, M., Tanaka, T. and Suzuki, Y., 1990. *Japanese Journal of Pharmacology*. 52: 513–521.
Ito, M., Shii, D., Segami, T., Kojima, R. and Suzuki, Y., 1992. *Japanese Journal of Pharmacology*. 59: 267–274.
Kashimura, H., Suzuki, K., Hassan, M., Ikezawa, K., Sawahata, T., Watanabe, T., Nakahara, A., Mutoh, H. and Tanaka, N., 1999. *Alimentary Pharmacology & Therapeutics Symposium Series*. 13: 483–487.
Ko, J. K. and Leung, C. C., 2010. *Journal of Gastroenterology and Hepatology*. 25: 1861–1868.
Laine, L., Takeuchi, K. and Tarnawski, A., 2008. *Gastroenterology*. 135: 41–60.
Matsukura, T. and Tanaka, H., 2000. *Biochemistry Biokhimiia*. 65: 817–823.
Matsumura, H., Nirei, K., Nakamura, H., Arakawa, Y., Higuchi, T., Hayashi, J., Yamagami, H., Matsuoka, S., Ogawa, M., Nakajima, N., Tanaka, N. and Moriyama, M., 2012. *Journal of Clinical Biochemistry and Nutrition*. 51: 178–184.
Milani, S. and Calabro, A., 2001. *Microscopy Research and Technique*. 53: 360–371.
Naito, Y., Yoshikawa, T., Yagi, N., Matsuyama, K., Yoshida, N., Seto, K. and Yoneta, T., 2001. *Digestive Diseases and Sciences*. 46: 845–851.
Nayeb-Hashemi, H. and Kaunitz, J. D., 2009. *Current Opinion in Gastroenterology*. 25: 537–543.
Odashima, M., Otaka, M., Jin, M., Konishi, N., Sato, T., Kato, S., Matsuhashi, T., Nakamura, C. and Watanabe, S., 2002. *Digestive Diseases and Sciences*. 47: 2799–2804.
Opoka, W., Adamek, D., Plonka, M., Reczynski, W., Bas, B., Drozdowicz, D., Jagielski, P., Sliwowski, Z., Adamski, P. and Brzozowski, T., 2010. *Journal of Physiology and Pharmacology*. 61: 581–591.
Palileo, C. and Kaunitz, J. D., 2011. *Current Opinion in Gastroenterology*. 27: 543–548.
Sakae, K., Agata, T., Kamide, R. and Yanagisawa, H., 2013. *Nutrition in Clinical Practice*. 28: 609–616.
Sakagami, M., Ikeda, M., Tomita, H., Ikui, A., Aiba, T., Takeda, N., Inokuchi, A., Kurono, Y., Nakashima, M., Shibasaki, Y. and Yotsuya, O., 2009. *Acta Oto-Laryngologica*. 129: 1115–1120.
Shao, L., Li, Q. H. and Tan, Z., 2004. *Biochemical and Biophysical Research Communications*. 324: 931–936.
Shimada, T., Watanabe, N., Ohtsuka, Y., Endoh, M., Kojima, K., Hiraishi, H. and Terano, A., 1999. *Journal of Pharmacology and Experimental Therapeutics*. 291: 345–352.

Sinha, M., Gautam, L., Shukla, P. K., Kaur, P., Sharma, S. and Singh, T. P., 2013. *Mediators of Inflammation*. 2013: 258209.
Suzuki, H., Mori, M., Seto, K., Miyazawa, M., Kai, A., Suematsu, M., Yoneta, T., Miura, S. and Ishii, H., 2001. *Alimentary Pharmacology & Therapeutics Symposium Series*. 15: 715–725.
Ueda, K., Ueyama, T., Oka, M., Ito, T., Tsuruo, Y. and Ichinose, M., 2009. *Journal of Pharmacological Sciences*. 110: 285–294.
Wada, I., Otaka, M., Jin, M., Odashima, M., Komatsu, K., Konishi, N., Matsuhashi, T., Horikawa, Y., Ohba, R., Itoh, H. and Watanabe, S., 2006. *Biochemical and Biophysical Research Communications*. 349: 611–618.
Wallace, J. L., Vong, L., Dharmani, P., Srivastava, V. and Chadee, K., 2011. *American Journal of Pathology*. 178: 1126–1133.
Watanabe, S., Wang, X. E., Hirose, M., Kivilioto, T., Osada, T., Miwa, H., Oide, H., Kitamura, T., Yoneta, T., Seto, K. and Sato, N., 1998. *Alimentary Pharmacology & Therapeutics Symposium Series*. 12: 1131–1138.
Watanabe, T., Ishihara, M., Matsuura, K., Mizuta, K. and Itoh, Y., 2010. *International Journal of Cancer*. 127: 1984–1990.
Watari, I., Oka, S., Tanaka, S., Aoyama, T., Imagawa, H., Shishido, T., Yoshida, S. and Chayama, K., 2013. *BMC Gastroenterology*. 13: 108.
Yoshikawa, T., Naito, Y., Tanigawa, T., Yoneta, T. and Kondo, M., 1991. *Biochimica et Biophysica Acta*. 1115: 15–22.
Yuan, Y., Padol, I. T. and Hunt, R. H., 2006. *Nature Clinical Practice Gastroenterology & Hepatology*. 3: 80–89.

Subject Index

Illustrations and figures are in **bold**. Tables are in *italics*.

AAPH (2,2′-azobis 2-amidino-propane dichloride) 177, 181–2, 221–2
Aβ peptides 419–21, 434, 452, 455, 463
absorption
  carnosine 11, 32–3, 91, 160–1, 382, 405
  D-carnosine 91
  glyoxal (GO) 142
  intestinal epithelial cells (IEC) 397–8
ABTS test 222–3, 234
accommodation (eyes) 319, **320**, 322
accuracy, LC-MS/MS 205–6
accurate mass 210
acetaldehyde 153, 360, 518, **519**
acetaminophen 514
acetate 518
N-acetyl-β-alanylhistamine *228*
N-acetyl carnosine
  analysis 209
  biosynthesis 29, 249
  carbonyl quenching 151
  chemical properties 12
  nitric oxide (NO) 416
  p*K* values *12*
  renal function 347
  therapeutic potential 62, 422–3, 458
  tissue distribution 12
N-α-acetyl-L-His (NAH) 118, 125, 127, 128, 129, 130
acetylcholinesterase inhibitors 437, 463
N-α-acetylhistidine 133
N-acetylhistidine deacetylase 266–7
acid dissociation constant (p*K*), definition 57
acidosis
  ischemia/reperfusion (I/R) injury 342
  skeletal muscle 280, 283, 285, 286, 287, 288, 304
Acipenseridae 30
acorbine 102
acrolein (ACR) 86, 140, **141**, 152, **153**, 158, 159
actin 56, 145–6, 328
actin *N*-methyltransferase 107
actinomyteces 11
actomyosin 56
acute kidney injury 497–8
acute myocardial infarction (AMI) 341, 347–8, 351
acute renal failure 351
acyl derivatives 229
adaptive responses 243
  see also dose-response
additive, definition 34
adenomatous polyposis coli (APC) 396, 407
adenosine 341

adenosine monophosphate (AMP) 303
adenosine triphosphate (ATP)
- acute kidney injury 497
- carnosine synthesis **100**, 105
- exercise 313
- hypoxia-inducible factors (HIFs) 376, 403
- ischemia/reperfusion (I/R) injury 341–2, 345, 346
- kidney 346
- liver 345, 511, 512
- mitochondria 341
- muscle tissue 56

S-adenosyl-L-homocysteine (SAH) **100**
S-adenosyl-L-methionine (SAM)
- anserine biosynthesis **100**, 107
- carnosine *N*-methyl transferases 109

adhesion molecules 343
adipose tissue 265, 272–3, 482
ADP-ribose 249
adrenal glands 261–2, 424–5
adrenal sympathetic nervous system (ANS) 261–2, 269, 270
adrenaline 259, 260
adriamycin 361
aerobic metabolism 55, 56
aflatoxins 358, 398
age-related macular degeneration 458
AGEs (advanced glycoxidation end products)
- AGEs-RAGE axis 146, 404, 416
- carnosine 88, 153, 223, 363, 364, 404, 455–6, 500
- cooking methods effect on meat levels 52–3
- definition 18–19, 56, 93, 139, 162–3, 368, 428, 524
- diabetes 88
- diabetic nephropathy *496*, 499
- disease pathology 18, 53, 416
- D-galactose overload 518
- meat products 26
- vegetarians 25

aggresome 458–60
aging
- animal models 240, 517–18
- antioxidants 172
- caloric restriction (CR) 247, 249–50
- carnosine effects 87, 357
- carnosine levels 24, 52, 297–8
- forkhead box, class O proteins (FOXOs) 381
- genotoxicity 358
- heat-shock response (HSR) 245, 247
- homocarnosine levels 24
- hormesis 243–4
- 4-hydroxy-*trans*-nonenal (HNE) 154
- inflammation 248
- lipid peroxidation 217
- liver injury 517–18
- metal-ion homeostasis 73
- methylglyoxal (MGO) 249–50
- microvascular disorders 71
- oxidative stress 248
- protein aggregation 249
- reactive oxygen species (ROS) 182
- skeletal muscle 297–8
- *in vitro* studies 495
- *see also* antiaging; apoptosis; longevity; vitagenes

Akt (serine/threonine-specific protein kinase) 375, 380, 400
Ala-His dipeptide 8
β-alanine
- anserine biosynthesis 107
- behavioural studies 476, **477–8**, 481, 482–3
- biosynthesis 298, 356, 481, 486
- blood–retinal barrier (BRB) 536, 539–40, 541
- carbonyl quenching 150

β-alanine (*continued*)
carnosine biosynthesis **100**, 107, **433**
carnosine synthase 102–3, 105
coenzyme A 493
oxidation inhibition 220, 222
paraesthesia 278–9, 288
retina 534
sources 289, 298
supplementation 278–80, 286, 287–8, 298, 299, 300–1
alanine transaminase (ALT) 346
β-alanyl-glycine 481
L-alanyl-L-histidine 482
β-alanyl-L-histidine *see* carnosine
β-alanyl-L-isoleucine 481
β-alanyl-L-leucine 481
β-alanyl-L-valine 481–2
β-alanyl-*N*-π-methyl-L-histidine *see* anserine
β-alanyl-*N*-τ-methyl-L-histidine *see* balenine
β-alanylhistamine 228
β-alanyllysine 104, 106
β-alanylornithine 102, 106
albumin 146, 226
alcohol dehydrogenases (ADH) 147, **148**, 518
alcoholic liver disease 146, 518–20
aldehyde dehydrogenases (ALDHs)
ethanol metabolism 518, **519**
functional overview 162
reactive carbonyl species (RCS) metabolism 147, **148**
aldehyde oxidase 518, **519**
aldehydes
carnosine 55, 223, 379, 413, 415, 495
D-carnosine 86
reactive 34, 495
see also acrolein; 4-hydroxy-*trans*-nonenal
ALDHs *see* aldehyde dehydrogenases
aldo-keto reductases 147
aldolase 378
aldose reductase 159, 162
ALEs (advanced lipoxidation end products)
carnosine 156
definition 19, 139, 163
disease pathology 18
algae, carnosine synthase homologs 102
allergic reactions 493
alphastat regulation 54, 226
Alzheimer's disease (AD)
AGEs (advanced glycoxidation end products) 416, 419, 455–6
amyloid pathology 62, 71, 146, 419, 434, 451–3
carnosine 87, 154, 218, 358, 417, 419–21, **422**, 434, 455–62, 528
drug screening 453–5
metal chelation potential 62
metal-ion homeostasis 73
overview 463
oxidative stress 421
prevalence 445
protein accumulation 417, 419
protein glycation 24, 416
Amadori products 34, 52, 142
amide derivatives 70–1
amido-carnosine 70
amine oxidase 140
amines 358
amino acids
blood–retinal barrier (BRB) 534–8
branched-chain, definition 486
aminoacylhistamine analogs 227–8
β-aminobutyrate 102
γ-aminobutyrate 102–3
γ-aminobutyric acid *see* GABA

γ-aminobutyryl-L-histidine *see* homocarnosine
β-aminobutyryllysine 104
aminoguanidine 151, 152
β-aminopeptidases 53
amphibians 30, 50
amygdala kindling 426
β-amyloid 93, 249, 419, 434, 452, 455, 456–60, 481
amyloid, definition 464–5
amyloid-β peptides 146
amyloid fibrils 89, 434, 451–3, 456, **457**, **459**, 461–3
amyloid precursor protein (APP) 452
amyotrophic lateral sclerosis 62
anaerobic glycolysis
  definition 322
  exercise 313, 314
  ischemia/reperfusion (I/R) injury 342
anaerobic metabolism 30–1, 54, 56, 225
analytical determination methods
  N-acetyl carnosine 209
  anserine 192–6, *196*, 198, 200, 201, **202–3**
  aqueous humour *194*, 208, 209
  balenine (ophidine) *194–5*, *196*, 200, 201, **202–3**
  carnosine 89–90, *193–5*, 196, *196*, 198, 200, 201, **202–3**
  cerebrospinal fluid *194*, *204*, 208
  EMADD potential for food 184
  fish 48, *52*
  homocarnosine *193–5*, 196, *196*, 200, 201, **202–3**
  meat 192, *193*, *204*, 206, 207–8
  plasma *194*, *204*, 208, 209
  urine *194*, *204*, 208, 209
androgens 298
aneuploidy 395
angiogenesis 326, **327**, 328, 337, 396, 398, 403
angiogenin 328
angiotensin-converting enzyme (ACE) 495
angiotensine II 494
Anguillidae 30, 122
animal models
  acute kidney injury 497–8
  aging 240, 517–18
  antigenotoxic properties 365
  behavioural studies 472, 486
  cancer 372, 373, 424–5
  D-carnosine octyl ester 72
  cerebral ischaemia 71
  circadian clock 269
  diabetes 258, 331–2, 499, 502, 521 *see also* Zucker rats
  gastric ulcers 556, 557
  hyperglycemia 259, 262
  hypertension 258, 497
  insulin resistance 521
  ischemia/reperfusion (I/R) injury 345, 346–7, 348, 349, 350
  liver injury 435, 514–17, 518, 519–20, 521–2
  metabolic syndrome 520
  neuroprotection 417–19, 434
  reactive carbonyl species (RCS) 154, **155**, 156, 159
  retina 528–9, 534–6
  testosterone 298
  trolox derivatives 68
  vascular dementia (VD) 438–9
  wound healing 329–32
  see also in vivo studies
Anolis carolinensis 109
anoxia tolerance 54
anserinase 266–7
  see also Xaa–methyl-His dipeptidase

anserine
analysis 192–6, *196*, 198, 200, 201, **202–3**
analytical determination methods 90
antioxidant activity 25, 29, 219–20, 221, *231*
bioavailability 209
biosynthesis 14, 29, **100**, 101, 106–7
brain 417
carbonyl quenching 14, 151, 152
carnosine 208
cooking methods effect on meat levels 28, 52
cyclodextrin derivatives 231
discovery 44, 170
functional overview 14
future research areas 110–11
hydrolysis 14
meat intake biomarker 53, 209
melting point *46*
metabolism 501–2
molecular mass *46*
nomenclature 43, 44
nuclear magnetic resonance (NMR) 48–9, *52*
p*K* values *12*, *46*, 191
reactive oxygen species (ROS) *181*, *182*, 186
seafood decomposition 130
sources 31–2, 52, 160
species variation 29, 30, 31
structure **8**, **44**, 118, **148**, **185**, **203**, **258**
tissue distribution 14, 208, 472
water solubility *46*
ANSN gene **123**, 128
antiaging
vitagenes 239–40
see also aging; longevity
anticancer agents
carnosine 62, 87, 265, 357, 361, 372–81
carnosine derivatives 383
carnosine drug potential 381–3
cyclo(His-Ala) dipeptide 9
cyclo(His-Gly) dipeptide 9
cyclo(His-Phe) dipeptide 9
gold-based complexes 7
tosylureido derivatives 69–70
antigenotoxic properties 358–65, **366**
antihyperalgesics 68
antimicrobial activity 9
antioxidants
acyl derivatives 229
amide derivatives 70–1
anserine 25, 219–20, 221, *231*
carcinine 220
carnosine 23, 24, 25, 26, 27, 55, 87, 219–21, 222, 344, 358, 404–5, 413–16, 494
carnosine derivatives 12
cyclodextrin derivatives *231*
cyclodextrins (CDs) derivatives 69
definition 18, 234, 428
dose-response 248
enzymes 218
histamine 344
histidine 344
histidine dipeptides 7
homocarnosine 221
imidazole dipeptides generally 171, 182–3
liver 511
meat products 26, 27, 28
muscle tissue 26
polyphenols 172
pseudopeptides 227–9
supplements 172, 248
α-tocopherol 26, 27, 494
urocanic acid 6
zinc (II)-carnosine complex 249, 556–7, *558*
see also EMADD (estimation method for antioxidants with DNA degradation)

anxiety 8
apoptosis
  cancer 377
  definition 351, 385, 447
  heat-shock proteins (HSPs) 244
  ischemia/reperfusion (I/R) injury 343, 349
  kidney *496*
  liver injury 511, 512, **513**, 516, 517
  neutrophils 495
  signal transduction pathways 396
apoptosis-inducing factor (AIF) 418
appetite 473
aquatic mammals 29
aqueous humour, analysis *194*, 208, 209
arginine 538
aromatic amines 358
ascorbic acid 27, 219, 514–15
  see also vitamin C
aspartate transaminase (AST) 346
aspirin 78, 398
aspirin-induced small-bowel injury 335, 549, *550*, 560
astaxanthin 180, 181
asthenopia 318–19, **320**, 321–2
astrocytes
  carnosine 418, 419, 432–3, 435, 441, **442**, 472
  function 445–6
  heat-shock proteins (HSPs) 249
  vascular dementia (VD) 437, 441, **442**
atherosclerosis
  carnosine 88, 89
  D-carnosine octyl ester 156
  reactive carbonyl species (RCS) 145, 161, 162
  Trp-His dipeptide 9
athletic performance *see* exercise performance
atomic force microscopy (AFM) 465
ATP *see* adenosine triphosphate
ATP-coupled ionic pumps 342
ATP-grasp domain-containing protein 1 (ATPGD1) 101, 104, 112, 265, 356
autistic spectrum disorders (ASD) 24, 382, 426
autonomic nervous system
  anserine 270
  carnosine 263
  circadian clock 269
  definition 272
  histamine 270
  histamine neurons 268
  overview 271
autophagy, definition 385
autosomal recessive inheritance disease 494
avidin 69, 76
axons 437

bacteria
  peptidases 119
  trehalose 79
balenine (ophidine)
  analysis *194–5*, *196*, 200, 201, **202–3**
  biosynthesis 14, 29, **100**
  carbonyl quenching 14
  discovery 44, 170
  functional overview 14
  future research areas 112
  hydrolysis 14
  isolation 202
  melting point *46*
  molecular mass *46*
  nomenclature 43, 44
  p*K* values *46*, 191
  sources 160
  species variation 29, 30, 32
  structure **8**, **44**, 118, **148**, **203**
  tissue distribution 14
  water solubility *46*
basement membrane 328
basic fibroblast growth factor (bFGF) 328, 555, 557, *558*
Bax (Bcl-2-associated X protein) 435, 518, 519, 523, 529

Bcl-2 (B-cell lymphoma 2) 435, 519–20, 523, 529
BCL2-associated athanogene 2 424
beef
  cooking methods 27, 28, 52
  imidazole dipeptide levels 31, *52*
  lipid peroxidation 26
  preservation 227
behavioural studies
  β-alanine 476, **477–8**
  animal models 472, 486
  anserine 473
  carnosine 472–81, 482–3, **484**, **485**
  carnosine derivatives 481–2
  histidine 476, **477–8**
bestatin 122–3, 266, 267
betaine 536
bicarbonate, gastric mucosa 553, 554
bilirubin 245
bioavailability
  anserine 209
  carnosine 11, 14, 33, 69, 91, 160, 239
  D-carnosine 91
  definition 93
biomarkers, definition 322
biosynthesis
  N-acetyl carnosine 29, 249
  β-alanine 298, 356, 481, 486
  anserine 14, 29, **100**, 101, 106–7
  balenine 14, 29, **100**
  carcinine 28–9
  carnosine 11, 25, 85, **100**, 101, 257, 265, **330**, 356, 432, **433**, 493
  dialdehydes 141, 142
  N-formiminoglutamic acid (FiGlu) **5**, 6
  glutamate 6
  histamine **5**, 7, 262, 347, 493
  homocarnosine 101, 103
  4-hydroxy-*trans*-nonenal (HNE) 93, 163, 511
  imidazole lactate **5**, 7
  imidazole pyruvate **5**, 7
  imidazolone-5-propionic acid **5**, 6
  keto-aldehydes 142–5
  malondialdehyde (MDA) 511
  reactive oxygen species (ROS) 171, 182, 413
  α,β unsaturated carbonyls 140
  urocanic acid **5**, 6
biotin
  derivatives **65**, 69, 76, 383
  function 78
birds
  anserine 50, 109
  carnosine 109
  HNMT-like protein 109
  longevity 29
  see also poultry meat
BLAST 113
blood–brain barrier (BBB) 71, 77, 413, 418, **420**, 471–2
blood–cerebrospinal fluid barrier (BCSFB) 531
blood clotting 326, **327**
blood glucose
  anserine 270
  carnosine 259–63, **261**, 269, *496*
  functional overview 271–2
  see also glucose metabolism
blood pressure
  carnosine 258, 263, **264**, *496*
  histamine 270
  suprachiasmatic nucleus (SCN) 269
  see also hypertension
blood–retinal barrier (BRB)
  structure 529, **530**, 542
  transport 531–42
Boc *see tert*-butyloxycarbonyl group
boiling, imidazole dipeptide levels 27, 28, 52
bone marrow 361–2

brain
 N-α-acetyl-L-His (NAH) 128, 129
 β-alanine 483
 amide derivatives 71
 animal models 68, 71, 417–19
 anserine 472
 cancer 423–5
 carnosine dipeptidases 267, **268**
 carnosine distribution 11, 208, 257, 432–3
 carnosine supplementation 471–2
 carnosine synthase/synthetase 103
 fetal 105
 histamine 268–9, 441
 homocarnosine 12, 103, 104, 208
 trolox derivatives 68
 see also neuroprotection; stroke/cerebral ischemia
branched-chain amino acids (BCAAs) 481–2, 483
breast cancer 87
breed variation 29, 30, 52
brown adipose tissue (BAT) 265, 269, 273
buffering 4, 54–5, 56
 see also proton buffering
buffering capacity
 β-alanine supplementation 282, 286
 carnosine 85, 280, 413
 definition 289, 307
 fatigue 301–2, 313–14
 myoglobin 55
tert-butyloxycarbonyl group (Boc) 70, 72

cachexia 337
Caco-2 cells 375, 405, 495, 531
cadmium 503, 515–16
Caenorhabditis elegans 380, 385, 495
caffeine 303
calcium
 cerebral ischemia 418
 ischemia/reperfusion (I/R) injury 342, 348
 mitochondria 341
 muscle tissue 56, 286, 302–4
calcium channels 9, 303
calcium release 303, 307
calcium sensitivity 286, 304, 307
caloric restriction (CR) 247, 249–50
cancer
 AGEs (advanced glycoxidation end products) 416
 apoptosis 377
 brain 423–5
 breast 87
 carnosine 62, 87, 265, 357, 361, 372–81
 carnosine drug potential 381–3
 cell cycle 376–7
 cell signaling 373–6
 cisplatin-induced nephrotoxicity *496*, 498
 colon 373, 393–406, 425
 cyclo(His-Phe) dipeptide 9
 drug side effects 361
 gastric adenocarcinoma 554
 gene regulation 373–6
 genotoxicity 358–9
 glycolysis 377–9, 435
 heat-shock response (HSR) 245
 hormesis 242
 inflammation 398–9
 lipid peroxidation 217
 liver 512, 522
 metal-ion homeostasis 73
 metastasis 377, 394, 396, 425
 nickel ions 222
 staging 394
 tosylureido derivatives 69–70
cap structure 407
capillary electophoresis (CE) 90
carbon tetrachloride (CCl_4) 362–3, **365**

carbonic anhydrases
  histidine function 3
  tosylureido derivatives 70
carbonyl quenching
  aging 249
  anserine 14
  balenine 14
  carnosine 10–11, 12, 17, 18, 19, 61, 149–59, 218, 249, 364, 495
  carnosine derivatives 12–14, 16, 18
  D-carnosine 14
  Gly-His dipeptide 13
  homocarnosine 13, 16
  O-methyl-β3-homotyrosine–D-His 13
  Michael addition 6, 10
  overview 9–11
  proteinogenic histidine dipeptides 10–11
  pseudopeptides 226
carboxyethyl-lysine (CEL) 142, **144**
S-(carboxymethyl)cysteine (CMC) 142, **143**
N-ε-(carboxymethyl)lysine (CML) 142, **143**
carboxypeptidase B 11
carcinine
  antioxidant activity 220
  biosynthesis 28–9
  homology modeling 14
  neuroprotection 13, 458
  p*K* values *12*
  structure **8**
  tissue distribution 13
carcinogens 243, 358–9
cardiac muscle *see* heart muscle
cardiovascular disease
  AGEs (advanced glycoxidation end products) 416
  carnosine 358
  lipid peroxidation 217, 218
  α-tocopherol 494
  see also myocardial infarction
cardiovascular system
  imidazole dipeptide function 24
  NO-donor derivatives 71
  Trp-His dipeptide 9
carnosinase
  N-acetyl carnosine 12
  anserine 14
  balenine 14
  carcinine 13
  carnosine derivatives 77, 382
  classification 119, 121, 266, 493
  deficiency 258, 493–4, 497
  functional overview 19, 163
  glioblastomas 423–4
  homocarnosine 12–13
  homology modeling 14, **15**
  inhibition 382
  NO-donor derivatives 229
  nomenclature 119, *120*, 121, 493
  research history 119–21
  serum levels 62, 160
  tissue distribution 11, 163
  wound healing 329–30
  see also carnosine dipeptidase I/II
carnosine
  absorption 11, 32–3, 91, 160–1, 382, 405
  aging effect on levels 24
  analysis 89–90, *193–5*, 196, *196*, 198, 200, 201, **202–3**
  antigenotoxic properties 359–65, **366**
  antioxidant activity 23, 24, 25, 26, 27, 55, 87, 219–21, 222, 344, 358, 404–5, 413–16 *see also* reactive oxygen species
  appearance 84
  bioavailability 11, 14, 33, 69, 91, 160, 239, 381
  biosynthesis 11, 25, 85, **100**, 101, 257, 265, **330**, 356, 432, **433**, 493

carbonyl quenching 10–11, 12, 17, 19, 61, 149–59, 161, 249, 364
carnosinase complex 14, **15**
chirality 84, 90, 92 *see also* D-carnosine
competitive inhibition 105–6
cooking methods effect on meat levels 27, 52
discovery 43–4, 149, 170, 412, 432
extraction 84
formula 84
FT-IR spectrum 45–6, **47**
functional overview 11, 19, 24, 61–2, 85–6, 238–9, 344, **357**, 494, 513, 549 *see also* antioxidants; carbonyl quenching; free radical scavenging; metal chelation; proton buffering
homeostasis 28–9, 91
hydrolysis 266, **330**, 412, 418, 493
lipophilicity 10
melting point *46*, 84
molecular mass *46*, 84
nomenclature 43, 84
nuclear magnetic resonance (NMR) 46–8, **49–50**
pharmacokinetic profile 11, 16, 62, 91
p*K* values *12*, *46*, 191
reactive oxygen species (ROS) *181*, *182*, 186
solubility 84
sources 24–5, 31–2, 50–2, 160, 549
species variation 29, 30, 31, 33
specific rotation 84
structure **8**, **44**, **85**, 118, **148**, **185**, **203**, **258**, **344**
supplementation 331, 347, 423, 434, 471–2, 499
synthesis 53, 84–5, 471
therapeutic potential 24, 62, 87–9, 248, 335–6, 357, 360, 363, 381–3, 419, 422–3, 455, 503
tissue distribution 11, 24, 25, 30, 33, 61, 85, 148, 208, 238, 257–8, 356, 412, 432, 549
transport 382, 530–4, **539**, 540, 541
UV-Vis absorption spectrum 45, **46**
vitagenes 248–50
water solubility *46*, 218

D-carnosine
absorption 91
bioavailability 62, 72, 91
carbonyl quenching 14, 156
derivatives 72, *73*, 162
homology modeling 14
inflammation 248
pharmacokinetic profile 91
p*K* values *12*
spinal cord injury therapy 62, 89
structure **85**
therapeutic potential 86–7

carnosine derivatives
acyl derivatives 229
amino group derivatization 63–70, 71–2
anticancer agents 383
antioxidant activity 12
bioavailability 69
carbonyl quenching 12–14, 161–2, 239
carboxylic group derivatization 70–2
D-carnosine derivatives 72, *73*
design 13–14, 16, 62, 161–2
metal complexes 73–6
NO-donors 71
reaction paths **67**
trolox derivatives 63–8, 232, 383

carnosine dipeptidase I (CN1)
  anserine 270
  function 126–7
  gene expression 124, 125
  homology modeling 125–6
  molecular identification 121
  nomenclature 119, *120*
  pH 125
  phylogenesis 130–1, **132**
  research history 119–21, 124
  specificity 125, *126*, 266
  tissue distribution 124, 267, **268**
carnosine dipeptidase II (CN2)
  anserine 270
  function 123
  gene expression 122
  inhibition 122
  molecular identification 121
  nomenclature 119, *120*, 122
  phylogenesis 130, **131**, **132**
  research history 119–21
  specificity 122, 266
  structure 267
  tissue distribution 121–2, 267, **268**, 269
carnosine *N*-methyl transferases
  anserine biosynthesis 14
  balenine biosynthesis 14
  future research areas 110–11
  gene encoding 107–8
  phylogenesis **111**
  research history 106–7
  specificity 109
  structure 112
D-carnosine octyl ester 14, 72, 156, 162, 163
carnosine-propanol 156, **157**, 159
carnosine synthase/synthetase
  ATP-grasp domain-containing protein 1 (ATPGD1) 265, 356
  gene expression 101–2, 296
  homologs 102
  regulation 105–6
  research history 100–1
  specificity 102–3, 106
  structure 101–2, 112
  tissue distribution 11, 85, 104–5
carnosinemia 133
carnosinuria 124, 133, 494
CARNS1 gene 101, 104–5, 106
L-carosine *228*
carp 30
caspases 345, 418, 512, 514, 516, 519
catalase (CAT)
  antioxidant activity 26, 218
  carnosine 365, 422
  diabetes 521
  function 367
  liver injury 514, 518, **519**, 520
catalytic constant (k_{cat}), definition 113
catalytic efficiency 113
cataract management 62, 89
catechin 180, *181*, *182*
cell cycle 376–7, 385, 401–2, **404**
cell death 171, 182, 511, 512
  see also apoptosis; ischemia/reperfusion (I/R) injury; necrosis
cell signaling *see* signal transduction pathways
cell-survival 244
cellular stress response 244–5
central nervous system (CNS)
  β-alanine 482–3
  anserine 14
  balenine 14
  carnosine 105, 347, 472, 498
  cyclo(His-Pro) dipeptide 8–9
  definition 486
  glioblastoma multiforme 386
  histamine receptors 347
  homocarnosine 12, 24
cerebral ischemia/stroke
  carnosine 417–19, **420**, 434, 435
  definition 352
  homocarnosine 417

ischemia/reperfusion (I/R) injury 341, 348–9
NO-donor derivatives 71
vascular dementia (VD) 437
cerebrospinal fluid
analysis *194*, *204*, 208
Aβ proteins 419
homocarnosine 12, 24
protein glycation 24
Cetacea see marine mammals
chaperones 87, 240, 244–5, 458, 464
see also heat-shock proteins
cheese 9
chemical shift, definition 56
chemokines 332, 516, 524
chemotherapy 243
chicken-breast extraction (CBEX™) 359
chicken extract 172, 173, 258, 259, 317, 321, 472
chickens
β-alanine 481
as brain models 472, 486
see also poultry meat
chirality, definition 92
Chlamydomonas reinhardtii 102
Chlorella variabilis 102
chlorogenic acid 180, *181*, *182*
cholesterol oxidation 27
see also lipid peroxidation
cholesterol synthesis 521
chromium 416
chromosomal instability (CIN) pathway 395
chromosome damage 361–3, 364, 365
chronic kidney disease (CKD) 494, *500*, 502–3, 504, 505
chronic obstructive pulmonary disease (COPD) 146
cigarettes 89, 515
cimetidine 418, 551
circadian clock 266, 269–70, 271, 272
circular dichroism (CD) spectroscopy 458
cirrhosis 512, **513**, 515, 518, 522
cisplatin *496*, 498
citrulline 480–1
Clupeidae 30
CNDP1 gene 121, **123**, 124–5, 127, 500–1, 502, 505
CNDP2 gene 121, 122, **123**, 501
cobalt 416
cod 30
coenzyme A 493
coenzyme Q10 180, *181*, *182*
cognitive defects, carcinine 13
cognitive function 382
collagen synthesis 326, **327**, 328, 330, 331, 493, 499
collagen type I 331, **333**, 512, **513**
collagenase 328
colon 549, *550*, 560
colon cancer
carnosine 265, 373, 399–406, 425
diet 397–8
genetics 395–6
mortality 393
pathogenesis 393
staging 394
types 394
comestibles, definition 337
comet assay 368
competitive inhibitors
carnosine 105–6
definition 133
connective tissue disease 361
constitutively expressed forms of NOS (cNOS) 479–80
cooking, carnosine levels 27–8, 52–3
copper (II) ions
carnosine 219, 360, 416, 421
carnosine derivative complexes 74–6, 231, **232**
chelation 7, 62, 85–6
DNA degradation 360, 365
dyshomeostasis 73
lipid peroxidation 69, 218, 231, 421
NO-donor derivatives 71, 229

Corbicula japonica 102
corneal diseases 335
correlation spectroscopy (COSY) 46, 48, **49**
corticosterone 474, 481
corticotropin-releasing factor 486
cortisol 315, **316**, 322
cosmetics 62
CpG island methylator phenotype (CIMP) 395
creatine phosphokinase (CK/CPK) 315, **316**, 322
CREB-binding protein 247
crocodiles 30
Crohn's disease 394, 398
α-crystallin 456
cyclin A protein 402
cyclin-dependent kinase (CDK) 402
cyclin-dependent kinase inhibitors (CK1) 402
cyclo-oxygenase-2 (COX-2) 343, 552, 554
cyclodextrins (CDs)
  definition 78
  derivatives **64–5**, **66–7**, 69, 71, 72, 74, 75, 229–31, **232**, 383
cyclodipeptides 8
cyclo(His-Ala) dipeptide **8**, 9
cyclo(His-Gly) dipeptide **8**, 9
cyclo(His-Phe) dipeptide 9
cyclo(His-Pro) dipeptide 8–9
cyclophosphamide (CP) 361, 367
CYP2E1 *see* cytochrome P450 2E1
cyprinids 30
cystic fibrosis 358
cytochrome P450 2E1 (CYP2E1) 511, 514, 515, 518, **519**, 520
cytokines
  cancer 398, 399, 400
  carnosine 248
  definition 407, 524
  ischemia/reperfusion (I/R) injury 343
  liver 511
  oxidative stress 184
  sepsis 516
  vascular dementia (VD) 437, 441
  wound healing 326, **333**
  see also interferon-γ; interleukin-1β (IL-1β); tumor necrosis factor alpha
cytosol nonspecific dipeptidase 121
cytotoxin inhibition 86, 88

DAF-16 385
trans-2,3-decadienal 223
trans-4-decanal 223
deficiency
  carnosinase 258, 493–4, 497
  folate 6–7
  histidinase 7
  urocanase 7
  zinc 549
dehiscence 331, 337
dementia
  carnosine 357
  definition 445
  see also Alzheimer's disease; vascular dementia
demyelination 437
2-deoxy-D-glucose (2DG) 259–60, **261**, 262, 269, 270, 472
3-deoxyglucosone 147
deoxyribose 379
derivatization reagents 196
desmosome 327, 337
detection limit (LOD) 206, 210
diabetes
  AGEs (advanced glycoxidation end products) 88
  animal models 258, 331–2, 499, 502, 521 *see also* Zucker rats
  carnosine 88, 154, 331–2, 335–6, 521
  D-carnosine 86
  CNDP1 gene polymorphism 161, 500–1
  nonalcoholic fatty liver disease (NAFLD) 521

reactive carbonyl species (RCS) 145, 162
vegetarians 25
wound healing 329, 331, 333–4, 335–6
diabetic nephropathy
carnosine dipeptidase I 126–7, 161
CNDP1 gene polymorphism *500*, 501
definition 505
pathology 498–9, 502
diabetic retinopathy 528
dialdehydes
biosynthesis 141, 142
carnosine 153–4
structure **141**
see also glyoxal; malondialdehyde
dicarbonyl melanoidins 52
α dicarbonyls 10
see also reactive carbonyl species
diene conjugate (DC) 350, 515, 517
diet
caloric restriction (CR) 247, 249–50
cancer 397
carnivorous 382
high-fat (HFD) 397, 520, 524
high fructose (HFr) 521
methionine- and choline-deficient diet (MCD) 521, 524
pyruvate dehydrogenase kinase expression 379
vegetarianism 25, 298–9
see also supplementation
dietary sources
β-alanine 298
antioxidants 172, 186
carnosine 24–5, 31–2, 549
histidine 6
imidazole dipeptides generally 50–3
zinc 549
dietary supplements *see* supplementation
digestion 33
dihydroxyacetone 379
dimeric copper complexes 75, 76
diopeter, definition 322
dipeptidases, classification 133
dipeptide, definition 92
dipeptidyl peptidase-4 (DPP-4) 9
diphenhydramine 418
diphenhydramine hydrochloride 262–3
diphthamide 11
DNA degradation
AGEs (advanced glycoxidation end products) 364
antioxidant effect 179–82, 358
carnosine 359–65, **366**
genotoxicity 358, 368
inflammation 398
reactive oxygen species (ROS) 171–2, 174–6, 177, **178**, 358, 363, 365, 367
DNA mismatch repair (MMR) 395, 407
L-dopa derivatives **64**, 68
dose-response
adipose tissue 265
β-alanine intake 278
blood pressure 263, **264**
hyperglycemia 259–60, 262
sympathetic nervous system 263, **264**, 265, 268
typical curve **241**, 242–3
see also hormesis
DPPH test 222, 232, 234
drug delivery, definition 78
drug uptake 408
dyslipidaemia 88, 89, 156, 519

ectotherm 133
edaravone 151
eels 30, 101, 122, **123**, 124–5, *126*
electrophilicity, definition 57

electrophoresis
  carbonyl quenching 150
  EMADD (estimation method for antioxidants with DNA degradation) 184
  reactive oxygen species (ROS) 176
electrospray ionization (ESI) 200, 206
Eleutherodactylus 30
EMADD (estimation method for antioxidants with DNA degradation) 171, 172–6, 184–5, 186
endogenous dipeptide, definition 337
endotherm 133
enolase 378
enzymes
  classification 133
  histidine function 3, 17
  housekeeping 133
  imidazole dipeptide function 24
  imidazole dipeptide synthesis 53
  molecular identification 112–13, 121
  nomenclature 119
  repair 104
  zinc-dependent 548
epidermal growth factor (EGF) 326, 555
epidermal growth factor receptor (EGFR) 373, 381, 396
epigallocatechin gallate (EGCG) 415
epilepsy 13, 426
epithelization **327**
equilibrative nucleoside transporter 2 (ENT2) 529
ergogenics 280, 283, 286, 289, 300–1, **301**
  see also exercise performance
ERKs *see* extracellular-signal-regulated kinases
Escherichia coli 79
esophagitis 398
estrogen-related receptors (ERRs) 379–80, 385
ethanol 518–20
eukaryotic elongation factor 1A (eEF1A) 246–7
eukaryotic translation initiation factor 4E (eIF4E) 375, 400, **401**, 407
evolution *see* phylogenesis
excitation–contraction coupling 301, 302–4, 307
excitin 482, 483, 486
excitotoxicity, definition 428, 447
exercise
  anaerobic glycolysis 313, 314
  carnosine release 266, 269, 305
  reactive oxygen species (ROS) 313
exercise capacity
  β-alanine supplementation 280–1, 283, 286
  definition 289
exercisc performance
  β-alanine supplementation 280–7
  anserine 314–15, **316**, 317–18
  carnosine 277, 280
  definition 289
  high-intensity exercise 281–5, 288, 296, 301, 302, 317
  proton buffering 226
  see also fatigue; skeletal muscle
cxpression sequence tag (EST) 113
extracellular matrix (ECM) 328, 498–9, 512
extracellular-signal-regulated kinases (ERKs)
  cancer 375, 381, 400, 402, **404**, 405, 406, 425
  kidney 495, 499
eye diseases
  N-acetyl carnosine 62, 209
  carnosine potential 24, 87, 88, 89, 209, 335

eyes
  accommodation 319, **320**, 322
  N-α-acetyl-L-His (NAH) 128, 129
  asthenopia 318–19, 321–2
  kinetic visual acuity 318, **319**
  Xaa–methyl-His dipeptidase 129
  see also retina

familial adenomatous polyposis (FAP) 394, 395, 396
Fas ligand 512, **513**
fatigue
  anserine 314–15, **316**, 317–19
  antioxidants 313–14
  buffering capacity 301–2, 313–14
  calcium sensitivity 304
  carnosine 171, 300
  excitation–contraction coupling 302–4
  high-intensity exercise 281, 286
  reactive oxygen species (ROS) 171
  see also exercise performance
fatty acid oxidation 518
fatty acid synthesis 520–1
Fenton reactions 182, 217, 218
ferrite nanoparticles 76
ferritin 183, 360, 422–3
ferulic acid 179, 180–1, *181*, *182*, 183, 186
fetal brain 105
fibrils
  α-crystalline 87, 89
  definition 465
fibroblast growth factors (FGFs) 326, **333**
fibroblasts 326, **327**, 328, 331, 417, 495
fibronectin 328, 494, 499
fibrosis (liver) 512, **513**, 515, 518, 522
fish
  analysis 48, *52*
  anserinase 266–7
  anserine 130, 314, 319
  carnosine dipeptidase I 125
  carnosine dipeptidase II 122
  carnosine synthase homologs 102
  HNMT-like genes 110
  imidazole dipeptide levels 30, 32, 50, *52*
  Xaa–methyl-His dipeptidase 127, 128, 130
flavour
  γ-Glu-His dipeptide 9
  lipid peroxidation effects 217
α-fluoromethylhistidine hydrochloride (FMH) 262–3
folate deficiency 6–7
food additives
  anserine 28
  carnosine 28, 423, 434
  chicken-breast extraction (CBEX™) 359
food analysis
  EMADD 184
  hams 29
  high-performance liquid chromatography (HPLC) 89–90
  liquid chromatography (LC) 192, *193*, *204*, 206, 207–8
  nuclear magnetic resonance (NMR) 48, *52*
  sodium chloride effect 207
  source comparison *52*
food degradation 217
food intake, behavioural studies 472–3
forkhead box, class O proteins (FOXOs) 379–80, 385–6
formaldehyde 153
N-formiminoglutamic acid (FiGlu) **5**, 6–7

Fourier transform infrared
spectroscopy (FT-IR) 57
free radical scavenging
carnosine 24, 239, 494
cyclodextrin derivatives 231
definition 57
imidazole dipeptides 55, 86, 221–3
free radicals
definition 34, 234
see also antioxidants; reactive nitrogen species; reactive oxygen species
frogs 30
fructose 379
FT-IR spectrum, carnosine 45–6, **47**

GABA (γ-aminobutyric acid)
carnosine dipeptidase I 126
functional overview 446
homocarnosine 12
retina 528, 538–9
GABA receptors 8, 476, 482, 540
GABA transporters 536
Gadidae 30
D-galactose 518
galactose 379
galectins 69
gastric adenocarcinoma 554
gastric muscosa defense mechanisms 553–5
gastric ulcers
carnosine 329, 335, 549
definition 561
pathogenesis 551–2
zinc (II)-carnosine complex 62, 331, 548–9, *550*, 556–60
gastritis 398, 554
gastrointestinal tract
carnosine 91
cyclo(His-Pro) dipeptide 9
gender differences in humans, carnosine levels 278, 297–8, 299
gender differences in meat 30, 52
gene expression
cancer 373–6
carnosine dipeptidase I 124, 125
carnosine dipeptidase II 122
carnosine *N*-methyl transferases 107–8
carnosine synthase/ synthetase 101–2
HNMT-like protein 109
sirtuins *241*
Xaa–methyl-His dipeptidase 128
see also vitagenes
genotoxicity, definition 358, 368
glial cells 432–3, 472
glial fibrillary acidic protein (GFAP) 441, **442**, 446, 529
glioblastoma multiforme 386
glioblastomas 62, 377–8, 379, 423–4
gliomas 425, 433
glomerular filtration rate (GFR) 498, 501
γ-Glu-His dipeptide, taste enhancement 9
glucagon
functional overview 272
hyperglycemia 259, 260
glucocorticoid receptor (GR) 379
glucocorticoids (GCs) 315, 386, 474, 517
gluconeogenesis 518
glucose *see* blood glucose
glucose-6-phosphate isomerase 378
glucose derivatives **65**, **66**, 69
glucose metabolism
carnosine 88, 269, 379
histidine dipeptides 9
see also blood glucose
glucose transporters (GLUTs) 378, 386, 529
glutamate
biosynthesis 6
excitotoxicity 419, **420**, 425, 434, 435, 447
retina 528

glutamate transporter-1 (GLT-1) 419, 434, 435, 441, 447
L-glutamylhistamine 228, 229
p-glutarylhistamine *228*
glutathione (GSH)
carnosine 358, 363, 365, 422, 514, 517, 518, 520
cellular stress response 245
functional overview 163, 367
heme oxygenase-1 (HO-1) 249
liver injury 514, 515, 516, 520
reactive carbonyl species (RCS) metabolism 147, **148**, 154, 155
reactive oxygen species (ROS) 181
sirtuins *241*
glutathione peroxidase (GPx)
antioxidant activity 26, 218
carnosine 358, 363, 365, **366**, 422, 514–15, 518, 520–1, 557, *558*
ethanol metabolism 518, **519**
function 367
α-tocopherol 521
glutathione reductase 346
glutathione transferases (GSTs) 147, **148**
Gly-His dipeptide
carbonyl quenching 13
metal chelation 7
p*K* values *12*
structure **8**
glycation
carnosine 363–4, 416
definition 34, 78–9, 428
glyceraldehyde-3-phosphate dehydrogenase 378
glycine receptors 476, 482
glycine transporter (GlyT1) 529
glycogen phosphorylase 262
glycogenolysis 259, 262
glycolysis
anaerobic 225, 313, 314, 322, 342
cancer 377–9, 435
Warburg effect 377–8, 385, 403, 408
glycosaminoglycan (GAG) mimetics 453, 455
glycosaminoglycans (GAGs) 465
glycosylation 133, 435, 502
glycoxidative stress 23
glycyl-β-alanine 483, **484**
glycyl-L-histidine 482
glyoxal 9
glyoxal (GO)
absorption 142
carnosine 153
metabolism 147
reactivity 142, **143**
sources 142
structure **141**
glyoxalases 147, **148**
GODIC 142, **143**
GOLD 142, **143**
gold, Gly-His dipeptide 7
gold nanoparticles 76
GPx *see* glutathione peroxidase
gradient heteronuclear multiple-bond correlation spectroscopy (gHMBC) 46, 48, **49–50**
granulation 327, 328, 329, 330, 331, 337
grilling (broiling), imidazole dipeptide levels 27, 28, 52
growth factors
gastric ulcers 555, 557, *558*
wound healing 326, 332–3
GSH *see* glutathione
GTPase 396
guanidinoacetic acid (GAA) 536, 541
guanine 365
Gulf War Illness 382

Haber-Weiss reactions 217
hair follicles 327
hazard assessment 242, 243, 359
HCDs (histidine-containing dipeptides) 4, 302

heart muscle
carcinine 13
carnosine 11, 257
imidazole dipeptide levels 28
see also cardiovascular system
heat-shock elements (HSEs) 245
heat-shock factors (HSFs) 245–7
heat-shock proteins (HSPs)
carnosine 248, 249, 415, 455, 460
cellular stress response 244–7
classification 563
cyclo(His-Pro) dipeptide 9
overview 465, 563
protein folding 458
vitagenes 239–40
zinc (II)-carnosine complex 460, 557, *558*, *559*
heat-shock response (HSR) 244–7, 250
heat shock RNA-1 (HSR-1) 247
heat stress *see* thermal stress
Helicobacter pylori 551, 552, 554, 558–9, 561
heme oxygenase-1 (HO-1) 239–40, 248, 249, 415, 556–7, *558*, *559*
hemoglobin 23
hemostasis 326, **327**, 337
hen egg-white lysozyme (HEWL) 458, **460**, 465
hepatic failure, definition 351
hepatic stellate cells (HSC) 510, 512, **513**, 522
hepatic sympathetic nerve activity 262, 269
hepatitis C 522, 549, *550*, 560
hepatocytes 510, 512, **513**
heptafluorobutyric acid (HFBA) 199
hereditary nonpolyposis colorectal cancer (HNPCC) 394, 395
herring 30
trans-2,4-hexadienal 223
hexameric copper complexes 74–5
hexanal 10, 219
trans-2-hexenal 150, 223
hexokinases 378
HIFs *see* hypoxia-inducible factors
high-intensity interval training (HIIT) 284, 286
high-performance liquid chromatography (HPLC) 89–90, 150, 296, 306
hippocampus 71, 124, 349, **443**, 480
His-Arg dipeptide 9
His-His dipeptide 7
histamine
antioxidant activity 344
autonomic nervous system 270
biosynthesis **5**, 7, 262, 347, 493
blood pressure 270
exercise 305
hyperglycemia 262–3
lipid peroxidation 220
neuroprotection 441
neurotransmission 272
structure **5**
wound healing 330
histamine *N*-methyltransferase (HNMT) 108, 109, **111**
histamine neurons 268–9, 272
histamine receptor antagonists 418, 434, 497, 529, 551, 557, 562
histamine receptors
carnosine 262, 263, 346, 347, 418, 498
definition 272
histargin
actinomyteces 11
carboxypeptidase B inhibition 11
structure **8**
histatins 4
histidinase 6
histidine
antioxidant activity 344
behavioural studies 476, **477–8**
blood–retinal barrier (BRB) 536–8, 539–40, 541

buffer function 4, 413
carbonyl quenching 150
carnosine biosynthesis **100**, **433**
carnosine synthase 102, 103, 105
carnosine synthesis 53
catabolism **5**, 6–7
dietary sources 6
function 3–4
hyperglycemia 262
imidazole ring 4–6, 17, **45**
ischemia/reperfusion (I/R) injury 347, 348
liver injury 520, 521
metabolic disorders 7
oxidation inhibition 222
phosphorylation 4
p*K* values 301–2
retina 534
structure **5**, **44**
wound healing 329–30

histidine decarboxylase (HDC)
function 7, 262, 330, 418–19, 446
skeletal muscle 305
tuberomammillary nucleus (TMN) 267, 269

histidine *N*-acetyltransferase 129
histidine-propanol **157**, 158
histidinemia 7
histidinomethylalanine 11
L-histidyl-β-alanine *see* rev-carnosine
histone-acetylases (HATs) 386
histone deacetylase 381
histone-deacetylases (HDACs) 386
histone-H2A 146
HNE *see* 4-hydroxy-*trans*-nonenal
HNMT-like protein 107–8, 109–10, **111**, 112
homeopathy 242
homeostasis
carnosine 28–9, 91
cellular 248, 249
extracellular matrix (ECM) 512
metals 426
mitochondria 341

homocarnosine
aging effect on levels 24
analysis *193–5*, 196, *196*, 200, 201, **202–3**
antioxidant activity 221
biosynthesis 101, 103
brain 103, 104, 417
carbonyl quenching 13, 16
carnosine dipeptidase I 126
cooking methods effect on meat levels 28, 52
copper (II) ions complexes 231, **232**
melting point *46*
metabolism 501–2
molecular mass *46*
pharmacokinetic profile 16
p*K* values *12*, *46*, 191
properties 12
species variation 31
structure **8**, 16, **44**, 118, **148**, **203**
tissue distribution 12, 24, 103, 104, 208
water solubility *46*

homodimer 133
homology modeling 14, 125–6, 465
hormesis 241–4, 248, 250
see also dose-response
housekeeping enzyme, definition 133
hPepT1 (human peptide transporter 1)
carnosine absorption 11, *12*, 91
carnosine derivatives *12*, 63, 70
D-carnosine 62, 72, 91
functional overview 19, 93, 163
homology modeling 14–16

HPLC *see* high-performance liquid chromatography
Huntington's disease 69, 358

hydralazine 151, 152
hydrocortisone 330
hydrogen peroxide (H_2O_2)
- antioxidants 86, 179, 180, **181**, 182, 183
- assay preparation 174
- carnosine 400
- DNA degradation 177, **178**, 360, 365, 398, **404**
- ethanol metabolism **519**
- iron reactions 217
- liver injury 515
- superoxide dismutase (SOD) 183, **519**
- transition metal ions 218

hydroperoxide radical $HO^{\cdot}$ 217, 220
hydrophilic interaction chromatography (HILIC) *197*, 198–200, 209, 210
hydrophilicity 192, 196, 494
hydroxide radical $OH^{\cdot}$
- antioxidants 172, 179, **180**, 181–2, 183
- assay preparation 174
- DNA degradation 177, **178**, 365
- ischemia/reperfusion (I/R) injury 344

3-hydroxy-3-methylglutaryl coenzyme A reductase 520
4-hydroxy-*trans*-nonenal (HNE)
- aging 154
- antigenic properties 146
- biosynthesis 93, 163, 511
- carboxylic group derivatives 71
- carnosine 150–2, 154–6, 223, **224**, 225, *230*
- carnosine derivatives *230*
- carnosine metabolites 158–9
- D-carnosine 86
- definition 79
- functional overview 163
- imidazole dipeptides generally 26
- liver fatty acid-binding protein (L-FABP) 146
- metabolic syndrome 154–5
- metabolism 147
- NO-donor derivatives 71
- structure **141**

8-hydroxyguanine 365
4-hydroxyhexanal (HHE) 140
hydroxyproline 515, 522
hyperactivity 474, **475**, 476–82, 483, 487
hyperglycemia
- anserine 270
- carnosinase 502
- carnosine 259–60, **261**, 270, 333, 472, 499
- definition 272
- histamine 262–3
- histidine 262
- suprachiasmatic nucleus (SCN) 269–70
- wound healing 328, 333–4

hyperoxia 361
hypertension
- animal models 258, 497
- carnosinase deficiency 258
- carnosine 156, 495–7
- definition 505
- renal sympathetic nerve activity (RSNA) 497
- vascular dementia (VD) 437

hypertriglyceridemia 521
hypnotic effects 482–3
hypoactivity 476, 482–3, 487
hypochlorous radical $ClO^{\cdot}$ 172, 173, 177, 179, 183, 314
hypothalamic pituitary adrenal axis 474, 487
hypothalamic–pituitary–adrenal (HPA) axis 517
hypothalamus 268–9, 272, 273, 472, 497
hypovolemic shock 341
hypoxanthine 341, 342
hypoxia
- cancer 376, **401**, 403
- definition 337

hyperglycemia 328
ischemia/reperfusion (I/R) injury 342, 346
sepsis 516
hypoxia-inducible factors (HIFs)
angiogenesis 326, 328
cancer 62, 375–6, 378, 386, **401**, 403, **404**, 405, 424
ischemia/reperfusion (I/R) injury 347
retina 529
hypoxia-responsive elements (HRE) **401**, 403

idiopathic taste disorder 549, *550*, 560
imidazole, structure **44**
imidazole-4-acetic acid 6
imidazole lactate **5**, 7
imidazole pyruvate **5**, 7
imidazole ring properties 4–6, 7–8, 12, 14, 17, **45**, 53, 85–6
imidazolone-5-propionic acid **5**, 6
imine 225
in silico studies, Alzheimer's disease (AD) 420–1, 456
in vitro studies
aging 495
β-alanine transport 536
Alzheimer's disease (AD) 419–20, 458
antigenotoxic properties 359, 360, 361–2
cancer 372–3, 378, 379
carbonyl quenching 150–4
carnosine transport 530–1
histidine transport 536–8
ischemia/reperfusion (I/R) injury 348
muscle contractile funciton *303*, 304
neuroprotection 415–16, 417, 418–19, 434, 435
wound healing 333–4, **335**
in vivo studies
amino acids transport 534–6
carbonyl quenching 154–9
see also animal models
inducible NO synthase (iNOS) 248, 343, 422, 435, 479–80
infertility 358
inflammation
acrolein 140
aging 248
cadmium 516
calcium 342
cancer 398–9
carnosine 154, 218, 239, 248, 399–400, **401**
definition 337
heat-shock response (HSR) 245
histamine 493
inducible NO synthase (iNOS) 248
ischemia/reperfusion (I/R) injury 343
liver injury 512, 514, 518, 520
NF-κB (nuclear factor kappa-light-chain-enhancer of activated B cells) 398
reactive carbonyl species (RCS) 146
scar formation 327
systemic inflammatory response syndrome 516
wound healing 326–7, 330
inflammatory bowel disease (IBD) 394, 397, 398, 407
inflammatory diseases
carnosine 358
lectins 69
pathology 239
inosine 341
inplane deformation, definition 57
insulin
carnosine 259, 260
functional overview 272
insulin-like growth factor-1 (IGF-1) 326, 331–2, 334, 557, *558*

insulin resistance, animal models 521
integrin receptors 328
interferon-α 522
interferon-γ (INF-γ) 248, 416, 435
interleukin-1β (IL-1β) 248, 435, 495, 556, *558*, *559*
interleukin-6 (IL-6)
 cancer 398
 carnosine 248, **333**, *558*
 diabetes 521
 liver injury 514, 515, 518, 520, 521
 wound healing 326
 zinc (II)-carnosine complex 556, *558*, *559*
interleukin-8 (IL-8)
 carnosine 184, 248, 400, 495, *558*
 hydrogen peroxide (H_2O_2) 375, 405
 zinc (II)-carnosine complex 556, *558*, *559*
intestinal epithelial cells (IEC) 397–8, 399–400, **401**, 405, 407
intestinal flora, histidine synthesis 6
invertebrates 50–1
ion trap-orbitrap 201, 207
ionized calcium-binding adapter-1 (Iba-1) 441, **442**
iron
 carnosine 86, 219, 344, 416
 DNA degradation 360, 365
 dyshomeostasis 73
 hydrogen peroxide (H_2O_2) 217
 ischemia/reperfusion (I/R) injury 342, 344
 oxidation 218
 transcription factors 403
ischemia
 definition 337, 351
 oxidative stress 249
ischemia/reperfusion (I/R) injury
 carnosine 61, 162, 344–50, 358, *496*, 497, 514, 528
 definition 341, 351
 pathology 341–3, 351
ischemic preconditioning 348–9, 352
isocitrate dehydrogenase *241*
isoketals
 reactivity **144**
 sources 144–5
isoleucine 481
isoleucyl-histidine 150, 223
L-isoleucyll-L-histidine 482
IUPAC nomenclature 44–5, 84

Janus kinase (JAK) 376

K^+ channels, histidine phosphorylation 4
k_{cat} *see* catalytic constant
keratinocytes 327, 331
keto-aldehydes
 biosynthesis 142–5
 carnosine 153–4
 structure **141**
 see also isoketals; methylglyoxal; 4-oxo-nonenal
Ki-67 518, 523
kidney
 carnosinase 500–2
 carnosine 494–5
 function 505
 ischemia/reperfusion (I/R) injury 346–7
kidney diseases
 carnosine 495–500, 504
 chronic kidney disease (CKD) *500*, 502–3, 504, 505
 see also diabetic nephropathy
kinetic visual acuity 318, **319**, 322
Kirsten rat sarcoma viral oncogene (KRAS) 396, 401–2, 403, **404**, 406, 407

K_M *see* Michaelis-Menten constant
kokumi molecules 9
Kupffer cells 345, 510, 511, 512, **513**, 516

L-type amino acid transporter 1 (LAT1) 538–9, 540, 541, 542
lactate 439
lactate dehydrogenase (LDH) 55, 349
lactic acid 225, 313, 342
β-lactoglobulin 8
lactose dervatives **65**, **67**, 69
lansoprazole 335
lectin 79
lectins 69
Leptodactylus 30
leucine 481
leucyl-histidine 150, 223
leukocytes 326
Lewy bodies 421, 434
lifespan *see* longevity
linearity, LC-MS/MS 205
lipid peroxidation
  acyl derivatives 229
  anserine 344
  antioxidants 26–7
  cadmium 516
  carnosine 25, 85–6, 87, 220, 344, 358, 365, 415, 441, 495, 515, 519, 520, 521
  definition 233, 523
  food degradation 217
  liposome model systems 219–21
  NO-donor derivatives 229
  nonalcoholic fatty liver disease (NAFLD) 520
  pathology 217, 421
  pH effects 85
  physiology overview 23, 26–7
  pseudopeptides 228
  reactive carbonyl species (RCS) 140, 141, 152
  α-tocopherol 521
  zinc (II)-carnosine complex 521, 557
lipoic acid derivatives **64**, 68
lipophilicity 10, 70–1, 77, *230*
lipopolysaccharides (LPS) 248, 416, 435
liposome model systems 219–21, 228, 233
liposomes 234
liquid chromatography (LC)
  carbonyl quenching **152**, 154
  definition 210
  elution order 192–6, *197*, 198
  high-performance (HPLC) 89–90, 150, 296, 306
  hydrophilic interaction (HILIC) *197*, 198–200, 209, 210
  reversed-phase chromatography (RP) *197*
  reversed-phase (RP) 90, 192–8, 210–11
  sample preparation *193–4*, 206, 207, 208
  tandem (MS/MS) 192, **196**, 200, 201–6
  ultraperformance (UPLC) 90, **196**, 198, 199, 211
liver
  β-alanine 481
  β-alanine biosynthesis 356, 481
  cancer 512, 522
  carnosine dipeptidases 267
  functional overview 510
  ischemia/reperfusion (I/R) injury 345–6
  sympathetic nervous system 261–2
  zinc (II)-carnosine complex 549, 560
liver fatty acid-binding protein (L-FABP) 146
liver injury
  carnosine 513–21
  pathogenesis 510–12, **513**, 523
livestock 34
locomotor effects 474, **475**

LOD *see* detection limit
longevity
birds 29
hormesis 243–4
vitagenes 240
see also aging; antiaging
LOQ *see* quantification limit
low density lipoproteins (LDL)
carnosine 87, 221–2
copper ion oxidation 69, 71, 218, 221–2
NO-donor derivatives 229
lymphocytes 360, 362–3, 435
Lynch syndrome *see* hereditary nonpolyposis colorectal cancer
lysine 538
lysozymes 458, **460**, 465

Maillard reactions 34, 52, 57
malic enzyme 520
malondialdehyde (MDA)
biosynthesis 141, 511
carnosine 153, 219, 220, 358, 363, 365, **366**, 418, 494, 514, 515, 516, 517
definition 367
nephrotoxicity 498
reactivity 141, **143**
structure **141**
mammalian target of rapamycin (mTOR) 373–5, 376, 378, 379, 386–7
mammals
buffering capacity 55
imidazole dipeptides distribution 50
manganese ions 266
mannitol 348
MAP kinases (MAPK) 247, 381, 495, 499
marine mammals
balenine 14, 32, 50
proton buffering 54–5
mass accuracy 210
mass spectrometry (MS)
carbonyl quenching 154
carbonyl quenching mechanisms 150–1, 152
carnosine *N*-methyl transferases 107
carnosine synthase/synthetase 101–2
high-resolution techniques 201, 209
HNMT-like protein 108
tandem (MS/MS) 90, 192, **196**, 200–6, 211
matrix effect
electrospray ionization (ESI) 206
salt 207
matrix metalloproteinases (MMPs)
brain 418, **420**, 434
cancer 377, 386, 425
liver 512, 521, 522
meat
analysis 192, *193*, *204*, 206, 207–8
breed variation 29, 30, 52
consumption levels measurement 53
extraction techniques 207
imidazole dipeptide levels 24–5, 33, *52*
pH levels 226
preservation 55, 227
quality 25–8, 34, 55
species variation 29, 31–2, 52
MEK *see* mitogen-activated protein kinases
melatonin 346, 514, 517
melting points *46*, 84
membrane transporter 542
MEROPS database 119
Met-His dipeptide *see* wheylin-1
metabolic disorders, D-carnosine 86
metabolic syndrome
animal models 86, 88, 154, **155**, 156, 520

4-hydroxy-*trans*-nonenal (HNE) 154–5
nonalcoholic fatty liver disease (NAFLD) 520
reactive carbonyl species (RCS) 161, 162
metabolism
ethanol 518, **519**
histidine 6–7
reactive carbonyl species (RCS) 147, **148**, 154, 155, 162
reactive oxygen species (ROS) 413
metal chelation
N-acetyl carnosine 12
carnosine 55, 62, 85–6, 218, 344, 357, 416, 418, 495
definition 18, 57, 428
imidazole dipeptides generally 7
metal homeostasis 426
metal-ion homeostasis 73
metastasis 377, 387, 394, 396, 425
methamphetamine 249
methionine 106
L-methionyl-L-histidine 482
O-methyl-β3-homotyrosine–D-His 13
N-methyl-D-aspartate (NMDA) receptor 437, 438
methylglyoxal (MGO)
aging 249–50
carnosine 153, 249–50
diabetes 499
metabolism 147, **148**
reactivity 142–4
sources 142, **144**
structure **141**
1-methylhistidine 53, 222
metmyoglobin 226, 227
Michael addition 6, 10, 12, 55, 79, 223
Michaelis-Menten constant (K_M), definition 113
microglia 445
micronuclei (MN) 368
microsatellite instability (MSI) 395
microvascular disorders 71
microwave cooking 27, 28, 52
milk products 11
Misgurnus anguillicaudatus 102
mitochondria
carnosine 62, 441
liver injury 511, 512
normal function 341
pyruvate 403
mitochondrial permeability transition (MPT) 512
mitogen-activated protein kinases (MAP kinases/MEK) 374, 375, 386, 396, **402**, **404**, 405, 406
model, definition 465
molecular chaperones *see* chaperones
molecular dynamics (MD) 421, 456, 458, 465
molecular mass *46*
molecular oxygen
definition 34
ischemia 342–3
monosaccharides 364
mTOR (mammalian target of rapamycin) 373–5, 376, 378, 379, 386–7
mucin glycoproteins 553, 554
muscle biopsy 289
muscle-fiber-type composition 296, 298, 307
muscle tissue
anaerobic metabolism 30–1, 56
antioxidants 26
carnosine synthase/synthetase 103, 105
pH 302
physiology overview 56
see also skeletal muscle
mutagenicity 359
myelin 437, 446
myelin base protein (MBP) 440, 441, **442**, 446

myeloperoxidase (MPO)
  acrolein biosynthesis 140
  gastric ulcers 557, *558*, *559*
  ischemia/reperfusion (I/R) injury 346
  liver injury 514, 516, 519
myocardial infarction 341, 347–8, 351
myofibrils 227
myofibroblasts 326, **327**, 328, 332
myoglobin 55
myosin 56

Na^+-dependent multivitamin transporter (SMVT) 529
Na^+/K^+ ATPase 342
nanoparticles 76–7, 79
necrosis
  definition 337
  ischemia/reperfusion (I/R) injury 342, 343
  liver injury 511, 512, 516
  nephrotoxicity 498
neovascularization 337
nephropathy
  definition 337
  see also diabetic nephropathy
nephrotoxicity *496*, 498, 505
nerve growth factor (NGF) 557, *558*, *559*
nervous system disease *see* neurodegenerative disorders
neuritic plaques 465
neurodegenerative disorders
  AGEs (advanced glycoxidation end products) 416
  definition 428
  DNA degradation 358
  heat-shock response (HSR) 245
  lipid peroxidation 218
  oxidative stress 249
  reactive carbonyl species (RCS) 145
  see also Alzheimer's disease; cerebral ischemia/stroke; vascular dementia
neuromuscular diseases 297
neuropathic pain control 68
neuropathy, definition 337
neuroprotection
  amide derivatives 71
  anserine 14
  balenine 14
  carcinine 13
  carnosine 86, 87, 218, 248, 249, 348–9, 413–17, 425–7, 434–5, 472
  cyclo(His-Pro) dipeptide 9
  definition 428
  histamine 441
  NO-donor derivatives 71
  in vitro studies 415–16, 417, 418–19, 434
neurotransmitters
  definition 446
  histamine 272
  hyperglycemia 262
  imidazole dipeptide function 24
neutrophils 326–7, 343, 345, 346, 495, 511
NF-κB (nuclear factor kappa-light-chain-enhancer of activated B cells)
  function 407
  heat-shock proteins (HSPs) 249
  inflammation 398
  RAGE activation 146
  zinc (II)-carnosine complex 556, *558*
nickel ions 222
nicotinamide adenine dinucleotide (NAD)-dependent deacetylases *see* sirtuins
nicotinamide adenine dinucleotide (NAD^+/NADH) 518
nipecotic acid 536, 541

nitric oxide (NO) 248, 345–6, 415–16, 476–81, 487
nitric oxide synthase 487
nitrosamines 358
nitrotyrosine 435
NO-donors 71, 76, 79, 229, *230*
NO synthase (NOS) 476–81
nociceptive pain 426
nomenclature
- carnosinase 119, *120*, 121, 493
- carnosine 84
- imidazole dipeptides 43, 44–5
- peptidases 119

nonalcoholic fatty liver disease (NAFLD) 520–1
nonalcoholic steatohepatitis (NASH) 520, 521
trans-2,3-nonenal 225, **225**
nonproteinogenic histidine dipeptides, overview 11–17
nonsteroidal anti-inflammatory drugs (NSAIDs) 398, 551, 559–60, 561–2
norepinephrine (noradrenaline) 260, 346, 474, 497
nuclear magnetic resonance (NMR)
- anserine 47–9, *52*
- Aβ peptides 421, 456
- carbonyl quenching mechanisms 151
- carnosine 46–8, **49–50**
- chemical shift definition 56
- definition 57

nucleophilicity, definition 57
nutrition, caloric restriction (CR) 247, 249–50

olfactory bulb 11, 104, 421, 433
oligodendrocytes 432–3, 435, 437, 441, 445, 472
oligomers 465
oligopeptide transporter 543
oligopeptide transporters (SLC15A) 531–2, 539
ontogenesis 113
ophidine *see* balenine
opthalmic therapies
- *N*-acetyl carnosine 62, 209
- carnosine potential 24, 87, 88, 89, 209, 335

oral mucosa 549, *550*
origin of life 8
ornithine 103, 538
orthologs 113
osteoarthritis 297
ovalbumin 153
oxi-imflammaging, definition 250
oxidation
- definition 217
- pro-oxidation activity 222, 233

oxidative stress
- aging 248
- Alzheimer's disease (AD) 421
- carnosine 155–6, 219, 239, 249, 358
- D-carnosine 89
- chromosomes 361
- definition 34, 427, 523
- D-galactose overload 518
- glutathione (GSH) 155
- hepatitis C 522
- imidazole dipeptide function 24
- ischemia/reperfusion (I/R) injury 349
- liver 511, 520
- microvascular disorders 71
- monosaccharides 364
- pathology 415
- reactive carbonyl species (RCS) 145, 161
- reactive oxygen species (ROS) 413, **511**
- sirtuins *241*
- α-tocopherol 494

4-oxo-nonenal (ONE)
- biosynthesis 144
- reactivity 144
- structure **141**

oxoaldehydes 147, **148**
oxygen
molecular 34, 342–3
singlet 182, 220, 344, 347, 348, 415, 494
oxygen-glucose deprivation (OGD) 428

p38 mitogen-activated protein kinases 375, 387, 495, 499
p53 mutation 396
pancreatic parasympathetic nerve 262
paracetamol 514
paraesthesia 278–9, 288, 289, 383
paralogs 113
parasympathetic nervous system 262
Parkinson's disease
carnosine 24, 62, 382, 416, 421–3, 434–5
L-dopa derivatives 68
metal-ion homeostasis 73
pathology 421
protein accumulation 417
penicillin 78
pentosidine 23
pepsin digestion 33
PepT1 (peptide transporter 1)
carnosine derivative design 162
functional overview 93, 160
see also hPepT1 (human peptide transporter 1)
peptidases
classification 119, 133
functional overview 134
nomenclature 119
peptide bond formation 8
peptide/histidine transporters (PHT1) 531, **532**
perches (fish) 30
Perkinsus marinus 102
peroxiredoxins 556, *558*, *559*
peroxisome proliferator activated receptors (PPARs) 379, 381, 387
peroxisome proliferator γ coactivator 1α (PGC-1α) 379–80, 381, 387
peroxisomes 518
peroxynitrite radical ONOO˙
antioxidants 172, 179, **180**, 183
assay preparation 174
DNA degradation 177, **178**
pH
carnosine dipeptidase I 125
lipid peroxidation 231
muscle tissue 302
physiological homeostasis overview 92–3
thiobarbituric acid reactive substances (TBARS) 226
see also acidosis; buffering
pH buffer, definition 307
phenotype stabilization 245
pheochromocytoma 424–5
phosphatidilocholine 221, 228
6-phospho-2-kinase/fructose-2,6 bisphosphatase 378
phosphofructokinase 1 378
phosphoglycerate kinase 378
phosphoglycerate mutase 378
phosphoinositide 3-kinase (PI3K) 374–5, 380–1, 387
phospholipases 343, 346, 497
phospholipids 140, 219, 226
phosphorylation
carnosine 380, 425
proteins 247, 400
phthaloyl-β-alanyle chloride 53
phylogenesis
carnosine dipeptidases 130–1, **132**
carnosine synthase/synthetase 102
definition 113
HNMT-like protein 109–10, **111**
p*K* values *12*, *46*, 54, 191, 199, 226, 301–2, 314
definition 57

placebo-controlled crossover trial 322
placebo-controlled parallel, double-blind clinical test 323
plasma
  analysis *194*, *204*, 208, 209
  exercise carnosine levels 266
plasmin 328
plasticity 243–4
platelet aggregation 9
platelet-derived growth factor (PDGF) 326, 327, 555, 557, *558*, *559*
pleiotropic molecule, definition 428
Pleuronectidae 30
polaprezinc *see* zinc (II)-carnosine complex
poly-ADP-ribose polymerase 1 (PARP-1) 435
polychromatic erythrocytes 361–2
polycyclic aromatic hydrocarbons 358
polyphenols 172
polyunsaturated fatty acids
  definition 34
  lipid peroxidation 26
polyunsaturated fatty acids (PUFAs) 140, 141, 219, 413
pork 26, 29, 31, *52*
postconditioning 243
POTs (proton oligopeptide transporters) 19
poultry meat
  analysis *193*, *204*, 207–8
  anserine 109
  antioxidants 26, 27
  breed variation 29, 30
  carnosine 109, 208
  cooking methods 28, 52
  imidazole dipeptide levels 31, *52*
  preservation 227
  quality 26
precision, LC-MS/MS 205
precolumn derivatization 210
preconditioning 243
primates, carnosine dipeptidase I 124
prion diseases 425–6
pro-oxidation 222, 233
procollagen α1 521
prodrugs
  N-acetyl carnosine 62
  D-carnosine 72
  definition 79
  L-dopa derivatives 68
prolinase *see* carnosine dipeptidase II
prolyl hydroxylases (PHD) 376, 403
L-prolylhistamine *228*, 229
prostaglandin E synthase (PGES) 554
prostaglandins 343, 349, 552, 554, *558*
proteasomes 495
protein aggregation
  aging 249
  Alzheimer's disease (AD) 455
  chaperones 245, 458–60
  heat-shock proteins (HSPs) 244
  reactive carbonyl species (RCS) 146–7
protein carbonyls (PCs) 350, 517
protein folding
  amyloid fibrils 451–2
  chaperones 458
  glioblastomas 62, 424
  heat-shock proteins (HSPs) 244
  process overview 464
  stress 240
protein function
  carnosine *415*
  histidine regulation 3–4
  reactive carbonyl species (RCS) 145–6
protein glycation 24, 435
protein kinase B (PKB) 380, 387

protein kinases
  cancer 374–5
  functional overview 375
  ischemia/reperfusion (I/R)
    injury 343
protein oligomerization 146–7
protein oxidation 226–7, 233
protein synthesis, tumour
  growth 375
proteinogenic histidine dipeptides,
  overview 7–11
proteinuria *496*, 502
proteostasis 249, 464, 465
proton buffering
  N-acetyl carnosine 12
  anserine 314
  capacity 54–5
  carnosine 3, 85
  definition 18
  imidazole dipeptides
    generally 225–6, 314
  imidazole ring properties 7
  "slyke" unit 54
proton-coupled peptide transporter
  (PePT1) 405
proton magnetic resonance
  spectroscopy (^{1}H-MRS) 289, 306
proton-pump inhibitors (PPI) 335,
  551, 557, 562
proton spectroscopy 296
pseudopeptides 226, 227–9
pyridoxal phosphate 151
pyridoxamine 152
pyrilamine 418
pyruvate dehydrogenase kinases
  (PDKs) 378, 379–81, 384–5, 403
pyruvate kinase 378

quadrupole 210
quadrupole-ion trap 201, 208
quantification limit (LOQ) 206, 210

Raf (rapidly accelerated
  fibrosarcoma) 375, 396, **402**, 406
RAGE 146, 404
rancidity 27, 34
ranitidine 551
rapamycin 373–4, 375, 386–7, 503
RAS family genes 396, 401, **402**,
  403, 406
rays (fish) 30
reactive aldehydes
  carnosine 495
  definition 34
reactive carbonyl species (RCS)
  assays 145
  classification 9, 139–40, **141**
  definition 18, 79, 139, 163
  functional overview 145–7
  glutathione (GSH) 249
  mercapturic acid
    derivatives 147, **148**, 154
  metabolism 147, **148**, 162
  pathology *140*, 145, 161
  reactivity 10
  see also dialdehydes;
    keto-aldehydes; α,β
    unsaturated carbonyls
reactive nitrogen species (RNS)
  358, 367
reactive oxygen species (ROS)
  aging 182
  anserine *181*, *182*, 183, 186
  biosynthesis 171, 182, 413
  brain 86
  carnosine *181*, *182*, 183, 186,
    239, 441, **443**, 515
  cell metabolism 363
  copper (II) ions
    complexes 231
  definition 79, 323, 367, 407–8,
    428, 445, 523
  DNA degradation 363, 365
  exercise 313
  functional overview 171, 217
  glutathione (GSH) 181, 365,
    514
  ischemia/reperfusion (I/R)
    injury 343, 345, 346
  ischemic preconditioning 348
  kidney 497
  liver 511, 514, 518

matrix metalloproteinases (MMPs) 418
measurement 35
mitochondria function 341
Parkinson's disease 68
rebamipide 554
renal brush-border membrane vesicles (BBMVs) 530
renal ischemia 346–7
renal sympathetic nerve activity (RSNA) 263, **264**, 268, 269, 346–7, 497
renoprotection 156, 347, *496*, 497–8, 504, 505
repair enzymes 104
reperfusion
definition 352
see also ischemia/reperfusion (I/R) injury
reptiles 30, 50
retina 528–9, 534, **539**
see also blood–retinal barrier (BRB)
retinal pigment epithelial cells (RPE cells) 529, **530**, 531, 539
retinal uptake index (RUI) 532, **533**, 534, 536, 541, 543
retinoid/retinoic receptors (RXRα/RARα) 379, 381
retinopathies, carcinine 13
rev-carnosine 483, **484**, **485**, 487
reversed-phase chromatography (RP) 90, 192–8, *197*, 210–11
ribosomal protein S6 kinases 1 and 2 (S6K1/2) 375
RNA polymerization 8

Saccharomyces cerevisiae 435
SAH *see* *S*-adenosyl-L-homocysteine
saliva, histatins 4
salmon 30, 48, *52*, 125, 319
salmon extract (SEAns) 314–21
Salmonella typhimurium 359
salsolinol 360, 422–3
SAM *see* *S*-adenosyl-L-methionine
sample stability 206
sarcoplasmic reticulum 303, 307
sardines 30
Sauria 30
scaffolding proteins 400, **401**
scar formation 327, 328
Schiff bases 13, 34, 55, 79, 141
schizophrenia 382, 425
SCN *see* suprachiasmatic nucleus
sedative properties 8, 476, 482–3
selected ion monitoring (SIM) 200–1
selective reaction monitoring (SRM) 201, **202**
senile plagues *see* neuritic plaques
sensitivity, LC-MS/MS 206
sepsis 516
Ser-His dipeptide 8
L-seryl-β-alanine 483
L-seryl-L-histidine 482
sharks 30
shelf life 27
short-chain dehydrogenases/reductases 147
sickle cell anemia 358
signal peptide 134
signal transducers and activators of transcription (STATs) 376, 378, 388
signal transduction pathways 373–6, 378, 384, 396, **402**, 406, 419
singlet oxygen 182, 220, 344, 347, 348, 415, 494
sirtuins 239–40, *241*, 247, 249–50
sister chromatid exchange 362, **363**, 368
skeletal muscle
acidosis 280, 283, 285, 286, 287, 288, 304
β-alanine supplementation 278
anaerobic metabolism 54–5, 225
calcium 56, 286, 302–4
carnosine biosynthesis 265
carnosine function 299–305

skeletal muscle (*continued*)
- carnosine levels 11, 25, 277, 278–80, 285, 295, 296–7, 299
- carnosine release 266, 269, 305
- histidine decarboxylase (HDC) 305
- imidazole dipeptide levels 28
- muscle-fiber-type composition 296, 298, 307
- *see also* exercise performance

skin 549
skinks 30
"slyke" unit 54
smoking 89, 515
smooth muscle actin 1 (SMA1) 331, 332–3
snakes 30
sodium bicarbonate 282, 283, 284, 285, 288, 289
sodium chloride, food analysis effect 207
sole (fish) 30
solubility 84
sources
- β-alanine 289, 298
- anserine 160
- antioxidants 172, 186
- balenine (ophidine) 160
- carnosine 24–5, 31–2, 160, 192, 549
- histidine 6
- imidazole dipeptides generally 50–3
- zinc 549

SP1 387–8
spatial memory 352
species variation 29, 302, 356
spermidine 140
spermine 140
spinal cord injury therapy 62, 89, 425
splenic sympathetic nerve 265
sports nutrition *see* ergogenics
stability of samples 206
standards for analysis 202–5
STAT3 signaling 376, 378, 419
steatosis *see* liver injury
sterol regulatory element binding proteins (SREBRs) 520–1
stomach cancer 554
streptavidin 69, 76
stress
- cellular stress response 244–5
- cortisol 315, **316**
- liver injury 516–17
- protein folding 240
- *see also* inflammation; oxidative stress

stretching, covalent bonds 45, 57
stroke/cerebral ischemia
- carnosine 417–19, **420**, 434, 435
- definition 352
- homocarnosine 417
- ischemia/reperfusion (I/R) injury 341, 348–9
- NO-donor derivatives 71
- vascular dementia (VD) 437

stromal-derived factor 1 (SDF-1) 331, 332–3
structure
- acrolein **141**
- activity relationship 15
- amyloid fibrils 452–3
- anserine **8, 44,** 118, **148, 185, 203, 258**
- balenine **8, 44,** 118, **148, 203**
- carcinine **8**
- carnosine **8, 44, 85,** 118, **148, 185, 203, 258, 344**
- carnosine derivatives **64–5, 66–7, 74**
- carnosine dipeptidase II (CN2) 267
- carnosine *N*-methyl transferases 112
- carnosine-propanol **157**
- carnosine synthase/synthetase 101–2, 112
- copper (II) complexes **74, 75**

copper (II) ions complexes **232**
cyclodextrin derivatives **232**
cyclo(His-Ala) dipeptide **8**
cyclo(His-Gly) dipeptide **8**
cyclo(His-Pro) dipeptide **8**
N-formiminoglutamic acid (FiGlu) **5**
Gly-His dipeptide **8**
glyoxal (GO) **141**
heat-shock factors (HSFs) 246
histamine **5**
histargin **8**
histidine **5, 44**
histidine-propanol **157**
homocarnosine **8, 44**, 118, **148, 203**
4-hydroxy-*trans*-nonenal (HNE) **141**
imidazole **44**
imidazole lactate **5**
imidazole pyruvate **5**
imidazolone-5-propionic acid **5**
malondialdehyde (MDA) **141**
methylglyoxal (MGO) **141**
4-oxo-nonenal (ONE) **141**
urocanic acid **5**
zinc (II)-carnosine complex 549, **551**
zinc (II)-carnosine complex (polaprezinc) **522**
sturgeon 30
sumoylation 247
superoxide anions 220, 229, 231, 342, 344, 413, 516, **519**
superoxide dismutase (SOD)
antioxidant activity 26
carnosine 358, 363, 365, **366**, 418, 422, 518
carnosine derivatives 76
ethanol metabolism **519**
function 218, 367
hydrogen peroxide (H_2O_2) 183
imidazole dipeptides 218, 514, 515
inflammation 249
sirtuins *241*
zinc (II)-carnosine complex 556, *558*, *559*
supplementation
β-alanine 278–80, 286, 287–8, 298, 299, 300–1
antioxidants 172, 248
carnosine 331, 347, 423, 434, 471–2, 499
compared to food intake 278
imidazole dipeptides 32–3, 34
sodium bicarbonate 283
α-tocopherol 26
suprachiasmatic nucleus (SCN) 269–70, 271, 272, 273, 497
sympathetic nervous system 261–2, 263, **264**, 271, 352
synthesis
carnosine 53, 84–5, 471
carnosine derivatives **67, 70**
endogenous 428
see also biosynthesis
α-synuclein 147, 421–2, 434
systemic inflammatory response syndrome 516
systemic lupus erythematosus (SLE) 146

tandem mass spectrometry (MS/MS) 90, 192, 200–6, 211
taste disorders 549, *550*, 560
taste (food)
γ-Glu-His dipeptide 9
lipid peroxidation effects 217
taurine transporter (TAUT) 296, 536, 538, 540, 541, 542, 543
tautomerism 4–6, 18, **45**, 57
TBARS *see* thiobarbituric acid reactive substances
temperature *see* thermal stress
termolysin 8
testicular germ cells 349, 352
testicular torsion 349–50

testosterone 298
tetrahydrofolate 6–7
TGF-β1 *see* transforming growth factor β 1
thermal stress 244, 245, 246
thioacetamide (TAA) 515, 522
thiobarbituric acid reactive substances (TBARS)
assay 25, 26, 145, 219, 220
carnosine 55, 219, 220, 227, 350
definition 35, 234
pH relationship 226
reactive carbonyl species (RCS) 145
zinc (II)-carnosine complex 557, *558*, *559*
thioflavin T 465
thioperamide maleate 263
thioredoxin/thioredoxin reductase system 240
threonine 140
thrombin 9
thymine 356, 481
tight junctions 397, 529, **530**
tissue inhibitors of metalloproteinases (TIMPs) 512, 521–2
α-tocopherol
antioxidant activity 26, 27, 55, 222, 344
carnosine 27, 494
liver injury 514–15, 521
tonofilaments 327, 337
TOR signaling pathway 499
tosylureido derivatives **65**, 69–70
toxicology 243, 359
TR-iBRB2 cells 543
training (exercise) 285–6, 287–8
tramiprosate **454**, 455
transaminase enzymes 7, 514, 516, 518, 519, 520
transcription factors 246, 343, 403, 520–1
transforming growth factor-a (TGF-a) 326, 555
transforming growth factor β 1 (TGF-β1)
carnosine *496*
diabetic nephropathy 498–9
function 505
gastric ulcers 555
liver 512
liver injury **513**, 521, 522
wound healing 326, 328, 331, 332
transition metal ions 182, 218, 226, 360, 365, 416
transplant patients 341
trefoil factor family (TFF) peptides 553, 554, 555
trehalose **65**, 69, 72, 75, 79, 383
trichostatin A (TSA) 381, 388
triose-phosphate isomerase 378
triple quadrupole (TQD) 200–1, 207, 208
trolox derivatives 63–8, 232, 383
troponin C 56, 304
Trp-His dipeptide
antimicrobial activity 9
cardiovascular system 9
TRPA1 (transient receptor potential cation channel, subfamily A, member 1) 68, 79
tuberomammillary nucleus (TMN) 267, 268–9, 273
tumor necrosis factor alpha (TNF-α)
carnosine 435
D-carnosine 248
diabetes 521
ethanol 518
liver injury 512, **513**, 514, 515, 520, 521
NF-κB 398
NO-donor derivatives 71
wound healing 326
zinc (II)-carnosine complex 521, 556, *558*, *559*
tumor necrosis factor-related apoptosis inducing ligand (TRAIL) 512, **513**

tuna fish 30, *52*
turkey meat *see* poultry meat
turtles 30
tyrosine hydroxylase 417
tyrosine kinase receptors 396

ubiquitin 152, 403
ulcerative colitis 394, 397, 398, 549, *550*, 560
ultraperformance liqud chromatography (UPLC) 90, **196**, 198, 199, 211
trans-2-undecanal 219
α,β unsaturated carbonyls
  carnosine 12, 150–2, 218, 223, 358
  metabolism **148**
  reactivity 140–1, **142**
  sources 140
  structure **141**
  see also acrolein; carbonyl quenching; 4-hydroxy-*trans*-nonenal; 4-hydroxyhexanal
uptake (of drug) 408
uracil 298, 356, 481
urine
  analysis *194*, *204*, 208, 209
  reactive carbonyl species (RCS) metabolites 147, **148**, 154, 156–9
urocanase deficiency 7
urocanic acid
  antioxidant function 6
  biosynthesis **5**, 6
  structure **5**
UV-Vis absorption spectrum, carnosine 45, **46**

vagal celiac nerve 261–2, 269
valine 481
L-valyl-β-alanine 483, **484**
valyl-histidine 150, 223
L-valyl-L-histidine 482
vanadium 416
vascular dementia (VD)
  animal models 438
  carnosine 438–44
  classification 435–6, 437
  pathology 436–7
  prevalence 436, 444
vascular endothelial growth factors (VEGFs) 326, 327, 328, 388, 555, 557, *558*, *559*
vasodilation 24
vasorelaxants 9
VDCCs (voltage-dependent l-type Ca^{++} channels) 9
vegetarianism 25, 298–9
venous stasis, definition 337
vision, carcinine 13
visual acuity 318, **319**
vitagenes
  carnosine 248–50
  definition 250
  function 239–40, 245
vitamin A 510
vitamin B7 *see* biotin
vitamin C 179, *181*, *182*, 183, 184, 186, 350
  see also ascorbic acid
vitamin E *181*, *182*, 515, 517
vitamin E derivatives **64**, 68, 87, 415, 426, 455
vitamin H *see* biotin
vitronectin 328
V_{max}, definition 113
voltage-dependent l-type Ca^{++} channels (VDCCs) 9
Volvox carteri 102
von Hippel–Lindau (VHL) protein 376, **401**, 403, 424

Warburg effect 377–8, 385, 403, 408
water solubility *46*, 218
wheylin-1 (Met-His dipeptide) 8
white adipose tissue (WAT) 265, 272–3, 482
Wnt signal transduction pathways 396, 408

wound healing
- abnormal 329
- β-alanylhistamine 228
- carnosine 88, 329–36
- hormesis 244
- process 326–9, 336
- zinc 548

wounds 325–6

Xaa–methyl-His dipeptidase
- function 129–30
- gene expression 128
- inhibition 129
- nomenclature 119, *120*, 127
- phylogenesis 130–1, **132**
- research history 127
- specificity *126*, 128–9
- tissue distribution 128
- *see also* anserinase

xanthine dehydrogenase 342

xanthine oxidase 342, 518, **519**

xantine oxidase/hypoxantine system 229

zinc
- bioavailability 551
- carnosine 62, 86, 358, 416, 418, 421, 455
- deficiency 549
- DNA degradation 365
- functional overview 549
- sources 549

zinc (II)-carnosine complex (polaprezinc)
- functional overview 563
- gastric ulcers 62, 331, 335, 548–9, *550*, 556–60
- heat-shock response (HSR) 249, 460, 557, *558*, *559*
- liver injury 521–2
- nanotapes 77
- structure **522**, 549, **551**
- wound healing 331, 333–4, 335

zolatidine 418

Zucker rats
- carnosine 72, 86, 88, 156
- reactive carbonyl species (RCS) 154, **155**